Lecture Notes in Networks and Systems

Volume 490

AF870096

The series "Lecture Notes in Networks and Systems" publishes the latest developments in Networks and Systems—quickly, informally and with high quality. Original research reported in proceedings and post-proceedings represents the core of LNNS.

Volumes published in LNNS embrace all aspects and subfields of, as well as new challenges in, Networks and Systems.

The series contains proceedings and edited volumes in systems and networks, spanning the areas of Cyber-Physical Systems, Autonomous Systems, Sensor Networks, Control Systems, Energy Systems, Automotive Systems, Biological Systems, Vehicular Networking and Connected Vehicles, Aerospace Systems, Automation, Manufacturing, Smart Grids, Nonlinear Systems, Power Systems, Robotics, Social Systems, Economic Systems and other. Of particular value to both the contributors and the readership are the short publication timeframe and the world-wide distribution and exposure which enable both a wide and rapid dissemination of research output.

The series covers the theory, applications, and perspectives on the state of the art and future developments relevant to systems and networks, decision making, control, complex processes and related areas, as embedded in the fields of interdisciplinary and applied sciences, engineering, computer science, physics, economics, social, and life sciences, as well as the paradigms and methodologies behind them.

Indexed by SCOPUS, INSPEC, WTI Frankfurt eG, zbMATH, SCImago.

All books published in the series are submitted for consideration in Web of Science.

For proposals from Asia please contact Aninda Bose (aninda.bose@springer.com).

Paramartha Dutta · Satyajit Chakrabarti · Abhishek Bhattacharya · Soumi Dutta · Celia Shahnaz
Editors

Emerging Technologies in Data Mining and Information Security

Proceedings of IEMIS 2022, Volume 2

Springer

Editors
Paramartha Dutta
Department of Computer and System Sciences
Visva-Bharati University
Santiniketan, West Bengal, India

Satyajit Chakrabarti
Department of Computer Science and Engineering
Institute of Engineering and Management
Kolkata, West Bengal, India

Abhishek Bhattacharya
Department of Computer Application and Science
Institute of Engineering and Management
Kolkata, West Bengal, India

Soumi Dutta
Department of Computer Application and Science
Institute of Engineering and Management
Kolkata, West Bengal, India

Celia Shahnaz
Department of Electrical and Electronic Engineering
Bangladesh University of Engineering and Technology
Dhaka, Bangladesh

ISSN 2367-3370 ISSN 2367-3389 (electronic)
Lecture Notes in Networks and Systems
ISBN 978-981-19-4051-4 ISBN 978-981-19-4052-1 (eBook)
https://doi.org/10.1007/978-981-19-4052-1

This Springer imprint is published by the registered company Springer Nature Singapore Pte Ltd.
The registered company address is: 152 Beach Road, #21-01/04 Gateway East, Singapore 189721, Singapore

Preface

This volume presents proceedings of the 3rd International Conference on Emerging Technologies in Data Mining and Information Security (IEMIS 2022), which took place at the Institute of Engineering and Management in Kolkata, India, from February 23 to 25, 2022. The volume appears in the series "Lecture Notes in Networks and Systems" (LNNS) published by Springer Nature, one of the largest and most prestigious scientific publishers, in the series which is one of the fastest-growing book series in their program. The LNNS is meant to include various high-quality and timely publications, primarily conference proceedings of relevant conference, congresses and symposia but also monographs, on the theory, applications and implementations of broadly perceived modern intelligent systems and intelligent computing, in their modern understanding, i.e., including tools and techniques of artificial intelligence (AI), computational intelligence (CI)–which includes data mining, information security, neural networks, fuzzy systems, evolutionary computing, as well as hybrid approaches that synergistically combine these areas—but also topics such as network security, cyber-intelligence, multi-agent systems, social intelligence, ambient intelligence, Web intelligence, computational neuroscience, artificial life, virtual worlds and societies, cognitive science and systems, perception and vision, self-organizing and adaptive systems, e-learning and teaching, human-centered and human-centric computing, autonomous robotics, knowledge-based paradigms, learning paradigms, machine ethics, intelligent data analysis, various issues related to network security, big data, security and trust management, to just mention a few. These areas are at the forefront of science and technology and have been found useful and powerful in a wide variety of disciplines such as engineering, natural sciences, computer, computation and information sciences, ICT, economics, business, e-commerce, environment, health care, life science and social sciences. The LNNS book series is submitted for indexing in ISI Conference Proceedings Citation Index (now run by Clarivate), EI Compendex, DBLP, SCOPUS, Google Scholar and Springer Link, and many other indexing services around the world. IEMIS 2022 is an annual conference series organized at the School of Information Technology, under the aegis of the Institute of Engineering and Management. Its idea came from the heritage of the other two cycles of events: IEMCON and UEMCON, which were organized by the Institute of Engineering and Management under the leadership of Prof. (Dr.) Satyajit Chakrabarti.

In this volume of "Lecture Notes in Networks and Systems," we would like to present the results of studies on selected problems of data mining and information security. Security implementation is the contemporary answer to new challenges in the threat evaluation of complex systems. Security approach in theory and engineering of complex systems (not only computer systems and networks) is based on a multidisciplinary attitude to information theory, technology and maintenance of the systems working in real (and very often unfriendly) environments. Such a transformation has shaped natural evolution in topical range of subsequent IEMIS conferences, which can be seen over the recent years. Human factors likewise infest the best digital dangers. Workforce administration and digital mindfulness are fundamental for accomplishing all-encompassing cybersecurity. This book will be of extraordinary incentive to a huge assortment of experts, scientists and understudies concentrating on the human part of the Internet and for the compelling assessment of safety efforts, interfaces, client-focused outline and plan for unique populaces, especially the elderly. We trust this book is instructive yet much more than it is provocative. We trust it moves, driving per user to examine different inquiries, applications and potential arrangements in making sheltered and secure plans for all.

The program committee of the IEMIS 2022 conference, its organizers and the editors of these proceedings would like to gratefully acknowledge the participation of all reviewers who helped to refine the contents of this volume and evaluated conference submissions. Our thanks go to all respected keynote speakers: Prof. Seyedali Mirjalili, Prof. Md. Abdur Razzak, Prof. Rafidah Md. Noor, Prof. Xin-She Yang, Prof. Reyer Zwiggelaar, Dr. Vincenzo Piuri and Dr. Shamim Kaiser and to our all session chairs.

Thanking all the authors who have chosen IEMIS 2022 as the publication platform for their research, we would like to express our hope that their papers will help in further developments in the design and analysis of engineering aspects of complex systems, being a valuable source of material for scientists, researchers, practitioners and students who work in these areas.

Santiniketan, India	Paramartha Dutta
Kolkata, India	Satyajit Chakrabarti
Kolkata, India	Abhishek Bhattacharya
Kolkata, India	Soumi Dutta
Dhaka, Bangladesh	Celia Shahnaz

About This Book

This book features research papers presented at the International Conference on Emerging Technologies in Data Mining and Information Security (IEMIS 2022) which was held at the Institute of Engineering and Management in Kolkata, India, on February 23–25, 2022.

Data mining is a current well-known topic in mirroring the exertion of finding learning from information. It gives the strategies that enable supervisors to acquire administrative data from their heritage frameworks. Its goal is to distinguish legitimate, novel, possibly valuable and justifiable connection and examples in information. Information mining is made conceivable by the very nearness of the expansive databases.

Information security advancement is an essential part to ensure open and private figuring structures. With the, in all cases, use of information development applications, affiliations are ending up more aware of the security risks to their benefits. Notwithstanding how strict the security techniques and parts are, more affiliations are getting the chance to be weak to a broad assortment of security breaks against their electronic resources. Network-intrusion area is a key protection part against security perils, which have been growing in rate generally.

This book comprises high-quality research work by academicians and industrial experts in the field of computing and communication, including full-length papers, research-in-progress papers and case studies related to all the areas of data mining, machine learning, Internet of things (IoT) and information security, etc.

About the Conference

Welcome to the 3rd International Conference on Emerging Technologies in Data Mining and Information Security (IEMIS 2022) which was held in Kolkata, India, on February 23–25, 2022. As a premier conference in the field, IEMIS 2022 provides a highly competitive forum for reporting the latest developments in the research and application of information security and data mining. We are pleased to present the proceedings of the conference as its published record. The theme this year is Crossroad of Data Mining and Information Security, a topic that is quickly gaining traction in both academic and industrial discussions because of the relevance of privacy-preserving data mining (PPDM model).

IEMIS is a young conference for research in the areas of information and network security, data sciences, big data and data mining. Although 2018 was the debut year for IEMIS, it has already witnessed significant growth. As evidence of that, IEMIS received a record of 610 submissions. The authors of submitted papers come from 35 countries and regions. The authors of accepted papers are from 11 countries.

We hope that this program will further stimulate research in information security and data mining and provide practitioners with better techniques, algorithms and tools for deployment. We feel honored and privileged to serve the best recent developments in the field of data mining and information security to you through this exciting program.

Dr. Satyajit Chakrabarti
President of IEM Group, India
Chief Patron, IEMIS 2022

Contents

Editors and Contributors

About the Editors

Dr. Paramartha Dutta is currently Professor in the Department of Computer and System Sciences in, Visva-Bharati University, Shantiniketan, India. He did Bachelors and Masters in Statistics from ISI, Kolkata, India. Subsequently, he did Master of Technology in Computer Science from ISI, Kolkata, India. He did Ph.D. (Engineering) from BESU, Shibpore, India. He is a co-author of eight authored books apart from thirteen edited books and more than 200 and 40 research publications in peer reviewed journals and conference proceedings. He is a co-inventor of 17 published patents. He is Fellow of IETE, Optical Society of India, IEI, Senior Member of ACM, IEEE, Computer Society of India, International Association for Computer Science and Information Technology and Member of Advanced-Computing and Communications Society, Indian Unit of Pattern-Recognition and AI, the Indian Affiliate of the International Association for Pattern Recognition, ISCA, Indian Society for Technical Education, System Society of India.

Dr. Satyajit Chakrabarti is Pro-Vice Chancellor, University of Engineering and Management, Kolkata and Jaipur Campus, India, and Director of Institute of Engineering and Management, IEM. As Director of one of the most reputed organizations in Engineering and Management in Eastern India, he launched a PGDM Programme to run AICTE approved management courses, toppers academy to train students for certificate courses, and software development in the field of ERP solutions. He was Project Manager in TELUS, Vancouver, Canada, from February 2006 to September 2009, where he was intensively involved in planning, execution, monitoring, communicating with stakeholders, negotiating with vendors and cross-functional teams, and motivating members. He managed a team of 50 employees and projects with a combined budget of $3 million.

Dr. Abhishek Bhattacharya is Assistant Professor at Institute of Engineering and Management, India. He has completed his Ph.D. (Engineering), BIT, Mesra. He is certified as Publons Academy Peer Reviewer, 2020. His research interests are data mining, cybersecurity, and mobile computing. She has published 25 conference and journal papers in Springer, IEEE, IGI Global, Taylor & Francis, etc. He has 3 book chapters in Taylor & Francis Group EAI. He is a peer reviewer and TPC member in different international journals. He was the editor in IEMIS 2020, IEMIS 2018 and special issues in IJWLTT. He is the member of several technical functional bodies such as IEEE, IFERP, MACUL, SDIWC, Internet Society, ICSES, ASR, AIDASCO, USERN, IRAN, and IAENG. He has published 3 patents.

Dr. Soumi Dutta is Associate Professor at Institute of Engineering and Management, India. She has completed her Ph.D. (Engineering), IIEST, Shibpur. She received her B.Tech. (IT) and M.Tech. (CSE) securing 1st position (gold medalist), from MAKAUT. She is certified as Publons Academy Peer Reviewer, 2020, and Certified Microsoft Innovative Educator, 2020. Her research interests are data mining, OSN data analysis, and image processing. She has published 30 conference and journal papers in Springer, IEEE, IGI Global, Taylor & Francis, etc. She has 5 book chapters in Taylor & Francis Group and IGI-Global. She is a peer reviewer and TPC member in different international journals. She was the editor in CIPR-2020, CIPR2019, IEMIS-2020, CIIR-2021, IEMIS-2018 special issues in IJWLTT. She is member of several technical functional bodies such as IEEE, ACM, IEEE, IFERP, MACUL, SDIWC, Internet-Society, ICSES, ASR, AIDASCO, USERN, IRAN, and IAENG. She has published 4 patents. She has delivered more than Keynote talks in Different International Conferences.

Dr. Celia Shahnaz received Ph.D. degree in electrical and computer engineering from Concordia University, Montreal, QC, Canada, in 2009. Currently she is serving as a Professor in the Department of Electrical and Electronic Engineering, BUET, from where she received her B.Sc. and M.Sc. degrees in 2000 and 2002, respectively. Dr. Celia is a senior member of IEEE, a fellow of IEB and has published more than 80 international journal and conference papers. Recently, she has been selected as the recipient of IEEE Member and Geographic Activities (MGA) Leadership award with citation " For leadership in engineering and technology driven innovative IEEE Women in Engineering activities for enhanced membership development and engagement in Region 10 and across the globe. Dr. Shahnaz was a recipient of the Canadian Commonwealth Scholarship and Fellowship for pursuing Ph.D. study in Canada. She is the recipient of Bangladesh-Academy of Science gold model for her contribution in science and Engineering in Bangladesh. Her research interests include the areas of speech analysis, speech enhancement, digital watermarking, biomedical signal processing, audio-visual recognition for biometric security, pattern recognition, multimedia communication, control system, robotics and signal processing and pattern recognition for power signals.

Contributors

Abbasi Nagma NextGen Life Sciences Pvt. Ltd., New Delhi, India

Abdullah A. Sheik School of Computer Science and Engineering, Vellore Institute of Technology, Chennai, India

Aditya St. Andrews Institutes of Technology and Management, Gurugram, India

Aggarwal Alok School of Computer Science, University of Petroleum and Energy Studies, Bidholi, Dehradun, India

Agrawal Jitendra Department of Computer Science, School of Information Technology, RGPV, Bhopal, India

Alshareef Abdulrhman M. Information System Department, FCIT, King Abdulaziz University, Jeddah, Saudi Arabia

Alsobhi Aisha Department of Information Systems, Faculty of Computing and Information Technology, King Abdulaziz University, Jeddah, Saudi Arabia

Ansari Farooq Asansol Engineering College, Asansol, West Bengal, India

Anubhuti GNDU Regional Campus, Gurdaspur, India

Anupama Damarla VIT-AP University, Vijayawada, Andhra Pradesh, India

Arif Abu Shamim Mohammad Computer Science and Engineering Discipline, Khulna University, Khulna, Bangladesh

Arora Prakriti HMRITM, Delhi, India

Ayve Khairun Nessa Computer Science and Engineering Discipline, Khulna University, Khulna, Bangladesh

Bakshi Gaytri School of Computer Science, University of Petroleum and Energy Studies, Bidholi, Dehradun, India

Bamba Inshita Manav Rachna International Institute of Research and Studies, Faridabad, Haryana, India

Banerjee Ayanita University of Engineering and Management, Kolkata, India

Banerjee Debanjan Sarva Siksha Mission, Kolkata, India

Banerjee Mallika Techno India University, Kolkata, India

Banerjee Priyanka Department of Computer Science, The Bhawanipur Education Society College, Kolkata, India

Baranwal Arpit Asansol Engineering College, Asansol, West Bengal, India

Baranwal Rahul Raj Rakuten, Bangalore, India

Basak Subhadip Computer Science and Engineering Department, JIS University, Kolkata, India

Batra Gunveen Manipal Institute of Technology, Manipal Academy of Higher Education, Manipal, India

Bhargav Rohan Vivekananda Institute of Professional Studies, New Delhi, India

Bharti Aditi University of Petroleum and Energy Studies, Dehradun, India

Bhati Irfan H. Google India, Bangalore, India

Bhattacharya Anuran Department of Computer Science, The Bhawanipur Education Society College, Kolkata, India

Bhoumik Debasmita Advanced Computing and Microelectronics Unit, Indian Statistical Institute, Kolkata, India

Bisht Muskan HMRITM, Delhi, India

Chakraborty Sanjay Computer Science and Engineering Department, JIS University, Kolkata, India

Chakravortty Somdatta Maulana Abul Kalam Azad University of Technology, Kolkata, West Bengal, India

Chandrakala C. B. Department of Information and Communication Technology, Manipal Institute of Technology, MAHE, Manipal, India

Chatterjee Biswajoy University of Engineering and Management, Jaipur, India

Chaudhary Riya Ajay Kumar Garg Engineering College, Ghaziabad, UP, India

Chauhan Kartikeya Singh Department of Electronics and Communications, Delhi Technological University, Bawana Road, New Delhi, Delhi, India

Chauhan Prashant University of Petroleum and Energy Studies, Dehradun, India

Chawla Pronika Manav Rachna International Institute of Research and Studies, Faridabad, Haryana, India

Chhibber Manasi Amity School of Engineering and Technology, Amity University, Noida, Uttar Pradesh, India

Choudhary Richa School of Computer Science, University of Petroleum and Energy Studies (UPES), Bidholi, Dehradun, Uttarakhand, India

Choudhury Tanupriya School of Computer Science, University of Petroleum and Energy Studies (UPES), Bidholi, Dehradun, Uttarakhand, India

Chowdhury Aritra Das Computer Science and Engineering Department, Future Institute of Engineering and Management, Kolkata, West Bengal, India

Dahiya Prachi Delhi Technological University (DTU), Delhi, India

Daniya T. GMR Institute of Technology, Rajam, AP, India

Das Sayan Computer Science and Engineering Department, JIS University, Kolkata, India

Das Srirupa RCC Institute of Information Technology, Kolkata, India

Deshmukh Rushali.A. Computer Engineering, JSPM'S Rajarshi Shahu College of Engineering, Tathawade, Pune, India

Dhall Garima Infosys, Bangalore, India

Dongre Swati Department of Computer Science, School of Information Technology, RGPV, Bhopal, India

Dutta Soumi Institute of Engineering and Management, Kolkata, India

Firdaus Hina St. Andrews Institutes of Technology and Management, Gurugram, India;
Faculty III, Human Computer Interaction, University of Siegen, Seigen, Germany

Freeda R. Adline Department of Computer Science, Hindustan Institute of Technology & Science (Deemed To Be University), Padur, Chennai, Tamil Nadu, India

Gaikwad Jyotsna S. Deogiri College, Aurangabad, India

Gaikwad Sonali S. Shri Shivaji Science and Arts College, Chikhli, Buldhana, India

Galety Mohammad Gouse College of Information Technology and Computer Science, Catholic University in Erbil, Erbil, Iraq

Gampala Veerraju Koneru Lakshmaiah Education Foundation, Vaddeswaram, Guntur, Andhra Pradesh, India

Ganguli Runa The Bhawanipur Education Society College, Kolkata, India

Ganguli Souvik Department of Electrical and Instrumentation Engineering, Thapar Institute of Engineering and Technology, Patiala, Punjab, India

Gautam Siddharth NSUT, Delhi, India

Ghosal Arijit St. Thomas' College of Engineering and Technology, Kolkata, India

Ghosh Jayasree Computer Science and Engineering Department, JIS University, Kolkata, India

Giri Parul MDU, Rohtak, India

Gupta Gopal ABES Engineering College, Ghaziabad, India

Gupta Meghna Amity Institute of Information Technology, Amity University, Noida, Uttar Pradesh, India

Gupta Neha Vivekananda Institute of Professional Studies, New Delhi, India

Gupta Romasha Amity Institute of Biotechnology, Amity University Uttar Pradesh, Noida, Uttar Pradesh, India

Gupta Vittesha Department of Computer Science and Engineering, Indira Gandhi Delhi Technical University for Women, Delhi, India

Hasanin Tawfiq Department of Information Systems, Faculty of Computing and Information Technology, King Abdulaziz University, Jeddah, Saudi Arabia

Jadhav Prachi Computer Engineering, JSPM'S Rajarshi Shahu College of Engineering, Tathawade, Pune, India

Jain Sarika Amity Institute of Information Technology, Amity University, Noida, Uttar Pradesh, India

Jain Vidit Vinay Manipal Institute of Technology, Manipal Academy of Higher Education, Manipal, India

Jaiswal Arunima Department of Computer Science and Engineering, Indira Gandhi Delhi Technical University for Women, Delhi, India

Jamshed Aatif ABES Engineering College, Ghaziabad, India

Jawale Rohan Computer Engineering, JSPM'S Rajarshi Shahu College of Engineering, Tathawade, Pune, India

Jayalalitha G. Department of Mathematics, School of Basic Sciences, Vels Institute of Science, Technology and Advanced Studies, Chennai, India

Joel J. P. C. Senac Faculty of Ceará, Fortaleza, CE, Brazil

Jose Deepa V. Christ Deemed to be University, Bangalore, India

Joseph Niju P. Department of Computer Science, CHRIST University, Bangalore, Karnataka, India

Kaleem Mohd. ABES Engineering College, Ghaziabad, Uttar Pradesh, India

Kalme Geetanjali A.P. Shah Institute of Technology, Thane, Maharashtra, India

Kalra Anushka Mahavir Swami Institute of Technology, Sonipat, India

Kapoor Manushi VMware, Bangalore, India

Kasturi Nivedita PES University, Bengaluru, Karnataka, India

Kataria Pooja Chandigarh University, Mohali, India

Kaur Gagandeep University of Petroleum and Energy Studies, Dehradun, India; Department of Electrical and Instrumentation Engineering, Thapar Institute of Engineering and Technology, Patiala, Punjab, India

Kaur Harjot GNDU Regional Campus, Gurdaspur, India

Khadidos Adil Department of Information Technology, Faculty of Computing and Information Technology, King Abdulaziz University, Jeddah, Saudi Arabia

Khadidos Alaa Omar Faculty of Computing and Information Technology, Department of Information Systems, King Abdulaziz University, Jeddah, Saudi Arabia

Khan Samadur Department of Computer Science and Engineering, Aliah University, Kolkata, India

Khatter Harsh KIET Group of Institutions, Delhi NCR, Ghaziabad, India

Kishore G. T. Tarun Department of Information Technology, Thiagarajar College of Engineering, Tamil Nadu, Madurai, India

Koneru Anu Priya Department of IT, Lakireddy Bali Reddy College of Engineering, Mylavaram, India

Kumar Shashwat School of Computer Science, University of Petroleum and Energy Studies (UPES), Bidholi, Dehradun, Uttarakhand, India

Kumar Vinod Delhi Technological University (DTU), Delhi, India

Kumar Rishav HMRITM, Delhi, India

Lavanya K. Department of IT, Lakireddy Bali Reddy College of Engineering, Mylavaram, India

Mahamud Sharder Shams Computer Science and Engineering Discipline, Khulna University, Khulna, Bangladesh

Mahdi Hussain Falih College of Engineering, University of Diyala, Baqubah, Iraq

Majumdar Ritajit Advanced Computing and Microelectronics Unit, Indian Statistical Institute, Kolkata, India

Malhorta Rishabh University of Petroleum and Energy Studies, Dehradun, India

Mallick Ankush Computer Science and Engineering Department, JIS University, Kolkata, India

Manas Shetty V. PES University, Bengaluru, Karnataka, India

Mangla Monika Department of Information Technology, Dwarkadas J Sanghvi College of Engineering, Mumbai, India

Manjunath Divya Bhanu Manipal Institute of Technology, Manipal Academy of Higher Education, Manipal, India

Manoj A. Department of Information Technology, Thiagarajar College of Engineering, Tamil Nadu, Madurai, India

Maram Balajee Chitkara University Institute of Engineering and Technology-CSE, Chitkara University, Baddi, India

Mathew Meera Associate Professor, School of Law, Christ (Deemed to be University) Delhi N.C.R, Delhi, India

Mazumdar Arpita Haldia Institute of Technology, Haldia, India

Mishra Anurag ABES Engineering College, Ghaziabad, India

Mitra Nayantara Institute of Engineering and Management, Kolkata, India

Mitra Sushmita Indian Statistical Institute, Kolkata, India

Mittal Harsh Department of Engineering Physics, Delhi Technological University, Bawana Road, New Delhi, Delhi, India

Mittra Ahana Computer Science and Engineering Department, Future Institute of Engineering and Management, Kolkata, West Bengal, India

Mohanty Sabyasachi Amity Institute of Biotechnology, Amity University Uttar Pradesh, Noida, Uttar Pradesh, India

Mohanty Sachi Nandan School of Computer Science & Engineering, VIT-AP University, Amaravati, India

Mohite Apeksha A.P. Shah Institute of Technology, Thane, Maharashtra, India

Mollah Ayatullah Faruk Department of Computer Science and Engineering, Aliah University, Kolkata, India

Mondal Arnab Computer Science and Engineering Department, JIS University, Kolkata, India

Mondal Arpan Computer Science and Engineering Department, JIS University, Kolkata, India

Mukherjee Manona Computer Science and Engineering Department, JIS University, Kolkata, India

Mukherjee Sabyasachi Asansol Engineering College, Asansol, West Bengal, India

Muniyal Balachandra Manipal Institute of Technology, Manipal Academy of Higher Education, Manipal, India

Munna Imran Computer Science and Engineering Discipline, Khulna University, Khulna, Bangladesh

Nag Tirthajyoti Computer Science and Engineering Department, JIS University, Kolkata, India

Nagavelli Umarani Dayananda Sagar Research Foundation, University of Mysore (UoM), Mysore, Karnataka, India

Naik Sakshi A.P. Shah Institute of Technology, Thane, Maharashtra, India

Narad Priyanka Amity Institute of Biotechnology, Amity University Uttar Pradesh, Noida, Uttar Pradesh, India

Narayan Yogendra Chandigarh University, Mohali, India

Negi Ashish HMRITM, Delhi, India

Nikam Ujwal Computer Engineering, JSPM'S Rajarshi Shahu College of Engineering, Tathawade, Pune, India

Pathak Nitish Bhagwan Parshuram Institute of Technology (BPIT), GGSIPU, New Delhi, India

Pati Soumen Kumar Maulana Abul Kalam Azad University of Technology, Nadia, West Bengal, India

Patil Dhanshri Computer Engineering, JSPM'S Rajarshi Shahu College of Engineering, Tathawade, Pune, India

Pawar Varsha Assistant Professor, Department of Computer Applications, CMR Institute of Technology, Bengaluru, India;
Research Scholar, Department of Computer Science, Christ Deemed to be University, Bengaluru, India

Phowakande Sayali A.P. Shah Institute of Technology, Thane, Maharashtra, India

Poddar Prerana G. Department of Electronics and Communication Engineering, BMS College of Engineering, Bengaluru, India

Prakhar Pratyay Manipal Institute of Technology, Manipal Academy of Higher Education, Manipal, India

Pramanik Sabyasachi Department of CSE, Haldia Institute of Technology, Haldia, West Bengal, India

Puri Akshita Manipal Institute of Technology, Manipal Academy of Higher Education, Manipal, India

Raghudathesh G. P. Manipal School of Information Sciences, MAHE, Manipal, India

Rahul Reddy K. PES University, Bengaluru, Karnataka, India

Rahul Mayur Department of Computer Application, UIET, CSJM University, Kanpur, India

Rai Rakesh Kumar IPEC, Ghaziabad, India

Raj N. Vandana Symbiosis Institute of Computer Studies and Research, Pune, India;
Symbiosis International (Deemed University), Pune, India

Rajendran P. Selvi Department of Computer Science, Hindustan Institute of Technology & Science (Deemed To Be University), Padur, Chennai, Tamil Nadu, India

Raji M. Department of Mathematics, School of Basic Sciences, Vels Institute of Science, Technology and Advanced Studies, Chennai, India

Rajput Arjun A.P. Shah Institute of Technology, Thane, Maharashtra, India

Rajput Sudhir Kumar School of Computer Science, University of Petroleum and Energy Studies (UPES), Dehradun, India

Rao B. Dinesh Manipal School of Information Sciences, MAHE, Manipal, India

Roy Siddhartha The Heritage College, Kolkata, Kolkata, India

Sachdeva Nitin Krishna Engineering College, Ghaziabad, India

Sahu Devanh EY, Gurgaon, India

Sahu Santhoshini GMR Institute of Technology, Rajam, AP, India

Saini Jatinderkumar R. Symbiosis Institute of Computer Studies and Research, Pune, India;
Symbiosis International (Deemed University), Pune, India

Samanta Debabrata Dayananda Sagar Research Foundation, University of Mysore (UoM), Mysore, Karnataka, India;
Department of Computer Science, CHRIST University, Bangalore, Karnataka, India

Sandhu Amanpreet Kaur University Institute of Computing, Chandigarh University, Mohali, Punjab, India

Sarkar Prasanta Department of Electrical Engineering, National Institute of Technical Teachers' Training and Research, Kolkata, West Bengal, India

Sarkar Prithish Computer Science and Engineering Department, Future Institute of Engineering and Management, Kolkata, West Bengal, India

Selvakumar S. Department of CSE, Visvesvaraya College of Engineering and Technology, Hyderabad, India

Sengupta Abhishek Amity Institute of Biotechnology, Amity University Uttar Pradesh, Noida, Uttar Pradesh, India

Shambharkar Prashant Giridhar Department of Computer Science and Engineering, Delhi Technological University, Bawana Road, New Delhi, Delhi, India

Sharma Ankita Chandigarh University, Chandigarh, India

Sharma Hitesh Kumar School of Computer Science, University of Petroleum and Energy Studies (UPES), Bidholi, Dehradun, Uttarakhand, India

Sharma Neelam Maharaja Agrasen Institute of Technology (MAIT), GGSIPU, New Delhi, India

Sharma Nonita Department of Information Technology, Indira Gandhi Delhi Technical University for Women, Delhi, India

Sharma Ritika Amity Institute of Biotechnology, Amity University Uttar Pradesh, Noida, Uttar Pradesh, India

Sharma Sanchit HMRITM, Delhi, India

Sharma Shivangi Computer Science and Engineering Department, Meerut Institute of Engineering and Technology, Meerut, India

Sharma Tripti Chandigarh University, Mohali, India

Shelar Sakshi Computer Engineering, JSPM'S Rajarshi Shahu College of Engineering, Tathawade, Pune, India

Shetty Nisha P. Manipal Institute of Technology, Manipal Academy of Higher Education, Manipal, India

Shrivastava Yashi University of Petroleum and Energy Studies, Dehradun, India

Shukla Rati GIS Cell, Motilal Nehru National Institute of Technology, Prayagraj, India

Shukla Vinod Kumar Department of Engineering and Architecture, Amity University, Dubai, UAE

Sidharth G. Department of Information Technology, Thiagarajar College of Engineering, Tamil Nadu, Madurai, India

Singh Ajay Kumar Computer Science and Engineering Department, Meerut Institute of Engineering and Technology, Meerut, 250005 India

Singha Anjani Kumar Aligarh Muslim University, Aligarh, India

Singhania Aditya Vikram Department of Computer Science, The Bhawanipur Education Society College, Kolkata, India

Singh Angad Manipal Institute of Technology, Manipal Academy of Higher Education, Manipal, India

Singh Isha HMRITM, Delhi, India

Singh Jahanvi Manav Rachna International Institute of Research and Studies, Faridabad, Haryana, India

Singh Juhi Department of Computer Science and Engineering, Amity University, Gurugram, Haryana, India

Sinha Sitesh Kumar Department of Computer Science and Engineering, Rabindranath Tagore University, Raisen, Madhya Pradesh, India

Srivastava Akhilesh Kumar ABES Engineering College, Ghaziabad, India

Sumathi D. VIT-AP University, Vijayawada, Andhra Pradesh, India

Sumukh R. R. PES University, Bengaluru, Karnataka, India

Sur Monoj Kumar Computer Science and Engineering Department, Future Institute of Engineering and Management, Kolkata, West Bengal, India

Tabasum Muskaan Department of CSIT, Siksha 'O' Anusandhan (DU), Bhubaneswar, Odisha, India

Thomas Benny Department of Computer Science, CHRIST Deemed to be University, Bangalore, Karnataka, India

Tiwari Namita Department of Mathematics, School of Sciences, CSJM University, Kanpur, India

Tiwari Pradeep Kumar Manipal University Jaipur, Jaipur, India

Tripathi Khushboo Department of Computer Science and Engineering, Amity University, Gurugram, Haryana, India

Tripathy Sushreeta Department of CA, Siksha 'O' Anusandhan (DU), Bhubaneswar, Odisha, India

Tulika Department of Electronics and Communication Engineering, BMS College of Engineering, Bengaluru, India

Tyagi Nitin HMRITM, Delhi, India

Tyagi Rajesh Kumar Department of Computer Science and Engineering, Amity University, Gurugram, Haryana, India

Uddin Abdul Hasib Computer Science and Engineering Discipline, Khulna University, Khulna, Bangladesh

Varma Neha Department of Computer Science and Engineering, Ajay Kumar Garg Engineering College, Ghaziabad, Uttar Pradesh, India

Vasra T. R. Abijeeth Department of Information Technology, Thiagarajar College of Engineering, Tamil Nadu, Madurai, India

Verma Akhilesh Ajay Kumar Garg Engineering College, Ghaziabad, UP, India

Vijay Suresh G. V. Department of CSE, Lakireddy Bali Reddy College of Engineering, Mylavaram, India

Vinay S. Department of Computer Science and Engineering, Rabindranath Tagore University, Raisen, Madhya Pradesh, India

Yadav Jay Kant Pratap Singh Department of Computer Science and Engineering, Ajay Kumar Garg Engineering College, Ghaziabad, Uttar Pradesh, India

Yadav Sunita Department of Computer Science and Engineering, Ajay Kumar Garg Engineering College, Ghaziabad, Uttar Pradesh, India

Yadav Vikash Department of Technical Education, Kanpur, Uttar Pradesh, India

Yashika Manav Rachna International Institute of Research and Studies, Faridabad, Haryana, India

Yogadisha S. PES University, Bengaluru, Karnataka, India

Data Science and Data Analytics

IOT Security: Recent Trends and Challenges

Prachi Dahiya and Vinod Kumar

Abstract The use of Internet of Things (IoT) devices is increasing day by day along with the increased use of advanced technologies in almost every field. IoT provides technologies like Wireless Sensor Networks (WSNs) and Radio Frequency Identification (RFID) which are used in several applications like forest fires, home automation, military equipment, traffic signals, medical sciences, agriculture, etc. Large amount of information is being obtained by the sensors and then sent to data centers through the transceivers for analysis and decision-making purposes. Based on these decisions, the actuators perform a particular tasks based on the applications. This large amount of data can be used by a third party for malicious purposes, so there is a need for information security almost at every level of IoT. IoT is the kind of technology that involves several fields of computer science at step. With the usage of multiple technologies like Artificial Intelligence, Machine Learning, Data Mining methods, etc., there is a need to provide proper security to the IoT system also. Several layers used in the IoT architecture need to be protected. Hence, there is a need to make a robust system in order to deal with the existing vulnerabilities, threats, attacks, and privacy and security challenges existing in IoT system. This paper discusses the recent trends in the IoT architecture along with acknowledging the threats present in different architecture layers. Along with this, the security threats and challenges are also discussed.

Keywords Internet of Things (IoT) · Threats and security · Sensors

P. Dahiya (✉) · V. Kumar
Delhi Technological University (DTU), Delhi, India
e-mail: prachidahiya_phdco11@dtu.ac.in

V. Kumar
e-mail: vinod.kumar@dtu.ac.in

P. Dutta et al. (eds.), *Emerging Technologies in Data Mining and Information Security*, Lecture Notes in Networks and Systems 490,
https://doi.org/10.1007/978-981-19-4052-1_1

1 Introduction

IoT or Internet of Things is defined as a network of interconnected physical devices which share the information over a secured network. These devices contain sensors, actuators, and several hardware and software technologies [1]. Various kind of technologies are connected with IoT such as embedded systems, Wireless Sensor Networks (WSNs), mobile computing, control systems, automation systems (home and office automated devices), security systems (cameras, alarm systems), and Radio Frequency Identifications (RFIDs) [2]. When several devices are used in a particular order and they all work like one complete machine or an entity, then such systems are called as smart cities, intelligent automobiles, smart homes, virtual power plants, etc., where no human help is required [3].

Sensors and actuators are the physical devices that are used to collect data and then act on the data provided by the system. They gather the data from the Internet also. Then, this data is acquired by the data acquisition systems (DAS) [4]. Then this data is sent to the next part for the processing through Wi-Fi or wired LANs or some other communicating medium. IoT generates a large amount of data in any application where it is being used, and hence, this data needs to be squeezed into the optimal size for further analysis [5]. In the end, the enhanced data analytics is done and then this data is being preprocessed through various Machine Learning (ML) and visualization techniques.

The technology is changing at such a fast rate that it has made it difficult for the human capability to absorb all the things happening around [6]. There are numerous vulnerabilities of the IoT system due to the involvement of a number of technologies and the participation of several hardware and software components. Due to the existence of several layers in the IoT architecture, every layer has its own security threats and problems. From sensors identification to the data analytics in the cloud storage, at every step there is a threat residing in IoT. There are several types of attacks like jamming, spoofing attacks, identification threats, malware like viruses, Trojan existing in the system, Denial-of-Service attack (DoS), etc. At every level of an IoT architecture, there is a threat that exists to a particular device or component or to the data [7].

This paper discusses the architecture and several hardware and software components [8] of an IoT system. Also, an enhanced literature survey is done about the threats related to IoT. It also summarizes various IoT security and threat issues into different groups and provides some counter measures on how to deal with such problems. The remaining paper is organized as follows: Sect. 2 provides a thorough literature survey about the latest research in IoT, several developments in IoT as well as the threats related to it. Then, Sect. 3 discusses the architecture of IoT which contains several layers and how they work in an interconnected system of IoT. Section 4 categorizes different threats and security issues into groups and discusses about them. In the end, Sect. 5 concludes the paper.

2 Literature Survey

A detailed research is being done about the security issues in IoT as well as how the architecture of IoT has evolved in the past few years in order to provide better performance to the users. It also discusses several research perspectives in the field of IoT.

IoT basically interacts with the people at the virtual level; thus, it forms several relationships among people such as professional relationships and social relationships [9]. In today's world, everybody is using the IoT technology in some form or another, and hence, it gives the idea of 'anywhere, anytime, anyone, anything, any media' concept. This leads to an improved user management, technology optimization according to the needs of the customer, enhanced data collection, and several advantages. This also leads to several security threats to the devices that are connected over the Internet. There is also complexity in the maintenance and deployment of the sensor nodes and connecting users with the cloud and to the system.

IoT is evolving day by day and with the increasing research which leads to the fusion of several technologies with IoT. This also leads to increasing threats and attacks on the connected devices like sensors which are associated with the system [10]. Also there is a threat to the data that is shared among devices over the Internet. Every layer present in the architecture deals with their own security problems existing in them. The attacks are shown in Fig. 1.

IoT not only provides several applications in order to communicate with the users but also provides users with several communication opportunities where the users can easily communicate with each other [11]. The information communicated over the Internet among the users is confidential at times, and hence, IoT provides with the information security techniques to protect the data of the users. It discusses the challenges faced by the IoT system related to the information security where

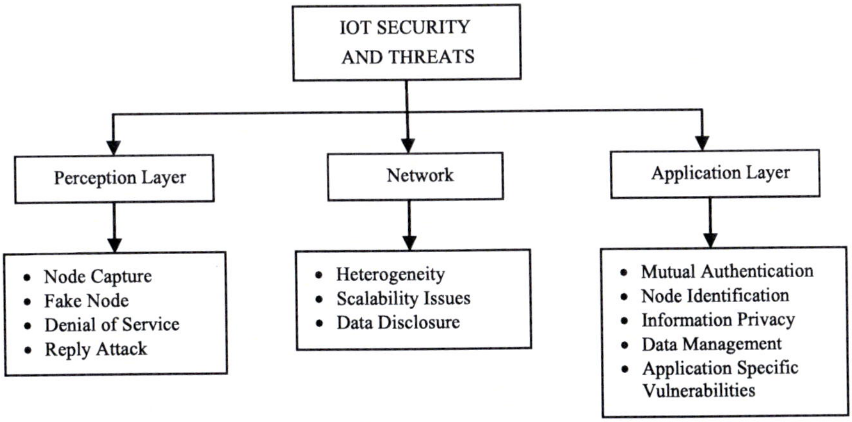

Fig. 1 Security threats of IoT

it provides some techniques to protect the shared data over the Internet among the users.

Computer-Aided Design (CAD) is a technology that helps the IOT systems in creation, manipulation, and modification of the design of the system [12]. Communications, documentations, and everything related to the design of the system are improved through CAD. There are several variations of CAD such as EDA (Electronic Design Automation) and MDA (Mechanical Design Automation). CAD helps IOT in analyzing the new and the latest technological trends used by the IOT systems. CAD helps in making more reliable IOT systems, and also, it helps in changing the designs of technological and engineering systems.

Data dissemination techniques are very important part of the IOT systems. Several dissemination techniques for latest IOT systems like Internet of Vehicles (IoVs), Internet of Drones (IoDs), and Internet of Battle Fields (IoBTs) are discussed in [13]. These dissemination techniques help in providing the security, confidentiality, privacy, authentication, authorization, and anonymity of the data disseminated through different devices. Hence, there is a need for secured data dissemination techniques such as efficient data aggregation techniques and cluster-based data dissemination. Several research opportunities are discussed in this paper, and there is need to further dig into this problem.

3 IOT Architecture

There is no specific architecture of IOT. Every system that uses IOT describes the architecture of IOT differently according to their needs and benefits. Different companies like IBM, Oracle, CISCO, etc., provide their own architecture for IOT [14, 15]. Basically, different layers have different purposes in the architecture starting from sensors and actuators to the data analytics part. Figure 2 shows a systematic architecture of IoT.

(1) Perception Layer: Sensors are used to collect the data and then convert that data into useful information such that it can be used for further analysis in the succeeding layers. Smart sensors are also used here which are mainly used for the computing and communication purposes. The type of sensors to be used mainly depends on the needs of the system like they can use analog or digital sensors based on the type of output they require, and scaler or vector sensors for the numerical or a directional output.

    There are various threats that are present in the perception layer such as node capture, Denial-of-Service attacks (DoS), fake node problems, and reply attack. In this layer, the attack is mainly on the nodes which might lead to the leakage of confidential information which lead to the security issues in the network. When a node is captured, it becomes malicious which can easily circulate the malicious codes to other devices which are connected over the Internet, thus infecting the whole IoT system.

Layer 1	Layer 2	Layer 3	Layer 4	Layer 5
Perception Layer (Sensors, Actuators, Smart Sensors, Wired/Wireless Devices, RFID tags)	**Transport Layer** (Internet Gateways, Data Acquisition/ Aggregation Systems, End-to-end data transfer)	**Network Layer** (Data Transfer Protocols- UDP, TCP, Routers, IPv4, IPv6, Mobile WiFi, Zigbee	**Data Processing Layer** (Edge IT analytics, Data Analysis Tools, pre processing)	**Application/Business Layer** Data Center/ Cloud Data, analytics. management, archive)
Control + Gather +Acquire	Enrich+ Stream+ Consolidate	Connect +Manage+ Transfer	Collect+ Assemble	Organize + Analyze

Fig. 2 IOT Architecture

(2) Transport Layer: The main purpose of this layer is to send and receive three important information from the perception layer. Transport layer acts as a communication medium that sends and receives data in the form of signals to the further layers. Several subsystem managements are used over here like the device management, access management, identity management, etc. These all subsystems receive the device's data. There are large number of transporting devices in this layer, and the vulnerability of these devices also increases and data can easily be stolen by the attacker.

(3) Network Layer: Several Data Acquisition Systems (DAS) are used here which connect the sensor network and then aggregate the output results. Many Internet gateways are used to monitor the data. They work through Wi-Fi and wired LANs and perform the further processing. Enormous amount of information is collected, and then, it is squeezed into the optimal size for the further analysis. Several conversions are made over here in terms of time and space. Network layer contains a lot of security threats like data integrity, data confidentiality, and availability.

(4) Data Processing Layer: Data collection, abstraction, accumulation, and aggregation are done in this layer. Although the data is already being squeezed and is converted into a simpler understandable format, this layer uses some enhanced analytics and preprocessing with the help of Machine Learning (ML) to make data more concise and efficient. Edge computing is an effective computing paradigm that mainly saves the data closer to the actual location where data is to be needed, and in this way, it also saves the bandwidth that is needed when the data is stored too far away from the actual location.

Large amount of data is present in the devices which work online, and hence, it becomes easy for the attackers to get access to this data. The attackers can easily leak this data which contains confidential information of the users such as their passwords and credit/debit card details. In the data processing layer, many auditing tools are used to provide the safety of the data.

(5) Application/Business Layer: The data which is routed from the devices and things is organized over here. It also includes collaboration from the people and business processes. An in-depth processing and a follow-up revision are done for getting a good feedback. The skills of IT and OT (Operational Technology) are required over here. Data not only from the sensors but also from the other devices can be included here in order to ensure a thorough analysis.

Node identification and authentication is an important security aspect in the application layer. This also helps in the security for the users and the information privacy. The data of the users is at a vulnerable area, and effective measures need to be taken in order to secure this data. Huge collections of data also pose a threat, and this data needs to be brief, concise, and accurate.

4 IOT Security Challenges

(1) IOT and Authentication

IoT authentication is basically a trust-building mechanism among the IoT devices and machines, and it is used to gain the proper access and control of the online data that is being transferred among the devices over a network. Authentication is a major aspect of IoT, and hence, authentication faces a lot of vulnerabilities with the security provisions in many applications. The authentication's main purpose is to protect the confidential data from going into the wrong hands but at one time, it protects the entire system from only one threat or vulnerability such as reply attacks or the Denial-of-Service attacks (DoS).

Cryptography has become an essential method in the past few years in order to secure the system which face the authentication issues and provide the best possible defense cryptographic keys and algorithms against the security loopholes present in the IoT applications. To create an effective defense mechanism against the prevalent attacks, one must build layered model with different layers to prevent the threat from penetrating into the systems. Hence, efficient cryptographic solutions are used by several companies who share their data online with their different systems and machines.

(2) IOT and Big Data

Since the usage of big data is done in the IoT devices, big data has created several problems which are results of increased online data which is sent and received among the IoT devices over the Internet. This confidential data is stored in the cloud that is present online, and it becomes easy for the attackers to hack this data. If the information is present on the local server, then it becomes easy to secure the data, but large amounts of data present online which are being shared by hundreds of devices and communicated over hundreds of channels cannot be secured easily.

Large number of security professionals is deployed to manage the growth of the online data such that the outgoing sources and the ingoing sources remain safe from the threats and attacks. The market of the online data security is fastly growing due to the increased threats. Their main emphasis is on the proactive data systems which integrate the data and provide router functions, network configurations, etc., to protect the online data. Data auditing can also be done by the third parties in order to ensure that the data is secured before sending to other parties.

(3) IOT and User Challenges

Data theft is the biggest security issue with the users who work on IoT devices as their personal and private information can easily be leaked by the attackers. Online purchasing records, browser histories, credit/debit card details, and username passwords can easily be hacked if a user is working online on their device. There are vulnerable devices which contain improperly secured vast amounts of data, these devices can be used by the attackers as gateways to enter into other secured devices also, and in this way, more sensitive information can be extracted. Sometimes, some devices are fit inside the body of the patients such as a pacemaker and heart monitors which can be controlled remotely by a doctor. So, if a user's medical records are leaked then one can harm them.

(4) IOT Malware

Malware is malicious software which is used to gain access and damage a system or a device. IOT technology works fully on the Internet, and hence, it is more vulnerable to the malware attacks as they are always online and lack security. One must know that which malware attacks your devices suffer, so that a detection algorithm can be used to locate the malware on the device. Botnets, Rootkit, Keylogger, Worms, Trojan horses, Ransomware, etc., are some of the most dangerous malware used to hack the IoT devices and steal the confidential data.

Several detection and analysis techniques are used in order to know which malware has attacked the system and to know what is the type of malware, its features, and what family it belongs to. Detection techniques include signature-based, heuristic-based, and specification-based techniques.

5 Conclusion

This paper discusses the recent trends in IOT and how it affects the current research scenario; different literatures have been studied in the literature review part focusing on the different techniques that are currently being used. One can see that the different layers of the architecture are exposed with different threats and challenges. IoT has provided with a promising techniques to share their data and use IoT systems in such a way that everything is automated and connected in today's world. This helps in making people's lives better and easier, improvement in the decision-making skills, etc. Today IoT works with several technologies like Big Data, Wireless Sensor Networks (WSNs), and these technologies help in making the work of IoT's systems faster and easier such that everyone can benefit from them. Although IoT has broaden the research areas for the people and has made difficult tasks easier for the people but the threats and vulnerabilities remain the same. Several scientists and researchers are still working toward a better IoT environment and engineer more complex security methods to protect the systems. The future work of this study will focus on a comprehensive review study of the recent literature with a review of these techniques.

References

1. Xiao L, Wan X, Lu X, Zhang Y, Wu D (2018) IoT security techniques based on machine learning. In: Signal processing and the Internet of Things, IEEE Signal Processing Magazine, 28 Aug 2018
2. Tawalbeh L, Muheidat F, Tawalbeh M, Quwaider M (2020) IoT privacy and security: challenges and solutions. Appl Sci, 15 June 2020
3. Boehm E (2020) The top Internet of Things (IoT) authentication methods and options, September 2020
4. Liao B, Ali Y, Nazir S, He L, Khan HU. Security analysis of IoT devices by using mobile computing: a systematic literature review. IEEE Access
5. Imran MA, Zoha A, Zhang L, Abbasi QH (2020) Grand challenges in IoT and sensor networks. Front Commun Netw
6. Romana R, Lopeza J, Mambo M (2018) Mobile edge computing, fog et al.: a survey and analysis of security threats and challenges. Fut Gener Comput Syst 78(3):680–698
7. Selvaraj SK, Sundaravaradhan S (2019) Challenges and opportunities in IoT healthcare systems: a systematic review. SN Appl Sci 2. Article number: 139
8. Bodkhe U, Tanwar S (2020) Secure data dissemination techniques for IoT applications: research challenges and opportunities. Wiley Online Library, February 2020
9. Tiwary A Dr, Mahato M, Chidar A, Chandrol MK, Shrivastava M, Tripathi M. Internet of Things (IoT): research, architectures and applications. Int J Fut Revol Comp Sci Commun Eng 4(3):23–27. ISSN: 2454-4248
10. Tewari A, Gupta BB (2018) Security, privacy and trust of different layers in Internet-of-things (IoTs) framework. Fut Gener Comp Syst
11. Zhang ZK, Cho MCY, Wang CW, Hsu CW, Chen CK, Shieh S (2014) IoT security: ongoing challenges and research opportunities. In: IEEE, fellow 7th international conference on service-oriented computing and applications
12. Xu T, Wendt JB, Potkonjak M. Security of IoT systems: design challenges and opportunities
13. Bodkhe U, Tanwar S (2020) Secure data dissemination techniques for IoT applications: research challenges and opportunities
14. Yao X, Farhaa F, Lia R, Psychoulab I, Chenb L, Ning H (2020) Security and privacy issues of physical objects in the IoT: challenges and opportunities. Digi Commun Netw. Science Direct, September 2020
15. Choo KKR, Yan Z, Meng W (2020) Editorial: blockchain in industrial IoT applications: security and privacy advances, challenges, and opportunities. IEEE Trans Ind Inform 16(6):4119–4121. ISSN: 1551-3203

Assistive Technology for Pedagogical Support and Application of Data Analysis in Neurodevelopmental Disability

Arpita Mazumdar, Biswajoy Chatterjee, Mallika Banerjee, and Irfan H. Bhati

Abstract Educational data mining (EDM) has been adapted in a variety of higher educational contexts. It can also play a significant role in the domain of special education of autism spectrum disorder (ASD) which is the third most common life-long neurodevelopmental disorder. Once in special school, children with ASD need user-friendly instructions and teaching equipments. A social and cultural appropriate educational intervention application may ease the process of learning. It may help to monitor the progress, to identify the learning pattern, and also to offer parental training. Moreover, machine learning models may be explored to diagnose the presence of autism in a child. Also, educators may be supported with educational assessment tool in form of mobile application to measure the performance parameters and thereby to observe and/or evaluate the learning progress of the child accurately and time efficiently. Computer-assisted pedagogical support may be finely tailored to serve children as well as educators across the disability spectrum in special education setup. In this paper, the authors aim to review the major existing research works carried out in this field of EDM at international and national levels. They also have proposed and implemented an integrated autism management platform and presented a brief overview of their ongoing project work.

Keywords Educational data mining · ASD · Machine learning · Educational intervention

A. Mazumdar (✉)
Haldia Institute of Technology, Haldia, India
e-mail: arpitamazumdar.hit@gmail.com

B. Chatterjee
University of Engineering and Management, Jaipur, India
e-mail: biswajoy.chatterjee@uem.edu.in

M. Banerjee
Techno India University, Kolkata, India

I. H. Bhati
Google India, Bangalore, India

P. Dutta et al. (eds.), *Emerging Technologies in Data Mining and Information Security*, Lecture Notes in Networks and Systems 490,
https://doi.org/10.1007/978-981-19-4052-1_2

1 Introduction

With the advent of technology, the use of Internet in education domain has created a new paradigm popularly known as e-learning through online mode in which huge amount of data about teaching–learning interaction are countlessly generated. From which meaningful interpretation is derived and in turn becomes useful to the education system. All these information provide an abundant source of educational data [1].

Educational data mining (EDM) describes a research field concerned with the application of data mining; machine learning, and statistics of information generated from educational settings. The application of EDM helps to analyze student learning process considering their interaction with the environment [2]. Several EDM methods and tools have been applied [3, 4] in mainstream educational system. Few common methods in EDM are enlisted in Table 1.

Special education sector may also be explored with EDM approach. According to the census 2011, in India, out of the 121-crore population about 2.68 crore persons suffer from disability. Autism spectrum disorder (ASD) is the third most common lifelong neurodevelopmental disorder that manifests within the first 3 years of the child. Children with ASD reveal unusual learning abilities and disabilities. In the disability spectrum, about 1% of the global population experience autism spectrum disorder (ASD). As "inclusive education" is considered as the right path toward sustainable development [5] dedicated initiatives for the welfare of persons with disability are indispensable.

Table 1 Common methods in EDM

Method	Target	Key applications
Prediction	To infer an output variable from some combination of other variable. Classification, regression, etc., are types of prediction methods	Predicting academic performance and detecting student behavior
Clustering	To identify groups of similar observations	Grouping students with similar learning and interaction patterns
Relationship mining	To study association among variables and to formulate rules. Association rule mining, correlation mining, sequential pattern mining are the major types	Identifying the relationships in learning behavior patterns and determining student difficulties
Distillation of data for human judgment	To represent data in comprehensible ways using summarization, visualization, and user-friendly interfaces	Assisting instructors to visualize and analyze the learning progress of the students

This research report describes few research-based applications for autism with a focus to add easiness in diagnosis process, to prepare a structured assessment procedure, and to make the educational intervention more effective. The report is ordered as follows.

Section 1 is the brief introduction of educational data mining and their common methods applied; Sect. 2 describes some significant work done in the domain for autism; Sect. 3 focuses on the problems faced by the beneficiaries in Indian society, wherein Sect. 4 illustrates a project work on an integrated autism management platform. Finally, future lines of the project work, relevance of the project in society, and policymaking are outlined in subsections A, B, and C, respectively.

2 Related Works

Some pioneer work in special education domain has been documented in Table 2.

3 Problems and Challenges Faced in Indian Society

- Lack of awareness of the problems encountered by a child with autism in both urban and rural areas.
- Inadequate knowledge of assessment procedure of a child with autism in both urban and rural areas.
- No "ALL INCLUSIVE" autism management platform available.
- Human prediction of future learning trends of a child with autism may be biased (not supported by data analysis).
- Existing application used/developed in foreign countries may not be socially and culturally appropriate.

4 An Integrated Autism Management Platform

In the present era, "inclusive education" is considered as the right path toward sustainable development as every student can learn just not on the same day or in the same way. Hence, initiatives for the welfare of persons with disability are absolutely necessary. Particularly, in India this field demands specialized research as the work in this field is very low.

The authors in their present research aim to develop an integrated autism management platform by which educating as well as controlling challenging behavior of the children with autism would be more manageable.

Table 2 A short review

Paper title	Abstract	Objective	Methodology	Result
"TOBY: Early Intervention in Autism through Technology—Venkatesh et al. Deakin—Curtin University, Australia., 2013 [6]".	Therapy outcomes by you (TOBY) a play pad application to teach students with Autism. It helps to monitor the rate of development and promotes parent education.	This early intervention application aims to act as a technical support during the stressful time between diagnosis and commencement of formal therapy and also during therapy itself.	TOBY is based on applied behavior analysis (ABA) therapy. The syllabus comprises of 52 foundation skills across four learning areas. It includes solo-on-screen activities, partner on-screen activities, natural environment activities. (NET tasks)	Based on the three trials, all participants could be assigned to three groups showing some evidence of learning. The children made successful task attempts, successful attempts with a high, medium, and low levels of prompting and failed attempts. NET tasks were both acceptable to parents and children.
"TOBY play pad application to teach children with ASD—A pilot trial—Dennis W Moore et al. Monash University, Melbourne Australia., 2015 [7]".	Therapy outcomes by you (TOBY) as an early intervention programming for children with ASD was acceptable to children and appreciated by parents.	To study the usage patterns and learning outcomes related with the use of TOBY play pad.	33 families having a child with a diagnosis of autism or pervasive developmental disorder were allowed to use TOBY for 4–6 weeks.	23 participants were engaged extensively and completed 17% to 100% of the curriculum.
"Data Mining of Intervention for Children with Autism Spectrum Disorder—Vellanki et al. Deakin University Australia, 2017 [8]".	Qualitative analysis and mapping of the progress of children with ASD.	Help analyze and build possible techniques toward personalized intervention for children with ASD.	Restricted Boltzmann machine model—mix variate modeling approach.	Observed how groups of similar features at the onset of the intervention responded in different manner to the same syllabus.

(continued)

Table 2 (continued)

Paper title	Abstract	Objective	Methodology	Result
"A Novel Approach to Predict the learning skills of Autistic children using SVM and Decision tree—MS Mythili et al. 2014 [9]".	Studied the problem faced in Autism, detect the degrees of autism with the help of data mining classification algorithm.	Design a classifier to classify the levels of autism based on performance of children with autism.	Based on various factors of learning skills like handwriting, spelling, read, write, language skill levels of autism are predicted using decision tree (J48) and support vector machine (SVM). The process was implemented in WEKA tool.	SVM better classifier than decision tree takes less time with better accuracy.
"Comparison of classification techniques for predicting the performance of students academic environment—M. Mayilavaganan et al. 2014 [10]".	Comparative performance of c4.5 algorithm, AODE, Naïve Bayesian classifier algo, multilabel—K-nearest neighbor algorithm to find the best-suited accuracy of classification.	Analyze the sample qualitative dataset collected from various educational Institutions.	c4.5 algorithm, Naïve Bayesian Classifier algo, AODE, multilabel—K-nearest neighbor algorithm.	Multilabeled—K-nearest Neighbor algorithm reported to produce the best accuracy.

Thorough field inspections have been carried out in several special schools, autism care centers. The conventional practice in the special education setup observed is as follows:

- When a child joins a special school for autism with some complaints, his/her activities are thoroughly observed by the special educators for few days (3–5 days). Student's cognitive development, behavior patterns, social development, etc., are evaluated as per the guideline of standard autism screening scales.
- To find the strength and weakness of a child various standard educational assessment tools are considered, namely Indian Portage guide developed by CBR network, behavioral assessment scales for Indian Children with mental retardation (Basic-MR), Functional Assessment Checklist for Programming (FACP) (Part A/B), etc.
- Individualized Education Plan (IEP) is designed by the educator for each student after baseline assessment.
- Student is supported with appropriate teaching–learning material.
- Repeated evaluation of their learning progress is carried out and revised IEP is outlined.

Students with autism may often show some unusual and challenging behaviors. And no single method of intervention works for all [11]. Therefore, it is necessary to develop a systematic plan for each student. It is even challenging for an educator to track his/her learning progress as it is time-consuming and tiring too. When teachers and other assistants are aided with proper information and technical support, they are better equipped to deal with challenging behavior.

An autism management platform projected by the authors aims to assist the special instructors, psychologists, and parents during the crucial period of diagnosis and early educational intervention. At first, the application software (in form of mobile application) screens a child as per the degree of autism. The next step is to assess him/her with appropriate educational assessment tool. The assessment application helps the instructor to visualize and analyze the learning progress data periodically which helps them to frame Individualized Education Plan (IEP) further. The basic educational intervention application based on pre-academic activities acts as a good teaching–learning material. The children and their parents can easily interact with such environments and learning becomes fun. It also eases the process of continuous monitoring and also helps to identify the learning pattern.

The workflow of the ongoing project is depicted in the following diagram (Fig. 1).

About 120 students have been screened by the autism screening tool application which is based on "Indian Scale for Assessment of Autism (ISAA)" [12]. Thirty-one of them have received various types of educational assessment with the help of assessment application. The comprehensive way of representation of the performance data using summarization, visualization, and user-friendly interfaces helps the instructors, psychologists to visualize and analyze the learning activities of the students. Three software trials have been conducted among the students with autism of age group 3–13 years. The impression of their learning trajectory is recorded, and subsequently, learning progress profile is framed for each one.

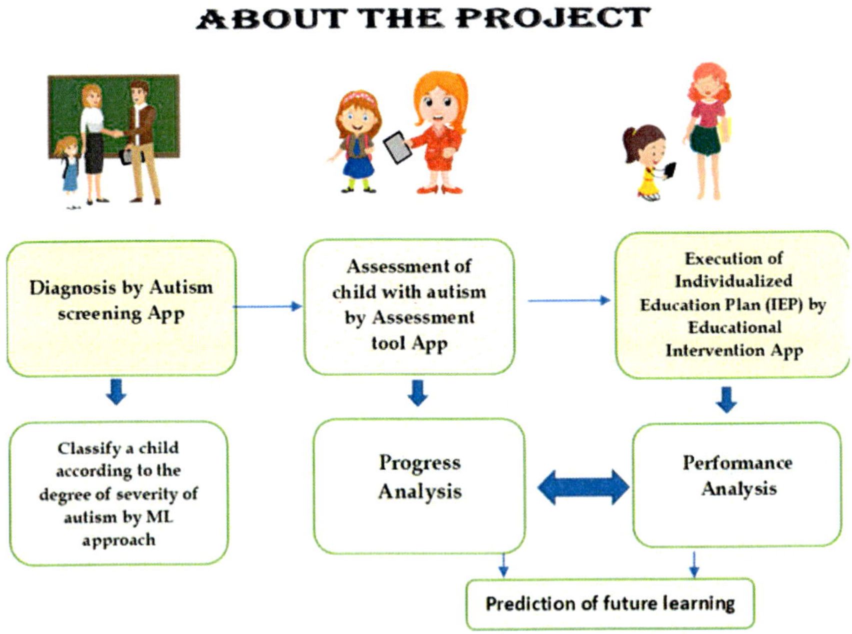

Fig. 1 Workflow of the proposed methodology

5 Projected/Future Work

The performance data of a child are to be analyzed for at least 3 years to measure ones learning progress at regular interval and also to facilitate the educator to develop tailored/customized intervention program. Subsequently, the pre-intervention profile is to be mapped to post-intervention profile (progress profile). It may help to predict an appropriate intervention path so as to orient the student to overcome their developmental inconsistencies. This study also aims to identify or predict suitable vocation of a student using machine learning approach based on performance and pre-vocational data.

A. Relevance of the project for society

- Build awareness about autism among rural as well as urban areas.
- Readiness for "inclusive education."
- Encourage parent to be a support system of their special child.

B. Relevance of the project for Policy making

The project outcome will facilitate the implementation of (RPWD Act 2016 and UNCRPD 2008) towards inclusive teaching–learning approaches for individuals with neurodevelopmental disabilities in an inclusive classroom settings and also boost up their self-confidence to be self-dependent.

Acknowledgements The authors express their sincere gratitude to the instructors, the children (Pradip Centre for Autism Management), and their parents who voluntarily participated in this research study and made the research possible.

References

1. Mostow J, Beck J (2006) Some useful tactics to modify, map and mine data from intelligent tutors. J Nat Lang Eng 12(2):195–208
2. Baker RSJD, Costa E, Amorim L, Magalhães J, Marinho T (2012) Mineração de Dados Educacionais: Conceitos, Técnicas, Ferramentas e Aplicações. Jornada de Atualização em Informática na Educação 1:1–29
3. García E, Romero C, Ventura S, de Castro C (2011) A collaborative educational association rule mining tool. Internet Higher Educ 14:77–88
4. Calvet Liñán L, Juan Pérez AA (2015) Educational data mining and learning analytics: differences, similarities, and time evolution. RUSC. Univ Knowl Soc J 12(3):98.https://doi.org/10.7238/rusc.v12i3.2515
5. Disabled persons in India a statistical profile 2016. Social Statistics Division, Govt. of India. Available from http://www.mospi.gov.in
6. Venkatesh S, Phung D, Duong T, Greenhill S, Adams B (2013) TOBY: early intervention in autism through technology. In: CHI 2013: changing perspectives: proceedings of the 31st annual conference on human factors in computing systems. Association for Computing Machinery, New York, N.Y., pp 3187–3196. https://doi.org/10.1145/2470654.2466437
7. Moore DW, Venkatesh S, Anderson A, Greenhill S, Phung D, Duong T, Cairns D, Marshall W, Whitehouse AJO (2015) TOBY play-pad application to teach children with ASD—a pilot trial. Dev Neurorehabil 18(4):213–217. https://doi.org/10.3109/17518423.2013.784817
8. Vellanki P, Duong T, Phung D, Venkatesh S (2016) Data mining of intervention for children with autism spectrum disorder. In: 2017 International summit on eHealth 360°, 2016—Budapest, Hungary, Duration: 14 Jun 2016 to 16 Jun 2016
9. Mythili MS et al (2014) A novel approach to predict the learning skills of autistic children using SVM and Decision tree. Int J Comp Sci Inf Technol 5(6):7288–7291
10. Mayilavaganan M et al (2014) Comparison of classification techniques for predicting the performance of students academic environment. In: International conference on communication and network technologies (ICCNT).https://doi.org/10.1109/CNT.2014.7062736
11. Teaching students with autism: resource guide for schools. Ministry of Education, British Columbia, 2000, pp 27–56
12. Ministry of Social Justice and Empowerment, Government of India 2009, New Delhi. ISAA. Report on assessment tool for autism: Indian scale for assessment of autism. Untitled-1 (thenationaltrust.gov.in)

Developing Smart ML-Based Recommendation System

Sakshi Naik, Sayali Phowakande, Arjun Rajput, Apeksha Mohite, and Geetanjali Kalme

Abstract Sometimes, music plays an important role in our life. Whether you are sad or happy, music plays an important factor as it expresses your mind, also the importance of music in your life will depend on your personal experience. This recommendation system is developed for those users who express their feeling and prefer listening as well as viewing music videos depending on their choice. The recommendation system will filter out the contents depending upon user choice with similar data. For this recommendation system, two techniques are used which are collaborative filtering and content-based filtering. Considering the issues for some users while searching, they can play music with help of a voice assistant.

Keywords Content-based filtering · Collaborative-based filtering · Hybrid filtering · Recommendation system · Music app · Smart system · Machine learning · Video recommendation · Voice assistant

1 Introduction

For some people, music has an important role in life. Music is one of the solutions for many problems as it enhances mood, can make excited, can also feel relaxing and

S. Naik (✉) · S. Phowakande · A. Rajput · A. Mohite · G. Kalme
A.P. Shah Institute of Technology, Thane, Maharashtra, India
e-mail: sakshinaik@apsit.edu.in

S. Phowakande
e-mail: sayaliphowakande@apsit.edu.in

A. Rajput
e-mail: arjunrajput@apsit.edu.in

A. Mohite
e-mail: atmohite@apsit.edu.in

G. Kalme
e-mail: grkalme@apsit.edu.in

P. Dutta et al. (eds.), *Emerging Technologies in Data Mining and Information Security*, Lecture Notes in Networks and Systems 490,
https://doi.org/10.1007/978-981-19-4052-1_3

calm [1–4]. Basically, it helps you from reducing stress, depression, pain. Considering the rapid development of mobile devices and the Internet has made it possible to get more closer to the music by music player system. The reason behind most portable music systems is that music can be played whenever and wherever. The increase in the number of music available exceeds the listening capacity of a single individual. Therefore, it is sometimes difficult to choose from millions of music. The solution for this issue is that there should be a good recommender system that can provide user music recommendations [5–7]. This system will basically recommend music to users according to analyzing the most popular, highly rated, and users preference, and this will help users to get personalized results. By considering all factors, we are developing a smart recommendation system that will be convenient for music listeners as well as there would be an increase of audience in the music field. This smart recommendation system is an advancement to the basic music player as it can recommend music to users by personalizing according to their choice and demands. Besides recommending music audio, this system also provides a recommendation of videos for 'now playing' music, i.e., while listening to the music, the music player screen while also provides a recommendation of music in video format. To make it more convenient, voice assistants would be integrated with this smart recommendation system to have a hand-free experience. This system even helps out those users who are unable to type manually by using voice assistant. Voice assistant is basically implemented to have a hands-free experience for users. Voice assistant would also result to recommend music on basis of the user's request.This would be beneficial for those are busy in other activities and wish to have music played simultaneously. In this recommendation system, the resulting recommendation is be provided based on audience preference or request. Besides this, the system will also recommend music based on categories like latest music, top hit music that will introduce users to new music. For processing, this recommendation machine learning algorithms need to implement. This algorithm helps to instruct the data set and provide desirable output depending upon the specific algorithm used. The most common approach toward recommendation systems has been the content-based technique and collaborative technique. Most of the recommendation systems use collaborative filtering techniques as recommended music based on a community of users, their preferences, and their browsing behavior. Whereas in content-based filtering is a technique in which music is recommended on the basis of users' similar preferences and by knowledge accumulated of the user by studying users' recent played, favorite music, search music, etc. Besides recommending music, this system will also provide video recommendations, i.e., while playing music, the system will also provide video option recommendations which will be useful for those users who want to play videos instead of audio, this can be implemented by navigating users to another page which will view video content. Another feature of the smart recommendation system is integrating the system with voice assistants. Voice assistant will also act as an interface for some users [8–11]. This feature would be useful for those users who are willing to work simultaneously while playing music, which means that requesting voice assistant through command would give results. Basically, voice assistance is integrated to have a hand-free experience.

2 Objective

2.1 *Compatibility*

To develop a cross platform application, i.e., developing single application that can be run on different operating system.

2.2 *Feasibility*

To build a hand-free mobile application by integrating it with voice assistant, which can make application more convenient.

2.3 *Regularity*

To keep a track of frequently played music by user.

2.4 *Usability*

To provide recommendations based on recorded information of users' preferences and suggesting video link of played music so that even videos can be watched.

2.5 *Serviceability*

To deliver a set of playlist from analyzing the current and future popularity of music, artist, and genres.

3 Literature Review

3.1 *Video Recommendation System Based on Human Interest*

In year 2019, authors [12] have published paper video recommendation system based on human interest. This system is develop for teenager user who are likely to watch videos on mobile. This paper proposes a video recommendation system that collects the reaction of the users for various videos which helps to know its relevance. Based on the viewers' watching history or browsing, the system is capable of recommending videos to the users. In this system, hybrid system is used which combination of content base filtering and collaborative filtering.

3.2 Music Recommendation Using Collaborative Filtering and Deep Learning

In year 2019, Anand Neil Arnold and Vaira Muthu S have published paper music recommendation using collaborative filtering and deep learning. This music recommendation system recommends user music as well as videos depending on user preference. Here, collaborative filtering is used in which existing history of the user and recommends music from other user's history which are similar.

3.3 Artificial Intelligence-Based Voice Assistant

In the year 2020, S Subhash, Prajwal N Srivatsa, S Siddesh, A Ullas Santhosh B publish paper artificial intelligence-based voice assistant. This system is basically an intelligent personalized assistant which can perform mental tasks like turning on/off smartphone applications with the help of the voice user interface (VUI) which is used to listen and process audio commands. For this, the PyCharm library was installed from Python packages. As a result, this system gives output on voice commands like playing songs on video, searching any location, and google search output.

3.4 Music Video Recommendation Based on Link Prediction Considering Local and Global Structures of a Network

In this literature, the authors Yui Matsumoto; Ryosuke Harakawa; Takahiro Ogawa; Miki Haseyama publish the paper in the year 2019 which was music video recommendation based on link prediction considering local and global structures of a network in this system, and they have implemented a novel method based on LP-LG SN for recommending music and videos. In this, they have the construction of a network by collaborative use of the multi-model feature. As a result, it can work well in real-world applications. In future, this application will introduce a framework to fuse prediction which can control the effect of local and global structure-based.

3.5 *FARMER'S ASSISTANT Using AI Voice Bot*

In the year 2021, authors Kiruthiga Devi M; Divakar M S; Vimal Kumar V; Martina Jaincy D E; Kalpana R A; Sanjai Kumar R M published a paper titled FARMER'S ASSISTANT using AI Voice Bot. The main purpose of this application is to develop a mobile application that can assist farmers depending on two techniques voice bot and suggestion bot. The multi-language response was generated depending on the farmer's queries. These queries were responded to by a multi-linguistic bot which was implemented using Google translator,pysttsx3, and Google search engines. This mobile application can improve increase in agriculture production and suggest farmer for progress in better farming practice.

4 Problem Definition

- Music industry has experienced a boom in recent years due to the rapid increase in music listeners.
- The number of music available exceeds the listening capacity of a single individual.
- It is sometimes difficult to choose from millions of music. However to manage, this user needs a recommendation system which can help their user to introduce new music by giving quality of recommendation.
- Our system is developed as a music recommendation system that can give recommendations based on similarity and rating features.
- Along with the music recommendation, a video link will also be provided for those users who are willing to even watch music in video format.
- To make more innovative, voice assistants are integrated which offers a productive and personalized experience for users (Fig. 1).

5 Proposed System Architecture

The main goal of our application is to recommend users with the latest, preferred, and previously played music along with the video link. This is can be implemented by applying machine learning filtering algorithms which are collaborative and content-based filtering. This algorithm provides music based on user history and by collecting other user preferences. The following are the modules which are been considered while implementing this application.

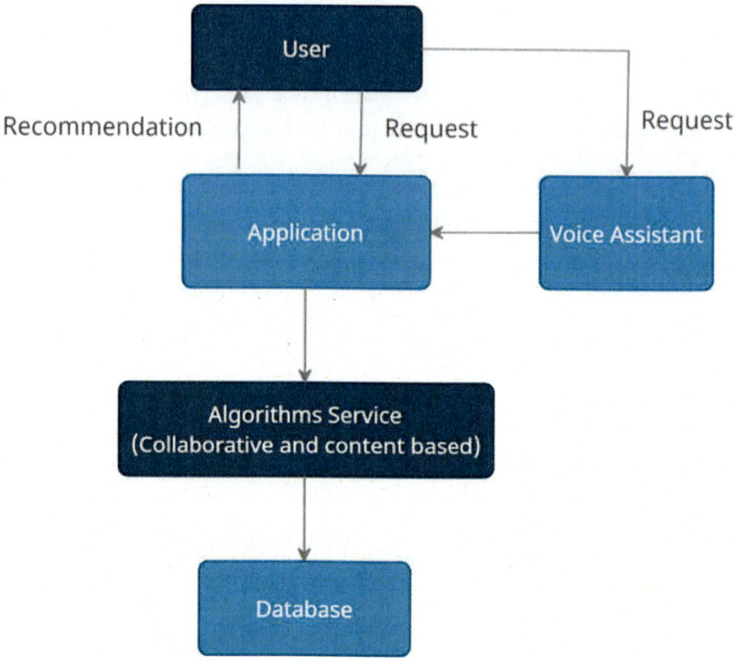

Fig. 1 Proposed system architecture

5.1 User

The user module is the targeted module that will request recommendations by interacting with the application or with a voice assistant to get music recommendations.

5.2 Application

This module is the main interaction with the user module which consists of a main application wherein music is recommended and played according to the user. This is a module where the user interacts the most to get recommendations.

5.3 Algorithms Service

This module consists of a machine learning mechanism which will recommend music by using algorithms like content-based and collaboration filtering.

5.4 *Database*

This module consists of a collection of user details and a music playlist which would be pushed toward user's dashboard depending upon the algorithm.

5.5 *Voice Assistant*

This module is implemented to perform hand-free use of application wherein user can command to assistant and assistant future send the request to the application.

6 Summary

By applying the knowledge and skill set, we are determined toward building a completely user interactive system that would be useful for every music listeners. This project will be implemented as a cross-platform application that will be compatible with multiple operating systems. So we have proposed a recommendation system with hybrid technique using ML. This system will recommend music to users depending upon preferences, recently played, and ratings of other users. Along with the music recommendation, a video link will also be suggested for those users who are interested to watch music in video format. To make it more innovative, voice assistants are integrated which a hand-free and personalized experience to users.

References

1. Jain S, Pawar T, Shah H, Morye O, Patil B (2019) Video recommendation system based on human interest. In: 2019 1st International conference on innovations in information and communication technology (ICIICT)(IEEE)
2. Kathavate S (2021) Music recommendation system using content and collaborative filtering methods. IJERT 10(02)
3. Arnold AN, Vairamuthu S (2019) Music recommendation using collaborative filtering and deep learning. Int J Innovat Technol Explor Eng (IJITEE) 8(7). ISSN: 2278-3075
4. Subhash S, Srivatsa PN, Siddesh S, Ullas S, Santhosh B (2020) Artificial intelligence-based voice assistant. In: 2020 Fourth world conference on smart trends in systems, security and sustainability (WorldS4)
5. Li X (2021) Research on the application of collaborative filtering algorithm in mobile e-commerce recommendation system. In: 2021 IEEE Asia-Pacific conference on image processing, electronics and computers (IPEC)
6. Matsumoto Y, Harakawa R, Ogawa T, Haseyama M (2019) Music video recommendation based on link prediction considering local and global structures of a network. IEE J
7. Singh J (2020) Collaborative filtering based hybrid music recommendation system. In: Third international conference on intelligent sustainable systems [ICISS 2020]

8. Chang S, Abdul A, Chen J, Liao H (2018) A personalized music recommendation system using convolutional neural networks approach. IEEE Int Conf Appl Syst Invent (ICASI) 2018:47–49. https://doi.org/10.1109/ICASI.2018.8394293
9. Sunitha M, Adilakshmi T (2016) Mobile based music recommendation system. Int Conf Invent Comput Technol (ICICT) 2016:1–4. https://doi.org/10.1109/INVENTIVE.2016.7830183
10. Wu X, Zhu Y (2016) T1—A hybrid approach based on collaborative filtering to recommending mobile apps. https://doi.org/10.1109/ICPADS.2016.0011
11. Jisha RC, Amrita JM, Vijay AR, Indhu GS (2020) Mobile App recommendation system using machine learning classification. In: Fourth international conference on computing methodologies and communication (ICCMC) 2020:940–943. https://doi.org/10.1109/ICCMC48092.2020.ICCMC-000174
12. Divakar MS, Kumar V, DE MJ, Kalpana RA, RM SK (2021) Farmer's assistant using AI voice bot. In: 2021 3rd international conference on signal processing and communication (ICPSC), pp 527–531. https://doi.org/10.1109/ICSPC51351.2021.9451760

Unsupervised Hybrid Change Detection Using Geospatial Spectral Classification of Time-Series Remote Sensing Datasets

Sriрupa Das and Somdatta Chakravortty

Abstract This paper presents a hybrid spectral unmixing-based change detection approach for the time-series remote sensing datasets. The proposed approach has an enormous capacity to act positively with the complex and heterogeneous background and distortion present within the datasets because of sensor errors and environmental hazards. The entropy and geospatial information have been adopted in the classification process followed by the smoothing operation, which improves the efficacy of the proposed method for time-series datasets. The derivative-based linear comparator with morphological operators has been applied on the binary classified datasets to estimate the changed elements and geospatial positions as the changed map. Ultimately, the efficacy and estimated outcomes of the hybrid classification-based change detection approach for the Yellow River Estuary of China have been validated using quantitative parameters, i.e., *overall error, percentage correct classification, kappa coefficient, and change map image.* The presented approach is thoroughly compared with the state of the art and advanced changed detection approaches and found that the proposed method is superior in numerous ways.

Keywords Multi-temporal · Fuzzy · Change detection · Time series · Geospatial

1 Introduction

The change identification and estimation within the remote sensing images have gained enormous attention in recent research fields and are widely employed in different real-life problems such as urban growth monitoring [1–4], environment

S. Das
RCC Institute of Information Technology, Kolkata, India

S. Chakravortty (✉)
Maulana Abul Kalam Azad University of Technology, Kolkata, West Bengal, India
e-mail: somdatta.chakravortty@makautwb.ac.in

P. Dutta et al. (eds.), *Emerging Technologies in Data Mining and Information Security*, Lecture Notes in Networks and Systems 490,
https://doi.org/10.1007/978-981-19-4052-1_4

monitoring [5], estimation of effects after natural calamities [6–9], agricultural monitoring like examining the growth of crops [10]. Among the remotely sensed time-series images, synthetic aperture radar (SAR) images are very popular [11–14] for change detection as they are less dependent on the atmospheric and weather conditions; therefore, they produce more accurate reflectance values. On the other hand, the images are prone to the induction of speckle noise, which makes the change detection task more challenging for the researchers. In the recent literature, lots of methods are reported to perform the operation of identification of change in multi-temporal images which include the algebraic methods through image ratio, regression-based analysis, absolute distance-based change estimation [15, 16], analysis of change vector [17], etc. Moreover, threshold, classification [18], and machine learning approaches have also gained popularity in recent literature. In the threshold-based methods, optimum threshold values are estimated to detect the changed pixels in the target images. In [12], a double-thresholding Kittler--Illingworth (DTKI) algorithm has been proposed to automatically estimate the decision thresholds using the objective function optimization to detect the changes in the log-ratio image. In classification-based change detection methods, the time-series images are done pre-classification individually and the classified images are analyzed for identifying the changes. Among the classification-based change detection methods fuzzy c-means is considered the most well-accepted classification method [19] but if the target image is corrupted with noise, then the chance of misclassification abruptly increases. To overcome this issue gradually, the spatial information is incorporated with the traditional FCM-based classification methods. In [20], a fuzzy c-means classification with spatial information has been proposed where the local features of the pixels are considered along with the pixel intensity values to reduce the noise effect in classification. The FCM with weighted local features has been proposed in [21], where a new weighted fuzzy factor and an adaptive kernel distance metric have been incorporated to accurately measure the damping extent of neighbors and reduce the effect of noise and outliers. The change detections using spectral unmixing are also reported [22, 23]. Advanced deep learning techniques are also grabbed the attraction in change detection of multi-temporal remote sensing images. In [24], the authors have proposed an unsupervised method for change detection incorporating the sparse autoencoder (SAE), CNN, and fuzzy-based classifier.

In this paper, a hybrid spectral unmixing-based change detection technique is presented for time-series remotely sensed images. The proposed approach has an enormous capacity to deal with the complex and heterogeneous background and distortion present within the datasets. The entropy and geospatial information have been adopted in the unmixing process followed by the smoothing operation, and the derivative-based linear comparator with morphological operators has been applied on the binary classified datasets to estimate the changed elements and geospatial positions as the change map. The outcomes of the proposed approach have been tested in the SAR image of the Yellow River Estuary of China in terms of overall error (OE), percentage correct classification (PCC), kappa coefficient (KC), and change map image and analyzed and compared thoroughly with the considered state-of-the-art change detection methods. The rest of the paper is organized as, Sect. 2 describes the proposed approach, Sect. 3 describes the results and discussions, and finally, Sect. 4 concludes the study.

2 Proposed Method

In this study, we have proposed an unsupervised spectral unmixing-based change detection method to efficiently estimate the changes in multi-temporal remote sensing datasets. An entropy-based spatial fuzzy c-means spectral unmixing approach (ESFCM) [25] has been adopted to unmix and classify the time-series remotely sensed datasets, where the local spatial correlation among the neighbors of the processing pixel is incorporated with the global membership measure estimated through Gaussian distribution to eliminate the effect of noise and reduce the chance of misclassification. The algorithm produces an efficient approach to reach the optimality by introducing entropy measure and local spatial information in the objective function, defined in Eq. 1.

$$J_{\mathrm{ESFCM}} = \sum_{j=1}^{C}\sum_{i=1}^{N} u_{ij}^{m} d_{ij} + \sum_{i=1}^{N} U(x_i) + \tau \sum_{i=1}^{N}\sum_{j=1}^{C} w_{ij} l_{jW_i} \tag{1}$$

$$\text{Constraints, } \sum_{i=1}^{N} U(x_i) \approx 0,\ \sum_{j=1}^{C} u_{ij} = 1,\ \sum_{j=1}^{C} w_{ij} = 1 \text{ and } \sum_{j=1}^{C} l_{jW_i} = 1.$$

where u_{ij} is the global membership function, defined in Eq. 2, d_{ij} is the Gaussian-based distance measure between the ith processing pixel and jth cluster, defined in Eq. 3, $w_{ij} l_{jW_i}$ is the weighted local spatial information, and $U(x_i)$ is the entropy measure of the ith processing pixel. The local membership function has been depicted in Eq. 4.

$$u_{ij} = \frac{1}{\sum_{k=1}^{C} \left(\frac{d_{ij}}{d_{ik}}\right)^{\frac{1}{(m-1)}}} \Big| G_{ij} = e^{-\left(\frac{x_i - v_j^2}{2\sigma_j^2}\right)} \tag{2}$$

$$d_{ij} = \left(\left(1 - G_{ij}\right) \Big| G_{ij} = e^{-\left(\frac{x_i - v_j^2}{2\sigma_j^2}\right)} \right) \text{ in } \mathbb{R}^2 \tag{3}$$

$$L_{\mathrm{ESFCM}} = \left\{ l_{jW_i} | x_k \in W_i, x_i \in (X \cap W_i) \forall k \right\}, \text{ in } \mathbb{R}^2 \tag{4}$$

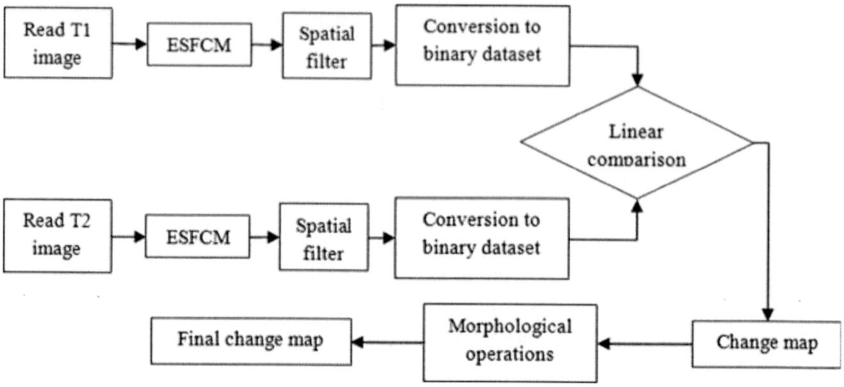

Fig. 1 Block diagram of the proposed method

where l_{jW_i} is the local spatial membership value for the ith processing pixel and $\sum_{j=1}^{C} l_{jW_i} = 1$ and W_i is the dynamic mask designed to incorporate the correlated neighbors of the processing pixel.

Algorithm 1: Hybrid ESFCM Classification-based Change Detection

Step 1: **Initialization** of the time-series remote sensing Yellow River Estuary in China datasets, classification method, thresholds, structuring element, and filters.
Step 2: The input images, T1 and T2 (noisy), are classified with the ESFCM classifier.
Step 3: The 'intraclasses' of classified datasets are smoothed using a standard spatial filter.
Step 4: The obtained smoothed 'intraclass' components are transformed to Binary datasets.
Step 5: The changed elements in each class are estimated by applying the first-order derivative on the results of **Step 4**.
Step 6: The final changed map is estimated based on the morphological operation, where the structural element (SE) is chosen as a (5 by 5) star.
Step 7: The registration of the estimated and ground truth changed maps is done to measure and analyze the quantitative validation parameters and qualitative visual results respectively.

After estimating the intraclass components from the target datasets, a smoothening operation on the estimated classes has been applied using a standard spatial filter and then the obtained intraclass components are transformed into binary datasets. Using the first-order derivative approach the changed elements from each obtained intraclass component are estimated. To generate the enhanced change map from the target time-series datasets, the morphological operation has been applied on the estimated changed elements using a structural element (SE) defined as a (5 by 5) star. For validating the estimated change map, it is co-registered with the ground truth change map and the pixel-to-pixel validation is done. The detailed steps of execution of the presented change detection approach have been illustrated in Fig. 1 and Algorithm 1.

3 Result and Discussions

The proposed method has been applied to the time-series SAR datasets of the Yellow River Estuary of China. The datasets were obtained by RADARSAT-2 sensor in June 2008 (T1) and June 2009 (T2), respectively. The datasets constitute 290 × 260 pixels with a spatial resolution of 8 m and severely corrupted speckle noise. The original time-series T1 and T2 images and the available ground truth change map have been depicted in Fig. 2. To validate the outcomes of the proposed method the overall error, percentage correct classification and kappa coefficient have been considered as the quantitative parameters and the visual change map has been considered as the qualitative assessment of the estimated outcomes. The proposed change detection method has been compared with the state-of-the-art classification-based and machine learning-based change detection algorithms, namely DTKI [12], FCM [19], FLICM [20], KWFLICM [21], and SAE + FCM + CNN [24].

Table 1 shows the comparative analysis of quantitative measures of the outcomes of the proposed method with the considered methods, where the proposed method shows superior performance over all the considered classification-based methods and comparable results with the machine learning-based methods. The qualitative comparison of performances of all the methods is shown in Fig. 3, where it has been seen that the visual change map of the proposed method is better concerning the other considered methods. The estimated per-pixel efficiency of the proposed method is 0.00418 s/pixel.

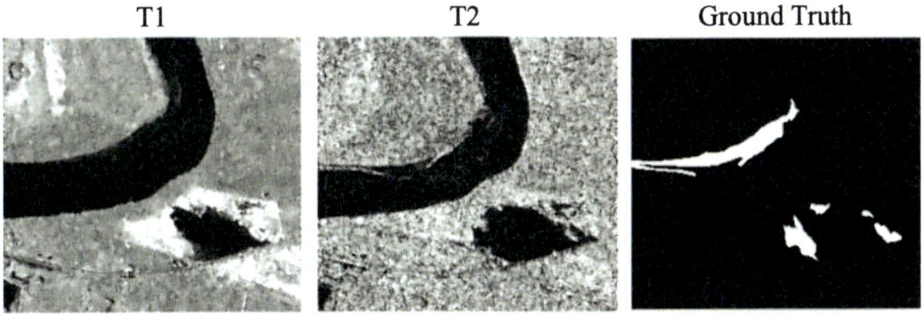

Fig. 2 Original T1 and T2 images of yellow river dataset

Table 1 Comparative analysis of the outcomes of all the methods

Methods	OE	PCC	KC
DTKI [12]	2001	0.9735	0.7276
FCM [19]	32,399	0.5703	0.1069
FLICM [20]	1453	0.9807	0.7913
KWFLICM [21]	31,268	0.5853	0.1153
SAE + FCM + CNN [24]	1013	0.9866	0.8389
Proposed Method	1041	0.9861	0.8338

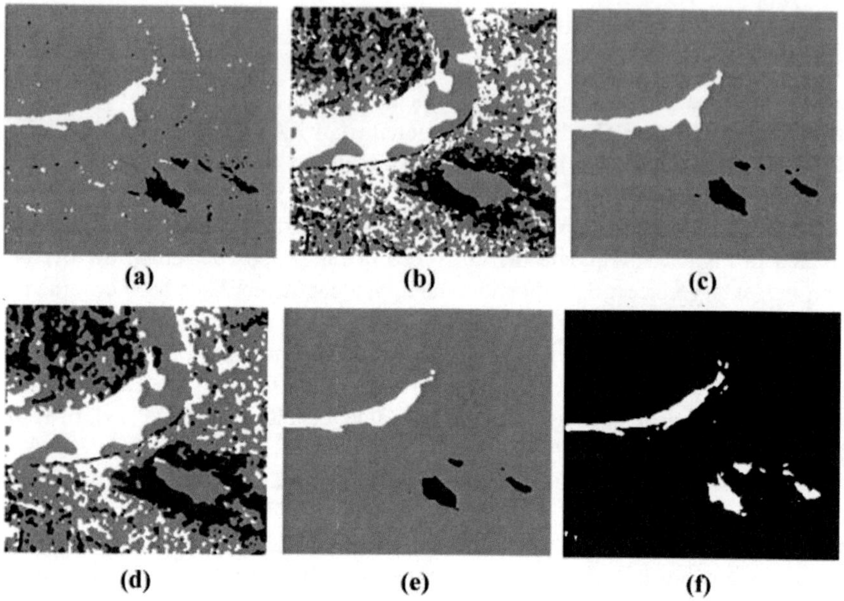

Fig. 3 Change maps generated by the methods **a** DTKI. **b** FCM. **c** FLICM. **d** KWFLICM. **e** SAE + FCM + CNN. **f** Proposed method

4 Conclusion

In this study, we have proposed a hybrid spectral unmixing-based change detection approach for the time-series remote sensing datasets. An entropy-based spatial fuzzy c-means-based spectral unmixing technique followed by a spatial filter has been employed to estimate the accurate intraclass elements from the target images. Then the first-order derivative-based method along with morphological operations has been used to estimate the change map. The method has been thoroughly compared with the available ground truth change map to measure the accuracy. The performance of the proposed method has been rigorously compared with the considered classification and machine learning-based change detection methods where the proposed method has shown its efficacy over the considered methods in terms of quantitative and qualitative measures.

References

1. Bouziani M, GoÃr´ta K, He DC (2010) Automatic change detection of buildings in urban environments from very high spatial resolution images using existing geodatabase and prior knowledge. ISPRS J Photogram Rem Sens 65:143–153
2. Ban Y, Yousif OA (2012) Multitemporal spaceborne SAR data for urban change detection in

China. IEEE J Sel Top Appl Earth Observ Rem Sens 5(4):1087–1094
3. Liu M, Zhang H, Wang C, Wu F (2014) Change detection of multilook polarimetric SAR images using heterogeneous clutter models. IEEE Trans Geosci Remote Sens 52(12):7483–7494
4. Gong M, Zhang P, Su L, Liu J (2016) Coupled dictionary learning for change detection from multisource data. IEEE Trans Geosci Remote Sens 54(12):7077–7091
5. Jin S, Yang L, Danielson P, Homer C, Fry J, Xian G (2013) A comprehensive change detection method for updating the national land cover database to circa 2011. Rem Sens Environ 132:159–175
6. Giustarini L, Hostache R, Matgen P, Schumann GJP, Bates PD, Mason DC (2013) A change detection approach to flood mapping in urban areas using TerraSAR-X. IEEE Trans Geosci Rem Sens 51:2417–2430
7. Yang H, Dou A, Zhang W, Huang S (2014) Study on extraction of earthquake damage information based on regional optimizing change detection from remote sensing image. In: Proceedings of IEEE geoscience and remote sensing symposium, pp 4272–4275
8. Li N, Wang R, Deng Y, Chen J, Liu Y, Du K, Lu P, Zhang Z, Zhao F (2014) Waterline mapping and change detection of tangjiashan dammed lake after wenchuan earthquake from multitemporal high-resolution airborne SAR imagery. IEEE J Sel Top Appl Earth Observ Rem Sens 7:3200–3209
9. Brunner D, Bruzzone L, Lemoine G (2010) Change detection for earthquake damage assessment in built-up areas using very high resolution optical and sar imagery. In: Proceedings of IEEE international geoscience and remote sensing symposium, pp 3210–3213
10. Rahman MR, Saha SK (2009) Spatial dynamics of cropland and cropping pattern change analysis using landsat TM and IRS P6 LISS III satellite images with GIS. Geo-spat Inf Sci 12:123–134
11. Bazi Y, Bruzzone L, Melgani F (2005) An unsupervised approach based on the generalized Gaussian model to automatic change detection in multitemporal SAR images. IEEE Trans Geosci Remote Sens 43(4):874–887
12. Bazi Y, Bruzzone L, Melgani F (2006) Automatic identification of the number and values of decision thresholds in the log-ratio image for change detection in SAR images. IEEE Geosci Rem Sens Lett 3(3):349–353
13. Gong M, Su L, Jia M, Chen W (2014) Fuzzy clustering with a modified MRF energy function for change detection in synthetic aperture radar images. IEEE Trans Fuzzy Syst 22:98–109
14. Parrilli S, Poderico M, Angelino CV, Verdoliva L (2012) A nonlocal SAR image denoising algorithm based on llmmse wavelet shrinkage. IEEE Trans Geosci Rem Sens 50:606–616
15. Du P, Liu S, Gamba P, Tan K, Xia J (2012) Fusion of difference images for change detection over urban areas. IEEE J Sel Top Appl Earth Observ Rem Sens 5(4):1076–1086
16. Hou Z, Li W, Li L, Tao R, Du Q (2021) Hyperspectral change detection based on multiple morphological profiles. IEEE Trans Geosci Rem Sens
17. Kwan C (2019) Methods and challenges using multispectral and hyperspectral images for practical change detection applications. Information 10(11):353
18. Yang X, Pavelsky TM, Bendezu LP, Zhang S (2021) Simple method to extract lake ice condition from landsat images. IEEE Trans Geosci Remote Sens
19. Bezdek JC (2013) Pattern recognition with fuzzy objective function algorithms. Springer Science & Business Media
20. Krinidis S, Chatzis V (2010) A robust fuzzy local information C-means clustering algorithm. IEEE Trans Image Process 19(5):1328–1337
21. Gong M, Liang Y, Shi J, Ma W, Ma J (2013) Fuzzy c-means clustering with local information and kernel metric for image segmentation. IEEE Trans Image Process 22:573–584
22. Liu S, Bruzzone L, Bovolo F, Du P (2016) Unsupervised multitemporal spectral unmixing for detecting multiple changes in hyperspectral images. IEEE Trans Geosci Remote Sens 54(5):2733–2748
23. Erturk A, Iordache M-D, Plaza A (2017) Sparse unmixing with dictionary pruning for hyperspectral change detection. IEEE J Sel Topics Appl Earth Observ Remote Sens 10(1):321–330

24. Gong M, Yang H, Zhang P (2017) Feature learning and change feature classification based on deep learning for ternary change detection in SAR images. ISPRS J Photogramm Remote Sens 129:212–225
25. Das S, Chakravortty S (2021) Efficient entropy-based spatial fuzzy c-means method for spectral unmixing of hyperspectral image. Soft Comput 25:7379–7397

Trust Based Resolving of Conflicts for Collaborative Data Sharing in Online Social Networks

Nisha P. Shetty, Balachandra Muniyal, Pratyay Prakhar, Angad Singh, Gunveen Batra, Akshita Puri, Divya Bhanu Manjunath, and Vidit Vinay Jain

Abstract Twenty-first century, the era of Internet, social networking platforms like Facebook and Twitter play a predominant role in everybody's life. Ever increasing adoption of gadgets such as mobile phones and tablets have made social media available all times. This recent surge in online interaction has made it imperative to have ample protection against privacy breaches to ensure a fine grained and a personalized data publishing online. Privacy concerns over communal data shared amongst multiple users are not properly addressed in most of the social media. The proposed work deals with effectively suggesting whether or not to grant access to the data which is co-owned by multiple users. Conflicts in such scenario are resolved by taking into consideration the privacy risk and confidentiality loss observed if the data is shared. For secure sharing of data, a trust framework based on the user's interest and interaction parameters is put forth. The proposed work can be extended to any data sharing multiuser platform.

Keywords OSN privacy · Multiparty authorization · Trust determination · Friend grouping

1 Introduction

The Internet has encroached the field of communication. Nowadays, telephones, letters, etc., are replaced by social networking platforms like Twitter, Facebook and Orkut where people share their thoughts, feelings and activities in the form of texts, multimedia posts and so on. Sites like Facebook and Twitter are the 3rd and 7th most visited sites globally, having millions of subscribers online. Not only communication sector but also many other areas such as employment exchange, law enforcement,

N. P. Shetty (✉) · B. Muniyal (✉) · P. Prakhar · A. Singh · G. Batra · A. Puri · D. B. Manjunath · V. V. Jain
Manipal Institute of Technology, Manipal Academy of Higher Education, Manipal 576104, India
e-mail: nisha.pshetty@manipal.edu

B. Muniyal
e-mail: bala.chandra@manipal.edu

P. Dutta et al. (eds.), *Emerging Technologies in Data Mining and Information Security*, Lecture Notes in Networks and Systems 490,
https://doi.org/10.1007/978-981-19-4052-1_5

advertising agencies, etc., rely on social media to glean information. It also serves as a platform for educational information exchange, political campaigns, etc. Many data breach scandals are occurring nowadays, the latest being Cambridge Analytica scandal [1] for which Facebook was fined a whooping sum of £500,000. Many organizations like GNIP [2], Epsilon, etc., harvest and sell data such as comments, likes, shares, etc., garnering huge profits. Henceforth, having control over data publishing in such platforms is the need of the hour.

Trust being a factor of authenticity and level of confidence amongst all parties involved can be evaluated in terms of interactions, relationships and profile validity in online social network (OSN). Some of the challenges faced in prediction of trust in OSN to determine access control are listed below [3, 4]:

- Almost all the existing data set in trust prediction lack in user-specified trust credentials. Meagre and sparse datasets in this domain make the research challenging.
- User A can trust the recommendations of a football player in sports, but not always likely in movies. Suitable context-specific data and background information are not available in OSN to make educated decisions in such scenarios.
- Relationships do not stay the same over time. XYZ may have a fallout with ABC in 2020 due to many factors such as change in interests and behaviours. Unless any explicit measure is taken in this context (like unfriending that particular user), there is no suitable technique to implicitly manage such changes.
- Users only have control over their virtual space. But, if any of the user's data reside over other's space, then they cannot govern who can access it. For example, in the event of untagging a photo, user only removes the link to his virtual space. However, his image in the photograph remains as long as the owner of the photograph keeps the photograph posted [5].

The proposed work introduces a novel model whose main contributions are listed below:

- An implicit trust framework depending on interest, demographic similarity and interpersonal connections which segregates the users' friend network into three categories [6].
- A systematic technique to resolve multiparty conflicts regarding sharing of a collaborative data.

2 Related Works

2.1 *Interest Generation and Interest Similarity Computation*

Lim and Datta [7] grouped communities in Twitter based on their interests by using the following relationship of Twitter. Their algorithm calculated the number of celebrities a user follows under a particular pursuit and ranked the user interest based on it. For example, if the followed more prominent movie stars than sports person, then his predominant interest was movies. The authors realized that the communities which share interests are often more connected. However, their method is solely based on availability of celebrities whose interests are well known in the user's profile and will fail otherwise. Over the period of time user's interest may vary from the celebrities they follow, so temporal factors must be taken into account while studying the evolution of such communities.

Nguyen et al. [8] computed similarity between the users by analysing their tweets and activities. Authors employed similarity metrics on feature set comprising of content, category, sentiments, groups involved and posting/commenting/liking activities. Most of the parameters of that study were incorporated in our work too. Major limitation of this work was that it gave a binary output value (similar /not similar) because of which it fails in the context of social media wherein the degree of similarity plays an important role affecting the user's judgement.

Yang et al. [9] perceived the similarity amongst users by comparing their video preference history, friendships and other such demographic information. The authors further established a correlation between these information and interest. Their work was effective in video recommendation and can be extended to any domain.

Ma [10] studied the parallels amongst the interests of friends in a social network. They voiced our views on improving privacy settings of the users by finding the closest friends of the user. However, their work is effective in a single-hop bidirectional framework. The authors fail to address the multi-hop relationships and following relationships as observed in networks like Twitter.

Zhou et al. [11] designed ISCoDe framework which takes user's interest in various topics as input, assess their similarities, and finally clusters the similar users together by generating the weighted graph of similar users. Two similarity metrics PS and InvKL were compared. Although PS was more susceptible to the changes in the community structure, InvKL was observed to offer better results amongst the nodes sharing more similarities. The authors plan on to accumulate a better feature set so as to improve upon their community detection module.

2.2 *Trust Computation*

Schwartz-Chassidim et al.'s [12] study corroborated our views on selective sharing of posts to enhance privacy in OSN. Their study on 181 Facebook users behaviour for a period of three months measured the frequency of users changing the intended audience for their post in multiple scenarios (0 if it remains same throughout else a positive value). They concluded that only about 30% of the people used the privacy settings option in the intended manner which showed its low success rate in protecting privacy. While the authors came to this conclusion by observing live user behaviour rather than the surveys, owing to the varied distribution and small sample size the method cannot be a conclusive one.

Almuzaini et al. [13] put forth a trust supervision study on WhatsApp to aid the users to make an edified verdict about the received message based on the trust value. The system computes each node's reputation weight rated by a significant number of reviewers to categorize if the node is malicious or not. Also, even at user level, provision is provided to make decisions in a decentralized manner based on past experiences. Since the current work in only deals with user-based trust management, the authors plan on to extend the scope to test the veracity of the messages as well.

Most of the existing crowdsourcing algorithms assign tasks to the users randomly leading to low precision and accuracy. Therefore, Zhang et al. [14] developed Social-Situ to compute the user's suitability for a task based on his qualifications and historical evidence. With this work as the base, the authors plan to combine machine learning algorithms with crowd sourcing approach to improve the trust prediction in OSN.

Baek and Kim [15] proposed a dynamic trust computation mechanism suitable for OSN. Their application computed trust dynamically and was able to withdraw permissions when encountered with a negative behaviour. Unlike static traditional methods, the proposed technique did not allow the trust transferal (same trust level to friend and friend of friends) and base the trust on relationships which changes all time in OSN. In so doing, the technique was able to overcome irresponsible data sharing on a better scale. As a future extension, the authors plan to experiment their methodology on real data to fine tune their parameters and develop OSN-friendly application.

Yassein et al. [16] developed a three-tier system which protects the users from malicious persons, URLs and abusive contents. Their method is an amalgamation of cryptography, feedback based and machine learning. It categorizes the entities as benign, risk and inappropriate. The major limitation observed was that the approach only dealt with textual content and not any multimedia data. Also, the method of categorizing the user as trustworthy or not before key sharing is not illustrated.

Zolfaghar and Aghaie [17] employed ensemble technique to check the trust amongst the users. However, their feature set involved direct trust values and user reviews to make a decision, which is not possible in an inherent OSN platform.

Saeidi [18] surveyed the OSN user opinion on parameters influencing trust scores in OSN. While computing the maximum trust between two nodes, the accuracies of artificial bee colony, ant colony and genetic algorithms were compared, amongst which artificial bee colony algorithm offered the most nominal computation time. However, the efficacy of the technique is impaired due to the small population size.

Son et al. [19] calculated the similarity amongst users by their assessment of certain items. Trust, its propagation value is calculated implicitly, and top recommendations are provided to the user accordingly. Since there is a sparsity of data online, future work needs to into account of that and predict with less data. Improvements in reliability measurements and belief propagation models in both directional and unidirectional networks can be a follow-up study in future.

2.3 *Multiparty Conflicts Resolution*

The work proposed by Ding and Zhang [20] in aims at striking a balance between privacy concern of patrons of communal item and the effect of social influence on their decision. This work allows the users to reach a unanimous binary decision on whether or not to share the data item under consideration. However, actions and penalties if shared are beyond the scope of this study.

Hu et al. [21] formulated many policies such as decision voting (either based on majority or sensitivity), and by enforcing other such strategies for data sharing or dissemination. However, the factors affected and their importance if the decision is "Deny" are not analysed. Any method, which inherently configures the privacy settings based on past data can elevate the time loss spent in configuring individually for all users.

Ali et al. [22] designed a middleware between OSN and coowners to regulate the data sharing process. Although this method seems effective, involvement of a third party again compromises the privacy as access rights must be given to a third party application. This method, however, provides security and confidentiality by means of encryption at the cost of computation and time overhead.

Ilia et al. [23] employed an access grant scheme wherein each collaborator can grant access or not to the requestor by deciding whether or not to share the secret key which facilitates the decryption of the data. Their method, however, is more complex due to multiple key management and such majority decision voting-based approach often compromises individual user's confidentiality.

Vishwamitra et al. [24] formulated a way to protect personal identifiable information of a user in a shared photo. Their application AppX automatically detects the face automatically and ensures if the photo needs to be shared by viewing the privacy settings of the user. Decision whether or not to share the photo is dependent on accord of the majority or sensitivity. The authors plan to incorporate machine learning algorithms to automate the system and reduce human interaction.

3 Methodology

3.1 *Trust Computation and Friend Grouping*

The methodology (as shown in Fig. 1) is grounded on the premise of using coarse online social network (OSN) data consisting of user interactions, interests and general information to group the existing friends into subsequent assorted categories.

- Data collection: The crawled Twitter data can be categorized into three categories [25].
- Tweets: These involve the top 100 latest tweets crawled per account.

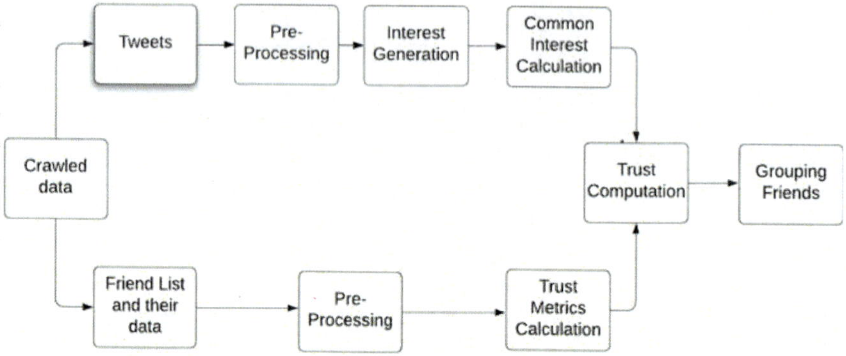

Fig. 1 Trust metrics calculation and friend grouping

- Demographic information: The information which is the part of profile such as name, age, sex, gender, marital status, location and other such basic personal information.
- Social relationship: Here, the follow/following/friend information is collected, and subsequently, those profiles were crawled for the above information.
- Data pre-processing: The collected data is improved in quality by removing incomplete/duplicate entries and noisy data. Text which is of less significance in detection process like stop words, special characters, emoji, etc., are removed. Ambiguous words are eliminated. Grammatical corrections like stemming and lemmatization are performed.
- Feature selection: Essential keywords are collected using CountVectorizer and TF-IDF.
- Interest Detection: The most frequent keywords are compared against the tag list of each topic, and the topic with the greatest number of common values with the keyword list is selected as the interest [26, 27].
- Common Interest Calculation: Similarity metrics such as cosine (for demographic data) and Jaccard (for interest data) are employed to compute the affinity between two users x and y [28, 29].
- Interest similarity computation can be broadly expressed using the following formula.

$$\text{Interest\_Sim}(x, y) = \text{In}x \cap \text{In}y / \text{In}x \cup \text{In}y \tag{1}$$

Demographic similarity can be expressed in the following manner

$$\text{Dem\_Sim}(x, y) = \frac{\sum_{i=1}^{D} \text{Dm}_{ix}.\text{Dm}_{iy}}{\left(\sqrt{\sum_{i=1}^{D} \text{Dm}_{ix}^2} \times \sqrt{\sum_{i=1}^{D} \text{Dm}_{iy}^2}\right)} \tag{2}$$

In social networks, users share a stronger intimacy with the people who belong to the same anthropology having similar interests. Intimacy amongst users can be computed using the following equation [28].

$$\text{si\_sim}(x, y) = \text{alpha} * \text{Dem\_Sim}(x, y) + \text{beta} * \text{Interest\_Sim}(x, y) \tag{3}$$

where

Dem_Sim(x, y)	demographic similarity between user x and y
si_sim(x, y)	social tie strength between user x and y
In_Sim(x, y)	interest similarity between user x and y
Dmi$x$/Dmi$y$	ith demographic attribute of user x/y.
Inx/Iny	interest of user x/y.

- Friend list data: Friend list data contains information such as friend/following information, demographic data, tweets, no. of likes, posts, shares, etc.
- Trust metrics calculation: The decentralized framework of OSN has often caused uncertainty and malicious misuse of data compromising integrity, reliability and confidentiality. In the context of a network environment, trust can be broadly defined as the competence of an individual to adhere to the security policies and enforce quality like dependability, clandestineness and integrity. As OSNs like Facebook, Twitter, etc., do not have an explicit trust rank mechanism, trust here is reflected through social tie strength and the amount of interactions between the users. In this work, raw OSN parameters such as demographic data, interests and interactions are mapped to the trust metrics in the following manner [28].
- Reliability is categorized as verifying if the user is genuine or not by assessing his presence on the platform. A genuine person has a significant number of followers and will be following a significant number of people.
- A person is said to be most popular if he has a significant amount of followers.
- Likes and comments to post ratio (LctP) is a metric used in our approach to determine the engagement rate of the user.
- Openness is the amount of posts posted by the user.
- Credibility is often perceived in three ways, i.e. as authenticity of the message or informant or the medium through which it is delivered. In case of OSN, credibility often befalls upon the legitimacy of the user which can be computed as the sum of openness and popularity. Demographic information such as expertise and age. enhances the credibility.

$$\text{Cred}(a) = \text{Open}(a) + \text{Pop}(a) \tag{4}$$

where

Cred(a)	credibility of user a
Pop(a)	popularity of user a
Open(a)	openness of user a

- Trust computation [28]: Trust here is computed as a facet based on common interests shared amongst users and the reputation earned by a user against OSN benchmarks. Amount of trust user y has on user x can be computed as

$$\mathrm{Ut}(x) = \mathrm{wt}_{\mathrm{si}} * \mathrm{si\_sim}(x, y) + \mathrm{wt}_{\mathrm{cred}} * \mathrm{Cred}(x) + \mathrm{wt}_{\mathrm{rel}} * \mathrm{Rel}(x) + \mathrm{wt}_{\mathrm{LctP}} * \mathrm{LctP}(x) \quad (5)$$

where

Ut(x)	trust of user x
Cred(x)	credibility of user x
si_sim(x, y)	social tie strength between user x and y
$\mathrm{wt}_{\mathrm{si}}$; $\mathrm{wt}_{\mathrm{cred}}$; $\mathrm{wt}_{\mathrm{rel}}$; $\mathrm{wt}_{\mathrm{LctP}}$	customized weights
LctP(x)	Likes and Comments to Post of user x
Rel(x)	reliability of user x

- Grouping friends: Based on the computed trust scores the friends of a particular user y are grouped into three levels via K-means, support vector machine, decision tree and K-nearest neighbour.

3.2 *Multiparty Authorization for Data Sharing*

Any OSN contains the profile (name, age, sex, area, interests, etc.), relationship (friend, following, group member/owner, etc.) and contents (tweets, photos and videos .). Currently OSN utilizes user to user relationship or group membership to allow access to any user's data. There is no regulation to restrict data sharing to a specific set of users for data residing in another's space. The methodology (as shown in Fig. 2) formulates a flexible way to collaboratively manage access to shared data, thereby resolving conflicts in a systematic manner.

3.2.1 Key Terminologies [5]

- Owner: User who shares a data item in OSN in his profile.
- Contributer: User who shares a data item in other person's profile.
- Stakeholder: Set of users tagged in a particular post.
- Disseminator: User who shares an item from someone else's space into his profile in OSN.
- Controller: Any of the above users who can modulate the access to the data.
- Negotiators: List of controllers and target users against whom the controllers have an apprehension to share the data.
- Target users: Users who are granted/ restricted the right to use the shared data item.

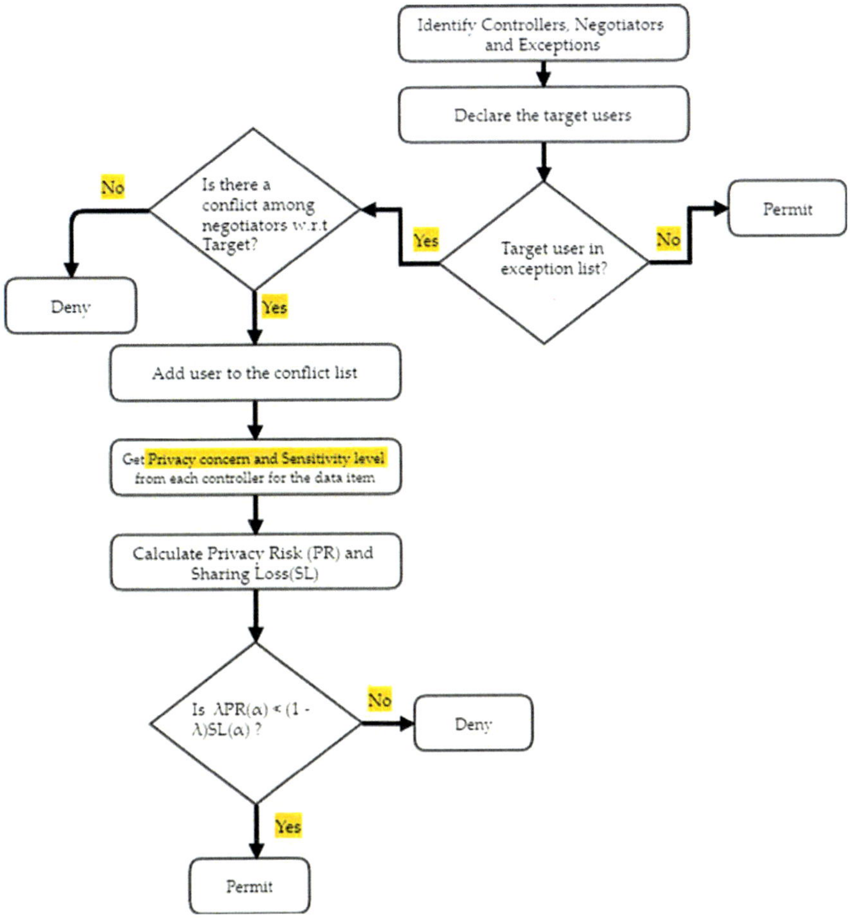

Fig. 2 Flowchart to resolve multi-party conflict in data sharing

The proposed algorithm takes the set of controllers (owner, contributor, stakeholder and disseminator), list of negotiating users and their exceptions, and target users as input. A list v is maintained to check individual decision whether or not (1 or 0) to share the data item of each of contributor is maintained. . For e.g., if the target user set is ["Adam", "Eve" and "Bob"] and there are two negotiating users Prat and Tim who do not want to give access to Bob and Eve, respectively, then v would be [[1, 1, 0], [1, 0, 1]] where 1 means that the access is granted and 0 means that the target user is an exception in sharing list for that particular negotiating user. A list c holds the conflict amongst the negotiating users whether or not to grant access. For e.g., if v is [[1, 1, 0], [1, 0, 1] and [1, 1, 1]] ; while matching the first elements of all three lists, no conflict is detected because all the elements are 1, and thus all the negotiating users have granted access to that target user. But, while iterating through the second element of the lists, we detect a conflict because the second negotiating user has not granted access to that particular target user, and thus target user is added to the list of conflicting users, i.e. c.

In any scenario, within the realms of OSN, privacy risk is a calculated measure of threat to the controller's privacy on sharing the contradictory data item. Sharing loss is the amount of loss observed by the controllers who were willing to share the data item to the target user if the ultimate decision was "Permit". Privacy concern is the access setting given by a controller concerning who can view a particular data item. For example, movie stars tend to keep certain photos private or only share amongst certain friends, not including the general public. It is an amalgamation of trust and visibility of the data item. Sensitivity is a measure of confidentiality associated with a data item.

It goes without saying that permitting the target users in conflicting sectors to access the data item will cause privacy risk and likewise rejecting the access to such users will result in sharing loss. The proposed approach aims to obtain a tradeoff between privacy risk and sharing loss [29]. In order to resolve the conflict, the privacy concern and sensitivity value (high, medium, low or none) for the data item as per each controller are gathered.

Sharing loss and privacy risk are calculated using equations

$$\mathrm{SL}(\alpha) = \sum_{i=1}^{\beta \in \mathrm{controllers}(\alpha)} (1 - pc\beta) \times (1 - sl\beta) \times \mathrm{len}[c(\alpha)] \tag{6}$$

$$\mathrm{PR}(\alpha) = \sum_{i=1}^{\beta \in \mathrm{controllers}(\alpha)} (pc\beta) \times (sl\beta) \times \mathrm{len}[c(\alpha)] \tag{7}$$

where

- $\mathrm{SL}(\alpha) \leftarrow$ sharing loss associated with data item α
- $\mathrm{PR}(\alpha) \leftarrow$ privacy risk associated with data item α
- $\mathrm{pc}\beta \leftarrow$ privacy concern expressed by controller β
- $\mathrm{sl}\beta \leftarrow$ sensitivity expressed by controller β
- $\mathrm{len}[c(\alpha)] \leftarrow$ total no. of conflicting target users for the data item α pertaining to each of the controllers in β.

Finally, to allow or dissuade the accessibility to the data item, decision is calculated and conveyed according to the following equation.

$$\text{Decision} = \begin{cases} \text{Deny} \quad \text{if } \lambda\mathrm{PR}(\alpha) \geq (1 - \lambda)\mathrm{SL}(\alpha) \\ \text{Permit if } \lambda\mathrm{PR}(\alpha) < (1 - \lambda)\mathrm{SL}(\alpha) \end{cases} \tag{8}$$

where λ is the custom weight assigned to the privacy risk and sharing loss with $0 \leq \lambda \leq 1$.

Table 1 Classifier accuracies for interest detection from tweets

Classifier	Accuracy
k-nearest neighbour (kNN)	88.22
Support vector machine (SVM)	88.63
Fuzzy C Means with naïve Bayes (NB)	82.49
Latent Dirichlet Allocation (LDA) with support vector machine (SVM)	80.9
Sequential minimal optimization (SMO)	90.44
naïve Bayes multinomial	85.7
random forests (RF)	92.34

Table 2 Classifier accuracies for friend grouping

Classifier	Accuracy
k-nearest neighbour	90
Support vector machine	82.0
k-means	86.66
Decision tree (DT)	89

4 Result Analysis

4.1 Trust Computation and Friend Grouping

The results obtained by comparing the classifiers based on a feature vector which contains structural and interaction properties mapped to the trust metrics are discussed below. Table 1 compares the accuracies of classifiers in detecting interests from tweets.

Table 2 compares the accuracies of classifiers in suitably grouping friends based on trust metrics.

4.2 Multiparty Authorization Conflict Resolution Framework for Data Sharing

A tradeoff between privacy risk and sharing loss has been achieved to put forth a unanimous decision as illustrated in the test cases as depicted in Table 3.

Table 3 Conflict resolution test cases

No. of controllers	No. of target users	Exception list	Conflict list	Sharing loss	Privacy risk	Decision
3	5	[[1, 1, 0, 1, 1], [1, 1, 1, 0, 1], [1, 1, 1, 1, 0]]	[1 1 1 1 1 1 0 1 1 1 0 1 1 1 0]	1.376	3.267	Deny
3	5	[[0, 1, 1, 1, 1], [1, 0, 1, 1, 1], [1, 1, 1, 1, 1]]	[0 1 1 1 0 1 1 1 1 1 1 1 1 1 1]	3.79	0.435	Permit

5 Conclusion and Future Work

Earlier, trust related works mostly dealt with explicit scores provided by users involved which is not usually recommended in OSN. It is better suited for business portals only. However, our framework takes into account homophily and context-based linkages to group the users into their assorted categories. Unlike prior binary trust recommendations (0 grant access; 1 no access), the current approach mainly focuses on granting access rights suitably to the users based on the computed trust levels. The importance and suitable mapping of trust metrics to various OSN parameters are presented in this paper. Since machine learning is gaining huge popularity in today's times, the application of these algorithms to predict trust is a promising scope for research. In the context of resolving conflicts regarding access rights to communal data, a multiparty access control policy is formulated which, unlike earlier majority-based decision approaches, provides an ingrained control to each decision maker. Each co-owner's privacy concern and sharing loss over the data item are taken into consideration to reach a consensus outcome.

A robust system which constantly monitors the trust equations between the individuals while changing the access rights accordingly needs to be developed. Newer algorithms on context detection, cryptography and data fusion to achieve better tradeoff between data loss and efficiency are the need of the hour. The proposed trust level methodology can be made more fine grained by incorporating deep learning and ensemble techniques provided more benchmark data sets on user's relationship, tweets and background information are made public. Crowdsourcing approach can be used in future to label more data sets. With technologies like facial recognition, etc., tags are generated automatically nowadays which can be a violation of privacy. No proper mechanism or rule to address means to control a particular user's data residing in other people's virtual space is present which can be a possible research venture.

References

1. Facebook–cambridge analytical data scandal (2018). https://en.wikipedia.org/wiki/Facebook%E2%80%93CambridgeAnalyticadatascandal
2. Meet gnip, the company that's using twitter's data to disrupt the online advertising industry (2018). https://www.businessinsider.com/gnip-2011-8?IR=T
3. Ghafari SM, Beheshti A, Joshi A, Paris C, Mahmood A, Yakhchi S, Orgun MA (2020) A survey on trust prediction in online social networks. IEEE Access 8:144292–144309
4. Liu S, Zhang L, Yan Z (2018) Predict pairwise trust based on machine learning in online social networks: a survey. IEEE Access 6:51297–51318
5. Hu H, Ahn GJ (2011) Multiparty authorization framework for data sharing in online social networks. In: Li Y (ed) Data and applications security and privacy XXV. Springer Berlin Heidelberg, pp 29–43
6. Wang J, Jing X, Yan Z, Fu Y, Pedrycz W, Yang LT (2020) A survey on trust evaluation based on machine learning. ACM Comput Surv 53(5). https://doi.org/10.1145/3408292
7. Lim KH, Datta A (2012) Finding twitter communities with common interests using following links of celebrities. In: Proceedings of the 3rd international workshop on modeling social media, ser. MSM'12. Association for Computing Machinery, New York, NY, USA, pp 25–32. https://doi.org/10.1145/2310057.2310064
8. Nguyen TH, Tran DQ, Dam GM, Nguyen MH (2018) Estimating the similarity of social network users based on behaviors. Vietnam J Comp Sci 5(2):165–175
9. Yang C, Zhou Y, Chiu DM (2016) Who are like-minded: Mining user interest similarity in online social networks
10. Ma H (2014) On measuring social friend interest similarities in recommender systems. In: Proceedings of the 37th international ACM SIGIR conference on research and development in information retrieval, ser. SIGIR'14. Association for Computing Machinery, New York, NY, USA, pp 465–474. https://doi.org/10.1145/2600428.2609635
11. Zhou K, Martin A, Pan Q (2015) A similarity-based community detection method with multiple prototype representation. Phys A: Stat Mech Appl 438:519–531. https://doi.org/10.1016/j.physa.2015.07.016
12. Schwartz-Chassidim H, Ayalon O, Mendel T, Hirschprung R, Toch E (2020) Selectivity in posting on social networks: the role of privacy concerns, social capital, and technical literacy. Heliyon 6(2):e03298. http://www.sciencedirect.com/science/article/pii/S2405844020301432
13. Almuzaini F, Alromaih S, Althnian A, Kurdi H (2020) Whatstrust: a trust management system for whatsapp. Electronics 9(12). https://www.mdpi.com/2079-9292/9/12/2190
14. Zhang Z, Jing J, Wang X, Choo KR, Gupta BB (2020) A crowdsourcing method for online social networks security assessment based on human-centric computing. Hum centric Comput Inf Sci 10:23. https://doi.org/10.1186/s13673-020-00230-0
15. Baek S, Kim S (2014) Trust-based access control model from sociological approach in dynamic online social network environment. Scien World J 2014
16. Yassein MB, Aljawarneh S, Wahsheh Y (2019) Hybrid real-time protection system for online social networks. In: Foundations of science, pp 1–30
17. Zolfaghar K, Aghaie A (2010) Mining trust and distrust relationships in social web applications. In: Proceedings of the 2010 IEEE 6th international conference on intelligent computer communication and processing, pp 73–80
18. Saeidi S (2020) A new model for calculating the maximum trust in online social networks and solving by artificial bee colony algorithm. Comput Social Netw 7:1–21
19. Son J, Choi W, Choi SM (2020) Trust information network in social internet of things using trust-aware recommender systems. Int J Distrib Sens Netw 16(4):1550147720908773. https://doi.org/10.1177/1550147720908773
20. Ding K, Zhang J (2020) Multi-party privacy conflict management in online social networks: a network game perspective. IEEE/ACM Trans Netw 28(6):2685–2698
21. Hu H, Ahn G, Jorgensen J (2013) Multiparty access control for online social networks: model and mechanisms. IEEE Trans Knowl Data Eng 25(7):1614–1627

22. Ali S, Rauf A, Islam N, Farman H (2017) A framework for secure and privacy protected collaborative contents sharing using public osn. Clust Comput 22:7275–7286
23. Ilia P, Carminati B, Ferrari E, Fragopoulou P, Ioannidis S (2017) Sampac: socially-aware collaborative multi-party access control. In: Proceedings of the seventh ACM on conference on data and application security and privacy, ser. CODASPY'17. Association for Computing Machinery, New York, NY, USA, pp 71–82. https://doi.org/10.1145/3029806.3029834
24. Vishwamitra N, Li Y, Wang K, Hu H, Caine K, Ahn GJ (2017) Towards pii-based multiparty access control for photo sharing in online social networks. In: Proceedings of the 22nd ACM on symposium on access control models and technologies, ser. SACMAT'17 Abstracts. Association for Computing Machinery, New York, NY, USA, pp 155–166. https://doi.org/10.1145/3078861.3078875
25. Khan J, Lee S (2018) Online social networks (osn) evolution model based on homophily and preferential attachment. Symmetry 10(11). https://www.mdpi.com/2073-8994/10/11/654
26. Michelson M, Macskassy SA (2010) Discovering users' topics of interest on twitter: a first look. In: Proceedings of the fourth workshop on analytics for noisy unstructured text data, ser. AND'10. Association for Computing Machinery, New York, NY, USA, pp 73–80. https://doi.org/10.1145/1871840.1871852
27. Krithika LB, Roy P, Jerlin MA (2017) Finding user personal interests by tweet-mining using advanced machine learning algorithm in r. IOP Conf Ser: Mater Sci Eng 263:042071, nov 2017. https://doi.org/10.1088/1757-899x/263/4/042071
28. Khan J, Lee S (2019) Implicit user trust modeling based on user attributes and behavior in online social networks. IEEE Access 7:142826–142842
29. Hu H, Ahn GJ, Jorgensen J (2011) Detecting and resolving privacy conflicts for collaborative data sharing in online social networks. In: Proceedings of the 27th annual computer security applications conference, ser. ACSAC'11. Association for Computing Machinery, New York, NY, USA, pp 103–112. https://doi.org/10.1145/2076732.2076747

Startup Profit Predictor Using Machine Learning Techniques

Manasi Chhibber

Abstract Startups focus on uplifting and transforming the old markets by bringing in new technologies that may possess the power to bring a revolution in the world. Hence, they find it very essential to keep their venture in a profitable position from beginning itself so that they can achieve their goals sustainably. Intelligent systems which use machine learning can process huge amount of statistical data and can be used to predict profits based on the startup's various expenses and other parameters. This can help them in regulating their expenses and grow quickly. The predictor makes use of four parameters, i.e., spend on R&D, administration, marketing (www.aitpoint.com, Last accessed 12 June 2021 [1]), and the location where the startup is based out of, and predicts an approximate value of the profit that it is most likely to make. On finding the perfect dataset, data preprocessing and data visualizations were done. The data was split for training and testing and the score of the model obtained on the test dataset was 0.9691. The startup profit predictor was successfully built using random forest regressor model and it can be used for making profit predictions with 96.91% accuracy.

Keywords Startup profit predictor · Random forest algorithm · Machine learning

1 Introduction

1.1 Need for Startup Profit Predictor

Startup costs are bound to incur whenever a new business begins to take shape. Be it spending on research, employees, advertising, promotions, or inventory, the initial costs are very high. Even a financial advisor comes at a hefty price. New businesses also lack experience [2]. An investment which might look good in the beginning may prove to be the reason for their decline in the future [3]. In such a scenario, relying on

M. Chhibber (✉)
Amity School of Engineering and Technology, Amity University, Noida, Uttar Pradesh, India
e-mail: manasichhibber@gmail.com

P. Dutta et al. (eds.), *Emerging Technologies in Data Mining and Information Security*, Lecture Notes in Networks and Systems 490,
https://doi.org/10.1007/978-981-19-4052-1_6

a human alone, might not be the best option. There is a high possibility of inaccurate results. This is the reason why there comes a need of something that is affordable, reliable, long-term, and intelligent enough to make smart, calculative decisions to minimize the expenses. Here are where artificially intelligent systems can be used which possess all these qualities.

1.2 Literature Review

The startup profit predictor should essentially take in parameters that are important for a startup's growth like its expenditure in various sectors and location where they are based in, history of founders, etc., as inputs and should predict an approximate value denoting the profit that the company can make. This can help the people associated with the startup to adjust their budget accordingly so that they can make reasonable choices and still align with their vision.

In the past, there has been a lot of research aimed at studying the trends of a startup and its profitability [4]. Fuertes-Callén et al. [5] analyzed the trend of certain startups for years and using the 'survival of the fittest' principle to predict startup survival probability using their first year's financial statements. There have been attempts to make a predictor too. Ünal [6] used multiple machine learning techniques and was able to predict profits with a maximum test accuracy of 94.5% using extreme gradient boosting algorithm. Ang et al. [7] attempted to predict a startup's post-money valuations using Bayesian optimization and gradient boosting regressor and were able to obtain an accuracy of 95% on test sets. Although, the results are promising but I have managed to achieve more accurate results while keeping in loop the main features which are responsible for a startup's growth.

1.3 Overview of Machine Learning

Machine learning aims at understanding data and then fitting it into models, so that it can be understood and made use of. Its primary focus is to learn from data and keep on improving its performance.

Machine learning is a bifurcation of artificial intelligence that covers the quantitative part of it. It enables computers to resolve problems by analyzing hundreds and thousands of examples, taking insights from them and then using that heuristic knowledge to solve the same problem with renewed terms [8]. Deep learning is further a peculiar field of machine learning wherein computers are trained to understand insights and make smart decisions on their own [9]. This involves a much deeper level of automation with respect to most of the ML algorithms.

Machine learning has a very strong impact on the society. Given ahead are some real-life examples. Netflix and Amazon provide recommendations for videos, movies, and TV shows to its users with the help of ML [10]. On the basis of their past views, machine learning produces suggestions that one might enjoy. This is very similar to how any person might recommend a TV show to someone, based on their learning of the types of shows that the person likes to watch. ML is used when banks are supposed to make a crucial decision of approving or disapproving someone's loan application. They predict the default probability for all the applicants and then grant or revoke the loan application on the basis of that probability. Companies that deal in telecommunication use their customers' data to segregate them and to predict whether they are going to unsubscribe the company's services or continue the same in the coming month. There are many other applications of ML as well that we can see every day in our day-to-day lives, like chatbots logging into phones. Even computer games use face recognition now. For all the above-mentioned applications, various machine learning algorithms are used.

Learning techniques in ML can be majorly divided into two types, supervised learning, and unsupervised learning [11]. In supervised learning technique, model is trained on the basis of labeled data. This means that some data is already tagged with the correct target value. Once fit on the training data, the model is provided with a new set of data so that it can produce an accurate outcome depending upon its learning. There are specifically two types of supervised learning methods or techniques. They are known as regression and classification techniques. Classification technique is used for predicting discrete class labels [8] or for predicting a category [12], whereas regression technique is for predicting continuous value type results. On the other hand, unsupervised learning technique is the one in which we allow the ML model to find structures within its input data which does not have labels of any kind. In this project, a supervised regression-based technique has been used because we took a dataset which had properly defined labels with a target variable which was not of categorical type.

There are many algorithms in ML which seek meaningful insights from data, interpret it, and can be used for making predictions. Based on what problem needs to be solved, any one of them could be chosen. Details on some of the algorithms are listed below (Table 1).

Table 1 Various machine learning algorithms

Algorithm	Learning technique	Application	Remarks
Linear regression	Supervised	Regression-based tasks	The data points are plotted, and a best fit line is chosen
Logistic regression	Supervised	Classification-based tasks	Logistic function when applied can be used for binary classification
Random forest	Supervised	Regression + classification-based tasks	Tree-based algorithm
K-means clustering	Unsupervised	Clustering-based tasks	Similar data points are grouped together

Most accurate results were provided by random forest regressor model, hence, it was used for building the startup profit predictor.

2 Random Forest Algorithm

Random forest algorithm is used in both classification and regression problems [13]. The algorithm is based on the notion of ensemble learning. Ensemble learning is the procedure of combining many classifiers to solve a single complex problem and hence heighten the accuracy of the model.

Random forest algorithm is very much dependent on decision trees. So, before moving on with the discussion of random forest regressor and random forest classifier, understanding of decision trees is important.

2.1 Decision Trees

Decision trees test an attribute and branch the cases based on the outcome [14] of the test. Each internal node represents a test, and each branch represents the outcome of the test. They are built using recursive partitioning and are used for classifying the data. It is important to find which attributes are the most [15] promising (more predictive) for splitting the data. The purity of leaves after splitting is based on this. A node in a tree is considered to be pure if in 100% of cases, and the node falls into a particular category of the target field. The recursive partitioning splits the training records into parts by minimizing the impurity at each and every step. The impurity level of nodes is calculated by calculating the entropy of the data in the node which is the amount of information which is in disarray in the data or the amount of randomness in it. It depends on the amount of data that [16] is present in that node and is calculated for each and every node. In decision trees algorithm, we look for trees having the least entropy in their respective nodes. It is also used for calculating homogeneity of samples in a particular node. If samples are totally homogeneous, then the entropy is zero, otherwise, if the samples are equally divided, then, it has entropy as one. As entropy decreases, the information gain increases, and reverse is also true. Hence, building a decision tree is all about looking for attributes that can return the best information gain. If we keep all these factors in our minds, then it becomes very easy to build a decision tree.

Random forest is the kind of classifier that consists of many decision trees based on [14] various groups formed from the given dataset, and it takes the average of all of them to improve the accuracy. Instead of relying on a single decision tree [14], this algorithm takes the prediction from every single tree and on the basis of majority votes of the predictions, it predicts the final result. More the number of trees in the forest, better the accuracy and it also prevents the problem of overfitting. It can perform both classification and regression and is capable of handling very large datasets having high dimensionalities. However, if a large number of trees are present, they can make the algorithm very slow and ineffective for the purpose of real-time predictions. Being an ensemble of decision trees, it often suffers from interpretability and sometimes fails to determine the importance of each variable in the dataset.

2.2 *Random Forest Regressor*

Individual decision trees have high variance, but there has been an observation that when all of them are combined together, the final variance becomes less as every decision tree gets very precisely trained on specific sample data and so the output does not depend on a single one rather multiple trees [17]. In case of regression problems, the final result is the mean of all the results. This is called as 'aggregation'. A random forest model performs various tasks with the help of multiple decision trees and by using a technique called 'bagging' which is a combination of 'bootstrap' and 'aggregation'. 'Bootstrap' is the part where we perform random row and feature sampling from the dataset. These act as sample datasets for every model or individual decision trees. Random forest is very effective if there exists a nonlinear or complex relationship between the features and the labels. The random forest algorithm is very robust as it uses a set of decision trees which are not correlated.

3 Other Algorithms Used

In contrast with using the random forest algorithm, some other algorithms were also taken into account for comparing the performance of different models based on the accuracy achieved.

Linear Regression. A linear regression model is used in cases [18], where the predictor attributes have a linear relationship with the target attribute. All the data points are [19] plotted in a scatter plot and a best fit line or hyper-plane is chosen in such a way that minimizes the sum of squared errors. The equation of this regression line is further used for making predictions.

XGBoost Algorithm. Similar to the random forest algorithm, XGBoost algorithm is also based on the concept of ensemble learning, but here, the trees are added one by one, and each time, they are fit to reduce the prediction errors given by the models preceding it. Theoretically, it is more efficient and effective than the random forest algorithm.

Decision Trees. An individual decision tree can also be used for making continuous value type predictions. Here, the mean of all the leaf nodes is calculated for generating the results.

Gradient Boosting Algorithm. The main aim of this algorithm is to form an ensemble model by making use of weak predictive models. The fitting of the model begins with a constant value like the mean of the values in the target attribute, and then, in each subsequent step, decision trees are fitted to compute the negative gradients in the samples. These gradients are updated in each iteration and a learning rate is set for minimizing the negative gradient from each following tree.

Light GBM Algorithm. It is just like the XGBoost algorithm, the only difference being in the manner of how the trees are spanned. In XGBoost, a level-wise tree growth occurs, whereas in light GBM, a leaf-wise tree growth occurs which reduces the losses. This algorithm has low space and time complexities and can support all parallel, distributed and GPU learning techniques.

Cat Boost Algorithm. In XGBoost and light GBM algorithms, numerous asymmetric trees are formed. The CatBoost algorithm follows an oblivious tree making approach. Here also, the weak learning trees are considered, and a greedy search algorithm creates a strong predictive model. This algorithm is efficient in terms of CPU utilization too [20].

4 Startup Profit Predictor

4.1 Acquiring Data and Preprocessing Data

For building the 'startup profit predictor,' the perfect dataset was found on Kaggle. It consisted of four attributes on the basis of which the profit could be predicted. These were: Research & development (R&D) spend, administration spend, marketing spend, and the state, where they are based in the USA, namely New York, California, and Florida. The total number of records in the dataset were fifty. The R&D spend, administration spend, and marketing spend attributes had float type data, so they did not require any encoding. On the flip side, the state attribute needed encoding. After label encoding, each state got assigned a unique integer from 0 to 2. Then, VIF or the variance inflation factor was checked to calculate the amount of multicollinearity between the attributes of the dataset. If the variance inflation factor is greater than 10, then it means that the attributes are highly correlated and because of it, the predictions for the desired attribute could be inaccurate. So, in case it is greater than 10 for any column, that attribute is dropped out of the dataset. On checking, the variance inflation factor for 'R&D spend,' 'administration,' 'marketing spend,' and 'state' attributes was found to be 8.386322, 4.815916, 7.674608, and 2.382637, respectively. As all of the attributes had variance inflation factor less than 10, so none of them was dropped. Moving on, outliers in the data were checked. The outliers result in the predictions being very off from the value that they actually should be. The 'seaborn' library was used for plotting box plots for each attribute. No outliers were found. With this, the

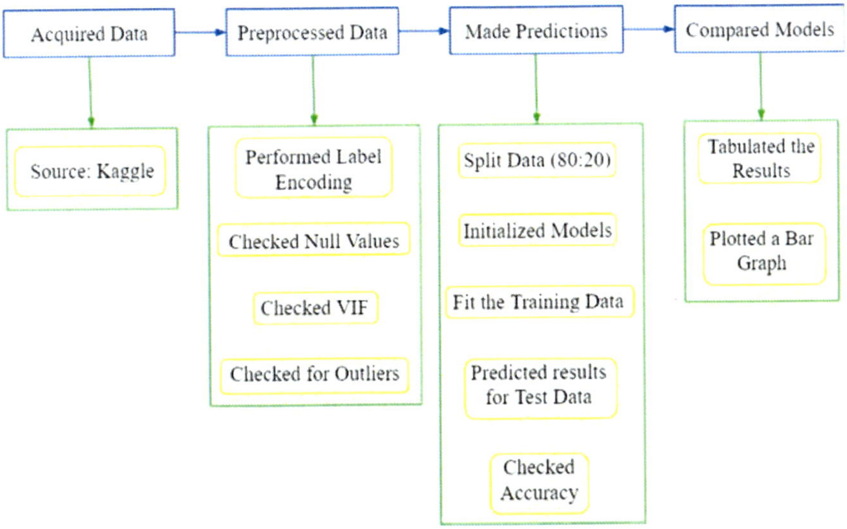

Fig. 1 Complete workflow

preprocessing of the data was complete. The dataset was clean, and predictions could be made accurately. This complete process is depicted in Fig. 1.

4.2 Building the Model

Before creating the model, it is essential to split the data for training and testing. The 'train_test_split' function was used from 'sklearn' library to perform the splitting. The data was split in the ratio of 80:20, where 80% was used for training the random forest regressor model and 20% was left for validation. The 'random_state' parameter are set to 0 so that I get the same results every time. This practice makes it easy to debug the code if required in the future. With this, the splitting was efficiently done. Next, the process of making the model was begun.

The random forest regressor model was initialized with parameter 'random_state' set to 0 as well. Then, the model was fit on training set which had 'X_train' and 'y_train' as parameters. The model was built successfully and now predictions on the test set could be made. The results were predicted on the test set, 'X_test,' then the inbuilt 'score' function was used for obtaining the accuracy score. For regressors, it is nothing but the R^2 score as denoted in Eq. 1 below. S_{res} is the sum of squares of the residual errors, and S_{tot} is the total sum of errors.

$$R^2 = 1 - \frac{S_{\text{res}}}{S_{\text{tot}}} \quad (1)$$

Table 2 Models' comparison summary

Algorithm used	Accuracy achieved ('score' value)	Time consumed (in seconds)
Linear regression	0.938686	1.28675
Random Forest	0.969131	0.119937
XGBoost	0.930025	0.889408
Decision tree	0.960979	0.00104856
Gradient boosting	0.939674	0.0304089
Light GBM	0.495692	0.0368373
Cat boost	0.80782	1.27958

The score came out to be 0.9691314428968053. Hence, it meant that the predictions were approximately 96.91% accurate.

4.3 *Comparing Results with Different Models*

A similar procedure was followed for making predictions using other regressor models. The results were tabulated using 'tabulate' library and compared with each other [21] using a bar plot built using 'matplotlib' library. It can be seen in Table 2 that the random forest regressor model provided the highest accuracy of 96.91% in 0.120 s. Using linear regression model, an accuracy of 93.86% was obtained in 1.287 s. Using XGBoost algorithm, 93.00% in 0.889 s, decision trees, 96.09% in 0.001 s, gradient boosting algorithm, 93.96% in 0.030 s, light GBM algorithm, only 49.56% in 0.037 s and CatBoost algorithm provided 80.78% accuracy in 1.280 s. If time constraints are there, then decision tree algorithm could be your best choice as it has a good accuracy and is time effective too. Accuracies achieved using different models are represented in Fig. 2.

5 Conclusion

Machine learning aims at giving computers the power to understand data and learn from it without being explicitly programmed. With the benefit of machine learning, we can extract insights and can train models so that it could be made efficient use of, by people and corporations. Supervised and unsupervised learning methods or techniques [22] are the principal ones which are used for training a machine learning model. Random forest algorithm is a kind of supervised learning technique. It is based on the idea of ensemble learning and uses decision trees as classifiers.

After taking several algorithms into consideration, the startup profit predictor was successively built using a random forest regressor model. All the data preprocessing and analysis resulted in a model that can provide 96.91% accurate results (Table 2).

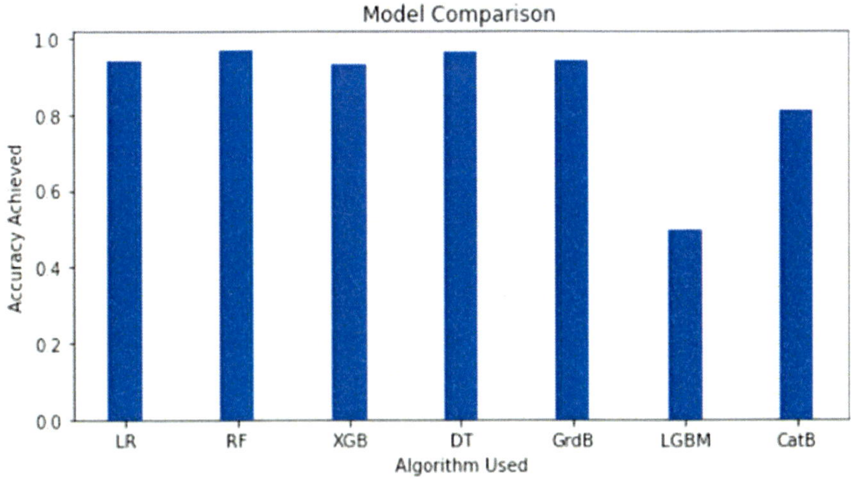

Fig. 2 Accuracies achieved using different models

In the future, a parameter could be formulated based on the accuracy of the models and their time complexity both, using which could be used to deduce the best possible algorithm and make the predictor with that. Also, to upgrade the project, a web or a mobile application could be made. It can be deployed so that the startup profit predictor is made accessible to everybody, everywhere.

References

1. AIT Point Website. www.aitpoint.com. Last accessed 12 June 2021
2. Hyytinen A, Pajarinen M, Rouvinen P (2015) Does innovativeness reduce startup survival rates? J Bus Ventur 30(4):564–581
3. Aggarwal R, Singh H (2013) Differential influence of blogs across different stages of decision making: the case of venture capitalists. MIS Q 37(4):1093–1112
4. Xiang G, Zheng Z, Wen M, Hong JI, Rosé CP, Liu C (2012) A supervised approach to predict company acquisition with factual and topic features using profiles and news articles on TechCrunch. In: ICWSM. AAAI, June 2012
5. Fuertes-Callén Y, Cuellar-Fernández B, Serrano-Cinca C (2020) Predicting startup survival using first years financial statements. J Small Bus Manage 58
6. Ünal C (2019) Searching for a unicorn: a machine learning approach towards startup success prediction. Master's thesis, Humboldt-Universität zu Berlin
7. Ang YQ, Chia A, Saghafian S (2020) Using machine learning to demystify startups funding, post-money valuation, and success. In: Post-money valuation, and success
8. GitHub. www.github.com. Last accessed 1 Oct 2021
9. IBM Community Page. https://community.ibm.com/community/user/ibmz-and-linuxone/blogs/subhasish-sarkar1/2020/03/01/ai-ml-and-deep-learning-introduction. Last accessed 6 July 2021
10. Irjlis Website. www.irjlis.com. Last accessed 12 July 2021

11. Dutta S, Ghatak S, Sarkar A, Pal R, Pal R, Roy R (2019) Cancer prediction based on fuzzy inference system. In: Tiwari S, Trivedi M, Mishra K, Misra A, Kumar K (eds) Smart innovations in communication and computational sciences. Advances in intelligent systems and computing, vol 851. Springer, Singapore. https://doi.org/10.1007/978-981-13-2414-7_13
12. Malik D, Munjal G (2021) Reviewing classification methods on health care. In: Intelligent healthcare. Springer, Cham, pp 127–142
13. Analytics Vidhya Website. https://www.analyticsvidhya.com/blog/2021/05/bagging-25-questions-to-test-your-skills-on-random-forest-algorithm/. Last accessed 22 June 2021
14. Section Website. www.section.io. Last accessed 9 June 2021
15. Covenant University Homepage. www.covenantuniversity.edu.ng. Last accessed 28 June 2021
16. Machine Learning Mastery Website. https://machinelearningmastery.com/how-to-develop-a-convolutional-neural-network-to-classify-photos-of-dogs-and-cats/. Last accessed 18 June 2021
17. GeeksforGeeks Website. www.geeksforgeeks.org. Last accessed 24 Aug 2021
18. Towards Data Science Website Homepage. www.towardsdatascience.com. Last accessed 10 Aug 2021
19. E-Scholarship Website. www.escholarship.org. Last accessed 5 Sept 2021
20. CatBoost Page, GitHub. https://github.com/catboost. Last accessed 13 Aug 2021
21. NWU Website. www.dspace.nwu.ac.za. Last accessed 26 Aug 2021
22. Worldwide Science Website. www.worldwidescience.org. Last accessed 15 Aug 2021

Right to Be Forgotten in a Post-AI World: How Effective is This Right Where Machines Do not Forget?

Gagandeep Kaur and Aditi Bharti

Abstract The right to be forgotten is the concept wherein an individual has the right to request for data deletion. While it is simple from human perspective, data deletion becomes complicated when an AI-based technology is involved as AI neither process the data nor do they 'forget' the way humans do. The laws on privacy have been enacted from the perspective of human memory, and therefore, their efficacy is in question when AI-based technologies process the data. Data protection laws are majorly about safeguarding the right to decide how the information is being used by the algorithm. This can become challenging when the data processing is done by an artificially intelligent entity, since it would be difficult to understand and explain how the information is correlated and used in the specific process. Moreover, the privacy rights can be exercised only when the individual is aware of the process details regarding the usage of the data. In this paper, the authors have analysed the right to be forgotten in the context of AI and explored legal provisions in the light of how far this right can be exercised with AIs.

Keywords Artificial intelligence · Machine learning · GDPR · Right to forget · Privacy rights

1 Introduction

The fact that AI is capable of processing huge amount of data at an extraordinary speed implies that it can also remember everything, unlike a human memory poses a challenge when it comes to forgetting that data, especially in a controlled manner [1]. Models that function on deep learning take decisions independently and therefore have relatively low transparency and are complex when compared to ordinary

G. Kaur (✉) · A. Bharti (✉)
University of Petroleum and Energy Studies, Dehradun, India
e-mail: gkaur@ddn.upes.ac.in

A. Bharti
e-mail: abharti@amity.edu

P. Dutta et al. (eds.), *Emerging Technologies in Data Mining and Information Security*, Lecture Notes in Networks and Systems 490,
https://doi.org/10.1007/978-981-19-4052-1_7

algorithm programmes. Online stores process a lot of data and information. With the enactment of GDPR, the companies are required to comply with the regulations before collecting the information. Pop-up consent banners are, therefore, the result of such compliance. The e-commerce apps are AI based, i.e. they record the information and suggests the consumers according to their search history or purchasing habits thereby, learning by using and storing the data. Extending data protection laws to such algorithms and deletion of such data are majorly about safeguarding the right to decide how the information is being used by the algorithm. This can become challenging when the data processing is done by an AI entity, since it would be difficult to understand and explain how the information is correlated and used in the specific process. GDPR grants autonomy to the individual to decide how his personal information is being used while imposing a duty on the user to maintain the privacy of the subject in the best possible manner [2]. Although, the right to control one's information over the Internet sounds justifiable, its implementation seems to be on the edge of impossibility. It is rather a challenging task for the legislature to address the issue of protecting personal information against any adverse usage but without compromising on the AI development. This paper examines the problem whether AI can actually forget? The paper would analyse the existing legal position on right to be forgotten on the basis of existing literature on the topic, followed by an analyses of the applicability of the privacy laws in the AI world. The authors, while concluding, would discuss the practical problems posed in enforcing right to be forgotten in AI [3–5].

2 Historical Footprints of the Right to Be Forgotten

The origin of this right can be traced back to French jurisprudence on the 'right to oblivion', which was to facilitate social re-integration of the offenders after serving their sentence. Drawing inspiration from this, EU made the right to be forgotten enforceable by enacting the 1995 European Union Data Protection Directive, which recognised the right of an individual to control, ratify, erase or block data related to them [6]. This right gained momentum in 2010 with a lawsuit filed against Google Inc. and a Spanish newspaper in Spain [7]. The citizen claimed violation of privacy rights by Google as their search results included information regarding his past legal liabilities. The court while upholding the right of privacy of the individual, stated that whoever market and promote their services in EU would be governed by data protection directives, and therefore, would have the duty to remove information associated with their customers, if requested. In yet another case, filed against Facebook, for not removing data, shadow profiling and storing excess personal data even though he had requested for removal of the information. Facebook had merely removed the links of the information that was 'deleted' by the user. Making the data invisible by removing the link to it was an interesting approach which was highlighted in this case [8]. This, again, raises a question that can an information be deleted from an AI?

3 Existing Regulatory Framework on Right to Be Forgotten

Due to ever-expanding nature of the Internet, the idea of privacy and subsequent threat to it has been one of the major concerns for legal and regulatory authorities. GDPR has recognised the right to erasure under Article 17 which casts a duty on the controller to remove the personal data without any undue delay when the data is no longer necessary; if the individual withdraws consent on which processing of data depends; when the individual objects to the data processing; when data has to be removed due to legal requirements; when the individual was a child at the time of data collection. Although, it is not an absolute right in itself [2].

Although, Article 17 have been incorporated to remove the information from the Internet, it fails to address the fundamental challenge posed by the basic nature of machine learning and its applicability in the industry [5, 6]. The development of an AI entity requires various types of data which might also include information collected for other purposes. Even though, using such information might be useful in giving out more accurate and useful analysis, it would nevertheless, amount to violation of privacy. For example, if an algorithm takes into account social media activity that determines if the person should get a loan from the bank or not.

4 Technical Analysis of AI Memory: Can AI Actually Forget?

The preliminary question to be asked in this reference is can we actually delete an information from an algorithm? The authors have tried to answer this issue with the help of an illustration based on database management system.

MySQL database have been taken to be the reference in respect of our example. Figure 1a is the state before data deletion. C1 to C5 represents the space for data storage as well as the start (I) and the end (S) of the database. C3 is the deleted record which is linked with ‘garbage-offset’, a collection of deleted and now free space. When a data is searched, the algorithm would look for the search tree where

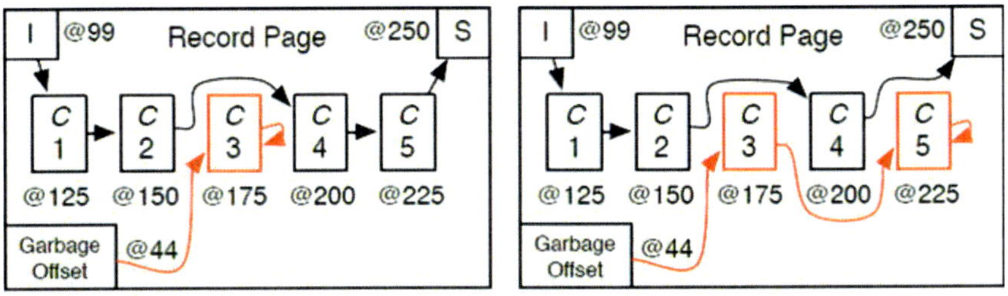

(a) One deleted record (b) Two deleted record

Fig. 1 Deletion in MySQL [9]

the data might be stored and would start from I and while following the path of arrows would try to locate the data. If the search ends at S with no result, no data was found. The task is removal of data stored in C5. The database would look for the data stored in C5 while navigating in the tree and would mark the space 'for deletion' once C5 is located. The bending row from C3 to C5 denotes that C5 is added to the garbage offset. It can be seen that even though the data has been superficially deleted, it is still stored in the database. This raises a question on the efficacy of right to be forgotten and what does the law actually mean by 'deletion of data'.

5 Conclusion

To conclude, implementing right to forget cannot be enforced in the same manner as in human environment. The laws have been framed with human perspective, and therefore, regulating a technology would pose challenges before the law. Moreover, the existing laws do not define what method of data deletion would satisfy the legal requirements under right to be forgotten. It will not be wrong to say that the current laws are ill-suited to address the problems raised by technological advancements. However, at the same time, it is imperative for the legal community to address this gap without compromising on the ability and providing an opportunity to AI to develop.

References

1. Esposito E (2017) Algorithmic memory and the right to be forgotten on the web. Big Data Society:1–11.https://doi.org/10.1177/2053951717703996
2. Regulation (EU) 2016/679 (General Data Protection Regulation)
3. Granter S, Beck A, Papke D (2017) AlphaGo, deep learning, and the future of the human microscopist. Arch Pathol Lab Med 141(5):619–621. https://doi.org/10.5858/arpa.2016-0471-ED
4. Getting the Future Right: Artificial Intelligence and Fundamental Rights. European Union Agency for Fundamental Rights (2020) Electronic resources. Retrieved 4 Nov 2021 from https://fra.europa.eu/sites/default/files/fra_uploads/fra-2020-artificial-intelligence_en.pdf
5. Tamò A, George D (2014) Oblivion, erasure and forgetting in the digital age. J Intellect Property, Inf Technol E-Commerce Law 5(71). Retrieved from https://www.jipitec.eu/issues/jipitec-5-2-2014/3997
6. Directive 95/46/EC of the European Parliament and of the Council of 24 October 1995 on the protection of individuals with regard to the processing of personal data and on the free movement of such data
7. Google Inc v. Agencia Española de Protección de Datos (AEPD), Mario Costeja González, Court of Justice of the European Union (2014) C-131/12. Retrieved from https://eur-lex.europa.eu/legal-content/EN/TXT/?uri=CELEX%3A62012CJ0131
8. The Data Protection Commissioner v Facebook Ireland Limited &Anr. ECLI: EU: C: 2020:559. Available at http://www.europe-v-facebook.org/sh2/HCJ.pdf
9. Fruhwirt P, Kieseberg P, Weippl E (2015) Using internal mysql/innodb btree index navigation for daINTETta hiding. In: IFIP international conference on digital forensics, pp 179–194

Age, Gender, and Gesture Classification Using Open-Source Computer Vision

Gaytri Bakshi, Alok Aggarwal, Devanh Sahu, Rahul Raj Baranwal, Garima Dhall, and Manushi Kapoor

Abstract In every country, there are a plethora of laws whose very foundation stands on the age of the concerned. Similarly, successful gender recognition is essential and critical for many applications in the commercial domains, like human–computer interactions: such as computer-aided physiological or psychological analysis of a human. In this work, a face and gesture detection and verification system are proposed that classifies them on the basis of gender while providing the most probable age range of the concerned face and also detecting the gesture of the hands using convolutional neural network architecture. The principal idea behind the system is to compare the image with the reference images stored as templates in the database and to determine the age and gender.

Keywords Age classification · Gender classification · Gesture classification · Open-source computer vision

G. Bakshi (✉) · A. Aggarwal
School of Computer Science, University of Petroleum and Energy Studies, Bidholi, Dehradun, India
e-mail: gaytri.b88@gmail.com

A. Aggarwal
e-mail: alok.aggarwal@ddn.upes.ac.in

D. Sahu
EY, Gurgaon, India

R. R. Baranwal
Rakuten, Bangalore, India

G. Dhall
Infosys, Bangalore, India

M. Kapoor
VMware, Bangalore, India

P. Dutta et al. (eds.), *Emerging Technologies in Data Mining and Information Security*, Lecture Notes in Networks and Systems 490,
https://doi.org/10.1007/978-981-19-4052-1_8

1 Introduction

Age is just a number', well that is not how law perceives it and neither does any other public authority. In every country, there are a plethora of laws whose very foundation stands on the age of the concerned [1, 2]. For example, there is a defined 'age of majority' in every country which legally demarcates childhood from adulthood and is usually set to 18. Furthermore, the age range of a person helps in the documentation and identification of a person as it is a reflection of the person's birth date, which most often is an essential field in the verification documents (Passport, driving license, etc.). Similarly, successful gender recognition is essential and critical for many applications in the commercial domains [3], like human–computer interactions: such as computer-aided physiological or psychological analysis of a human.

Documents verification for the identification of a person on prime security check-points [4] like airports, railways and metro stations, border check points, etc., and verification of the identity of a person in exam centers rely heavily on the written documentation provided by the person himself. Furthermore, the age restriction in movie theaters and similar online restrictions depend upon the word of the person rather than a tangible proof of age. Age and gender classification using open-source computer vision is an application of deep learning applied to faces. Here, the convolutional neural network architecture is being used which is a deep neural network (DNN) used for the processing and recognition of images. The proposed work deals with two problems: gender classification and age detection. Both of them are framed as classification problems. It is obvious that gender determination is a classification problem, but in the case of age detection, it seems as if it should be approached as a regression problem given that we are expecting a real number as an output. However, even humans determine only an age approximation of a new face instead of an accurate age prediction. Therefore, it is better to frame the problem as a classification problem to estimate an age range of a new face. For example, age in range (48–53) is one class, while age in range (60–100) is another.

In this work, a face and gesture detection and verification system is proposed that classifies them on the basis of gender while providing the most probable age range of the concerned face and also detecting the gesture of the hands using convolutional neural network architecture. The principal idea behind the system is to compare the image with the reference images stored as templates in the database and to determine the age and gender. Objectives of the proposed work are detecting faces using the function getFaceBox in C++, predicting gender of the concerned face, by choosing the max of the probabilities of two classes (i.e., male and female), predicting the age range (by choosing one of the predefined classes of age) of the concerned face, detecting the hands and gestures of the concerned person, by choosing the maximum probability among various gesture classes, and displaying the output of the network on the face-image.

Rest of the paper is organized as follows. Few major works done by earlier researchers is summarized in Sect. 2. Section 3 gives the proposed methodology.

Obtained results are discussed in Sect. 4. Finally, work is concluded in Sect. 5 with conclusion and limitations of the work.

2 Related Work

Kalansuriya and Dharmaratne [5] have considered gender classification in according to the geometric difference of primary features in male and female, and in case of age detection of those faces, the proposed approach classifies them into different age groups. Eidinger et al. [6] proposed a system of age and gender detection from the images of faces acquired in challenging or in the wild conditions that would help in making it more accurate when facial features are clearly visible. The myriad variety of faces (faces from different cultures, countries, and of different colors) can make the system more efficient in recognizing a face or the origin of a certain facial structure. Yaman et al. [7] provided important insights in understanding the importance of geometric features of face not only in the gender classification but also in the age detection of a face. Only, ear images have been used to detect gender and age of a person, and hence, this outlines the importance of facial features in age and gender detection. A database of human face images has been described in [8] which is designed as an aid in studying the problem of unconstrained face recognition. This database aids in studying the unconstrained, face recognition problem. Toews and Arbel [9] proposed a novel framework for detecting, localizing, and classifying faces in terms of visual traits, e.g., sex or age, from arbitrary viewpoints and in the presence of occlusion.

3 Proposed Methodology

A convolutional neural network architecture is used which has three convolutional layers, two fully connected layers, and a final output layer (Fig. 1). The first convolutional layer has 96 nodes of kernel size 7. Second convolutional layer has 256 nodes of kernel size 5. Third convolutional layer has 384 nodes of kernel size 3. The two fully connected layers have 512 nodes each. The complete process is divided into following four parts: detecting the face, predicting the gender, determining the age, and detecting the count of fingers. Finally, results are displayed.

The first part requires detecting something in the picture which is the closest in resemblance to a human face. This has been done using deep neural network (DNN) along with the function getFaceBox in C++. Detecting face using DNN face detection is shown in Fig. 1.

Once the face is detected, gender network will be loaded to the memory, and the detected face will be passed through the network. The system will generate forward passes which will give the probabilities of two classes (male and female); maximum of the two probabilities will be selected as output of the gender classification.

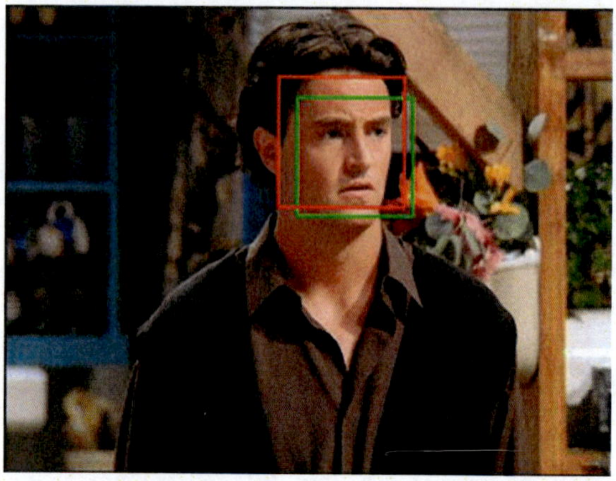

Fig. 1 Detecting face using DNN face detection

After this, the system can proceed with age range determination. Since the age determination problem is also a classification problem, same network architecture as that of the gender network has been used. Hence, the age network will be loaded to the memory, and finally, one of the following 'age range' class will be selected as the output of the age detection: (0–2), (4–6), (8–12), (15–20), (25–32), (38–43), (48–53), (60–100) (i.e., 8 nodes in the final softmax layer).

A. *Algorithm for Face Detection and Gender Classification.*

Face Detection

DNN Face Detector—OpenCV's deep learning face detector is based on the single-shot multibox detector (SSD) framework. In object detection, the task is to determine both the location and class of the object: the 'localization task' and 'classification task'. Deep learning methods for object detection use convolutional neural network-based approaches to achieve this. The earliest methods of object detection separated these tasks into two logical steps:

(1) Propose candidate regions using a region proposal network.
(2) Classify them using a classifier.

Faster R-CNN and its predecessors are two-stage networks. While these have very high accuracy, they are also relatively heavy since there are two neural networks used to do the detection task. Thus, they are termed 'two-stage' networks. Newer methods complete both tasks in one neural network, which is why they are called single-shot detectors. The basic idea behind a single-shot detector is that it extracts features, proposes regions, and classifies them all in one forward pass.

MultiBox: This is the name of a technique for bounding box regression. SSD is good for large scale and real-time detections. SSD also has different aspect ratios built in, which can be useful. For example, giraffes are shaped very differently than cars. Blob has been used to store images.

Gender Classification

Gender network will be loaded into memory, and the detected face will be passed through the network. The forward pass gives the probabilities or confidence of the two classes. Maximum of the two outputs has been taken and used as the final gender prediction.

Age Range Classification

After this, the system can proceed with age range determination; since the age determination problem is also a classification problem, it has the same network architecture as that of the gender network. Hence, the age network will be loaded to the memory, and finally, one of the following 'age range' class will be selected as the output of the age detection: (0–2), (4–6), (8–12), (15–20), (25–32), (38–43), (48–53), (60–100) (i.e., 8 nodes in the final softmax layer) as represented in the network architecture in Fig. 2.

B. *Gesture recognition algorithm*

Basically, the first task is to detect hand by background subtraction and HSV segmentation in the video frame. Once the hand is segmented, later, the number of fingers raised will be detected. The proposed method is to first find the largest contour in the image which is assumed to be the hand. Then, the convex hull and convexity defects will be found which are most probably the space between fingers. Finally, the system will display the output on the concerned image, which is done using the 'imshow' function in C++. The flow diagram in Fig. 3 describes the steps of the algorithm.

4 Results and Discussions

Various results obtained from the proposed approach are given below. Detecting something in the picture which is closest in resemblance to a human face is shown in Fig. 4. Predicting gender of the concerned face, by choosing the max of the probabilities of two classes (i.e., male and female), is shown in Figs. 5 and 6, respectively, for female and male faces. Predicting the age range (by choosing one of the predefined classes of age) of the concerned face is shown in Fig. 7. Figure 8 shows the detection of fingers on the concerned hand, by choosing the maximum probability among various gesture classes. Figure 9 displays the output of the network on the face image.

5 Conclusion and Limitations of the Work

In this work, a face and gesture detection and verification system is proposed that classifies them on the basis of gender while providing the most probable age range

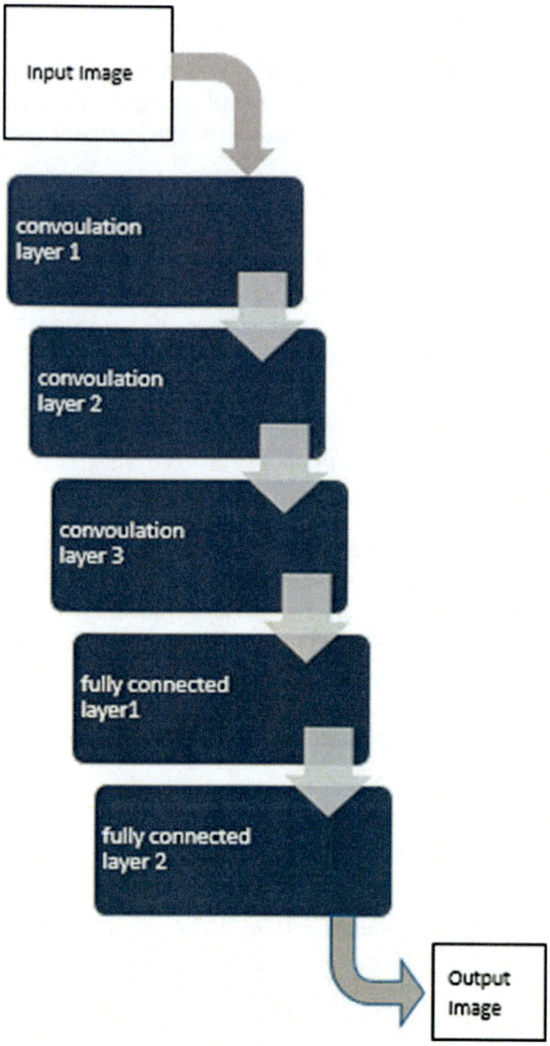

Fig. 2 Schematic diagram of the network architecture

of the concerned face and also detecting the gesture of the hands using convolutional neural network architecture. The principal idea behind the system is to compare the image with the reference images stored as templates in the database and to determine the age and gender. Following objectives are achieved: detecting faces using the function getFaceBox in C++, predicting gender of the concerned face, by choosing the max of the probabilities of two classes (i.e., male and female), predicting the age range (by choosing one of the predefined classes of age) of the concerned face, detecting the hands and gestures of the concerned person, by choosing the maximum probability among various gesture classes, and displaying the output of the network on the face image. Major limitations of the proposed work are as follows: Poor

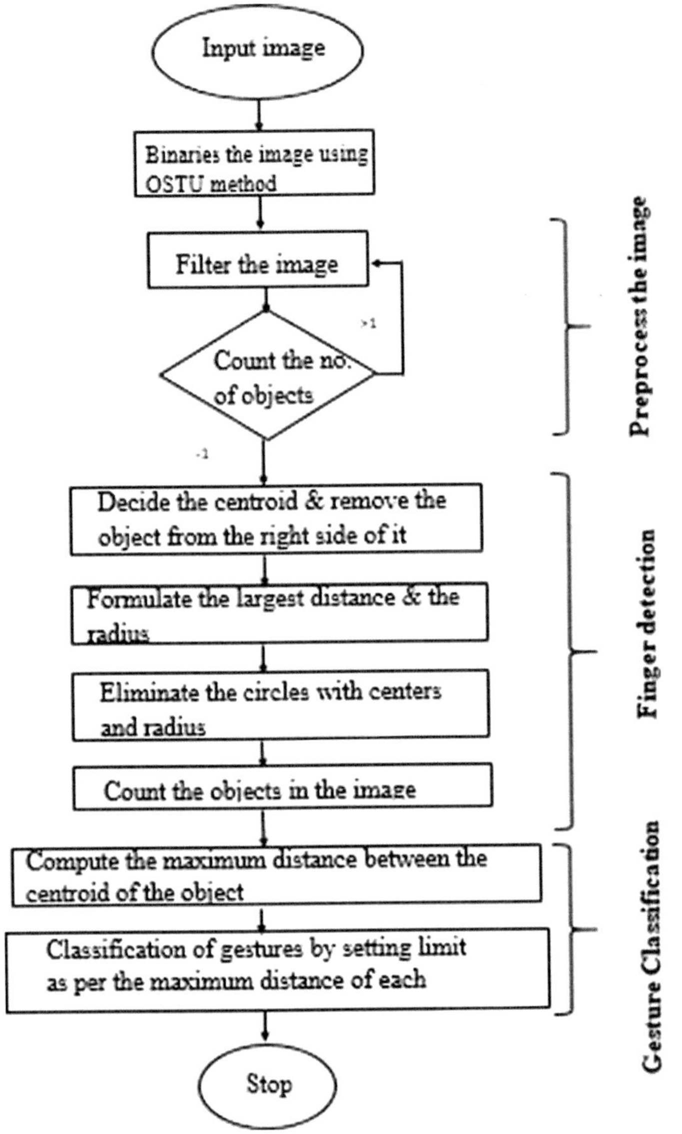

Fig. 3 Flowchart for hand gesture recognition algorithm based on finger counting

image quality limits facial recognition's effectiveness; small image sizes make facial recognition more difficult, and different face angles can throw-off facial recognition's reliability.

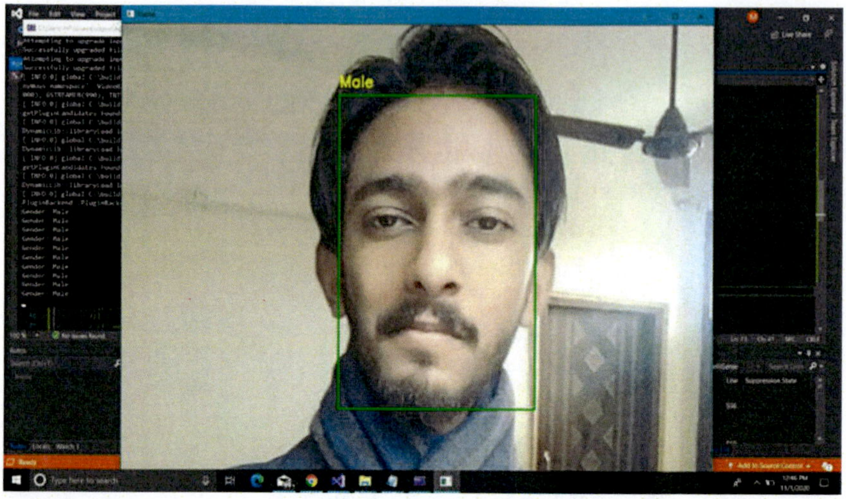

Fig. 4 Detecting faces with computer vision

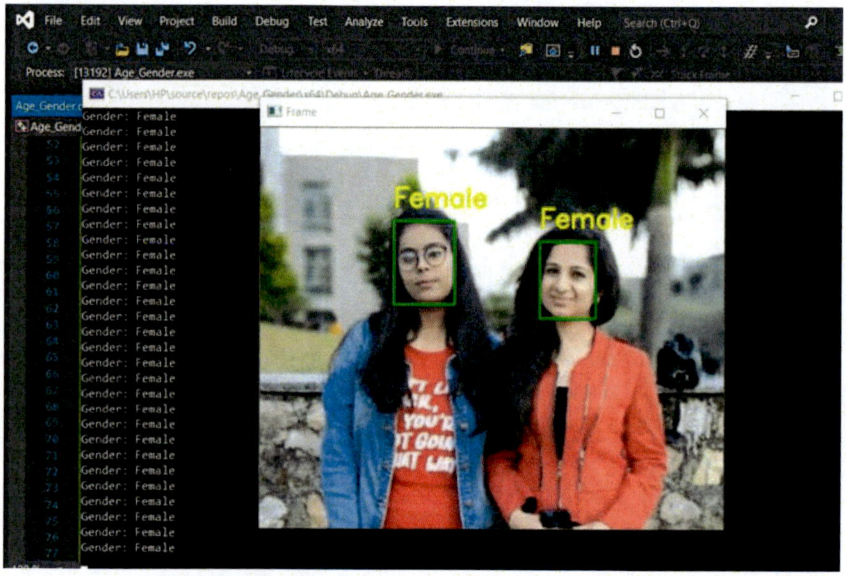

Fig. 5 Gender detected (female) after face detection

Fig. 6 Gender detected (male) after face detection

Fig. 7 Age range assignment after face detection

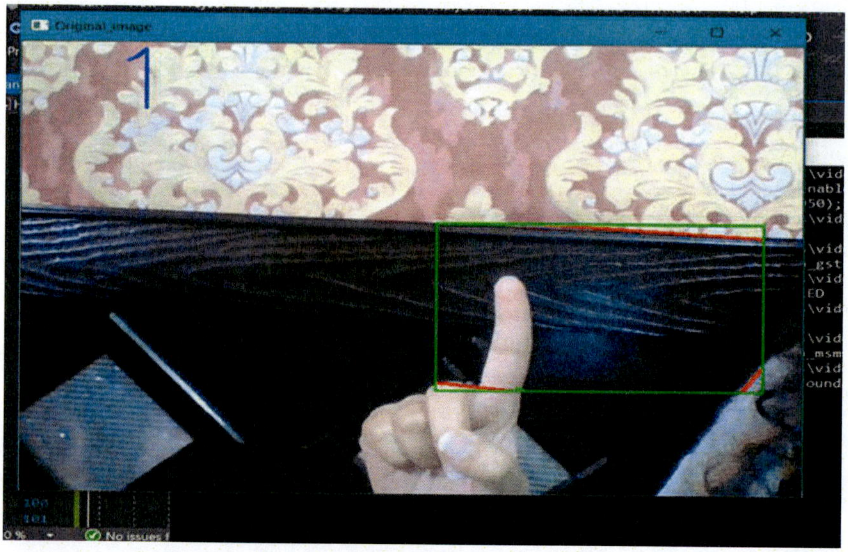

Fig. 8 Number of fingers detection after hand detection

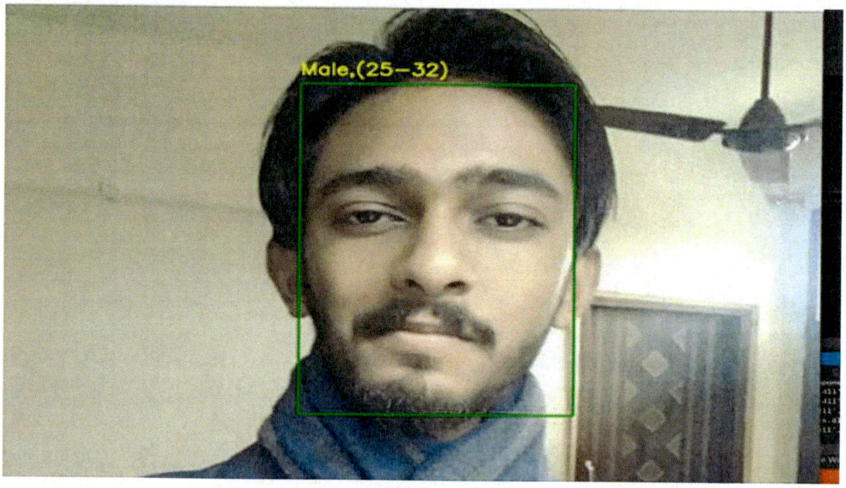

Fig. 9 **a** Gender detected (male) and age range assigned after face detection. **b** Gender detected (female) and age range assigned after face detection

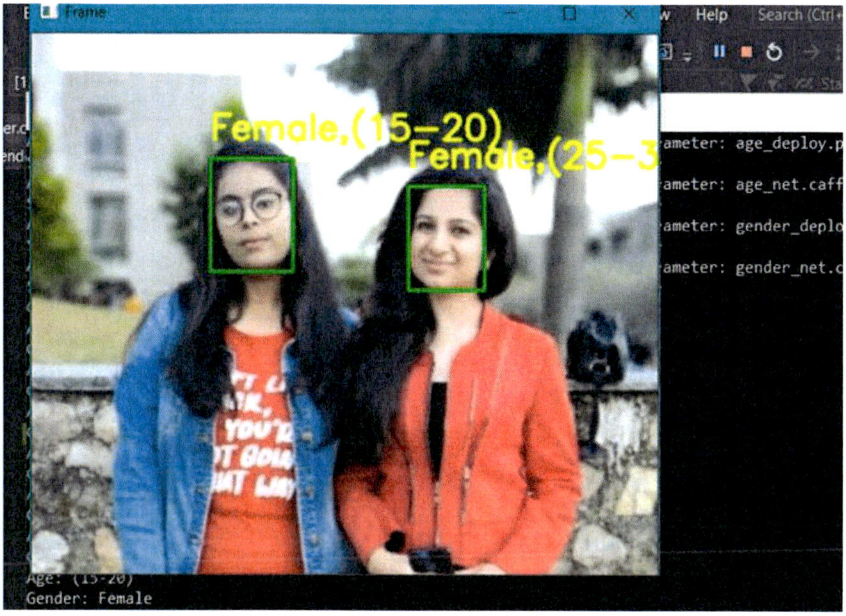

Fig. 9 (continued)

References

1. Hanmer L, Elefante M (2016) The role of identification in ending child marriage
2. Dahan M, Sudan R (2015) Digital IDs for development
3. Lin F, Wu Y, Zhuang Y, Long X, Xu W (2016) Human gender classification: a review. Int J Biometrics 8(3–4):275–300
4. Kumar VN, Srinivasan B (2012) Enhancement of security and privacy in biometric passport inspection system using face, fingerprint, and iris recognition. Int J Comp Netw Inf Secur 4(8)
5. Kalansuriya TR, Dharmaratne AT (2015) Neural network-based age and gender classification for facial images. Int J Adv ICT Emerg Regions (ICTer) 7(2):1–10. https://doi.org/10.4038/icter.v7i2.7154
6. Eidinger E, Enbar R, Hassner T (2014) Age and gender estimation of unfiltered faces. IEEE Trans Inf Forensics Secur 9(12):2170–2179. https://doi.org/10.1109/TIFS.2014.2359646
7. Yaman D, Eyiokur FI, Sezgin N, Ekenel HK (2018) Age and gender classification from ear images. In: 2018 International workshop on biometrics and forensics (IWBF), Sassari, 2018, pp 1–7.https://doi.org/10.1109/IWBF.2018.8401568
8. Huang GB, Ramesh M, Berg T, Learned-Miller E (2007) Labeled faces in the wild: a database for studying face recognition in unconstrained environments. Technical report, Technical Report 07-49, University of Massachusetts, Amherst, 2007
9. Toews M, Arbel T (2009) Detection, localization, and sex classification of faces from arbitrary view points and under occlusion. Trans Pattern Anal Mach Intell 31(9):1567–1581

Estimated Time of Arrival for Sustainable Transport Using Deep Neural Network

Aditya and Hina Firdaus

Abstract Time of arrival estimation is a challenge in a world where swiftness and accuracy are a new normal. During the year 1997, a summit in the United Nations showed concern for sustainable transport. This sustainability is for rural–urban linkage, pollution-free environment, health, etc. We understood the importance of estimating the time of arrival for the sustainable transportation and formulated a deep neural network model which increases our estimation of the time of heavy vehicles from source to destination. In our proposed research paper, we created a grid-based network dataset from a raw GPS truck data stored in a form of Data Lake. We introduced a vanilla neural network with three-layered architecture which works well with our grid-based dataset. Gradually, the velocity and volume of data increased and the dataset obtained was with labeled attributes of continuous type. The traditional machine learning algorithms failed or took more time to train on rapidly increasing GPS-dataset. The vanilla deep neural network in a three-layered architecture outperforms various deep learning method with an accuracy of 85%.

Keywords Neural network · Grid learning · Random forest · Logistic regression · Neural network · Sustainable transport

1 Introduction

Machine learning models are evolving rapidly, and so the data are increasing in velocity and volume. The industrialization is in revolution 4.0 where automation is mostly computerized. This situation [1] was well understood by United Nations in the year of 1997 and raised an issue for sustainable transport. The machine learning and big data can help us to understand the performance of exponentially increasing

---

Aditya · H. Firdaus (✉)
St. Andrews Institutes of Technology and Management, Gurugram, India
e-mail: hina.firdaus@student.uni-siegen.de

H. Firdaus
Faculty III, Human Computer Interaction, University of Siegen, Seigen, Germany

P. Dutta et al. (eds.), *Emerging Technologies in Data Mining and Information Security*, Lecture Notes in Networks and Systems 490,
https://doi.org/10.1007/978-981-19-4052-1_9

data size in various real-life applications such as health care, education, government, transport, etc.

1.1 Vanilla Neural Network

The concept of multilayer perceptron is based on forward and backward traversing [2]. The model is generally divided in three phases primary phase is input layer and tertiary layer is output. The output layer can have single to n number of probabilistic output. The middle layer is the hidden which encapsulates activation function and optimizers. This algorithm works well in the labeled dataset and falls between the [3] supervised learning and [4] unsupervised learning termed as [5] semi-supervised learning.

1.2 Ideation of Problem Statement

Literature of Albeaik et al. [6] have proposed deep neural network and deep reinforcement learning on a heavy truck archival dataset, where the performance of heavy truck is analyzed in various dynamics. The deep models performed better to understand the functionality of longitudinal truck structure. Nezafat et al. [7] also worked on the truck body dataset using transfer learning where the deep neural network hidden layer is treated to re-model the obtained image dataset of the truck. Here the emphasis was on pre-trained CNN models which improved the accuracy for classification of truck. Another paper we considered was by He et al. [8] here the bus journey travel time is predicted. The bus travel time includes traffic and halting time. The dataset they used was archival nature so long short term memory (LSTM) model was an able to predict accurately the arrival time of bus between sources to destination.

2 Methods and Materials

There were various challenges in the proposed project; firstly collection of heavy vehicle data precisely trucks. We contacted the local truck drivers with global positioning system (GPS). Raw data were collected in data lake, these data are stored in centralize location in the form of 'as is'. Data Lake is more advance and smooth transitioning between raw and processed data. We can store N numbers of structured, semi-structured, and unstructured data type of audio, videos, images, and texts. Despite of having lots of benefit using Data Lake, there are few challenges which we have faced, as the data size was increasing handling metadata got difficult. Second challenge we faced was ability to update and delete data, which make it less secure.

Table 1 Attributes of initial data collected

S. No.	Parameters	Total values
1	GPS_DATE	2,037,096
2	VEH_EXIT	
3	ACTUAL_ARRIVAL	
4	TIME_TAKEN	
5	IN_SEC	
6	PRD	
7	MONTH	
8	DAY	
9	HOUR	
10	DATEOFWEEK	
11	FROM	
12	TO	
13	DISTANCE	
14	TIME_DIFF	
15	HOUR	

The parameters in which our data were stored are in Table 1, where the attributes are GPS_DATE (date obtained from truck GPS), VEH_EXIT (vehicle exit time), ACTUAL_ARRIVAL (time of truck arrival and continuous type of data), TIME_TAKEN (encoded value in 0, 1, 2, 3, 4, 5, discrete type), IN_SEC (GPS data from truck in seconds discrete type), PRD (continuous type), MONTH (time series type), DAY (time series type), HOUR (time series type), DATEOFWEEK (time series type), FROM (continuous type), TO (continuous type), DISTANCE (continuous type), TIME_DIFF (continuous type), and HOUR (continuous type).

The major challenge in front of us is to predict the total duration of the trip done by the truck vehicle at the start of trip. We also need to re-calculate these estimated time of arrival predictions at every step of the process. This challenge is important from the technical point of view and also for the sustainability of the environment. If our solution is in a way where time of travel is less during the duration of journey which make the estimation of petrol/CNG/diesel price more accurate which help the truck owners to know the total estimation price of the journey in the beginning state itself.

Keeping these challenges in our mind, we made a huge deal focusing on arranging and pooling our raw data. These data are utilized in two ways, i.e., batch preprocessing and near real-time processing [9]. In one of the research paper, the collected GPS data from vehicle is only focused to the local roads, environment, and weather. These kinds of data do not work in general situation. The dataset we made will work for any individual or a company situated across the globe. That is why, we have not considered the route of roads from Google Maps or other source as the route changes a lot in due to development which can be external factors outside of our control.

Table 2 Attributes obtained after preprocessing

S. No.	Attributes	Total value
1	FROM_LAT	2,021,023
2	FROM_LON	
3	TO_LAT	
4	TO_LON	
5	MONTH	
6	DAYOFMONTH	
7	HOUR	
8	DAYOFWEEK	
9	DISTANCE	
10	ACTUAL_HOURS	
11	PREDICTED_HOURS	
12	DIS_CLUSTER	

As the data were from multiple streams, we divided the data processing into two different stages first was batch pre-processing. From Table 1, you can see the volume of our data was 2,037,096. With batching, we attained a refined dataset as shown in Table 2, FROM_LAT (latitude of source as continuous data type), FROM_LON (longitude of source as continuous data type), TO_LAT (latitude of destination as continuous data type), TO_LON (longitude of destination as continuous data type), MONTH (discrete type of data), DAYOFMONTH (discrete type of data), HOUR (discrete type of data), DAYOFWEEK (discrete type of data), DISTANCE (obtained attributes from source and destination distance as continuous type), ACTUAL_HOURS (obtained data of continuous type), PREDICTED_HOURS (obtained attributes of continuous type), and DIS_CLUSTER (obtained attributes of continuous type).

In the later stage, the preprocessed data are in near real-time processing, which help us analyzing the everyday data and evaluating them in almost real time.

The project was developed in Python programming language, ML library using TensorFlow library. For backend, we used structured data and we used oracle, and for unstructured, we used MongoDB. We have used eight core processor Linux servers.

For analyzing the data, we have introduced grid-based learning. Where the whole dataset is divided in grids and all the other grids are connected with each other. We were familiar with the real world problem of transport when heavy vehicles have to halt and take unusual turns as they may create an outliers in our analysis.

3 Model Development Evolution

In the previous section, we have seen the challenges of dataset creation and utilization. Now in this section, we will elaborate you about the process of model evaluation. The obtained dataset was first evaluated using the machine learning model, as you know the dataset was already obtained labeled. We choose supervised learning model like logistic regression to analyze the continuous type of data, but the algorithm could not improve the accuracy of more than 65% of total predicted time. We also found the model is not performing well if the batch size of data is increasing with higher volume, accuracy of the model started decreasing.

Aftermath, we choose random forest which showed an accuracy of 78% which was better than that of the logistic regression. Challenge we started facing when the size of data started increasing which made prediction to decline and training time was very high. Carefully, analyzing the machine learning model and their failure, we opted for neural network [10]. This model helped as training model with more than with 80% accuracy also with high volume of data. Refer Fig. 1, here we have modeled the whole process.

Primary stage of development is data collection and preprocessing which we have already explained in the previous section. Later stage is training of model in various machine learning and vanilla neural network. After that, we evaluated the model in the metrics of accuracy. The accuracy is maintained using MAPE metric model evaluation and for testing, we used A/B testing. Most researchers do not show the phase of deployment and monitoring of the model. Here we have deployed our trained model for automation in a regularly trained model set using the Rest API and hosted in Linux server. The use of live monitoring of the near real-time processing and trained data helped us to gain insight of the data. As in maintenance phase, we live compared the time of arrival with the previous accuracy metrics we obtained judging on various factors like halt, turns, etc.

In the process of creating the training model in vanilla neural network, we created a three-layered architecture. Each layer had its importance; first layer is input with the trained model second layer we had activation function with an optimizer [11]. Adam optimizer with [12] gradient descent optimizer. The third layer is output layer; we have a single output because the dataset we used is of continuous type. As shown

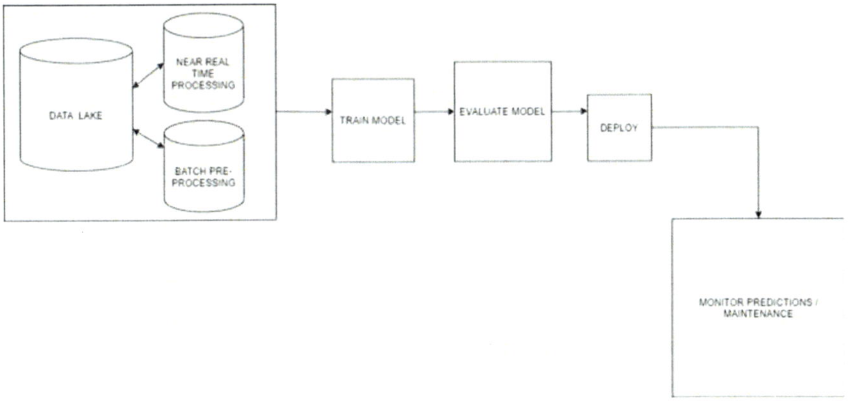

Fig. 1 Model for estimation of arrival time of truck

in the Fig. 2, the first input layer constitute of three major units that are, 1. truck live location (x_i), 2. distance between to and from location, (d_i). 3. time of arrival (a_i), and 4. truck type. The arrival of vehicle $x_i = [x_1, x_2, x_3, \ldots, x_n]$ each input calculates a speculated vehicle arrival by

$$x_i = a_i - d_i$$

where

a_i actual time of arrival
d_i departure time of the vehicle.

In the hidden layer, we have optimizers in a form of Adam optimizers and gradient descent optimizer with the activation function. Adam optimizer is referring

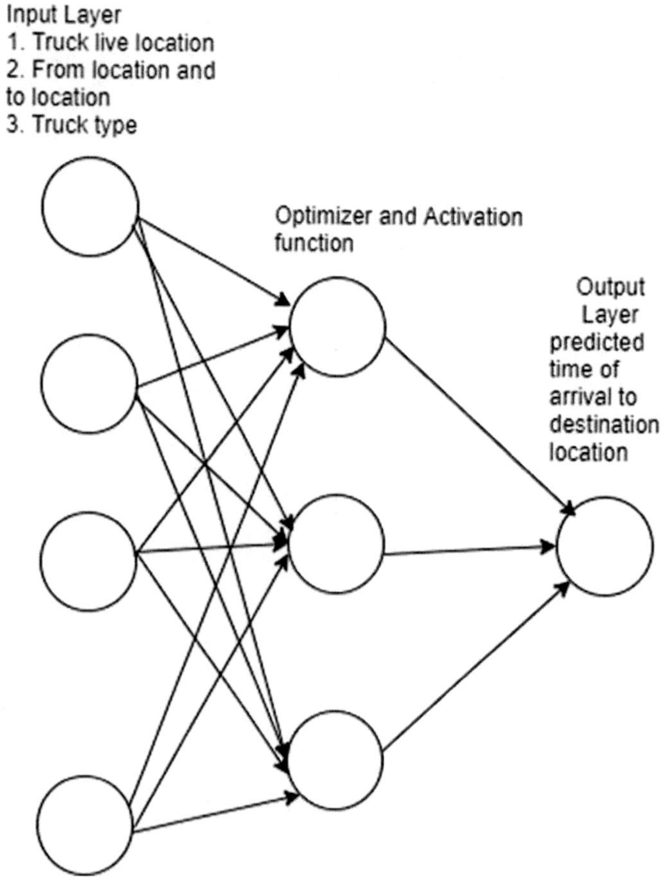

Fig. 2 Three-layered vanilla neural network architecture

as learning technique which updates rate for individual parameters. It uses successive and previous gradients and updates weight of each neural network. For the calculation of momentum, gradient descent algorithm is taken into consideration. This can be calculated as

$$W_{t+1} = W_t - LG_t$$
$$G_t = MG_{t-1} + (1 - M)[\delta f/\delta W_t]$$

where

G_t	aggregate of gradients at time t
$G_{t\text{-}1}$	aggregate of gradients at time $t - 1$
W_t	weights at time t
W_{t+1}	weights at time $t + 1$
L	learning rate at time t
M	moving average parameter constant 0.9
δf	derivative of loss function
δW_t	derivative of weights at time t.

After the calculation of momentum, it is important to understand the formulation of adaptive gradient which we define as root mean square prop aka RMSprop.

$$W_{t+1} = W_t - L_t/(S_t + \varepsilon)^{1/2}[\delta f/\delta W_t]$$
$$S_t = MS_t + (1 - M)[\delta f/\delta W_t]^2$$

where

G_t	aggregate of gradients at time t
W_t	weights at time t
W_{t+1}	weights at time $t + 1$
L	learning rate at time t
M	moving average parameter constant 0.9
δf	derivative of loss function
δW_t	derivative of weights at time t
ε	a small positive constant (10^{-8})
S_t	sum of square of past gradients.

The obtained Adam optimizers from the above calculation is

$$G_t = M_1G_{t-1} + (1 - M_1)[\delta f/\delta W_t]S_t = M_2S_t + (1 - M_2)[\delta f/\delta W_t]^2$$

M_1 and M_2 are decay rates of average gradients in the both methods where $M_1 = 0.9$ and $M_2 = 0.999$.

It is evident the Adam optimizer works fast in real-time fast moving or changing data because the rate of training is very low and performance measurement is very high.

The dataset we have used is of continuous in nature. The logistic regression traditionally should have performed well but due to the volume of data we are training regression is unable to perform. The third layer is output layer which act as regression model producing single output of the predicting the time of arrival for the truck at the location.

In the evaluation, we used simple A/B testing because we need to make a model which can enhance the prediction in every new input. This testing method works well in controlled random scenarios. In the A part of the test, we can keep the instances unchanged; on the other hand, the B part can be a new set of inputs. This is a hypothetical testing method so in various scenarios were considered and the best scenario is finally adapted for the model.

To curb any error in the model, we have used mean absolute percentage error (MAPE) which is used to calculate the exact forecast as an absolute average of error in percentage for each time period.

$$M = 1/N \sum_{t=1}^{N} [A_t - F_t/A_t]$$

where

N number of exact points
A_t actual elements
F_t forecast elements.

The algorithm we curate considering all these models is like this

Input:

Live locations of truck
Distance between from location and to location
Type of truck.

Output:

Predicted arrival time to the destination location

Step 1: Get the Current Location of the vehicle.
Step 2: Get the body type of a vehicle.
Step 3: Obtain the current Vehicle location Time(UTC/Date-time format).
Step 4: Split the datetime into 'Month', 'DayofMonth', 'Hour', 'Dayofweek'.
Step 5: Get the geo-location of To Branch.
Step 6: Label encoding body-type of vehicle.
Step 7: Calculate Distance between locations using Haversine formula.
Step 8: Train the data in vanilla neural network.

Table 3 Table for model accuracy

S. No.	Models	Accuracy (%)
1	Logistic regression	65
2	Random forest	78
3	Three-layered vanilla neural network	85

Step 9: Output will be the predicted hours as the estimated time of truck arrival.

4 Observation

The experiment we have performed was for a live project we did for the local truck vendors. The biggest challenge we have faced in the whole experimental phase was obtaining and storing the data. Another challenge was increasing the prediction of accuracy. Refer Table 3 for comparison between various models that we have used.

We can make an observation despite of using the best performing and state-of-the-art supervised learning model which generally outperforms in continuous type of data, failed in the scenario of our dataset [13] Which make it true the no lunch theorem in data science is more accurate which enable a thought to use more state-of-the-art algorithm on more complicated dataset to analyze the formulation and performance of the model. The deep learning is kind of semi-supervised learning which we have tried in our dataset. The addition of our own route and start time does not created issue in the accuracy of the model, which is a proof that the vanilla neural network being a classical gradient descent approach can work well in continuous huge volume data.

The monitoring and maintenance phase of our modulation was another challenge. Due to change in the route, the rest API use to get stuck and also the real time updating due to traffic was also not a factor that we have considered initially in our project. As we were using the GPS, the halting time in traffic or the other cases were purely neglected. This could be a challenge which we would like to take up in our rest research.

5 Concluding Remarks

In the proposed research work, we have used the entire classical machine learning model and re-evaluated their performance based on the dataset we have used. In the future, we are going to integrate the IoT and instead of near real-time processing, we will upgrade the proposed research for real-time locations. The biggest challenge is availability of hardware, because the biggest challenge for any machine learning model or neural network is type of dataset and volumes. We need more research and quality work where obtained dataset is mostly formulated in state-of-the-art algorithm to understand the behavior of an algorithm based on these factors. In the future work, we would like to use convolutional neural network, reinforcement neural network with our dataset.

References

1. Sustainable Transport UN (2021) Sustainable development knowledge platform. In: Sustainabledevelopment.un.org. https://sustainabledevelopment.un.org/topics/sustainabletransport. Accessed 6 Nov 2021
2. Hastie T et al (2009) The elements of statistical learning: data mining, inference, and prediction. Springer, New York, NY
3. Caruana R, Niculescu-Mizil A (2006) An empirical comparison of supervised learning algorithms. In: Proceedings of the 23rd international conference on machine learning, pp 161–168
4. Barlow HB (1989) Unsupervised learning. Neural Comput 1(3):295–311
5. Zhu X, Goldberg AB (2009) Introduction to semi-supervised learning. Synth Lect Artif Intell Mach Learn 3(1):1–130
6. Albeaik S et al (2019) Deep truck: a deep neural network model for longitudinal dynamics of heavy duty trucks. In: 2019 IEEE intelligent transportation systems conference (ITSC). IEEE, pp 4158–4163
7. Nezafat V et al (2019) Transfer learning using deep neural networks for classification of truck body types based on side-fire lidar data. J Big Data Anal Transp 1:71–82. https://doi.org/10.1007/s42421-019-00005-9
8. He P et al (2019) Travel-time prediction of bus journey with multiple bus trips. IEEE Trans Intell Transp Syst 20:4192–4205. https://doi.org/10.1109/tits.2018.2883342
9. Wang D et al (2018) When will you arrive? Estimating travel time based on deep neural networks. In: Thirty-second AAAI conference on artificial intelligence
10. Wasserman PD, Schwartz TJ (1988) Neural networks. II. What are they and why is everybody so interested in them now? IEEE Expert 3:10–15
11. Kingma DP, Ba J (2014) Adam: a method for stochastic optimization. arXiv preprint arXiv:1412.6980
12. Ruder S (2016) An overview of gradient descent optimization algorithms. arXiv preprint arXiv:1609.04747
13. Firdaus H, Hassan S (2020) Unsupervised learning on healthcare survey data with particle swarm optimization. In: Learning and analytics in intelligent systems, pp 57–89.https://doi.org/10.1007/978-3-030-40850-3_4

Texture Feature Based on ANN for Security Aspects

S. Vinay and Sitesh Kumar Sinha

Abstract Centered on ELM, to achieve high-performance picture steganography, we suggested an innovative OEPF model. An updated ELM algorithm is used in this method to set up the supervised computational framework to evaluate the optimized hiding picture position with a minimum distortion. The ELM is trained on a part of the image and checked in regression methodology to pick the optimum message hiding spot. It has permitted the supreme outcomes of the expected measurement indicators to be obtained. Training is carried out on the basis of a collection of extracted textures, statistical characteristics and their related visible imperceptibility metrics that use a portion of the graphic. For output enhancement, the learned model is further utilized. To surpass the current innovative frameworks, the proposed model is shown.

Keywords OEPF model · ELM · ANN · Texture feature

1 Introduction

By avoiding the message in an image, more conventional steganographic approaches hide the secret information into a picture. The importance of the spatial characteristics of the photograph nevertheless, the message homogeneity characteristics of the image blocks critically assess the determination of the best embedding [1] place. A spot with the least image distortion is known to be the optimal one. Every type of distortion in the image must be used to shield the embedding mechanism from steganalysis [2].

After the payload [3] is inserted, in addition, both physically and objectively, the cover image and stego image have to be nearly similar. The primary factors that cause the distortion are the chosen region and the embedding process. ELM is recommended for locating the superior hidden position based on the OEPF model. It should be noted that because of its generalized estimation ability, ELM [4] is advantageous, that makes speedy learning with a better over-fitting evasion compared

S. Vinay (✉) · S. K. Sinha
Department of Computer Science and Engineering, Rabindranath Tagore University, Raisen, Madhya Pradesh, India
e-mail: vinaygnanesh@yahoo.com

P. Dutta et al. (eds.), *Emerging Technologies in Data Mining and Information Security*, Lecture Notes in Networks and Systems 490,
https://doi.org/10.1007/978-981-19-4052-1_10

to other contemporary Neural Network-oriented techniques. To train any separate-secret-level Neural Network consisting of a differing quantity of neurons, a modified ELM is thus used [5].

2 Literature Review

Increased steganographic results, including ANN [6] and genetic algorithm (GA) [6] or both depending on the spatial domain was used to eliminate distortion and in order to speed up training, GA and ANN are used. In order to increase the embedding power, the ANN frequency domain is used. In order to achieve good approximation capability, rapid convergence and stable surface performances, spatial sector-related ANN is used. In addition to increasing estimation capability and minimizing distortion, ANN form is used. The ANN is often used for the embedding of a message with steganography, which presumes the hidden message represented in an image. The steganography was therefore able to openly modify the message details provided that visual information is maintained. This presumption, however, does not extend to text data [7, 8]. ANN is also utilized to validate the picture for digital watermarking, where a hidden message is not needed. ANN is used to optimize power, to detect steganographic contents, to classify the hidden information in a picture while adding to steganalysis or as a classification, and to determine the upper and lower bound of the integrative capacity. GAs was also used for various applications of steganography. For quest and optimization [4, 9], GA is used in modeling the steganography dilemma. In addition, the GAs can be optimized for minimally skewed used when a stego picture is collected similar to the cover picture. This introduces Discrete Cosine Transform with Markov as identification and classification related to pictures. Numerous methods for hiding in combined areas of spatial [10, 11] and frequency domain. Recent use of the learning resources of NNs is made of standard methods for data hiding to improve optimization. ANN is used in steganography either to identify the stego file, or to detect hidden information [12, 13] in a picture. The authors are working to minimize the distortion [14] of the stego file, by choosing the image position for hiding messages as adequately as possible. Theoretically, at incredibly high learning speeds, an ELM shows strong generalization efficiency and universal approximation. It may be used for the purposes of either grouping or regression. Inspired by some noteworthy ones we recommend an ELM-related supervised framework for image steganography also known as the Optimal Embedding Position Finder (OEPF). Moreover, during training in regression mode, an enhanced fusion criterion (fusion1) [15] is implemented to realize the best output metric for steganography. To assess the results, another novel fusion metric (fusion2) [16] has been created. And to the best of our understanding, we are using the unit for the first time.

3 Preparation of Data Sets

Prior to ELM training [17] and research, it shows the schematic structure in the development of the learning data array including the functional entity. For the construction of learning data array, texture attribute extraction, the metric measurement and the hiding are done. The hiding and the feature extrication process are customary to quickly Wavelet Dependent Embedding Transform. As previously mentioned, for each squared frame for the data set, the message must be inserted in the corresponding square window. Utilizing the hiding procedure and the estimation of the resulting visual imperceptibility measurements, data is derived from the raw data collection. The following steps will be taken to fulfill these objectives.

4 Extraction of Texture Feature

Using the following steps, the texture features are extracted:

1. For each squared frame whose sub-blocks are utilized to insert the information bits, the co-occurrence table is constructed.
2. In a square frame [18], the feature function [(energy (E), contrast (D), homogeneity (Homo), correlation (CRel), mean (M), entropy (Ent), standard deviation (SD)] for the co-occurrence table is determined. The expression in versatility gives

$$\text{Features} = (C, \text{Enr}, H, \text{Ent}, \text{Corr}, M, \text{Std})$$

5 Extreme Learning Machine Learning

The matrix structure is defined by

$$Y = (g_{1i}, g_{2i}, \ldots, g_{7i}; f_{ij}), \text{ with } i = 1, \ldots, n$$

where n is the quantity of squared frames, $g_{1i}, g_{2i}, \ldots, g_{7i}$ are the extracted characteristics, f_{ij} is the related performance metrics, and the Correlation, MSE, SSIM, and fusion1 relevant to $j = 1, 2, 3, 4$.

To predict f_i, a neural network of n' concealed neurons are constructed and instructed on a portion of Y. In addition, before implementing the ELM-based model, the training and the testing stages are undergone by the use of RMSE. We are now shifting our attention to deciding the appropriate percentage of preparation and the possible quantity of neurons.

6 Result and Discussion

See Figs. 1 and 2.

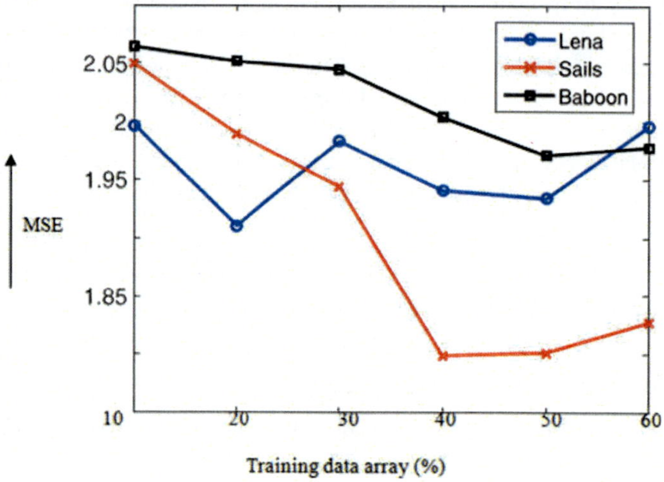

Fig. 1 Lena, sails, and baboon photos, training data collection percentage-based variance of correlation

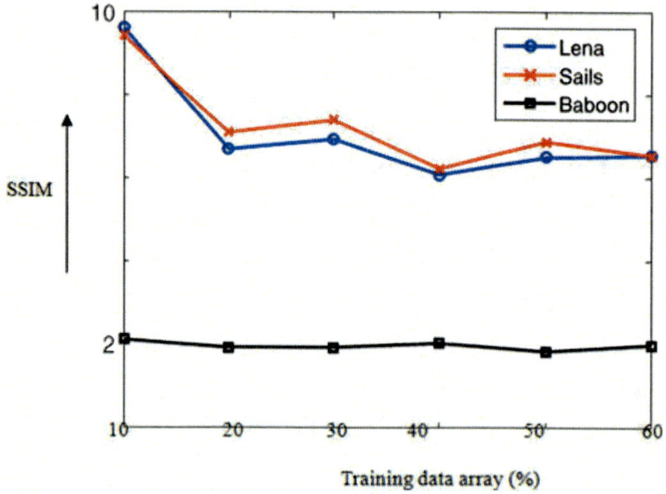

Fig. 2 Lena, sails, and baboon images, training data set the percentage-based variance of SSIM

7 Conclusion

The structure of the suggested OEPF framework that is accomplished utilizing the sequences below, is schematically outlined in Figs. 1 and 2. For instruction as well as validation, the data collection is partitioned into 50% for training and 50% for checking. Focused on the training data array, the ELM regression model is designed to be split into 80% for tutoring and 20% for verification. The Extreme Learning Machine regression framework is additionally utilized to anticipate the optimal square frame related to the fusion2 metric. In order to produce the stego image, the data hiding process is carried out to embed the hidden information into the defined optimized square frame.

References

1. Pandey KK, Rangari V, Sinha SK. An enhanced symmetric key cryptography algorithm to improve data security. Int J Comp Appl 74(20)
2. Aleya KF, Samanta D (2013) Automated damaged flower detection using image processing. https://www.semanticscholar.org/paper/AUTOMATED-DAMAGED-FLOWER-DETECTION-USING-IMAGE-Aleya-Samanta/11f8ebd4082acef98b7329cecc81601b6ec20bc8
3. Saxena AK, Sinha S, Shukla P (2017) General study of intrusion detection system and survey of agent based intrusion detection system. In: 2017 International conference on computing, communication and automation (ICCCA), pp 471–421. https://doi.org/10.1109/CCAA.2017.8229866
4. Khadri SKA, Samanta D, Paul M (2014) Approach of message communication using Fibonacci series: in cryptology. Lect Notes Inf Theory 2(2):168–171. https://doi.org/10.12720/lnit.2.2.168-171
5. Kureethara V, Biswas J, Samanta D, Eapen NG. Balanced constrained partitioning of distinct objects. In: International journal of innovative technology and exploring engineering. ISSN: 2278-3075(Online). https://doi.org/10.35940/ijitee.K1023.09811S19
6. Dhakar BS, Sinha SK, Pandey KK (2013) A survey of translation quality of English to Hindi online translation systems (Google and Bing). Int J Scien Res Publ 3(1):2250–3153. January 2013
7. Anwar Z, Banerjee S, Eapen NG, Samanta D (2019) A clinical study of hepatitis B. J Crit Rev 6(5):81–84.https://doi.org/10.22159/jcr.06.05.13
8. Samanta D et al (2020) Distributed feedback laser (DFB) for signal power amplitude level improvement in long spectral band. J Opt Commun, Apr 2020. www.degruyter.com. https://doi.org/10.1515/joc-2019-0252
9. Biswal AK, Singh D, Pattanayak BK, Samanta D, Yang MH (2021) IoT-based smart alert system for drowsy driver detection. Wirel Commun Mob Comput 2021, Article ID 6627217, 13 p. https://doi.org/10.1155/2021/6627217
10. Samanta D, Paul M, Khadri SKA (2013) Message communication using phase shifting method (PSM). Int J Adv Res Comp Sci 4(11):9–11. https://doi.org/10.26483/ijarcs.v4i11.1936
11. Maheswari M, Geetha S, Selva Kumar S, Karuppiah M, Samanta D, Park Y. PEVRM: probabilistic evolution based version recommendation model for mobile applications. IEEE Access. https://doi.org/10.1109/ACCESS.2021.3053583
12. Mukherjee M, Samanta D (2014) Fibonacci based text hiding using image cryptography. Lect Notes Inf Theory 2(2):172–176. https://doi.org/10.12720/lnit.2.2.172-176

13. Gomathy V, Padhy N, Samanta D et al (2020) Malicious node detection using heterogeneous cluster based secure routing protocol (HCBS) in wireless adhoc sensor networks. J Ambient Intell Human Comput 11:4995–5001. https://doi.org/10.1007/s12652-020-01797-3
14. Khadri SKA et al (2016) Message encryption using Pascal triangle multiplication: in cryptology. Asian J Math Comp Res 262–270, Sept 2016
15. Jaferi F, Saeid KT, Borah L, Samanta D (2016) Recognition of potential drug-drug interactions in diabetic's patients in hospital pharmacy. Int J Control Theory Appl 10(2017)(9(2016)):481–487. ISSN: 0974-5572
16. Kuchy SA, Ahmed SK, Khadri MM, Samanta D, Le D-N (2017) An aggregation approach based on elastic search. J Eng Appl Sci 12:9451–9454. https://doi.org/10.36478/jeasci.2017.9451.9454
17. Manu MK, Roy S, Samanta D (2018) Effects of liver cancer drugs on cellular energy metabolism in hepatocellular carcinoma cells. Int J Pharma Res 10(3). ISSN: 0975-2366, July–Sept 2018. https://doi.org/10.31838/ijpr/2018.10.03.079
18. Sivakumar P, Nagaraju R, Samanta D et al (2020) A novel free space communication system using nonlinear InGaAsP microsystem resonators for enabling power-control toward smart cities. Wireless Netw 26:2317–2328. https://doi.org/10.1007/s11276-019-02075-7

ML-Based Prediction Model for Cardiovascular Disease

Umarani Nagavelli, Debabrata Samanta, and Benny Thomas

Abstract In this paper, the prediction of cardiovascular disease model based on the machine learning algorithm is implemented. In medical system applications, data mining and machine learning play an important role. Machine learning algorithms will predict heart disease or cardiovascular disease. Initially, online datasets are applied to preprocessing stage. Preprocessing stage will divide the data from baseline data. In the same way, CVD events are collected from data follow-ups. After that, data will be screened using the regression model. The regression model consists of logistic regression, support vector machine, naïve Bayes, random forest, and K-nearest neighbors. Based on the techniques, the disease will be classified. Before classification, a testing procedure will be performed. At last from results, it can observe that accuracy, misclassification, and reliability will be increased in a very effective way.

Keywords Cardiovascular disease (CVD) · Logistic regression · Naïve Bayes · Random forest

1 Introduction

Heart disease characterizes a scope of conditions that influence human heart. The name "heart disease" is regularly utilized ordinarily with the name "cardiovascular sickness". Heart disease is a term that permits to an enormous number of clinical conditions identified with heart. These clinical conditions describe the unpredictable medical issue that straightforwardly influences the heart and every one of its parts [1]. In the most part of the body, heart permits mainly different conditions. For an

U. Nagavelli · D. Samanta (✉)
Dayananda Sagar Research Foundation, University of Mysore (UoM), Mysore, Karnataka, India
e-mail: debabrata.samanta369@gmail.com

D. Samanta · B. Thomas
Department of Computer Science, CHRIST Deemed to be University, Bangalore, Karnataka, India
e-mail: frbenny@christuniversity.in

P. Dutta et al. (eds.), *Emerging Technologies in Data Mining and Information Security*, Lecture Notes in Networks and Systems 490,
https://doi.org/10.1007/978-981-19-4052-1_11

heart failure mainly, valves and veins will plays very important role. Studies on bradyarrhythmias, tachyarrhythmias, atrial, and ventricular arrhythmias were included in the cardiac arrhythmias category. For each study, at least two investigators worked independently to retrieve data. Through consensus, the extracted data were compared and reconciled. We manually estimated positive and negative cases in case studies that did not provide them using conventional equations and statistics found in the manuscripts or provided by the authors. If the data of interest was not reported in the publications, we contacted the authors. Based on the heart's beats, valves, and muscle, the condition of heart is enabled and decided which type of heart disease is obtained. Cardiovascular diseases are various types in that mostly popular are Coronary Artery Disease (CAD) and heart failure (HF) [2]. Due to the supply routes in the coronary don't flow properly, the heart failure will be obtained. The main intent of coronary veins is to supply the blood to heart. Data mining is nothing but a data which is valuable to obscure and performs the complex calculations. The data will be recorded in the huge amount based on big data technology. Information mining is fundamentally a movement of noticing the examples in the information which is important and with specific data by utilizing big data [3]. The helpful examples with covered up designs and obscure connections are scientifically taken care of educated choice through this big data investigation measure [4].

Earlier 25 years back, the heart disease rate has been increased very fastly compared to previous years. According to the World Health Organization (WHO) report, 17.5 million total world deaths result from heart attacks and strokes. Computer science & technology is used for diagnosing heart disease in bioinformatics and biomedicine [5]. Machine learning is further guided using heart disease where datasets will be gathered. Repositories will record the datasets in this world. Only this need to apply some classifiers of machine learning techniques to signify the heart disease in a human. In this paper, it is surveyed the research papers to compare the accuracy of various machine learning algorithms about heart disease based on the datasets given and their attributes. By using detection processed, the heart diseases are detecting very fast [6, 7]. The death rate has been decreased by using this effective detection process. One of the human body's most indispensable organ is the heart. Respiratory failures are the most widely recognized heart condition in India. The heart pumps through the body's circulatory framework, siphons blood. Oxygen is appropriated through the circulatory arrangement of the body in the blood, and if the heart doesn't work accurately, the whole circulatory arrangement of the body will fall flat [8]. So if the heart doesn't work as expected, it could even prompt demise.

Chest pain: One of the indications of a respiratory failure is chest torment. This happens for the most part due to the blockage of the plaque of the coronary conduit of the body. Arm torment: In this, the heart pumps blood from most of the left part. Low oxygen: The degree of oxygen diminishes in the body due to the plaque which actuates tipsiness and loss of equilibrium. Tiredness: The reason for weariness proposes that it becomes hard to perform essential undertakings. Over the top perspiring: Sweating is another normal symptom [9]. Diabetics: For this situation, patients have a pulse of 100 BPM and surprisingly a pulse of 130 BPM seldom. Bradycardia: The patient might have a slow beat of 60 BPM in this cycle. Cerebrovascular infection: The patient will typically have a high pulse of 200 BPM better than expected and may cause a coronary failure higher. Hypertension: The pulse of the patient normally differs from 100–200 BPM in the present circumstance. A great deal of work has been completed utilizing the UCI machine learning dataset to foresee coronary disease [10]. Utilizing various information mining strategies, different degrees of precision have been accomplished. Regularly, heart cannot push the important measure of blood to different spaces of the body to fulfill the typical working of the body in this disease, and along these lines, cardiovascular breakdown in the long run happens [11]. The predominance of coronary disease is exceptionally high in the US. Manifestations of coronary disease incorporate windedness, actual body exhaustion, enlarged feet, and sluggishness with related signs, for example, expanded jugular venous pressing factor and fringe edema due to practical or then again non-utilitarian cardiovascular inconsistencies. The beginning phase examination approaches used to identify heart disease have been troublesome, and the subsequent trouble is one of the key elements influencing the way of life. Concluded treatment of coronary disease is undeniably challenging, particularly in non-industrial nations, attributable to the uncommon accessibility of indicative instruments and the deficiency of specialists and different administrations influencing the legitimate forecast of heart disease. The exact and right location of coronary disease is essential to decrease the related danger of genuine heart entanglements and to further develop heart wellbeing nature [12].

2 Literature Survey

In [13], the researchers introduced for heart disease patients based on data mining technique. Random forest, KNN, artificial neural network, and support vector machine techniques are used in this paper. By using diagnosing of heart disease, high accuracy is obtained from artificial neural network. In [14], the researchers introduced hybrid machine learning techniques using heart disease prediction. By using hybrid machine learning methodology, effective results are obtained. Hybrid approach is nothing but the random forest and linear method combination. This will collect the prediction of attributes and dataset base on subsets. The subset is the attribute for data of preprocessed knowledge in cardiovascular disease. After preprocessing stage, hybrid techniques is applied to the cardiovascular disease. In [15], the researchers introduced heart failure prediction based on sequential data modeling. This model is designed based on the neural network. To predict the heart disease, Electronic Health Record (EHR) is predicted from data of real world. This will analyze the data of clinical records which are sequential in nature. In [16], the researchers introduced heart disease prediction based on Evolutionary Rule Learning. This will extract the information and eliminate the manual task from the electronic records. Frequent pattern growth association is used to generate the strong association rules on patient's dataset. In [17], the researchers introduced Improved Heart Disease Detection for optimized random forest model and random search algorithm. To select the factors, random search algorithm (RSA) is introduced. To diagnose the cardiovascular disease, random forest model is introduced. Grid search algorithmic program is optimized in this model [18].

3 Cardiovascular Disease Model Based on Machine Learning

Algorithm

Figure 1 shows the flowchart of cardiovascular disease model based on machine learning algorithm. Initially, online datasets are applied to preprocessing stage. Preprocessing stage will divide the data from base line data. In the same way, CVD events collected from data follow-ups. After that, data will be screened using regression model. Regression model consists of K-nearest neighbors, support vector machine, random forest, logistic regression, and Naïve Bayes. Based upon the techniques, the

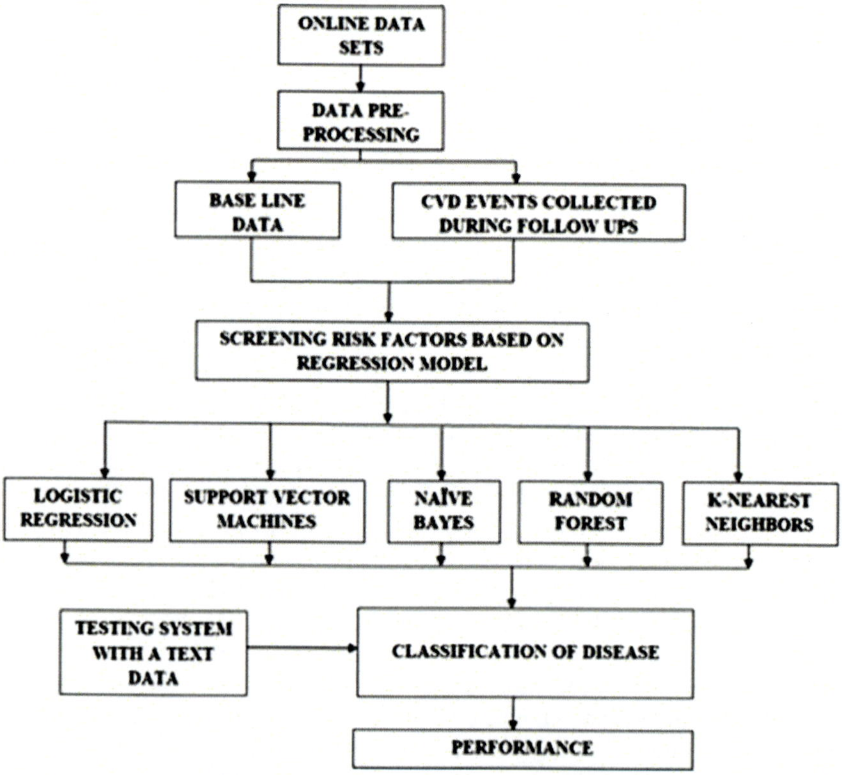

Fig. 1 Flowchart of cardiovascular disease model based on using machine learning algorithm

disease will be classified. Before classification, testing procedure will be performed [19].

By changing the Comma Separated Document format (CSV) to Excel File, cardiovascular dataset is set up. There are various requirements which will record the missing data, copy the records, and evaluate pointless information. There are no missing data values for different attributes. Logistic regression is nothing but data which is added in straight line ,and it is estimated to have value more than 0 and 1. By using objective class, the data is utilized based on the events. This logistic regression will fit the data in straight line by using logistic curves [20, 21]. The method P(X|Y) is known as Bayes Rule which will prepare and discover the dataset. By using hypothesis of Bayes', Naïve Bayes calculations are arranged. Here the data is autonomous to one another by using indicators. Generally, the data is assumed similar to the class which is autonomous.

Random forest is taken from the supervised learning. This is done based on the combination of classifications and regression. This is introduced to overcome the issues of classification. Here basically, forest consists of trees and large number of trees represents a robust forest [22]. So in the same way, the forest algorithm will make the choice on trees to sample the data. Similarly, supervised machine learning algorithm is taken as K- nearest neighbors (KNN) algorithm. This is used for both classification and regression. This is introduced to overcome the issues of classification and regression problems. There is no training phase in the K-nearest neighbors algorithm. While performing the classification of data, training will be performed automatically. At last, this is a nonparametric learning algorithm which will calculate the fundamental data [23].

4 Results

Figure 2 shows the comparison of accuracy for cardiovascular disease model and cardiovascular disease model using machine learning algorithm. From this, it can observe that accuracy for cardiovascular disease model using machine learning algorithm is increased.

Figure 3 shows the comparison of reliability for cardiovascular disease model and cardiovascular disease model using machine learning algorithm. From this, it can observe that reliability for cardiovascular disease model using machine learning algorithm is increased.

Figure 4 shows the comparison of misclassification for cardiovascular disease model and cardiovascular disease model using machine learning algorithm. From this, it can observe that misclassification is reduced for cardiovascular disease model using machine learning algorithm. Table 1 represents comparison part.

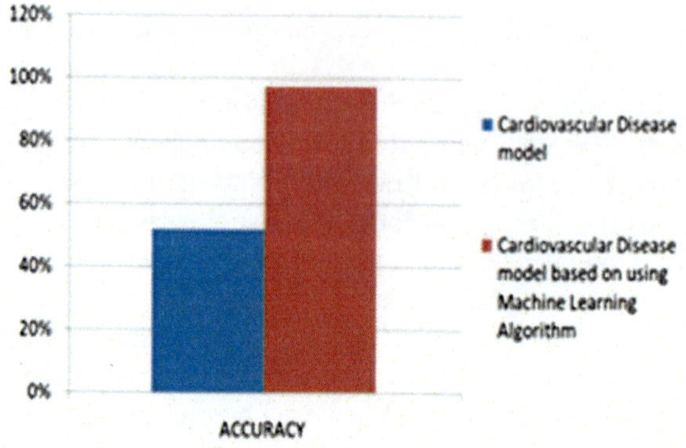

Fig. 2 Comparison of accuracy

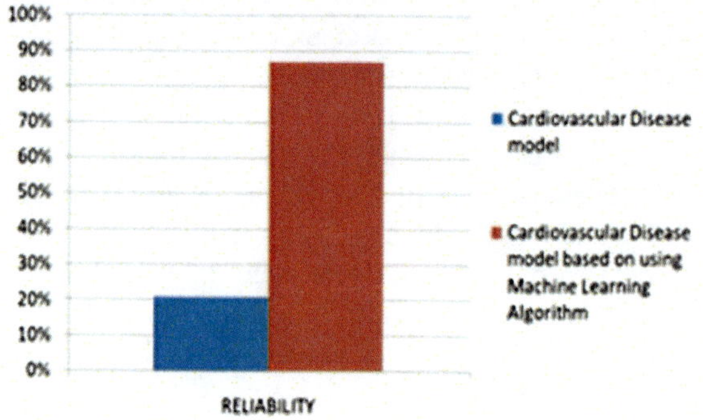

Fig. 3 Comparison of reliability

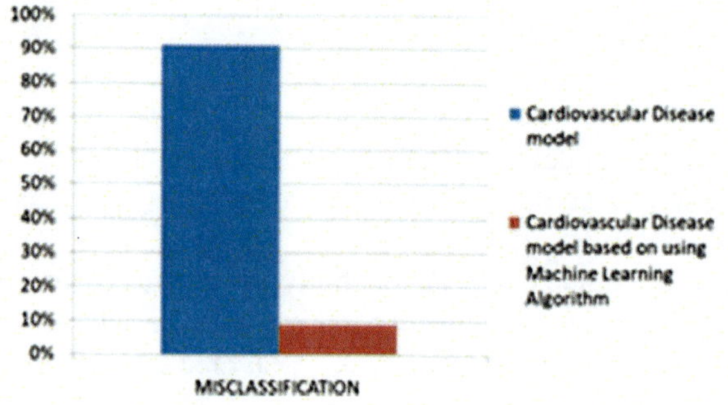

Fig. 4 Comparison of misclassification

Table 1 Comparison table

S. No.	Parameter	Cardiovascular disease model	Cardiovascular disease model using machine learning algorithm
1	Accuracy	Low	High
2	Reliability	Low	High
3	Misclassification	High	Low

5 Conclusion

In this paper, prediction of cardiovascular disease model based on machine learning algorithm was implemented. Machine learning algorithm will predict the heart disease or cardiovascular disease. At last from results, it can observe that accuracy, misclassification, and reliability will be increased in effective way. Additional population-based investigations of the cardiovascular disease prediction model described in this article are needed, with a larger sample size, longer follow-up period, and coverage of more locations in India, as well as external validation.

References

1. Rogers Aaron J, Miller Jessica M, Ramaswamy K, Palaniappan S (2019) Cardiac tissue chips (ctcs) for modeling cardiovascular disease. IEEE Trans Biomed Eng 66(12):3436–3443
2. Tang L, Bian C, Fang L, Xiong Y (2020) Study on the changes of cardiovascular disease influencing factors of pilots in china, pp 57–60
3. Mostafa N, Azim MA, Kabir MR, Ajwad R (2020) Identifying the risk of cardiovascular diseases from the analysis of physiological attributes, pp 1014–1017
4. Ji N, Xiang T, Bonato P, Lovell NH, Ooi SY, Clifton DA, Akay M, Ding XR, Yan BP, Mok V, Fotiadis DI (2021) Recommendation to use wearable-based mhealth in closed-loop management of acute cardiovascular disease patients during the covid-19 pandemic. IEEE J Biomed Health Inf 25(4):903–908
5. Deepika P, Sasikala S (2020) Enhanced model for prediction and classification of cardiovascular disease using decision tree with particle swarm optimization, pp 1068–1072
6. Rajasekaran C, Jayanthi KB, Sudha S, Kuchelar R (2019) Automated diagnosis of cardiovascular disease through measurement of intima media thickness using deep neural networks, pp 6636–6639
7. Deviaene M, Borzée P, Buyse B, Testelmans D, Van Huffel S, Varon C (2019) Pulse oximetry markers for cardiovascular disease in sleep apnea, pp 1–4
8. Peng C-C, Lai Y-C, Huang C-W, Wang J-G, Wang S-H, Wang Y-Z (2020) Cardiovascular diseases prediction using artificial neural networks: a survey, pp 141–144
9. Sinha A, Gopinathan P, Chung Y-D, Shiesh S-C, Lee G-B (2019) An aptamer based sandwich assay for simultaneous detection of multiple cardiovascular biomarkers on a multilayered integrated microfluidic system, pp 1075–1077
10. Heydari Z, Moeinvaziri F, Agarwal T, Pooyan P, Shpichka A, Maiti TK, Timashev P, Baharvand H, Vosough M (2021) Organoids: a novel modality in disease modeling. Bio-Des Manufact 4(4):689–716

11. Kaseke T, Opara UL, Fawole OA (2021) Novel seeds pretreatment techniques: effect on oil quality and antioxidant properties: a review. J Food Sci Technol 58(12):4451–4464
12. dos Santos LR, de Sousa Melo SR, Severo JS, Beatriz Silva Morais J, da Silva LD, de Paiva Sousa M, de Sousa TGV, Henriques GS, do Nascimento Marreiro D, (2021) Cardiovascular diseases in obesity: what is the role of magnesium? Biol Trace Element Res 199(11):4020–4027
13. Jayanta B, Pritam K, Debabrata S (2021) Reducing approximation error with rapid convergence rate for non-negative matrix factorization (NMF). Math Statist 9(3):285–289
14. Kirschner A, Koch SE, Robbins N, Karthik F, Mudigonda P, Ramasubramanian R, Nieman ML, Lorenz JN, Rubinstein J (2021) Pharmacologic inhibition of pain response to incomplete vascular occlusion blunts cardiovascular preconditioning response. Cardiovasc Toxicol 21(11):889–900
15. Khamparia A, Singh PK, Rani P, Samanta D, Khanna A, Bhushan B (2021) An internet of health things-driven deep learning framework for detection and classification of skin cancer using transfer learning. Trans Emerg Telecommun Technol 32(7):e3963
16. Tatsunori T, Reiko T, Risa T, Shigehiro U, Hiroyuki K, Yuji S, Shouichi F (2021) Impact of polypharmacy on all-cause mortality and hospitalization in incident hemodialysis patients: a cohort study. Clin Exp Nephrol 25(11):1215–1223
17. Dudenkov DV, Mara KC, Maxson JA, Thacher TD (2021) Serum 25-hydroxyvitamin D values and risk of incident cardiovascular disease: a population-based retrospective cohort study. J Steroid Biochem Molecular Biol 213:105953
18. Samanta D, Galety MG, Shivamurthaiah M, Kariyappala S (2020) A hybridization approach based semantic approach to the software engineering. TEST Eng Manage 83:5441–5447
19. Harjutsalo V, Pongrac Barlovic D, Groop P-H (2021) Long-term population-based trends in the incidence of cardiovascular disease in individuals with type 1 diabetes from Finland: a retrospective, nationwide, cohort study. Lancet Diabetes Endocrinol 9(9):575–585
20. Kaze AD, Santhanam P, Erqou S, Bertoni AG, Ahima RS, Echouffo-Tcheugui JB (2021) Microvascular disease and cardiovascular outcomes among individuals with type 2 diabetes. Diabetes Res Clin Pract 176:108859
21. Kumar R, Kumar R, Samanta D, Paul M, Kumar V (2017) A combining approach using dft and fir filter to enhance impulse response, pp 134–137
22. Bin W, Zhiyun Z, Shanshan L, Shuangyuan W, Chen Y, Xu Y, Xu M, Weiqing W, Guang N, Mian L, Tiange W, Yufang B (2021) Impact of diabetes on subclinical atherosclerosis and major cardiovascular events in individuals with and without non-alcoholic fatty liver disease. Diabetes Res Clin Pract 177:108873
23. Frisoli A Jr, Paes AT, Kimura AD, Azevedo E, Ambrosio V (2021) Measuring forearm bone density instead of lumbar spine bone density improves the sensitivity of diagnosing osteoporosis in older adults with cardiovascular diseases: data from SARCOS study. Bone Rep 15:101134

Sentiment Analysis in Airlines Industry Using Machine Learning Techniques

Neha Gupta and Rohan Bhargav

Abstract With the increasing power of Internet, businesses get a huge number of customer feedbacks through: their business website, social media page, business listings, etc. Majority of business do not know how to use this information to improve themselves. However, unstructured feedback on Facebook/Instagram/Twitter is where the volume lies. But the problem is these feedbacks are unstructured and there is no aggregated sentiment that we may conclude from them. To analyze these unstructured customer feedbacks at scale, machine learning is used. In this work we present a survey on various machine learning techniques that have been used in past eight years for analysis of tweets/comments related to airline industry.

Keywords Sentiment analysis · Natural language processing (NLP) · Naïve Bayes · Logistic regression · Deep learning · CNN · LSTM

Abbreviations

LR	Logistic Regression
RF	Random Forest
NB	Naïve Bayes
SVM	Support Vector Machine
DT	Decision Tree
KNN	K-Nearest Neighbor
CNN	Convolutional Network
DL	Deep Learning
XGB	Extreme Gradient Boost
RNN	Recurrent Neural Network

N. Gupta (✉) · R. Bhargav
Vivekananda Institute of Professional Studies, New Delhi, India
e-mail: neha.gupta@vips.edu

P. Dutta et al. (eds.), *Emerging Technologies in Data Mining and Information Security*, Lecture Notes in Networks and Systems 490,
https://doi.org/10.1007/978-981-19-4052-1_12

1 Introduction

Traveling is one of the important aspects of life. For traveling various means of transportation are there. But the most preferred mode is Air Transportation. Airline industry is very competitive as many companies are into it. In such a competitive market consumers opinion and feedback matters a lot. With increasing power of Internet, consumers are using social media platforms such as Tweeter/Facebook/Instagram to review brands and share their outlook about various facilities. Consumer's review guides companies to select the top marketing policies to increase their revenue and how they can improve the quality of services. Satisfaction level of consumer for any airline company can be obtained by doing sentiment analysis of comments/tweets by people.

Sentiment analysis aims to conclude the outlook of a person related to some topic or the contextual polarity of a text. With recent breakthroughs in deep learning algorithms' capacity to analyses text has greatly increased. Sentiment analysis is the process of detecting and understanding the sentiments expressed majorly in textual language via natural language processing techniques. A fundamental job in sentiment analysis is to detect and classify a sentence's or document's polarity as positive, negative, or neutral. These categories may be expanded to more complex emotion words such as "happy," "sad," and "angry," which are commonly used in social media text analytics and hate speech identification. Machine learning algorithms have a major role in sentiment analysis of comments/reviews. A lot of research work in this direction has been done using supervised techniques such as LR, NB, KNN, RF, SVM [1–3] and unsupervised techniques such as K-mean clustering, Birch algorithm, neural network, Apriori algorithm [4–6]. Recently, deep learning techniques that are based on neural networks have worked exceptionally in computer vision and natural language processing. Many researchers [5, 7–10] have used deep learning techniques such as CNN, RNN, LSTM, UPNN, and DE noising Auto encoder for sentiment analysis of customers in Airline industry. In this work we present a survey on machine learning techniques that have been utilized for sentiment analysis of customers tweet for airline industry. We have also discussed general difficulties that people face in applying sentiment analysis on Airline data. Further, Sect. 2 consists of process of evaluation of airline tweets/comments, machine learning techniques used or it and the scores that are calculated to evaluate results. In Sect. 3 literature has been reviewed and analysis of review is available in Sect. 4. Finally, conclusion is in Sect. 5.

2 Background

2.1 Process for Evaluation of Airline Tweets/Comments

Evaluation of airline tweets consists of various steps that need to be followed properly for best results. Steps involved in the process are:

(1) **Data collection**: Data related to flights can be mined from various online platforms such as Google Flight, Sky Scanner, Twitter, Facebook, and Trip advisor or an existing dataset from Kaggle may be used. Dataset represents the opinion of passengers in the form of tweets/comments. Figure 1 depicts airline dataset that consist of various columns such as tweet_id, airline_sentiment, negative reason, and airline name. Generally a dataset contains many columns that tell about data.

(2) **Preprocess data**: After collecting data, we pre preprocess it and clean by removing punctuation such as "@, &, ?, ., ;" and stop words. Unnecessary words such as "To, it, how, and" are also removed. Next we convert all letters into lowercase so that all words are unified. Finally stemming is done to know the origin of the word. E.g., "flew" is converted into "fly". After cleaning the comments, words need to be represented in feature vector. For that various word encoding processes such as TF-IDF, Word2vec, Glove, N-gram may be used depending on the type of machine learning technique used.

(3) **Sentiment analysis training**: Now selected machine learning technique need to be trained with the prepared training datasets. The data is generally split into 70% training and 30% testing dataset.

```
pd.DataFrame(airline_dataset,columns=['tweet_id','airline_sentiment','negativereason','airline'])
```

	tweet_id	airline_sentiment	negativereason	airline
0	570306133677760513	neutral	NaN	Virgin America
1	570301130888122368	positive	NaN	Virgin America
2	570301083672813571	neutral	NaN	Virgin America
3	570301031407624196	negative	Bad Flight	Virgin America
4	570300817074462722	negative	Can't Tell	Virgin America
...				
14635	569587686496825344	positive	NaN	American
14636	569587371693355008	negative	Customer Service Issue	American
14637	569587242672398336	neutral	NaN	American
14638	569587188687634433	negative	Customer Service Issue	American
14639	569587140490866689	neutral	NaN	American

14640 rows × 4 columns

Fig. 1 Airline dataset

(4) **Sentiment prediction**: Any comment/tweet is given as input and sentiment analysis in term of "negative", "positive," and "neutral" is done by considered machine learning model.

(5) **Comparison with other classifier/machine learning technique**: Comparison of results from different techniques may be done in terms of accuracy, precision, recall, and f1-score and we can select the best technique.

2.2 Formula/Score Used for Analysis of Results

$$P = \frac{\text{tp}}{\text{tp} + \text{fp}}$$

- **Precision**: It is the percentage of accurate estimates produced by a classifier. Its range is between 0 and 1. High value of precision means better classification.

$$R = \frac{\text{tp}}{\text{tp} + \text{fn}}$$

- **Recall**: It is ability to find all the relevant cases within a data set. Mathematically, it is defined as the number of true positives divided by the number of true positives (tp) and the number of false negatives (fn). Recall value lies between 0 and 1. Higher value of recall means better classification.
- ***F*1-Score**: It is a combination of precision and recall. Recall is given more importance while calculating the score. Its range is also between 0 and 1. High value of $F1$-score means good classification.

$$F1 = \frac{2 * P * R}{P + R}$$

- **Accuracy**: It is defined as the percentage of correctly classified positive and negative reviews.

2.3 Classification of Machine Learning Techniques

Supervised Learning: Supervised learning begins with a recognized dataset in which data is already classified. Supervised learning is envisioned to find patterns in data that may be used in analytics process. Data has labeled features that tell the significance of data. For example, we can train learning algorithm to differentiate between animals on the basis of provided pictures and explanations.

Unsupervised Learning: It is used when we have unlabeled data. For example, applications, such as Twitter, Facebook and Instagram have large amounts of unlabeled

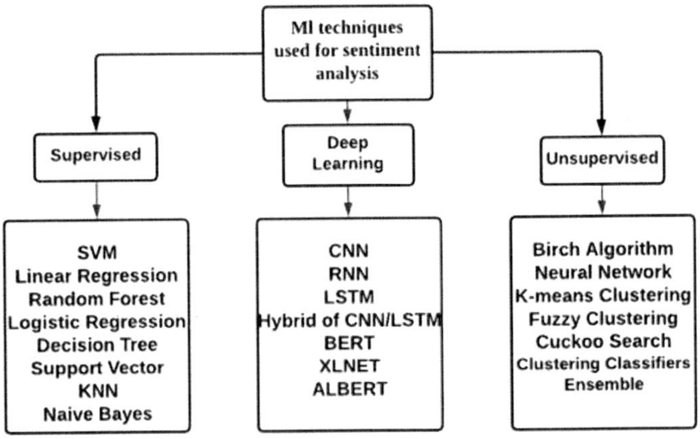

Fig. 2 Classification of machine learning techniques

data. Unsupervised algorithms classify the data on the basis of patterns or clusters it discovers. It uses a repetitive process to analyze data without human interference.

Deep Learning: Deep learning uses neural networks in consecutive layers to learn from data in an iterative way. It is used when we are trying to learn information from unstructured data. Deep learning neural networks are aimed to imitate the human brain. So that computers can be taught to solve poorly defined problems. A five-year-old child can easily identify the dissimilarity between his teacher's face and guard's face. In contrast, the computer needs lot of computation to find out who is who (Fig. 2).

3 Literature Survey

This section consists of study of various research works that have used supervised, unsupervised, and deep learning techniques for sentiment analysis of airline tweets/comments.

3.1 Supervised Learning Techniques

Al-Qahtani [1] has used two supervised (LR, NB) and four deep learning (CNN, BERT, XLNET, and ALBERT) techniques for prediction of user sentiments from US airline tweets. For encoding the words in tweets, Bag of Word, TF-IDF with Bigram, Trigram, Word2Vec, and Spacy has been utilized. They have applied BERT and ALBERT techniques with binary classes and multiple classes. On the basis of accuracy, precision, recall, and F1-score, BERT and ALBERT methods outdid all other techniques.

A comparative study on six US based airline companies has been done by Rane and Kumar [2] using DT, RF, Gaussian NB, SVM, KNN, LR, and AdaBoost. Classifiers were trained on 80% of the data and rest data was used for testing. Results depict that LR, AdaBoost, RF, and SVM performed with an accuracy of 80%. RF, LR, KNN, NB, DT, XGB, and AdaBoost have been utilized by Veerakumari and Prajna [3] on US Airline dataset. The results provide a comparison of bagging and non-bagging classification approaches. The proposed ensemble bagging classifiers outperform better accuracy than the non-bagging classifiers.

Rustam et al. [4] used LR and stochastic gradient descent classifier and LSTM in US Airline dataset to make the final prediction. For extraction they used word2vec and TF-IDF. The proposed approach performed better and achieved an accuracy of 0.789 and 0.791 with TF and TF-IDF, respectively. Results depict that LSTM performed poorly on the chosen dataset.

Ankit and Saleena [5] have used NB, RF, SVM, and LR for analysis of consumers review. Proposed approach was compared to a number of classic sentiment analysis approaches as well as the most widely used majority voting ensemble classification system. Proposed classifier outperforms stand-alone classifiers and the commonly used majority vote ensemble classifier.

The Machine learning methods such as SVM and NB have the greatest accuracy and may be regarded baseline learning methods. Kharde and Sonawane [6] have applied NB, Max Entropy, and SVM and compared with lexicon-based approaches. Results show that lexicon-based approaches are extremely successful in some situations and take less effort in human-labeled documents.

Saad [11] has used six supervised techniques that are SVM, LR, RF, XgBoost, NB, and DT to classify tweets. SVM beat other classifiers with an accuracy of 83.31%, according to the conclusions of the researcher. They also compared their findings to those of others, and they concluded that this improvement would produce a remarkable addition to the field of sentiment analysis.

Sinha and Sharma [7] have used NB, SVM, Decision tree, K-means, and RF classifier. The most accurate classifier looks to be the random forest classifier which has given 9% more accuracy than decision tree. The key conceptual difference of random forest is that it is a collection of many decision trees. KNN + NB classifier also shows improvement in accuracy as against the cases where the classifiers were used individually.

3.2 *Unsupervised Learning Techniques*

Tiwari et al. [8] have used Birch Clustering algorithm and Association rule mining to analyze the twitter dataset. They compared their proposed approach with traditional algorithms in terms of accuracy (DT 63%, K-neighbors' 67%, NB 69%, and AdaBoost 74%). This is far higher than the accuracy of prior classification systems.

A meta-heuristic method called CSK has been utilized by Pandey [9]. It is based on cuckoo search (CS) and *k*-means (*K*). Because clustering is important in analyzing the perspectives and feelings in user tweets, the study presents a method for determining the best cluster head from the Twitter dataset. Proposed approach has been compared with particle swarm optimization, differential evolution, cuckoo search, improved cuckoo search, gauss-based cuckoo search, and two *n*-grams methods. Al-Sharuee et al. [12] used *k*-means Classifier and Clustering classifiers Ensemble which is a collection of clustering classifiers using voting mechanism with different weight schemes. The base classifier is a modified *k*-means algorithm. The technique combines an automated contextual analysis with a clustering classifier ensemble. The proposed ensemble method is distinguished from prior work in the literature by its dependability.

Fuzzy logic with unsupervised tools has been used by Batista et al. [13] for sentiment analysis. They build a Fuzzy Sentiment model from social networks data in order to combine sentiment analysis with standard dimensions in a multidimensional model. They also compared the proposed fuzzy process with six machine learning algorithms in terms of accuracy.

Suresh and Raj [14] have used fuzzy clustering method and Simple *K*-means on real dataset for analyzing twitter feeds on a certain brand's thoughts. The Simple *K*-means Partitioning clustering approach is efficient than the Expectation–Maximization clustering, with a precision of about 75.5%.

3.3 Deep Learning Techniques

CNN and BERT Algorithm are used by Heidari and Rafatirad [15] on US Airline dataset. They choose the BERT model over the traditional sentiment classification model (Text blob, TF-IDF) as a robust transfer learning model. They used Glove, BERT, and Google Flight information to feed the CNN model. Result is a model with excellent prediction accuracy.

Barakat et al. [16] used CNN and LSTM in aviation business, more particularly to the airport sector, with the purpose of measuring ASQ. Models in this study that were trained with aviation related data performed better than those trained with general purpose data.

LSTM, Simple RNN, and Stacked LSTM models have been used by Manchikanti and Madhurika [17] on a sentiment analysis dataset. They utilized a basic vanilla LSTM model with one embedding layer, one LSTM layer, and one output layer in this dataset. For word encoding they extract the bag of words transformations of all the words in the dataset. They observed that LSTM has given better accuracy than Simple RNN, and Stacked LSTM has given slightly better accuracy than Vanilla LSTM with 91% with stacked.

Dang et.al [18] used DNN, CNN, RNN to solve sentiment polarity in airline data. For transforming the tweets into words they applied word embedding and TD-IDF. The studies indicated that CNN outdid other models, offering a solid combination of accuracy and CPU runtime. With most datasets, RNN reliability is somewhat greater than CNN reliability, but its computational time is significantly longer.

A novel Attention-based Bidirectional CNN-RNN Deep Model (ABCDM) is proposed by Basiri et al. [10] for sentiment analysis. Reason for using CNN in ABCDM is that it allows the model to extract local features in addition to those retrieved by the LSTM and GRU layers. ABCDM exploits publicly available pre-trained Glove word embedding vectors as the initial weights of the embedding layer.

Jain et al. [19] implemented hybrid of (CNN-LSTM) model for sentiment analysis. They calculated various parameters, such as accuracy, precision, recall, and F1-measure, to measure the model's performance. In future they want to implement other hybrid models.

SVM, ANN, and CNN have been used by Kumar and Zymbler [20] to develop classification model that maps the tweet into positive and negative category. Features were extracted from the tweets using word embedding with Glove dictionary approach and *n*-gram approach. It provides a drastic improvement in the performance of classifications model.

4 Analysis

In this section we have analyzed the considered research work on the basis of machine learning technique and word encoding process utilized for sentiment analysis. As well as we have also considered that on which dataset approach has been implemented. Complete analysis is available in Table 1. After analysis it may be concluded that for word encoding, word2vec gives the most promising results as it overcomes the difficulties faced in assessing the comments from bag of words and TF-IDF.

For machine learning techniques it may be concluded that supervised and deep learning techniques are widely used and unsupervised learning is less used by authors for sentiment analysis of airline reviews. Deep learning techniques such as LSTM, CNN and hybrid of CNN and LSTM give most promising results in terms of accuracy, precision, and $F1$-score.

4.1 Difficulties Faced in Airline Sentiment Analysis

Following are some of the difficulties that may be faced in analysis of consumer sentiments in airline industry.

1. When consumer want to show their anger, they may use sarcastic comments. Sarcastic comment use positive words but actually its meaning is negative. Thus, it becomes difficult sometimes for machine learning techniques to predict such comments properly.

Table 1 Analysis of considered research works

Paper	ML technique	Word Encoding Process	Dataset	Result
[1]	LR, NB, CNN, BERT, ALBERT, and XLNET	TF-IDF Word2vec Glove	US Airline dataset from Kaggle	BERT and ALBERT methods with binary sentiment tasks outdid all other algorithms
[2]	DT, RF, Gaussian Naïve Bayes, SVM, KNN, LR, and AdaBoost	Doc2vec	US Airline	Sentiment count was visualized combining all six airlines
[3]	RF, LR, KNN, Naïve Bayes (NB), DT, XGB, and AdaBoost	Bagging Approach	Twitter US Airline	The proposed ensemble bagging classifiers outperformed in terms of accuracy than the non-bagging classifiers
[4]	LR, Stochastic gradient descent classifier (SGDC) and LSTM	TF-IDF and word2vec	US Airline	The findings demonstrate that LSTM performs poorly on the chosen dataset
[5]	NB classifier, RF, SVM, and LR	Bag of Words	Tweet Airline Dataset	The proposed classification system was compared with classic sentiment analysis approaches and voting ensemble classification system
[6]	NB, Max Entropy, and SVM	Word embedding	Tweet Airline Dataset	The machine learning methods such as SVM and naive Bayes have the greatest accuracy, whereas lexicon-based approaches take less effort in human-labeled documents
[11]	SVM, LR, Random RF, XgBoost, NB, and DT	Word2vec	Twitter airline-sentiment	SVM beat other classifiers with an accuracy of 83.31%

(continued)

Table 1 (continued)

Paper	ML technique	Word Encoding Process	Dataset	Result
[7]	NB, SVM, DT, *K*-means, and Random Forest classifier	Bag of Words	US Airline	K-nearest neighbor + Naïve Bayes classifier shows improvement in accuracy as against where the classifiers were used individually
[8]	BIRCH Clustering and Association Rule Mining	Bag of Words	US Airline	Birch clustering that was used to understanding about positive opinion
[9]	Cuckoo search (CS) and *k*-means (K)	TF-IDF	Twitter Dataset	Proposed approach performed better than particle swarm optimization, differential evolution, cuckoo search
[12]	*K*-means Classifier and Clustering Classifiers Ensemble	TF-IDF		The technique combines an automated contextual analysis with a clustering classifier ensemble
[13]	Fuzzy-based sentiment analysis	Bag of words	Social Network Tweets	The results show that FSD provides analysis with a wider range of query possibilities in a multidimensional model
[14]	Fuzzy Clustering Method	TF-IDF	Twitter Sentiment Analysis	The suggested model has accuracy of 76.4 and takes less time to develop
[15]	CNN, BERT	TF-IDF	US Airline	CNN performed better than BERT
[16]	CNN and RNN	Word embedding	Airline Service Quality (ASQ)	This research successfully applied CNN and LSTM to the airport sector, with the purpose of measuring ASQ

(continued)

Table 1 (continued)

Paper	ML technique	Word Encoding Process	Dataset	Result
[17]	RNN, LSTM and Stacked LSTM	Bag of words	Kaggle Airline dataset	LSTM has given better accuracy than Simple RNN, Stacked LSTM gives better accuracy than Vanilla LSTM
[18]	DNN, CNN, RNN	Word Embedding and TD-IDF		CNN outdid other models, offering a solid combination of accuracy and CPU runtime
[10]	DNN and LSTM	Glove Word embedding	Twitter dataset	CNN-RNN Deep Model (ABCDM) performed best among all models
[19]	CNN, LSTM	Keras Word Embedding	Twitter Airline Dataset and Airline quality	CNN-LSTM hybrid classifier performs better than individual model
[20]	CNN, SVM, ANN	Word embedding with glove Dictionary	Twitter Airline Dataset	CNN improves the performance of classification

2. A word can have subjective or objective meaning. It is important to analyze that which case must be considered because depending on case meaning will change.
3. Sometimes sentences may be thwarted, i.e., in many sentences only some part of sentence will conclude the polarity of comment, e.g., Emirates should be amazing. It has world class amenities, attendants and services. This comment may be considered as positive but actually it is a negative comment.
4. Length of comments is mostly limited; thus, it makes pattern recognition more difficult.
5. Sometimes comment about two entities may be made in one sentence. It is important to separate text related to each entity otherwise it may not predict correct sentiment.

   For example, I like Terminal 3 but I do not like airlines that go from there to Iraq. It is positive for T3 and negative for airlines that operate from there for Iraq. But it may be predicted as neutral by bag of words process.

5 Conclusion

As customer repeat visits are one of the most essential requirements for future service success, a number of previous research have studied both the motives for and the barriers to their return visits. Furthermore, this survey used users' feedback comments and airline service satisfaction rankings as possible resources to predict their repeat visits to the service by utilizing machine learning methods. This article discusses several ways to sentiment analysis and how machine learning techniques were considered to be the most effective way to evaluate sentiment. After analysis we found that mostly supervised and deep learning techniques are used in sentiment analysis of airline tweets/comments. Among deep learning techniques, CNN and LSTM are widely used techniques. Type of word encoding process used also has impact on the results. We have identified difficulties that are faced in sentiment analysis of airline comments. And from the analysis it may also be concluded that difficulties are associated with the type of encoding process used. Word2vec process is more reliable compared to Bag of words and TF-IDF process for encoding words.

References

1. Al-Qahtani R (2021) Preprint predict sentiment of airline tweets using ML models. EasyChair
2. Rane A, Kumar A (2018) Sentiment classification system of twitter data for US airline service analysis. In: Proceedings of 42nd IEEE computer software and applications conference, COMPSAC 2018, Tokyo, Japan, pp 769–773
3. Veerakumari M, Prajna B (2021) Collaborative classification approach for airline tweets using sentiment analysis. Turk J Comp Math Educ 12(3)
4. Rustam F, Ashraf I, Mehmood A, Ullah S, Choi GS (2019) Tweets classification on the base of sentiments for US airline companies. Entropy 21(11)
5. Ankit, Saleena N (2018) An ensemble classification system for Twitter sentiment analysis. Procedia Comp Sci 132:937–946
6. Kharde VA, Sonawane SS (2016) Sentiment analysis of Twitter data: a survey of techniques. Int J Comp Appl 139(11)
7. Sinha A, Sharma P (2020) Comparative analysis of machine learning classifiers on US Airline Twitter dataset. In: International research journal of engineering and technology. www.irjet.net
8. Tiwari P, Yadav P, Kumar S, Mishra BK, Nguyen GN, Gochhayat SP, Singh J, Prasad M (2019) Sentiment analysis for airlines services based on Twitter dataset. In: Social network analytics, pp 149–162
9. Pandey AC, Rajpoot DS, Saraswat M (2017) Twitter sentiment analysis using hybrid cuckoo search method. Inf Process Manag 53:764–779
10. Basiri ME, Nemati S, Abdar M, Cambria E, Acharya UR (2021) ABCDM: an attention-based bidirectional CNN-RNN deep model for sentiment analysis. Futur Gener Comput Syst 115:279–294
11. Saad AI (2020) Opinion mining on US Airline Twitter data using machine learning techniques. In: 16th International computer engineering conference, ICENCO 2020, pp 59–63
12. AL-Sharuee MT, Liu F, Pratama M (2017) An automatic contextual analysis and clustering classifiers ensemble approach to sentiment analysis
13. Gutiérrez-Batista K, Vila MA, Martin-Bautista MJ (2021) Building a fuzzy sentiment dimension for multidimensional analysis in social networks. Appl Soft Comput 108

14. Suresh H (2016) An unsupervised fuzzy clustering method for twitter sentiment analysis. In 2016 Int Conf Comput Syst Inf Technol Sustain Solutions (CSITSS) (pp. 80–85). IEEE
15. Heidari M, & Rafatirad S (2020) Using transfer learning approach to implement convolutional neural network model to recommend airline tickets by using online reviews. In 2020 15th Int Workshop on Semantic and Social Media Adaptation and Personalization (SMA) (pp. 1–6). IEEE
16. Barakat H, Yeniterzi R, Martín-Domingo L (2021) Applying deep learning models to twitter data to detect airport service quality. J Air Transp Manage 91
17. Manchikanti K, Madhurika B (2020) Airline tweets sentiment analysis using RNN and LSTM techniques. Int J Adv Trends Comp Sci Eng 9(5):8197–8201
18. Dang NC, Moreno-García MN, de la Prieta F (2020) Sentiment analysis based on deep learning: a comparative study. Electronics (Switzerland) 9(3)
19. Jain PK, Saravanan V, Pamula R (2021) A hybrid CNN-LSTM: a deep learning approach for consumer sentiment analysis using qualitative user-generated contents. ACM Trans Asian Low-Resour Lang Inf Process 20(5):1–15
20. Kumar S, Zymbler M (2019) A machine learning approach to analyze customer satisfaction from airline tweets. J Big Data 6(1). https://doi.org/10.1186/s40537-019-0224-1

Facial Recognition to Detect Mood and Play Songs Using Machine Intelligence

S. Yogadisha, R.R. Sumukh, V. Manas Shetty, K. Rahul Reddy, and Nivedita Kasturi

Abstract Face recognition technology has gotten a lot of press because of its vast range of applications and market possibilities ranging from gaming security to entertainment Kim et al. [1]. It is used in a variety of fields, including security systems, digital video processing, and so on. When there are hundreds of songs, it is difficult for music listeners to manually create and segregate the playlist. The system's overall concept is to determine the song's mood based on the presence of tags in the lyrics, and then, using the MobileNet model, to detect the user's emotion and play the song on the music player according to the user's preferences. To create the required system, we employed two separate models: the music mood prediction model and the mood detection model utilising face recognition. We applied Tf-Idf on a variety of classifiers, with a particular focus on the random forest classifier. For detection of the face, Haar cascade method was employed. The FER dataset was utilised to train the model using the MobileNet model.

Keywords Music · Recognition · OpenCV · Haar cascade · MobileNet · Tf-Idf · Random forest · Face recognition · Mood detection · Emotion classifier

1 Introduction

Music is interconnected with emotion of human beings [2]. Music's remarkable power to generate emotions ensures its central position in human culture and daily life. Music is frequently appreciated and sought after for its capacity to elicit or transmit emotions [3]. People have a tough time making and arranging the playlist on their own once they have many songs. Users do modification and updating for every song in their playlist occasionally. Facial observation to detect mood and recommend song consequently helps overcome time and effort required in using a music player. Music recommendation systems exist however the suggestion of songs continues to

S. Yogadisha · R.R. Sumukh · V. Manas Shetty · K. Rahul Reddy · N. Kasturi (✉)
PES University, Bengaluru, Karnataka, India
e-mail: niveditak@pes.edu

P. Dutta et al. (eds.), *Emerging Technologies in Data Mining and Information Security*, Lecture Notes in Networks and Systems 490,
https://doi.org/10.1007/978-981-19-4052-1_13

be a tedious task, and hence, this is known to be a hurdle in the enjoyable process of listening to songs. High-performance personal computers have become very much ordinary as the information society's technology improves on daily basis. As a result, human–computer interaction is rapidly evolving into a bidirectional interface, and developing human–machine interaction systems require a greater understanding of human emotions. Byun and Lee [4, 5].

The rest of the work is organised in the following sections: Related work is summarised in Sect. 2, problem statement discussed in Sect. 3, design methodology in Sect. 4 has elaboration on the various methods used for our findings and results Sect. 5 consists of results and conclusion and future work, and Sect. 6 presents conclusion and future aspects of the work.

2 Related Work

The key problem in hand is to detect the mood of the users and suggest and play songs accordingly to them. To overcome this challenge, various models have been proposed. We have referred to four papers that describe the techniques that can be used to approach this problem and we have chosen the best.

1. Heart Player
   A smart music player involving emotion recognition, expression, and recommendation in this paper focuses on two technologies, namely when a song is being played, an animation figure expresses the emotion of the song by certain facial expressions and also generates a series of analysis results, including the user's preference, today's mood and music personality. Fan et al. [6]. Calculating programme is responsible for calculating the emotion of the music synchronously when the music is being played. A model is built to calculate the emotion score via these music features. The classification is based on a refinement of the Tellegen-Watson-Clark mood model. The score is then calculated by the computer, and if the score falls into one of the intervals, the emotion of the song is identified.
2. Mood-Based Music Player
   Anuja Arora, Aastha Kaul, and Vatsala Mittal in their paper mood-based music player depict the usage of various algorithms for the classification, and they have used KNN, SVM, and random forest for prediction of each song into different mood, and they have used HAAR cascade classifier and Fisher face algorithm to detect user's mood through their facial expressions. Arora et al. [7]. They combine these two to generate a user customised music playlist. The Haar cascade and Fisher face algorithm together give an accurate result of the user's emotion with an accuracy of 92%. Using the audio files lyrics, the accuracy of the model is around 81% for detecting the mood of the song using SVM. The three most faced problems are the presence of unidentified elements like glasses or beard, quality of static images, and unidentifiable facial gestures. Large posture alter-

ations have a tendency to wreak havoc on pre-existing algorithms. To reduce this standard image, input format (.jpeg) is taken.

3. Facial Expression-Based Music Player
   The paper facial expression-based music player is about the proposed work of a facial expression recognition system to play a tune based on the detected expression It extracts features using the PCA method, and it classifies phrases using the Euclidean distance classifier. Kamble and Kulkarni [8]. PCA is a statistical strategy used in pattern recognition and signal processing to reduce the number of variables in facial recognition techniques. In this work, real images (user images) are captured utilising the in-built camera. The final result shows the accuracy level obtained is up to 84.82%.
4. Person-Specific Face Tracking with Online Recognition
   In the paper person-specific face tracking with online recognition in order to ensure the precision of identity recognition with different poses, they used the canonical correlation analysis (CCA) technique to project the extracted characteristics of faces into a latent space, and then incrementally trained these projected features using an online SVM (LASVM). Cai et al. [9]. Person-specific face tracking surpasses various state-of-the-art face trackers, according to test results. The behaviour of a certain individual can be continuously studied with the help of person-specific facial tracking.

3 Problem Statement

Because of its massive application price and market potential, face recognition technology has gotten a lot of press. It is being enforced in varied fields like security systems, digital video process, etc. Music is that type of art that is thought to own a bigger reference to a person's feelings. People have a hard time manually maintaining their playlists. Songs users like or prefer might not be given priority and others non preferred might have an upper priority. The overall construct of the system is to acknowledge facial feelings and suggest songs expeditiously supporting the user's mood.

4 Design Methodology

The major solution is divided into two parts: the first is to predict the mood of each song using a Tf-Idf model with a random forest classifier, and the second is to detect the user's emotion using openCV, the Haar cascade technique, and the MobileNet Tensorflow model. The same model was done in other models before such as convolutional neural networks and fuzzy logic approach [3, 10].

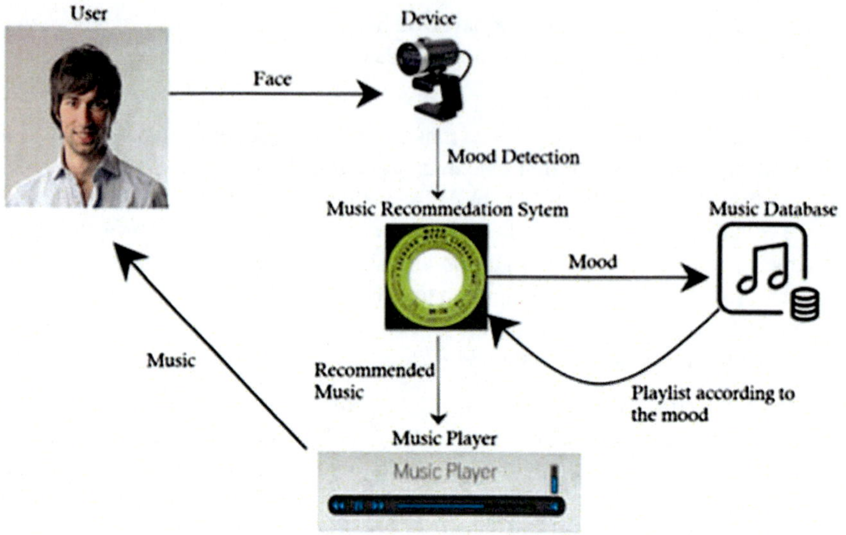

Fig. 1 Design of the system

Figure 1 discusses the model's design is self-explanatory; it starts with a simple webcam to capture the image of the user's face [11], which is then used to detect the user's mood in the music recommendation system, and this emotion data is fed to any content-based recommendation engine (here our database) [12], the matched songs are then made into a playlist, and the songs are played using the music player for the user to enjoy.

4.1 Pre-processing

In the music mood prediction model, we have performed 4 types of pre-processing on dataset

1. Tokenization-Taking a text or set of text and breaking it into different individual words.
2. Stop-word removal-Stop words like (the, a, in, an, etc.) have been eliminated.
3. Punctuation removal-Eliminates punctuation from lyrics
4. Stemming-Inflected words are reduced to a slender wordstem or root form.

Although our training dataset had already been cropped and cleaned, we needed to clip our webcam photos for our testing dataset so that the model would not learn the image's background noise. To speed up detection, we shrunk the image by 1/4, then constructed coordinates around the face, which were detected using Haar cascade, and a test frame was prepared to manually check if the cropping was operating properly. This was the most important stage in the emotion detection model's pre-processing.

4.2 Model Building

Following are the steps used in the process of model building:

1. Music Mood Prediction Model
   The dataset we begin with is a subset of the Million Song dataset containing 10,000 songs. PyLyrics package was used for getting lyrics for all of the songs. Our model is being built using English lyrics. Last.FM was used to extract tags for the remaining songs in our collection. Genre, mood, artist type, and other factors can all be used to categorise tags. These four tags were chosen from Last.FM tags because they represent the most common and basic human moods.
   (a) Happy Tags: cheerful, cheer up, festive, jolly, gleeful, etc.
   (b) Sad Tags: sadness, unhappy, melancholic, depress, etc.
   (c) Angry Tags: anger, angry, choleric, fury, outraged, rage, etc.
   (d) Relaxed Tags: calm, comfort, quiet, serene, mellow, etc.

   Using these tags as a guide, we attempted to categorise each song into a specific mood based on the presence of tags in the song's lyrics. On the dataset, we used pre-processing techniques such as tokenization, stop-word removal, punctuation removal, and stemming. Because we cannot supply words as input to models, feature engineering is required for the lyrics column. For this, we employed the word-level Tf-Idf NLP model. The Tf-Idf model delivered the best accuracy for feature engineering when using the random forest classifier.
2. Mood Detection Model Using Face Recognition.
   (a) Facial emotion recognition (FER) dataset from kaggle: Greyscale photographs of faces at a resolution of 48 $\times$ 48 pixels make up the data. The faces have been automatically registered to be roughly in the middle of each image and to occupy roughly the same amount of space [13]. The purpose is to categorise each face into one of seven groups based on the emotion portrayed in the expression (0 = Angry, 1= Happy, 2 = Neutral, 3 = Sad) [14].
   (b) OpenCV: OpenCV is a big open-source library for computer vision, machine learning, and image processing that is utilised in real-time activities. In images and movies, it can recognise objects, faces, and even human handwriting. When Python is used in conjunction with other modules such as NumPy, the OpenCV array structure can be examined. We employ vector space and execute mathematical operations on these features to recognise visual patterns and their various features.

(c) Haar Cascade algorithm: It is a face detection method that works with photos or real-time video. This algorithm takes into account three features: edge features to locate the corners of the face, line features to locate the forehead, nose, and brow, and centre-surround features to locate the cheekbones, chin, and mouth. The Haar cascade classifier employs AdaBoost, which distributes larger weights to successfully categorised features, and the procedure is repeated until the face is correctly recognised. It outperforms several algorithms when it comes to detecting faces in low-light or dark environments.

(d) MobileNet pre-trained model in Tensorflow has been used by us for retraining our network. It is a practical CNN architecture that is been applied in real-world scenarios. It can categorise photographs into up to 1000 different categories. MobileNet is much smaller than VGG 16 yet has the same accuracy. This is because VGG has a lot more undesirable connections, which increases the network's complexity and size, which can only be decreased through pruning. MobileNet has a low latency, operates quickly, and uses very little battery power, making it perfect for use in mobile applications. It also has excellent accuracy. Our model was tested using the validation set for the ImageNet Large-Scale Visual Recognition Challenge 2012 (ILSVRC2012). In the Large-Scale Visual Recognition Challenge 2012, the ImageNet dataset (10,000,000 tagged images representing 10,000+ object types) is used (ILSVRC2012). With 1000 categories and 1.2 million images available for training, ImageNet is a subset of ImageNet.

5 Results and Discussion

We used particular tags in lyrics to forecast the correct mood, which was then compared to the mood of the real song for the music mood prediction algorithm. We trained the emotion detection model using Kaggle's FER dataset, which already had four emotion shots cut, and then tested it on a real-time dataset that included images of all group members divided into four emotions. The following findings were achieved using the methods indicated above:

1. Music mood prediction model
   The main goal was to create a model that could analyse a song based on its lyrics and detect its mood. Random forest, multinomial Naive Bayes, ensemble bagging and boosting, and SVM were used to perform word-level Tf-Idf. The Tf-Idf model, in combination with the random forest classifier, has the maximum accuracy of 74.6% . We are getting lyrics for all of the songs using the PyLyrics package, which uses the LyricWikia.com API to acquire lyrics for songs, using the artist name and song title. We are using English lyrics to build our model. As a result, all song lyrics not written in English are removed from the dataset. Tags will now be retrieved using the Last.FMAPI. Tags can be categorised by

Classifier	Countvectorizer	Word Level Tf-idf	Ngram Vectors
Naive Bayes	57.1	45.5	59.2
Random forest	69.7	**74.1**	67.2
Logistic Regression	63.1	62.6	59.2
Boosting	52.3	54.0	37.3
Bagging	**65.2**	63.9	60.5
SVM	64.3	**66.5**	61.3

Fig. 2 Results obtained on music mood prediction model

genre, mood, artist type, and other factors.We create the class labels for moods in our dataset by matching the tags we found on Last.FM with the tag groups we defined. Pre-processing processes include tokenization, stop-word removal, punctuation removal, and stemming.

2. Emotion detection model On the training dataset, which was acquired from the FER dataset, this model had a 98.1% accuracy. Each image has a mood associated with it, and if the MobileNet model correctly detects that mood, it is considered a success. On the testing dataset, which consisted of pictures of various types of team members, around 98 images out of 100 were correctly classified, and 92.4% on the training dataset. When given a real-time dataset, testing was done to assess how accurate the model was, and 92 out of 100 photographs were properly identified. Figure 2 shows the results of the models which is used for predication of the mood.

6 Conclusion and Future Work

The system begins with clicking the photograph of a person in order to recognise their present mood which is beneficial in saving time. It can help a person by being quick and accurate in suggesting the songs a person would most likely want to listen to at the moment by matching the mood detected with the database. Haar cascade can identify images in dark spaces as well as in sharp edges. This system recognises a single mood at a time and cannot detect multiple emotions simultaneously. This proposed system can be used efficiently in order to escalate our lifestyles by using in various health care, mobile, and entertainment purposes. To implement the proposed system on mobiles and cars as an app/system with a perfect user interface [15]. It can enable the recommendation for different TV series and movies depending on the emotion.

References

1. Kim K, Kim G, Choi H-II (2008) Face detection using the improved feature tracker. In: Fourth international conference on networked computing and advanced information management. https://doi.org/10.1109/NCM.2008.230
2. Parikh DP (2021) Emotion based music recommendation system. Int J Res Appl Sci Eng Technol 9:1674–1680
3. Barthet M, Fazekas G, Sandler M (2013) Music emotion recognition: from content- to context-based models. From Sounds to Music and Emotions 228–252
4. Byun S-W, Lee S-P (2020) Human emotion recognition based on the weighted integration method using image sequences and acoustic features. Multimedia Tools Appl. https://doi.org/10.1007/s11042-020-09842-1
5. Yu Z, Zhao M, Wu Y, Liu P, Chen H (2016) Research on automatic music recommendation algorithm based on facial micro-expression recognition. In: 39th Chinese control conference (CCC). https://doi.org/10.23919/CCC50068.2020.9189600
6. Fan S, Tan C, Fan X, Su H, Zhang J (2011) HeartPlayer: a smart music player involving emotion recognition, expression and recommendation. Lect Notes Comp Sci 483-485
7. Arora A, Kaul A, Mittal V (2019) Mood based music player. In: International conference on signal processing and communication (ICSC). https://doi.org/10.1109/ICSC45622.2019.8938384
8. Kamble SG, Kulkarni AH (2016) Facial expression based music player. In: International conference on advances in computing, communications and informatics (ICACCI). https://doi.org/10.1109/ICACCI.2016.7732105
9. Cai Z, Wen L, Cao D, Lei Z, Yi D, Li SZ (2013) Person-specific face tracking with online recognition. In: 10th IEEE international conference and workshops on automatic face and gesture recognition (FG). https://doi.org/10.1109/FG.2013.6553730
10. Li X, Lang J (2018) Simple real-time multi-face tracking based on Convolutional neural networks. In: 15th Conference on computer and robot vision (CRV). https://doi.org/10.1109/CRV.2018.00054
11. Mohaghegh M, Pang Z (2018) A four-component people identification and counting system using deep neural network. In: 5th Asia-Pacific world congress on computer science and engineering (APWC on CSE). https://doi.org/10.1109/APWConCSE.2018.00011
12. Ayata D, Yaslan Y, Kamasak ME (2018) Emotion based music recommendation system using wearable physiological sensors. IEEE Trans Consumer Electron 64:196–203. https://doi.org/10.1109/TCE.2018.2844736
13. Kim K, Kim G-Y, Choi H-I (2008) Automatic face detection using feature tracker. In: International Conference on Convergence and Hybrid Information Technology. https://doi.org/10.1109/ICHIT.2008.203
14. Li H, Wen G (2019) Sample awareness-based personalized facial expression recognition. Appl Intell 49:2956–2969. https://doi.org/10.1007/s10489-019-01427-2
15. Çano E, Coppola R, Gargiulo E, Marengo M, Morisio M (2017) Mood-based on-car music recommendations. Lect Notes Inst Comput Sci Soc Inf Telecommun Eng 154–163

Banana Leaf Diseases and Machine Learning Algorithms Applied to Detect Diseases: A Study

Meghna Gupta and Sarika Jain

Abstract The world is changing continuously, nothing is permanent. What we think is new today becomes obsolete in a few days with better versions. All things have become computerized day by day. People are opting for technical methods to deal with the changes rather than following traditional methods. Agriculture is not an exception. Since in India, 70% of the population is dependent on agriculture and also agriculture has a 20.5% share in India's GDP, i.e. 17–18% of India's income comes from agriculture, farmers have started opting for new methods to increase the productivity of crops. Researchers are working on Artificial Intelligence-based technologies to increase the life of crops by which crop diseases can be predicted in their early stages. India is a land of agriculture and there are varieties of crops available. Since there are different climatic conditions, depending on which soil also changes its behaviour. Pests are also a major problem. Image Processing has evolved as an effective thing for the early analysis and detection of plant disease. Several algorithms have used to analyze the diseases at the early stage that results in minimum loss to the farmers and good quality of crops. This paper is presenting a study on the diseases found in banana crop along with their solutions available. This paper is different from other survey papers because it has focused mainly on banana crop whilst study had done on multiple crops earlier in a single paper.

Keywords Panama wilt · Sigatoka disease · Bract mosaic virus · Deep learning models · InceptionV2 · ResNet50 · MobileNetV1

M. Gupta (✉) · S. Jain
Amity Institute of Information Technology, Amity University, Noida, Uttar Pradesh, India
e-mail: meghna.gupta6@s.amity.edu

S. Jain
e-mail: sjain@amity.edu

P. Dutta et al. (eds.), *Emerging Technologies in Data Mining and Information Security*, Lecture Notes in Networks and Systems 490,
https://doi.org/10.1007/978-981-19-4052-1_14

1 Introduction

Banana is amongst one of the most popular and favourite fruit crop grown in India as it is available around the year, affordable, nutritious, comes in various ranges and has medical importance [1]. It grows in near about 120 countries and India leads the world in its production with 14.2 million tonnes. Banana is mainly grown in humid tropical regions of South East Asia [2]. It is the second most favourite fruit of India after Mango; however, several banana diseases and pests have caused major harm to these crops. Some Banana diseases are Panama Wilt, Leaf Spot, Leaf Streak or Sigatoka Disease, Anthracnose, Cigar End Tip Rot, Crown Rot, Stem-end Rot, Pseudostem Heart Rot, Head Rot, bacterial Wilt or Moko Disease, Banana Bunchy Top Virus, Banana Streak Virus, Mosaic Virus, Banana Bract Mosaic Virus, etc.[3]. Early detection of these diseases is the first crucial step to save these crops. Although farmers follow the traditional approaches, these approaches are not fruitful, as a lot of information must consider by a farmer to detect these diseases and save the crops from a big loss. In addition, traditional methods are fruitful only for small farms where the farmers can keep an eye on all plants along with proper knowledge to identify diseases and the necessary control measures to prevent them. But when the size of farms increases, keeping an eye on all plants is a time-consuming process, requires more labour and may result in the fast-spreading of diseases if not properly cared for [4]. Today, Artificial Intelligence emerged as a rescue and increased the productivity of crops using several image processing and deep learning models [2].

Diseases of Banana

Several diseases may affect banana plants. Some of these diseases and their symptoms are mentions in this section.

A. **Panama Wilt**: It is also known as Fusarium wilt (FW). It is due to the presence of a fungi Fusarium oxysporum f. sp. cubense (foc) that is found in soil. This disease was first to report in banana plants in Australia and spread worldwide including areas in India, Lebanon, Pakistan, Mozambique, Asia and Oman with the casual trading of planting material and the spore-bearing soil movement [5, 6]. The symptoms found in most banana plants are the drying and light yellow colour of the lower leaves, especially near the margins that turn to dark yellow colour with time results in dead leaves. (Fig. 1).
B. **Leaf Streak or Sigatoka Disease**: It is also known as a black streak or black Sigatoka and caused by the pathogen Mycosphaerella fijiensis. Its symptoms start emerging in one-month-old leaves. The pathogen causes light yellow spots or streaks initially, which later turned browned or light grey centred with few centimetres in size [7]. The spots size increases with time and the tissues around this turn yellow and spread in the whole leaf resulting in brown colour and the leaf ultimately dies hampering the fruit maturity (Fig. 2).
C. **Banana Streak Virus**: It is called a badnavirus. The symptoms are the yellow-coloured lines that start from the middle of the leaf to the margins. These fine

Fig. 1 Banana leaf affected from panama wilt

Fig. 2 Banana leaf affected by sigatoka disease

streaks may be broken or continuous and turn brown or black with the time that results in small or weak plants (Fig. 3).

D. **Banana bract Mosaic Virus**: It is known as a banana bract mosaic or Kokkan disease, and it first appeared in the Philippines in 1979. The virus can affect at any stage of the plant. Green or red-brown streaks appear on the leaf stalks and occasionally on the centre portion of young banana leaves as symptoms (Fig. 4).

E. **Banana Bunchy top disease**: It is considered the most damaging banana virus disease and the virus responsible for it is the Banana bunchy top virus, from

Fig. 3 Banana leaf affected from banana streak

Fig. 4 Banana plant affected by banana virus bract mosaic virus

which this disease has taken its name. It is restricted to certain African and Asia—Pacific countries. The dark green streaks on the bottom half of the leaf's midrib, and subsequently on the secondary veins, are the first signs. The streak consists of dost dash patterns, the most diagnostic symptom of the bunchy top, the so-called 'Morse code'. Streak signs become more visible on the leaf blade as the infection progresses. Dark green hook-like vein extensions may also be observed in the tiny, light green zone between the midrib and the lamina (Fig. 5).

2 Research Methodology

Different types of leaves images are considered identifying the diseases of plants and then passed through different classification techniques to differentiate between an

Fig. 5 Banana plant affected from banana bunchy top virus

affected and unaffected area of leaves [8]. Image Processing methods help develop automated systems for identifying plant diseases under various constraints. The colour of banana leaf is the main criteria to indicate a healthy plant. The green colour of the leaf indicates a healthy plant whereas yellow, black or brown colour indicates that any bacteria, fungi or plant virus [9] infects the leaf. There are many layers formed whilst determining whether the leaf is infected or not. We can summarize these layers as mentioned in Fig. 1. Plant Diseases can be identified by following the five major steps:

A. **Capture the image with proper adjustment**: It is the process of capturing an image through an image-capturing device with require resolution for better results called Image acquisition [10, 11]. The application determines how an image database is constructed. The picture database is responsible for the classifier's improved efficiency, which determines the algorithm's resilience [3]. This layer should ensure that all the caught leaf pictures are taken at a legitimate point, at an appropriate distance and at an appropriate lighting condition to get high precision during handling. In this progression, we can download images from any certifiable site of plant pictures, containing various kinds of contaminated leaf pictures or could be taken through a computerized camera [11]. All solid and sick pictures are saved in RGB shading structure in picture data set and these are found physically with given interesting names and numbers [9, 11].
B. **Remove disturbance from Image**: This offers accurate details about a picture with proper visual comprehension in terms of radiance, dimensions, signal level, etc. called image pre-processing [11]. For localization purposes, it adjusts the dynamic range of particular features but does not modify the default information in the leaf image [8, 12]. The purpose is to remove any kind of disturbance that can be in terms of insects, dew drops, dust on plant leaves [4, 12]. It can be achieved by converting the image from RGB to grayscale for precise output. To be precise image pre-processing is the process of image enhancement, removal of disturbance and colour conversion [7, 12].

C. **Extracting region of interest:** The process of segregating unwanted regions from the diseased region is called Image Segmentation [4, 10]. Components like infected area, desired leaf area, background region that are required to be processed are extracted in this step [11]. The background regions are extracted to get it subtracted from the input image to get the more clear understanding of the desired area [12]. Thresholding, histogram, watershed, edge detection, colour image segmentation are some commonly used methods used for segmentation in the identification of Banana leaf diseases [3, 10–12].

D. **Extracting new features from the image**: It is the process of extracting the desired portion of the image based on colour, size, shape, texture, etc. called Feature Extraction [8]. Because of the large number of pixels and the possibility of irregular leaf placement, input picture pixels cannot be used directly for categorization. The input picture is converted into a numerical array of characteristics that may be utilised to represent the picture uniquely in this phase [11]. Today, researchers are focussing more on the texture of leaves to identify plant diseases [3, 12, 13].

E. **Remove Redundancy, categorize features and predict accuracy**: As the dataset for training and testing is considered to be large for good results, it might raise the cases of redundancy. Redundancy can directly affect the classification results [11]. Machine learning algorithms like support vector machine, Back Propagation neural Networks, K-Nearest neighbour, Convolutional Neural Network, Artificial neural network, etc. [3, 10] can be used to identify the disease infection in the extracted region of leaves. These algorithms of statistical classifiers will help farmers for diagnosing plant diseases in their early stage. The steps have been summarized in the below-mentioned Fig. 6.

3 Related Work

Deep Learning Models helps to identify diseases by plant's condition and image processing with great accuracy. Michel et al. used the three different Convolution Neural Network (CNN) architecture that is InceptionV2, ResNet50 and MobileNetV1 and forms 18 different classes (by grouping diseases based on plant parts) by collecting 18,000 images from different parts of the banana plant and got accuracy between 70 and 99% in most of the models [2]. Karthik and Praburam [14] used the Economic Threshold Level Algorithm by setting a threshold value and comparing this value with the pixel values of the captured images to check the presence of the banana streak virus. For this, they have used a camera interfaced in an Embedded Linux board to capture the images and an HSV algorithm to detect plant diseases. Vipinadas and Thamizharasi [15] have used the image pattern classification using MATLAB and identify diseases based on shape, texture and colour. The work was done for detecting Black Sigatoka disease in banana. They started by capturing the image in RGB format and then convert the image into YCbCr colour

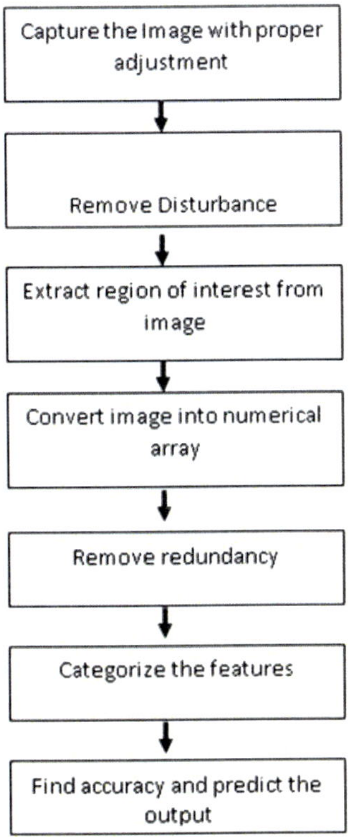

Fig. 6 Steps to identify leaf diseases

space. Later they extracted the Y component from the greyscale image, classified the images using a support vector machine, and got a good result. Gomez Selvaraj et al. [16] have used advanced machine learning models on the high-resolution satellite imagery data and UAV data, connected through mobile apps to detect and classify banana plants. They used the Random Forest model and got an accuracy of 99.4 and 92.8% for banana bunchy top disease (BBTD) and Xanthomonas Wilt of Banana (BXW). They also applied the model to detect healthy banana cluster and individual banana plants and got an accuracy of 93.3% and 90.8%, respectively. Escudero et al. have used the LeNet CNN topology for Banana Sigatoka Leaf disease detection and used 70% data for training and 30% for testing. They created their own database and sampled images from banana harvests of Risaralda department. They identify the infection in different segments of the leaf along with whole banana leaf. A comparative study was done between SVM and CNN and results showed that CNN performs better than SVM. All the mentioned have been summarized in Table 1.

Table 1 Summary of leaf diseases of banana and algorithms applied with the data set

Diseases	Algorithm	Dataset	Accuracy	Challenges
Xanthomonas wilt, Bunchy top disease, Black Sigatoka, Yellow Sigatoka, Fusarium wilt [2]	InceptionV2, ResNet50 and MobileNetV	Approx. 18,000 images have been collected, 12,600 used for study, 70% are used for training, 20% for testing and 10% for validation	Inception V2 and ResNet 50-Based Models performed better with more than 90% accuracy	Background variations due to field surroundings, dried leaves and overlapping leaves
Banana streak virus [14]	HSV algorithm	Not mentioned	88% accuracy for unaffected leaves, 81.6% accuracy for moderately affected leaves, 84.8% accuracy for fully affected leaves	Accuracy of detecting the disease is not high
Black sigatoka [15]	Support vector machine	180 images	Accuracy is not mentioned	Experiment is only limited to whether the leaf is diseased or not but not giving the amount of infection in leaf
Black sigatoka and black speckle [17]	Convolutional neural network	3700 Images	92–98%	Background, illumination, resolution, pose, size
Xanthomonas Wilt of Banana (BXW), banana bunchy top disease (BBTD), healthy banana clusters, and individual banana plants [16]	Random forest	8272 images	Banana Bunchy Top disease—99.4%, Xanthomonas Wilt of Banana—92.8%, Healthy Banana Cluster—93.3%, individual banana plant—90.8%	Using open source satellites with medium resolution, accurate mapping of banana is difficult
Black sigatoka and panama wilt [18]	Support vector machine and anfis classifier	50 videos sample of healthy and diseased banana leaves out of which 7537 frames had been captured	SVM accuracy is 92% and ANFIS Classifier accuracy is 100%	Not mentioned
Black sigatoka diseases [3]	Support vector machine	799 Images	96%	Needs improved segmentation and feature extraction for better recognition

(continued)

Table 1 (continued)

Diseases	Algorithm	Dataset	Accuracy	Challenges
Black sigatoka [19]	CNN Inception V3 and SVM	4244 images	CNN—90% and SVM—86%	Not mentioned

4 Conclusion and Future Scope

Many Image processing algorithms have been used to identify banana diseases and results in high accuracy. This paper has discussed various banana diseases, study the algorithms applied to those diseases, and conclude that most algorithms have shown results with more than 80% accuracy. Most of the researchers have applied the SVM algorithm on different diseases of banana and got good results. Along with it, Convolutional Neural Networks is an emerging field where we can train the data set using Machine Learning algorithms. Future work can be done by training the CNN model for banana disease recognition by capturing the images from different UAV's or designing a framework for the sustainability of banana crop.

References

1. Dita M, Barquero M, Heck D, Mizubuti ESG, Staver CP (2018) Fusarium wilt of banana: current knowledge on epidemiology and research needs toward sustainable disease management. Front Plant Sci 9:1468. https://doi.org/10.3389/fpls.2018.01468
2. Selvaraj MG et al (2019) AI-powered banana diseases and pest detection. Plant Methods 15(1):1–11. https://doi.org/10.1186/s13007-019-0475-z
3. Upadhyay A, Oommen NM, Mahadik S (2021) Identification and assessment of black sigatoka disease in banana leaf. Lect Notes Netw Syst 135:237–244. https://doi.org/10.1007/978-981-15-5421-6_24
4. Saranya N, Pavithra L, Kanthimathi N, Ragavi B, Sandhiyadevi P (2020) Detection of banana leaf and fruit diseases using neural networks. In: Proceedings of 2nd international conference inventive research in computing applications, ICIRCA 2020, pp 493–499. https://doi.org/10.1109/ICIRCA48905.2020.9183006
5. Ploetz RC (2015) Fusarium wilt of banana. Phytopathology 105(12):1512–1521. https://doi.org/10.1094/PHYTO-04-15-0101-RVW
6. Ye H et al (2020) Recognition of banana Fusarium wilt based on UAV remote sensing. Remote Sens 12(6):1–14. https://doi.org/10.3390/rs12060938
7. Prabha DS, Kumar JS (2014) Study on banana leaf disease identification using image processing methods. Int J Res Comput Sci Inf Technol 2(2):2319–5010
8. Devaraj A, Rathan K, Jaahnavi S, Indira K (2019) Identification of plant disease using image processing technique. In: Proceedings of 2019 IEEE international conference communication and signal processing, ICCSP 2019, pp 749–753. https://doi.org/10.1109/ICCSP.2019.8698056
9. Sakhamuri S, Kompalli VS (2020) An overview on prediction of plant leaves disease using image processing techniques. IOP Conf Ser Mater Sci Eng 981(2):10–16. https://doi.org/10.1088/1757-899X/981/2/022024
10. Patel A, Agravat S (2021) Banana leaves diseases and techniques: a survey. Lect Notes Data Eng Commun Technol 52:209–215. https://doi.org/10.1007/978-981-15-4474-3_24

11. Jogekar R, Tiwari N (2020) Summary of leaf-based plant disease detection systems: a compilation of systematic study findings to classify the leaf disease classification schemes. In: Proceedings world conference smart trends in systems, security and sustainability, WS4 2020, pp 745–750. https://doi.org/10.1109/WorldS450073.2020.9210401
12. Swain S (2020) A review on plant leaf diseases detection and classification based on machine learning models. IX(VI):5195–5205
13. Singh J, Goyal G, Gupta S (2019) FADU-EV an automated framework for pre-release emotive analysis of theatrical trailers. Multimed Tools Appl 78(6):7207–7224. https://doi.org/10.1007/s11042-018-6412-8
14. Karthik G, Praburam N (2016) Diseases from banana plant using embedded linux board
15. Vipinadas MJ, Thamizharasi A (2016) Banana leaf disease identification technique. Int J Adv Eng Res Sci 3(6):236756
16. Gomez Selvaraj M et al (2020) Detection of banana plants and their major diseases through aerial images and machine learning methods: a case study in DR Congo and Republic of Benin. ISPRS J Photogramm Remote Sens 169(Sept):110–124. https://doi.org/10.1016/j.isprsjprs.2020.08.025
17. Amara J, Bouaziz B, Algergawy A (2017) A deep learning-based approach for banana leaf diseases classification. Lecture notes informatics (LNI), Proceeding—Ser. Gesellschaft fur Inform., vol 266, pp 79–88
18. Vipinadas MJ, Thamizharasi A (2016) Detection and grading of diseases in banana leaves using machine learning. Int J Sci Eng Res 7(7):916–924
19. Escudero CA, Calvo F, Bejarano A (2021) Black Sigatoka classification using convolutional neural networks. 11(4). https://doi.org/10.18178/ijmlc.2021.11.4.1055

Covid-19 Prediction Analysis Using Machine Learning Approach

Prithish Sarkar, Ahana Mittra, Aritra Das Chowdhury, and Monoj Kumar Sur

Abstract The unforeseen outbreak of Covid-19 which resulted in a global pandemic posed a threat to the human civilization. The entire world is trying their best to combat against the outspread of the disease. The rapid spread of the disease has put governing bodies under pressure and made it difficult to confront the situation. The RT-PCR which is the test confirming if a person has Covid-19 infection, is restricted by the shortfall of reagents, time taking, high cost and need for dedicated labs with trained pathologists. With the sudden rise in daily cases, there were large queues for Covid-19 tests, stressing the medical laboratories with many such laboratories facing shortage of kits for testing. Hence, there is a requirement for cost effective and quick diagnostic model to determine positive and negative cases of Covid-19. This paper aims to predict Covid-19 infection in an individual person from initial symptoms and information like fever, cough, sore throat using machine learning algorithms. The study includes working with six predicting models, MLP, GBC, Decision tree, SVM, Logistic Regression and Random forest with highest accuracy of 92.94% achieved in logical regression. The results can help in the initial diagnosis of Covid-19, especially when there is a shortage of RT-PCR kits, specialized laboratories and to screen large number of patients.

Keywords Covid-19 · Machine learning · Logistic regression · Classification

---

P. Sarkar · A. Mittra · A. D. Chowdhury · M. K. Sur (✉)
Computer Science and Engineering Department, Future Institute of Engineering and Management, Kolkata, West Bengal 700150, India
e-mail: monojdada@gmail.com

A. Mittra
e-mail: ahana.mittra.fiem.cse19@teamfuture.in

A. D. Chowdhury
e-mail: aritra.das.chowdhury.fiem.cse19@teamfuture.in

P. Dutta et al. (eds.), *Emerging Technologies in Data Mining and Information Security*, Lecture Notes in Networks and Systems 490,
https://doi.org/10.1007/978-981-19-4052-1_15

1 Introduction

In early December of 2019, a novel coronavirus (COVID-19) was discovered in Wuhan province of the People's Republic of China [1]. Within a short span of time, the disease spread from China to over 100 countries worldwide resulting to Covid-19 pandemic. As of 8th November 2021, the total number of global cases stood at 62,875,460 with 1,4681,763 deaths across the globe [2]. The infectious disease threatens the human race, limit economic and social growth and endanger stability. The economic and social disturbance caused by the pandemic is disastrous. Tens of million people are on the risk of falling into acute poverty and number of undernourished people could count up to 132 million by the end of 2020 [3]. This pandemic carries on to put stress on medical infrastructure and systems throughout the world in many aspects, the sudden rise in need of hospital beds, ICUs, scarcity of medical equipment and health workers getting infected by the disease. So, effective and prompt clinical decision of medical resources is decisive to handle the situation. The most legitimate confirming test for Covid-19 is RT-PCR (Reverse Transcriptase Polymerase Chain Reaction). Many developing countries has been facing a shortage of RT-PCR kits, which resulted in delay in testing individuals for covid-19 infection, and resulted in increased infection rate and delay in taking critical precautions. The alternative to RT-PCR test can be Chest CT-Scan [4] and Chest X-rays. Still, the two mentioned methods cannot always be used for screening patients because of high cost and low availability of devices.

Machine Learning [5] refers to the automatic detection of valid patterns in the dataset by using the data and algorithms emulating the way humans learn and gradually enhance its accuracy. The algorithms enable to build models that aids in pattern recognition, classification and prediction. The training data is the input to the learning algorithm and the output is the expertise which can perform a task. Contemporary biology has the comfort of benefitting from the advancements in the area of machine learning [6].

The artificial intelligence (AI) and machine learning models have been widely used and carried out trials on various healthcare sector. The recent spread of Covid-19 infections demands the need of using such techniques to identify, predict and help in prioritizing RT-PCR test with speed and efficiency, which will aid in effective screening of Covid-19 infections and the reduce the load on healthcare departments.

Studies have stressed on Chest X-rays and CT images using deep learning algorithm to predict Covid-19 infection [7]. But it is difficult to screen large number of patients due to high cost, radiation doses and inadequate number of available devices. Thus, the task to differentiate between covid positive and negative in a cost effective and more speedy way remains as a challenge. Previous works had the limitations and future scope of comparing and bringing out the best suited algorithm for the prediction [8]. In this study we aim to predict covid-19 infection positive and negative cases by knowing eight basic information(features) and compare different six models.

With the aim of helping medical specialist across the globe in screening patients where healthcare resources are limited. The Covid-19 infection had sign and symptoms of cough [9], fever [10] and shortness of breath [11]. Along with these symptoms, total five symptoms along with three basic information have been used. The Israeli Ministry of Health published data of every individual who tested Covid-19 positive by RT-PCR test [12]. The dataset translated to English [13] and is available to be accessed [14]. Here, we train six different machine learning algorithms on the same dataset and predict the possibility of an individual being Covid positive or not by having basic knowledge of the person. Next we will describe the Dataset, Methodology, Result and Discussion and lastly conclusion.

2 Dataset

In the current experiment, we have used the dataset mentioned in [12]. The dataset contains a total of 2,742,596 cases. Table 1 reflects the number of positive, negative cases, and tests with no results. The data has been recorded from 2020/03/11 to 2020/11/12 and the date is in yyyy/mm/dd format. The dataset consists of records on a daily basis of all the residents who tested for covid-19 confirmed by RT-PCR test.

The following list describes the dataset's features used by six machine learning model:

1. Basic information:
   (a) Sex (male/female)
   (b) Age $\geq$60 (true/false)
2. Symptoms:
   (a) Cough (true/false)
   (b) Fever (true/false)
   (c) Sore throat (true/false)
   (d) Shortness of breath (true/false)
   (e) Headche (true/false)
3. Other information:
   (a) Known contact with an individual confirmed to have COVID-19 (true/false)

We convert the true/false value to 1 and 0, and male/female to 1 and 0 values respectively. The collected data was transformed from the categorical values to binary

Table 1 Details of the database used in the experiment

Total number of tests	Total covid-19 positive results	Total covid-19 negative results	Total tests with no result
2,742,596	220,975	2,480,403	41,218

Table 2 Information of features with number of samples

Features	Status	No. of samples	Total
Cough	Yes (1)	2,631,258	2,742,596
	No (0)	111,338	
Fever	Yes (1)	2,645,600	2,742,596
	No (0)	96,996	
Sore throat	Yes (1)	2,712,512	2,742,596
	No (0)	30,084	
Shortness of breath	Yes (1)	2,731,579	2,742,596
	No (0)	11,017	
Headache	Yes (1)	2,682,655	2,742,596
	No (0)	59,941	
Age 60 and above	Yes (1)	1,908,553	2,287,838
	No (0)	286,399	
	NAN	547,644	
Gender	Yes (1)	1,278,266	2,742,596
	No (0)	1,371,444	
	NAN	92,886	
Test indication	Other (0)	2,547,559	236,255
	Contact with confirmed (1)	170,742	
	Abroad (2)	24,295	

values using label encoders. We created a dictionary where the label is treated as the key and in the value is assigned by enumerating the labels starting at 0. Then the mapped dictionary is used to transform the class labels to integers. The column containing the dates are dropped and the Nan values removed. The dataset contains the detailed information of used eight features with number of samples shown in Table 2.

3 Methodology

The methodology used in the present experiment for the prediction of COVID-19 is mentioned in the following Fig. 1. The dataset collected from [14] first checked for missing values and null values. The unnecessary information like date is dropped. Then the null/missing data is identified and handled. The features with null or missing values are dropped. In the next step, features with categorical data is identified. The identified data is represented into numerical values. The resultant feature file contains the features cough, fever, shortness of breath and early symptoms of headache and sore throat [15]. Basic information of gender and age are also considered as feature

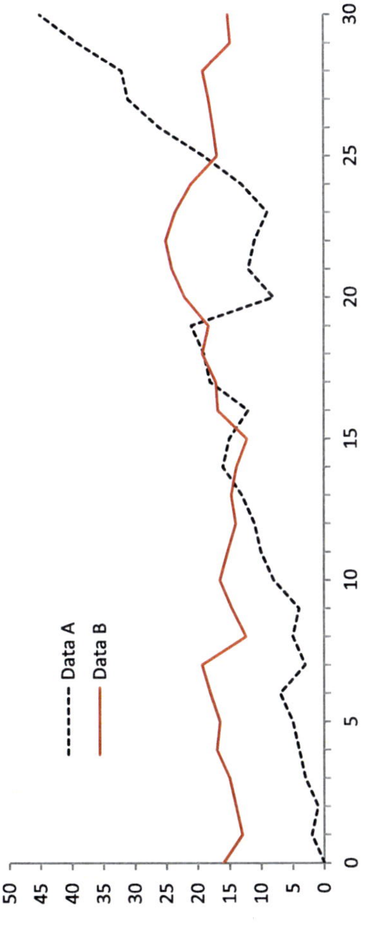

Fig. 1 Flowchart of the proposed covid-19 prediction approach

for the prediction of COVID-19. Next, the dataset has been splited into train and test data. 70% of the dataset has been used as the train set and rest 30% used as the test set. Lastly, six popular machine learning algorithms (MLP, GBC, Decision tree, SVM, Logistic Regression and Random forest) are applied on the feature set for the prediction purpose.

4 Result and Discussion

The six mentioned machine learning algorithms are applied on the previously stated feature set to generate training model. The test set is then fed to the corresponding build models to find the prediction and thereby finding the strength of the various models by evaluating the metrics like precision, recall, F1 Score, and achieved accuracy. Table 3 reflects the outcomes observed by applied six machine learning approaches.

From Table 3 it can be observed that Logistic Regression provides the best outcome over other approaches and is highlighted by marking with blue color. The confusion matrix obtained through best performing model has been shown in Fig. 2.

5 Discussion and Conclusion

The worldwide pandemic of covid-19 has become a major threat to many countries in terms of health, security as well as financial instability. The newly recognized infection, covid-19 spreads throughout the world, which brings a challenge to get readily available data and information. The availability of data is crucial for the timely training and evaluation of models which predict the disease. There were shortage of RT-PCR test kit in under developed to developing countries. The RT-PCR test whose alternative is using CT and X-Ray images [4]. These tests are not very suitable due to the lack of available devices, high cost, time consuming and radiation doses. All

Table 3 Results obtained from applied six machine learning approaches

Model	Precision	Recall	Accuracy
Support vector machine	0.86	0.90	89.60
Random forest	0.88	0.91	90.28
Decision tree	0.90	0.92	91.93
Multi-layer perceptron	0.91	0.92	92.49
Gradient boosting	0.91	0.93	92.79
Logistic regression	0.90	0.93	92.94

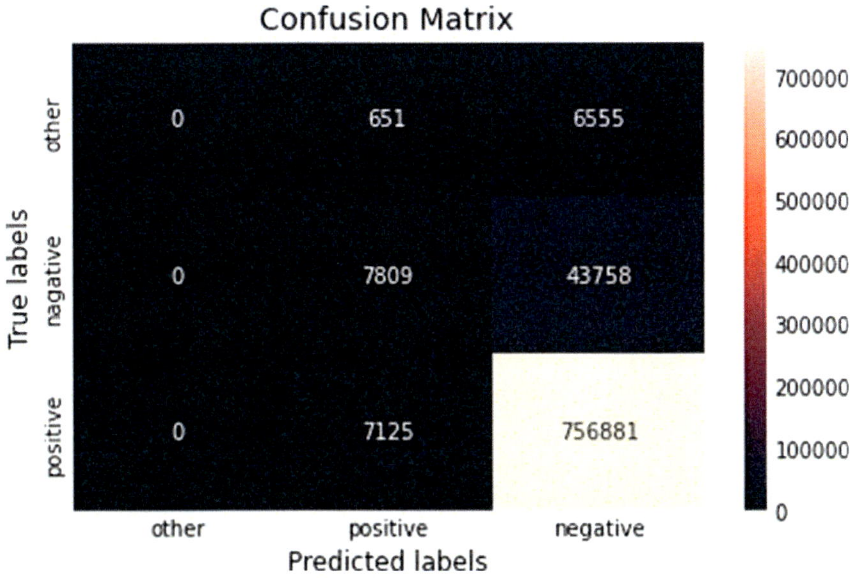

Fig. 2 Logistic regression

these factors led to a long queue of patients awaiting covid test, and thus resulting in increased infection rate. With a solution to screen the patients on priority of covid test, we propose few models to predict the covid-19 positive and negative cases using machine learning classification algorithm on few clinical features. The collected dataset had six five clinical symptoms and two basic information. Using this information, we trained five predictive models (i.e. Random Forest, Logical regression, MLP, GBC, Decision Tree). The models were validated and evaluated using confusion matrix, accuracy, precision, recall, F1 score and support. The logistic regression outperformed all other models with an accuracy of 94.37%. Every other models performed well with achieving an accuracy of 89+%. Thus, the models can be an effective tool to screen large number of patients. The dataset was limited to one country, Israel. The outbreak of Covid-19 may not be repeated everywhere [16] as reported and observed by many. Hence, our work is limited to a particular country with unavailability of data from other countries. For future research, applying the models on data of different country and on a larger data with feature as image would enhance the standard and accuracy of the prediction.

References

1. Zhu N, Zhang D, Wang W, Li X, Yang B, Song J, Zhao X, Huang B, Shi W, Lu R, Niu P (2020) A novel coronavirus from patients with pneumonia in China, 2019. New England J Med
2. https://www.patientcareonline.com/view/covid-19-updates-us-vaccinations-booster-doses-global-data-as-of-november-8-2021

3. https://www.who.int/news/item/13-10-2020-impact-of-covid-19-on-people's-livelihoods-their-health-and-our-food-systems
4. Majidi H, Niksolat F (2020) Chest CT in patients suspected of COVID-19 infection: a reliable alternative for RT-PCR. Am J Emergency Med 38(12):2730–2732
5. Shalev-Shwartz, S. and Ben-David, S., 2014. Understanding machine learning: From theory to algorithms. Cambridge university press
6. Tarca AL, Carey VJ, Chen XW, Romero R, Drăghici S (2007) Machine learning and its applications to biology. PLoS computational biology 3(6):e116
7. Ai T, Yang Z, Hou H, Zhan C, Chen C, Lv W, Tao Q, Sun Z, Xia L (2020) Correlation of chest CT and RT-PCR testing for coronavirus disease 2019 (COVID-19) in China: a report of 1014 cases. Radiology 296(2):E32–E40
8. Ardabili SF, Mosavi A, Ghamisi P, Ferdinand F, Varkonyi-Koczy AR, Reuter U, Rabczuk T, Atkinson PM (2020) Covid-19 outbreak prediction with machine learning. Algorithms 13(10):249
9. Lu R, Zhao X, Li J, Niu P, Yang B, Wu H, Wang W, Song H, Huang B, Zhu N, Bi Y (2020) Genomic characterisation and epidemiology of 2019 novel coronavirus: implications for virus origins and receptor binding. The Lancet 395(10224):565–574
10. Chan JFW, Yuan S, Kok KH, To KKW, Chu H, Yang J, Xing F, Liu J, Yip CCY, Poon RWS, Tsoi HW (2020) A familial cluster of pneumonia associated with the 2019 novel coronavirus indicating person-to-person transmission: a study of a family cluster. The Lancet 395(10223):514–523
11. Shi Y, Yi Y, Li P, Kuang T, Li L, Dong M, Ma Q, Cao C (2003) Diagnosis of severe acute respiratory syndrome (SARS) by detection of SARS coronavirus nucleocapsid antibodies in an antigen-capturing enzyme-linked immunosorbent assay. J Clin Microbiol 41(12):5781–5782
12. COVID-19-Government Data (2020) . https://data.gov.il/dataset/covid-19
13. Gupta R, Pandey G, Chaudhary P, Pal SK (2020) Machine learning models for government to predict COVID-19 outbreak. Dig Gov Res Pract 1(4):1–6
14. https://github.com/nshomron/covidpred
15. Kotsiantis SB, Zaharakis I, Pintelas P (2007) Supervised machine learning: a review of classification techniques. Emerg Artif Intell Appl Comput Eng 160(1):3–24
16. Remuzzi A, Remuzzi G (2020) COVID-19 and Italy: what next? The lancet 395(10231):1225–1228

Smart City Driven by AI and Data Mining: The Need of Urbanization

Sudhir Kumar Rajput, Tanupriya Choudhury, Hitesh Kumar Sharma, and Hussain Falih Mahdi

Abstract In the modern world of urbanization, the needs of any smart city are to manage the various important wheels of the city like water and electricity, urban transportation and traffic system, pragmatic approach to manage the solid waste, centralized management of information, better disaster management for the city, control over crime, an active emergency response system, renovating the heritage monuments and making the city beautiful, etc., for the people living in the city. The city authorities are engaged to make their city smart by means of implementing different solutions vide different schemes of the Central or State Government or of combined effort, integrating solutions and generated data which is huge can be utilized to discover the gap, improve operation and services for citizens of the city. This research paper highlights the use of artificial intelligence (AI) and data mining in implementing the sustainable solutions for a smart city.

Keywords Smart city intelligence · Artificial intelligence · Data mining

1 Introduction

As the city grows, the need of electricity and water, better and efficient transportation, collecting solid waste generating from the houses, consistent efforts to make the city beautiful, need to provide safety to citizens, control and reduce the crime and need of an effective disaster management for emergency response keeps on growing to maintain and sustain the city. For maintenance and sustainability of the city and

S. K. Rajput (✉) · T. Choudhury · H. K. Sharma (✉)
School of Computer Science, University of Petroleum and Energy Studies (UPES), Dehradun, India
e-mail: skrcnb15@gmail.com

H. K. Sharma
e-mail: hkshitesh@gmail.com

H. F. Mahdi
College of Engineering, University of Diyala, Baqubah, Iraq
e-mail: hussain.mahdi@ieee.org

P. Dutta et al. (eds.), *Emerging Technologies in Data Mining and Information Security*, Lecture Notes in Networks and Systems 490,
https://doi.org/10.1007/978-981-19-4052-1_16

Fig. 1 Basic needs of the citizens

fulfilling the needs of the people living in the city, smart city have become the need of today's world, and this concept has been useful to improve the quality of life of the city residents and is being adapted by different countries in the world. The paper presents basic needs of citizens (Fig. 1) and data mining techniques along with some algorithms. The literature review part covers the study of the papers worked on use of artificial intelligence in smart cities, while the case study presents one of the smart city implementation in India along with the ideas of using AI and data mining that can be further useful.

The idea of smart city came from developing the cities with the aim of improving the quality of services offered to the people living in the city and at the same time promoting economic growth of the city. Thereby smart city works on developing smart solutions, integrating with existing system, improving the operational efficiency of the government services for the citizens of the city and sharing the information with them. To achieve this, the smart city solutions use the information and communication technologies (ICT) for acquisition, processing and sharing the data.

Smart city is envisaged to have four pillars: Social infrastructure, physical infrastructure, institutional infrastructure (including governance) and economic infrastructure. City residents are the centre of attention for each of these pillars. From these four pillars, basic needs of the citizens that any city should fulfil are needs of sewage and waste management, basic infrastructure needs like roads, bus and railway stations, water and electricity supply needs, urban transportation and traffic management, educational needs and employment needs which make the city livable and sustainable [1, 2].

The smart city projects are being implemented using information communication technologies (ICT) in which lot of data is generated from implementation to daily routine operations of the projects. The use of AI and data mining techniques on the generated data can further bring a transformational change in planning, improvement

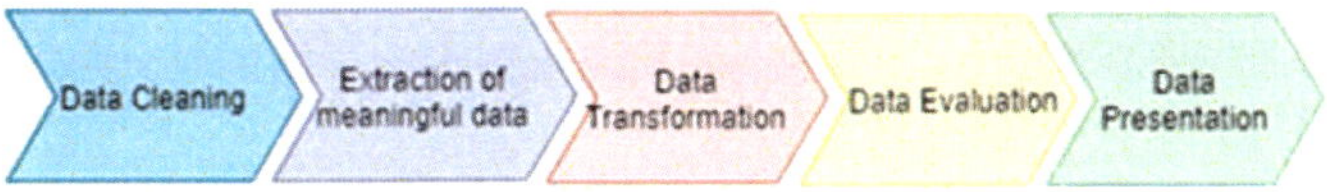

Fig. 2 Data mining process

and strengthening the operations. The process of the data mining includes the steps of data cleaning, extraction of meaningful data, data transformation, pattern evaluation and data presentation (Fig. 2).

The important data mining techniques useful in finding the pattern evaluation or data modelling are regression, classification, association, clustering and prediction. Regression analysis is used to predict the numeric attributes or continuous values. Classification is used to find the important information and features from the dataset and helps in classifying the dataset. Classification can be categorical (example gender and marital status .) or numerical (example, age, temperature, etc.). Association rule helps to find the interesting associations and relationships between items of large dataset. The association rule also provides the detail about how frequently an itemset is occurring in a transaction. For example, if there are two items *X* and *Y* purchased by the customers frequently, then it is better to keep these items together in the store or provide some discount offer on second item if first one is purchased. This can results in increase of the sales. This is known as market based analysis wherein association rule allows retailers to find the relationships between the items that people tend to buy together frequently. For example, customers purchasing bread and butter can also purchase milk. The two terms used in association rule mining are support and confidence. The support tells about usefulness and certainty of rules. About 10% support means total 10% of transactions in database follow the rule.

$$\text{Support}(X \rightarrow Y) = \text{Support\_count}(X \cup Y)$$

A confidence of 70% means that 70% of the customers who purchased bread and butter also bought milk.

$$\text{Confidence}(X \rightarrow Y) = \text{Support\_count}(X \cup Y)/\text{Support\_count}(X)$$

A strong rule satisfies both, i.e. minimum support and minimum confidence [3]. The word 'Clustering' has come from clusters which are formed using unlabelled dataset. The clusters are formed in such a way that each dataset belongs to only one group of categories that have similar type of features or attributes. Clustering is helpful in discovering the categories of groups in the unlabelled dataset by itself without the need for training. Prediction is a futuristic event that will occur which is predicted based on analysis of past events in right sequence. It uses a combination of the data mining techniques like trends, sequential patterns, clustering, classification, etc. (Table 1).

Table 1 Data mining techniques

Clustering	Classification	Prediction	Association	Regression
It is a technique to club together related data in different groups	Categories data in different categories based on the attributes	Finding unknown facts and based on that find future actions	Relation between two or more different variable in dataset	Finding impact of one variable value on other variable

Most important algorithms used in data mining techniques are decision tree algorithm for classification and regression, Apriori algorithm for association and prediction and K-means algorithm for clustering. These are briefed below:

1.1 Decision Tree Algorithm

Decision tree algorithm is a well-known supervised learning algorithm, used for classification and regression models. The decision tree helps to visualize the decisions which make it easy to understand. The classification or regression models are built using training data, and then the accuracy of the model is checked, and then the model is tested for new data. The decision tree follows the top-down approach where the top region presents all the observations at a single place called as root node. The root node splits into two or more branches that further split into leaf node and terminal node. Due to its focus on current node rather than future nodes, this approach is also termed as greedy approach. The algorithm runs till the stop criteria such as the minimum number of observations are met or satisfied. After completing a decision tree, many nodes may represent outliers or noisy data which are removed using tree pruning method. This avoids the overfitting and also improves the accuracy of the model [4].

1.2 Apriori Algorithm

Before proceeding to Apriori algorithm, let us understand the itemset and frequent itemset. A set of items together is known as an itemset. An itemset consists of two or more than two items. If an itemset has *n*-items, then it is known as *n*-itemset. An itemset occurring frequently is known a frequent itemset. Apriori algorithm is used to find out the frequent itemsets in the dataset. The algorithm is called Apriori because it uses previous knowledge of frequent itemset properties. An iterative approach or level-wise search is applied where *n*-frequent itemsets are used to find $n + 1$ itemsets. This data mining algorithm follows join and prune steps iteratively till the most frequent itemset is achieved. A minimum support threshold is provided in the problem or it is assumed by the user [5].

1.3 K-Means Clustering Algorithm

K-means is an algorithm from category of unsupervised learning and it is used to solve the clustering related problems. K in K-means algorithm presents the number of pre-defined clusters that are required. For example, if $K = 2$ or 5, then there will be two clusters or five clusters, respectively. Every cluster has a centroid which is used to calculate the distance of a data point from it. The main purpose of this algorithm is to minimize the sum of distances between the data point and their corresponding centroids, resulting into division of dataset into K-number of clusters. The process is repeated until it finds the best clusters having homogenous data and without overlapping.

2 Literature Survey

The various papers from different researchers are studied to understand the development of smart cities and their migration towards using AI and data mining. The detailed review of the papers is presented (Table 2).

Based on thorough study of different papers, Sujata et al. [6] in their paper presented that the development of the smart cities lies basically on six factors which are social, management, economic, legal, technology and sustainability (SMELTS). The projects of smart cities should consider the citizens of the city as core, and there should be transparent policies for everyone including the management. The use of ICT is another important requirement for implementing various smart cities projects. The development can only be said sustainable if it is done without disrupting the environment and integrating the infrastructures for optimized use of services.

Mathur and Modani [7] in their paper pointed that AI is the need of the smart cities to improve citizen's quality of life and making a robust society. These objectives of smart city can be achieved by implementing AI in public transportation to maximize the capacity, to make an effective electricity supply system with smart grid, smart metres and smart appliances, roboticbased waste management system to separate recyclable items, using AI for doctors in the healthcare workforce, using AI in safety and security systems for surveillance and analysis. To create smart energy cities, importance of technologies like Internet of Things (IoT) and data mining have been depicted by Horban [8] in her paper. She also mentioned that the governance body of smart cities which are city municipalities should present the key performance indicator (KPI) based on their cities vision, objectives, opportunities, and challenges, etc. Shahrour et al. [9] discussed about the digital components for smart cities and also developed the platform for sharing data and information to be used by the management of SunRise smart city in France. The architecture of platforms consisted of data collection, data storage, data analysis and system management, data display, visualization, user interface and stakeholder interaction. The platform was also connected to the professional data management tools such as OsiSoft, ARCGIS

Table 2 Synopsis of literature review on smart cities from 2016 to 2021

S. No.	References	Author and findings of the paper	Year of publication
1	Ref. [6]	In the review paper presented by Sujata et al. [6], it is stated that smart cities are developed on six factors of management, economic, legal, technology and sustainability (SMELTS)	2016
2	Ref. [7]	Mathur and Modani [7] outlined how use of AI in public transportation, electricity supply, waste management, healthcare, safety and security can benefit to the society and citizens	2016
3	Ref. [8]	In order to estimate the energy consumption and analysis of related data, Horban [8] in her paper presented the concept of using data mining technology and big data analytics	2016
4	Ref. [9]	Shahrour et al. [9] in their paper presented the architecture of the smart city platform for implementing the digital solutions	2017
5	Ref. [10]	For the data generated by intelligent transportation system, research paper of Chen [10] is about using spatial clustering algorithm-based data mining technique. For mining algorithm-based solution for traffic jam and improving operational efficiency, he has proposed to study early peak and late peak in the traffic flow data and to construct the traffic flow model library	2017
6	Ref. [11]	The paper of Srivastava et al. [11] emphasizes on implementation of smart solutions using AI which also raises the privacy concerns to the human life and therefore proposed the solutions based on human intervention	2017
7	Ref. [12]	As per Yang et al. [12], big data techniques and tools can also be used to help and provide relief to the citizens of the city suffering from natural disasters	2017
8	Ref. [13]	Navarathna et al. [13] mentioned about the use of AI, machine learning, deep learning, big data analytics and IoT for creating solutions to various problems faced by citizens	2018
9	Ref. [14]	Dias et al. [14] mentioned about the challenges faced by researchers to collect smart city. To compensate this, they proposed a data management plan and also developed ontology for description of smart cities data	2018
10	Ref. [15]	The paper from Nageswari Amma [15] focuses on the need to protect the sensitive data generated from smart environment and proposed the possible encryption algorithm	2018

(continued)

Table 2 (continued)

S. No.	References	Author and findings of the paper	Year of publication
11	Ref. [16]	In this research work, the author described the law and policy requirement in implementation of smart city, and how AI is useful for getting proper insight for converting a normal city to smart city	2019
12	Ref. [17]	In this paper, author has presented the smart city infrastructure level requirement and suggested a framework to efficient implementation of smart city concepts in a normal city	2019
13	Ref. [18]	Laadan et al. [18] provided an analysis of violations like missing fire hydrants and demonstrated the potential of data-driven approaches for quickly locating and addressing them	2020
14	Ref. [19]	Based on data mining techniques, Mystakidis and Tjortjis [19] proposed a methodology for predicting traffic congestion in the smart cities to inform the drivers about congested routes and take decision for alternate routes	2020
15	Ref. [20]	The research article by Huang and Nazir [20] presents that devices connected to IoT in smart cities produce a massive data, and therefore, there is risk of breaching security and privacy. The researchers presented the analytic network process (ANP) for evaluating the smart cities IoT based on use cases	2021

and MATLAB. Srivastava et al. [11] highlighted about already implemented solutions like video surveillance, drones and cyber security in the smart cities. The use of AI-based techniques has benefit of making the cities smarter, but on other side, these techniques are also becoming a threat to the privacy of people and their control of life. Therefore, the solutions being implemented using AI should also consider human intervention.

As per Navarathna et al. [13], most of the problems related to the traffic management, parking, water and waste management, educational and industrial sectors can be resolved efficiently using AI techniques like machine learning, deep learning and big data analytics. For resolution of the problems of smart cities, pattern recognition, image recognition, IoT, ICT and big data could emerge as solutions. Dias et al. [14] realized the need of the data for the researchers, and therefore, to support the researchers proposed a data management plan which will help researchers from the point of collecting data up to the point of publishing the data. The data management plan (DMP) was developed for Porto smart city in Portugal. DMP also included a set of guidelines for data management and elaboration of ontology for smart cities was one of the important guideline. The paper by Nageswari Amma [15] presented the need to protect the sensitive data generated from smart solutions developed for smart cities and addressed the issue of protecting the sensitive data as well as classification of the sensitive data. The protection of private data was proposed using homomorphic encryption which allows to use the encrypted data for other purposes without decrypting, while the classification of data was proposed using naive Bayes algorithm. Founoun et al. [16] in their paper analyse the existing regulations and present the tools for regulation support in the implementation of the smart city in Morocco and to know the transformation level the city is ready for. For this purpose, the textual evaluation method based on the similarity analysis is proposed. They also addressed the assessment of the orientation of the regulatory efforts within the cities through the analysis of diagnosis results. Mystakidis and Tjortjis [19] proposed a methodology for predicting traffic congestion, and if the traffic is severe, the drivers can choose to adopt alternate route. The methodology was implemented using data mining and big data techniques along with Python, SQL, and GIS technologies and same was tested on data originated from the most traffic congested streets in Thessaloniki (2nd most populated city in Greece). The outcome showed that the quality and size of the data play an important role in achieving the accuracy for the model. Huang and Nazir [20] mentioned that with use of IoT, there are number of devices in smart cities remain connected for providing or utilizing different services, and these IoT devices are increasing day by day due to increase of users. These IoT devices in the smart cities produce a huge volume of data which also imposes the risk of breaching security and privacy. The researchers presented the approach of analytic network process (ANP) for evaluating the smart cities IoT based on use cases.

Table 2 presents the summary of above-mentioned papers.

The research papers from different researchers provide insights, challenges and risks in implementation of smart cities world-wide. This paper provides the use of AI and data mining in implementation of smart solutions for developing the smart cities in Indian perspective.

3 Case Study: Implementation of Jaipur Smart City (2017–18)

More than hundred cities are being developed as smart cities with the initiative of Indian Government. As the needs of the cities are same, the smart cities have been provided a bouquet of smart solutions based on the size of the city. Jaipur smart city in India [2] has been presented as a case study to implement the solutions along with the proposed role of AI and data mining.

3.1 Water and Electricity

The water and electricity are the two basic needs of any city and should be given high priority. The authority managing the water supply in the city shall make sure the city has sufficient water as per the consumption and a proper distribution network for supply of water in the houses and commercial buildings. To reduce the wastage of water and increase the revenue, the authority responsible for water and electricity supplies are installing smart metres in the premises of the users [21–24]. Role of AI and data mining: All over the world, shortage of water is being experienced and hence there is need to reduce and minimize the wastage of the water. The AI can play a big role in achieving the goal of an intelligent and resilient water management system which can become reality upon integrating SCADA with AI. The SCADA system updates the data from the smart metres. Since the SCADA system for electricity and water stores lot of data, the data mining technique will also be helpful in discovering the areas usages and wastages and deploying the strategic solutions accordingly.

3.2 Sewage and Solid Waste Management

Any sustainable city should have good sewage system and solid waste management system to manage the solid waste it generates on daily basis. The techniques to recycle the waste are compulsory to avoid creating the mountains of solid wastes. In the recycling technique of the solid waste, there is requirement to separate the solid waste into dry and wet. This also requires the collection in the same manner from the houses. The main components of solid waste management are waste collection, transfer and transport of solid waste, onsite handling, storage and processing, and waste recovery and final disposal. The role of AI in waste management begins with smart bins which are fitted with the sensors to monitor the fullness of waste in the bin, sending the alarm to the monitoring centre to collect the waste and empty the bin. Smart bins are also paired with a mobile application that lets collecting agency to know the status of bins and preventing it from overflow and also the users to know the location of the nearest available waste bin for disposal of the wastes. After collection, the waste is sorted manually at the waste management facilities before sending it at the recycling centre(s).

Role of AI and data mining: The use of AI at waste management facilities can help in automated sorting of the waste. This way, the waste management facilities can also be integrated with the recycling centres and optimizing the process of solid waste management.

3.3 ITS in Urban Transportation

Apart from electricity and water, the transportation is another basic need in any of the city. The authorities of the city are putting their efforts to provide better transportation to the city residents. The mini buses, city buses and metro are some of the transport means where the authorities are working for implementation of ITS. The ITS system helps in smoothing the operations and provides better information to the users. Like metro, to make the city bus transport services reliable, the city buses and bus stops need to be implemented with automatic vehicle location (AVL) and passenger information system (PIS). Each bus stop shall display the expected time of arrival (ETA) of the buses and any delay, if expected to the users. The monitoring centre will keep track of the bus locations and their schedule for adherence. Further, the integration of the operations of the metro and city bus services will not only make the public transport available for the entire city and reliable to the users, but will also improve their operational efficiency.

Role of AI and data mining: ITS in transportation is used for scheduling, optimizing service, information sharing with the commuters, improve operational efficiency and brand building. Use of ITS in the city bus service, bus rapid transit (BRT), light rail transit (LRT) and mass rapid transit (MRT) generates a lot of data. This data can be cleaned for thorough study applying data mining algorithms to discover the history of the traffic patterns and jams and improved operational efficiency of the transportation system. The data mining can also be used to study the demand pattern and add the service or increase the frequency during different periods of demand. This way, data mining can help a lot in ITS system for transportation system and smoothing the traffic in the city.

3.4 Digitization of Police Stations and Crime Control

The safer city attracts the tourists and the business as well. The police stations are responsible for controlling the crime and making a city safer and keep the records of the complaints and criminals. The digitization of the police stations does not only make the records accessibility faster, but also available from anywhere. This data can help in increased transparency and controlling over the crime and criminals. The mobile applications are developed to reduce the crime in the city, faster response in emergency and build the confidence between the city residents and the police system. The mobile application has the panic button to use it and a 24 $\times$ 7 monitoring centre connected with the patrolling vehicles and police stations. The city residents are motivated to use the mobile application for using it in emergency situation like theft, robbery, molestation, etc.

Role of AI and data mining: The initial steps required are related with digitization like filing and acceptance of complaint on the computer and digitization of investigation record which will make the record of the criminals and victims as well. Then all the police stations in the city need to be connected to the data centre location in the city. The data centre and the data shall be in high security. To improve the operational efficiency, use of AI through image recognition can help in identifying accessing and finding the criminal records, if available in the database.

3.5 *Common Payment System*

The common payment system in the city is being implemented with help of nationalized bank which will provide a prepaid card to the city residents, and the entire ecosystem is connected with a central clearing house (CCH). To penetrate the use of prepaid common payment card, the different authorities of the city need to be agreed for integrating the card for payment of different services like transportation (city buses, MRT or LRT), heritage site's or monuments entry fees or even purchasing the goods at the shops. CCH ensures to distribute the amount amongst the different authorities.

Role of AI and data mining: AI is already being used in the banking and financial institutions, and it is playing an important role in fraud detection in credit card systems. Apart from thatt, AI is also used to enhance customer service and provide recommendation on loans. Though the common payment cards are essential for the transport services without the need of the PIN, they can also be integrated to carry out the e-commerce transaction. With use of AI, the system is able to learn a card user's normal behaviour over time and spot outliers, and notifies to the user in case of suspect transaction. Further, the AI-based chatbots driven customer care centres should be developed to solve the user's query or help them and enhance the user's experience to motivate the use of the card.

4 Centralized Management of Information

As seen above, there is need to have central data acquisition and monitoring agency for proper functioning of different solutions required to be implemented to make a city smart. For data acquisition, the data centre needs to be build up locally or it can be set up based on the cloud computing technologies. The central system is also a point of coordination between amongst different authorities. The disaster recovery site is also required to secure the data. The authorities can share the data amongst them and also use the data to implement any other solution required for the city.

Role of AI and data mining: A platform based on AI should be developed to share the data amongst the authorities which can be studied and help in identify the projects to be launched in the city.

5 Conclusion and Future Work

The use of artificial intelligence and data mining has become the need of the present situation, where everything revolves around the data. Using AI and data mining, many projects in the smart cities can see a transformational change in their urbanization in a sustainable manner. This research work has been focussed from the perspective of implementing solutions using artificial intelligence and data mining in the smart cities of India. The research on world-wide implementation of smart cities can be done as the future work.

Acknowledgements I would like to express my deep gratitude to Dr. Tanupriya Choudhury, Professor at School of Computer Science, UPES for his guidance and continuous support towards completion of this research paper.

References

1. https://www.twi-global.com/technical-knowledge/faqs/what-is-a-smart-city. Accessed on 9 June 2021
2. Smart cities mission: a step towards smart India. National Portal of India. Accessed on 9 June 2021
3. https://towardsdatascience.com/association-rule-mining-be4122fc1793. Accessed on 16 July 2021
4. https://www.softwaretestinghelp.com/decision-tree-algorithm-examples-data-mining/. Accessed on 16 July 2021
5. https://www.geeksforgeeks.org/apriori-algorithm/. Accessed on 16 July 2021
6. Sujata J, Saksham S, Tanvi G, Shreya (2016) Developing smart cities: an integrated framework. In: 6th International conference on advances on computing and communications, ICACC 2016, 6–8 Sept 2016, Cochin, India
7. Mathur S, Modani US (2016) Smart city—a gateway for artificial intelligence in India, In: 2016 IEEE students' conference on electrical, electronics and computer science
8. Horban V (2016) A multifaceted approach to smart energy city concept through using Big Data analytics. In: IEEE first international conference on data stream mining and processing, 23–27 Aug 2016
9. Shahrour I, Ali S, Maaziz Z (2017) Comprehensive management platform for smart cities. IEEE
10. Chen Y (2017) Intelligent transport decision analysis system based on Big Data mining. In: 7th International conference on education, management, information and computer science (ICEMC 2017)
11. Srivastava S, Bisht A, Narayan N (2017) Safety and security in smart cities using artificial intelligence-a review. In: 2017 7th International conference on cloud computing, data science and engineering—confluence
12. Yang C, Su G, Chen J (2017) Using Big Data to enhance crisis response and disaster resilience for a smart city. In: 2017 IEEE 2nd international conference on Big Data analysis
13. Navarathna PJ, Malagi VP (2018) Artificial intelligence in smart city analysis. In: International conference on smart systems and inventive technology (ICSSIT 2018)
14. Dias P, Rodrigues J, Aguiar A, David G (2018) Planning and managing data for smart cities: an application profile for the UrbanSense project
15. Nageswari Amma NG (2018) Privacy preserving data mining classifier for smart city applications. In: Proceedings of the international conference on communication and electronics systems (ICCES 2018)
16. Founoun A, Hayar A, Haqiq A (2019) The textual data analysis approach to assist the diagnosis of smart cities initiatives. In: 5th IEEE international smart cities conference (ISC2 2019)
17. Heaton J, Parlikad AK (2019) A conceptual framework for the alignment of infrastructure assets to citizen requirements within a smart cities framework. Cities 90(2019)
18. Laadan D, Arviv E, Fire M (2020) Using data mining for infrastructure and safety violations discovery in cities. IEEE 2020

19. Mystakidis A, Tjortjis C (2020) Big Data mining for smart cities: predicting traffic congestion using classification. IEEE 2020
20. Huang C, Nazir S (2021) Analyzing and evaluating smart cities for IoT based on use cases using the analytic network process. Hindawi Mob Inf Syst 2021, Article ID 6674479
21. Kshitiz K et al (2017) Detecting hate speech and insults on social commentary using nlp and machine learning. Int J Eng Technol Sci Res 4(12):279–285
22. Praveen P, Shaik MA, Kumar TS, Choudhury T (2021) Smart farming: securing farmers using block chain technology and IOT. In: Choudhury T, Khanna A, Toe TT, Khurana M, Gia Nhu N (eds) Blockchain applications in IoT ecosystem. EAI/Springer innovations in communication and computing. Springer, Cham. https://doi.org/10.1007/978-3-030-65691-1_15
23. Bhasin S, Choudhury T, Gupta SC, Kumar P (2017) Smart city implementation model based on IoT. In: 2017 International conference on big data analytics and computational intelligence (ICBDAC), pp 211–216.https://doi.org/10.1109/ICBDACI.2017.8070836
24. Kumar M, Singh TP, Choudhury T, Gupta SC (2019) ICT—the smart education system in India. In: 2019 International conference on contemporary computing and informatics (IC3I), pp 279–282. https://doi.org/10.1109/IC3I46837.2019.9055562
25. Biswas R et al (2012) A framework for automated database tuning using dynamic SGA parameters and basic operating system utilities. Database Syst J III(4/2012)
26. Sharma HK (2013) E-COCOMO: the extended cost constructive model for cleanroom software engineering. Database Syst J 4(4):3–11
27. Kumar S, Dubey S, Gupta P (2015) Auto-selection and management of dynamic SGA parameters in RDBMS. In: 2015 2nd International conference on computing for sustainable global development (INDIACom), pp 1763–1768
28. Khanchi I, Agarwal N, Seth P (2019) Real time activity logger: a user activity detection system. Int J Eng Adv Technol 9(1):1991–1994
29. Wieclaw L, Pasichnyk V, Kunanets N, Duda O, Matsiuk O, Falat P (2017) Cloud computing technologies in smart city projects. In: The 9th IEEE international conference on intelligent data acquisition and advanced computing systems: technology and applications, 21–23 Sept, 2017

Machine Learning for Speech Recognition

Rakesh Kumar Rai, Parul Giri, and Isha Singh

Abstract Machine learning paradigms that are used in automatic speech recognition (ASR) have been improved in the past decade. The improvement in speech recognition has been developed to help make the technology more efficient by dealing with various challenges affecting speech recognition such as speaker identification, capitalization, correct formatting, domain-specific terminology, background noise, the timing of words, and punctuation placement. Some other issues that have come up in speech recognition include data security and privacy, deployment, and language coverage. Any speech recognition system must have a noise-removal feature to perform in the best way possible. This paper gives a brief about the machine learning techniques that can be used in speech recognition. A better understanding of the models will help understand the systems, and it may help improve the technology even further for the benefit of society.

Keywords Machine learning · Speech recognition · Technology · Algorithms

1 Introduction

Machine learning paradigms that are used in automatic speech recognition (ASR) have been improved in the past decade [1]. Speech recognition is used in applications that include dictation and transcription. However, there are several issues that affect the technology, and they have to be dealt with in the best way possible. The issues include recognizing the voices in the noisy environments, carrying out multimodal speech recognition, and multi-lingual recognition. Automatic speech recognition is

R. K. Rai (✉)
IPEC, Ghaziabad, India
e-mail: rakeshnitn25@gmail.com

P. Giri
MDU, Rohtak, India

I. Singh
HMRITM, Hamidpur, India

P. Dutta et al. (eds.), *Emerging Technologies in Data Mining and Information Security*, Lecture Notes in Networks and Systems 490,
https://doi.org/10.1007/978-981-19-4052-1_17

undergoing a paradigm shift in such a way that the emerging concepts such as big data and the Internet of Things have been assimilated [2, 3]. The fact that speech recognition gives the people a chance to communicate with machines makes it one of the technology that is used in many applications. Speech recognition has made it possible for people to carry out their day-to-day activities in the best way possible [4]. The improvement in speech recognition has been developed to help make the technology more efficient by dealing with various challenges affecting speech recognition such as speaker identification, capitalization, correct formatting, domain-specific terminology, background noise, the timing of words, and punctuation placement. Some other issues that have come up in speech recognition include data security and privacy, deployment, and language coverage [1, 5]. This paper highlights and explains the recent advancements made in machine learning for speech recognition.

Speech Recognition Terminology: Recognition of discourse is an innovation that empowers a gadget to get the words expressed by a human into an amplifier. These words are subsequently handled through discourse acknowledgment, and eventually, the framework yields perceived words. The discourse acknowledgment process comprises of various advances that are talked about individually in the accompanying areas [1, 6]. Our task breaks down the language hindrance with the goal that people can collaborate with one another in their favored language [7, 8]. Discourse acknowledgment frameworks can be arranged by their capacity to comprehend the terms and arrangements of words they have in various gatherings. An advantageous condition in the discourse acknowledgment process is the place where the verbally expressed word is heard. The acknowledgment motor regards all words verbally expressed by an individual; in any case, by and by, the discourse acknowledgment motor's effectiveness relies upon an assortment of elements [9]. The key factors that are considered for a discourse acknowledgment motor are phrasing, simultaneous clients, and uproarious settings.

Speech Recognition Process: The correspondence of which means from one language (the source) to another language (the objective) is interpretation. Essentially, discourse combination is utilized for two principle reasons. The PC sound card creates the comparing computerized portrayal of gotten sound through receiver input. The strategy for making an interpretation of the simple sign into advanced structure is digitization. Testing changes a persistent sign into a discrete sign; in this manner, quantization is characterized as the strategy for approximating a persistent arrangement of qualities [10].

2 Architecture and Modeling of Speech Recognition

Any speech recognition system must have a noise-removal feature to perform in the best way possible (Fig. 1). Various noise-removal mechanisms are available, including speech enhancement techniques such as Wiener filtering, windowing, spectral amplitude estimation, and spectral subtraction in objective and subjective

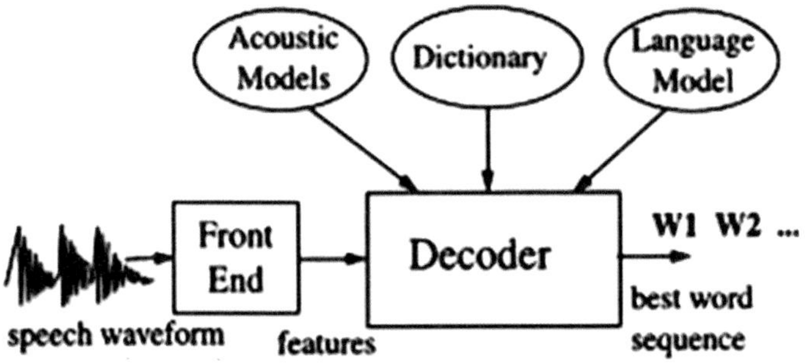

Fig. 1 Speech recognition architecture

manners. The input speech signal enters the system through an auditory front end that will then pre-process the movement and as a result produce spectral-like features. The system then passes the features to a phone likelihood estimator. The decoder uses the phone likelihoods, HMM models, and n-gram language models (LM) to decode the speech. The decoder then sends the output words to the parser that will convert them into a form that can be read by humans [1].

The acoustic preprocessing unit that is found in the auditory front end works to reduce the influence of the unwanted components of the speech, such as noise. Thus, the unit transforms the speech signal to speech frames, from which the requisite speech vectors are generated to assist characterize the input speech signal [1, 11].

2.1 Sub-decoder

The statistical formulation of speech recognition is shown below. If the acoustic observation $A = a_1, a_2, \ldots, a_k$, then have to find the word sequence $W = w_1, w_2, \ldots, W_k$ to maximize the probability $P(W|A)$. Bayes' rule will help us in defining the model as shown below:

$$P\left(\frac{W}{A}\right) = \frac{P\left(\frac{A}{W}\right)P(W)}{P(A)} \tag{1}$$

$P(AW)$ is the acoustic model; $P(A)$ is the constant for a complete sentence, whereas $P(W)$ represents the LM [1].

2.2 *Acoustic Model*

The acoustic model is used to extract the acoustic elements from the speech frames by modeling the auditory input using the sequence of the states that represent the phone likelihoods. It calculates the chances of a given acoustic sentence in a *W* in a given word sequence that has been entered into the system. Each phone's representation could be utilized in conjunction with the pronunciation lexicon; the words could be mapped, or *W* could be described in terms of feature vectors to characterize the spectral variabilities, and the speech signals could be estimated using a neural network or a GMM [1, 6, 7].

2.3 *Language Model*

It is an a priori probability that is independent of the acoustic sequence that is employed in the observation of word sequence *W*. The probability that a word *wk* will be given the preceding *n* – 1 words, as estimated using n-gram LMs, is as follows.

$$P(W) = II\left(\frac{wk}{wk-1}, wk-2\right) \tag{2}$$

3 Machine Learning for Speech Recognition

Machine learning has led to the development of speech recognition, which has led to the development of voice assistants such as Amazon Echo, Siri, Google Home, Cortana, and others. Machine learning came from the desire to have computers learn several skills such as facial features, handwriting, and speech. These applications get their data from a clip of spoken words, which are then turned into lessons then the computer acts on the terms appropriately. Concisely, they can also be referred to as speech-to-text algorithms. Applications such as Siri can interpret the spoken words then answer or respond by carrying out the action they have been asked to. Machine learning makes the concept of speech recognition possible [6].

A system has to be trained for speech recognition to work in the best way possible. Speech recognition training makes it possible for the AI models to understand the unique inputs that have been presented in the form of recorded audio data. It is important to note that there is a long way to go to achieve perfection in many cases as far as speech recognition is concerned. Algorithms used in speech recognition include deep neural networks, PLP features, Viterbi search, and discrimination training, among many others. Some of the algorithms that are used in speech recognition are available in an open-source format. The speech recognition algorithms convert sound waves into valuable data that can be processed to give information. A large amount of data is required to make these algorithms function better and produce effective results [12].

Some machine learning techniques such as Gaussian mixture, artificial neural networks, support vector machines, and the hidden Markov models have been helpful in automatic speech recognition.

3.1 Supervised Learning

It is a process that involves training the machine with a labeled dataset where each class or output response is known. It rides on the assumption that a hypothesis will show better performance, provided that the available training data are large enough. The curve-fitting problem is an excellent example of supervised learning. In the application, the professionals trained the machines to make the best-curved surface to fit the training dataset in the best way possible. They also design the device to interpolate new collected data over the curved surface during the testing process. This category is made up of constrained MLPs, perceptron's, and multilayer perceptron's.

3.2 Unsupervised Learning

In this case, the computer is expected to pick the patterns in the unlabeled dataset then learning them without getting any feedback from the environment. The challenge in this case includes finding the patterns in the dataset provided then partitioning them accordingly.

3.3 Semi-supervised Learning

In this case, the machine is provided with both the labeled and unlabeled data in the training process. It typically involves using a small portion of labeled data with a large portion of the labeled data. This method is mainly used in cases where getting the labeled data is very expensive, and the training process has to go on in the best way possible. Semi-supervised learning is therefore appropriate in some instances since it helps in one way or the other.

3.4 Active Learning

In this case, the algorithm demands the provision of the labels used for the example. The user therefore has the task of providing the required labels. The user, therefore, plays a vital role in the process of labeling the data. It is mainly applied in cases where there is plenty of labeled data, but labeling it is costly. Therefore, it is a way to save costs, especially when a person is working on a very tight budget.

4 Proposed Methodology

In this exploration, the work depends on the flowchart underneath (Fig. 2). The models outlined already are comprised of millions of boundaries, from which the guidance corpus should be learned. We utilize extra data where proper, for example, text that is firmly connected to the discourse we are going to decipher [6]. It is feasible to compose this text in the source language, the objective language, or both.

This paper serves researchers, clinicians, innovation makers, and purchasers as an excellent investigation of arising innovations in many fields, for example, conduct science, brain research, transportation, and medication.

Voice recognition with constant prescient voice interpretation gadget streamlining utilizing multimodal vector wellsprings of data and usefulness was introduced. The critical creation and responsibility of this work is the way in which outside data input is utilized to increment the framework's precision, in this way permitting a prominent improvement, contrasted with the cycles of nature. In expansion, another drive has been dispatched from an insightful point of view, while staying a reasonable one, and was talked about. The framework we need changes Hindi over to English, according to our conversation and arranging, as well as the other way around.

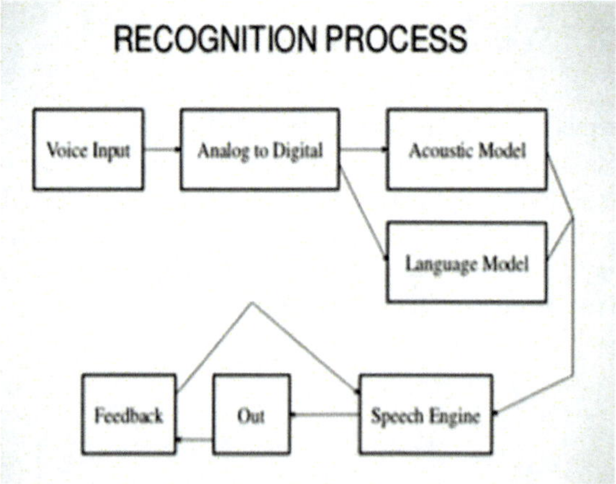

Fig. 2 Working model of speech recognition

5 Simulation Result

The encoder–decoder model's fundamental function is to condense the information grouping into a single fixed-length vector from which each result time step is decoded. Center is proposed as a methodology for both modification and deciphering with long groupings, and this issue is recognized to be more worrisome. The consideration model creates a freely separated setting vector for each outcome time step rather than storing the input arrangement as a solitary fixed-setting vector. The technique is expanded to a machine interpretation issue, as with the encoder–decoder text, and utilizes GRU units rather than LSTM memory cells [8].

Worked with FC, or upheld composing, is a logically ruined method that endeavors to help correspondence by individuals with mental imbalance or other correspondence inabilities, and who are non-verbal (Fig. 3).

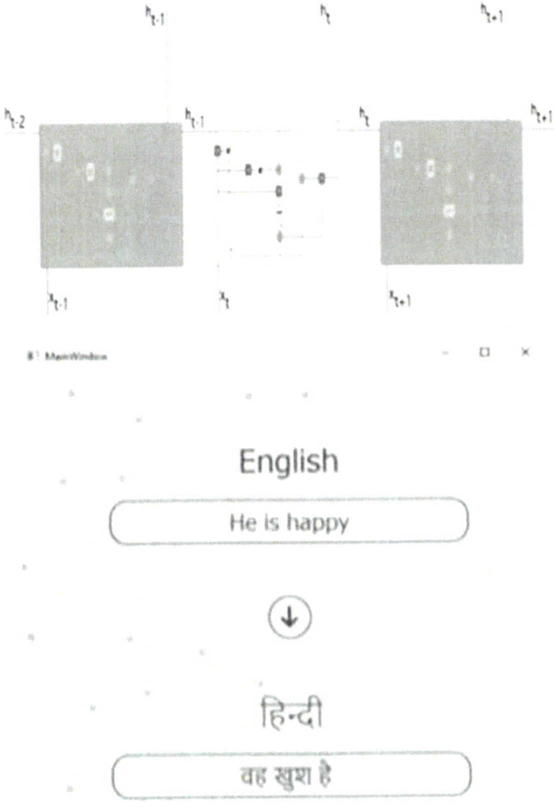

Fig. 3 First translation from English to Hindi

It is executed on the last pivot of course; however, we need to carry out it on the first pivot here, as the score structure is as per the following: group size, worst case scenario length, secret size. Support vector = total (loads of concentration * EO, hub = 1). The equivalent clarification as above applies for a hub choice of 1.

Inserting yield = the information is moved through an inserting layer to the decoder. Incorporated vector = idea (inserting yield, setting vector).

6 Conclusion

This paper gives a brief about the machine learning techniques that can be used in speech recognition. A better understanding of the models will help understand the systems, and it may help improve the technology even further for the benefit of society. Machine learning paradigms that are used in automatic speech recognition (ASR) have been improved in the past decade. Any speech recognition system must have a noise-removal feature to perform in the best way possible. Various noise removal mechanisms are available, including speech enhancement techniques such as Wiener filtering, windowing, spectral amplitude estimation, and spectral subtraction in objective and subjective manners.

References

1. Padmanabhan J, Johnson Premkumar MJ (2015) Machine learning in automatic speech recognition: a survey. IETE Tech Rev 32(4):240–251
2. Agarwalla S, Sarma KK (2016) Machine learning-based sample extraction for automatic speech recognition using dialectal Assamese speech. Neural Netw 78:97–111
3. Ding Jr I, Yen CT, Hsu YM (2013) Developments of machine learning schemes for dynamic time-wrapping-based speech recognition. Math Problems Eng
4. Alhawiti KM (2015) Advances in artificial intelligence using speech recognition. Int J Comput Inf Eng 9(6):1432–1435
5. Matarneh R, Maksymova S, Lyashenko V, Belova N (2017) Speech recognition systems: a comparative review
6. Akçay MB, Oğuz K (2020) Speech emotion recognition: emotional models, databases, features, preprocessing methods, supporting modalities, and classifiers. Speech Commun 116:56–76
7. Tsontzos G, Diakoloukas V, Koniaris C, Digalakis V (2013) Estimation of general identifiable linear dynamic models with an application in speech characteristics vectors. Comput Stand Interfaces 35(5):490–506
8. Wu Y et al (2016) Google's neural machine translation system: bridging the gap between human and machine translation, pp 1–23. arXiv preprint arXiv:1609.08144
9. Peng S, Lv T, Han X, Wu S, Yan C, Zhang H (2019) Remote speaker recognition based on the enhanced LDV-captured speech. Appl Acoust 143:165–170
10. Varghese AA, Cherian JP, Kizhakkethottam JJ (2015) Overview on emotion recognition system. In: 2015 international conference on soft computing and networks security (ICSNS), Coimbatore, pp 1–5

11. Maran V, Keske-Soares M (2021) towards a speech therapy support system based on phonological processes early detection. Comput Speech Lang 65:101130
12. Ault SV, Perez RJ, Kimble CA, Wang J (2018) On speech recognition algorithms. Int J Mach Learn Comput 8(6):518–523

Analysis of Road Accidents Prediction and Interpretation Using KNN Classification Model

Santhoshini Sahu, Balajee Maram, Veerraju Gampala, and T. Daniya

Abstract Roadway accidents are very common and cause a great threat to developing and developed countries. Some basic knowledge is very important for everyone because safety of road accident and its prediction is more important in present days. Most of the variables influence the frequent accidents, and these are as road features, weather conditions, type of accident, road condition. These parameters or influential components are used in selecting the effective model for evaluating the accident reasons. This paper presents a model system for analysis of road accidents prediction and interpretation using KNN classification model. By using this described method, the best and effective performance of the road accident prediction model and their reasons is discovered with K-nearest neighbor classification. The described method is compared with the previous methods like logistic regression (LR) and Naïve Bayes (NB). According to the performance parameters, it can be clear that best model of road accident prediction is discovered. Therefore, the government takes the suggestive results from the model and improves the road safety measurements.

Keywords Road accident · K-nearest neighbor (KNN) · Traffic safety · Road features

1 Introduction

Many societies and individuals lives are enhanced by the motorization, but with the money, these benefits have come. In road accidents, too many people are losing

S. Sahu · T. Daniya
GMR Institute of Technology, Rajam, AP, India
e-mail: daniya.t@gmrit.edu.in

V. Gampala
Koneru Lakshmaiah Education Foundation, Vaddeswaram, Guntur, Andhra Pradesh 522502, India

B. Maram (✉)
Chitkara University Institute of Engineering and Technology-CSE, Chitkara University, Baddi, India
e-mail: maram.balajee@chitkarauniversity.edu.in

P. Dutta et al. (eds.), *Emerging Technologies in Data Mining and Information Security*, Lecture Notes in Networks and Systems 490,
https://doi.org/10.1007/978-981-19-4052-1_18

their life even in the high-income countries which indicates the downward trend. At the same time, road traffic injury burden is also increasing in terms of societal and economic costs [1].

In developing countries, because of road traffic accidents (RTA), deaths and injury rates are increased widely. From road traffic accidents (RTA) more than 90% of persons with disabled road traffic injuries and 85% of all deaths happened [2]. Around the globe, eighth leading cause of death is because of the road traffic accidents [3]. Road traffic accidents (RTAs) causing injuries and fatalities costs are highly influenced the development of socioeconomic and societal well-being. In accidents, every year, more than 150,000 persons are died, and around 400 injuries are happened per day in India. To the traffic, more than 1 million vehicles are added averagely for every year.

In the world, there is rise in traffic and related issues which has disadvantageous effect on the people of the community property and lives [4]. Energy consumption and rising fuel, urban pollution, wasting millions of hours a day in traffic jams, wasting national asserts, and community facilities are the conditions for accidents occurrence which results ultimately death, injury, and property loss or damage. These are the poor traffic conditions and facilities results [5]. It provides the study, reduction in number of accidents and improving the safety measurements of road by the investigation of accident severity affecting factors [6].

In recent years, traffic safety is the major evaluating issue with the different studies of using multiple logistic regressions [7]. Several engineering studies use machine learning techniques and metaheuristic algorithms [8], especially in transportation problems which are required more accurate predictive results than statistical methods. This is achieved because of having the capacity for handling the difficult classification problems and functions. By using optimized prediction tasks, the analysis of pattern recognition tools can be processed in present days [9]. In addition, this different datasets-based several engineering problems use different prediction models. Traditional methods like regression models together with machine learning approaches are given the high accuracy predictive models. One of those models is ANN-based accident severity prediction [10]. Therefore, severity of accidents and type can be easily investigated by employing the machine learning models with higher precision. In order to understand the effects of the determining factors in an activity and predict incidents in the future, nonlinear links between variables with different forms of ANN may be modeled.

2 Traffic Accident Detection Techniques

With reference to severity analysis, three main indicators are influenced as property damage, number of injuries and number of facilities based prediction systems are developed in previous studies in accordance with one or two of three aspects or indicators. For example, one independent variable with four alternatives-based accident severity is taken by the Mannera and Wunsch-Ziegler; these are named as severe injury, fatal, property damage, and light injury. Based on the property damage indicator, Milton et al. investigated as injury and possible injury [11]. One or two of the three aspects of severity-based considerations are developed by many researches in previous studies.

Road accidents using Data Mining Technique: Transportation department uses the data mining techniques and producing the results as generating the trends and patterns for future prediction in [12]. Road accidents influencing parameters correlation using the dataset are analyzed previously which only concentrates for existing roads and does not take any consideration for roads under construction and newly planned roads. From transportation department, required data are collected. The imputation method cleans the collected data. Data selection is used to reduce the dataset complexity. Data reduction uses the discretization technique [13]. The relation between the points which influence the accidents is obtained by using the data mining techniques as clustering and associate rule mining. In order to reduce the time for correlation process, map reduce programming is used for correlation. Accident-prone zones are predicted by applying the geospatial predictive modeling, Naive Bayes classification, and SVM machine learning techniques for the newly planned roads.

Traffic Accident detection using Random Forest Classifier: Constant car crash recognition strategy, joins vehicle to vehicle correspondence procedures with AI techniques. Utilizations procedures of managed AI calculations, for example, ANN, SVM, and random forests are executed on traffic information to build up a model to recognize mishap cases from ordinary cases [14]. The proposed framework utilizes reenacted information gathered from vehicular specially appointed systems (VANETs) in view of the rates and arranges of the vehicles, and afterward, it sends traffic cautions to the drivers. In addition, the presented strategy can give evaluated topographical area of the conceivable mishap.

Traffic Accident detection using regression trees: It presents a probabilistic models of crash injury seriousness. A person on foot is killed at regular intervals and harmed like clockwork in car accidents. Classification regression tree procedure is mainstream information extracting strategy that needn't bother with a particular utilitarian structure. Study has recorded a relapse tree investigation for analyzing factors that impact crash seriousness in crashes including a person on foot. The regression tree gave agreeable outcomes. Results were instinctive and predictable with the consequences of past investigations on person on foot crashes that utilized other diagnostic strategies. The data gave in the examination should help TxDOT to more readily focus on their endeavors for decreasing the number and seriousness of passer by crashes.

Accident duration and severity-based independent studies are developed previously even though these two are found that correlate between each other [15]. In addition from three aspects, only, one or two aspects are investigated by using previous methods or studies, and these three aspects are property damage, number of injuries, and number of facilities. Both the accident duration and severity are estimated in the present work of model system development, and it is the aim of this work. Then, accident severity three indicators as property damage, number of injuries, and number of facilities are investigated, respectively.

3 Road Accidents Prediction and Interpretation Using KNN Classification Model

Figure 1 shows the architecture of the road accident prediction and interpretation using KNN classification model analysis. Property damage, number of injuries, and number of fatalities are mainly three aspects which are included in the severity analysis. Along with these three main parameters, some more influencing parameters are also existed as emergency services (emergency medical services, tow services, fire and rescue services, and police services), accident characteristics (crash type, vehicle fire, number of lanes affected, and accident occurrence time), accident duration, vehicle characteristics (disabled vehicles involved, hazardous material involved, debris involved, and vehicle type involved), road conditions (roadway surface condition, road geometrics, pavement condition, and number of lanes), environmental factors (visibility distance and weather conditions). Only, three prediction models based on severity prediction will be developed. Then, accident severity output is taken as input of accident duration modeling input.

Dataset

From the official Web site of weather forecast and accident reports, the dataset is collected. The described system is executed by using the road and weather conditions which are extracted from the government data. The collected data can be grouped according to the accident occurrence time as regular hours, early hours and peak hours, and so on. The collision avoidance system (CAS) records the previous events data frames which are equated with an hour-based similar strategies. From 2012 to 2019 period, happened accidents data are included in the collected dataset. Total 19 variables with 31,661 road accidents records which are the accident causing variables are present in the chosen dataset.

Data preprocessing

In order to develop a model with machine learning techniques, there is a significant stage of data preprocessing for eliminating the impurities present in the dataset. In the dataset, column mean estimation replaces the missing esteem with all records. Nominal values are here by the conversion of all categorical values. Attribute selection, normalization, transformation, splitting, and cleaning are several steps in data preprocessing.

Accidents are classified into four classes:

1. Fatal
2. Grievous injury

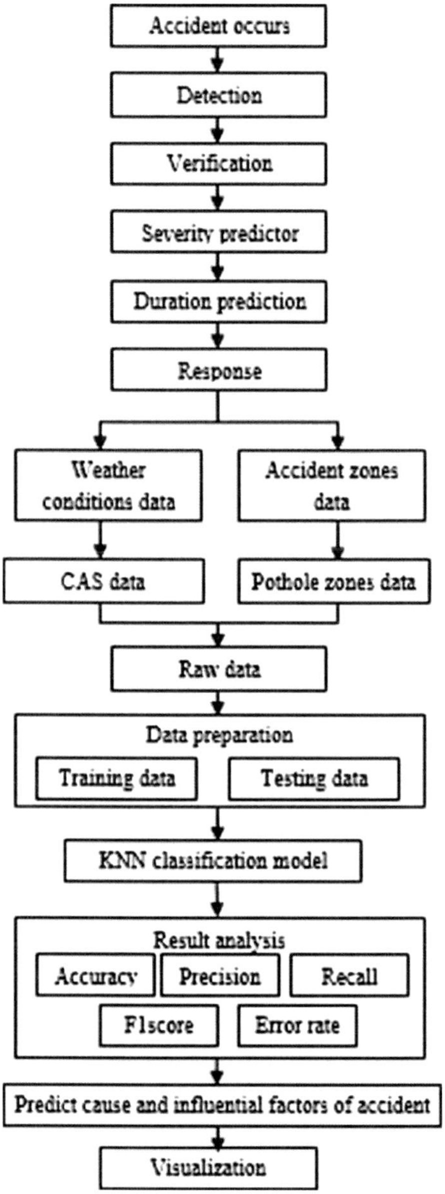

Fig. 1 Architecture of road accidents prediction

3. Minor injury
4. Non injury

Important properties selection for the model advancement or improvement is called as feature selection or attributes selection or variable choice. The selected attributes are most powerful in order to do the experiment. So, the attributes are selected in different edges and apply the procedure for every edge with the calculation of accuracies and adopts the best accuracy producing attribute edge selection.

Training and Testing

To get the efficient system, it has to be trained in different circumstances. 60% of the data is divided for training, and 40% of the data is dividing for testing. Therefore, there is no inconsistency. Then, the system is ready to take any progressive improvements within the system that means if any requirement of continuous preprocessing then old dataset is combined with the preprocessed one for further use.

K-Nearest Neighbors (KNN)

Supervised and unsupervised are two models which were dividing from the data mining techniques. One of the popular unsupervised data mining techniques is clustering. Regression and classification models are built by using the KNN model. Each processing case is stored with in this model by the set of rules and takes the majority votes from their neighbors for characterizing the new instances. Clustering important task is data objects grouping process into clusters. Therefore, same group containing objects are similar than another group objects. Numerical data mostly use the k-means clustering algorithm. In the form of k clusters, the data are grouped by using k-means clustering algorithm. Data type and nature influences the selection of suitable clustering algorithm.

Procedure for k-means Clustering Algorithm

Let us consider the data points set as x and set of centers as v.

i. From the calculation of elbow point, k random points are selected which is best and efficient.
ii. In this step, cluster centers to each datum distances are calculated.
iii. The minimum distance-contained data points from cluster centers are grouped as clusters.
iv. By using below formula, new cluster center is acquired.

$$V_i = \frac{1}{C_i} \sum_{i=1}^{C_i} x_i$$

where an ith cluster number of data points are represented with the C_i.
v. Again, calculate the new distances from new obtained cluster centers to data points.
vi. Repeat the process from step 3 until no datum was reassigned.

The presented KNN classification-based road accident prediction and interpretation performance is evaluated by using the performance parameters, and these are as accuracy, precision, recall, F1-score, and error rate.

4 Result Analysis

From the platform of Open Government, information is collected for result analysis. High potholes and high accident-prone areas are identified as more dangerous for the accidents, especially in morning hours with particular conditions. Low visibility with over speeding in the early hours for pothole severities and medium accident-prone areas are having the high chances of accident occurrence. Atmospheric conditions and human errors are causing parameters for accidents in low accident-prone areas.

Performance parameters

The described road accident prediction and interpretation using KNN classification model performance is evaluated by using the parameters like accuracy, precision, recall, F1-score and RMSE (root mean square error). Different machine learning techniques such as logistic regression (LR), naïve bayes (NB), K-nearest neighbors (KNN) based road accident prediction performance is evaluated with the different parameters and these values are shown in Table 1. Four terms as true negative (TN), true positive (TP), false negative (FN), and false positive (FP)define the above performance metrics.

Accuracy

The ratio of correct predictions to total predictions is defined by the parameter called accuracy and its expression as follows:

$$\text{Accuracy} = \frac{(\text{TP} + \text{TN})}{(\text{TP} + \text{TN} + \text{FP} + \text{FN})}$$

Precision

The ratio of classification algorithm predicted accurate positive scores to the total positives scores is defined by the precision or positive predictive value.

$$\text{Precision} = \frac{\text{TP}}{(\text{TP} + \text{FP})}$$

Table 1 Different machine learning methods comparative analysis

Parameters	(LR)	(NB)	(KNN)
Accuracy	94	93	97
Precision	92	90	96
Recall	95	89	97
F1-measure	93	90	98
RMSE	0.45	0.5	0.14

Recall

The ratio of accurate positive scores to the number of true positives and false negatives is defined by the recall parameter. It is also named as sensitivity or true-positive rate.

$$\text{Recall} = \frac{\text{TP}}{(\text{TP} + \text{FN})}$$

F1-measure

The classifiers performance is measured with statistics with the help of F-score. It requires the precision and recall values of classifiers and calculates the value in between 0 to 1 and classifiers indicatives. According to this, classifiers are arranged from lowest to highest performance. F1-score can be computed as:

$$F1\text{Score} = 2 \times \frac{\text{Precision} \times \text{Recall}}{\text{Precision} + \text{Recall}}$$

RMSE (root mean square error)

RMSE determined on the weather conditions for which KNN, with most minimal RMSE of 0.14 and it is better than Logistic Regression algorithm with 0.45 and Naive Bayes algorithm with 0.5.

The graphical representation of different machine learning techniques for the road accident detection model accuracy and precision parameters is represented in Fig. 2.

The graphical representation of different machine learning techniques for the road accident detection model recall and F1-measure parameters is represented in Fig. 3.

The graphical representation of different machine learning techniques for the road accident detection model RMSE (root mean square error) value is represented in Fig. 4.

From above results, it is clear that presenting KNN machine learning model road accident prediction model is very efficient than other classifications.

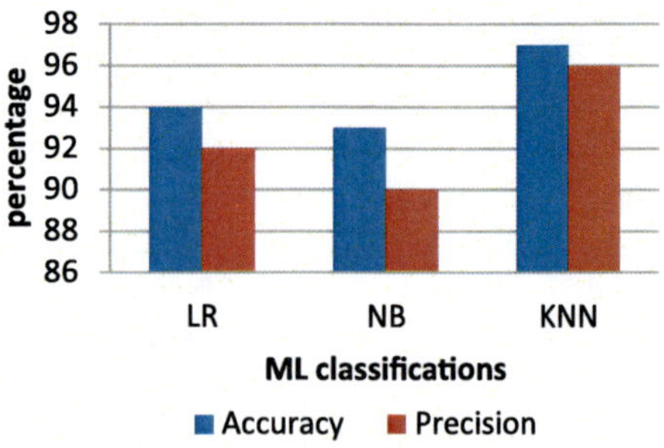

Fig. 2 Comparative analysis of different Ml methods in terms of accuracy and precision

Fig. 3 Comparative analysis of different Ml methods in terms of recall and F1-measure

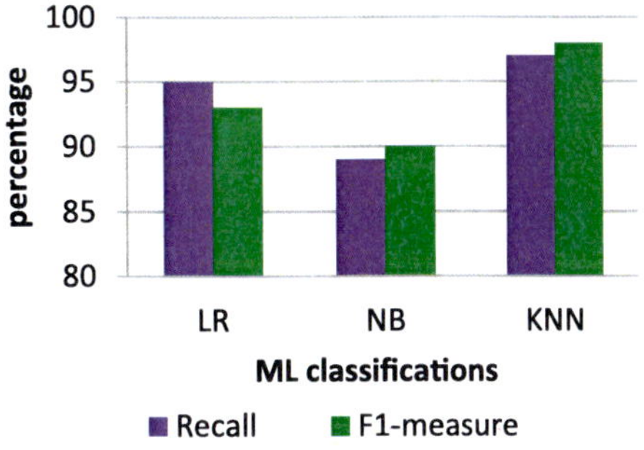

Fig. 4 Graph for RMSE values

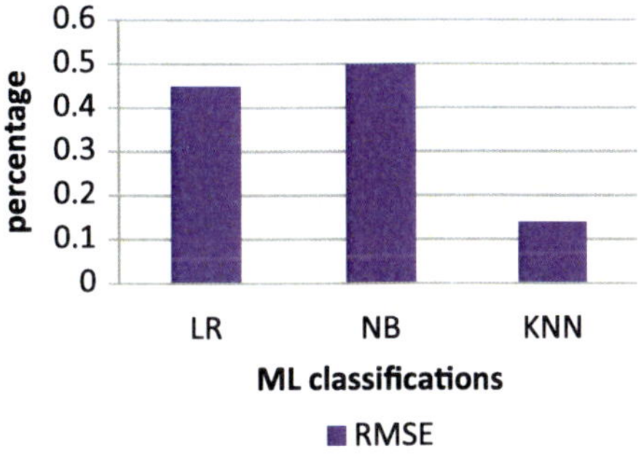

5 Conclusions

This paper presented a model system for analysis of road accidents prediction and interpretation using KNN classification model. Many problems can be solved by employing the machine learning methods. Several parameters influence the occurrence of the road accidents. Property damage, number of injuries, and number of fatalities are enclosed with the accident severity and accident duration. From results, it is clear that, accuracy, precision, recall, F1-score values are high and as well as RMSE value is progressively low for KNN classification-based road accident prediction and interpretation. By using this investigation, how every accident variables influenced by the every component and used in knowing the safe driving recommendations for decreasing or reducing the number of accidents.

References

1. Rana S, Faysal MRH, Saha SC, Noman AA, Shikder K (2021) Road accident prevention by detecting drowsiness & ensure safety issues. In: 2021 2nd international conference on robotics, electrical and signal processing techniques (ICREST)
2. Zahran ESMM, Tan SJ, Yap YH, Tan EH, Pena CMF, Yee HF, Uddin MR (2019) An investigation into the impact of alternate road lighting on road traffic accident hotspots using spatial

analysis. In: 2019 4th international conference on intelligent transportation engineering (ICITE)
3. Tai W-K, Wang H-C, Chiang C-Y, Chien C-Y, Lai K, Huang T-C (2018) RTAIS: road traffic accident information system. In: 2018 IEEE 20th international conference on high performance computing and communications; IEEE 16th international conference on smart city; IEEE 4th international conference on data science and systems (HPCC/SmartCity/DSS)
4. Ki Y-K, Kim J-H, Kim T-K, Heo N-W, Choi J-W, Jeong J-H (2018) Method for automatic detection of traffic incidents using neural networks and traffic data. In: 2018 IEEE 9th annual information technology, electronics and mobile communication conference (IEMCON)
5. Blagoiev M, Gruicin I, Ionascu M-E, Marcu M (2018) A study on correlation between air pollution and traffic. In: 2018 26th telecommunications forum (TELFOR)
6. Xia X-L, Nan B, Xu C (2017) Real-time traffic accident severity prediction using data mining technologies. In: 2017 international conference on network and information systems for computers (ICNISC)
7. Zhao G, Li J, Zhou W (2015) Regression analysis of association between vehicle performance and driver casualty risk in traffic accidents. In: 2015 international conference on transportation information and safety (ICTIS)
8. Dib O, Manier M-A, Caminada A (2015) A hybrid metaheuristic for routing in road networks. In: 2015 IEEE 18th international conference on intelligent transportation systems
9. Divyashree V, Sumathi S (2015) Extension theory based partial discharge pattern recognition using statistical operators. In: 2015 international conference on power and advanced control engineering (ICPACE)
10. Raut S, Karmore S (2015) Review on: severity estimation unit of automotive accident. In: 2015 international conference on advances in computer engineering and applications
11. Gschwendtner K, Lienkamp M, Kiss M (2014) Prospective analysis-method for estimating the effect of advanced driver assistance systems on property damage. In: 17th international IEEE conference on intelligent transportation systems (ITSC)
12. Shanthi S, Geetha Ramani R (2012) Gender specific classification of road accident patterns through data mining techniques. In: IEEE-international conference on advances in engineering, science and management (ICAESM -2012)
13. Shehzad K (2012) EDISC: a class-tailored discretization technique for rule-based classification. IEEE Trans Knowl Data Eng 24(8)
14. Hamner B (2010) Predicting future traffic congestion from automated traffic recorder readings with an ensemble of random forests. In: 2010 IEEE international conference on data mining workshops
15. Lee Y, Wei C-H (2008) A computerized feature reduction using cluster methods for accident duration forecasting on freeway. In: 2008 IEEE Asia-Pacific services computing conference

Comparative Analysis of Machine Learning Classification Algorithms for Predictive Models Using WEKA

Siddhartha Roy and Runa Ganguli

Abstract The world is becoming progressively more dependent on technology. Artificial intelligence (AI), part of such technology is the simulation and emulation of human intelligence by machine and computer systems. Machine learning, one of the most significant branches of AI, makes it possible for machines to learn from experience and historical data using simple and complex algorithms. Predictive algorithm is a scientific idea of empirically establishing a relationship between the historical set of data which can then be used to make future decision in an attempt to solve some real-life problems. The action of predictive algorithm on big data results into predictive modeling which has been applied by various researchers to solve numerous problems including prediction of weather condition, rate of disease spread, birth rate, death rate, rate of road accident, and in population prediction too. In this paper, we have presented a comparative analysis on the performance of four major machine learning classification algorithms, namely K-nearest neighbor, naïve Bayes, random forest, and SVM on three case studies of predictive modeling using WEKA tool. The case studies selected for this paper are station-wise rainfall prediction, life time of a car prediction, and detection of breast cancer.

Keywords AI · Machine learning · Predictive model · K-nearest neighbor · Big data · naïve Bayes · Random forest · SVM

1 Introduction

Artificial Intelligence (AI) is one of the major areas not only in the field of computer science but also in other discipline such as medicine, engineering, natural sciences, applied sciences, agriculture, and language that has been used to solve various types of

S. Roy (✉)
The Heritage College, Kolkata, 994 Chowbaga Road, Anandapur, Kolkata 700107, India
e-mail: siddhartha.roy@thc.edu.in

R. Ganguli
The Bhawanipur Education Society College, 5 L.L.R. Sarani, Kolkata 700020, India
e-mail: runa.ganguli@gmail.com

P. Dutta et al. (eds.), *Emerging Technologies in Data Mining and Information Security*, Lecture Notes in Networks and Systems 490,
https://doi.org/10.1007/978-981-19-4052-1_19

real-life problems. Machine learning, a field of AI, provides statistical tools to explore and analyze data. To have a functioning ML algorithm, there must be sufficient historical data from where the algorithm can acquire series of experience to solve similar problems in the future. Humans naturally make enquiries and predict likely future occurrences. After decades of research and development, computer science provides predictive algorithm [1] as one of the areas of ML which is focused to the area of forecasting or predicting the likelihoods of future events making use of the series of available data (big data), which are made available through modern technologies deal with only predictive analysis. Predictive analytics deals an unknown event of interest by learning from historic data to predict a future event. It should be noted that data can exist in both structured format like age, salary, occupation, marital status, etc., and also in unstructured format including textual data from social media content, or any open text forms. Predictive models work on such data to discover patterns and relationships which helps different organizations to gather constructive information rather than having mere assumptions. The process of predictive analytics begins with problem identification followed by objective determination. After data collection from multiple sources, data is pre-processed to same format which is used by machine learning algorithms. Data predictive analytics model using naive Bayes algorithm provides probabilistic classification based on Bayes' theorem with independent assumptions between the features. Naive Bayes approach sometime produces optimal results for large datasets, especially for big data. Support vector machine (SVM) [2] which is mostly used in classification, significantly reduces the need for labeled training instances which helps in text and hypertext classification. It has been found through research that SVM attains higher search accuracy [3] compared to traditional query processing schemes after just three or four iterative rounds. Modified version of SVM uses privileged approach for image segmentation systems. The k-nearest neighbor algorithm also called lazy-learning algorithm, uses less time during classification [4]. The purpose of k-NN algorithm is to find set of k objects from training dataset which are closest to input data instance. This is done by distance calculation (mostly Euclidean distance) and assigning maximum voted classes out of these neighboring classes [5]. The choice of an optimal value of k is significant here as the performance of the classifier depends on the selection of k. Random forest algorithm is one of the widely used techniques in machine learning which consists of a large number of trees. In random forest, each individual tree is constructed by some approach of creating single decision tree. The class with the maximum votes will be preferred in model's prediction. One of the key points of random forest is that a large number of less correlated models (trees) operating as a committee outperform any of the individual constituent model and has a lower error rate [6].

In this paper, we have studied few real-life case studies of predictive modelling, applied four machine learning classification algorithms, namely K-nearest neighbor, naïve Bayes, random forest and SVM on various data sets, and presented few observations regarding the comparative performance analysis of these algorithms. For analyzing the predictive accuracy of the above models, WEKA tool has been used. After the introductory part in Sect. 1, a brief report on the related work is presented in

Sect. 2. In Sect. 3, we mention about the general model evaluation and performance measurement techniques. Section 4 highlights the three use cases of our research work and its detailed simulation result using WEKA tool. Class-wise detailed analysis, comparative performance study along with some key observations is listed in Sect. 5. Finally, we conclude the paper in Sect. 6 mentioning the future scope of this research work.

2 Related Work

There are two main machine learning methods existing in supervised model; one is the technique of predicting continuous-valued output from input, i.e. input data is given in the form of features and output variable is continuous. Another technique is classification, where each data member is assigned to predefined sets. The goal of a classifier is to learn the relationship between input variables and categorical output. For last few decades, many people have worked with machine learning algorithms to compare various aspects of these algorithms. The learning methods have good performance if the predictions are calibrated after training [7]. Different data sets and parameters on accuracy are taken into consideration to compare three algorithms, namely naive Bayes, random forest, decision tree, and SVM. Vrushali [8] investigated various algorithms like J48, naïve Bayes and random forest classifiers and to enhance the algorithm performance evaluated them by fertility index. Archana et al. [9] emphasized on the major category of classification algorithms which includes k-nearest neighbor classifier, Naive bayes, and SVM. This paper provides an inclusive survey of various classification algorithms and their merits and demerits. Xindong et al. [10] have given the descriptive study of various classification algorithms which includes SMO (sequential minimal optimization which is used for training SVM) and Naïve bayes. In their paper, they have discussed the impact of the algorithms on different datasets. Different classification algorithms were applied and tested on various datasets from University of California, Irvine (UCI) using WEKA tool in the research work by Kharche et al. [11]. Performance evaluation has been carried out for each algorithm with respect to accuracy and execution time. Sharma [12], in her paper, used WEKA tool to carry out a comparative analysis of various decision tree classification algorithms. Potter [13] conducted comparison of classification algorithms to perform the diagnosis of the patients on breast cancer dataset. A survey of naive Bayes classifier by Vidhya et al. [14] highlights the importance of this machine learning approach for text document classification. Various feature selection approaches have been discoursed and associated along with the metrics involving text document classification.

3 Dataset Splitting and Performance Measuring Metric

In this section, we will be discussing different ways of dataset splitting or partitioning into training and testing sets followed by some of the performance measuring metrics.

3.1 Dataset Splitting

Selection of a proper way of dataset splitting into training and testing sets is one of the key strategies in the area of machine learning. There are four ways of splitting that exists in WEKA, namely training dataset, supplied dataset, percentage split, and cross validation. In training dataset, preparation of the model is based on the entire dataset and then evaluation is done on the same. In supplied dataset, the model is prepared on the entire training dataset, but for testing or evaluation of the performance of the model, separate set is used. This technique is good for large dataset. In percentage spilt, the dataset is randomly divided into training and testing partitions where some portion of the data set is used for training purpose and the rest for testing. But, the main limitation of this technique is that the accuracy depends on sample training set. Biased sample used for training will not produce accurate prediction. For a particular sample selection, bias reduction can be done by repeating the whole process several times across different sample groups for training and testing data. Cross validation is used for this purpose. In cross validation, a dataset is fragmented into k partitions or folds. A model is trained on all of the $(k - 1)$ partitions except one that is held out as the test set, and then this process is iterated for k times by selecting different $(k - 1)$ partitions and one test set, so that at least each of the partition has chance of being held out as the test set. To get unbiased result, the average performance is calculated of all k partitions.

3.2 Performance Measuring Metrics

After proper splitting of data into training and testing set, next task is to analyze the performance metrics. Performance metrics help to decide, if one algorithm is better or worse than another. These are mainly classification accuracy, accuracy by class, and confusion matrix. Accuracy by class contains measuring fields such as TP rate, FP rate, precision, recall, F-measure, MCC, ROC area, and PRC area. MCC and ROC are measurement metrics which give an idea of how the classifiers are performing in general.

4 Case Study and Simulation Results

In this paper, we have taken three datasets from Kaggle; Rainfall_data_Bangladesh (1948 to 2014), Lifetime_of_a_car and breast_cancer_data. We have applied various classification algorithms, namely naïve Bayes, K-nearest neighbor, support vector machine, and random forest on the above-mentioned datasets using WEKA tool. We first generate stratified cross validation summary to compare the performance for each dataset and then present detailed analysis class wise. We have used 10-fold cross validation using WEKA.

4.1 Use Case 1: Rainfall Prediction

The climate of Bangladesh in the central-north is subtropical, whereas tropical in the south. However, the place experiences an enjoyably warm and sunny winter from the month of November to February, while a short spring and summer between March to May, and a monsoon from June to October. This flat country is spanned by the huge Ganges-Brahmaputra delta, and hence get exposed to floods and storm surges whenever cyclones hit the Bay of Bengal. The rainfall in Bangladesh varies, depending upon various season and location. Winter is very dry and annual rainfall is less than 4 for only less than 4% of the annual rainfall. Rainfall in this season is less than 4% which varies from 20 mm in the west and 40 mm in the northeast, which is caused by the western disturbances coming from the north-western part of the country. Rainfall for Bangladesh for last 40 years are collected from Kaggle dataset. The objective is to predict station-wise (region) rainfall distribution using four popular classification algorithms and compare their prediction accuracy. Illustration of class-wise detailed analysis using WEKA tool is shown in Fig. 1. (a) KNN; (b) Naïve Bayes; (c) Random Forest; (d) SVM.

4.2 Use Case 2: Lifetime of a Car

Life time of a car or more specifically performance of an automobile engine depends on various factors such as fuel consumption and emissions, temperature, and moisture. To use the engine, in its optimum range it is important to check those factors in a regular basis. An engine's burn rate depends on the variations in moisture influence. High-moisture content suppresses the burn rate and extends the combustion duration. This results in low thermal efficiency. Moreover, with the increase in humidity, engine is more likely to misfire. Couple of things that get affected due to cylinder temperature are engine fuel efficiency, knocking, exhaust gas temperature, and mainly NO_2 emission. Higher air temperature results in higher NO_2 emission, hence, heat transfer and knocking tendency. In this case study, we have developed some predictive model

TP Rate	FP Rate	Precision	Recall	F-Measure	MCC	ROC Area
0.732	0.010	0.720	0.732	0.726	0.716	0.870
0.665	0.012	0.658	0.665	0.661	0.650	0.830
0.762	0.010	0.723	0.762	0.742	0.733	0.879
0.655	0.009	0.698	0.655	0.676	0.666	0.827
0.653	0.011	0.657	0.653	0.655	0.644	0.828
0.757	0.009	0.736	0.757	0.746	0.738	0.879
0.780	0.008	0.763	0.780	0.771	0.763	0.887
0.719	0.007	0.781	0.719	0.749	0.742	0.859
0.778	0.009	0.710	0.778	0.742	0.735	0.893

(a)

TP Rate	FP Rate	Precision	Recall	F-Measure	MCC	ROC Area
0.998	0.000	1.000	0.998	0.999	0.999	1.000
1.000	0.000	1.000	1.000	1.000	1.000	1.000
1.000	0.000	1.000	1.000	1.000	1.000	1.000
0.998	0.000	1.000	0.998	0.999	0.999	1.000
1.000	0.000	0.998	1.000	0.999	0.999	1.000
1.000	0.000	1.000	1.000	1.000	1.000	1.000
1.000	0.000	1.000	1.000	1.000	1.000	1.000
1.000	0.000	1.000	1.000	1.000	1.000	1.000
1.000	0.000	1.000	1.000	1.000	1.000	1.000
1.000	0.000	1.000	1.000	1.000	1.000	1.000

(b)

TP Rate	FP Rate	Precision	Recall	F-Measure	MCC	ROC Area
1.000	0.000	1.000	1.000	1.000	1.000	1.000
1.000	0.000	1.000	1.000	1.000	1.000	1.000
1.000	0.000	1.000	1.000	1.000	1.000	1.000
1.000	0.000	1.000	1.000	1.000	1.000	1.000
1.000	0.000	1.000	1.000	1.000	1.000	1.000
1.000	0.000	1.000	1.000	1.000	1.000	1.000
1.000	0.000	1.000	1.000	1.000	1.000	1.000
1.000	0.000	1.000	1.000	1.000	1.000	1.000
1.000	0.000	1.000	1.000	1.000	1.000	1.000
1.000	0.000	1.000	1.000	1.000	1.000	1.000
1.000	0.000	1.000	1.000	1.000	1.000	1.000

(c)

TP Rate	FP Rate	Precision	Recall	F-Measure	MCC	ROC Area
0.988	0.013	0.731	0.988	0.840	0.844	0.994
0.458	0.004	0.808	0.458	0.585	0.599	0.982
0.812	0.033	0.456	0.812	0.584	0.592	0.978
0.448	0.012	0.536	0.448	0.488	0.475	0.982
0.244	0.002	0.764	0.244	0.370	0.423	0.988
0.969	0.024	0.575	0.969	0.722	0.736	0.987
0.346	0.002	0.855	0.346	0.492	0.535	0.969
0.969	0.028	0.545	0.969	0.698	0.716	0.986
0.000	0.000	?	0.000	?	?	0.983
0.973	0.034	0.496	0.973	0.657	0.682	0.982
0.303	0.003	0.738	0.303	0.429	0.463	0.990

(d)

Fig. 1 Class-wise detailed analysis of rainfall prediction using WEKA tool: **a** KNN. **b** naïve Bayes. **c** random forest. **d** SVM

for life time of a car influenced by various issues like intake air's effect of variations in moisture level, pressure, and temperature on engine performance based on four service providers. Total sample taken here is 1000. Illustration of class-wise detailed analysis using WEKA tool is shown in Fig. 2. (a) KNN; (b) naïve Bayes; (c) random forest; (d) SVM.

TP Rate	FP Rate	Precision	Recall	F-Measure	MCC	ROC Are
0.298	0.228	0.290	0.298	0.294	0.069	0.535
0.390	0.204	0.394	0.390	0.392	0.187	0.593
0.361	0.211	0.382	0.361	0.371	0.153	0.575
0.595	0.144	0.569	0.595	0.582	0.445	0.726

(a)

TP Rate	FP Rate	Precision	Recall	F-Measure	MCC	ROC Area
0.336	0.217	0.327	0.336	0.331	0.118	0.592
0.409	0.253	0.355	0.409	0.380	0.149	0.530
0.338	0.196	0.385	0.338	0.360	0.148	0.659
0.335	0.194	0.355	0.335	0.345	0.144	0.720

(b)

TP Rate	FP Rate	Precision	Recall	F-Measure	MCC	ROC Area
0.534	0.167	0.500	0.534	0.516	0.359	0.809
0.531	0.111	0.619	0.531	0.572	0.443	0.827
0.541	0.177	0.526	0.541	0.533	0.361	0.816
0.587	0.148	0.559	0.587	0.573	0.432	0.859

(c)

TP Rate	FP Rate	Precision	Recall	F-Measure	MCC	ROC A
0.113	0.049	0.422	0.113	0.179	0.113	0.624
0.445	0.319	0.322	0.445	0.374	0.115	0.548
0.316	0.315	0.267	0.316	0.289	0.001	0.556
0.421	0.222	0.378	0.421	0.398	0.193	0.690

(d)

Fig. 2 Class-wise detailed analysis of lifetime of a car using WEKA tool: **a** KNN. **b** naïve Bayes. **c** random forest. **d** SVM

TP Rate	FP Rate	Precision	Recall	F-Measure	MCC	ROC Area	
0.943	0.028	0.952	0.943	0.948	0.917	0.956	0.920
0.972	0.057	0.967	0.972	0.969	0.917	0.956	0.955

(a)

TP Rate	FP Rate	Precision	Recall	F-Measure	MCC	ROC Area	
0.896	0.056	0.905	0.896	0.900	0.842	0.976	0.956
0.944	0.104	0.939	0.944	0.941	0.842	0.977	0.983

(b)

TP Rate	FP Rate	Precision	Recall	F-Measure	MCC	ROC Area	
0.925	0.020	0.966	0.925	0.945	0.913	0.991	0.989
0.980	0.075	0.956	0.980	0.968	0.913	0.991	0.991

(c)

TP Rate	FP Rate	Precision	Recall	F-Measure	MCC	ROC Area	
0.948	0.003	0.995	0.948	0.971	0.955	0.973	0.963
0.997	0.052	0.970	0.997	0.983	0.955	0.973	0.969

(d)

Fig. 3 Class-wise detailed analysis of breast cancer detection using WEKA tool: **a** KNN. **b** naïve Bayes. **c** random forest. **d** SVM

4.3 Use Case 3: Breast Cancer Detection

One of the most hazardous and common reproductive cancers affecting mostly women worldwide is breast cancer. Increase in the number of breast cancer cases is an alarming public health issues in modern society. However, early diagnosis of breast cancer can help in improving survival chances significantly, as it can aid in timely medical treatment to patients. Also proper diagnosis and accurate classification of benign tumors can avoid patients undergoing needless treatments. This brings to the need of an important research area of correct diagnosis of breast cancer and classification of patients into malignant (M) and benign (B) groups. This is a classic case of Machine learning problem which includes cancer pattern classification and forecast modeling from complex breast cancer datasets. We have taken various features to compute for each cell nucleus such as radius, texture, perimeter, area, smoothness, compactness, concavity, concave points, symmetry, and fractal dimension. In this paper, the main objective is to classify the breast cancer type to be either benign or malignant. To achieve this, we have taken 569 sample data and used four machine learning classification algorithms to fit a function. Illustration of class-wise detailed analysis using WEKA tool is shown in Fig. 3. (a) KNN; (b) naïve Bayes; (c) random forest; (d) SVM.

5 Comparative Analysis of Various Use Cases Using WEKA Tool

In the previous section, we have obtained the simulation results using WEKA tool. In Table 1, we have analyzed various data obtained from the simulation results using

Table 1 Case-wise comparative analysis of the classification algorithms

	KNN	Naïve Bayes	Random Forest	SVM
	Accuracy percentage	Accuracy percentage	Accuracy percentage	Accuracy percentage
Case1 (Rainfall Prediction)	0.74	0.99	0.99	0.61
Case2 (Life Time of a Car)	0.41	0.36	0.55	0.33
Case3 (Breast Cancer Detection)	0.96	0.92	0.96	0.98

performance measuring metrics. In this table, we have listed the accuracy percentage and kappa statistics for each classifier algorithms based on the use cases.

From Table 1, we have the following observations:

1. In both rainfall prediction and life time of a car, random forest gives the best. In case of breast cancer detection, SVM gives us the best accuracy percentage as well as kappa statistics.
2. In case of rainfall prediction and breast cancer detection, accuracy is not significantly different. But, in case of life time of a car, accuracy significantly differs.
3. Among all cases, life time of a car gives us the worst prediction result for all classifier algorithms.
4. In case of rainfall prediction, both naïve Bayes and random forest give the best prediction result (99%).

Interpretation of the above observations

It is observed that no single algorithm dominates when choosing a machine learning model. It depends hugely on the type of data. Some perform better with large datasets, while some works best when number of features or dimensions in data is high.

Observation 1: Rainfall prediction has a very large dataset whereas in case of life time of a car, the number of dimensions is high. This is why, in both the cases, random forest gives better accuracy than the other algorithms as random forest works well with large dataset or with high dimensionality. On the other hand, SVM gives better prediction when a dataset is nonlinear and consists of a huge number of features. It is noticed that the dataset used for breast cancer detection consists of a large count of features as well as it is nonlinear. Thus, the performance of SVM for breast cancer detection is obvious.

Observation 2: Accuracy of classifier algorithms depends on various factors such as record set (more training data means more accuracy in testing data), high correlation among attributes of a dataset, balancing of a dataset, etc. Now for life time of a car, the dataset is very small (only 1000 instances). Each attribute of this dataset, such as temperature or pressure is not correlated to each other. Moreover, the dataset is purely unbalanced. Hence, the performance is poor with respect to any classifier model.

Observation 3: In general, naive Bayes gives us best performance when we have a small training data with relatively small number of features. In the rainfall prediction model, features are very less. Hence, naïve Bayes classifier model gives us the best prediction result along with random forest.

From Table 2, it is clear that none of the accuracy percentage and F-Measure are significantly different. In fact, for most of the cases, the value is equal. Then the question should arise that which one is better; F1-score or accuracy percentage to decide the best performance measuring metric. Accuracy can be computed by taking into consideration a significant count of true negatives which has very less focusing points with respect to most business circumstances. Business houses mostly focus on false negative and false positive because usually it has significant associated business costs. So, it is better to use F1 score or F measure, if we need to maintain a balance between precision and recall as well as cases of uneven class distribution (high count of actual negatives). Thus, in most general cases, we should use F1 score as a measure of performance accuracy of prediction model.

6 Conclusion

In this paper, we carried out an experimental work to compare four popular machine learning classification algorithms on three real-life use cases, namely rainfall prediction, life time of a car, and breast cancer detection based on different performance metrics. The measuring attributes play a vital role for prediction in these cases. We have observed that random forest produces best result for rainfall prediction and life time of a car, whereas SVM gives us the best predictive results for breast cancer detection. The level of accuracy and prediction highly depends on the data being used as input, like data size, dimension, correlation among attributes, and uneven class distribution. We have observed that in most cases F1 score does not significantly differ from the accuracy percentage. Every algorithm has its advantages and limitations, it is difficult to select the best algorithm for a particular dataset. In our paper, we have carried out simulation using diverse datasets and have drawn inference from various angle of measurement. A limitation of this paper is that we have made the entire performance analysis based on the simulation results obtained from WEKA tool. Consequently, how the simulation results using WEKA tool differ from program implemented classifier model need to be tested. In fact, to increase the prediction accuracy, developing a hybrid prediction model where multiple machine learning algorithms that can work together can give optimal results. Many real-world problems consist of a large number of input features, termed as high dimensions which can be reduced by dimensionality reduction and feature selection. This is another future scope of our paper where the analysis can be extended to work on other diversified and high dimensionality datasets.

Table 2 Case-wise comparison between accuracy and F-measure for the classification algorithms

	KNN		Naïve Based		Random Forest		SVM	
	Accuracy percentage	F-measure	Accuracy percentage	F-measure	Accuracy percentage	F-measure	Accuracy percentage	F-measure
Case 1	0.743	0.742	0.999	0.999	0.995	1.000	0.613	0.757
Case 2	0.41	0.409	0.355	0.355	0.548	0.547	0.326	0.311
Case 3	0.961	0.948	0.922	0.962	0.959	0.968	0.978	0.977

References

1. Kumar V, Garg ML (2018) Predictive analytics: a review of trends and techniques. Int J Comput Appl (0975–8887) 82(1):31–37
2. Bartlett P, Shawe-Taylor J (1999) Generalization performance of support vector machines and other pattern classifiers. Advances in Kernel methods—support vector learning, pp 43–54
3. Burges C (1998) A tutorial on support vector machines for pattern recognition. Data Min Knowl Disc 2:121–167
4. Zhao S, Rui C, Zhang Y (2013) MICkNN: multi-instance covering kNN algorithm. Tsinghua Sci Technol 18(4):360–368
5. Kotsiantis SB, Zaharakis I, Pintelas P (2007) Supervised machine learning: a review of classification techniques. Emerg Artif Intell Appl Comput Eng 160:3–24
6. Horning N (2010) Random forests: an algorithm for image classification and generation of continuous fields data sets. In: Proceedings of the international conference on geoinformatics for spatial infrastructure development in earth and allied sciences, vol 911, Osaka, Japan (2010, December)
7. Caruana R, Niculescu-Mizil A (2006) An empirical comparison of supervised learning algorithms. In: Proceedings of the 23rd international conference on Machine learning, pp 161–168 (2006, June)
8. Bhuyar V (2014) Comparative analysis of classification techniques on soil data to predict fertility rate for Aurangabad district. Int J Emerg Trends Technol Comput Sci 3(2):200–203
9. Archana S, Elangovan K (2014) Survey of classification techniques in data mining. Int J Comput Sci Mobile Appl 2(2):65–71
10. Wu X, Kumar V, Quinlan JR, Ghosh J, Yang Q, Motoda H, Zhou ZH (2008) Top 10 algorithms in data mining. Knowl Inf Syst 14(1):1–37
11. Kharche D, Rajeswari K, Abin D (2014) Comparison of different datasets using various classification techniques with weka. Int J Comput Sci Mobile Comput 3(4)
12. Sharma P (2014) Comparative analysis of various decision tree classification algorithms using WEKA. Int J Recent Innov Trends Comput Commun 3(2):684–690
13. Potter R (2007) Comparison of classification algorithms applied to breast cancer diagnosis and prognosis. In: Industrial conference on data mining-posters and workshops, pp 40–49 (2007, July)
14. Vidhya KA, Aghila G (2010) A survey of Naive Bayes machine learning approach in text document classification. Int J Comput Sci Inf Secur 7(2)

Cataract Detection on Ocular Fundus Images Using Machine Learning

Vittesha Gupta, Arunima Jaiswal, Tanupriya Choudhury, and Nitin Sachdeva

Abstract Cataract is a very common condition affecting the elderly. The onset of cataract is slow and is therefore not detected until it starts to obstruct the vision of the affected patient. One way to detect cataract is through study of the ocular fundus images. Machine learning techniques trained on available datasets offer effective and fast methods of detecting abnormalities in ocular fundus and can be used for identifying patients affected by cataract. In this paper, we trained and tested various machine learning techniques in order to perform binary classification for cataract detection on ocular fundus images. We implemented techniques such as support vector machines, random forest, decision tree, logistic regression, naïve Bayes, k-nearest neighbors, XGBoost, light gradient boosting, and voting classifier. Improved results were obtained through light gradient boosting.

Keywords Cataract · Ocular fundus · Machine learning · Eye diseases

1 Introduction

The undiagnosed slow onset of cataracts has made it the cause of a large number of visual impairment cases globally [1]. With age, the lens behind the iris of the eye starts to become foggy. This fogginess in the lens is called a cataract. A cataract is an obstruction in the path of light which enters the eye and therefore causes problems

V. Gupta (✉) · A. Jaiswal
Department of Computer Science and Engineering, Indira Gandhi Delhi Technical University for Women, Delhi, India
e-mail: vittesha110btcse19@igdtuw.ac.in

A. Jaiswal
e-mail: arunimajaiswal@igdtuw.ac.in

T. Choudhury
University of Petroleum and Energy Studies UPES, Dehradun, Uttarakhand, India

N. Sachdeva
Krishna Engineering College, Ghaziabad, India

P. Dutta et al. (eds.), *Emerging Technologies in Data Mining and Information Security*, Lecture Notes in Networks and Systems 490,
https://doi.org/10.1007/978-981-19-4052-1_20

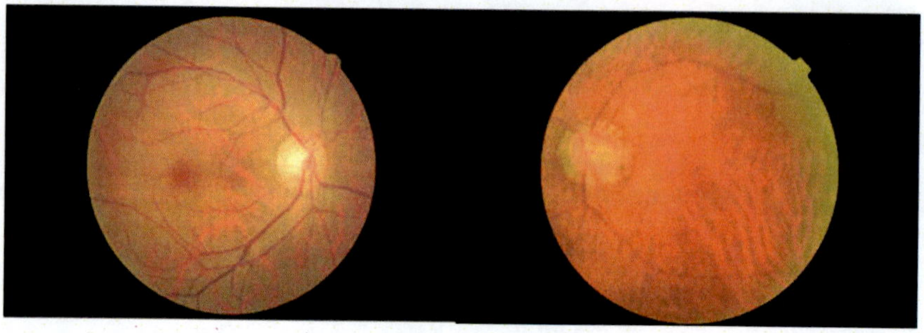

Fig. 1 Ocular fundus images for normal (left) and cataract affected (right) eyes

in vision (see Fig. 1) [2]. If left untreated, it can result in blindness. Timely detection of cataracts through the correct tests is a very important step for the prevention of extreme cases. Cataracts can be diagnosed with the help of several tests performed and analyzed by ophthalmologists.

For developing, rural and remote areas [3], this poses a huge challenge. This gap between the need and availability of expert doctors can be bridged using artificial intelligence (AI). Machine learning (ML) techniques can be used to train computers to detect abnormalities in ocular fundus images, analyze them, and categorize these abnormalities into potential diseases. These trained computers do not require any human intervention and can therefore be used in remote areas where expert help cannot reach. Motivated by this, we aim to determine the best performing technique for cataract detection to deliver an improved diagnosis.

In this paper, we trained various ML techniques such as K-nearest neighbors (KNN), naïve Bayes (NB), random forest (RF), support vector machine (SVM), decision tree (DT), logistic regression (LR), XGBoost (XGB), LightGBM (LGB), and voting classifier (VC) on images of ocular fundus. The evaluation has been performed on the publicly available kaggle dataset cataract detection [2] which consists of ocular fundus images for normal and cataract affected patients. We performed binary classification on the cataract dataset, classifying the fundus into normal and cataract cases. All the ML techniques have been assessed using accuracy, precision, recall, f1 score, and Matthews correlation coefficient (MCC) as efficacy.

The organization of the paper is as follows: In Sect. 2, tests for clinical examination such as slit lamp tests and the existing work done in application of AI and deep learning (DL) in healthcare have been discussed. Section 3 consists of explanation of the various ML techniques used for analysis. In Sect. 4, the results obtained have been discussed and observations have been made based on these results. Section 5 discusses the future scope of AI in this direction.

2 Related Work

Cataracts are diagnosed clinically through a series of optical tests performed by ophthalmologists [4]. Researches have explored use of AI in various healthcare and biomedical applications [5]. The challenges associated with used of AI systems in the medical area along with the social, legal, and economic implications have been studied in detail. Use of ensemble techniques such as stacking and majority voting for detection of cataract has been studied where texture, wavelet and sketch-based features have been extracted from fundus images for analysis [6]. Application of deep learning (DL) technique and convolutional neural network (CNN) for training and testing on available slit-lamp images has also been studied [7]. There has been some research in the development and validation of DL algorithms [8] for diagnosis of age-related diseases such as diabetic retinopathy and related eye diseases [9]. Ocular fundus images have been used for the detection of eye diseases by many researchers [10]. Traditionally, cataract was diagnosed using slit-lamp tests for clinical and automated examinations. There has been research in the use of SVM for automatic detection of nuclear cataract from images obtained through slit lamps [11]. New methods have been proposed by researchers for grading of cataract using the images of slit-lamp tests [12]. Study of use of DL methods [13–16] and transfer learning [17] in cataract classification has also been done.

3 Application of Techniques

In this work, we have implemented various traditional ML techniques based on supervised learning such as KNN, SVM, DT, NB, and LR. We also analyzed ensemble techniques like VC along with other ensemble techniques based on one of the better performing traditional methods like RF, XGB, and LGB [18–21]. These techniques are shown in Table 1. The performance measures used for comparison are accuracy scores, the precision score, recall score, f1 score, and the MCC.

These techniques have been implemented using python 3.

4 Result and Discussion

In this work, we have used the publicly available dataset —Cataract dataset which consists of 300 normal and 100 cataract images. We have used 100 images of ocular fundus of normal and 100 images of cataract affected eyes to prevent bias raised from imbalanced data. For the training and testing data, a 75–25 split has been performed where 75 images of each category are used for training and 25 images of each category are used for testing purposes (see Fig. 2).

Table 1 Implemented ML techniques

Technique	Details
KNN	This model plots the data points into space and uses the nearest k points to assign a class to the selected data point. This label is decided as the most common label appearing in its neighboring points
SVM	This model plots data points into a k-dimensional (here k = 2) space. It finds a hyperplane to draw a definitive boundary between points to distinguish them into distinct classes
NB	This model creates a probability distribution based on various features for all classes present in the dataset. The final label is decided using this probability distribution for every data point
DT	This model uses answers obtained on questioning different data points of the dataset till a label is decided and constructs a tree-like structure made of nodes and directed edges
RF	This model uses majority voting of the labels obtained through a collection of DT models to decide on a final label for the input data
LR	This model works by calculating the probability of each class for every data point lying between 0 and 1 to assign the final label to the input data
XGB	This model uses DT as a base estimator. Initially, equal weightage is given to all the data points. In the next iteration, correctly labeled points have their weights reduced and wrongly labeled have their weights increased
LGB	This model has DT as the base estimator and is quite similar to the XGB, but it gives a better performance than XGB as XGB builds the DT greedily instead of fully
VC	This model works by assembling the findings of various ML models. It assigns the label which has the highest probability. For this analysis, we have used LR, RF, SVM, DT, KNN, and NB models

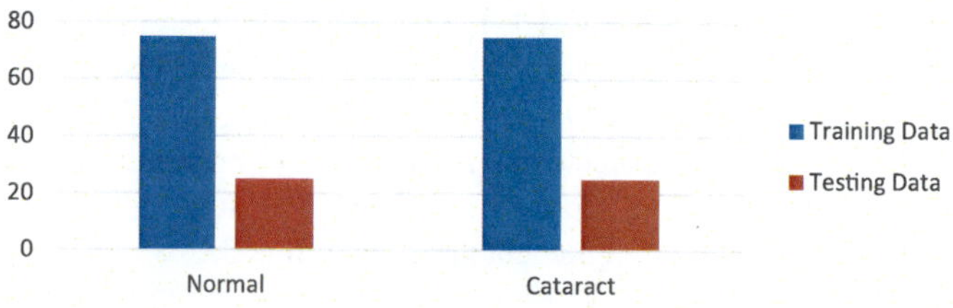

Fig. 2 Distribution of cataract dataset into training and testing data for analysis

During preprocessing (see Fig. 3), the images are resized into 32 by 32 size and then flattened into vectors. These vectors are used as features which are then made into a data frame with normal and cataract label. One hot encoding is done, and the final dataset is split into training and testing data.

After application of the aforementioned ML techniques in Sect. 3 on the cataract dataset, accuracy (see Fig. 4), precision, recall, f1 score (see Fig. 5) and MCC (see Fig. 6) scores were compared and analyzed. It was observed that the highest

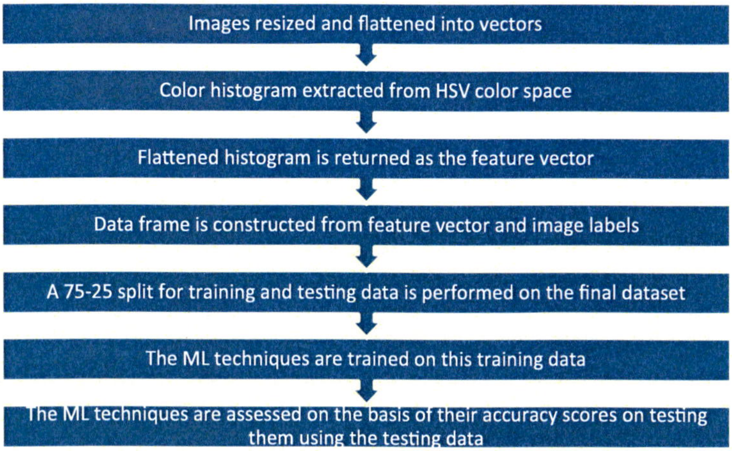

Fig. 3 Steps followed during analysis of ML techniques on the cataract dataset

Fig. 4 Accuracy scores obtained from testing ML and DL techniques on cataract dataset

performing ML technique in all parameters was LGB, with the second-best performance by RF, which was followed by XGBM, and then the remaining ML techniques were followed in this order—DT, VC, NB, KNN, SVM, and LR. Motivated by this, we implemented a DL technique, namely—Convolutional neural network (CNN). CNN is a type of regularized multilayer perceptron [22]. It is most commonly used for image classification problems. Neurons of one layer are linked with each of the other layers to fully connect the layers. It was observed that CNN gave an accuracy score of 0.875. On comparing accuracy scores of ML and DL techniques, we observed that DL-based technique, namely CNN gave superlative results.

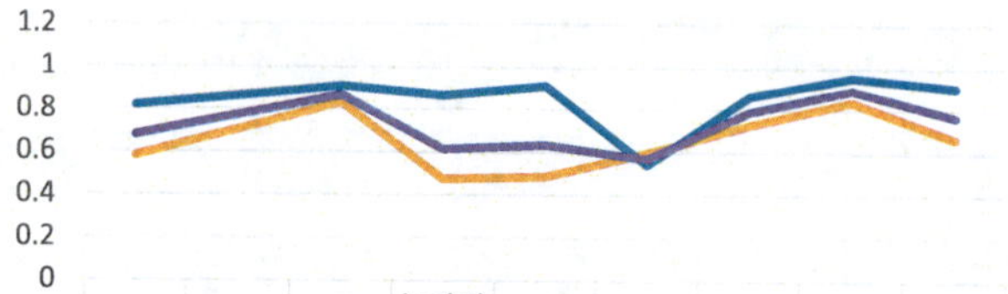

	Naïve Bayes	Decision Tree	Random Forest	Logistic Regression	SVM	KNN	XGBoost	LightGBM	Voting Classifier
Precision	0.581	0.704	0.833	0.475	0.488	0.6	0.731	0.84	0.667
Recall	0.818	0.864	0.909	0.864	0.909	0.545	0.864	0.954	0.909
F1 Score	0.679	0.775	0.869	0.613	0.635	0.571	0.792	0.894	0.769

Fig. 5 Precision, recall, and F1 scores obtained from testing ML and DL techniques on this cataract dataset

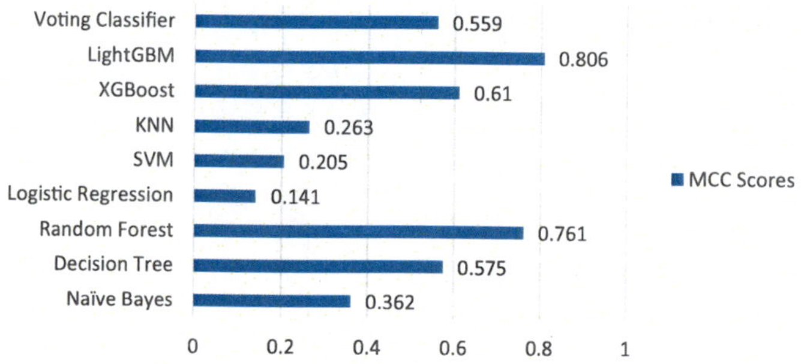

Fig. 6 MCC Scores obtained from testing ML and DL techniques on this cataract dataset

5 Conclusion and Future Scope

Cataract is one of prevalent diseases that affect a large number of people worldwide. It is one of the major causes of visual imparity and can lead to serious issues such as blindness if left untreated. ML and DL techniques can be used for analysis of ocular fundus test results to detect cataract at an early stage and take the correct preventive measures or cure. In this paper, we assessed various ML techniques such as RF, NB, LR, DT, SVM, VC, KNN, XGB, and LGB, and we also compared one DL technique CNN. We observed that DL technique (CNN) gave the highest accuracy of 0.875, followed by RF with score of 0.86. The lowest accuracy was observed with SVM with a score of 0.58. Future enhancement can be done through application of other soft computing techniques including hybrid methods and transfer learning for cataract detection on other datasets as well.

References

1. Flaxman SR, Bourne RRA, Resnikoff S et al (2017) Global causes of blindness and distance vision impairment 1990–2020: a systematic review and meta-analysis. Lancet Global Health 5:e1221–e1234
2. https://www.kaggle.com/jr2ngb/cataractdataset
3. Goh JHL, Lim ZW, Fang X, Anees A, Nusinovici S, Rim TH, Cheng C-Y, Tham Y-C (2020) Artificial intelligence for cataract detection and management. Asia-Pacific J Ophthalmol 9(2):88–95
4. Chylack LT Jr, Wolfe JK, Singer DM et al (1993) The lens opacities classification system III. The longitudinal study of cataract study group. Arch Ophthalmol 111:831–836
5. Yu KH, Beam AL, Kohane IS (2018) Artificial intelligence in healthcare. Nat Biomed Eng 2:719–731
6. Yang J-J, Li J, Shen R, Zeng Y, He J, Bi J, Li Y, Zhang Q, Peng L, Wang Q (2016) Exploiting ensemble learning for automatic cataract detection and grading. Comput Methods Programs Biomed 124:45–57. ISSN 0169-2607
7. Qian X, Patton EW, Swaney J, Xing Q, Zeng T (2018) Machine learning on cataracts classification using SqueezeNet. In: 2018 4th international conference on universal village (UV), pp 1–3. https://doi.org/10.1109/UV.2018.8642133
8. LeCun Y, Bengio Y, Hinton G (2015) Deep learning. Nature 521:436–444
9. Ting DSW, Cheung CY, Lim G et al (2017) Development and validation of a deep learning system for diabetic retinopathy and related eye diseases using retinal images from multiethnic populations with diabetes. JAMA 318:2211–2223
10. Gulshan V, Peng L, Coram M et al (2016) Development and validation of a deep learning algorithm for detection of diabetic retinopathy in retinal fundus photographs. JAMA 316:2402–2410
11. Li H, Lim JH, Liu J et al (2009) An automatic diagnosis system of nuclear cataract using slit-lamp images. Conf Proc IEEE Eng Med Biol Soc 2009:3693–3696
12. Xu Y, Gao X, Lin S et al (2013) Automatic grading of nuclear cataracts from slit-lamp lens images using group sparsity regression, pp 468–475. Berlin, Heidelberg: Springer Berlin Heidelberg
13. Dai W, Tham YC, Chee ML et al (2020) Systemic medications and cortical cataract: the Singapore epidemiology of eye diseases study. Br J Ophthalmol 104:330–335
14. Dong Y, Zhang Q, Qiao Z, Yang J (2017) Classification of cataract fundus image based on deep learning. In: 2017 IEEE international conference on imaging systems and techniques (IST), pp 1–5
15. Ran J, Niu K, He Z, Zhang H, Song H (2018) Cataract detection and grading based on combination of deep convolutional neural network and random forests. In: 2018 international conference on network infrastructure and digital content (IC-NIDC), pp 155–159
16. Zhang L, Li J, Zhang I et al (2017) Automatic cataract detection and grading using deep convolutional neural network. In: 2017 IEEE 14th international conference on networking. sensing and control (ICNSC), Calabria, pp 60–65
17. Pratap T, Kokil P (2019) Computer-aided diagnosis of cataract using deep transfer learning. Biomed Signal Process Control 53:101533
18. Rayyan M, Jaiswal A (2015) Robotics the roadmap to a digital life. In: 2015 international conference on green computing and Internet of Things (ICGCIoT), pp 1240–1244. https://doi.org/10.1109/ICGCIoT.2015.7380653
19. Kumar A, Jaiswal A, Empirical Study of Twitter and Tumblr for Sentiment Analysis using Soft Computing Techniques. In: Proceedings of the World Congress on Engineering and Computer Science, vol 1, iaeng.org
20. Jaiswal A, Malhotra R (2018) Software reliability prediction using machine learning techniques. Int J Syst Assur Eng Manag 9(1):230–244

21. Kumar A, Jaiswal A (2020) Particle swarm optimized ensemble learning for enhanced predictive sentiment accuracy of tweets. In: Singh P, Panigrahi B, Suryadevara N, Sharma S, Singh A (eds) Proceedings of ICETIT 2019. Lecture Notes in Electrical Engineering, vol 605. Springer, Cham
22. Kumar A, Jaiswal A (2020) Deep learning based sentiment classification on user-generated big data. Recent Adv Comput Sci Commun 13(5)

Plant Species Classification from Bark Texture Using Hybrid Network and Transfer Learning

Abdul Hasib Uddin, Imran Munna, and Abu Shamim Mohammad Arif

Abstract Every plant species has unique characteristics for leaf shape, vein structure, bark texture, fruit shape, or others. These characteristics identify plant species distinctly. In this paper, we have classified plant species from two bark image datasets, Bark-101 and Trunk12. We developed residual learning-based extended DenseNet201, where the DenseNet201 part of the network was pre-trained with the ImageNet dataset. The proposed extended part of the model is comprised of 14 layers with one residual block. We then compared the performances against benchmark results. For the Bark-101 dataset, our hybrid model achieved 53.88%. In the case of the Trunk12 dataset, we applied 10-fold cross-validation and our model gained 96.56% on average. This research will significantly help plant researchers and others to classify particular plants easily.

Keywords Plant species · Bark-101 · trunk12 · DenseNet201 · Residual network · Transfer learning

1 Introduction

It is quite difficult to estimate the species of a tree solely by looking at its bark or leaf. However, the species can be classified by using the texture of the bark or the picture of the leaves using machine learning approaches. An important visual signal for many types of textures is natural pictures. Texture analysis has received great importance in a variety of application domains, including object recognition and content-based image retrieval [1]. Numerous authors have provided tools for developing methods to help automate the process of cross-disciplinary plant identification between botany and computer vision [2–8]. The bark of a particular plant organ provides rich textural information and has recently attracted the interest of the researchers [4, 9–11]. A bark is mainly characterized by its structure [12]. The main disadvantage of bark texture

A. H. Uddin (✉) · I. Munna · A. S. M. Arif
Computer Science and Engineering Discipline, Khulna University, Khulna, Bangladesh
e-mail: abdulhasibuddin@gmail.com

P. Dutta et al. (eds.), *Emerging Technologies in Data Mining and Information Security*, Lecture Notes in Networks and Systems 490,
https://doi.org/10.1007/978-981-19-4052-1_21

analysis is that it deals with irregular texture structures and high inter-class similarities. Four benchmark color texture databases are generally considered, namely VisTex, BarkTex, OuTex, and Curet. Unlike the first three databases, each class in the Curet database is defined by a single color texture. These databases have been used several times in the framework of the supervised classification of images. This is necessary to manage supervised color texture classification and to extract training and validation image subsets from the database. During a supervised learning stage, the training subset is used to create a discriminatory feature space.

Firstly, training images are coded in one or more color spaces [13]. The color and texture properties are then calculated from the coded images. Thus, each image is presented in a characteristic space, where a classifier works at a decision stage to determine which category the image belongs to. Thus, a valid subset is used to evaluate the performance of the proposed classification at this decision stage [14]. Learning algorithms commonly used in the structure of color texture classification include local algorithms, where classifications such as k-nearest neighbor (KNN) and support vector machine (SVM) classifiers are considered [15].

We have used the convolutional neural network (CNN) algorithm in our proposed study. And, we were able to achieve significantly better results than the existing works.

2 Datasets

Table 1 holds the summary of the two datasets for our purpose. We used Bark-101 dataset developed by Ratajczak et al. [16]. This dataset contains a total of 2592 bark images against 101 classes/plant species. Each class has a minimum of 2 to a maximum of 138 images. Image widths range between 69 and 800 pixels, while image heights range within 112–804 pixels. The dataset is split into 50% training and 50% testing sets. We considered 20% of the training data as the validation set.

Trunk12 dataset was constructed by Švab [17]. There are a total of 393 bark images for 12 different species. Each class contains from 30 up to 45 images. The dimensions of these images are 1000 × 1334p.

Table 1 Dataset description [16]

Dataset information	Bark-101	Trunk12
Classes	101	12
Total images	2592	393
Images per classes	2–138	30–45
Image size	(69–800) × (112–804)	1000 × 1334
Train/Test splits	50/50	-

3 Implementation

3.1 Model Description

The model (Fig. 1) consists of a total of 721 layers, among which the last 14 layers are custom built and configured for bark texture analysis. The input shape of our proposed architecture is 256 × 256p. Hence, at first we resize each image to 256 × 256. We configured the input layer to take RGB images. Then, the 3-channel RGB images are fed into ImageNet pre-trained DenseNet 201, which automatically extracts useful features from each image. The final output of this stage is 8 × 8 × 1920. This means that on the last layer of DenseNet 201, there are 1920 filters for final feature extraction and the extracted features are 8 × 8p. Next, all of the 1920 features are passed through a 2D convolution layer. This layer has a rectified linear unit (ReLU) activation function. Similar to the previous stage of feature extraction, this layer also applied 1920 filters and outputs an 8 × 8 matrix for each filter. After that, a batch normalization layer is added to provide some regularization, reduce generalization error, and accelerate training. The output shape of this layer is also 8 × 8 × 1920. The next layer is another 2D convolution layer without any activation function. The output of the layer is then added with the initially extracted features from DenseNet 201 stage to preserve useful information.

The resultant matrix goes through an activation layer, which applies ReLU activation on the inputs. Again, another batch normalization is applied, and the output is then flattened. The flatten layer takes the input of size 8 × 8 × 1920 and flattens it into a vector of size 122,880.

After the flatten layer, there are three dense layers, each layer is followed by a consecutive dropout layer. The output of each of these six layers is 1024. We used exponential linear unit (ELU) activation function and L2 regularizer in those dense layers. The value of the regularizer was set to 0.001. About 50% of neurons were dropped randomly in each of the three dropout layers. Finally, the output layer was again a dense layer with 44 neurons for the Bark-101 dataset and 12 neurons for the Trunk12 dataset. We used the softmax function as the classifier.

3.2 Implementation on Bark-101 and Trunk12

In Fig. 2, validation accuracy and train accuracy curves for the Bark-101 dataset can be seen from the dual-line graph. The trend of train accuracy is upwards from the 1st to the 25th epoch. Since then, the trend has not fluctuated too much, which is the upward trend. On the other hand, the trend of validation accuracy goes up to the 16th epoch, but then the trend fluctuates a lot, which does not increase the accuracy much. Graphs of train accuracy and validation accuracy intersect at approximately the 26th epoch. Model validation occurs when validation accuracy is not good, and

Fig. 1 Proposed model

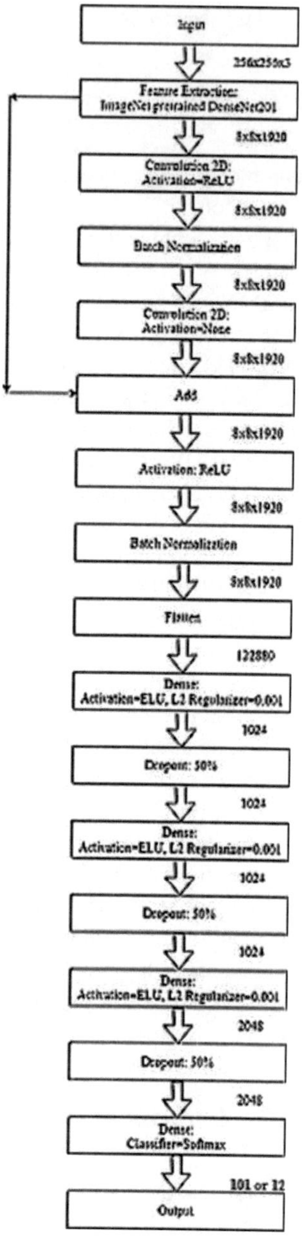

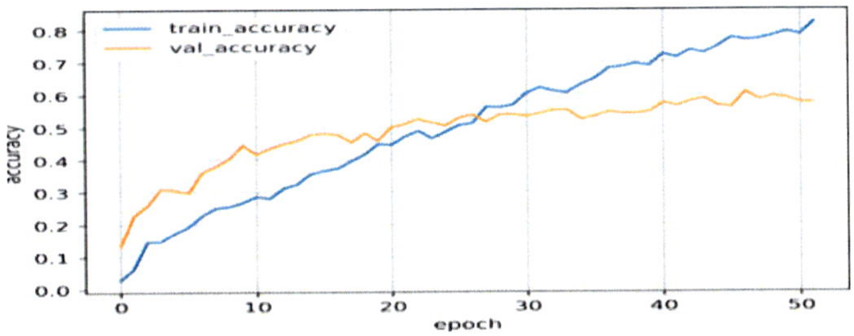

Fig. 2 Train accuracy versus validation accuracy plot for Bark-101 dataset

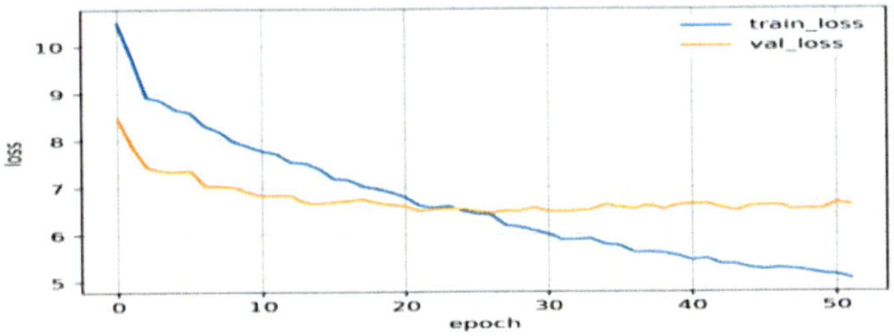

Fig. 3 Training loss versus validation loss plot for Bark-101 dataset

at the same time, train accuracy is good. The model memorizes the data set instead of learning it.

The graph in Fig. 3 shows that both train loss and validation loss for training our proposed model on the Bark-101 dataset are downward. The performance of training is good, while the performance of validation is not. This phenomenon is called overfitting. Validation data is not given to the model during training. The model does not know about the data set beforehand, and the anonymous data that is finally tested after the model is trained is called test data. We used validation data set to determine if the model is learning properly during training.

Trunk12 dataset is not partitioned into train-validation-test set by the authors. Hence, we applied 10-fold cross-validation on this dataset considering the average accuracy as final. Figure 4 shows that on the third fold, both train accuracy and validation accuracy go up by 20 epochs. The trend has not changed much from 20 to 50 epochs; it has ended in the same situation. Train data, validation data, and test data are all separated here.

Again, Fig. 5 demonstrates that on the third fold of training on the Trunk12 dataset, both the training loss and validation loss of the trend are downward. Train loss from 1 to 10 epoch is rapidly downward. Then, from the 10 to 50 epoch, the trend does not change much. Similarly, the validation loss also goes down from 1 to around 22 epochs, and since then the same trend of validation loss has continued to decrease slowly. Train data, validation data, and test data are not separated in Trunk 12.

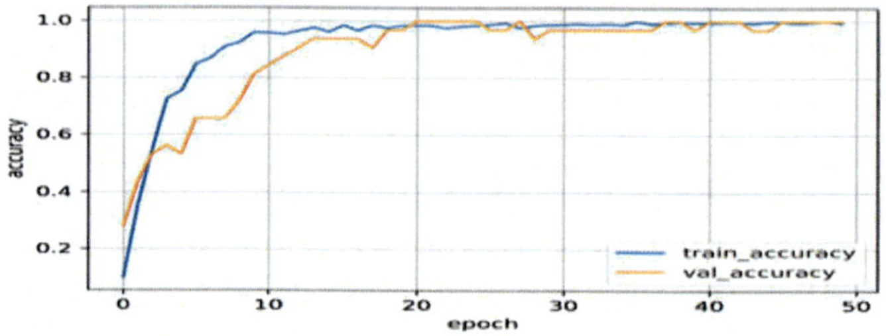

Fig. 4 Train accuracy versus validation accuracy plot for Trunk12 dataset on third fold

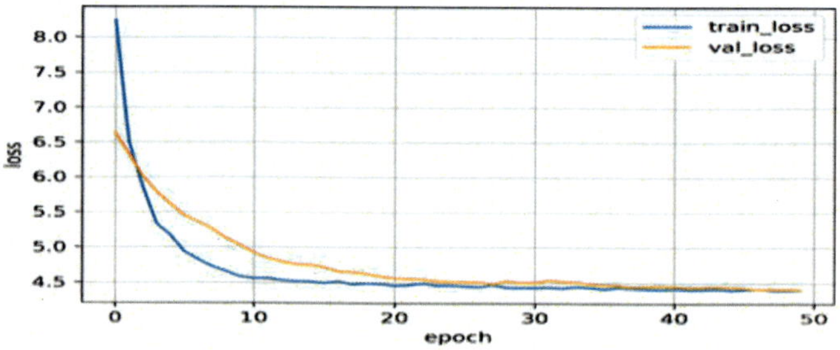

Fig. 5 Train loss versus validation loss plot for Trunk12 dataset on third fold

4 Results and Discussions

Table 2 shows the performance comparison between our proposed and existing methods for Bark 101 dataset. The existing methods are of two types, such as KNN and SVM. The top-1 accuracy of the H30, H180, LCoLBP, LS-LCoLBP, LS-LCoLBP / H30, LS-LCoLBP/H180, GWs/H30, and GWs/H180 descriptors for k-nearest neighbor (KNN) method are 19.1%, 22.2%, 34.2%, 28.3%, 27.6%, 27.8%, 28.2%, and 31.8%, respectively. On the other hand, the corresponding top-1 accuracy for the aforementioned descriptors for support vector machine (SVM) classifiers are 20.4, 20.9, 41.9, 30.1, 32.1, 31.0, 31.7, and 32.2%, accordingly. In contrast, the top-1 accuracy of our proposed architecture is 53.88% which is better than any other descriptors mentioned previously.

Figure 6 shows the results for 10-fold cross-validation applied on Trunk12 dataset. The accuracy of the corresponding folds is 93.75, 100, 100, 96.88, 93.75, 93.75, 96.88, 96.88, 100, and 93.75%. The average accuracy is 96.56%.

Table 3 represents a comparison of the results for the Trunk12 dataset. The accuracy of the descriptors MSLBP, H30, H180, CLBP, and LS-CLBP is around 63% to 70%. On the other hand, SNBP, LCoLBP, CLBP/H30, CLBP/H180, GWs/H30, GWs/180, LS-LCoLBP, LS-LCoLBP/H30, and LS-LCoLBP/H180 have accuracy in between 70 and 80%. Again, LCoLBP/H30, LCoLBP /H180 hold accuracy around 84%. Finally, our proposed descriptor successfully achieved an accuracy level of

Table 2 Top-1 accuracy comparison for Bark-101 dataset (in comparison with the performances reported in [16])

Descriptor	Size	Top-1 accuracy (%)		
		KNN	SVM	CNN
Porebski18	10,752	–	–	–
Wang17	267	–	–	–
Sandid16	3072	–	–	–
H30	30	19.1	20.4	–
H180	180	22.2	20.9	–
LCoLBP	240	34.2	41.9	–
LS-LCoLBP	90	28.3	30.1	–
LS-LCoLBP/H30	120	27.6	32.1	–
LS-LCoLBP/H180	270	27.8	31.0	–
GWs/H30	151	28.2	31.7	–
GWs/H180	301	31.8	32.2	–
Proposed	721	–	–	53.88

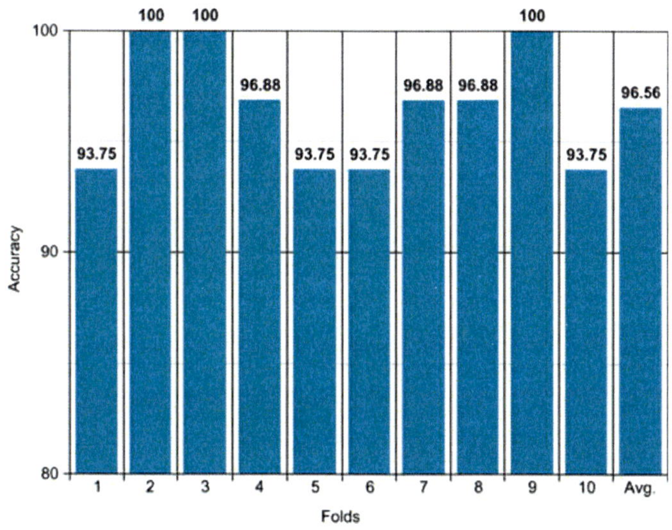

Fig. 6 Results for 10-fold cross-validation applied on Trunk12 dataset

96.56%, which is the best performance for the Trunk12 dataset classification problem in Table 3.

Table 3 Top-1 accuracy comparison for Trunk12 dataset (in comparison with the performances reported in [16])

Descriptor	Size	Top-1 accuracy (%)
MSLBP	2816	63.3
SMBP	10,240	71.0
H30	30	64.4
H180	180	69.0
LCoLBP	240	77.1
LCoLBP/H30	270	84.2
LCoLBP/H180	420	84.2
CLBP	66	70.0
CLBP/H30	96	77.4
CLBP/H180	246	78.1
GWs	121	39.9
GWs/H30	151	74.3
GWs/H180	301	76.1
LS-LCoLBP	90	74.6
LS-LCoLBP/H30	120	80.7
LS-LCoLBP/H180	270	80.7
LS-CLBP	30	70.0
LS-CLBP/H30	60	77.4
LS-CLBP/H180	210	78.1
Proposed	721	96.56

5 Conclusion

In this manuscript, we introduced an extended version of the DenseNet201 model for plant species classification using bark texture images. We deployed the ImageNet pre-trained DenseNet201 model for automatic feature extraction instead of manual descriptor selection. We then structured a 14-layer classifier model, which sits on top of the dense network for texture classification. We applied 10-fold cross-validation for the Trunk12 dataset and calculated the average. For both Bark-101 and Trunk12 datasets, our proposed approach outperforms all the other methodologies by a significant margin.

Acknowledgement This work is funded by the division of Information and Communication Technology (ICT), Ministry of Posts, Telecommunications and Information Technology, Government of the People's Republic of Bangladesh.

References

1. Mirmehdi M (2008) Handbook of texture analysis. Imperial College Press
2. Belhumeur PN, Chen D, Feiner S, Jacobs DW, Kress WJ, Ling H, Lopez I, Ramamoorthi R, Sheorey S, White S et al (2008) Searching the worlds herbaria: a system for visual identification of plant species. In: European conference on computer vision, pp 116–129. Springer
3. Nilsback M-E, Zisserman A (2010) Delving deeper into the whorl of flower segmentation. Image Vis Comput 28(6):1049–1062
4. Wendel A, Sternig S, Godec M (2011) Automated identification of tree species from images of the bark, leaves and needles. In: 16th computer vision winter workshop. Citeseer, p 67
5. Mouine S, Yahiaoui I, Verroust-Blondet A (2013) A shape-based approach for leaf classification using multiscaletriangular representation. In: Proceedings of the 3rd ACM conference on International conference on multimedia retrieval. ACM, pp 127–134
6. Mouine S, Yahiaoui I, Verroust-Blondet A, Joyeux L, Selmi S, Goeau H (2013) An android application for leaf-based plant identification. In: Proceedings of the 3rd ACM conference on International conference on multimedia retrieval. ACM, pp 309–310
7. Mouine S, Yahiaoui I, Verroust-Blondet A (2013) Plant species recognition using spatial correlation between the leaf margin and the leaf salient points. In: ICIP 2013-IEEE international conference on image processing. IEEE
8. Mzoughi O, Yahiaoui I, Boujemaa N, Zagrouba E (2016) Semanticbased automatic structuring of leaf images for advanced plant species identification. Multimedia Tools Appl 75(3):1615–1646
9. Sulc M, Matas J (2013) Kernel-mapped histograms of multi-scale lbps for tree bark recognition. In: Image and vision computing New Zealand (IVCNZ), 28th international conference of. IEEE, pp 82–87
10. Sixta T (2011) Image and video-based recognition of natural objects. Ph.D. dissertation, Diploma Thesis. Czech Technical University in Prague. Faculty of Electrical Engineering
11. Boudra S, Yahiaoui I, Behloul A (2017) Statistical radial binary patterns (srbp) for bark texture identification. In: International conference on advanced concepts for intelligent vision systems, pp 101–113. Springer
12. Wojtech M (2011) Bark: a field guide to trees of the Northeast. University Press of New England
13. Porebski A, Vandenbroucke N, Macaire L (2013) Supervised texture classification: color space or texture feature selection? Pattern Anal Appl 16(1):1–18. https://doi.org/10.1007/s10044-012-0291-9
14. Vandenbroucke N, Alata O, Lecomte C, Porebski A, Qazi I (2012) Color texture attributes, chapter 6, Digital Color Imaging, ISTE Ltd/Wiley
15. Hable R (2013) Universal consistency of localized versions of regularized kernel methods. J Mach Learn Res 14:153–186
16. Ratajczak R, Bertrand S, Crispim-Junior C, Tougne L (2019) Efficient bark recognition in the wild. In: International conference on computer vision theory and applications (VISAPP 2019)
17. Švab M (2014) Computer-vision-based tree trunk recognition. PhD diss., Bsc Thesis (Mentor: doc. dr. Matej Kristan), Fakulteta za racunalništvo in informatiko, Univerza v Ljubljani

Tomato Leaf Disease Recognition with Deep Transfer Learning

Sharder Shams Mahamud, Khairun Nessa Ayve, Abdul Hasib Uddin, and Abu Shamim Mohammad Arif

Abstract Transfer learning has introduced us to a new aspect of sensitive image classification tasks, such as disease recognition from both flora and fauna. It enables us to achieve significant performance faster and more effectively. On the other hand, plant disease recognition is undoubtedly important from nutritional and financial aspects. In this manuscript, we have deployed ImageNet pre-trained ResNet152V2 along with a custom 10 layers densely connected neural network for automatic disease classification from ten types of tomato leaf images, nine diseased classes, and healthy class. Our approach demonstrates significant performance against other recent works on tomato leaf disease classification. Our model performed well with almost 97% accuracy on the testing set.

Keywords Tomato leaf disease · Transfer learning · ResNet152V2 · DenseNet · Disease classification

1 Introduction

Plant diseases detection in time plays an important induction in the agricultural sector. Crop diseases pose a significant risk to maintenance protection, but evidence of their rapid differentiation remains a problem in many parts of the world due to the lack of presence on critical grounds. In declining food production, toxic pathogens, poor disease control, and harsh climate change are some of the major factors that are growing [1].

Tomatoes are the principal notable alter in our nation and all through the world. However, they besides make pests and illnesses in plants bring about pulverization of plants or segment of the plant causing in declined yield of harvests which prompts food disturbance. Besides, the mindfulness about the administration of bugs or infectious prevention is not much in different pieces of the country [2].

S. S. Mahamud (✉) · K. N. Ayve · A. H. Uddin · A. S. M. Arif
Computer Science and Engineering Discipline, Khulna University, Khulna, Bangladesh
e-mail: info.shamsbd@gmail.com

P. Dutta et al. (eds.), *Emerging Technologies in Data Mining and Information Security*, Lecture Notes in Networks and Systems 490,
https://doi.org/10.1007/978-981-19-4052-1_22

Deep learning is an artificial intelligence science that copies the processes of the human brain in information handling and generation of designs for utilizing in choice making [3]. Deep learning may be a modern slant in machine learning and it accomplishes the state-of-the-art comes about in numerous investigate areas, such as computer vision, drug design, and bioinformatics [4]. The advantage of deep learning is the capacity to abuse straightforwardly crude information without utilizing the hand-crafted features [4, 5]. As of late, the utilize of deep learning has given great outcomes in both scholastic and mechanical areas, much appreciated for two primary reasons [4]. Firstly, huge sums of information are created ordinary. Consequently, this information can be utilized in arrange to train a deep model. Besides, the control of computing given by graphics processing unit (GPU) and high-performance computing (HPC) makes conceivable the preparation of deep models.

2 Proposed Methodology

In this part, we discuss the proposed deep neural network pre-trained architecture which is used for preprocessing, feature extraction, and classification.

ResNet152V2 is some kind of pre-trained neural network architecture that is used for low-level feature extraction from the image set [6]. Then these features are used to feed dense and fully connected softmax layer for the classification of tomato leaf diseases.

2.1 Data Augmentation and Preprocessing

Data preprocessing of tomato leaf disease images is an essential part of the neural network. Sometimes, as a result of less or small data, the neural network can be overfitted. The neural network can immensely perform well in overfitting but perform less than expected in testing data. Techniques of data augmentation have been applied using different neural network architectures to minimize overfitting that exacerbates the size of dataset and accomplished geometric conversion to the image dataset.

In our work, an image data generator which is a built-in function in Keras and TensorFlow is used for preprocessing, and image pixel values are divided by 255 for rescaling. Image data was augmented in different ways such as horizontal flip, vertical flip, and zoom range for further research, and this can generate dataset on multiple cores in real time. It was fed into the neural network.

2.2 Automatic Feature Extraction Using Pre-trained Architectures

Image data has so many features like shape, color, and texture. These should be extracted manually. Then the part feature selection had come. This is a time-consuming process. A deep neural network helps us to avoid the manual feature extraction process. Former extraction of feature is carried out by particular CNN-powered architectures where dense network usage and which is combined into a fully integrated layer. These features are important for classification task that illustrates completely variant behavior of features like visibility, shrink and the roundness of the lesion. Within the planned model, four effective and up-to-date pre-trained CNN model, ResNet152 is taken into account as a characteristics extractor for disease detection from the tomato plant disease image dataset. This design was earlier trained on completely different image disease identification founded on the thought of transfer learning.

2.3 Transfer Learning

Transfer learning could be a machine learning strategy where a model created for an assignment is reused as the beginning point for a model on a secondary task. It is the process of transferring knowledge that is learned from one domain and used in another domain for extracting feature purpose and image classification tasks [7]. Within the deep learning see, TL is carried out by employing a deep CNN show which was prepared prior on an expansive dataset. The pre-trained CNN model is assisting prepared (fine tuning) on a modern dataset with littler numbers of preparing pictures comparable to the already trained datasets [7]. It alludes to the situation whereby what has been learned in one setting is misused to make strides optimization in another setting [8].

2.4 ResNet152

ResNet presents a structure called the remaining learning unit to lighten the corruption of profound neural systems. A residual block makes the utilize of skip associations to address the debasement issue. Residual networks (ResNet) were introduced as a family of different deep neural systems with comparable structures but distinctive profundities. This unit's structure may be a feedforward arrange with an easy route association that includes modern inputs into the network and creates unused yields [9]. Connections that skip one or more layers are easy route associations (moreover known as skip connections) [10]. Over a long time, deep convolutional neural systems have arranged breakthroughs within the area of picture identification and classification. Going deeper to illuminate more complex errands and to make strides classification or acknowledgment exactness has ended up a trend [11].

2.5 *Framework for Tomato Disease Classification*

In our classification task, our model took red, green, and blue (RGB) tomato leaf disease images as the input. The images had fed into the ResNet152V2 layer with a pre-trained ImageNet weight, and the final layer is a dense layer with a fully-connected softmax function. The dense layer consists of a total ten layers which have nine input layers and one output layer. Every input dense layer has 1024 units, used regularizer was L2 regularizer, used activation function was ELU, used optimizer was Adam, and the final activation function is softmax. Figure 1 indicates the overview of our model.

The hyperparameters were tuned in different values, but the final selected value is,

- Optimizer: Adam ($\beta 1 = 0.9$, $\beta 2 = 0.999$ and $\varepsilon = 10E–4$)
- Number of epoch: 25

Fig. 1 Block diagram of our proposed model

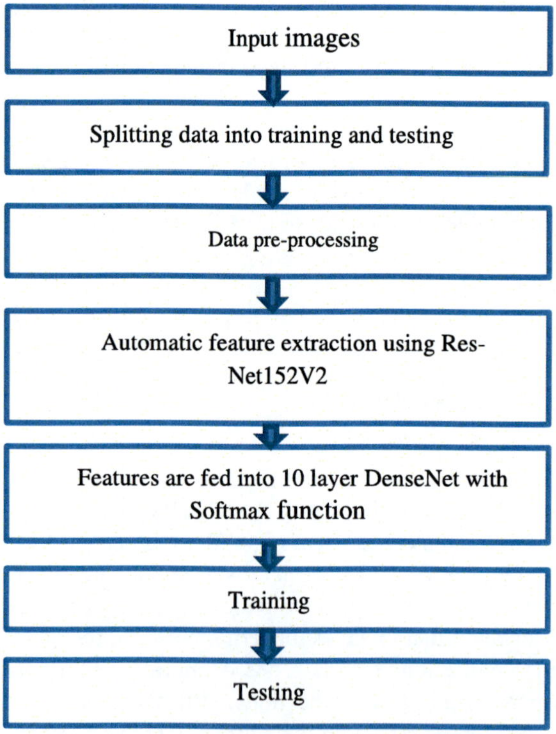

- Batch size: 128
- Regularizer: L2
- Number of layers: five (four input layer and one output layer)
- Initial activation function: ELU
- Final activation function: Softmax
- learning rate: 0.00001
- Dropout: 50%
- Loss: categorical_crossentropy

3 Experimental Results and Comparison

3.1 Dataset Collection

We collected tomato leaf images for disease prediction from https://www.kaggle.com/kaustubhb999/tomatoleaf. It is a public access repository where we find images of tomato leaf diseases from different ten classes. The dataset folder is divided into two subfolders. One is the training set folder, and the other is the validation set folder. Each folder has ten classes, the training folder consists of 1000 images, and the validation folder consists of 100 images for each class (in Table 1).

Table 1 Number of classes and images of our dataset

#	Name of diseases	No of Training set images	No of validation set images
1	Bacterial spot	1000	100
2	Early blight	1000	100
3	Late blight	1000	100
4	Leaf mold	1000	100
5	Septoria leaf spot	1000	100
6	Twospotted spider mite	1000	100
7	Target spot	1000	100
8	Yellow leaf curl virus	1000	100
9	Mosaic virus	1000	100
10	Healthy	1000	100
	Sub total	10,000	1000

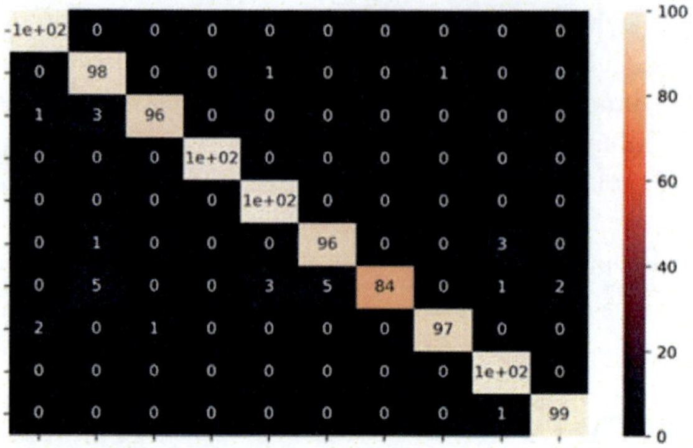

Fig. 2 Confusion matrix for our proposed method

Table 2 Model performance

Metrics	Performance (after 10 epochs)
Training accuracy (%)	98.80
Validation accuracy (%)	98.49
Test accuracy (%)	**96.98**
Training loss	3.4093
Validation loss	3.3813
Test loss	3.6833

3.2 *Experimental Result*

The confusion matrix in Fig. 2 has been used for evaluating classifier performance for disease classification. We had evaluated the experimented result in the terms of accuracy, f1-score, recall, and precision. Accuracy and loss metrics were measured during the training of the dataset. For both training and validation data, these metrics were measured.

Our dataset consists of a total of 10,000 images belonging to ten different classes. The data was spilled into 90 and 10%, which means for training 90% (9000 images) data was used and for testing 10% (1000 images) data was used. About 1000 images data were used for validation.

Table 2 was tabulated both training and validation accuracy and training and validation loss. See Fig. 3 for training, validation and testing accuracy and loss for our proposed model and Fig. 4 is our proposed ten layer DenseNet model.

3.3 *Performance Comparison*

A lot of work had been done before in this area. In recent years, for image classification, deep learning model is mostly used. We compared some of the state-of artwork

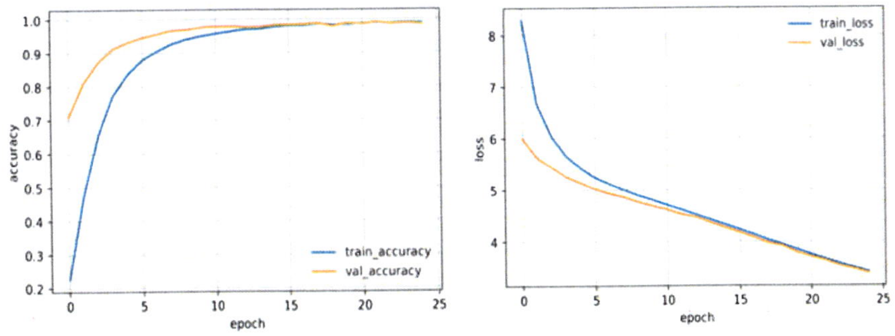

Fig. 3 Graph of accuracy and loss in training and validation

Fig. 4 Proposed ten layer DenseNet model

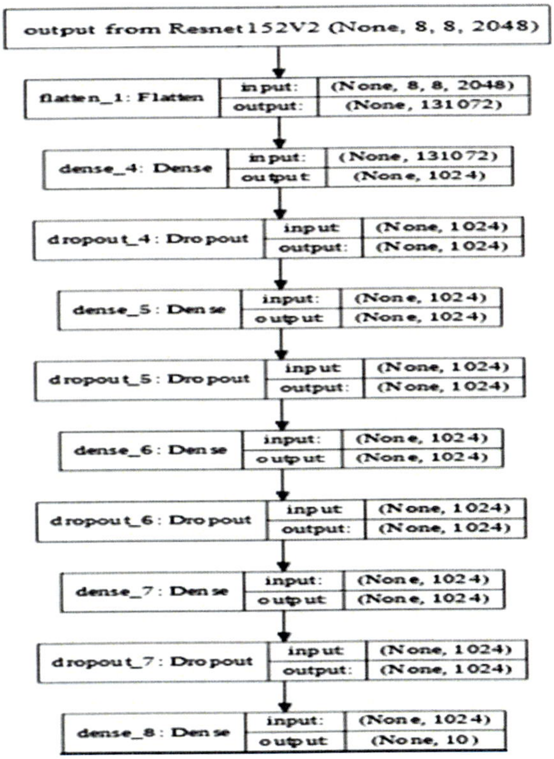

(in Table 3) in this area, and our model performed well with almost 97% accuracy on the testing set.

Table 3 Comparison of some deep learning model for tomato leaf disease classification

	Dataset	Number of images	Applied model	Performance	Number of classes
Zhang et al. [12]	AI Challenger platform	11,476		Validation accuracy	11
			Residual Attention Network	88.52%	
			SE-ResNet	88.83%	
			Sufflenet_V2	86.49%	
Gadekallu et al. [13]	Plant–village dataset	6000		Test accuracy	10
			DNN (using PCA with dimensionality reduction)	90%	
			DNN (using PCA and WOA with dimensionality reduction)	94%	
Jiang et al. [14]	AI Challenger platform	3000	ResNet 50 with 3 experiment	Test accuracy	3
			ReLU,7 × 7	95.7	
			L-ReLU,7 × 7	97.3	
			L-ReLU,11 × 11	98.0	
Proposedmodel	Kaggle dataset	10,000	ResNet 152V2	Test accuracy	10
				96.98%	

4 Conclusion and Future Work

In our model, features were automated extracted from tomato leaf images using pre-trained deep neural network model ResNet 152V2, which take advantage of making better the accuracy, precision, recall, and classification of tomato diseases. The outcome of our model correctly classified tomato leaf diseases which helps to take appropriate steps to reduce production loss. The use of transfer learning for tomato disease classification has made a better result even without any upgraded hardware like TPU's. In the future work, our proposed network could be trained with the dataset containing more data and could be implemented on different types of datasets of diseases for proactive predictions in different territories affecting our human life.

References

1. Maniyath SR, Vinod PV, Niveditha M, Pooja R, Prasad Bhat N, Shashank N et al (2018) Plant disease detection using machine learning. In: Proc. int. conf. design innov. 3Cs comput. commun. control (ICDI3C), pp 41–45, Apr 2018
2. Basavaiah J, Anthony AA (2020) Tomato leaf disease classification using multiple feature extraction techniques. Springer Science+Business Media, LLC, part of Springer Nature 2020
3. Alajrami MA, Abu-Naser SS (2019) Type of tomato classification using deep learning. Int J Acad Pedagogical Res (IJAPR) 3(12). ISSN: 2643-9603
4. Al Hiary H, Bani Ahmad S, Reyalat M, Braik M, ALRahamneh Z (2011) Fast and accurate detection and classification of plant diseases. Int J Comput Appl 17:31–38. https://doi.org/10.5120/ijca
5. Mokhtar U, El-Bendary N, Hassenian AE, Emary E, Mahmoud MA, Hefny H, Tolba MF, Mokhtar U, Hassenian AE, Emary E, Mahmoud MA (2015) SVM-based detection of tomato leaves diseases. In: Advances in intelligent systems and computing
6. Khamparia A, Singh PK, Rani P, Samanta D, Khanna A, Bhushan B (2020) An internet of health things-driven deep learning framework for detection and classification of skin cancer using transfer learning. Trans Emerging Tel Tech e3963
7. Deniz E, Şengür A, Kadiroğlu Z et al (2018) Transfer learning based histopathologic image classification for breast cancer detection. Health Inf Sci Syst 6:18
8. Hussain M, Bird JJ, Faria DR (2019) A study on CNN transfer learning for image classification. In: Lotfi A, Bouchachia H, Gegov A, Langensiepen C, McGinnity M (eds) Advances in computational intelligence systems. UKCI 2018. Advances in intelligent systems and computing, vol 840. Springer, Cham. https://doi.org/10.1007/978-3-319-97982-3_16
9. Nguyen LD, Lin D, Lin Z, Cao J (2018) Deep CNNs for microscopic image classification by exploiting transfer learning and feature concatenation. IEEE international symposium on circuits and systems (ISCAS)
10. Kumar V, Arora, Sisodia J (2020) ResNet-based approach for detection and classification of plant leaf diseases. In: 2020 international conference on electronics and sustainable communication systems (ICESC), pp 495–502. https://doi.org/10.1109/ICESC48915.2020.9155585
11. Sai Bharadwaj Reddy A, Sujitha Juliet D (2019) Transfer learning with ResNet-50 for malaria cell-image classification. In: International conference on communication and signal processing, April 4–6, 2019, India
12. Zhang T, Zhu X, Liu Y, Zhang K, Imran A (2020) Deep learning based classification for tomato diseases recognition. Earth Environ Sci (IOP)
13. Gadekallu TR, Rajput DS, Reddy MPK et al (2020) A novel PCA–whale optimization-based deep neural network model for classification of tomato plant diseases using GPU. J Real-Time Image Proc
14. Jiang D, Li F, Yang Y, Yu S (2020) A tomato leaf diseases classification method based on deep learning. Chin Control Decis Conf (CCDC) 2020:1446–1450. https://doi.org/10.1109/CCDC49329.2020.9164457

A Machine Learning Supervised Model to Detect Cyber-Begging in Social Media

Abdulrhman M. Alshareef

Abstract The misuse of social media leads many of the users to fall into crimes that they are not aware of it especially financial ones. Recently, cyber-begging has grown significantly on social platforms, impacting users and actual beneficiaries. The effects of this threat are serious and harmful that could lead to financial crime charges. With the spread of social media in recent times, this threat has become a globally important issue due to the economic and security implications. Most of the researchers discussed different types of cybercrimes to discover them and develop solutions to them. However, the number of research that focuses on cyber-begging in social media is insufficient. In this paper, we used machine learning to automatically identify cyber-begging in social media. The proposed model identifies cyber-begging using the Naive Bayes (NB) classification algorithm by training and testing the classifier using a real dataset collected from Twitter.

Keywords Supervised model · Machine learning · Cyber-begging · Social media · Cybercrimes detection

1 Introduction

The tremendous development of technology is an important motivation for the appearance of social media platforms. This contributed to accelerating the pace of communication between individuals and societies without the obstacles of location and time. Thus, this led to an expansion in the field of delivering information and obtaining it quickly and easily. Despite the effectiveness and positiveness of social media, yet some negative aspects that affect the positive use of it. Social media platforms have become a fertile environment for fraud, scam, and beggary for several factors. Among those factors affecting, there are economic factors, social factors, or political factors.

A. M. Alshareef (✉)
Information System Department, FCIT, King Abdulaziz University, Jeddah, Saudi Arabia
e-mail: amralshareef@kau.edu.sa

P. Dutta et al. (eds.), *Emerging Technologies in Data Mining and Information Security*, Lecture Notes in Networks and Systems 490,
https://doi.org/10.1007/978-981-19-4052-1_23

Nowadays, most social media users are either a witness or a victim of cyber-begging. Cyber-begging is an online form of the traditional begging, which asks strangers for a donation to fulfill instant needs such as food and money [1]. Begging relies on the use of persuasion tactics to bring money. These methods affect people because of their tendency to do charitable work and help others. The phenomenon of cyber-begging is an actual threat that affects societies in terms of both financial and security aspects. Indeed, vulnerable users are increasingly exposed to messages of solicitation and requests for assistance. Cyber-begging includes posting plea messages, which range from asking for bills or debts to be paid to transferring money for charitable causes through publishing stories that may or may not be true. In contrast to traditional begging, cyber-begging is not restricted to certain locations and times yet instead can occur anytime and anywhere. In addition, the beggar is considered anonymous in matters of name, real status, and location. This type of threat leads to uncontrolled losses, including financial losses, security problems, and even file legal charges. The spread of the this phenomenon worldwide in general, and in Arab world in particular, is due to reasons related to that most religions motivated followers to help the needy people. Recently, the warning against cyber-begging through news Websites has increased due to the negative effects of the phenomenon and the potential legal and security consequences on Arab societies. Although this phenomenon has a negative effect on social network users and societies, yet few researchers study the cyber-begging phenomenon. Most of the prior works detect cyber-begging in English. This paper focuses on detecting cyber-begging in Arabic.

This paper proposes a supervised learning model to integrate attributes to detect cyber-begging in social media. We will collect dataset from social media platform (Twitter). This dataset will be labeled by an expert in Arabic language to identify the begging keyword. Then, we use the labeled dataset to build the supervised learning model. Later, the model will be used detect cyber-begging tweets. The contributions of this paper come to address the following goals: First, identify and categorize the cyber-begging tweets. Second, build a dataset for a begging keyword. Third, building supervised learning model using a labeled dataset. Fourth, detect cyber-begging tweets based on the likelihood of the associated begging keyword. Accordingly, the proposed model can benefit the following applications: fraud detection, cyber-criminal finance analysis, and money laundering analyzes.

To achieve the first and second goals, we employ the natural language processing (NLP), the Twitter API to build labeled dataset. We retrieve and categorize the tweets collection. The tweets collection is retrieved and classified based on common terms usually used by the traditional begging. The common term will be extracted from literature review. Then, the dataset will be labeled by expert to identify the begging tweets. Later, a begging keywords dataset will be built. To achieve the third goal, we employ the Naive Bayes classifier to build a supervised learning model. The paper uses Naive Bayes classifier to identify begging terms in a tweet. The main goal is to determine the probability weight of begging keywords in the tweets in compere to non-begging keywords. In order to build the detection model, we use a legacy labeled training dataset in order to build the supervised learning model. To fulfill the fourth goal, we utilize the supervised learning model to detect the cyber-begging tweets.

This paper organized as follows. Section 2 presents some work that is related to this matter. Section 3 illustrates the proposed model. Section 4 evaluates the proposed model and discusses the results. Section 5 concludes the paper by identifying some limitations and future work.

2 Literature Review

The rapid development in the daily lifestyle and rapid economic changes prompted people to find different means to earn money quickly, whether legal or illegal [2]. Cyber-begging considers one of the recent methods used by fraudsters to receive easy money online out of pity. Researchers studies the cyber-begging phenomenon from social and linguistics perspective [1–3]. However, the lack of studies that aim to detect cyber-begging on social media motivates us to establish this paper. This paper focuses on using machine learning techniques to detect cyber-begging.

Ariyo [1] and Alabi et al. [3] discuss the linguistics bases used to establish a cyber-begging dialog. The studies illustrate different examples of communication phrases that are used for begging from linguistics perspective. The main target for both studies was to study the link between the linguistic features and the emotional effects that were intended to be reflected in the dialog. Thus, it will affect the individual reaction toward cyber-begging requests. Although they provide a linguistics basis for the cyber-begging phenomenon, yet their studies need to be implemented technically to detect cyber-begging.

Among the applications of artificial intelligence (AI) is machine learning. It is used to give the application the automation capability to learn from given information to improve the outcome results [4]. That will help the computer to make an independent decision based on legacy information. Consequently, it facilitates the computer work to complete sophisticated jobs that required plenty of time from individuals [5, 6]. Machine learning divided into two main types: supervised learning and unsupervised learning [4]. The different between them is based on the type on the used legacy information to construct the learning model. Supervised learning model is using pre-labeled data to construct the learning model, while the unsupervised learning used pre-unlabeled data [7, 8]. The main goal for both types is to construct a structured model to classify the new coming data based on the given legacy data. Despite the challenges of identifying cyber-begging, yet machine learning systems help to overcome these challenges significantly. As it discussed early, cyber-begging mostly uses text-based message to communicate with different user over social media. Although the lack of paper discussing detecting cyber-begging from a technical perspective, yet different papers discussed the use of machine learning algorithms to detect similar text-based phenomena. Djedjiga et al. [8] propose a methodology based on analyzing social media data to detected cyberbully in Arabic. Their analysis is prepared to identify existing of cyberbully action based on collected data from different social media platform. They use traditional Naive Bayes techniques as supervised learning technique to detect cyberbully messages in the social media platform. Nandakumar et al. [6] study the proposed methods to detect cyberbully activities using two different supervised learning technique. The paper compeer between Naive Bayes classifier and support vector machine based on precision and run time complexity. The results show that using Naive Bayes classifier enhances the outcome result in compeer to support vector machine. In addition, it reduces the required time to complete the tasks. As a result, it shows that Naive Bayes classifier overcomes support vector machine on both precision and run time complexity. Fatahillah et al. [9] proposed an approach to classify tweets that contain optimistic and pessimistic dialogs. The authors collected Indonesian language tweets using Twitter API. The authors used machine learning classifiers (Naive Bayes) to classify tweets dialogs. Tseng et al. [10] used Naive Bayes classifier algorithm to classify tweets in order to offer suggestions to the users based on the trending topics. The classifier takes each Twitter messages and calculate the probabilities to classify the messages to one of the four classes based on the common sense of trending tweets. Santoshi et al. [11] propose a methodology to evaluate the tweets and categorize them into "spam and ham" considering the used terms. They used Naive Bayes classifier algorithm to identify terms and categorize them into spam and non-spam terms. The classifier breaks each tweet into separate terms, then estimate the likelihoods to identify the tweets to each category based on the used terms. The paper shows that the used classifier provides better results with minimum load.

This paper proposed a supervised learning algorithm to detect cyber-begging. The Naive Bayes classifier will be used due to its sufficiency and simplicity [10, 12, 13]. It is due to that most studies targeting similar text-based supports the use of Naive Bayes classifier [14]. In addition, the lack of research that using supervised learning algorithm to detect cyber-begging on social media.

3 Proposed Model

This paper aims to identify cyber-begging act in a given text. Figure 1 illustrated the system architecture of the proposed detection model. Naive Bayes (NB) classifier will be used to construct the detection supervised learning model based on the collected data. The data required to build the model will be collected from most used social media platform. This research relies on tweet data from Twitter. The system applies the following five steps:

1. Searching and retrieving the collection of relevant tweets from Twitter platform.
2. Prepare and clean the tweet information to create clean dataset.
3. Classify the cleaned dataset to label the tweets data to one of the following classes: begging tweets or non-begging tweets.
4. Construct the supervised learning model using the training data from the labeled dataset.
5. Validate the supervised learning model using the testing data from the labeled dataset.
6. Predict the new entered data using the constructed learning model.

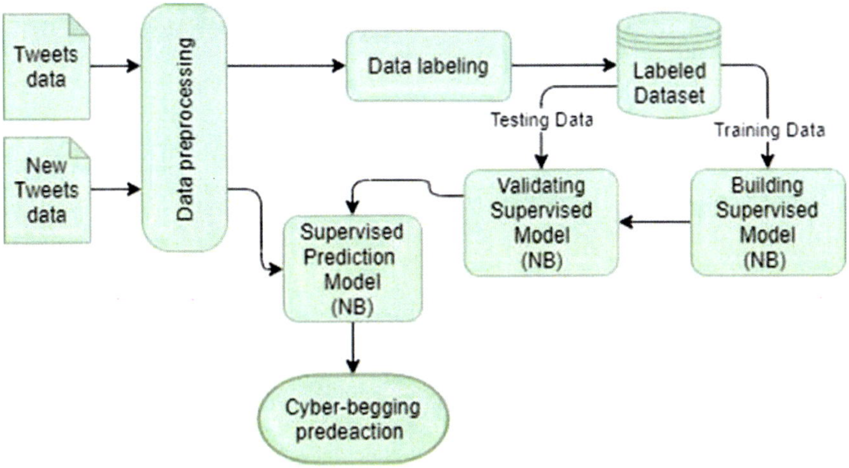

Fig. 1 The system architecture of the proposed detection model

3.1 Data Collection

To construct the detection supervised model, we need to collect data from social media platform. Recently, Twitter platform considered the most used one among different social media platforms. The tweet dataset contains text data that will be used to construct the detection model. In order to fulfill the first and second steps, Twitter API will be used to collect dataset randomly using relevant begging keyword. Then, we prepare the collected dataset through applying different preprocessing steps to avoid incomplete data and misconducted interpretation. Preprocessing steps include remove noise data, cleaning the tweets form irrelevant information, normalization the text to smooth the data processing. The cleaned dataset then we be labeled in the next step.

3.2 Data Annotation

Data annotation usually begins with asking people to make decisions about a specific part of unlabeled data. The main idea behind labeling the dataset is to construct a staple supervised model. In order to complete this task, a common standard needs to be set. The annotation process should follow this standard to label this dataset. The used stander rules are:

1. The tweets will be classified into two classes as follows:
    - Begging: The tweet contains a keyword for begging, or the context of the sentence indicates that.
    - Non-begging: The tweet does not contain any begging relevant keyword.

2. Request from individuals to annotate the given tweets with respect to the aforementioned classes specification.
3. The dataset will be divided into two subsets for training and testing.

The supervised learning model uses individuals provided labeled data to learn the original patterns in a training process as shows in Table 1a. The trained model can be used to dedicate the cyber-begging on new data.

3.3 Supervised Learning Model Construction

This paper aims to propose a supervised learning model to classify given tweets. The Naive Bayes classifier will be used due to its sufficiency and simplicity. It relies on Bayes theorem [8, 15]. The theorem aims to calculate the fitting probability of the given data to one of the designated classes. The Naive Bayes classifier uses prior knowledge to build the learning model. The prior knowledge extracted from the given training data that has been labeled by one of the predefined designated classes.

The main idea in this model is to estimate the probability of a term mentioned in each class per given tweets. In order to do that, Bayes theorem will be used as a base of Naive Bayes classifier construction. We calculate the probability of given terms $t_1, \ldots, t_i$ from a specified tuple tweet T belongs to class C as in Eq. (1). The classifier will predict that T fits in class that have the highest posterior probability. As $P(t_1, \ldots, t_i)$ is constant for all classes which is the probability of any given keyword, only $P(t_1, \ldots, t_i|C)P(C)$ need to be maximized.

$$P(C \mid T) = \frac{P(C)P(t_1, \ldots, t_i \mid C)}{P(t_1, \ldots, t_i)} \tag{1}$$

The size of each tweet can reach up to 280 words. Thus, $P(t_1, \ldots, t_i \mid C)$ calculation could be computationally expensive due to the size of each tweet. Therefore, the naive assumption of class-conditional independent will be used to decrease the computation process. The concept assumes that each attribute value is independent of the other values in the tuple of attributes for an object in a given class.

Thus, the estimation of the probability for attributes given class calculated in Eqs. (2) and (3)

$$P(C \mid T) = P(C)\prod_{i=1}^{n} P(t_i \mid C) \tag{2}$$

where:

$$\prod_{i=1}^{n} P(t_i \mid C) = P(t_1 \mid C) \times P(t_2 \mid C) \times \cdots \times P(t_i \mid C) \tag{3}$$

The Probability of a term t_i in T given class C calculate as in Eq. (4):

Table 1 **a** An example of Arabic labeled tweets and extracted keywords, **b** an example of the probabilities of Arabic tweets classification based on used keywords

Label	Tweets	Keywords
Begging	حالة إنسانية ، محتاجين مساعدة لسداد فاتورة كهرباء 2500 ريال.	محتاج، سداد، فاتورة، إنسانية كهرباء، مساعدة، ريال
Begging	الراتب مو قد الحاجة، المديونة عالية، محتاجين مساعدة.	راتب ، ديون، مساعدة ، محتاج
Non-Begging	الحمد لله مو محتاج راتب، دخل عالي ، وبدون ديون.	محتاج ، دخل ، عالي، ديون، راتب.
Non-Begging	محتاج مساعدة في سداد فاتورة الكهرباء ,ماعرف أدخل الموقع.	محتاج، سداد، فاتورة، كهرباء، مساعدة، موقع.
Begging	تكفون ، فاتورة كهرباء 1500 ريال محتاجة سداد ، دخل العائلة لايكفي.	محتاج، سداد، فاتورة، دخل، كهرباء ، ريال

(a)

	Tweets	Begging	Non-Begging
1	الراتب مو قد الحاجة، المديونة عالية، محتاجين مساعدة.	0.0096	0.0032
2	الحمد لله مو محتاج راتب، دخل عالي ، وبدون ديون.	0	0.00064
3	محتاج مساعدة في سداد فاتورة الكهرباء ,ماعرف أدخل الموقع.	0	0.000128
4	تكفون ، فاتورة كهرباء 1500 ريال محتاجة سداد ، دخل العائلة لايكفي.	0.001536	0.000128

(b)

$$P(t_i \mid C) = \frac{\text{Number of } t_i \text{ in } T \text{ belong to } C}{\text{Number of } T \text{ belong to } C} \tag{4}$$

The Probability of a term t_i given class C calculate as in Eq. (5):

$$P(C) = \frac{\text{Number of } T \text{ belong to } C}{\text{Number of } T \text{ in all data}} \tag{5}$$

The main idea here is to predict the class of given tweets based on the likelihood of using terms that has been previously classified in one of the designated classes: begging or non-begging. The contracted learning model using the labeled data helps to decide on a new data entry by comparing them with the legacy data.

4 Experimental Evaluation and Results

To construct and validate the supervised learning model, the dataset has been collected from one of the social media platforms. The collected data have been crawled using Twitter API. We consider using Twitter platform as a main source of data due to its popularity in the Arabic world and the information richness of the collected data. The dataset was collected concerning Arabic tweets for our experiment. The tweets are mainly about asking for help or money, requesting to pay bills, and courts' debt verdicts. Tweets on these matters are generally considered as a rich source for cyber-begging. Since publishers are pleading with users to donate by explaining the suffering. Furthermore, the dataset contains a set of tweets that is unlikely to categorized as begging tweets to validate our experiments. The collected a dataset is around 35,890 tweets which are stored in CSV files.

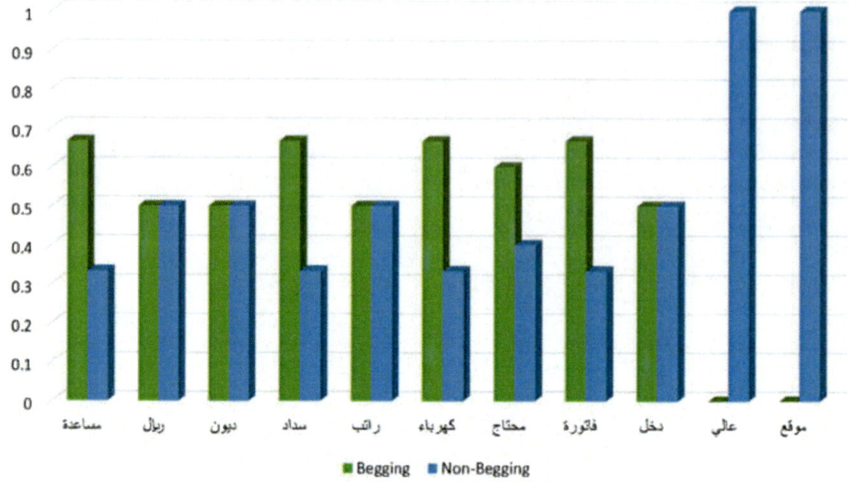

Fig. 2 The probabilities of some Arabic keywords from the dataset

After cleaning and labeling the dataset, it was divided into two subsets: a randomly withheld 20% of the dataset as the test set and the remaining 80% as the training set, then conducted five runs using different training-test distributing datasets to avoid the possibility that the test set was biased. Table 1a illustrates an example if begging keywords of the training dataset that has been collected from the tweets. The dataset will be used to train the classifier system. Naive Bayes algorithm will be used to calculate the key word likelihood for keyword to classify the testing dataset. Figure 2 illustrates the probabilities of some keywords from Table 1a based on that dataset sample. As shown, the keyword مساعدة has 0.67 likelihood to consider the tweets as a begging phrase and a probability of 0.33 to consider the tweets as a non-begging phrase. Thus, any tweets that have a keyword مساعدة has higher possibility to be classified as begging phrase. Furthermore, the keyword ديون has a probability of 0.5 to consider the tweets as a begging and the same probability for non-begging, this indicates that the keyword ديون has the same possibility to be a begging and non-begging keyword. However, Table 1b shows the probabilities calculation for some tweets have keyword مساعدة . Although the keyword has higher probabilities to classify the tweets contains it as begging phrase, yet tweet number 3 has been classified as non-begging phrase. This due to that tweet number 3 include keyword موقع which has no likelihood to consider the as a begging keyword. Thus, the phrase contains keyword موقع will be classified as non-begging phrase.

Four types of evaluation metrics were adopted: average precision (P), average recall (R), average F-measure (F_1), and Accuracy (Acc) to assess the performance of proposed classifier model to determine the cyber-begging tweets [8, 14]. The used metrics are calculated as following:

Table 2 An example of confusion matrix representing the classifier performance

Classifier performance	Predicted begging	Predicted non-begging	Total
Actual begging	4740	168	4908
Actual non-begging	198	2072	2270
Total	4938	2240	7178

$$P = \frac{\text{TP}}{\text{TP} + \text{FP}} \tag{6}$$

$$R = \frac{\text{TP}}{\text{TP} + \text{FN}} \tag{7}$$

$$F_1 = 2\frac{P \cdot R}{P + R} \tag{8}$$

$$\text{Acc} = \frac{\text{TP} + \text{TN}}{\text{TP} + \text{TN} + \text{FP} + \text{FN}} \tag{9}$$

where true positives (TP) represent the positive tuples (actual positives) that were correctly labeled (predicted positives) by the classifier. While true negatives (TN) represent the negative tuples (actual negatives) that were correctly labeled (predicted negatives) by the classifier. In addition, false positives (FP) represent the negative tuples (actual positives) that were incorrectly labeled (predicted negatives) by the classifier. While false negatives (FN) represent the positive tuples (actual negatives) that were incorrectly labeled (predicted positives) by the classifier. Table 2 illustrated an example of confusion matrix representing the classifier performance for one of the testing runs regarding identifying the correct category. The table will be used to calculate the precision, recall, F-measure, and accuracy using Eqs. (6)–(9).

In this example, Naive Bayes has been used to classify the testing dataset which was over 7000 tweets. About 96.5% out of 4908 tweets were classified accurately as begging phrases, and about 91.3% out of 2270 tweets were classified accurately as non-begging phrases. Figure 3 summarized the average of precision, recall, accuracy, and F-score metrics that aim to evaluate the performance of proposed classifier model. It demonstrates the performance efficiency of the proposed model in term of these metrics. This indicates a positive influence on detect cyber-begging.

5 Conclusion and Future Work

In this paper, a supervised learning model proposed to deduct cyber-begging. We take advantage of the Naive Bayes classification algorithm to estimate the probability

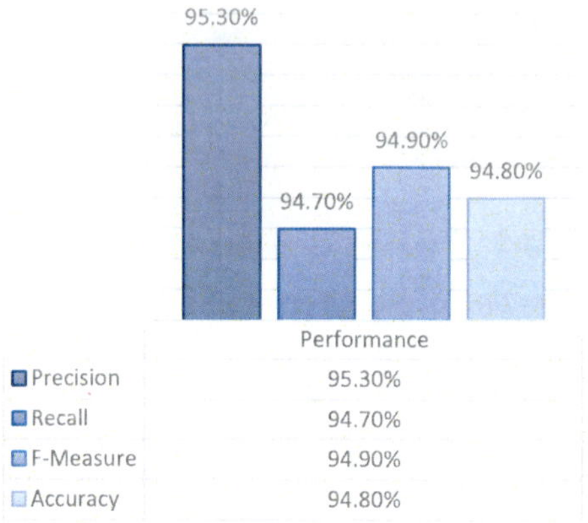

Fig. 3 The performance metrics summarizing

cyber-begging in social media. The detection model has been utilized to classify Arabic tweets in Twitter platform. Twitter API has been used to collect the data to facilitate the analysis cyber-begging in Arabic world smoothly. We conducted an experiments on a subset of Twitter dataset to evaluate the performance of the proposed supervised learning model. The experiment results show that the proposed supervised learning model obtained 94.8% in accuracy as one of the main performance indicators in machine learning. Researchers can benefit from the supervised learning model to determine cyber-begging in social media platform. Potential enhancements will be addressed in regrades by implementing different machine learning techniques to enrich the detection model such as SVM, K-means, and regression. In future, different attribute such as user id, location, and time will be considered to evaluate the cyber-begging in s social media platforms.

References

1. Samuel AK (2013) Analytical study of discourse strategies in internet begging relating to financial incapacity. Int J Lang Learn Appl Linguist World 36
2. Alhashlamoun RMA (2021) Electronic begging and its social and economic impact on Jordanian society from the point of view of a sample of Facebook users. Hum Soc Sci 5(4):60–76
3. Alabi O, Tshotsho B, Cekiso M, Landa N (2017) An examination of emotive style in online begging discourse. Gender Behav 15(2):8631–8641
4. Diwakar D, Kumar R, Gour B, Khan AU (2019) Proposed machine learning classifier algorithm for sentiment analysis. In: 2019 16th international conference on wireless and optical communication networks (WOCN). IEEE, 2019, pp 1–6
5. Wang J (2020) Using machine learning to identify movie genres through online movie synopses. In: 2020 2nd international conference on information technology and computer application (ITCA). IEEE, 2020, pp 1–6
6. Nandakumar V, Kovoor BC, Sreeja M (2018) Cyberbullying revelation in twitter data using naïve bayes classifier algorithm. Int J Adv Res Comput Sci 9(1)

7. Samal B, Behera AK, Panda M (2017) Performance analysis of supervised machine learning techniques for sentiment analysis. In: 2017 3rd international conference on sensing, signal processing and security (ICSSS). IEEE, 2017, pp 128–133
8. Mouheb D, Albarghash R, Mowakeh MF, Al Aghbari Z, Kamel I (2019) Detection of Arabic cyberbullying on social networks using machine learning. In: 2019 IEEE/ACS 16th international conference on computer systems and applications (AICCSA). IEEE, 2019, pp 1–5
9. Fatahillah NR, Suryati P, Haryawan C (2017) Implementation of Naive Bayes classifier algorithm on social media (twitter) to the teaching of Indonesian hate speech. In: 2017 international conference on sustainable information engineering and technology (SIET). IEEE, 2017, pp 128–131
10. Tseng C, Patel N, Paranjape H, Lin T, Teoh S (2012) Classifying twitter data with Naïve Bayes classifier. In: 2012 IEEE international conference on granular computing. IEEE, 2012, pp 294–299
11. Santoshi KU, Bhavya SS, Sri YB, Venkateswarlu B (2021) Twitter spam detection using Naïve Bayes classifier. In: 2021 6th international conference on inventive computation technologies (ICICT). IEEE, 2021, pp 773–777
12. Zulfikar WB, Irfan M, Alam CN, Indra M (2017) The comparation of text mining with Naïve Bayes classifier, nearest neighbor, and decision tree to detect Indonesian swear words on twitter. In: 2017 5th international conference on cyber and IT service management (CITSM). IEEE, 2017, pp 1–5
13. Erşahin B, Aktaş Ö, Kılınç D, Akyol C (2017) Twitter fake account detection. In: 2017 international conference on computer science and engineering (UBMK). IEEE, 2017, pp 388–392
14. Rahat AM, Kahir A, Masum AKM (2019) Comparison of Naïve Bayes and SVM algorithm based on sentiment analysis using review dataset. In: (2019) 8th international conference system modeling and advancement in research trends (SMART). IEEE 2019, pp 266–270
15. Abd DH, Sadiq AT, Abbas AR (2020) Political Arabic articles classification based on machine learning and hybrid vector. In: 2020 5th international conference on innovative technologies in intelligent systems and industrial applications (CITISIA). IEEE, 2020, pp 1–7

Retinal Optical Coherence Tomography Classification Using Deep Learning

Hitesh Kumar Sharma, Richa Choudhary, Shashwat Kumar, and Tanupriya Choudhury

Abstract The execution of medical decision-making or clinical diagnosis in medical or clinical imaging generally involves difficulties with reliability, robustness, and interpretability of data. In this project, we setup an analytic and predictive system dependent on a deep learning technique for the screening of patients with treatable blinding retinal illnesses, specifically related to choroidal neovascularization (CNV), diabetic macular edema, DRUSEN, or normal retina. We use the power of transfer learning in our neural network, which prepares our deep-learning model to perform better on X-ray retinal images. When a dataset of optical coherence tomography (OCT) pictures is given to a model, we are able to exhibit results comparable to that of human specialists in characterizing and segregating age-related illnesses of the retina, especially corresponding to choroidal neovascularization, macular degeneration, and diabetic macular edema. Transfer learning leverages the power of pre-trained weights and biases to allow a deep learning neural network to find better patters and classify more efficiently. This system may assist in the conclusion and reference of treatable retinal diseases and eye-related conditions, consequently allowing early detection and treatment of illnesses, bettering medical treatment of patients.

Keywords Transfer learning · Deep learning · Neural network · Choroidal neovascularization · Diabetic retinopathy · Diabetic macular edema · Optical coherence tomography · Retinal OCT · Eye disease · Treatment · Medical deep learning

H. K. Sharma (✉) · R. Choudhary · S. Kumar · T. Choudhury
School of Computer Science, University of Petroleum and Energy Studies (UPES), Energy Acres, Bidholi, Dehradun, Uttarakhand 248007, India
e-mail: hkshitesh@gmail.com; hksharma@ddn.upes.ac.in

T. Choudhury
e-mail: tanupriya1986@gmail.com; tanupriya@ddn.upes.ac.in

P. Dutta et al. (eds.), *Emerging Technologies in Data Mining and Information Security*, Lecture Notes in Networks and Systems 490,
https://doi.org/10.1007/978-981-19-4052-1_24

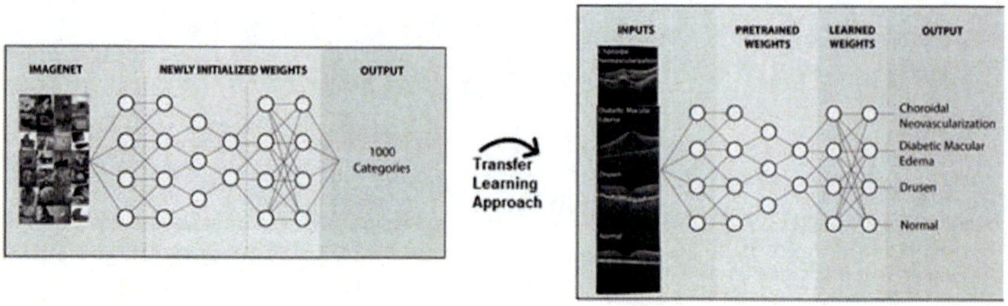

Fig. 1 Application of transfer learning on Retinal OCT images

1 Introduction

Machine learning and deep learning have the potential to transform disease assessment and treatment by conducting multiclass or binary classification that are hard for humans or even medical practitioners to accomplish and by examining massive amounts of data quickly. Notwithstanding its potential, ML's medical correctness and practical application are still issues that need to be resolved [1–3]. A special device can capture a high-resolution image of the back of the eye where these diseases typically manifest. We believe we can use these optical coherence tomography images to simultaneously provide rental screening to areas without proper axes while reducing the patient burden in populated urban centers [4–6].

Using an accurately labeled dataset of OCT images, we construct a deep, convolutional neural network to train using a transfer learning algorithm that helps the network learn faster by adapting weights pre-trained on a different data set. This neural network is designed to require little of the pre-processing that traditional hand-engineered algorithms required simply in neural networks use their learned weights to predict a classification for an input image [7, 8].

We show that (Fig. 1) using a dataset of optical coherence tomography images, we can characterize age-related macular degeneration and diabetic macular edema in a way that is comparable to that of human specialists. Transfer learning makes use of pre-trained weights and biases to help a deep learning neural network uncover better patterns and classify them more quickly. This approach may aid in the diagnosis and referral of treatable retinal diseases and eye-related ailments, enabling for early detection and treatment of illnesses and therefore improving patient medical care [9–12].

2 Literature Review

Diabetic, macular, DEMA, and age-related macular degeneration affect tens of millions of individuals worldwide. Numerous people may suffer irreparable eyesight loss if they do not seek medical help right away. Owing to a dearth of competent professionals in their region, many persons with these eyesight disorders are ignorant of their situation. At about the same time, many people in urban centers are unable to receive prompt medical attention due to large patient numbers and a scarcity of licensed specialists to match demand. Optical coherence tomography is an imaging system that has shown to be quite useful for assessment and monitoring blinded rental properties in recent years [13, 14]. When components of a pre-trained deep learning model are repurposed in a new deep learning model, this is known as transfer learning. Generalized information can be transferred across the two models if they are designed to accomplish similar tasks. This method of deep learning development saves time and money by reducing the amount of labeled data needed to train new models [15–17]. It has become a more significant aspect of deep learning's evolution, and it's progressively being employed as a development tool. Deep learning is getting more and more prevalent in today's environment. In a variety of industries, deep learning algorithms are being employed to execute hard jobs. Advertising campaigns can be fine-tuned for a higher return on investment, network performance could be improved, and speech recognition systems can be advanced. The continuous improvement of these models will rely heavily on transfer learning. There are many various types of machine learning algorithms, but supervised machine learning is among the most prominent [18, 19]. To train models, this class of machine learning algorithm utilizes tagged training data. Accurately labeling datasets necessitates expertise, and training algorithms [20–22] are frequently resource-intensive and time-consuming. Transfer learning offers a solution to this issue, and as a result, this is becoming a popular deep learning technique.

3 Proposed CNN Model for Facial Expression Recognition

A convolutional neural network, deep neural network, and transfer learning-based model are being used for our project. It, along with, specifications of dataset used and the architecture of neural network designed, is discussed as follows.

3.1 Convolutional Neural Network (CNN)

Convolutional neural networks (CNN) are one of the kinds of deep neural networks that are frequently used for analysis and perusal of visual data (Fig. 2). Analysis of clinical images for diagnosis, computer vision techniques for image and video perusals, image classification, recommendation systems, NLP or natural language processing, and financial or time series data analysis are all areas where CNN can be used. Retinal Optical Coherence Tomography is a deep learning approach for image clasification. It is based upon CNN and Transfer Learning methods. In this method, an image as input can be given to a model, assign trainable weights and biases to the various parts of the deep neural network, and distinguish one illness from another.

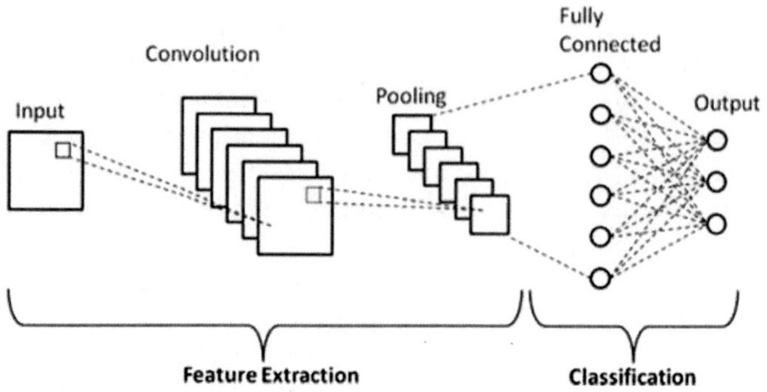

Fig. 2 Image representing a convolutional neural network [19]

CNN generally outperforms other deep learning algorithms both in terms of time taken for training and efficiency of output.

3.2 *Transfer Learning*

Transfer learning is the process of using attribute embeddings from a previously trained model to avoid having to train a new model from the ground up (Fig. 3).

Pre-trained models are typically trained on large datasets, which are a common standard in the field of deep learning and computer vision. The weights calculated by the algorithms can be used to various image processing applications.

These algorithms can be used to make accurate predictions on assigned challenges directly or as part of the training process for a new model. When pre-trained models are used in a new model, the training time and generalization error are reduced. When

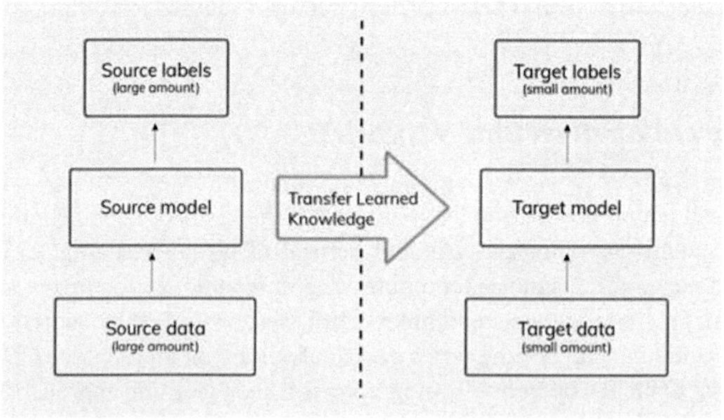

Fig. 3 Transfer learning approach

you just have a tiny training dataset, transfer learning comes in handy. We can utilize the weights from the pre-trained models to generate the weights of the new model in this situation.

3.3 Dataset Specifications

The dataset (Table 1) used for the classification of retinal optical coherence tomography images uses the data from Mendeley data archive. The dataset has the following directory specification (Refer to GitHub link). Dataset is available in a directory named "OCT2017".

```
import os

for root, dirs, files in os.walk("OCT2017"):
    print(f"There are {len(dirs)} directories and {len(files)} files in {root}")

There are 3 directories and 0 files in OCT2017
There are 4 directories and 0 files in OCT2017/train
There are 0 directories and 26315 files in OCT2017/train/NORMAL
There are 0 directories and 11348 files in OCT2017/train/DME
There are 0 directories and 8616 files in OCT2017/train/DRUSEN
There are 0 directories and 37205 files in OCT2017/train/CNV
There are 4 directories and 0 files in OCT2017/test
There are 0 directories and 242 files in OCT2017/test/NORMAL
There are 0 directories and 242 files in OCT2017/test/DME
There are 0 directories and 242 files in OCT2017/test/DRUSEN
There are 0 directories and 242 files in OCT2017/test/CNV
There are 4 directories and 0 files in OCT2017/val
There are 0 directories and 8 files in OCT2017/val/NORMAL
There are 0 directories and 8 files in OCT2017/val/DME
There are 0 directories and 8 files in OCT2017/val/DRUSEN
There are 0 directories and 8 files in OCT2017/val/CNV
```

Dataset is distributed as follows:
The dataset is unbalanced at the DRUSEN class.
Here, the abbreviations are as follows:

CNV: Choroidal neovascularization.
DME: Diabetic macular edema.
DRUSEN: Drusen accumulation in retina.
NORMAL: Normal retinal without any disease.

Some of the images of the dataset is shown as below (Fig. 4):

Table 1 Dataset description

Sets available →	Training set	Testing set	Validation set
Classes of images ↓			
CNV	37,205	242	8
DME	11,348	242	8
DRUSEN	8616	242	8
NORMAL	26,315	242	8

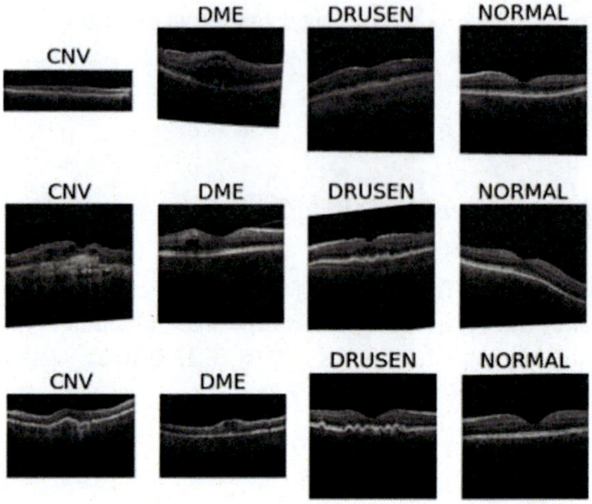

Fig. 4 Example images from the dataset for all classes

Table 2 Model description

Total params	17,673,823
Trainable params	0
Non-trainable params	17,673,823

3.4 Proposed Model Architecture

The model used is a transfer learning model that uses a pre-trained model named EfficientNetB4 which will have the following specifications (Table 2):

The model is non-trainable because it is a pre-trained classification model. We are using this model's weights and biases and adding on to it extra layers for training to customize it for our model.

The pre-trained model "EfficientNetB4" is used as a layer in our model. The architecture of our model is shown below:

4 Implementation of Proposed CNN Model

The model architecture is shown as follows:

The input of the model is a (224, 224, and 3) vector which is an image vector, resized to match the input of the network, and the output is a (1, 4) vector representing the one-hot encoded value of the classification, i.e. the output class (Fig. 5).

The base model used (EfficientNetB4) has the following structure. (Being an extremely big model, only a part of it is shown).

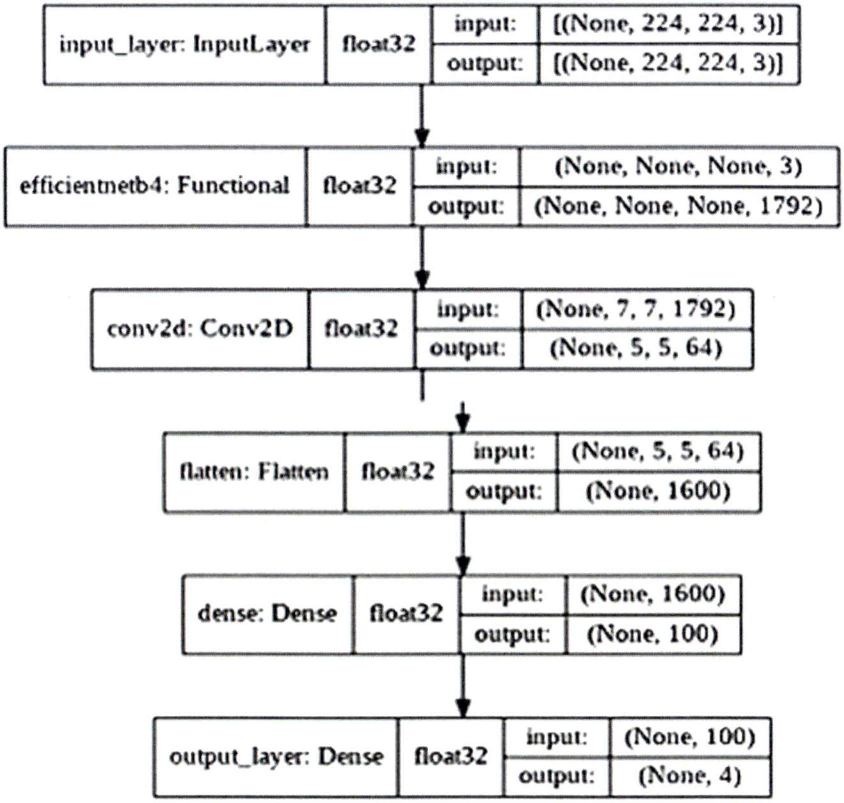

Fig. 5 Huge model of EfficientNetB4

5 Experimental Results

All the parameters use the concept of true positives, true negatives, false positives, and false negatives. These values are determined after plotting the confusion matrix of the evaluation or model's predictions. The confusion matrix of our transfer learning model is shown below (Fig. 6). Vertical columns represent predicted results based on the input images given, and horizontal rows represent actual label of the image given as an input.

The classification report of the model is shown below (Table 3) and (Fig. 7, 8, and 9).

The loss and accuracy curves of the model are shown below:

The precision and recall curve of the model are shown below:

From the images, its clear that the training accuracy of the model is reaching 98% while the validation accuracy of the model is nearly 1. It also shows that the training loss of the model is quite low and that of the validation loss is nearly zero. The graphs of precision and recall show a steep increase of both in training and validation which

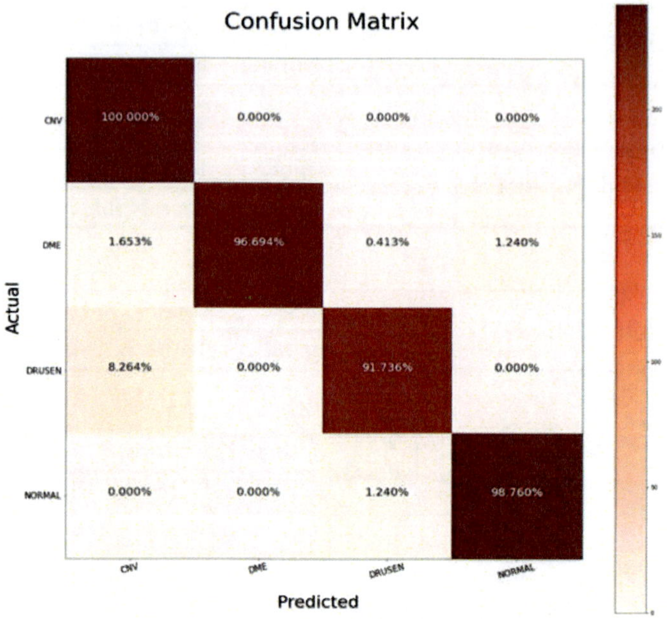

Fig. 6 Confusion matrix for the predictions on the testing set

Table 3 The F-1 scores of the classification model are shown below for different classes

	Class names	F1-scores
3	NORMAL	0.987603
1	DME	0.983193
0	CNV	0.952756
2	DRUSEN	0.948718

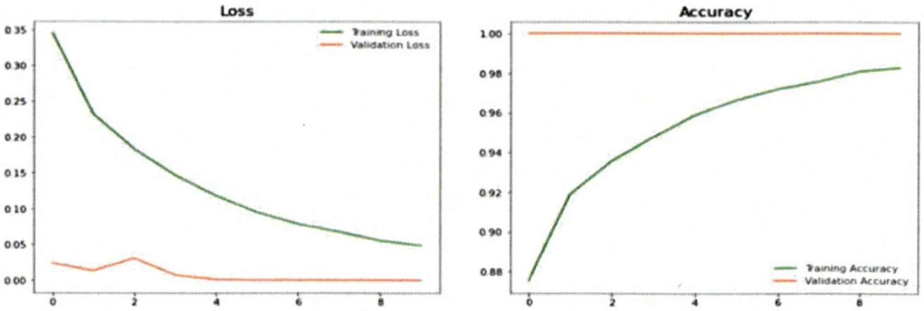

Fig. 7 Training and validation—accuracy and loss

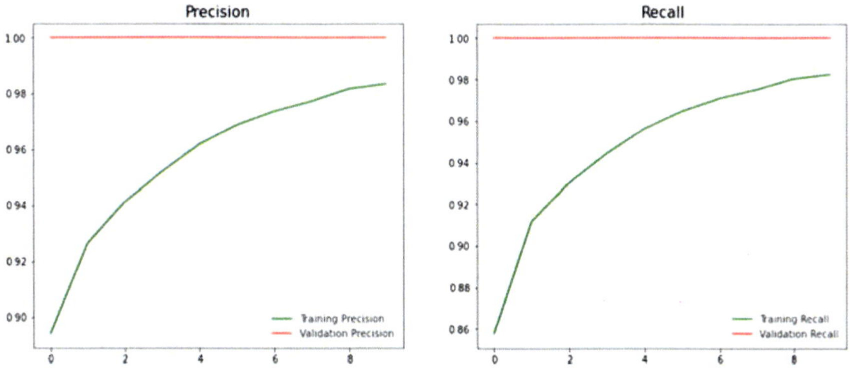

Fig. 8 Training and validation—precision and recall

Fig. 9 Prediction by model on an unseen image

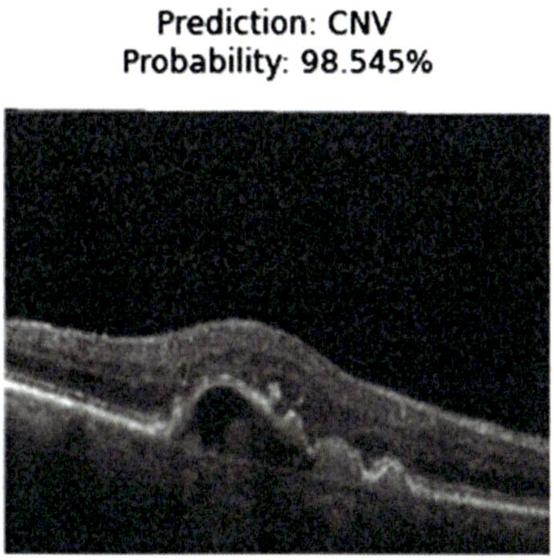

means that the model is quite good in its job. A prediction is done on the model (real-time) shows that it is performing correctly on the required input.

6 Conclusion

Retinal disease classification is a way of identifying the type of retinal disease a patient is afflicted with and the probability of the occurrence of this disease. Our model gives a training accuracy of 98.3% and a testing accuracy of 96.8%. Additionally, the model gives a training precision of 98.34% and a training recall of 98.24%. Testing the model leads us to get a testing precision of 96.9% and a testing recall of 96.8%. These numbers are encouraging and let us safely assume that the model is capable enough to give good results in the production and real life. Furthermore, the darker primary diagonal of the confusion matrix gives us more confidence in the built model. We hope to achieve greater accuracy and better attributes by doing further research and experimentation. We hope to employ many different types of datasets of the eye diseases which will help us attain better accuracy and provide reliable identification of more diseases. We also hope to build a robust model to get better results also on noisy images.

References

1. Kermany DS, Goldbaum M, Cai W, Anthony Lewis MXia H, Zhang K (2018) Identifying medical diagnoses and treatable diseases by image-based deep learning
2. Bressler NM, Cabrera DeBuc D, Burlina P (2019) Deep learning based retinal OCT segmentation. https://arxiv.org/pdf/1801.09749.pdf
3. Duan, Xi D, Zhu Y, Zhu H, IEEE, Hui Xiong, Fellow, IEEE, and Qing He
4. Kshitiz K et al (2017) Detecting hate speech and insults on social commentary using nlp and machine learning. Int J Eng Technol Sci Res 4(12):279–285
5. Pan SJ, Yang Q (2010) A survey on transfer learning. IEEE Trans Knowl Data Eng 10:1345–1359
6. Wang Z, Dai Z, Poczos B, Carbonell J (2019) Characterizing and avoiding negative transfer. In: Proc. IEEE conference on computer vision and pattern recognition. Long Beach 11293–11302
7. Kumar S, Dubey S, Gupta P (2015) Auto-selection and management of dynamic SGA parameters in RDBMS. In: 2015 2nd international conference on computing for sustainable global development (INDIACom), pp 1763–1768
8. Huang J, Smola AJ, Gretton A, Borgwardt KM, Sch¨olkopf B (2006) Correcting sample selection bias by unlabeled data. In: Proc. 20th annual conference on neural information processing systems, Vancouver, pp 601–608
9. Sugiyama M, Suzuki T, Nakajima S, Kashima H, Bnau P, Kawanabe M (2008) Direct importance estimation for covariate shift adaptation. Ann Inst Stat Math 60(4):699–746
10. Tian J, Varga B, Tatrai E, Fanni P, Somfai GM, Smiddy WE, DeBuc DC (2016) Performance evaluation of automated segmentation software on optical coherence tomography volume data. J Biophotonics 9(5):478–489
11. Klein R, Klein BEK (2013) The prevalence of age-related eye diseases and visual impairment in aging: current estimates. Invest Ophthalmol Vis Sci 54(14)
12. Biswas R et al (2012) A framework for automated database tuning using dynamic SGA parameters and basic operating system utilities. Database Syst J III(4)
13. Kumar SH (2013) E-COCOMO: the extended cost constructive model for cleanroom software engineering. Database Syst J 4(4):3–11
14. Bird AC, Bressler NM, Bressler SB, Chisholm IH, Coscas G, Davis MD, De Jong PTVM, Klaver CCW, Klein BE, Klein R et al (1995) An international classification and grading system for age-related maculopathy and age-related macular degeneration. Survey Ophthalmol 39(5):367–374
15. Bressler NM (2004) Age-related macular degeneration is the leading cause of blindness... JAMA 291(15):1900–1901
16. Khanchi I, Agarwal N, Seth P (2019) Real time activity logger: a user activity detection system. Int J Eng Adv Technol 9(1):1991–1994
17. Bhushan A, Rastogi P (2017) I/O and memory management: two keys for tuning RDBMS. In: Proceedings on 2016 2nd international conference on next generation computing technologies, NGCT 2016 7877416, pp 208–214
18. Jaffe GJ, Martin DF, Toth CA, Daniel E, Maguire MG, Ying G-H, Grunwald JE, Huang J, Comparison of Agerelated Macular Degeneration Treatments Trials Research Group et al (2013) Macular morphology and visual acuity in the comparison of age-related macular degeneration treatments trials. Ophthalmology 120(9):1860–1870

19. https://www.upgrad.com/blog/basic-cnn-architecture/
20. Khanna A, Sah A, Choudhury T (2020) Intelligent mobile edge computing: a deep learning based approach. In: Singh M, Gupta P, Tyagi V, Flusser J, Ören T, Valentino G (eds) Advances in computing and data sciences. ICACDS 2020. Communications in Computer and Information Science, vol 1244. Springer, Singapore. https://doi.org/10.1007/978-981-15-6634-9_11
21. Sille R, Choudhury T, Chauhan P, Sharma D (2021) A systematic approach for deep learning based brain tumor segmentation. Ingénierie des Systèmes d'Information
22. Vashishtha P, Choudhury T (2020) Procedia computer science, 2020, DMAPS: an effective and efficient way for the air purification of the outdoors:(Deep-mind Air Purification System for a smart city)

Deep Learning-Based Emotion Recognition Using Supervised Learning

Mayur Rahul, Namita Tiwari, Rati Shukla, Mohd. Kaleem, and Vikash Yadav

Abstract Facial emotion recognition using deep learning is getting lot of attention by the researchers in the past years. Since the images are nonlinear and noisy in nature, it is very difficult to develop an intelligent system capable of higher accuracy to recognize facial emotion. In this research, we introduced an intelligent framework for facial emotion recognition that used deep learning for feature extraction and random forest for classification. The extracted features obtain from the deep learning are used further for the classification to identify different types of emotions. The results are obtained from the three publicly available datasets: AffectNet, CK+, and EMOTIC. We have also shown the effectiveness of our framework in comparison with the research where deep learning is used as feature extractor to recognize emotions. The results prove that the introduced technique can able to improve the emotion recognition rate on the different size datasets.

Keywords Brain-computer interface · Emotion recognition · Deep learning · Random forest · CK+ · EMOTIC

M. Rahul
Department of Computer Application, UIET, CSJM University, Kanpur, India

N. Tiwari
Department of Mathematics, School of Sciences, CSJM University, Kanpur, India

R. Shukla
GIS Cell, Motilal Nehru National Institute of Technology, Prayagraj, India

Mohd. Kaleem
ABES Engineering College, Ghaziabad, Uttar Pradesh, India

V. Yadav (✉)
Department of Technical Education, Kanpur, Uttar Pradesh, India
e-mail: vikas.yadav.cs@gmail.com

P. Dutta et al. (eds.), *Emerging Technologies in Data Mining and Information Security*, Lecture Notes in Networks and Systems 490,
https://doi.org/10.1007/978-981-19-4052-1_25

1 Introduction

Understanding and reading emotion is an important piece of human edification. Computerized facial emotions identification has great implementations such as health care, human–robot interaction (HRI), behavioral analysis, driver emotion recognition, virtual assistance, emotion sensing, and smart home appliances [1]. Various modalities for emotions recognition like body gestures, face, speech, and biological signals [2]. Facial emotion recognition plays an important role in direct interpersonal communication. Automatic facial emotion recognition is based on important features which are extracted from the human faces. Appearance-based features and geometric features are the two main categories of facial features. Geometric features are related to some places of face landmarks, whereas appearances are basically texture variance of facial image. Deep learning and machine learning are also very popular field in the recognition of facial emotions. The goal is to achieve recognition similar to human beings. But, it is very difficult to overcome many challenges like selection of feature extraction, perceiving and conveying emotions, and some mixed emotions [3].

The main challenge in the human facial emotions recognition is to perform reliable and fast in wild environment. Emotion recognition is very complex due to mixed and dynamic nature of vast range of expression that human able to express that are different from seven basic facial emotions (Fig. 1) like disgust, anger, fear, joy, surprise, sadness, and neutral. The main focus of maximum research in this field is on these seven facial emotions recognition. The existence of various complex and mixed are very challenging task nowadays. The other problems like variations in head pose, illumination conditions, partial occlusions, inaccuracy in face authorization,

Fig. 1 Six basic facial expressions

subject identity, and camera lens resolution or distortions. The above-mentioned factors produce variations among classes. These factors will affect the performance of machine learning and deep learning algorithms. The large and dynamic set of data is required to learn all these variations by the given model [4].

This research attempts to produce solution to various problems related to some barriers in feature extraction from human faces like illumination variation, pose, and partial occlusion. To introduce a model based on deep learning and supervised learning for facial emotion recognition that works well with the problems like various head poses, illumination, and partial occlusion. Although large dataset is used for the purpose of training, still these problems may causes lot of problems for the researchers. Deep learning is used to overcome these problems. Capability of deep learning to adapt this type of situation is very good.

Deep learning is an emerging area in the field of computer science and its accuracy to recognize facial emotion recognition is significantly very good. Deep learning is the field of machine learning that used the concept of artificial neural networks. Deep learning is comparable to how peoples learn from the past examples. This can be obtained learning form the past similar type of tasks. The traditional base learning models are also called base learners, give a hypothesis that based on constant inductive bias. This type of problem, the search space for the hypothesis space is limited to the fixed bias. Deep learning is normally good for these types of situations. The deep learning refines and analyzes the base learners by creating changes in learning techniques. This is obtained with the help of optimization variables that are capable to train the base learners. The model will be easily adapted to similar type of task. Deep learning is very commonly used in the human facial emotion recognition. This is very useful in classification problem in general.

The problem we discuss so far is that the goal of our research is to recognize facial sentiments from face images with different illumination, head posture, and partial occlusions. This is very problematic for artificial intelligence to classify emotions under these problems unless they are not trained by large dataset. Our research consists of framework that used deep learning as feature extraction and random forest for classification because it can produce good results in emotion recognition without creating overfitting. This work is done to achieve significantly good accuracy in different datasets like AffectNet, CK+, and EMOTIC.

Assessment of our proposed model was processed on AffectNet, CK+, and EMOTIC. All the experiments are conducted on these datasets based on partially independent subject way. The meaning is that they contain around 54 subjects in common and their emotions as well in test and train sets. Assessment of each model is done with each datasets which is completely independent.

The main findings of this paper are as follows:

1. A method is introduced to recognize emotion from different facial images with pose, illumination, and occlusion. From the past research, no such research has been done for the facial emotion recognition using deep learning and supervised learning.

2. Despite of training is done in the dataset for still head poses and illuminations, our model is able to adapt all the variations like illumination, contrast, color, and head poses. That is, our model is able to give better results than traditional machine learning models.
3. Our model is also able to produce good results with less training datasets in the publicly available datasets like AffectNet, CK+, and EMOTIC.
4. Our model is able to detect emotion recognition with high accuracy and able to tag each of it.

The rest of the remaining paper is organized as follows: Section 2 discusses the related work in field of facial emotion identification, Sect. 3 explains the proposed model and its implementation, Sect. 4 presents the experiments and results of our proposed model, and finally concluded in Sect. 5 with some future suggestions.

2 Related Works

Previous systems for facial emotion identification from images with different problems like illumination, head posture, and partial occlusion are described in this part.

Robust principal component analysis (RPCA) is powerful to outliers and missing data that was used in emotion identification from images with partial occlusion [5]. After obtaining the extracted features using RPCA, geometric and gabor wavelet features were extricated. They assessed their features from three publicly available datasets such as MUG, JAFFE, and CK+. The black rectangles have been set in the face images due to absence of partial occluded images in the datasets. They got the accuracy of 90% for occluded images with the CK+ datasets with KNN, LDA, and PCA with the fusion of features from HOG, gabor wavelets, and HOG. They are able to consider only ERMOPI occluded naturally images but unable to recognize natural occlusion. Further, they were able to explain that RPCA is powerful for missing information only but not pose and illumination variation.

Cotter et al. introduced a framework based on sparse representation by combination of linear training images for each class of test image [6]. They were created the artificial occlusion using black and white box set on face images for experiment. Their main finding is that the accuracy depends on the box color set for the occlusion of the face images. Their results were not included in the presence of natural occlusion. Further, they were also not considered the variation poses and illumination in the face images. Ali et al. proposed the system based on the transformation after projection of occluded images at some specified angle to 1D functional signal [7]. They were able to extract occluded features by using second-order statistics or high-order spectra (HOS) called bi-spectra. These bi-spectra were able to extract features from contour and texture information of the face image. They used the CK+ dataset for their experiments. Their technique got the accuracy of 91.3% on upper part of the occlusion of face images [8]. Mao et al. introduced a Bayesian framework for multi-pose FER. Their method got the accuracy of 90.24% using CMU multi-pie dataset [9].

Deep learning-based systems give very favorable results for emotion recognition in real world problems. Mollahosseini et al. introduced a framework based on deep learning using two convolutional and four inception layers to obtain the result of 94.7% on CMU multi-pie database [10]. The given inception layer is used to reduce the problem of overfitting. They were also able to increase the computational requirements using sparse networks.

Lai et al. proposed a system based on generative adversarial network (GAN) for the formulation of face in multi-view images for emotion modeling and joint pose [11]. They used stacked convolutional auto encoder (SCAE) to monitor illumination and pose invariance. They used the SCAE for trained rigorously minimize facial pose and maximize image luminance. They used the various publicly available datasets to pre-train the model. Their model is able to obtain the accuracy of 79.5% in unseen pictures in the real-world environment.

Palaniswamy et al. introduced a framework based on various CNN having attention technique to overcome the problem of illumination, age variance, and simultaneous pose [12]. They applied the technique of transferring observation from occluded face area to clear face region. They evaluate their model in different wild databases like AffectNet and RAF-DB also in laboratory database like MMI, CK+, SFEW, and Oulu-CASIA. Their result is depending on facial landmark of the model. They reported their accuracy in 54.84% in AffectNet and 80.54% in RAF-DB for occluded images. Ngo et al. proposed a system to monitor the issue of data imbalance in facial images [13]. They fuse the two loss function in order to pre-trained dataset. They used the AffectNet dataset with eight emotions, obtain 60.70% accuracy.

Although machine learning models give very good result, their performance requires more specimens for training. Hence, in given introduced research in emotion identification in natural world problems where finite sets are present, deep learning is favorable as described in next section.

3 Proposed Methodology

We introduced a deep learning system based on CNN network to recognize the emotion in facial images. The improvement in deep learning network is based on the addition of more layers. More layers will responsible for the flow of gradient factors in the given network or improve regularization, especially for the classification issue in large number of classes. Although, for emotion recognition, we are able to show that only few convolution network is enough for few number of classes and able to obtain higher accuracy. We are also able to present better accuracy compare to existing accuracies.

The given face images, every part is not necessary for the efficient emotion recognition form the datasets. Our goal is to focus on some specific region of the facial images to get the better accuracy. For this reason, we have decided to incorporate CNN to focus on this particular region.

The feature extraction process of the system consists of six convolution layers with three layers followed by max-pooling layer and activation function of rectified linear unit. Further, it is followed by dropout layer and fully-connected layer. The localization network comprises of three convolution layer and two fully-connected layers. After the regression of given transformation parameters, all the inputs are translated in the form of grid data in warped form. The given spatial transformer part is very important part to focus on the most appropriate region of the face. We use the most popular affine transformation technique for warping of data from input to output.

The given system is trained by loss function using alternating direction method of multipliers (ADMM). The loss function in this research is combination of cross entropy and regularization term. These two after addition are used to train our model even if the dataset is small like CK+ and EMOTIC. We train each model for every datasets used in this model. We have also tried it with more layers, approximately 20, but still there are no changes found in the results. So, we have tried it with the less layers, i.e. six layers in this research.

4 Experiments and Results

In this research, we have analyzed the results obtained from various facial expression datasets includes AffectNet [14], EMOTIC [15], and CK+ [16]. We will give a brief introduction to all datasets used in this research.

AffectNet: AffectNet is the biggest database for facial affect in motionless images which covers both dimensional and categorical models. The database is composed with the help of 1250 expression-related labels in six desperate languages that are Farsi, Portuguese, German, Spanish, Arabic, and English. The database comprise of more than one million facial images and extracted face landmark regions.

CK+: The Extended Cohn-Kanade Dataset (CK+) is a public available database for emotion identification and action units. The dataset contains a total of 5876 tag images of 123 persons, where the given sequences range from peak to neutral expression. Images in the CK+ dataset are all taken with identical backgrounds, mostly 640×490 pixels and grayscale.

EMOTIC: EMOTIC is a database of facial images with persons in real environments, elucidate with apparent expressions. The EMOTIC database combines two distinct types of expression representation that contain a set of 26 separate classes, and with continuous dimensions dominance, arousal, and valence. The database comprise of approximately 34,320 annotated persons and 23,571 images.

The experiments have been conducted on the above datasets to present the performance of our model. For every dataset, we divide the entire dataset in train set, test set, and validation set. The three datasets are divided as 80% for train set, 10%

for test set, and 10% for validation set. We trained model for each datasets in our experiments, but we have maintain all the parameters and same in all the datasets.

We initialized some of its parameters as: Gaussian random parameters with SD as 0.007 and mean as 0, alternating direction method of multipliers (ADMM) as 0.003 and L2 regularization as 0.002. The average time to take training process is around 1.5–2 h. The performance of our model in all datasets is depicted in Tables 1, 2, and 3.

Table 1 Confusion matrix using AffectNet dataset

	Angry	Disgust	Fear	Happy	Contempt	Sad	Surprise
Angry	66.3	2.1	2.3	3.6	4.7	14.2	7.2
Disgust	5.4	57.5	15.4	4.6	3.2	3.7	10.2
Fear	5.2	5.1	63.7	6.7	4.9	7.3	6.1
Happy	2.5	3.1	2.1	80.3	7.3	2.1	2.6
Contempt	4.9	5.2	6.7	10.9	58.80	6.6	6.9
Sad	12.9	6.2	6.3	3.1	3.0	65.1	3.4
Surprise	4.4	6.2	10.3	5.6	5.2	5.9	62.4

Table 2 Confusion matrix using CK+ dataset

	Angry	Disgust	Fear	Happy	Contempt	Sad	Surprise
Angry	90.2	0	0	0	5.4	4.4	0
Disgust	0	100	0	0	0	0	0
Fear	0	0	92.1	3.1	0	0	4.8
Happy	0	0	2.6	97.4	0	0	0
Contempt	0	0	0	5.9	84.7	10.4	0
Sad	2.7	0	0	0	6.8	90.5	0
Surprise	0	0	0	0	2.6	0	97.4

Table 3 Confusion matrix using EMOTIC dataset

	Angry	Disgust	Fear	Happy	Contempt	Sad	Surprise
Angry	75.6	5.3	1.4	14.7	0.6	0.8	1.6
Disgust	8.4	79.4	5.2	2.1	2.3	2.0	0.6
Fear	9.1	2.5	80.3	1.1	0.6	3.8	2.6
Happy	3.5	5.8	14	68.1	3.1	2.8	2.7
Contempt	2.3	3.3	8.6	6.2	71.8	5.7	2.1
Sad	3.1	4.1	1.1	0.8	3.4	86.3	1.2
Surprise	2.4	3.6	3.1	3.5	4.9	4.2	78.3

5 Conclusion and Future Works

In this paper, a method is introduced to recognize emotion from different facial images with pose, illumination, and occlusion. From the past research, no such research has been done for the facial emotion recognition using deep learning and supervised learning. Despite of training is done in the dataset for still head poses and illuminations, our model is able to adapt all the variations like illumination, contrast, color, and head poses. That is, our model is able to give better results than traditional machine learning models. Our model is also able to produce good results with less training datasets in the publicly available datasets like AffectNet, CK+, and EMOTIC. Our model is able to detect emotion recognition with high accuracy and able to label each of them. The performance of our model for CK+ dataset is best as compare to AffectNet and EMOTIC datasets. In the future, we will incorporate more deep learning techniques to improve the results and also conduct experiments on other available datasets.

References

1. Yadav V, Mishra D (2020) Home automation system using Raspberry Pi Zero W. Int J Adv Intell Paradigms 16(2):216–226. https://doi.org/10.1504/IJAIP.2018.10017087
2. Singh S, Yadav V (2020) Face recognition using HOG feature extraction and SVM classifier. Int J Emerging Trends Eng Res 8(9):6437–6440. https://doi.org/10.30534/ijeter/2020/244892020
3. Rahul M, Yadav V (2019) Zernike moments based facial expression recognition using two staged hidden markov model. In: Advances in computer communication & computational sciences, proceedings of IC4S 2018, vol 924, pp 661–670, May 2019
4. Martinez B, Valstar MF (2016) Advances, challenges, and opportunities in automatic facial expression recognition. In: Advances in face detection and facial image analysis. https://doi.org/10.1007/978-3-319-25958-1_4
5. Cornejo JYR, Pedrini H (2017) Emotion recognition based on occluded facial expressions. I: Lecture notes in computer science (Including Subseries Lecture Notes in Artificial Intelligence and Lecture Notes in Bioinformatics). https://doi.org/10.1007/978-3-319-68560-1_28
6. Cotter SF (2010) Sparse representation for accurate classification of corrupted and occluded facial expressions. In: ICASSP. In: IEEE international conference on acoustics, speech and signal processing—proceedings. https://doi.org/10.1109/ICASSP.2010.5494903
7. Ali H, Hariharan M, Zaaba SK, Elshaikh M (2018) Facial expression recognition in the presence of partially occluded images using higher order spectra. In: Regional conference on science, technology and social sciences (RCSTSS 2016). https://doi.org/10.1007/978-981-13-0074-5_15
8. Tiwari N, Padhye S (2013) Analysis on the generalization of Proxy Signature. Secur Commun Netw 6:549–556
9. Mao Q, Rao Q, Yu Y, Dong M (2017) Hierarchical Bayesian theme models for multipose facial expression recognition. IEEE Trans Multimed. https://doi.org/10.1109/TMM.2016.2629282
10. Mollahosseini A., Chan D, Mahoor MH (2016) Going deeper in facial expression recognition using deep neural networks. In: 2016 IEEE winter conference on applications of computer vision, WACV 2016. https://doi.org/10.1109/WACV.2016.7477450

11. Lai YH, Lai SH (2018) Emotion-preserving representation learning via generative adversarial network for multi-view facial expression recognition. In: Proceedings—13th IEEE international conference on automatic face and gesture recognition, FG 2018. https://doi.org/10.1109/FG.2018.00046
12. Palaniswamy S, Tripathi S (2018) Emotion recognition from facial expressions using images with pose, illumination and age variation for human-computer/robot interaction. J ICT Res Appl 12(1):14–34. https://doi.org/10.5614/itbj.ict.res.appl.10.5614/itbj.ict.res.appl.2018.12.110.5614/itbj.ict.res.appl.2018.12.1.2
13. Ngo QT, Yoon S (2020) Facial expression recognition based on weighted-cluster loss and deep transfer learning using a highly imbalanced dataset. Sensors (Switzerland) 20(9):2639. https://doi.org/10.3390/s20092639
14. Mollahosseini A, Hasani B, Mahoor MH (2017) AffectNet: a new database for facial expression, valence, and arousal computation in the wild. IEEE Trans Affect Comput
15. Kosti R, Álvarez JM, Recasens A, Lapedriza A (2019) Context based emotion recognition using emotic dataset. IEEE Trans Pattern Anal Mach Intell (PAMI)
16. Lucey P, Cohn JF, Kanade T, Saragih J, Ambadar Z, Matthews I (2010) The extended cohn-kanade dataset (ck+): a complete dataset for action unit and emotion-specified expression. In: Proceedings of the IEEE computer society conference on computer vision and pattern recognition workshops (CVPRW), San Francisco, CA, USA, 13–18 June 2010

Comprehensive Review of Learnable and Adaptive Recommendation Systems

Swati Dongre and Jitendra Agrawal

Abstract Due to the dynamic and heterogeneous nature of the Web, it is increasingly difficult for users to choose amongst large amounts of data. For this reason, the modelling of users and access to personalised information becomes essential. Recommendation systems stand out as systems that aim to provide the most appropriate and efficient services by offering personalised recommendations to users. Traditional recommendation systems provide suggestions to their users with static approaches and do not include user preferences that change over time in suggestion strategies. In this study, a comprehensive review and comparison of recommendation systems that can adaptively develop suggestion approaches according to changing user preferences and learn user preferences are presented.

Keywords Collaborative filtering · Classification technique · Recommendation system · Social media · User preferences

1 Introduction

Recommendation systems can be defined as programmes that aim to offer the most suitable products or services for specific users by predicting the items that users may be interested in based on their interactions with each other and with the items in the system [1]. The main goal is to deal with information overload issues by removing personalised services from large amounts of data. The most crucial feature of recommendation systems is that whilst creating personalised recommendations for a user, they analyse the behaviour of other users and generate predictions for user preferences [2].

S. Dongre (✉) · J. Agrawal
Department of Computer Science, School of Information Technology, RGPV, Bhopal, India
e-mail: dswati31@gmail.com

J. Agrawal
e-mail: jitendra@rgtu.net

P. Dutta et al. (eds.), *Emerging Technologies in Data Mining and Information Security*, Lecture Notes in Networks and Systems 490,
https://doi.org/10.1007/978-981-19-4052-1_26

Recommender systems offer suggestions based on the preferences of users whilst browsing the Web. A suggestion is made by creating user profiles based on the past behaviour of users. Although recommendation systems are widely used in e-commerce platforms, their domains become increasingly common [3]. Recommender systems can be created using collaborative filtering [4], content-based filtering [5], and hybrid methods.

The collaborative filtering method generates suggestions based on the similarities between user preferences [4]. The collaborative filtering process works to identify users with similar preferences and present items that have not yet seen to new users, based on the evaluations of these users for items. Content-based filtering uses information in item contents to predict items that users may be interested [6]. Hybrid methods are applied by combining these approaches to eliminate the disadvantages of collaborative filtering and content-based filtering methods.

In the era of purposive Web mining, i.e. Web 4.0, recommendation systems widely used collaborative filtering method. The effectiveness of the collaborative filtering method depends on the algorithm used to identify users similar to the current user [7]. Various data mining methods and information retrieval methods are used for this purpose. The adaptation of recommendation systems to changing user needs is an evolving issue. In a constantly changing and evolving Web environment, presenting the relevant content at the right timing through a recommendation mechanism is one of the most critical factors. It is possible to obtain high-prediction performance with various model-based methods used in recommendation systems. However, the disadvantage of such model-based methods is that they require long training periods. Besides, user profiles should be updated adaptively according to the changing user evaluations due to the static nature of model-based methods [8].

With the development of recommendation system technologies, systems have been developed for different application areas. Most of the recommendation system applications created using hybrid systems where the advantages of different suggestion submission approaches are discussed. Since the hybrid systems used are developed with static approaches, they cannot change their strategies at runtime and according to user behaviour. Since user behaviours and preferences may vary, recommendation systems need to update their user profiles and provide suggestions dynamically at runtime [9].

This paper presents a comprehensive analysis regarding the state of the art of adaptive learning via recommendation system over Web. The rest of the article organised as follows: Sect. 2 presents an overview of the recommendation system, Sect. 3 covers literature survey over different domains . Section 4 presents detail description of adaptive recommendation system, and subsequently, Sect. 5 presents parameter of adaptive learning, and Sect. 6 covers application of adaptive recommendation system, and finally, Sect. 6 concludes the recent research and potential gap for upcoming research.

2 Literature Survey

In recent studies on recommendation systems, researchers use content-based filtering and collaborative filtering techniques to analyse user preferences with data mining methods and conduct studies on personalised information provision systems [10]. The clustering approach, one of the data mining methods, is widely used in recommendation systems. Clustering algorithms enable the creation of clusters with objects with similar features according to the determined variables and maximise the difference between the clusters formed by objects with different properties [11].

In the study conducted by [12], a recommendation system was designed that offers users the appropriate products according to the repetitive purchasing model. The number of iterations in the purchase for each product per user was used as a recommendation criterion in the system. The system has been implemented using user-based filtering, product-based filtering, and analysis of products purchased simultaneously with association rules. Using a company's sales data for a year and a half, the proposals' performance was analysed. Better results were obtained from the product-based collaborative filtering method than the user-based collaborative filtering method.

Troussas et al. [13] present an hybrid proposal approach was developed that offers mobile product and service recommendations by combining fuzzy clustering techniques with user-based and product-based filtering techniques. It aims to deal with the sparseness problems experienced in recommendation systems, increase the accuracy of estimates, and eliminate the uncertainties in customer data. The product-based collaborative filtering method was used to create user-product matrices, and recommendations were made with user-based collaborative filtering methods based on this matrix. It is aimed to solve the sparseness problem with the completed user-product matrix. Fuzzy clustering techniques are used to generate suggestions using linguistic variables and ambiguous information that define user preferences. The developed method is applied to Fuzzy-based Telecom Product Recommender System (FTCP RS) software. The results of the analysis showed that the suggestions offered by the FTCP RS application to the users after the proposed approach gave more effective outcomes.

Li et al. [14] present an LORI system was developed based on location-based user evaluations. Whilst traditional recommendation systems do not consider the positional characteristics of the users, the proposed method uses an approach that offers recommendations based on the location information of the users. The LARS system presents its recommendations by combining them under a single system using three types of location-based evaluation. These evaluations are provided by the user, location, evaluation score, and product quartet. The first type of evaluation is those whose position information is known, but the product information is unknown-for example, users who evaluate books at home. The second is the evaluations where the location is not precise-for example, user reviews for an unidentified restaurant. The third is the evaluations where location and product information are available. These evaluations are defined by the user, user location, assessment, product, and product location five. For example, location-specific user reviews for a restaurant whose location is known. The system was designed using location-based user and product information and content-based filtering methods. Analysis results using the four square social network and MovieLens movie recommendation system data showed that the proposed system offers nearly twice as successful suggestions.

Carmel et al. [15] present an interactive hybrid recommendation system that generates suggestion estimates using collaborative filtering methods with information obtained from multiple social and semantic Web sources such as Wikipedia, Facebook, and Twitter is presented. In the developed system, an interactive user interface is presented to explain the suggestion process and reveal the preferences of end-users, unlike traditional recommendation systems. In order to determine the quality of the recommendations, evaluations regarding the comparison results of the new system using interactive and non-interactive hybrid strategies and social and semantic Web APIs are presented.

Wu and Gou [16] present an comprehensive survey of neighbourhood-based methods, which are amongst the collaborative filtering approaches that maintain their popularity due to their simplicity, efficiency and providing accurate and personalised recommendations, are presented to find solutions to the problems they encounter in product recommendations. The main features, as well as the advantages of such methods, are described. Finally, issues of lack of information about users and limited coverage issues often observed in large business advice systems are discussed, and solutions are offered to overcome these problems. It has been observed that dimension reduction methods eliminate the sparseness and coverage problems in recommendation systems.

Hao and Cheang [9] present a recommendation system to reveal the importance and effectiveness of tag and timing information to predict user preferences and create a practical resource recommendation model. In the study, a recommendation model was developed using tagging and timing information in online social systems to obtain user evaluation. Source suggestions were made using the tags that users have created on social platforms in the past and user tags that change according to time. The study aims to achieve better performance results by adding time and tag information to collaborative filtering methods to provide personalised recommendations on platforms that offer social tagging.

3 Recommendation System

Due to the development of Web technologies and the increase in the amount of information available on the Web, it becomes an important issue to support people's decisions with suggestions. It can be an excellent strategy to give users this support in the Web environment according to the choices made by other users who think like them. Recommendation systems support users in selecting the most useful from the growing body of information and facilitating access to information [1, 3, 5] .

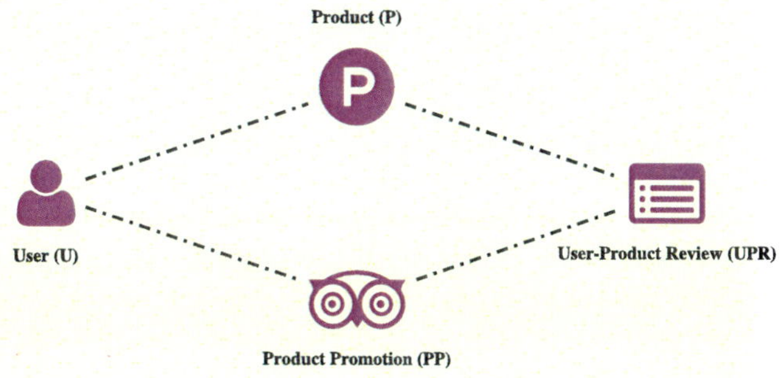

Fig. 1 Recommendation system

With the development and widespread use of Internet technologies, recommendation systems have become systems that we encounter in most areas of our lives. Suggestion advertisements on Websites and product suggestions offered on online shopping sites can be given as examples of recommendation systems. Recommendation systems have recently become an essential field of study as they can deal with information overload problems [17].

The recommendation systems are primarily aimed at users who have no experience with the recommendations to be presented. Popular online stores such as amazon.com offer personalised recommendations to their users in this context. Since the suggestions are personalised, users or user groups can receive different suggestions. On the other hand, magazines or newspapers can often provide more simple impersonal recommendations about current events. Although they can be valuable and effective in some situations, non-personalised systems are not addressed in recommendation systems research.

In its simplest form, personalised recommendations are presented with various lists of items. In this ranking, recommendation systems try to predict the most suitable products based on the user's preferences and constraints. These estimates are derived from the voting values that recommendation systems receive from users or interpretation of users' behaviour.

In general, each recommendation system follows specific processes, as shown in Fig. 1, to produce product recommendations. Suggestion submission approaches can be classified according to the source of information used by recommendation systems. Three possible sources of data can be considered as input to the recommendation process. Available sources can be user data (demographic characteristics), item data (keywords, product categories), and user-product reviews [8].

4 Adaptive Learning System

Although various recommendation system algorithms have been proposed to provide users with recommendations, some challenges still need to be resolved for recommendation systems to work efficiently. The recommendation systems have a cold start problem because there are no indicators in new users and new products [18]. Besides, recommendation systems are seen as black boxes because they do not provide any information about the logic or justification of offering suggestions to the user. This prevents comprehension of the results suggested by the recommendation systems and can lead to trust issues when the recommendations fail. Also, this approach does not allow users to feedback. Since it is difficult to predict the changing interests of the users, different approaches that can guide the process need to be developed. These approaches promise to solve high-risk applications such as increasing product variety and innovation of recommended results and adapting adaptively to changing user preferences [19].

Recommendation systems are systems that offer recommendations using the information that users obtain based on their interactions. The concept of learning and adaptability in recommendation systems, on the other hand, refers to the ability to provide recommendations based on independent parameters and user interactions. Adaptive recommendation systems can directly or indirectly record changes in the domain of users and update their recommendations [20].

From the perspective of human psychology, user dynamics should be taken into account to develop advice systems that can offer recommendations to each other in real life. Since the human mind has a very complex and challenging to interpret the structure, characteristics that will reveal various personality traits must be adapted to recommendation systems. Psychological factors such as transparency and reliability should not be ignored for the recommendations to be as realistic as real-life recommendations [21].

In systems where user interactions exist, the ultimate goal is user satisfaction. The increase in the amount of information obtained from user interfaces about users makes user modelling much more accessible. For example, in systems that directly ask users for their evaluations about products with a simple interface, the number of users making evaluations is deficient. In modern systems that acquire user interactions dynamically, the system experience is increased with the user data obtained indirectly, such as the click information of the users on the Web page and the pages they have viewed. Another factor affecting user satisfaction in recommendation systems is the accuracy of the recommendations presented. The accuracy of the suggestions offered provides solutions that can meet user needs and expectations for each user. Suggestion accuracy is provided by an efficient algorithm and enriched data [22].

Adaptive advice systems record content changes resulting from user interactions on the system side and make updates and changes. Such changes affect the recommendations offered to the users and the applicability of the system. However, concepts such as the context, diversity, and innovation of the items to bent as suggestions directly affect the relevance of the recommendation systems [23].

5 Parameter for Adaptive and Learnable Advice Systems

Collaborative filtering, content-based filtering, and hybrid methods used in classical recommendation systems are based on estimating the liking of different users for different products. Content-based filtering methods create user profiles based on the items that users have previously evaluated [18]. The disadvantage of this method is that the suggestions offered are limited to things similar to items that users liked before. The collaborative filtering method presents users' preferences with similar tastes as suggestions based on the users' past evaluations of the items. It is aimed to increase the quality of the recommendation by combining the beneficial aspects of the hybrid methods and collaborative and content-based filtering methods under different strategies [19]. Since these approaches used in traditional recommendation systems cannot update user models according to content information and time, they lack different types of recommendations. Besides, traditional methods are limited to the existing user data and the algorithm used since they are built on the assumption that users' interests will not change over a certain period. However, this assumption becomes invalid in the rapidly developing Web environment and fast-paced life. Since traditional recommendation systems cannot keep up with changes in user preferences, dynamic parameters should be included in the proposal submission process. These parameters are temporal context, innovation, diversity, dynamic environment, and material characteristics.

5.1 *Temporal*

Temporal changes mean that the preferences of the recommendation system users regarding the products change over time. The product perception of the users and the popularity of existing products lead to the emergence of new choices [24]. Likewise, customer tendencies may also develop in the direction of re-examining the effects they have seen before. For this reason, the modelling of temporal dynamics is essential to increase the accuracy of the recommendations offered by the recommendation systems and to develop general customer preference profiles. However, material changes pose various challenges in platforms where multiple users and products intersect. Changing user dynamics affect each other, and changes occurring on data related to each other necessitate updating user profiles [25].

5.2 *Real Time*

The developing Web technologies, institutions, and organisations can directly offer more than one digital content to their users. However, the biggest challenge faced by institutions and organisations is optimising the recommendations to be presented by identifying the appropriate users rather than the lack of content. Personalised suggestions should be presented with an approach that can attract all users to increase customer satisfaction and increase user loyalty. With personalised recommendations, it is aimed to collect information about site users, manage existing content, analyse user behaviour and interactions, and provide users with the right content [26]. However, this situation causes cold start problems in systems that offer news suggestions as an example, due to rapidly changing contents and changing new users.

Various hybrid recommendation systems and adaptive advice systems have been developed to overcome these problems stemming from real-time dynamics. Cold start problems experienced by collaborative filtering methods for new users and items can be resolved using content-based filtering methods or updating user profiles adaptively. Although hybrid methods eliminate some of the limitations of collaborative filtering and content-based strategies, they fall short in dynamic content and user profile modelling. Dynamic range means adding or deleting new items to the system and means that the popularity of the content changes over time with user interactions. For example, the newsworthiness of a newly added content to a Website offering news suggestions is determined by the rate of users clicking on this content over time [9].

5.3 Contextual

Contextual features such as days, weather, location, and seasons can influence user choices. Traditional recommendation systems do not consider contextual features at the time of submitting suggestions and focus only on users and elements. However, in systems offering movie suggestions as an example, it is not sufficient to consider only users and items. Still, also contextual information should be included in the proposal submission process. In the recommendation systems that offer holiday suggestions using the concept of temporal context, the suggestions offered vary from season to season. For example, users can choose to read the stock market news on weekdays and read reviews about movies on weekends. Context information is also frequently used in COVID-19 for mapping recommendation in patient [27].

5.4 Diversity

The concept of diversity refers to the offer of different products with similar features to the users. Diversity is the degree to which the items in a group differ from each other. The diversity of the things in the recommendation systems and presented as suggestions facilitate user interests and adaptive profile modelling. Studies in the literature provide the balance between accuracy and a variety of recommendations [11].

5.5 Innovation

Traditional recommendation systems can generate ineffective recommendations for users looking for new items, as they generally offer the most liked or popular products as recommendations. Collaborative filtering methods suggest fewer new items than content-based filtering methods but have a higher perception of user quality. Since there is no previous user evaluation of newly added items in traditional recommendation systems, these items are unlikely to be presented as suggestions. In adaptive advice systems, a match is made between the contents of newly added items and continuously updated user models. In this way, it is aimed to increase the quality and diversity of the suggestions with freshly added items [28].

6 Application of Adaptive Recommendation System

Adaptive learning via recommends systems aim to extract demandable items that may interest users amongst the vast amounts of data as suggestions to their users. Recommendation systems can be used in many adaptive learning areas such as e-commerce, e-business, e-learning, e-tourism, and e-resource services.

6.1 E-commerce

Recently, many e-commerce systems have been developed to guide individual customers. E-commerce systems commonly receive feedback from users about their products with the help of scoring scales. For example, iTunes receives a score between 1 and 5 for songs or albums purchased from users. These evaluation data can then be used to make recommendations to users [6].

Another method used to establish user-product relationships in e-commerce systems is tagging. For example, the MovieLens Website expects its users to provide brief views and comments about the movies. Large commerce sites such as Amazon and eBay also use referral systems to assist their customers with their purchases. Websites like this benefit from the demographic characteristics of the users or their past purchasing behaviour [29].

Wang et al. [30] presents music recommendation system was developed using social media information and acoustic-based content information of music. Hypergraph structure is used to model the relationships between music pieces and users. Whilst a connection can be established between each edge and two elements in classical graphs, it is aimed to establish a connection between the item sets in the proposed method.

6.2 E-business

E-business systems can be divided into individual customer-oriented systems (B2C) and systems that aim to offer products and services to business users (B2B). E-business systems use recommendation systems to maintain current business methods more quickly and efficiently. It aims to provide business partners or users with news about their services and products, make product comparisons, and offer product and service recommendations to consumers [31]. Zhang et al. [32] present a fuzzy logic-based telecom products recommendation system was developed. With the developed system, customers are offered service plans and recommendations as packages, and at the same time, explanations are made regarding the provided suggestions. In particular, a hybrid system has been developed using product-based collaborative filtering method, user-based collaborative filtering method, and fuzzy clustering methods to eliminate uncertainties in customer data.

6.3 E-learning

Developed based on traditional e-learning systems and e-learning advice systems have become increasingly popular for educational institutions since the early 2000s. These systems generally aim to assist students in selecting courses, subjects, and learning materials that interest students and create a classroom environment and online discussion [10]. User-oriented e-learning advice system was developed to provide personalised support for Web-based education systems. The developed system has been designed by considering student needs with the help of field experts in the e-learning cycle. Data such as student needs, age, and class status were used to generate personalised recommendations through knowledge-based methods [7, 33].

6.4 E-tourism

E-tourism systems aim to offer the most suitable offers to the increasing holiday and tourism options. Advisory systems in this area have been carried out for transportation, restaurants, accommodation, and places to visit. These systems offer recommendations depending on the demographic characteristics of the users, past preferences, and temporal variables [8]. Trip planning recommendation system was developed to provide personalised tour plans for tourists who want to visit the Portuguese city of Oporto. A hybrid system has been designed to overcome scalability, rarity, and new users and products in recommendation systems. The proposed approach has been achieved by combining collaborative filtering and content-based filtering methods with clustering algorithms, relational classification algorithms, and fuzzy logic [34].

6.5 E-source

E-source systems are systems where movies, music, videos, Web pages, documents, and contents uploaded by users are presented as suggestions. E-resource recommendation systems offer content shared by other users as suggestions to users according to their relevance [10]. Nguyen et al. [19] in 2013, a new method was presented to provide better Web page suggestions by integrating Websites' domain and Web usage information through semantic development methods. In the developed system, ontology structure was used to represent domain knowledge. Also, a conceptual prediction model has been designed to integrate domain knowledge and Web usage information. The proposed method was compared with the results obtained from Web use mining, and it was seen that higher performance was obtained.

7 Conclusion

Recommender systems aim to help users choose products that may be useful to them amongst large amounts of data. Different methods such as collaborative filtering, content-based filtering, and demographic filtering are used in recommendation systems. Most of these methods use static approaches in the proposal stage. Traditional recommendation systems used in real-world applications are based on bilateral relationships between products and users. User profiles are created through direct or indirect feedback received from users about the products, and suggestions are made based on the choices they have made in the past. It is based on the assumption that user preferences will not change over time. However, the interaction of the users with the system and each other, the contents and contexts of the products to be offered as suggestions should be included in the process.

In this paper, recommendation systems based on updating existing user profile models according to dynamic changes in user preferences and learning user preferences were examined. It has been demonstrated that the hybrid methods based on learning user profile models according to changing dynamics, adaptive updating of the close user set used in classical collaborative filtering methods, and dynamically updating prediction strategies achieve higher success rates. The studied studies have been hybridised using classical methods and data mining, machine learning, and artificial intelligence techniques. It has been observed that the advisory systems examined have primarily eliminated the problems experienced by traditional systems such as sparseness, scalability, and cold start, and the recommendations presented have higher accuracy. accuracy.

References

1. Patel K, Patel HB (2020) A state-of-the-art survey on recommendation system and prospective extensions. Comput Electron Agric 178:105779
2. Li Z, Cheng B, Gao X, Chen H, Chen G (2021) A unified task recommendation strategy for realistic mobile crowdsourcing system. Theor Comput Sci 857:43–58
3. Stöckli DR, Khobzi H (2021) Recommendation systems and convergence of online reviews: the type of product network matters! Decis Support Syst 142:113475
4. Bhalse N, Thakur R (2021) Algorithm for movie recommendation system using collaborative filtering. Mater Today Proc
5. Kulkarni S, Rodd SF (2020) Context aware recommendation systems: a review of the state of the art techniques. Comput Sci Rev 37:100255
6. Hwangbo H, Kim YS, Cha KJ (2018) Recommendation system development for fashion retail e-commerce. Electron Commerce Res Appl 28:94–101
7. Santos OC, Boticario JG, Pérez-Marín D (2014) Extending web-based educational systems with personalised support through user centred designed recommendations along the e-learning life cycle. Sci Comput Program 88:92–109. (Software Development Concerns in the e-Learning Domain)
8. Kolahkaj M, Harounabadi A, Nikravanshalmani A, Chinipardaz R (2020) A hybrid context-aware approach for e-tourism package recommendation based on asymmetric similarity measurement and sequential pattern mining. Electron Commer Res Appl 42:100978
9. Hao PY, Cheang WH, Chiang JH (2019) Real-time event embedding for poi recommendation. Neurocomputing 349:1–11
10. Aher SB, Lobo LM (2013) Combination of machine learning algorithms for recommendation of courses in e-learning system based on historical data. Knowl Based Syst 51:1–14
11. Berbague CE, Karabadji NE, Seridi H, Symeonidis P, Manolopoulos Y, Dhifli W (2021) An overlapping clustering approach for precision, diversity and novelty-aware recommendations. Expert Syst Appl 177:114917
12. Tatiana K, Mikhail M (2018) Market basket analysis of heterogeneous data sources for recommendation system improvement. Procedia Comput Sci 136:246–254. (7th International Young Scientists Conference on Computational Science, YSC2018, 02-06 July2018, Heraklion, Greece)
13. Troussas C, Krouska A, Sgouropoulou C (2020) Collaboration and fuzzy-modeled personalization for mobile game-based learning in higher education. Comput Edu 144:103698
14. Li J, Liu G, Yan C, Jiang C (2019) Lori: a learning-to-rank-based integration method of location recommendation. IEEE Trans Comput Soc Syst 6(3):430–440
15. Vaisman CL, Gonen I, Pinter Y (2018) Nonhuman language agents in online collaborative communities: comparing hebrew wikipedia and facebook translations. Discourse Context Media 21:10–17
16. Wu Y, Gou J (2021) Leveraging neighborhood session information with dual attentive neural network for session-based recommendation. Neurocomputing 439:234–242
17. Amer AA, Abdalla HI, Nguyen L (2021) Enhancing recommendation systems performance using highly-effective similarity measures. Knowl Based Syst 217:106842
18. Da'u A, Salim N, Idris R (2021) An adaptive deep learning method for item recommendation system. Knowl Based Syst 213:106681
19. Wang S-L, Wu C-Y (2011) Application of context-aware and personalized recommendation to implement an adaptive ubiquitous learning system. Expert Syst Appl 38(9):10831–10838
20. Chen G, Zeng F, Zhang J, Lu T, Shen J, Shu W (2021) An adaptive trust model based on recommendation filtering algorithm for the internet of things systems. Comput Netw 190:107952
21. Zhang W, Wang B, Zhou L (2021) Analysis of text feature of English corpus with dynamic adaptive recommendation algorithm fused with multiple data source English language. Microprocess Microsyst 104075
22. Wang Y, Han L (2021) Adaptive time series prediction and recommendation. Inf Process Manage 58(3):102494

23. Moonen L, Binkley D, Pugh S (2020) On adaptive change recommendation. J Syst Softw 164:110550
24. Yuen MC, King I, Leung KS (2021) Temporal context-aware task recommendation in crowdsourcing systems. Knowl Based Syst 219:106770
25. Guo S, Wang Y, Yuan H, Huang Z, Chen J, Wang X (2021) Taert: triple-attentional explainable recommendation with temporal convolutional network. Inf Sci 567:185–200
26. Safran M, Che D (2017) Real-time recommendation algorithms for crowdsourcing systems. Appl Comput Inf 13(1):47–56
27. Lotfi T, Stevens A, Akl EA, Falavigna M, Kredo T, Mathew JL (2021) Getting trustworthy guidelines into the hands of global decision-makers and supporting their consideration of contextual factors for implementation: recommendation mapping of covid-19 guidelines. J Clin Epidemiol
28. George S, Lathabai HH, Prabhakaran T, Changat M (2021) A framework for inventor collaboration recommendation system based on network approach. Expert Syst Appl 176:114833
29. Wang K, Zhang T, Xue T, Lu Y, Na SG (2020) E-commerce personalized recommendation analysis by deeply-learned clustering. J Visual Commun Image Represent 71:102735
30. Wang R, Ma X, Jiang C, Ye Y, Zhang Y (2020) Heterogeneous information network-based music recommendation system in mobile networks. Comput Commun 150:429–437
31. Abumalloh RA, Ibrahim O, Nilashi M (2020) Loyalty of young female arabic customers towards recommendation agents: a new model for b2c e-commerce. Technol Soc 61:101253
32. Zhang Z, Lin H, Liu K, Wu D, Zhang G, Lu J (2013) Hybrid fuzzy-based personalized recommender system for telecom products, services. Inf Sci 235:117–129. (Data-based Control. Decision, Scheduling and Fault Diagnostics)
33. De Medio C, Limongelli C, Sciarrone F, Temperini M (2020) Moodlerec: a recommendation system for creating courses using the moodle e-learning platform. Comput Hum Behav 104:106168
34. Zhang X, Yu L, Wang M, Gao W (2018) Fm-based: algorithm research on rural tourism recommendation combining seasonal and distribution features. Pattern Recogn Lett

Autoencoder: An Unsupervised Deep Learning Approach

Sushreeta Tripathy and Muskaan Tabasum

Abstract With the advent of science and technology, it has been observed that autoencoder plays a vital role in unsupervised learning and in deep architectures for many tasks like transfer learning and other tasks. Learn efficient data coding in an unsupervised manner, a variety of artificial neural network is used, which is known as autoencoder. Here we present an entire overview of an autoencoder and learning about what actually autoencoder is and how are they used. Here we will learn about the architecture of autoencoder, their implementation, types of autoencoder like denoising autoencoder, sparse autoencoder, the use cases, and a lot more things.

Keywords Autoencoder · Unsupervised learning · Denoising encoder · Sparse encoder

1 Introduction

A special kind of neural networks wherever output is almost same as its input is known as autoencoder. They restrict the data inside a code of lower-dimensional form and then produced the output as its original form. The code has compact "summary" or "compression" of the input, which is called as the latent-space narration. Encoder, code and decoder are the three basic parts of autoencoder. The encoders compress the data and generate the code. Afterwards, the decoder regenerates the output by only using the same principle [1–3] (Fig. 1).

S. Tripathy (✉)
Department of CA, Siksha 'O' Anusandhan (DU), Bhubaneswar, Odisha, India
e-mail: sushreetatripathy@soa.ac.in

M. Tabasum
Department of CSIT, Siksha 'O' Anusandhan (DU), Bhubaneswar, Odisha, India

P. Dutta et al. (eds.), *Emerging Technologies in Data Mining and Information Security*, Lecture Notes in Networks and Systems 490,
https://doi.org/10.1007/978-981-19-4052-1_27

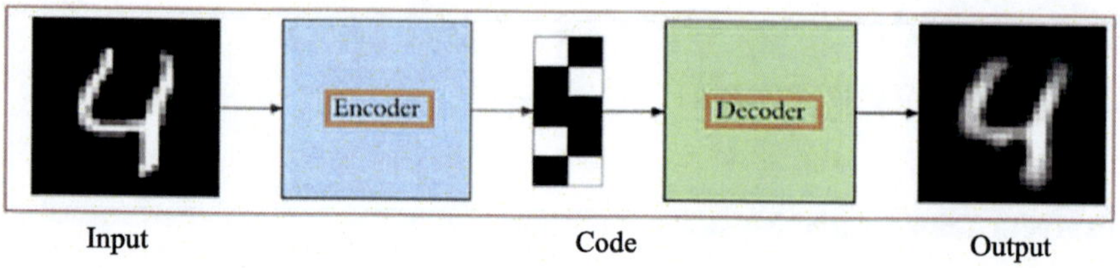

Fig. 1 Shows the first image is input, middle image is the code and the last image is the output. Three things play a very vital role in construction or making of an autoencoder, those are an encoding approach, decoding approach and to match the output with the target, a loss function is needed. Autoencoders are mainly an algorithm used for reducing and compressing of dimensions with some of the important properties [4–7]

1.1 Unsupervised

In this phase just we have to give the raw input data to train an autoencoder. Generally autoencoders are supervised itself because they produced their labels from the practise data. Hence, it is considered as an unsupervised learning approach because it do not need explicit labels to train on [8, 9].

1.2 Lossy

The output of the autoencoder will be a close but slightly degraded representation of the input. But if we want a lossless compression then we need to go for something else rather than an autoencoder.

1.3 Specific on Data

Autoencoders can meaningfully compress data. So we cannot forecast an autoencoder prepared on transcribed digits to compress photos. Since they acquire characteristics distinct to the given training data, they are separate from a conventional data compression algorithm like gzip. Dimensionality compression or feature learning was the most traditional application, but recently the concept of autoencoder has become broadly used for training generative models of data. Some of the most potent artificial bits of intelligence during the 2010s included sparse autoencoders accumulated inside of deep neural networks [10].

For this purpose, we begin in Sect. 2 by describing different types of autoencoder along with its working principle. In Sect. 3, we review the deep architecture and in Sect. 4 discuss its application area. Finally, in Sect. 5, we conclude the theory of deep architectures.

2 Types of Autoencoder

See Fig. 2.

In this type of neural network an autoencoder is copies its input to the output. Encoder learn silent features when it doing so. It learns useful properties of its input. It has basically two different properties, i.e.

Undercomplete autoencoder:

- h has lower dimension than x
- f or g has lower capacity
- Most silent features of input learn by hidden layer h
- By minimizing the loss function we achieved learning
- (Represent the difference between x and reconstructed x)

$$L(x, g(h))$$

Overcomplete autoencoder:

- h has higher dimension than x
- The region behind it to copy input to its output. It can be solved by using regularization:

$$L(x.g(h)) + \Omega(h) \quad \text{where } \Omega(h) \text{ is a regularizer}$$

A reconstructing side is also learned, along with its reduction side, wherever the autoencoder attempts to produce from the compressed encoding for description as near as practicable to its primary data, this causes autoencoders to determine significant attributes being in the information.

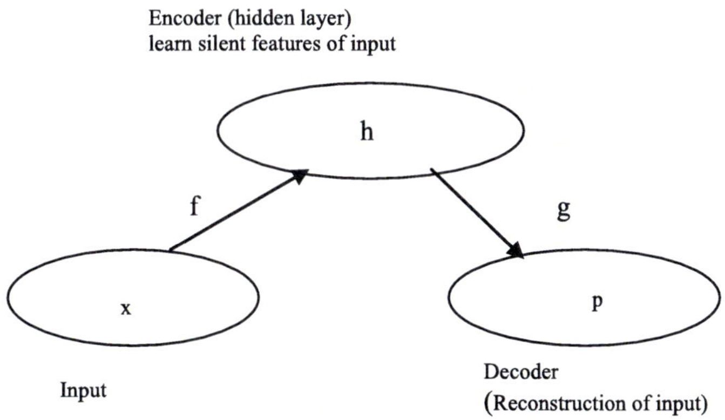

Fig. 2 The basic working principle of autoencoder. Overalls it shows three different nodes, i.e. input, hidden layer encoder and decoder [7]

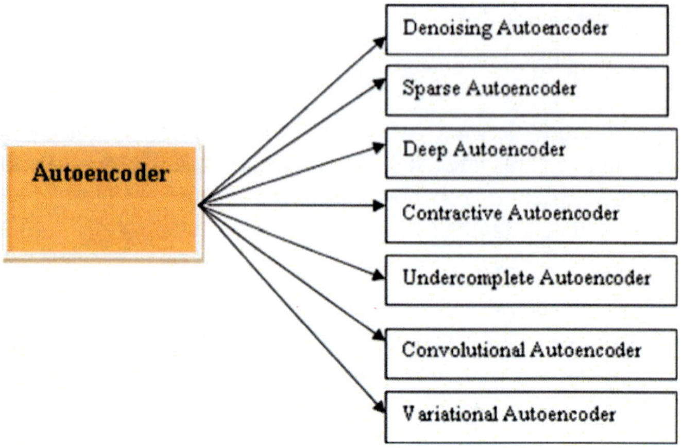

Fig. 3 Different types of autoencoder

A description acknowledges a proper restoration about its input, later it becomes much of the information existing in the data. Newly, the autoencoder theory has grown more popularly used for studying generative representations of data. There are, primarily, seven types of autoencoders.

Finally, autoencoders act by compressing the data into a possible space design and then building the product from this design. This sort of network is of two parts:

- **Encoder:** Here is the section regarding the network that compresses the into possible space design. It can represent an encoding function $h = f(x)$.
- **Decoder:** That part tries to restore the data from the possible space design. It can represent a decoding function $p = g(h)$.

Figure 3 displays different types of autoencoder.

3 Architecture

The simplest form of an autoencoder is a feed-forward, non-recurrent neural network. Both the encoder and decoder connect feed-forward neural networks. Autoencoders exercises with only a single layer encoder and a single layer decoder but practising deep encoders and decoders grants many benefits.

- Depth can exponentially lessen the computational cost.
- Depth can exponentially reduce the quantity of exercise data required to learn any elements.
- Experimentally, deep autoencoders allow more reliable compression compared to shallow or linear autoencoders.

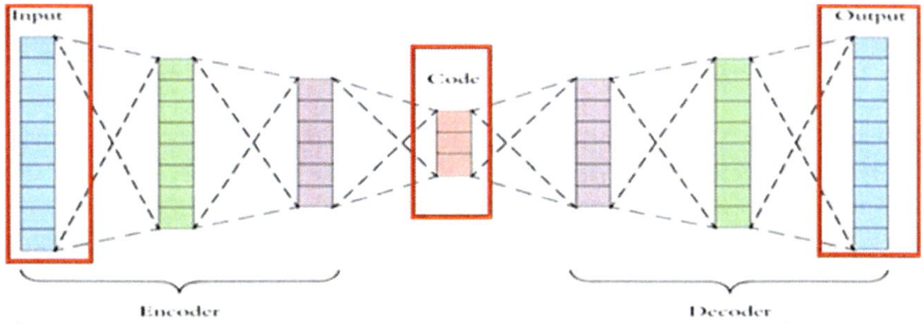

Fig. 4 Represents architecture of an autoencoder

Code is a separate layer of an Artificial Neural Network (ANN) with the dimensionality from our selection [11, 12]. The quantity of nodes in the core layer, which draws the language size, is a *hyperparameter* that initiated ere practise the autoencoder.

Initially, the data reaches through the encoder, which is a wholly connected ANN, to generate the code. The decoder, which has a related ANN structure, then makes the product only practising the code. The intention is to arrange a result indistinguishable with the data. Perceive that particular decoder design is the reflection about the encoder. The only thing one needs to keep in mind is that the dimensionality of the data and product requires to be identical. Before training an autoencoder, there are 4 hyperparameters that we need to set:

Code size: It points to the number of nodes in the intermediate layer, which is a more petite size decisions in more compression.

The number of layers: The autoencoder can be as deep as we want it to be. Externally considering the input and output, in Fig. 4, shows two layers in both the encoder and decoder.

Nodes per layer: The autoencoder architecture is the layers accumulate one over another. Naturally accumulated autoencoders resemble alike a box of pringles. The quantity of nodes by layer diminishes with every successive layer of the encoder and progress back in the decoder. Here we can view that the decoder does symmetric to the encoder in courses of the layered structure.

Loss function: They map an assortment of parameter values for the network onto a scalar value that symbolizes how well that parameter can accomplish the task which the **system** signifies to do. If the data contents are in the range [0, 1], then we typically manage cross-entropy. Contrarily, we use the mean squared error.

4 Experimental and Theoretical Application

The autoencoder mainly focuses on the field where the specific object and its precise boundary detection are essential. Some of its vast areas of application include medical, pattern recognition, visualization, image compression [13].

With the increasing use of medical imaging approaches such as MRI, pathology research, and surgical guidance, picture segmentation has become an important tools for medical image analysis. However, due to the ambiguity of the medical image itself and the complexity of the human anatomy, how to effectively remove the corresponding human tissue is still a challenging task. Autoencoder is a common technique which is used in image segmentation, and it is also a well-known application in the field of preprocessing of digital pictures [14–16].

5 Conclusion

Autoencoder form a very interesting group of neural network architectures. It applies many areas like computer vision, natural language processing and other fields. The goal of the autoencoder is to get a compressed and significant representation. We would like to have a design that is significant to us, and at the same time fitting for restoration. In that trade-off, it is essential to find the strategies which serve all demands.

References

1. Baldi P (2012, June) Autoencoders, unsupervised learning, and deep architectures. In: Proceedings of ICML workshop on unsupervised and transfer learning, pp 37–49
2. Bank D, Koenigstein N, Giryes R (2020) Autoencoders. arXiv preprint arXiv:2003.05991
3. Tripathy S, Swarnkar T (2020) A comparative analysis on filtering techniques used in of mammogram image. In: Advanced computing and intelligent engineering, pp 455–464. Springer, Singapore
4. Tripathy S, Swarnkar T (2020) Performance observation of mammograms using an improved dynamic window based adaptive median filter. J Discrete Math Sci Crypt 23(1):167–175
5. Tripathy S, Swarnkar T (2020) Unified preprocessing and enhancement technique for mammogram images. Procedia Comput Sci 167:285–292
6. Tripathy S (2019) Performance evaluation of several machine learning techniques used in the diagnosis of mammograms. Int J Innov Technol Exploring Eng 8:2278–3075
7. Zhang G, Liu Y, Jin X (2020) A survey of autoencoder-based recommender systems. Front Comput Sci 14:430–450. https://doi.org/10.1007/s11704-018-8052-6
8. Tripathy S, Swarnkar T (2019) Imaging & machine learning techniques used for early identification of cancer in breast mammogram. Int J Recent Technol Eng 8:7376–7383
9. Tripathy S, Hota S, Satapathy P (2013) MTACO-miner: modified threshold ant colony optimization miner for classification rule mining. Emerg Res Comput Inf Commun Appl 1–6
10. Tripathy S, Hota S (2012) A survey on partitioning and parallel partitioning clustering algorithms. In: International conference on computing and control engineering, vol 40

11. Tripathy S, Swarnkar T (2021) Application of big data problem-solving framework in health-care sector—recent advancement. In: Intelligent and cloud computing, pp 819–826. Springer, Singapore
12. Tripathy S, Swarnkar T (2020) Investigation of the FFANN model for mammogram classification using an improved gray level co-occurances matrix. Int J Adv Sci Technol 29(4):4214–4226
13. Tripathy S, Singh R (2021) Convolutional neural network: an overview and application in image classification. In: Advances in intelligent systems and computing, Springer, Singapore
14. Tripathy S (2021) Detection of Cotton leaf disease using image processing techniques. J Phys Conf Series 2062 (IOP Publishing)
15. Mohanty SS, Tripathy S (2021) Application of different filtering techniques in digital image processing. J Phys Conf Ser 2062 (IOP Publishing)
16. Kadam VJ, Jadhav SM, Vijayakumar K (2019) Breast cancer diagnosis using feature ensemble learning based on stacked sparse autoencoders and softmax regression. J Med Syst 43(8):1–11

Stock Price Prediction Using Principal Component Analysis and Linear Regression

Rushali.A. Deshmukh, Prachi Jadhav, Sakshi Shelar, Ujwal Nikam, Dhanshri Patil, and Rohan Jawale

Abstract To determine the future stock value of a company is the main purpose of stock price prediction there is a continuous change in the price of stocks which is affected by different industries and market conditions. The high dimensionality of data is a challenge for machine learning models because highly correlated dimensions/attributes may exert influence on precision of the model. PCA is used to reduce dimensionality to fit linear regression algorithm for future stock price prediction. Linear regression algorithm is used prior to and subsequent to implementation of Principal Component Analysis on the Tesla stock price data. Results manifest that production of machine learning models can be boosted by PCA, reducing the correlation and appropriate selection of principal components for high redundancy of data. Root mean square value and R-square value is used for assessment.

Keywords Principal component analysis · Linear regression · Root mean square error · R-square value

1 Introduction

The stock price prediction is one of the popular topics in financial field. Economic behaviors, social economy gets affected directly due to trends of the stocks. To erect a forecast model traditional stock method is used like ARIMA model to construct an autoregressive model to forecast stock price. Non-linear forecasting model requires huge financial data based on rough sets, suggest a decision tree system. The advantages of rough sets and decision trees are mixed in this method, but overfitting occurs. The trend will be distressed because of huge amount of noise in the datasets. To overcome this, an artificial neural network is preferred as it deals with non-linear data

Rushali.A. Deshmukh · P. Jadhav · S. Shelar · U. Nikam · D. Patil (✉) · R. Jawale
Computer Engineering, JSPM'S Rajarshi Shahu College of Engineering, Tathawade, Pune, India
e-mail: dhanashri2509@gmail.com

Rushali.A. Deshmukh
e-mail: radeshmukh_comp@jspmrscoe.edu.in

P. Dutta et al. (eds.), *Emerging Technologies in Data Mining and Information Security*, Lecture Notes in Networks and Systems 490,
https://doi.org/10.1007/978-981-19-4052-1_28

models, so it is widely used in time series prediction. In practical application ANN is used to the local optimal problem and SVM is used to reduce it. SVM improves the generalizability of the model. Random forest gives better results than SVM after optimization of parameters. Layered feature representation is mainly analyzed by DNN with complex, deep non-linear relationships. Aim is to predict trend of stock price with the help of liner regression as a classification model. PCA is used to optimize the results with accurate predicted values by reducing dimensionality and redundancy.

2 Literature Survey

Shah et al. [1] examined the results of news. Specifically, they advanced a dictionary-based sentiment analysis model and examined the results of news sentiments on stocks for the pharmaceutical market. They achieved a directional correctness of 70.59% Mehtab [2]. In this paper, regression models consist of NIFTY 50 of certain periods. Afterward, they used deep learning-based regression models using LSTM for prediction NekoeiQachkanloo [3]. A trading system that serves as an artificial counselor for us. Support vector regression helps to predicts future prices. For the recommendation part, two methods were used, first Markowitz's portfolio theory, second a fuzzy investment counselor.

Mehtab and Sen [4] developed eight regression and classification methods. For augmenting a predictive framework, they bring public sentiments, and as the two inputs, a fuzzy neural network-based SOFNN algorithm is used. Focus is the ability of various neural networks. Solving the accurate problem in the model by using Genetic Algorithm by Chen [5]. Used two-stage price prediction model GA-LSTM with the help of feature selection and GA-LSTM Algorithm reaches a minimum and 0.0039 MSE score. Using crossover and mutation rate, and several factor combined improves the performance. Reinforcement learning, agent-based learning apart from supervised and unsupervised learning. Q-learning is a model-free learning algorithm were used Sathya [6] Buying and Selling actions can be done with layered neural network. There is method called expReplay when memory gets full.

Identifying the trends by data mining and developing time series data to forecast periodical changes by Sorna Shanthi [7] using regression analysis model. The Gaussian process recognizes the trend in stock data. The algorithm will be learning geometrical shapes. Generative adversarial network, predicts the price and CNN as discriminator used with LSTM and RNN.

The patterns are recognized this model in historical market. Chacón et al. [8] Empirical mode decomposition, improved the correctness of financial time series. EMD and sample entropy to reduce complexity of forecasting accuracy. LSTM model is the best suited model for forecasting. Gaussian distribution with parameters (u = sigma), used for noise removal. First, two IMF's will have lower prediction accuracy than other cases. Stock price trend prediction (classification) and stock price forecast (regression) are forecasting types. To predict yesterday's or 5-day closing price, by using short-term. Support Vector Machine (SVM), used to maximize the minimum interval, the algorithm altered into an optimization problem.

Preprocessing of missing data is essential, can be obtained by the fill method implemented by Yuan [9]. The best suited for the stock price trend is the RF model. Ketter et al. [10] used time series, interval forecasting, financial services industry, least-square support vector regression, sliding window method. Maio et al. developed its forecasting uses ANN. Later, Xiong et al. introduced multiple output support vector regression methods used for predicting stock price value. This system provides guessing performance in the financial field. The prediction mainly uses tweet mining, machine learning, sentiment analysis, model stacking, stock movement direction prediction, and textual features extraction. To utilize Twitter data as information of six well-known NASDAQ companies. To predict stock market prediction Chou et al. [11] need a sophisticated way of sentiment analysis. Li [12] proposed a deep neural network for prediction, incorporates the news articles as hidden.

For the prediction, Cho et al. [13] used finance, equality research report, natural language processing, stock market, investment strategy, and binary classification. Using binary classification, dependent on NLP-elements, framework was explored. Therefore, they predict positive stock prices. Shen [14] by using Recursive feature elimination for feature engineering and LSTM model for forecasting the price of the trend of the market included financial aspects as input data. The binary accuracy is 93.25%. The limitation is REE is not suitable for the long term. Scanning the significant acutance between markets and experts. Calculated the Kendall-Tau rank-order correlation and Mean Absolute Percent Error to analyze the market behavior and elasticity on trading performance. Blohm et al. [15] conclude that Price elasticity affects the market design on trading performance thus, the difference in trading performance will be large for moderate elasticity settings and smaller for low and high elasticity settings.

3 Techniques

1. **SUPPORT VECTOR MACHINE:** SVM is the flexible algorithm. Kernel function with VM solves the linear indivisibility problem. The chief purpose is to find a hyperplane. Small design, non-linear and pattern recognition for that solving SVM is involved. It gives better prediction than neural networks.
2. **RANDOM FOREST ALGORITHM (RF):** It is based on a decision tree algorithm in which record branched into small features and specially used for classification and regression analysis. In this, the random data of the stocks are taken and then train the models to achieve better accuracy.
3. **ARTIFICIAL NEURAL NETWORKS (ANN):** For non-linear models, ANN mostly requires. This algorithm is similar to the human brain. Input, output the hidden layer are used. The input to the model is the features data and the output is obtained by calculation of hidden layers in between. This algorithm contains nodes and is calculated by the error backpropagation algorithm which contains

various steps like Parameter Initialization, Forward propagation, calculate the total error, and error backpropagation.

4. **EMPIRICAL MODE DECOMPOSITION (EMD):** Development in the accuracy and reduced complexity EMD is used. This method depends on Fourier transformation that decomposes signal based on time scale and finding the regular pattern in time series ARIMA used.
5. **LONG SHORT-TERM MEMORY (LSTM)**: It is the most useful algorithm and is used in many projects of the stock market [10, 12]. In this, certain operations are performed and then we train the LSTM model SVM is involved in solving small design, non-linear and pattern recognition. It gives better prediction than neural networks.

4 Methodology

Tesla's closing price is predicted for the system. Using various machine learning algorithms to prediction the future price of stocks we train the machine from the numerous data points of historical values to make a future price prediction (Fig. 1).

1. **DESCRIPTION OF DATA:** The dataset includes 4 years data from 2010-06-29 to 2020-02-03 of Tesla. High, Low, Open, Close, Adjacent close, and Volume are the attributes of data. The extracted prices are of day-wise closing.
2. **DATA PRE-PROCESSING:** It is a very important step that our model will outcome important details from data. Every time we do not get the clean and formulated data Therefore before assign the data in our model, we pre-process

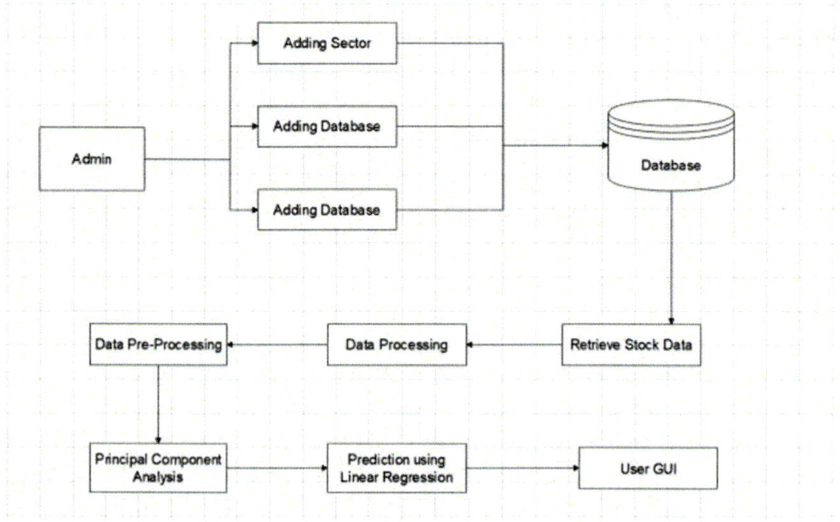

Fig. 1 System Architecture

and removing outliers, noises, missing values. Data pre-processed by normalization. Normalization changes the numerical values without making differences in the ranges of values.

3. **PRINCIPAL COMPONENT ANALYSIS:** Dimensions are reduced of various variables from given data set. The principal component is to outcome uncorrelated variables from correlated variables by using the mathematical procedure. Reduce the extension of convert data, only the first few element well-advised as the first principal component have the largest conflict in data. Less important variables are still present in the most valuable parts of all the variables so it is important as new variables after PCA are independent of one another this is helpful for the linear model requires independent variables should be independent of one another.
4. **LINEAR REGRESSION:** Supervised technique to fit best line depends on slop and intercept by the Independent variable X and another scalar on dependent the variable Y on X. Insert the input in the form of (X_i, Y_i), we have to predict $Yn + 1$ where i = 1 to n, and for a new point $Xn + 1$. Equation is:

$$\text{i. } \hat{y} = a + b_o x \tag{1}$$

$$\text{ii. } y = b_o + b_1 x_1 + b_2 x_{12} + b_2 x_{13} + \cdots + b_n x_{1n} \tag{2}$$

5 Implementation and Results

First, we upload the data set, i.e., Tesla stock data once the dataset is uploaded then clean the dataset and check for null values. In the dataset there are total 4 input features including Open, High, Low, volume. Hence, the absolute value of feature is less significant than the change of a feature with respect to time. The main motive is to get a highest correlation between features so that principal components extraction for PCA can become easier. Accuracy score and root means square error is used as estimation criteria to judge the performance of regression on given datasets, which are defined as:

RMSE = Regression score = the coefficient of determination R^2 of the prediction.

The coefficient R^2 is defined as (1 − SSE/SST), where SSE is the residual sum of squares ((y_ actual - predicted)** 2).sum() and SST is the total sum of squares. ((y_actual – y_actual.mean()) **2).sum().

6 Results

The high correlation in data causes data redundancy because the X variables values are very close to each other and the high dimensionality. Hence there is an overfitting problem faced to fit the regression model.

R-squared value without PCA: 0.9996 (Figs. 2 and 3).

After implementation of PCA

R-squared value: 0.9725.

After implementing we can see the lower values of R-square. There is standardization and normalization are done with the data values and got the two principal components so it is more feasible to fit the regression model after the PCA. After implementing PCA, outcome is close to predicted values (Fig. 4 and 5).

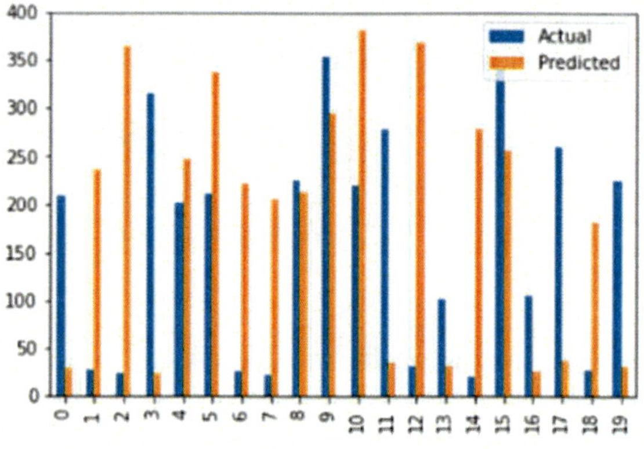

Fig. 2 Actual and predicted values before implementation of PCA

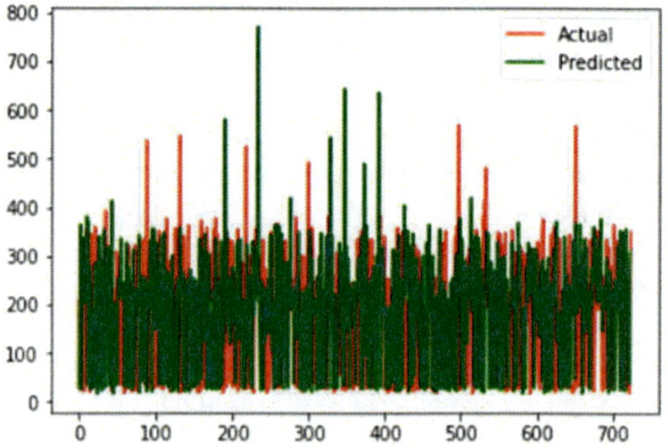

Fig. 3 Variation between actual and predicted values before implementation of PCA

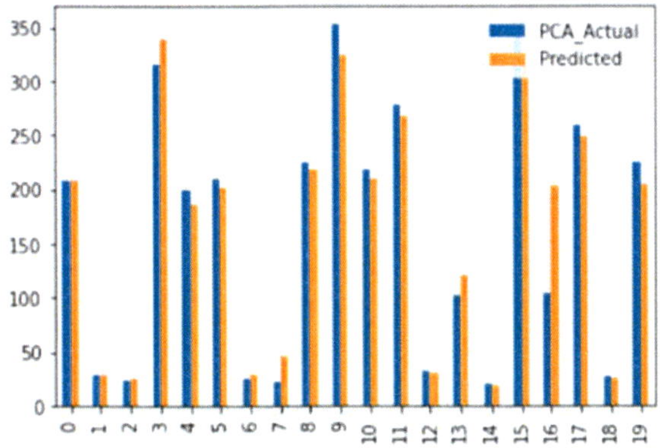

Fig. 4 Actual and predicted values after implementation of PCA

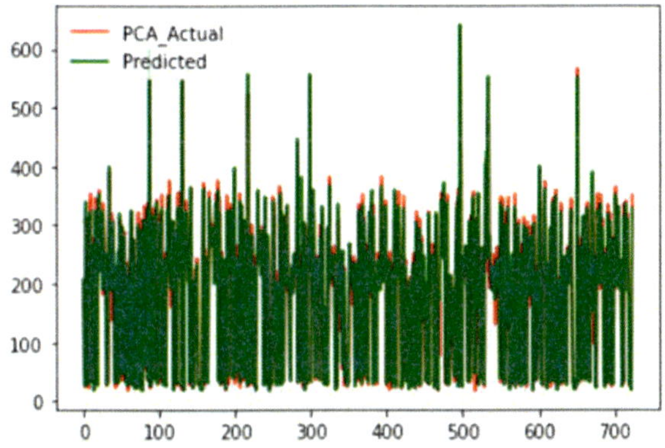

Fig. 5 Actual and predicted values are closer after implementation of PCA

7 Conclusion and Future Scope

Stock market investment is a fascinating and trending topic. However, there are many factors involved in this making decision a complex task. A clever prediction system will help investors make investments accurately and profitably with information of future prediction direction of stock prices. The high-dimension data is an improvement in the ML performance in the classification of PCA. For PCA the impact of features aided, has been investigated. This paper is helpful for researchers to acknowledge the effect of implementing PCA on highly correlated datasets and helpful for selecting various optimum principal components, used while scrutinize the automatic selection of parameters for techniques, which are k for KNN, kernel parameters in SVM, and the number of PCs. Sentiment analysis has a high impact on future prices. Thus, the investigation can produce a highly efficient prediction of the mixture. For stock price prediction is not able to predict dynamic and rapid changing patterns in stock price movement a disadvantage of the existing propositions. Many studies estimate their machine learning model by using one market and one time period without considering whether the system will be effective in other situations.

References

1. Shah D, Isah H, Zulkernine F (2018) Predicting the effects of news sentiments on the stock market. In: 2018 IEEE international conference on big data (Big Data) 2018 Dec 10, pp 4705–4708. IEEE
2. Mehtab S, Sen J, Dutta A (2020) Stock price prediction using machine learning and LSTM-based deep learning models. In: Symposium on machine learning and metaheuristics algorithms, and applications. Springer, Singapore, pp 88–106, 2020 Oct 14
3. NekoeiQachkanloo H, Ghojogh B, Pasand AS, Crowley M (2019) Artificial counselor system for stock investment. In: Proceedings of the AAAI conference on artificial intelligence 2019 Jul 17, vol 33(01), pp 9558–9564
4. Mehtab S, Sen J (2019) A robust ,Predictive model for stock price prediction using deep learning and natural language processing. Available at SSRN 3502624, 2019 Dec 12
5. Chen S, Zhou C (2020) Stock prediction based on genetic algorithm feature selection and long short-term memory neural network. IEEE Access 24(9):9066–9072
6. Sathya R, Kulkarni P, Khalil MN, Chandra Nigam S (2020) Stock price prediction using reinforcement learning and feature extraction. IJRTE 8(6)
7. Sorna Shanthi D, Aarthi T, Bhuvanesh AK, Chooriya Prabha RA (2020) Pattern recognition in stock market
8. Chacón HD, Kesici E, Najafirad P (2020) Improving financial time series prediction accuracy using ensemble empirical mode decomposition and recurrent neural networks. IEEE Access 25(8):117133–117145
9. Yuan X, Yuan J, Jiang T, Ain QU (2020) Integrated long-term stock selection models based on feature selection and machine learning algorithms for China stock market. IEEE Access 24(8):22672–22685
10. Ketter W, Collins J, Gini M, Gupta A, Schrater P (2005) A computational approach to predicting economic regimes in automated exchanges, 2005 Jan 1, pp 147–152
11. Chou JS, Truong DN, Le TL (2020) Interval forecasting of financial time series by accelerated particle swarm-optimized multi-output machine learning system. IEEE Access 10(8):14798–14808
12. Li X, Li Y, Yang H, Yang L, Liu XY (2019) DP-LSTM: Differential privacy-inspired LSTM for stock prediction using financial news. arXiv preprint arXiv:1912.10806. 2019 Dec 20
13. Cho P, Park JH, Song JW (2021) Equity research report-driven investment strategy in Korea using binary classification on stock price direction. IEEE Access 22(9):46364–46373
14. Shen J, Shafiq MO (2020) Short-term stock market price trend prediction using a comprehensive deep learning system. J Big Data 7(1):1–33
15. Blohm I, Riedl C, Füller J, Köroglu O, Leimeister JM, Krcmar H (2012) The effects of prediction market design and price elasticity on the trading performance of users: an experimental analysis. arXiv preprint arXiv:1204.3457, 2012 Apr 16

An Extensive Survey on ICT-Based English Language Teaching and Learning

Nayantara Mitra and Ayanita Banerjee

Abstract Learning and teaching a language is perceived to be a difficult task. Advanced teaching aids should be available to make such intricacies easier, as it is the need of the hour. The rapid growth of the Internet has ushered in a technological revolution in all aspect of our lives, including teaching and learning. The use of information and communication technology (ICT) has a noteworthy impact on the quality and quantity of the teaching–learning process. ICT can improve teaching and learning by providing dynamic, interactive, and interesting content, as well as actual opportunity for personalized education.). Due to technological advancements, today's classroom atmosphere differs significantly from the previous-traditional arrangement. This paper, on the other hand, focuses on ICT tools that aid in the development of English language learning and teaching methods in order to demonstrate how technology affects second/foreign language education and how it can be effectively used in the English as a Foreign Language (EFL) classroom; these tools include both web-based and non-web-based tools, as well as how they can be used in the classroom.

Keywords Information Communications Technology (ICT) · Foreign language · Teaching and learning · ELT · ESL · etc.

1 Introduction

ICT is an acronym for information and communication technology. It is defined as a collection of technical tools and resources for communicating, creating, disseminating, and managing information. We live in the technological age, and it has impacted every aspect of human life. Technological improvements and innovations

N. Mitra (✉)
Institute of Engineering and Management, Kolkata, India
e-mail: nayantara.mitra@iemcal.com

A. Banerjee
University of Engineering and Management, Kolkata, India
e-mail: ayanita.banerjee@uem.edu.in

P. Dutta et al. (eds.), *Emerging Technologies in Data Mining and Information Security*, Lecture Notes in Networks and Systems 490,
https://doi.org/10.1007/978-981-19-4052-1_29

have had a significant influence on academic achievement and administration in the field of education. For a huge percentage of students, traditional ways of teaching higher education have become less compelling. The use of ICT tools increases the scope of teaching–learning process. The use of information and communication technology (ICT) broadens the breadth of the teaching–learning process.

It gives us access to fresh learning materials. To be consistent with the globalized world, English language learning and instruction must adapt and reinvent it. The goal of this research is to emphasize the beneficial benefits of ICT in order to keep up with modernized communities in today's digital world. In reality, the use of technology gives learners with previously unimagined chances to practice English and immerse themselves in authentic language situations [1]. For example, they can use Skype Chat for interaction or social media sites like Facebook or Twitter for writing practice [2, 3]. Furthermore, the usage of ICTs increases learners' motivation due to multimedia capabilities such as visual aids, audios, and videos [4].

ICT can be characterized as scientific, technological, and engineering-based management techniques used in information storage and communication mechanisms with optimal time and space consumption, as opposed to other outdated methodologies. Any communication equipment or application, such as a computer, a mobile phone, a radio, a television, or a satellite system, is included in the term "ICT." Today's teachers can employ a variety of technological tools to make their lessons more imaginative and engaging.

One of the most essential languages in the globalization and knowledge explosion processes is English. It is the most widely used method of communication worldwide. As a result, it is known as a Link Language, a Global Language, and a Lingua Franca. It is referred to as English as a Second Language) (ESL) in India . The ability to communicate in English has become increasingly important for learning and earning.

As a result, beginning in elementary school, English must be taught and pupils' English language skills must be developed. At many levels, the government, NGOs, and educational institutions are collaborating and taking steps to improve English Language Teaching (ELT) and increase English language abilities among students. In our country, numerous approaches and methods are employed to teach English and build English language abilities. However, the majority of them are outdated, uninteresting, ineffective, and not motivating.

As a result, it is vital to apply modern Information and Communication Technology (ICT) methodologies and tools to improve students' understanding and learning of basic English language abilities, such as Listening, Speaking, Reading, and Writing (LSRW). ICT has a lot to offer teachers and students in terms of expanding their vocabulary and improving their English language skills. The study's goal is to determine the strategies for using ICT in English language teaching and learning, as well as their effectiveness and practicality for use in regular classes. This study also discusses the various ICT tools and their benefits and drawbacks. For the goal of language learning, the students' perceptions of ICT and their expectations have been investigated.

The remaining pieces of paper are placed as follows. In Sect. 2, a detailed literature survey on the benefits of incorporating ICT in the classroom is briefly depicted, whereas Sect. 3 depicts the proposed approaches and tools of ICT, Sects. 4 and 5 discuss the benefits and drawbacks of using these tools in a regular classroom environment, and Sect. 6 discusses the conclusion followed by the current study's future work.

2 Literature Survey

Teaching English using technology is not a new for teachers all around the world [5], particularly for non-native English speakers. ICT is mostly employed to offer learners with real content. Students can use these materials to improve their speaking, listening, reading, and writing skills. Jurich et al. [5] lists several benefits of using ICT into the teaching and learning process, particularly in ELT classes. The first is to provide multisensory stimuli that can quickly improve English language skills [6]. The second point to consider is motivation. For those who are interested, technology has the potential to be a great eucational tool, and it should be introuced and embraced at an young age. As a result, students who use technology are more likely to remain focused on their work for longer periods of time. Collaborative learning is the third thing. In English learning, ICT can be used in a variety of ways. Collis et al. [7] divided ICT applications into three categories: "learning resources," which includes educational software, online resources, and video resources, and "instructional organization of learning," which includes software and technology tools for classroom lecturing, course management, and so on.

As follows, there are various technical fields that could potentially help to the field of education. The first is called Extended Learning, and it involves using modern communication tools or social networking sites like Facebook, Twitter, blogs, wikis, and instant messaging to augment traditional teaching and learning. Teaching and learning are no longer limited to the classroom; they are now reinforced beyond the classroom through social networking sites where students may participate in a platform for communication that allows collaborative conversation, exchange of ideas, and critical thinking.

The second category is Ubiquitous Wireless, which encourages students to learn on the go using portable or mobile devices such as laptops, tablets, cell phones, and other similar devices. The third category, Intelligent Searching, enables learners to more effectively search, organize, and retrieve material. Educational Gaming, which consists of games and simulations, is the fourth category. It is seen as a learning tool that can improve motivation, communication, critical thinking, and problem-solving abilities [8].

3 Methodology

3.1 ICT Tools in Teaching and Learning

Digital infrastructures such as computers, laptops, desktops, data projectors, software applications, printers, scanners, and interactive teaching boxes are examples of information communication technology equipment.

ICT tools are the most up-to-date technology, equipment, and concepts utilized in information and communication technology interactions between students and teachers (e.g., flipped classroom, mobile apps).

Web-based learning and non-web-based learning are the two primary groups of ICT technologies. Learning that is not based on the Internet.

1. Radio and television are effective language learning aids. Both instruments provide low-cost access to a variety of enrichment activities. The rapidity of current affairs broadcasts ensures that learners' language exposure is current and rooted in native speakers of the real world. Teachers might use radio to get their pupils to listen to lectures given by prominent and distinguished speakers. The other key technical medium utilized by language teachers is television, which appeals to both the eyes and ears. Television gives a complete audio-visual simulation that is active and realistic. Along with the face emotion, it conveys language expressiveness.
2. Films—In the hands of a knowledgeable and clever instructor, films are the most potent tool. Films entice students, pique their interest, and keep it for longer. Films are an effective way to present facts, abilities, and background information. The function of the speech organs and pronunciation pique the curiosity of elementary school kids.
3. Language Lab—One of the most modern technology teaching aids is the language lab. Students can listen to audios and understand the various accents in use, they can talk, and they can even record their voices in the language lab. Listening to standardized materials could help pupils improve their pronunciation. The language lab is solely focused on achieving results and enriching the English language learning experience. In recent years, lab materials have incorporated not only audios but also films, games using a flash interface, and the Internet. A language lab provides a much more relaxed environment than a regular classroom.

3.2 Web-Based Education

One of the fastest growing sectors is web-based learning, often known as e-learning, technology-based learning, remote learning, and online education. It enables the creation of well-designed, learner–centered, cost–effective, interactive, officiate, and adaptable e-learning environments (khan, 2005). Thousands of English language web-based programs are available that provide training in a number of basic language abilities such as Learning, Speaking, Reading, and Writing, and are interactive in a variety of ways. The following are some of the most common technologies used to promote education:

1. YouTube—YouTube is a video-sharing tool that you may use in your classroom to find and share authentic video content.
2. Email—By creating a free personal email account, students may contact with native speakers of the target language via email (g-mail, yahoo, etc.). Students can send their homework to the teachers who are responsible for it and have it rectified. The teacher can also provide modifications, feedback, and recommendations for improving each piece of work before returning it.
3. Blogs—A blog is a frequently updated personal or professional journal intended for public consumption. Blogs allow students to upload and link files, making them ideal for use as online personal journals. According to Pinkman (2005), blogging becomes communicative and interactive when participants take on several roles in the writing process, such as readers/reviewers who respond to other authors' postings and writers-readers who respond to criticism of their own posts.
4. Mobile phone—Learners can use the dictionary feature on their phones to look up new words and expand their vocabulary. They can double-check the spelling, pronunciation, and usage of the word they're looking for. They can also use Short Message Service (SMS) to ask questions to their professors and get answers from them.
5. IPods—IPods are a type of multimedia gadget that allows users to create, deliver, and exchange texts, images, audio, and video scripts as needed. Students can read and respond to text messages that are sent by their teachers. Students can also record and listen to their lectures, poems, news, short stories, and other presentations. As a result, iPods provide an opportunity for English language learners to improve their listening, pronunciation, vocabulary, grammar, and writing skills.
6. Video conferencing software—With built-in features like chat, screen sharing, and recording, video conferencing software provides online communication for audio meetings, video meetings, and seminars. These programs are used to facilitate long-distance or international communication, collaborate more effectively, and save money on trip. Employees at various levels of a firm can utilize video conferencing solutions to arrange or attend virtual meetings with co-workers, business partners, or customers, regardless of their physical location.

4 The Benefits of Using ICT in ELT

1. Materials—Using ICT, students can quickly access resources or information by just pressing a button. Sites such as Google, Wikipedia, Yahoo enables students to get relevant results within a matter of seconds.

2. Long-distance learning—Many students have employed computer technologies to learn from anywhere; computer gadgets such as iPads can connect to the Internet from anywhere, allowing students to attend remote classes. Many universities and colleges have included online education as part of their curriculum, and students from all over the world are enrolled.
3. Motivating Factor—For many students, the Internet may be a motivating tool. Young people are enthralled by technology. Educators must take use of this curiosity, excitement, and enthusiasm for the Internet in order to improve learning. If your students are already keen learners, you can use the Internet to give them with additional learning activities that are not available in the classroom.
4. Cooperative Learning—The Internet makes cooperative learning easier, fosters conversation, and makes the classroom more interesting. A class listener, for example, can participate in class discussions via email in a way that is not conceivable within the four walls of the classroom.

5 Disadvantages of Using ICT for Education

The use of the Internet for educational purposes is burdened with issues. As a result, one should expect the issues that arise from using the Internet in the classroom to evolve as well. There are some drawbacks of using ICT in education and learning:

1. **Plagiarism**—Though some of the Web sites claim to assist students in writing term papers, there have been multiple instances of students obtaining information from the Internet and handing it in for grades. This difficulty can be mitigated by requiring students to credit research sources. Plagiarism.org (http://www.plagiarism.org/) is such a free internet service that helps us reduce plagiarism in the classroom. This service also claims to avoid plagiarism by determining whether or not a term paper has been plagiarized from the web.
2. **Student Privacy**—When students are online, criminals, marketers, and other unscrupulous people can readily obtain information about them. These could endanger children' lives or perhaps result in a lawsuit against the institution. To avoid this issue, children should be taught about the dangers of sharing personal information with strangers via the internet. Parents and instructors must monitor their children's online activities.
3. **Low-Income Groups**—According to the US Department of Education, over half of public schools with a high minority enrollment had a lower percentage of Internet connection than public schools with a low minority enrollment in 1997. The same may be true of such schools' classrooms. Additionally, students from low-income families may not have access to computers at home or may have PCs with restricted Internet access. As a result, students from low-income homes may face challenges. To reduce the influence of social or economic standing, we should set Internet assignments that students can easily do while at school. If required, schools may need to keep computer laboratories open.

Schools might need to maintain computer labs available if necessary. Computer use in public libraries should be encouraged as well.

4. **Preparatory Time**—Using the Internet for educational purposes effectively demands a substantial amount of planning time. In addition to creating Internet-based lesson plans, we may need to search the Internet for lesson plans and modify them to meet the curriculum objectives, or visit websites to find ones that are acceptable for schools. We do not have a choice but to prepare to help pupils become responsible Internet users.
6. **New Administrative Responsibilities**—Teachers and school administrators face additional administrative challenges as a result of Internet-based education. These tasks include developing and implementing an appropriate policy, as well as providing training, developing new assessment criteria as needed, and resolving parental concerns.

6 Data Analysis

6.1 Findings

According to the survey, 83.6% of respondents use ICT for educational purposes, with 94.2% agreeing that ICT is extremely helpful in language learning. When it came to ICT usage per day, 40% of the participants said they spent more than three to five hours per day, as seen in Table 1. Nearly a quarter (24.2%) said they spent six to eight hours a day, while 20.2% said they spent zero to two hours and 8.6% said they spent more than nine hours. The same has been represented in the form of bar diagram for better understanding. Figure 1 shows the Graphical Presentation Daily use of ICT in hours.

The question on which learning reasons the students used ICT was included in the survey. Table 2 demonstrates that the most common reason for using ICT is for Google Classroom (87.5%), followed by YouTube (85.7%), Google for study materials (78.4%), and finally mailing for educational purposes (67.2%). ICT was used by half of them for online vocabulary and reading exercises (51.7% and 53.8%, respectively). Only 33.6% of people utilize blogs to learn. Figure 2 shows the use of ICT for learning purposes.

The study also included a question on how the learners utilized ICT for non-learning activities. Table 3 reveals that ICT is primarily used for social media

Table 1 Daily use of ICT in hours

ICT use in hours per day	Response rate (%)
0–2	20.2
3–5	40.08
6–8	24.5
More than 9	8.6

Fig. 1 Graphical presentation daily use of ICT in hours

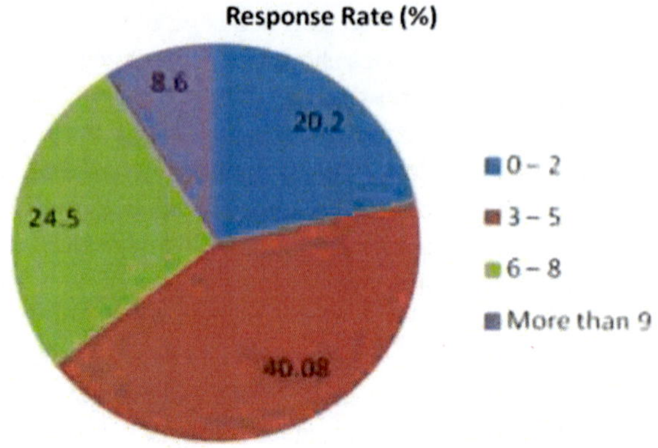

Table 2 Use of ICT for learning purposes

Activity	Response
Email	67.2
YouTube	85.7
Online Vocabulary exercise	51.7
Online Reading exercise	53.8
Google Classroom	87.5
Google search for study material	78.4
Blogs	33.6

Fig. 2 Graphical Representation of Use of ICT for learning purposes

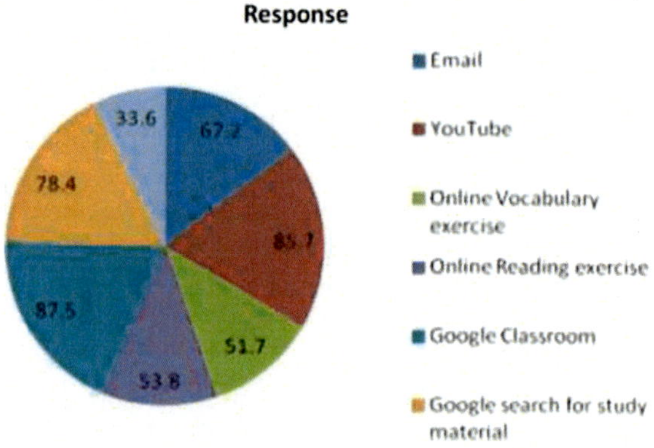

(81.4%), listening to music (76.2%), writing letters (74.5%), and watching movies (74.5%). (72.8%). Other common uses include chatting apps (71.1%), online shopping (68.1%), Google maps (63.7%), and news (53.4%). Figure 3 displays Graphical Representation of Use of ICT for learning purposes.

Table 3 Usage of ICT can be used for non-learning reasons

Activity	Response rate (%)
Unofficial emails	74.5
Social Media	81.4
Google Map	63.7
Watching news	53.4
Watching movies	72.8
Online shopping	68.1
Chat application	71.1
Listening to music	76.2
Other purposes	28.4

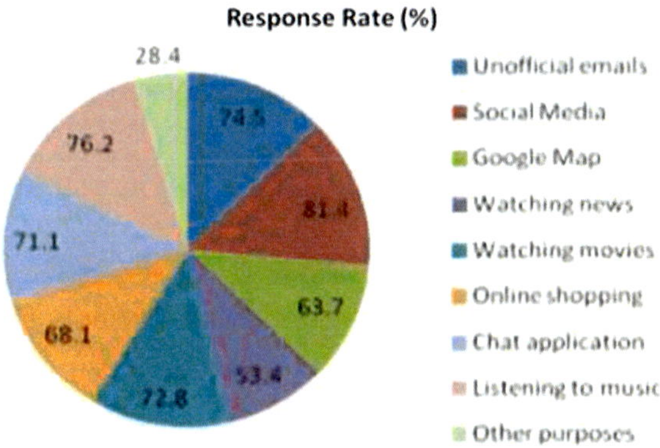

Fig. 3 Graphical representation of usage of ICT can be used for non-learning reasons

7 Conclusion

To summarize, not all lessons may be accessed via the Internet. We must persuade students that utilizing the Internet contributes something fresh, something of real worth to their learning. However, pupils should be taught how to make the most of the technology accessible to them. We should communicate with other professors in the system since cooperation and mutual understanding are crucial, especially when the institution has a limited number of Internet accounts.

The student can communicate with or work with other students or experts in the subject from all around the world via the Internet. They can also join a news group focused on a certain topic. In terms of communication, what's most amazing about the Internet is that it is race, age, place of origin, and gender agnostic. Students can also utilize the Internet to make their project outcomes public so that their classmates all around the world can see them. This may offer some students with the drive they need to do their assignments on time and in proper English. As a consequence, any student may benefit from a Net communication project.

Many educational institutions use ICT as a core instrument. The use of information and communication technology (ICT) broadens the area of instruction. It provides high-quality learning materials and promotes learning autonomy. Students must have English communicative abilities in addition to academic brilliance in order to have a bright future. Technological aids must be included in curriculums to make them more user-friendly. Learners can share their work, which can help to encourage cultural variety, increase motivation, and boost self-esteem.

References

1. Kramsch C, Thorne SL (2002) Foreign language learning as global communicative practice. In: Block D, Cameron D (eds) Globalization and language teaching, pp 83–100. Routledge, London and New York
2. Dalton ML (2011) Social networking and second language acquisition: exploiting Skype (TM) chat for the purpose of investigating interaction in L2 English learning. Master's thesis, Iowa State University, the USA. Retrieved February 2, 2014 from http://lib.dr.iastate.edu/etd/10221/
3. Cheng HY (2012) Applying Twitter to EFL reading and writing in a Taiwanese college setting. doctoral dissertation, Indiana State University, the USA. Retrieved March 25, 2014 fromhttp://scholars.indstate.edu//handle/10484/4574
4. Altiner C (2011) Integrating a computer-based flashcard program into academic vocabulary learning. Doctoral Dissertation, Iowa State University, the USA. Retrieved March 10, 2014 from http://lib.dr.iastate.edu/cgi/viewcontent.cgi?article=1122&context=etd
5. Rank T, Warren C, Millum T (2011) Teaching English using ICT: a practical guide for Secondary School Teachers. London: Continuum
6. Jurich S (2001) ICT and teaching of foreign languages. Retrieved October 24, 2015 from http://www.techknowlogia.org/TKL_Articles/PDF/335.pdf
7. Garimella S, Srinivasan V (2014) A large scale study of the effectiveness of multi-sensory learning technology for learning English as a second language. Retrieved November 23, 2015 from http://www.englishhelper.com/whitepaper.pdf
8. Collis B, Moonen J (2001) Flexible learning in a digital world. Experiences and expectations. Kogan Page, London

A Study on Using AI in Promoting English Language Learning

Nayantara Mitra and Ayanita Banerjee

Abstract The transformation in the way we live, work, and learn is characterised by the fast proliferation of technology and digital applications. It is a revolution propelled by the convergence and amplification of coming achievements in artificial intelligence, automation, and robotics, and amplified by the widespread connectedness of billions of people with mobile devices that provide unparalleled access to data and information. When AI is brought up in a classroom setting, it may be rather daunting. AI is commonly thought to relate to the collection and use of data. Whilst this is true, the purpose of this article is to demystify the background and demonstrate the most practical ways in which language instructors may employ AI in the ELT classroom. The goal of this research is to show how successful AI-based instructional programmes are in teaching and learning English. It is impossible to prove effectiveness without putting it to the test in the real world. Digitalization has become an integral element of our daily lives. It is regarded as the world's unavoidable feature and driving force. Due to technology advancements and their applications, the world has achieved significant progress in a multitude of fields. Education, being one of the most important industries, has adopted a variety of techniques over time. In teaching and learning, trial and error are constantly present. In ELT classes, Information and Communications Technology (ICT)-based technology and gadgets have already been utilised. The most important and user-friendly AI-based strategies may help teachers make English language education more fascinating, interactive, dynamic, and joyful. It is certain that Artificial Intelligence would affect the entire educational system's view in the future (Bernard Marr and Matt Ward, Artificial Intelligence in Practice, UK, 2019, Print).

Keywords Artificial intelligence · English teaching · English learning · Digitization · Education · Trial and error · ICT · Virtual reality

N. Mitra (✉)
Institute of Engineering and Management, Kolkata, India
e-mail: nayantara.mitra@iemcal.com

A. Banerjee
University of Engineering and Management, Kolkata, India
e-mail: ayanita.banerjee@uem.edu.in

P. Dutta et al. (eds.), *Emerging Technologies in Data Mining and Information Security*, Lecture Notes in Networks and Systems 490,
https://doi.org/10.1007/978-981-19-4052-1_30

1 Introduction

Machines demonstrate artificial intelligence, but humans and other animals display natural intelligence (NI). Artificial Intelligence is defined as any device that detects its surroundings and takes activities to improve its chances of attaining its objectives. Artificial Intelligence is defined as a machine that can do the cognitive activities of a human brain, such as learning, communication, and problem solving. Artificial intelligence is a development of human intellect that is currently one of the most advanced technologies. It has also achieved a number of educational accomplishments. Artificial intelligence (AI) is a computer simulation of human intelligence. Education is the engine that propels and supports societal progress. Nowadays, English is one of the most widely spoken languages on the planet. As a result, artificial intelligence, machine learning, and intelligent search may be used to successfully boost English teaching and learning reform [2].

Artificial intelligence's rise is not only a useful method of English teaching and learning, but also a significant demonstration of science and technology's ability to create societal change. When artificial intelligence (AI) is used in language learning and teaching, the term Computer Assisted Language Learning (CALL) takes on a new meaning. The role of AI in language teaching and learning is that it allows students to acquire a language more quickly. A typical ESL class has between 30 and 35 students with varying levels of ability. ESL teachers have a wide variety of obligations in this situation. As a result, the goal of this research is to see how artificial intelligence (AI) may be employed in second language classes to boost productivity and minimise the obligations of ESL teachers. When a student utilises Siri, Apple's iOS operating system, to talk in English, this is an example of AI being utilised for language acquisition. This is especially important for English as a Second Language (ESL) and English as a Foreign Language (EFL) students (English as a Foreign Language). More precisely, it aids a group of students in improving their speaking skills, particularly their pronunciation and listening abilities.

The study's goal is to discover the tactics and techniques for using AI to teach/learn English, as well as the success of these strategies from the perspective of university students and the practicality of implementing them. The project also tries to figure out how to use AI applications in the creation of English teaching and learning programmes [3].

2 Literature Survey

A successful education system has the twin role of developing a nation's most valuable resource, its human resource. Younger generations must also be taught so that they are 'ready for life' and can contribute positively to their country's progress and enrichment. They must be exposed to such learning environments with the support of current tools and knowledgeable trainers in order for their learning outcomes to be optimised and tailored to each learner's ability. In order to do this, contemporary education should establish the objective of preparing students to be 'AI Ready.' AI is at the heart of the world's many applications; it incorporates and develops a wide range of capabilities that may be used in a variety of fields of research and operations. The picture below in Fig. 1, depicts some of the most critical AI abilities that share substantial characteristics and links with those in other disciplines of study.

A close examination of the graph above leads us to assume that many of the technologies, as well as the underlying principles that each of them follows, have a significant link to the teaching and learning processes at the secondary and post-secondary levels. As a result, it is critical that AI not only be taught as a topic in schools, but also as a connection to the teaching of other disciplines at all levels. Many AI-based applications are now accessible to help a learner study at his or her own speed and in his or her own fashion. Figure 2 represents five ways AI capabilities would assist instructors in achieving desired learning outcomes.

Once AI technologies are in place, teachers will be able to spend more time in the classroom since they will have more free time. As a result, she or he may now concentrate on her pupils' individual learning methods. She or he can then focus appropriately on the problem of acquiring language processing, reasoning, and cognitive modelling skills, having assumed AI capabilities.

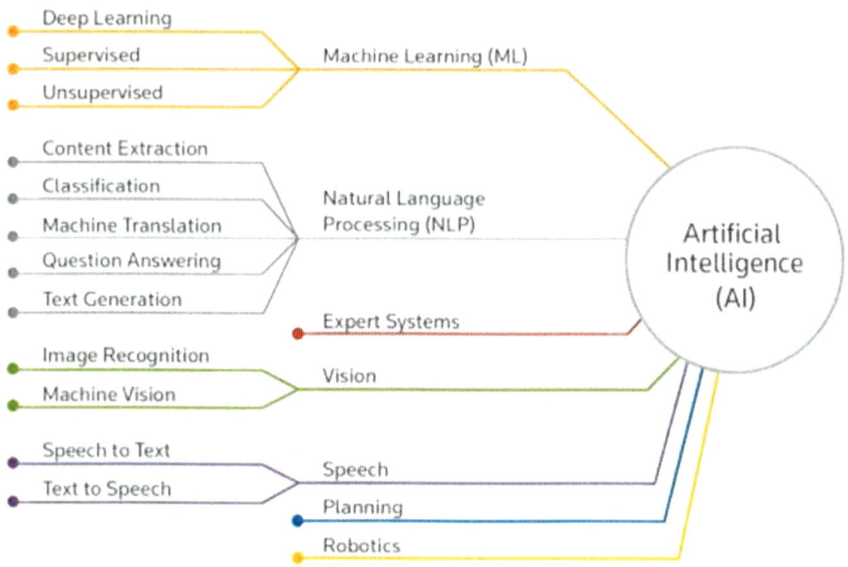

Fig. 1 Competencies of AI with teaching learning process

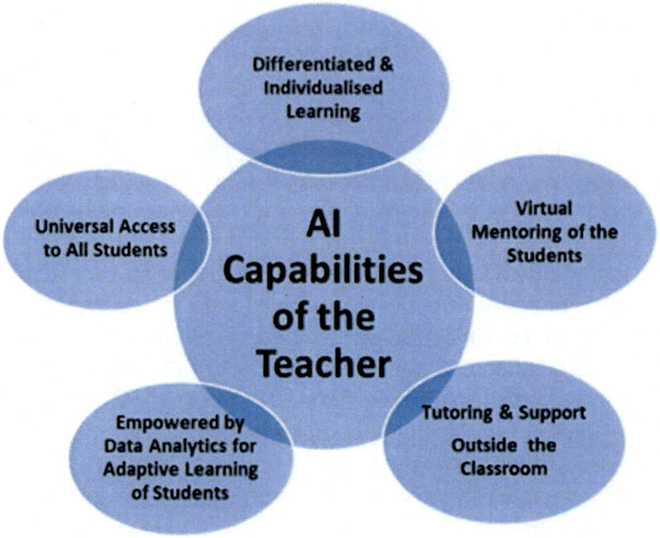

Fig. 2 Desired outcomes that can be achieved by AI in an ESL classroom

3 Benefits of AI in Language Learning

It is nearly hard for a teacher in a classroom of 25 or more pupils to find the proper approach for everyone. However, when employing artificial intelligence to learn a new language, the demands of each individual learner may be taken into account. Educators may acquire a lot of data on students, their interests, their talents, and so on if AI is integrated into the learning process. This information, when examined, can pave the path for tailored instruction [4].

Learners may work at their own speed using AI-powered language learning tools, repeating themes and stressing items they struggle with, engaging them with the activities they excel at, appealing to their interests, and taking into consideration aspects like cultural background. Teachers may also use data to figure out what's going on in their pupils' heads and anticipate their future performance.

1. *Providing immediate feedback*—In language learning, AI gives pupils quick feedback. It is hard to wait for results after putting in a lot of effort for a big test. When the mistakes are pointed out after a week, you may not remember how or why you committed them. The AI language learning platform can grade examinations and even analyse compositions after they've been submitted, highlighting errors and suggesting ways to avoid them in the future. This allows pupils to correct their faults right away and, as a result, score better on following examinations.
2. *No fear of failure*—Making mistakes is the best way to learn. Students are self-conscious when they make mistakes, obtain lower grades, or fail to answer questions, worrying that their teacher's impression of them may suffer. AI does

not criticise or condemn pupils in language learning, nor does it tell them they are not smart enough in front of their peers, nor does it threaten them with complaints to their parents or a principle visit. This strategy allows learners to be assessed without being judged.

3. *Deeper engagement in the learning process*—Students will be able to study at their own pace, set their own goals, and follow a tailored curriculum from anywhere on the planet when AI is used to learn a new language. Teachers will not have to go over the same material every year thanks to a personalised learning technique that vary from student to student. In addition, AI will aid in the creation of fun games, quizzes, and other learning and exploration activities that connect academic programmes with students' interests.

4 Major Applications of AI in Language Learning

1. **Placement exam**—Provide learners with a well-crafted placement test at the outset of their study to match the difficulty level of the course to their real talents. Every subsequent question in the AI-based placement test is adjusted in difficulty based on your previous response. Following a learner's correct response, a more difficult question is presented. In the event that you make a mistake, the next question will be much easier. A placement exam is used to assess a learner's knowledge level and identify shortcomings that should be addressed first.
2. **Chatbots (text and voice)**—Another beneficial use of AI-powered language learning apps is chatbots for communication practise. It simulates real-life discussion to help students enhance their communication abilities. AI algorithms are used by language learning chatbots to comprehend the context of the discussion and respond in a unique fashion. It implies that for a comparable question, each student receives a distinct response. AI chatbots that use speech recognition to teach proper pronunciation are the newest trend. In real time, an AI-powered speech chatbot converses with a student. The subjects discussed are from day-to-day living. The student merely speaks into the smartphone's microphone, and the AI listens and analyses pronunciation in real time.
3. **Tailored learning route**—In order to satisfy and keep users, the app must provide a personalised learning path. Each learner's skills and weaknesses are assessed by the AI system, which then generates a tailored collection of learning materials. Due to a more immersive and customised approach to learners' demands, using AI in language learning helps to cut learning time. The period when a piece of information is updated is adjusted using an AI-powered spaced repetition approach dependent on the difficulty level of the learned material. The following is an example of how AI-driven spaced repetition could appear to users. There are five degrees of difficulty for each topic. The topic is deemed done after all five levels have been completed (highlighted with gold). However,

after a certain amount of time (determined by AI), the icon becomes broken, indicating that the content needs to be evaluated.

4. **Automated assessment**—The Duolingo English Test is the most sophisticated example of using AI for automated assessment. This exam establishes a language level based on an international benchmark. Machine Learning and Natural Language Processing are two AI technologies that are used to:
   - generate hundreds of assessment tests automatically;
   - analyse and grade complicated responses;
   - align test results with CEFR language competence levels (A1 - C2);
   - collect all data to calculate the final score.
5. **Proactive smart alerts**—Notifications urging users to practise skills keep them on track and help the software retain users. App notifications may be made more effective with the help of artificial intelligence. To do so, you must choose when to send notifications and what to tell each learner. With a daily activity log in the app, AI determines the optimal moment to deliver the reminder and what to say in the reminder to reintroduce the learner. Duolingo observed significant increases in the number of individuals returning after implementing AI-enabled alerts [5].

5 Top 5 AI-Powered Language Learning Apps

1. **Duolingo**—A language learning website and smartphone app based in the United States. It is the world's most popular language learning platform. Duolingo has 106 distinct language courses available in 38 different languages. It comes with AI algorithms to provide each student with a personalised learning experience. Duolingo's language chatbots allow users to learn a new language without the shame of conversing incorrectly with a native speaker. Duolingo's AI features include:
   - Chatbots to mimic real-life contact with native speakers;
   - Placement tests to determine a student's present ability level;
   - Proactive reminders to alert learners to the need to begin studying;
   - Personalised lessons to adjust the difficulty level of exercises for a specific learner;
   - The business operates on a freemium strategy. Although the information is completely free, Duolingo does charge a price for its premium service.
2. **Memrise**—Memrise is a British language learning tool that focuses on boosting information retention through spaced repetition of flashcards. In 2017, the app was named Best App Winner at the Google Play Awards. Memrise can assist you in learning 23 different languages. More than 50 million people from 189 countries are amongst their users. The app's unique feature is real time item recognition. This approach entails taking a snapshot of an object and feeding it

to the app, which will then determine the object's name in the target language. Memrise employs the following AI features:

- Speech recognition feature to practise pronunciation
- Text chatbots to practise learner's vocabulary in conversational circumstances
- Individualised learning activity to modify the difficulty level and material relevancy for each user

  Memrise began as free language learning software with paid premium features. Later, the firm changed its business plan to become a full-fledged language app for a fee. The early learning courses and material are remain free, allowing learners to test the app without having to purchase.

3. **Rosetta Stone**—Users may study 30 languages with this language learning programme. This software use the TruAccent speech engine to ensure proper articulation, and it teaches vocabulary and grammar through spaced repetition using graphics, text, and sound. The business collaborated with the US Army to develop a unique military version of Arabic to aid troops in the Middle East in learning the language for crucial military discussions and phrases. Rosetta Stone has also received several accolades from non-profits, periodicals, and the software business.

   Rosetta Stone's AI characteristics include:

- Real time object recognition to identify items surrounding a student and learn new words;
- Speech recognition to examine learner's pronunciation of new words and offer quick feedback.

4. **Busuu**—Busuu offers over 100 million learners throughout the world online and mobile courses in 12 different languages. Busuu's distinguishing characteristic is language learning based on a combination of AI-powered learning content, peer engagement, and one-on-one live tutoring with expert teachers. Busuu for Organisations is a service provided by the firm for colleges and corporations. Organisations may provide Busuu Premium access to their students or staff, measure their development, and utilise the app over time. Furthermore, businesses receive customised training that incorporate teachings on specific situations. Uber, for example, receives training on circumstances and words that a driver can encounter with passengers. Busuu's AI features include:

- Tailored learning to give information that is appropriate to the learner's interests, objectives, and skills;
- Grammar training to review rules on-demand and practise weaknesses;
- Vocabulary training to teach vocabulary in a timely manner for greater retention.

  Learners may sign up for a free account or upgrade to a Premium account to have access to additional features such as advanced grammar courses,

Offline Mode, McGraw-Hill Education accreditation, and our adaptive Vocabulary Trainer.

5. **Babel**—Babel is a language learning platform that provides courses in 14 different languages. Babel learning employs interactive discussions to focus on real-life scenarios. Curated review sessions reintroduce previously learned content in new circumstances, reinforcing it.

   Babbel's AI features include:

   - Voice recognition technology to improve pronunciation;
   - Tailored learning activities to keep students engaged;
   - A review manager tool that uses the spaced repetition strategy to improve information retention.

Bebel's founders improved the app a year after introducing the original free version, opting for paid content over advertising and a mixed-finance model (freemium). Learners can pick from four subscription levels based on the length of the course.

The features of the various AI-based programmes for learning English are summarised in the table below in Table 1 [6].

6 Everyday Tools for Language Learners

1. **Google Docs speech recognition**—Google Docs now allows you to type with your voice. This is a free and mobile-friendly speech recognition function. It can help with conversational tasks. It may be used by students to assess the intelligibility of their own speech.
2. **Using Google Assistant** to practise speaking and listening skills—Google Assistant is an excellent tool for practising speaking and listening abilities. A language student can ask the assistant simple queries such, "What's the current news?" "How is the weather?" "What are the greatest TV series to watch for foreigners?" and so on. This is an excellent technique to assess pronunciation intelligibility as well as listening and comprehension abilities.
3. **Practising directions with Google Maps**—Tools with built-in AI are useful not only for practising pronunciation but also for improving speaking abilities. Google Maps is a navigation and mapping programme that is not designed for language learning. However, utilising this app to learn directions, names of different places, and how to go someplace is quite beneficial.

To begin, a student can study the fundamental instructions such as turn right/left, cross the street, next to, move straight ahead, opposite to, and so on. They can also make a list of neighbouring sites.

After then, students can practise their speaking skills by describing how to go from their current location to the destinations specified.

Table 1 Comparative study of 5 main language learning applications

Application	Pros	Cons
Duolingo	• Makes use of a lot of pictures and symbols to help you remember what you have learnt • Duolingo is really simple to use, and you have a good understanding of the information and possibilities • Learned phrases or terminology are always uttered aloud. Recording activities are also available to help you enhance your speaking abilities • You may obtain the meaning/translation of a term in a translation job if you are unclear about it. In almost every activity, grammar rules are mentioned • You may retry particular workouts or refine previously gained abilities, which help to reinforce what you have learned • Whole app is free of charge	• There is no information on how the entire course works at the start • If you wish to learn a certain topic, you will not be able to do so unless you complete all of the prior tasks • The terminology and sample phrases are occasionally odd and would not be used in everyday situations
Memrise	• Strong language study content for beginners • More than just flashcard learning • Good customization options in settings • High quality	• No guarantee on quality of user-generated content
Rosetta Stone	• Outstanding user experience • Sophisticated interface • Highly instinctive • Mobile apps offered • Optional live teaching sessions • Bountiful bonus content • Strong tech backing	• Expensive • Absences of translations provided or cultural information • No assignment for students with previous language contact • Repetitive
Busuu	• The layout is engaging and easy to use • The conversation lessons are very useful • The social feature is awesome • Busuu offers useful cultural tips • You can create a study plan that's relevant to your life	• Limited Language Options • Some exercises do not have translations • The grammar explanations and practise have a lot of space for improvement • The Chinese lessons have poor quality • You Can Receive Incorrect Corrections • Lack of Entertainment

(continued)

Table 1 (continued)

Application	Pros	Cons
Babel	• A good overview of your progress, different assignments, and other possibilities is provided by Babbel • Babbel uses a lot of attractive graphics to help you remember vocabulary and phrases • Babbel uses a lot of appealing pictures to help you remember vocabulary and sentences the layout and style is highly inviting and enjoyable • Beginners, middle school courses, grammar, business English, talk and listen, read and write, nation and people, extras, words and phrases, and more are all covered	• There is no information regarding how the course is structured at the start (e.g. how many questions are there? variety of tasks?) • To begin, you only have access to one part. If you wish to access additional tasks, you must subscribe for one, three, six, or twelve months • Some applications are designed to be played like a game, but Babbel is designed to be used like conventional education and might get tedious after a time

7 The Prerequisites for Successful AI Learning Implementation

AI has the ability to provide a tailored environment in which adult learners develop English skills using all of their senses at the same time, based on their present level of English, occupational needs, or hobbies. However, keep in mind that the use of an AI robot's synthetic voice should be restricted to instructions and recommendations, with the majority of lessons and drills being recorded by native speakers, as the emotional component of a human voice are required for the subconscious training of correct pronunciation. Adult learners should have access to AI classes that are based on real content and recorded by native speakers. AI could hardly be expected to produce an original answer when applied to a static field of knowledge. Even expert English teachers have had a high failure rate in the past because traditional techniques of teaching a foreign language do not address the following issues:

- How to avoid a dreadful forgetting curve for the subject taught.
- How to overcome the ingrained tendency of thinking in one's native language and cross-translating, both of which are seen as major roadblocks to fluency.
- How to develop intuitive grammar, which functions as a sense of right and wrong. Whilst compared to the formal grammar rules that are memorised in the original language but are impractical when speaking, this is the case.
- How each student may use a mobile application to construct their own lessons and use simultaneous repetition to the new courses?

Adaptive or customised learning using AI, in my opinion, is a fantastic notion, but the success of its implementation will be determined by the learning approach used. Without modifying the learning methodology and introducing a new, more effective methodology, using AI to traditional ways of teaching English that have a success rate of roughly 5% would not drastically boost the success rate.

8 Conclusion

If the AI algorithm is built on new learning approaches, such as subconscious English skill training, AI might become the finest instrument for learning English. Subconscious training is seven times quicker than traditional conscious learning and belongs to one of the mind's systems. The AI tools will provide an active and interesting learning environment for the students here, allowing them to access personalised lessons, view their development records and frequent errors, communicate with the teacher and others when they need clarifications, and learn the lesson at home if they are absent. To begin, however, both ESL teachers and students need have a solid understanding of how to use a computer, courses should have computers for all students with internet access, and the system should be maintained by specialists. Artificial intelligence research and development will continue. We should not only pay attention to the practise and application of AI in teaching in modern English teaching and learning, but also fully integrate AI and English teaching activities and interact effectively, optimise the effect and mode of English teaching, and promote the healthy development and reform of English education.

References

1. Marr B, Ward M (2019) Artificial Intelligence in Practice. UK, Print
2. Dodigovic M (2005) Artificial intelligence in second language learning—raising error awareness. Cromwell Press Ltd. UK, Print
3. Gogate L, Hollich G (1962) Theoretical and computational models of word learning: trends in psychology and artificial intelligence. Information Science Reference, The United States of America, Print
4. Swartz ML, Yazdani M (1992) Intelligent tutoring systems for foreign language learning. Springer Science and Business Media, Print
5. https://www.telc.net/en/about-telc/news/detail/is-artificial-intelligence-the-future-of-language-learning.html
6. https://www.intellias.com/how-ai-helps-crack-a-new-language/(6)

Pattern Recognition

Block-Based Discrete Cosine Approaches for Removal of JPEG Compression Artifacts

Amanpreet Kaur Sandhu

Abstract Image compression plays an important role in different fields, such as medical imaging, digital photography, multimedia, interactive Television, mobile communications. The main function of image compression is to reduce the size of image, transmission cost and also occupy less storage space. It maintains the perceptual quality of an image without ant loss of important information. Therefore, space and time are two main components of image compression. Moreover, Image compression is a successful and popular technique which reduces the size of image and display tangible information of given data. Lossless, lossy and hybrid compression are the different types of techniques which used to reduce the size of images and videos. In this paper, advantages and efficiency of image compression have been discussed. There are two types of blocking artifacts filters that are discussed along with types of blocking artifacts.

Keywords Blocking artifacts · Image compression · Filter · Outliers

1 Introduction

The fundamental development in image compression is the establishment of various standards such as Joint Picture Expert Group (JPEG) for still images and Moving Picture Expert Group (MPEG) for videos. For example, using JPEG standard, an 8 bits/pixels image can be compressed between 1 and 2 bits/pixels with minimum visual artifacts. The amount of compression can be measured by the bandwidth requirement of an application. The application having small bandwidth available needs high compression ratios. Block-based Discrete Cosine Transform (BDCT) is used to reduce the size of images as well as videos with high energy compaction feature. In BDCT, images are divided into number of blocks with size N × N, whereas N × N is the size of sub-image block [1]. As the compression increase, compression techniques yield visually blocking artifacts which degrades the perceptual quality of

A. K. Sandhu (✉)
University Institute of Computing, Chandigarh University, Mohali, Punjab, India
e-mail: aman123.brar@gmail.com

P. Dutta et al. (eds.), *Emerging Technologies in Data Mining and Information Security*, Lecture Notes in Networks and Systems 490,
https://doi.org/10.1007/978-981-19-4052-1_31

an image. Blocking artifacts are most annoying distortion and destroy the relationship among two adjacent blocks. Therefore, blocking artifacts are alleviate with transform coded techniques especially at low bit rates. Moreover, discontinuity commonly exists in the block boundaries of 8 × 8 blocks because of independent transformation and quantization process of 8 × 8 DCT coefficients [2].

There are generally three kinds of blocking artifacts namely, stair noise, grid noise and corner outliers. These artifacts become more noticeable in the smooth regions of image. To maintain the perceptual quality of image and to achieve higher bit rate, post-processing methods provides an excellent solution. To alleviate such blocking artifacts several post-processing techniques have been designed in last decades [3]. These techniques only generate satisfactory performance with unreasonable assumption that quantization noise has been produced or generated. The block-based techniques using low pass filtering to alleviate different types of blocking artifacts. However, these filtering techniques are not adaptive in nature and these are able to suppress the blocking artifacts and therefore output generated as over-smoothing as well as over blurring of reconstructed images especially at low bit rates.

1.1 Image Compression Fundamentals

The primary objective of image compression is to reduce the size of data and transmission cost and maintaining the perceptual information. There are three basic types of data redundancies in image compression [4]:

(i) **Coding Redundancy**: The coding redundancy is also called statistical redundancy which is used to represent the set of events as well as information. The code word is assigned to an event or every piece of information, called code symbols. The number of code symbols in each code word is determined as its length. The 2-D intensity arrays are represented by 8-bit codes which consist of extra bits than are required to exhibit the intensity.

(ii) **Spatial Redundancy/Interpixel Redundancy**: In spatial redundancy, the gray values of adjacent pixels are highly correlated as well as may be practically predicted from each other. In this every pixel of 2-D intensity array is dependent on or similar to surrounding pixels and the information is unessential replicated in representation of dependent or similar pixels.

(iii) **Irrelevant Information/Psychovisual Redundancy**: Irrelevant information is based on Human Visual System (HVS) of image information. 2-D intensity arrays variations may be perceived in domain of constant intensity. The HVS does not respond with equal sensitivity to all visual information. Therefore, in normal visual processing certain information is not very important than other information.

1.2 Advantages of Image Compression

There are the following advantages of image compression [5]:

(a) It reduces the communication costs as well as data storage requirements while transmitting the large volume of data over the network with the help of an effective bandwidth.
(b) It improves the quality of various multimedia presentation systems via limited data transmission rates.
(c) The speed of input–output functions in the computing device is extremely increased because of smaller representation of the data. For example, in the hierarchy storage system, the principle of data compression makes faster storage level, while decreasing the burden on input–output communication channels.
(d) It also reduces the cost of backup and recovery of data in the computing device by saving the large volume of databases into compressed forms.

1.3 Image Compression Efficiency

Compression efficiency is commonly determined as the following way, CR

$$\text{CR} = \frac{\text{Total data size of orignal input image}}{\text{Total data size of compressed image}}$$

where CR compares the local data size of an original input image with the total size of compressed image [6]. The compression efficiency can be also determined in the form of average bit rate, bits per pixel (bpp)

$$\frac{\text{Total number of bits in encoder input}}{\text{Total number of bits in decoder output}}$$

The idea behind image compression is to reduce or remove the redundancies that occupy more storage space as well as transmission time. The data redundancies do not carry any new type of information and thus have no contribution in improving the image quality. In general, image compression can be classified into two parts namely lossless and lossy image compression [1].

(i) **Lossless Compression**: Lossless compression approach performs a perfect reconstruction and it does not loss any information during compression-decompression process. Therefore, it allows error free and reversible compression. In other words, it does not loss the perceptual quality of an image as well as preserves the image contents and allows lower compression ratios such as 1–3:1.
(ii) **Lossy Compression**: Lossy image compression technique is irreversible in nature which allows higher compression ratios such as 30–100:1. The lossy

compression generates imperceptible difference between original and reconstructed data. The lossy compression algorithms are generally applicable where small loss of information is tolerable by human visual system (HVS) such as graphics, speech, images and videos.

(iii) **Hybrid Compression**: Hybrid compression is the combination of both lossy as well as lossless compression within a single encoding block. The primary feature of hybrid compression is that it has advantages of both lossless and lossy compression. The hybrid compression ratio lies between lossless and lossy compression ratios.

2 Existing Filtering Methods

A large number of deblocking techniques have been proposed for detection and reduction of blocking artifacts in compressed images. After reviewing the literature of techniques, it has been found that these deblocking approaches can be classified into two categories [7]:

(i) Post filtering method
(ii) Loop filtering method.

The post filtering method is performed after the image decompressed. It improves the visual quality of an image without any changes in encoding or decoding procedures. It makes compatible with JPEG and MPEG coding standards. The preprocessing/loop filtering is performed within the coding loop. In other words, blocking artifacts are alleviating at encoding side. However, loop filtering-based techniques are difficult to integrate with aforesaid standards.

The majority of post-processing filtering methods are implemented in spatial as well as frequency domain techniques.

3 Blocking Artifacts Overview

It is well known that BDCT coded images suffers from so-called blocking artifacts because of independent quantization/transformation of each block. In BDCT, an image is divided into blocks of size $N \times N$ ($N = 8$) where $N \times N$ is the size of sub-image blocks. Sub-image blocks can be of size 2×2, 4×4, 8×8, 16×16, 32×32 pixels. This has been proposed [1] that the Mean Square Error (MSE) increases as in sub-image size decreases, maximum for 2×2 sub-image. The MSE is almost minimum for the block size of 8×8 and becomes constant as the block size is further increased. The image and video coding standards are widely used to code images and frames on a block-by-block basis in which each block is transformed, quantized and coded separately [8, 9].

When compression rate increases, the correlation between adjoining pixels of adjacent block decreases and reconstruction of more pixels becomes poor. Therefore, an artificial discontinuity appears along with the horizontal as well as vertical block boundary named as blocking artifacts, which becomes more visible as compression ratio increases. Such artifacts are very disturbing to the viewer as it severely reduces the image quality. Increasing bandwidths and bit rate beyond a certain limit is not possible and too costly. Consequently, it becomes very important to eliminate different types of blocking artifacts from compressed images especially at low bit rate [10].

The primary drawback of BDCT compression schemes is that it may results in visible artifacts near the block boundaries (horizontal/vertical) due to the rough quantization mechanism of DCT coefficients. Blockiness is the most serious defect in compression artifacts. Blocking artifacts eliminate several high frequency DCT coefficients in every segmented macro-block from an original image and also generate serious quantization errors to low frequency coefficients. These artifacts become more noticeable in the smooth regions of image.

(a) **Qualitative Definition**

Blockiness is qualitatively defined as "distortion of an image characterized by presence of critical block encoding framework". It cannot be defined mathematically in any standard. However, the blockiness may be represented mathematically as artificial boundary discontinuities in amplitude at the block boundary [11]. The higher value of blockiness determines that the higher visibility of block structure. Figure 1a illustrates that how blocking artifacts are visible in JPEG compressed image by using 8×8 block procedure of DCT. Figure 1b shows the zoomed part near the eye area along with blocking effects.

(b) **Quantitative Definition**

Blockiness is quantitatively defined as "an intensity jumps at block boundaries". It can be determined in units of grayscale intensity. Whenever, the size of a block is known, the value of intensity gradients at each block boundary

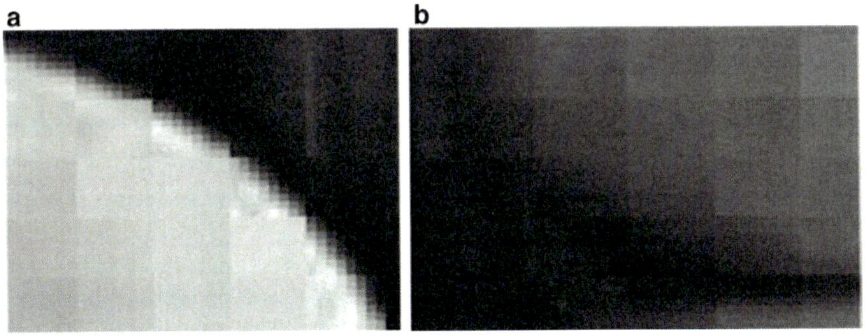

Fig. 1 **a** Stair noise, **b** grid noise

may be evaluated. The shape of a block is commonly square and rectangular and the gradients values are determined in both horizontal as well as vertical directions.

3.1 *Observations of "Blocking Effects"*

There are the following two primary observations of blocking effects that are determined in transformed based coding [12]:

(a) In terms of human visual system (HVS), the human eyes are more sensitive to the flatter regions of an image with different complexity. Therefore, blocking effects are more noticeable in smooth regions than in complex regions.
(b) The deblocking approaches can alleviate some high frequency discontinuities among the block boundaries. Although, it may blur the true edges of an original image.

3.2 *Need for Detection and Reduction of Blocking Effects*

Detection and reduction of blocking artifacts in JPEG compressed images are important as:

(i) It reduces the computational load as well as time of deblocking techniques.
(ii) It severely reduces the perceptual quality of an image, making it unpleasant to the viewer. After reduction of compression artifacts, the subjective picture quality improves.

3.3 *Types of Blocking Artifacts*

There are the following types of blocking artifacts which visually annoying in JPEG decompressed images:

(a) **Staircase Noise**: Staircase noise is one of the most dominant blocking artifacts which appear along curved or diagonal edges. When an image divided into 8×8 DCT block includes an edge, an edge is degraded and block boundary appear to be edge. Figure 1a shows staircase noise [13].
(b) **Grid Noise**: Grid noise artifacts occur in monotone area of an image. When a slightly change of image intensity gradients along with 8×8 block boundaries and these artifacts are noticeable in monotone area. Although, this change is determined as grid noise artifacts. Figure 1b shows grid noise.
(c) **Corner Outlier**: A corner outlier traditionally appears at the corner point of 8×8 DCT block, where corner point is much larger or smaller than surrounding pixels. Moreover, the corner outliers are commonly appearing

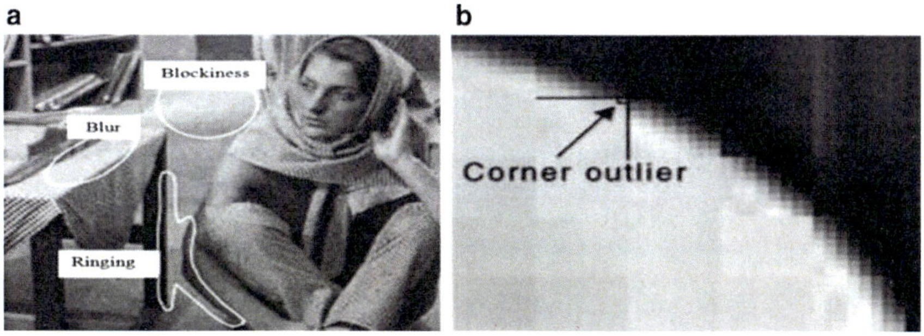

Fig. 2 **a** Blocking artifacts, **b** corner outlier

around the vertices of a block with no size less than 8 × 8. Figure 2a, b shows blocking artifacts and corner outliers [14].

(d) **Ringing Artifacts**: Ringing artifacts occur when an image is transformed into frequency transform domain with the help of compression techniques. The ringing effects are mainly caused by truncation or coarse quantization of high frequency DCT components (Gibbs phenomenon) that make decompressed image as noisy pattern called mosquito noise or ringing artifacts near the image edges [15].

(e) **Blurring Artifacts**: Blurring artifacts means that an image is much smoother than original image. When a high spatial frequency DCT components are lost at low bit rate data, it may result in un-sharpness or fuzziness or blurring artifacts. Moreover, a moderate degree of blur is not observed by HVS. Blockiness is also noticeable at low bit rate JPEG images, multimedia, MPEG videos and DCT coded images [16, 17].

3.4 Blocking Artifacts Techniques

There are following detection techniques for blocking artifacts (Table 1).

4 Conclusion

The post-processing techniques are classified into POCS, Estimation theoretic-based deblocking methods, wavelet-based methods, Adaptive deblocking methods and Frequency domain methods. The compressed images are affected by mainly three types of artifacts namely blocking artifacts, ringing artifacts and corner outliers. These artifacts can be suppressed by using post-processing approaches so that further analysis may be performed effectively. The removal of blocking effects is an important aspect especially at low bit rates in image and video coding. Further, the key

Table 1 Comparison of various deblocking techniques

Projection on convex set (POCS)-based deblocking methods	Estimation theoretic-based deblocking methods	Wavelet-based deblocking methods	Adaptive deblocking filtering methods	Frequency domain techniques
The POCS-based algorithms are recursive in nature which require high computational burdened from DCT/IDCT in every iterative step	These methods are iterative in nature and takes more time	Wavelet-based methods are non-iterative in nature	The main issue of adaptive deblocking filtering is the trade-off between sufficient smoothing and maintaining details of an image. The term trade-off is accounted by the adaptive nature of an algorithm	Transform domain techniques are fast in nature as compared to spatial domain methods. The limitation of spatial domain techniques generate output as over-smoothing and over blurring of an image due to its low pass filtering nature
These methods are very complex in nature and generated the result as worse performance that is not acceptable	These techniques are not easily applicable on real time applications	Various statistical characteristics used for block discontinuities	There are still some problems occur, especially in some complex regions of an image. Ringing artifacts are introduced during the coding of strong edges	Sometimes, a low pass filtering approach introduces blurring while removing blocking artifacts
Algorithms based on POCS need abundant computational burden. Requires numbers of iterations to achieve convergence	High computational complexity as well as execution speed of algorithms is slow	Sometimes, wavelet-based techniques show poor performance in edges areas	Sometimes, the computational load is very high due to multiple reapplications of JPEG compression	These types of algorithms fail to remove artifacts in diagonal area as well as curved edge areas

feature of post-processing approaches is that these techniques are adaptive in nature, images can be made more uniform to avoid blur and preserve the image contents. After reviewing the literature, it has been found that there are some drawbacks that exist in existing post-processing techniques.

References

1. Kaur A, Sidhu JS, Bhullar JS (2018) Artifacts reduction based on separate modes in compressed images. J Intell Fuzzy Syst 35(2):1645–1656
2. Sabbavarapu SR, Gottapu SR, Bhima PR (2021) A discrete wavelet transforms and recurrent neural network based medical image compression for MRI and CT images. J Ambient Intell Humaniz Comput 12:6333–6345
3. Kaur A, Sidhu JS, Bhullar JS (2021) Adaptive deblocking technique based on separate modes for removing compression effects in jpeg coded images. Int J Comput Appl 43(6):501–513
4. Saravanan S, Sujitha Juliet D (2021) A hybrid approach for region-based medical image compression with nature-inspired optimization algorithm. Innovations in Computer Science and Engineering, Lecture Notes in Networks and Systems 171. https://doi.org/10.1007/978-981-33-4543-0_24
5. Xin G, Fan P (2021) A lossless compression method for multi-component medical images based on big data mining. https://doi.org/10.1038/s41598-021-91920-x
6. Chaudhary AK, Mehrotra R, Ansari MA, Tripathi P (2021) Novel scheme for medical image compression using Huffman and DCT techniques. In: Agrawal R et al (eds) Advances in smart communication and imaging systems. Lecture Notes in Electrical Engineering 721. https://doi.org/10.1007/978-981-15-9938-5_28
7. Taimori A, Razzazi F, Behrad A, Babaie-Zadeh AAM (2021) A part-level learning strategy for JPEG image recompression detection. Multimed Tools Appl 80:12235–12247
8. Eerenberg O, Kettenis J, Peter HN (2013) Block-based detection systems for visual artifact location. IEEE Trans Consum Electron 59(2):376–384
9. Ahmed ST, Sankar S (2020) Investigative protocol design of layer optimized image compression in telemedicine environment. Procedia Comput Sci 167:2617–2622
10. Kaur A, Sidhu JS, Bhullar JS (2018) Artifacts reduction based on separate modes in compressed images. J Intell Fuzzy Syst 35:1645–1656. https://doi.org/10.3233/JIFS-169702
11. Kumar R, Patbhaje U, Kumar A (2019) An efficient technique for image compression and quality retrieval using matrix completion. J King Saud Univ Comput Inf Sci
12. Zhang G, Wang J, Yan C, Wang S (2019) Application research of image compression and wireless network traffic video streaming. J Vis Commun Image R 59(2019):168–175
13. Han J, Saxena A, Melkote V, Rose K (2012) Jointly optimized spatial prediction and block transform for video and image coding. IEEE Trans Image Process 21(4):1874–1884
14. Chen Y (2017) Variational JPEG artifacts suppression based on high-order MRFs. Signal Process Image Commun 52:33–40
15. Kim J (2009) Adaptive blocking artifacts reduction using wavelet- based block analysis. IEEE Trans Consum Electron 55(2):933–940
16. Li J (2013) An improved wavelet image lossless compression algorithm. Optik 124(11):1041–1044
17. Brahimi T, Laouir F, Boubchir L, Chérif AA (2017) An improved wavelet-based image coder for embedded greyscale and colour image compression. Int J Electron Commun 73:183–192

ODNN-LDA: Automated Lung Cancer Detection on CT Images Using an Optimal Deep Linear Discriminate Learning Model

Alaa Omar Khadidos

Abstract Lung cancer is classified as one of the most common forms of cancer that cannot be ignored and can lead to death if treated late. According to the American Cancer Society, CT scans can currently diagnose lung cancer in its earliest stages. However, there are various occasions in which the doctors' experience diagnosing lung cancer can cause some complications. Human survival rates can be improved by early identification. The average survival rate for persons with lung cancer increases from 14 to 49% if the disease is detected early. CT is significantly superior to X-ray in terms of accuracy, but a complete diagnosis needs numerous imaging modalities that work together to support one another. There is much evidence that deep learning is a popular and effective tool for medical imaging diagnosis. This paper proposes a new method of automatic diagnosis classification that has been developed for the scanned CT images of the lungs. A implementation of the Optimal Deep Neural Network (ODNN) in addition to the Linear Discriminate Analysis (LDA) was used in this work to analysed lung images obtained by CT scan. Malignant or benign lung nodules can be classified by obtaining in-depth features from the CT images of lungs and the redundancy of the dimensionality of the feature with Linear Dimensionality Reduction (LDR). The proposed method uses the Modified Gravitational Search Algorithm (MGSA) to shows an improvement of the ODNN's classification of lung cancer. In terms of sensitivity, specificity, and accuracy, the proposed classifier scored 96.2%, 85.37%, and 93.33%, respectively.

Keywords Lung cancer · Deep learning · CT images · Optimization · Classification · Image processing

A. O. Khadidos (✉)
Faculty of Computing and Information Technology, Department of Information Systems, King Abdulaziz University, Jeddah, Saudi Arabia
e-mail: aokhadidos@kau.edu.sa

P. Dutta et al. (eds.), *Emerging Technologies in Data Mining and Information Security*, Lecture Notes in Networks and Systems 490,
https://doi.org/10.1007/978-981-19-4052-1_32

1 Introduction

Medical image evaluation has exceptional dominance in the health sector, mainly in the non-invasive treatment and clinical testing [1]. For an accurate diagnosis, X-ray, ct, MRI, and ultrasound pictures are employed. To record images on film using attractive fields, CT scans are one of the filtering methods used by medical imagers. One-of-a-kind lung cancer is responsible for 1.61 million fatalities per year. Indonesia has a third most common cancer in the world, with most cases occurring in MIoT centres [2]. Getting a lung cancer diagnosis early is not easy because more than 80% of patients with cancer are diagnosed correctly. There are four fundamental phases of lung cancer described in this research. Lung cancer is considered the third most common cancer amongst women, which comes following to breast and the colorectal cancer. In image processing, the feature extraction procedure is one of the easiest and most effective ways to reduce dimensionality. CT imaging's non-intrusive nature is one of its most noticeable properties.

According to the American Lung Association, CT scan is the most effective approach to detect lung nodules since it can provide a three-dimensional (3D) image of the chest, resulting in higher resolution of nodules and tumour pathology. The use of a computer-processed CT scan to aid in lung nodule diagnosis is common in clinics. For this reason, CADx can discriminate between benign and malignant pulmonary nodules based on texture, form and growth rate because the likelihood of malignancy depends on geometric size, shape, and appearance. This means a CADx system's success may be judged by its diagnostic accuracy, speed, and automation level [3, 4].

Deep learning algorithms have showed potential in tumour histopathology evaluations during the past few years. Deep learning algorithms are particularly suited for labour-intensive tasks such as ROI detection or segmentation, element quantification, and visualisation. AI has also been used to handle experience-dependent problems, such as histology grading, classification or subclassification, and prognostic inference. It was found to be viable to conduct imaging genomics research, including biomarker prediction or discovery, as well as tumour microenvironment (TME) characterisation [5] using digital histopathology slides.

Convolutional neural network (CNN) models with single or multiple convolutional neural network (CNN) layers have been successful in histological categorization of lung cancer [6]. For viewing, annotating, and data mining of full slide images, computational methods have been created (WSIs). However, QuPath, DeepFocus, ConvPath, HistQC, and ACD Model [7] are referred to in the article just as generic WSI analysis tools, and not as lung cancer specific techniques. Several pioneering research have also studied the link between DNA genotypes and morphological traits. Whilst breakthroughs were made in NSCLC, a limited number of cases or a single cohort, there is still much work to be done before they have clinical impact. In addition, cases of pulmonary tuberculosis (PTB) with non-typical radiographic signs require surgical inspections to distinguish them from cancer in terms of potential infectiousness, according to the report. Organisation pneumonia (OP) and bronchogenic cancer are difficult to identify, hence patients with a high suspicion of malignancy receive surgical removal.

Extracted or selected features-set could extract a relevant information from input data [2]. For training and testing, the reduced features are assigned to an SVM. Using a neural network models In addition to the binarization image processing increasing the possibility of categorization of lung cancer images. For lung cancer categorization, a neural network model was used, which had an accuracy rate of about 80%. SVM, KNN, and ANN are some of the classifiers that have been studied for lung cancer classification. Based on statistical learning hypotheses, the SVM is a universally beneficial learning tool. Although these techniques are expensive, they are able to diagnose lung cancer at an advanced stage, resulting in a low survival rate.

2 Literature Review

Xie et al. [8] proposed a Fuse-TSD method in 2018. For the heterogeneity of the nodules, a Fourier-shape descriptor is used, and a DCNN is used for the training of the characteristics of the nodes. Using CNN, Chougrad et al. [9] investigated a breast cancer classification system based on the CAD framework. In general, deep learning algorithms requires large datasets to develop systems, whereas transfer learning characteristics only takes a small number of medical images as a data source. CNNs were optimally trained using transfer learning. 98.94% accuracy was achieved by CNN. When paired with wavelet transform and principal component analysis [10]. Van Ginneken et al. [11] and colleagues suggest using transfer learning from OverFeat [12], an object detection network previously trained for natural photos. For each nodule candidate, they begin by extracting 2-D patches from the CT scan. In order to achieve a grayscale image of 221 by 221 pixels, we employed Hounsfield units for scaling and linear interpolation for each patch. These 865 scans from the LIDC dataset were included, as well as nodules 3 mm and larger that were declared positive by three or four radiologists. A section thickness of more than 2.5 mm is excluded, along with images that include inconsistent or inaccurate DICOM information [13, 14]. To do this, the three systems x, y, and z were used in conjunction with the CAD system's score. In the second stage, they utilise a linear SVM classifier to evaluate the chance that a nodule has been detected. By itself, the CAD system came up with 37,262 potential places. Seventy-eight percent of the candidates had genuine nodule. According to Kumar et al. [15], an autoencoder-based CAD system can classify lung nodules based on deep features. Patients' CT images are combined with diagnostic data from the LIDC datasets to arrive at this result. It is estimated that 157 people have been diagnosed with nodules at two levels after undergoing biopsy, surgical excision, or analysing radiological images (the patient level and the nodule level). Diagnoses were used since it was the only way to determine the likelihood of cancer. It begins with a 2-D CT image annotated with nodules. This is followed by a five-layer denoising autoencoder trained using L-BFGS [16]. Fourth-layer features that have been learned are taken from the fourth layer. It was decided to build 200-dimensional feature vectors for every instance using the features from the 4th layer (instance meaning one slice containing nodules). The vector is then used to classify nodules using a binary decision tree. The radiologists' annotations are used to extract characteristics from the autoencoder as long as the nodule is less than 3 millimetres in diameter. According to the size of the nodules, they constructed a rectangle window that adapts. This is followed by resizing each rectangular section to a predetermined size, creating a fixed length input for the autoencoder. Treatment of malignant nodules is reserved for those having a grading of 0, 2, or 3. It was 75.01% accurate with a sensitivity of 0.8325 at a scan rate of 0.39 frames per second. Hua et al. [17] Convolutional layers are configured to 2 dimensions in the CNN to capture local spatial patterns. With the help of max-pooling and sigmoid activation functions, the dimensionality is minimised. For the first layer, they utilise four feature maps, followed by six feature maps, and then an FC layer to classify the nodule. To get a preliminary model, the DBN is first trained unsupervised. For classification tasks, it is then fine-tuned under supervision. Each layer of the DBN is trained one at a time, starting at the bottom and working upward. The maximum log-likelihood is approximated using the stochastic gradient descent method and the counteractive divergence procedure [18]. An LDA summarization approach based on Euclidean norm was developed by Hao Wang and colleagues in 2016 to overcome the drawbacks of conventional LDA. To execute step classification, a multi-class SVM is coupled. When it comes to accuracy and scalability, this method outperforms all others.

3 Methodology

The suggested approach for classifying CT images of the human lung includes processes such as pre-processing, reduction, feature extraction, and classification. First, CT scans were analysed for image quality. Then, employing methods, a feature extraction technique was employed to extract some main features such as histogram, texture, and wavelet from the images. Dimensionality reduction strategies are used to reduce the number of features used in the classification process after features have been retrieved. LDA is a feature reduction method. The LDA-based reduction was used in the proposed classification approach to reduce computation time and cost. It takes longer to compute and uses more memory when you employ the maximum number of attributes. Phase of classification: On the basis of extracted features during this stage, the scanned CT lung images were classified normally, benign, and malignant. To train and test a classifier, it is typical to use training data. As a result of this categorization technique, it is possible to determine whether or not the photos contain lung cancer regions. The ODNN classifier was used in the current work, as proposed in [19], and MGSA optimisation was applied to optimise the structure [19]. Figuring out how to classify CT lung images using this method is really simple, requiring little effort in both training and testing.

3.1 Filtering Phase and Contrast Enhancement

As a result of this noise, the images from the medical datasets were distorted in some way. This filter replaces pixel respect with the middle value if the selected image is noisy and the targeted pixels' neighbouring worth is between 0 and 255. Adaptive histogram equalisation is a contrast enhancement procedure that is performed after removing the noise from the datasets.

In order to obtain the histogram, which existing in the primary position of each line, subtract the trailing column, which includes the new leading row. So that it can distinguish the image's grey level and adaptively adjust its dispersion of two adjacent grey levels in a new histogram, CT images' complexity will need to be raised and limited.

3.2 Feature Extraction

Feature extraction approach, in general, was designed to characterise the image in its unique and compact form of single values or matrix vectors. Features are extracted from an image to reduce its dimensionality in image processing so that it may be utilised for the classification use. It could involves minimising the input data to a smaller set of features that are reflective of it. As a result, the features are used to help classifiers determine what they represent. By estimating positive qualities, the goal of the feature extraction is to reduce the data processing. Texture features, Histogram features, and wavelet features were all recovered from distinct bands of CT images in the current study.

3.3 Features of Histogram

Images are represented as pixels in histograms. Each power value is represented by a histogram, which shows the amount of pixels in the image at each power level. By varying the histogram power settings, it is feasible to approximate a predefined histogram. A total range of Grey levels are decisived from the image input using the histogram approach. In the provided example, a 256 Gy levels ranging from 0 to 255 was determined. Their common characteristics are mean, skewness, kurtosis, and variance as illustrated in Fig. 1.

Mean: As a rough idea of power, the mean gives the average Grey level of each region, but it has no bearing on the image texture.

Standard Deviation: Image contrast variance, or standard deviation, is defined as the square root of variance in the image. High and low variance values are used to determine the image contrast level. An image with strong contrast has a large variance, whilst an image with low contrast has little volatility.

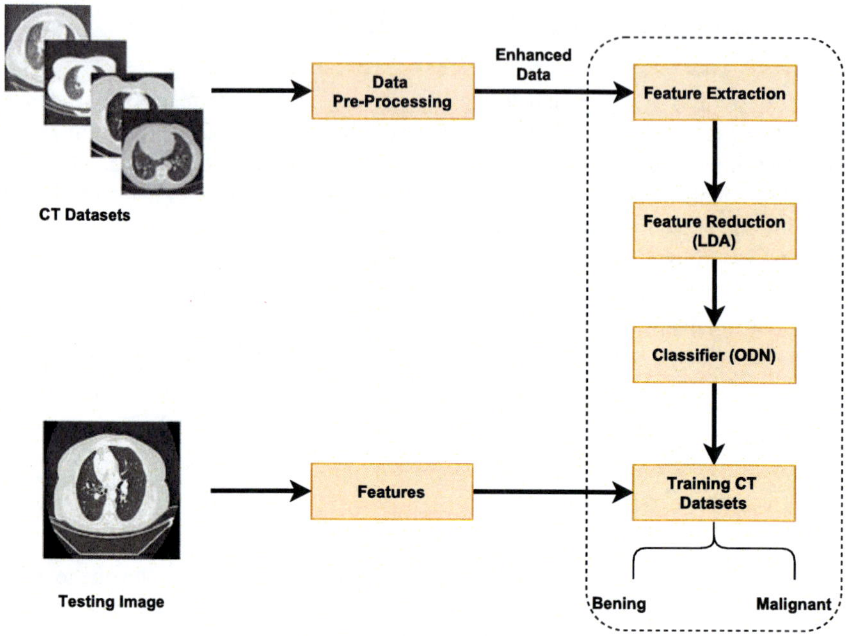

Fig. 1 Proposed CT image block diagram classifier

Skewness: According to the tail of the histogram, the skewness of the image is computed. It is possible to classify the histogram tail value into positive and negative values.

Kurtosis: Measuring of real-valued random variable's distribution shows the anomalous of the image. This determines the shape of the distribution, statisticians utilise Kurtosis and Skewness.

3.4 Lung CT Images Classification

In this study, CT scans were classified using a deep neural network (DNN). In contrast to the typical NN structure, the DL structure incorporates more hidden layers between the input and output layers into the system architecture, resulting in more unexpected and nonlinear connections. DNN is used to classify the output component vector once features have been selected. A deep DBN and RBM classifier are used by this classifier. In the provided model, MGSA optimisation is suggested in order to improve classification performance.

3.5 *Deep Belief Network*

An advanced neural network called a DBN is used during the training stage. It has multiple hidden layers. In the DBN paradigm, the system's hidden unit states are used to determine the system's conviction. There are two main parameters in a DBN: layer bias and weights amongst layer units. Setting up the system to train DNN help the constrained restricted RBM [17] which is a difficult undertaking.

4 Results and Discussion

During the implementation of the proposed CT lung images categorization models, The MATLAB 2020 was used. To classify cancer images, a typical CT dataset was employed. The proposed model was compared to current used classifiers such as NN and SVM.

This work uses a collection of 50 low-dose lung cancer CT imaging datasets [5] to detect lung cancer [7]. CT scan images with a slice thickness of 1.25 millimetres were obtained in one breath. Using a radiologist's expertise, the radiologist identified nodules and recorded them in the dataset. Figure 2 shows the test photos considered for the proposed work.

Classification models are often evaluated using the following techniques. Table 1 shows the training and testing images for an ODNN-based classification algorithm. For training, a total of 70 scanned photos were employed. The remaining 30 scanned images were used to put the lung image research technique through its paces.

Point operations were used to illustrate the ODNN model's results. It was determined that a malignant or non-cancerous lung image could be detected by the proposed model. The programme is able to forecast a new patient's lung health based on the testing data.

As shown in the illustration, the LDA was used to reduce the complexity of the framework during the feature reduction period. Dimension of feature vector has decreased over time. By comparing them, it is clear that the suggested technique requires less computation time whilst keeping high classification accuracy and this is because of LDA-based feature reduction model. As a result, the LDA should maintain the same biases and weightings unless the images' properties have significantly

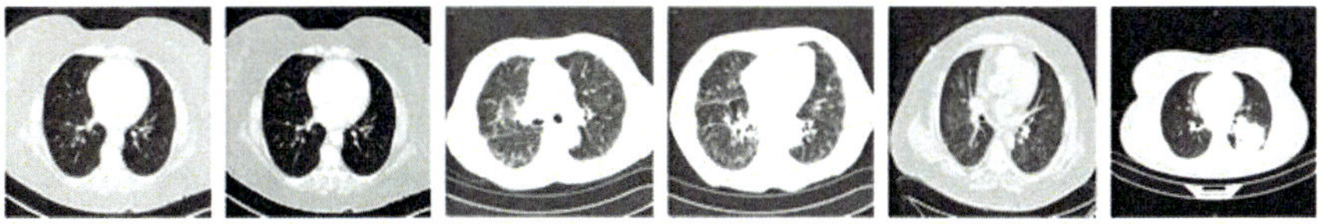

Fig. 2 Sample datasets

Table 1 Training analysis and testing analysis

Phase	Target images	Proposed classifier			
		Normal	Malignant	Benign	Total images
Training	Normal	23	2	3	28
	Malignant	3	19	3	25
	Benign	2	9	1	12
	Total images	28	30	7	65
Testing	Normal	7	1	3	11
	Malignant	2	10	2	14
	Benign	1	12	1	14
	Total images	10	23	6	39

Table 2 The proposed CT lung image classification scan with pre-processed results

	Normal		Bening		Malignant	
	CT Image	Enhanced	CT Image	Enhanced	CT Image	Enhanced
	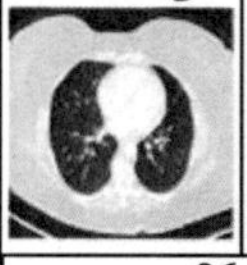	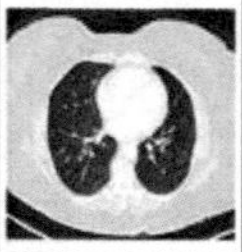	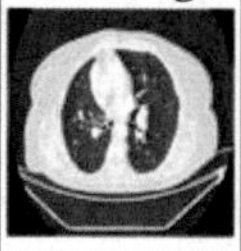	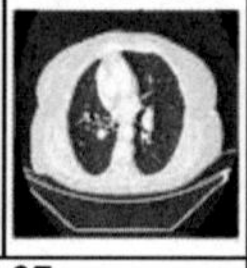	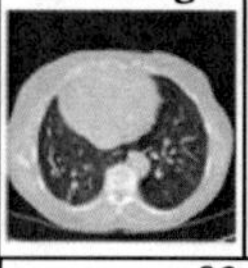	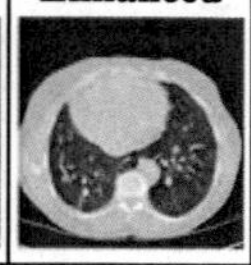
Accuracy	96.22		86.37		93.33	
Sensitivity	93.4		89.43		81.2	
Specificity	87.3		81.2		94.1	

changed. The precision rates increase when the element vectors are restricted to the part selected by the LDA.

For the proposed approach, the accuracy level of lung cancer image categorization is shown in Table 2. These two classifiers use supervised machine learning to categorise CT images as benign or malignant (normal or benign). To illustrate its superiority, the proposed ODNN method was compared with current classifiers. The tabulated data suggest that the evolutionary classification technique eliminates sensitivity to initial clustering parameters. To improve categorisation, texture and colour factors are taken into consideration when grouping CT lung cancer datasets. The validation analysis can reveal the kernel functions with the highest accuracy and the ones with the largest variation in accuracy.

5 Conclusion

In comparison to existing classification algorithms, the proposed ODNN with feature reduction outperformed them in lung CT image categorisation. The automatic lung cancer categorisation method reduces manual labelling time and removes the chance of human error during the labelling process. Using machine learning approaches, the researchers intended to enhance the precision and accuracy of distinguishing between normal and diseased lung images. With values of 96.22, 95.2, and 95.24%, the proposed approach is effective in classifying lung images. Since the proposed algorithm has a high level of accuracy, it is clear that it is capable of identifying cancerous areas in CT images. In this study, the categorisation results indicate the advantages of this strategy: it is fast, easy to use, non-invasive, and inexpensive. A multi-classifier-based cancer detection approach will be used in the future with high dose CT lung images and appropriate feature selection.

References

1. Szegedy C, Vanhoucke V, Ioffe S, Shlens J, Wojna Z (2016) Rethinking the inception architecture for computer vision. In: Proceedings of the IEEE conference on computer vision and pattern recognition, pp 2818–2826
2. Kumar PR, Sarkar A, Mohanty SN, Kumar PP (2020) Segmentation of white blood cells using image segmentation algorithms. In: 2020 5th international conference on computing, communication and security (ICCCS). IEEE, 2020, pp 1–4
3. Chollet F (2017) Xception: deep learning with depthwise separable convolutions. In: Proceedings of the IEEE conference on computer vision and pattern recognition, pp 1251–1258
4. Huang G, Liu Z, Van Der Maaten L, Weinberger KQ (2017) Densely connected convolutional networks. In: Proceedings of the IEEE conference on computer vision and pattern recognition, 2017, pp 4700–4708
5. Dauphin YN, Fan A, Auli M, Grangier D (2017) Language modeling with gated convolutional networks. In: International conference on machine learning. PMLR, 2017, pp 933–941
6. Jenuwine NM, Mahesh SN, Furst JD, Raicu DS (2018) Lung nodule detection from ct scans using 3d convolutional neural networks without candidate selection. In: Medical imaging 2018: computer-aided diagnosis, vol 10575. International Society for Optics and Photonics, 2018, p 1057539
7. Zhu W, Liu C, Fan W, Xie X (2018) Deeplung: deep 3d dual path nets for automated pulmonary nodule detection and classification. In: 2018 IEEE winter conference on applications of computer vision (WACV). IEEE, 2018, pp 673–681
8. Xie Y, Zhang J, Xia Y, Fulham M, Zhang Y (2018) Fusing texture, shape and deep model-learned information at decision level for automated classification of lung nodules on chest ct. Inf Fusion 42:102–110
9. Chougrad H, Zouaki H, Alheyane O (2018) Deep convolutional neural networks for breast cancer screening. Comput Methods Programs Biomed 157:19–30
10. Mohsen H, El-Dahshan E-SA, El-Horbaty E-SM, Salem A-BM (2018) Classification using deep learning neural networks for brain tumors. Future Comput Inf J 3(1):68–71
11. Van Ginneken B, Setio AA, Jacobs C, Ciompi F (2015) Off-the-shelf convolutional neural network features for pulmonary nodule detection in computed tomography scans. In: IEEE 12th international symposium on biomedical imaging (ISBI). IEEE 2015, pp 286–289
12. SermanetP E, FergusR LO (2018) Integrated recognition, localization and detection using convolutional networks. *arXiv preprint* arXiv:1312.6229
13. Food US, Drug US food and drug administration home page. (Online). Available: https://www.fda.gov
14. Burges CJ (1998) A tutorial on support vector machines for pattern recognition. Data Mining Knowl Discov 2(2):121–167
15. Kumar D, Wong A, Clausi DA (2015) Lung nodule classification using deep features in ct images. In: 2015 12th conference on computer and robot vision. IEEE, 2015, pp 133–138

16. Liu DC, Nocedal J (1989) On the limited memory BFGS method for large scale optimization. Math Program 45(1):503–528
17. Hua K-L, Hsu C-H, Hidayati SC, Cheng W-H, Chen Y-J (2015) Computer-aided classification of lung nodules on computed tomography images via deep learning technique. OncoTargets Therapy 8
18. Hinton GE, Osindero S, Teh Y-W (2006) A fast learning algorithm for deep belief nets. Neural Comput 18(7):1527–1554
19. Lakshmanaprabu S, Mohanty SN, Shankar K, Arunkumar N, Ramirez G (2019) Optimal deep learning model for classification of lung cancer on CT images. Future Gener Comput Syst 92:374–382

Enhancement of Low-Resolution Images Using Deep Convolutional GAN

Tulika and Prerana G. Poddar

Abstract Neural networks have expanded the scope of machine learning to unlock new opportunities across industrial dimensions. Convolutional GAN is a recent technique which has achieved promising results in the areas of image enhancement and classification. In this paper, deep convolutional GAN (DCGAN) is implemented, and the network is trained on the fashion-MNIST dataset. The implementation of DCGAN is done using leaky ReLU activation functions and sequential modeling. The proposed implementation has resulted in successful data retrieval from grayscale images for relevant apparel categories. The collective PSNR and SSIM results achieved in this work are better when compared with the other contemporary image enhancement techniques available in literature.

Keywords Deep learning · Deep convolutional generative adversarial network (DCGAN) · Leaky rectified linear activation function (Leaky ReLU) · Sequential modeling

1 Introduction

Deep neural networks have shown remarkable performance in recognition and classification. The fast growth in neural networks and advancement in system capabilities such as processing speed, power consumption and computation have broaden the aspects of computer-vision applications. In addition, all these advance techniques have achieved great results and cost effectiveness in image processing tasks such as object recognition, image enhancement, detection, and classification when compared to traditional techniques. All neural networks which are used in supervised and unsupervised learning are trained not programmed, so the applications following this

Tulika (✉) · P. G. Poddar
Department of Electronics and Communication Engineering, BMS College of Engineering, Bengaluru 560019, India
e-mail: tulika.ldc19@bmsce.ac.in

P. G. Poddar
e-mail: preranagp.ece@bmsce.ac.in

P. Dutta et al. (eds.), *Emerging Technologies in Data Mining and Information Security*, Lecture Notes in Networks and Systems 490,
https://doi.org/10.1007/978-981-19-4052-1_33

approach require less proficiency and fine tuning. Deep neural network is best suited when massive datasets are available for training along with high-end processing devices; the network is unaware of the feature outcome; and when domain proficiency is less or not available. Neural networks can utilize supervised and unsupervised learning. The huge, highly interpreted datasets used in neural networks have played a great role in this implementation.

In supervised learning [1] approach, the network is trained using labelled data. Training is done on datasets in which the data is already tagged with the correct answer. Just like a child learns in the supervision of elders, the network learns from labeled data and then analyzes the other input data based on its knowledge. Unsupervised learning [1] technique deals with unlabeled data and is used where the process does not require supervision. As compared to supervised learning, unsupervised learning process deals with more complicated data and the model can discover features and information on its own. Sometimes this technique is unpredictable, and the trainer does not know which features the model will select or discard.

Generative adversarial network proposed by Goodfellow is one such unsupervised and semi-supervised learning technique [2, 3], which is quite useful in image processing and computer vision. GAN is a type of neural network which trains generative models using adversarial technique. These techniques have the capability to learn from unlabeled data, collect important relevant features from those data, and then transform those samples into meaningful images. There are different circumstances in real world where data gets corrupted or lost due to technical failures, device failures, and network failures. GAN can be used in such circumstances for data recovery. In case of unsupervised learning, GANs have emerged as a favorable technique and have shown great capabilities in processing simple images. This technique has recently gained attraction from researchers all over the world and a variety of derived GAN models have been proposed to overcome the shortcomings, for example.

Contemporary to the research in GANs, there are several image enhancement algorithms being investigated by authors, such as U-Net algorithm, optronic convolutional neural networks, and fourier ptychography. These algorithms are also based on neural networks techniques and have achieved prominent results. In optronic convolutional neural networks [4, 5], the computational operations are performed in optics and the remaining data transmission, error control, etc., are done in electronics. Along with convolutional layers, other layers like down-sampling layers, fully connected layers, and non-linear activation functions are implemented in optics using optical 4f system. To implement down-sampling, stride convolutions are used, and for nonlinear activation functions, CMOSs camera modulation function have been used with the property of fourier optics. But, the architecture does not give a flexible layering structure. The layers and their order are fixed in this approach. Fourier ptychography [6] is an image reconstruction method in which phase retrieval optimization is used. In this method, multiple images are captured using a lens and then all the images are combined using phase retrieval process to get a high-resolution image. The choice of LEDs, the distance between them, and the numerical aperture used—all these factors affect the reconstruction quality. The limitation is that large number of attempts are required for training with different numerical aperture values. Similarly, U-Net algorithm [7] is used for feature extraction by removing residual noise from the images.

Motivated by the available literature, we have implemented deep convolutional generative adversarial network (DCGAN) approach for enhancement of low-resolution images, and we provide a performance comparison with available techniques when identical datasets are used as input. The paper is organized as follows: Sect. 1 introduces the problem domain, Sect. 2 provides the architecture and mathematical working of DCGAN, the derived GAN model, Sect. 3 discusses the proposed implementation of DCGAN method, Sect. 4 gives detail of all the definitions and parameter settings, stepwise results, and comparison with literature, and finally Sect. 5 concludes the paper.

2 Deep Convolutional GAN

The deep convolutional generative adversarial network is a derived model of the GAN, which uses the convolutional and transpose convolutional layers. The DCGAN was first described by Radford et al. [8] and it differs from the Vanilla GAN proposed by Goodfellow in 2014. It uniquely replaces the fully connected layer with the de-convolution layer in the generator. It is one of the most common CNN, and it allows training of a pair of deep convolutional generator and discriminator networks. This technique has received immense success in image restoration, detection, and recognition tasks. A major application is the generation of synthetic images like the dataset during training. The technique is often used for editing photographs such as human faces and postures, for translating images such as winter images to summer images, generating cartoon characters, or emojis . GAN can also be used in detecting malicious component in an image and prevent adversarial hacking. It provides an innovative way to improve healthcare facilities in detecting tumors, and in drug discovery.

2.1 Architecture and Working

Figure 1 shows the basic DCGAN architecture. There are two modules: generator G and discriminator D. The generator network aims to generate synthetic data which are possibly the best fit of the original data distribution. That is why the G is trained on the original data x. Based on the training, it generates synthetic images from random noise vector z. $G(z)$ are the synthetic samples (multidimensional vectors) obtained from the generator by mapping the noise vector z to a new data space using generator. The discriminator D is trained on mixed image (original + synthetic) samples and learns to distinguish between original and synthetic samples.

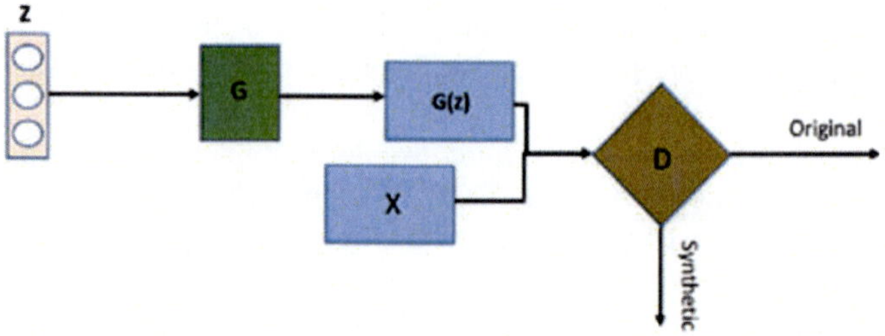

Fig. 1 DCGAN architecture

As training proceeds, the generator tries to fool the discriminator by sending synthetic images. It purposefully marks the synthetic image as original to confuse the discriminator. If the discriminator gives a correct decision by identifying the correct label of the sample, the training is assumed to be successful, and the feedback is sent to the generator to create more realistic samples. If the discriminator fails, the optimal state for generator is reached, implying that generator has learnt the original data distribution. In this model, discriminator needs to be well synchronized with the generator while training. Generator should not be trained alone without updating the discriminator.

Let us assume, $P_{\text{data}}(x)$ as original data distribution, $P_g(x)$ as generator-distribution over data x, and input-noise-distribution as $P_z(z)$. $D(x)$ depicts expectation that x comes from original sample and not from P_g. The loss function derived in the original paper by Ian Goodfellow [3] for GAN networks is derived from the binary cross-entropy loss formula in (1):

$$L(\hat{x}, x) = [x \cdot \log\ \hat{x} + (1 - x) \cdot \log(1 - \hat{x})] \tag{1}$$

where $x =$ original data, $\hat{x} =$ reconstructed data.

Discriminator Loss: While training discriminator, $P_{\text{data}}(x)$ is $x = 1$ (original data) labeling for original sample is 1 and $\hat{x} = D(x)$, substituting this in (1) will give:

$$L(D(x), 1) = \log(D(x)) \tag{2}$$

Similarly, $x = 0$ (synthetic data), label for synthetic sample is 0 and $\hat{x} = D(G(z))$ because synthetic data is obtained from generator. Substituting this in (1) will give:

$$L(D(G(z)), 0) = \log(1 - D(G(z))) \tag{3}$$

Since, D aims to distinguish between synthetic/original data. For this, (2) and (3) are maximized and the final loss function is given by:

$$L(D) = \max[\log(D(x)) + \log(1 - D(G(z)))] \tag{4}$$

Generator Loss: Since the generator is trying to compete with the discriminator, the module aims to minimize (4) and the loss function is given as:

$$L(G) = \min\left[\log(D(x)) + \log(1 - D(G(z)))\right] \quad (5)$$

Thus, the generator loss is dependent on discriminator prediction on whether the data generated by the generator is original.

Combined Loss: Combining (4) and (5) for single dataset will give the expectation as:

$$L = \min{}^{G}\max{}^{D}\left[\log(D(x)) + \log(1 - D(G(z)))\right] \quad (6)$$

For the entire dataset, the expectation of the above (6) is given by:

$$\min{}^{G}\max{}^{D}(V)(D, G) = \min{}^{G}\max{}^{D}\left(E_{x\sim p_{\text{data}}(x)}\left[\log(D(x))\right] + E_{z\sim P_z(z)}\left[\log(1 - D(G(z)))\right]\right) \quad (7)$$

3 Proposed Implementation

In this work, the DCGAN network shown in Fig. 1 has been implemented using Python, one of the widely used programming languages [5]. Many libraries and in-built module such as NumPy, OpenCV, Tensor Flow, and Pillow/PIL facilitate the process of implementation of desired image processing functionality. The tensor flow module in Python is used to load datasets.

The steps below are repeated for training the DCGAN model for each epoch output.

- Step 1: Generating random noise vector.
- Step 2: Images are generated by the generator using the random noise.
- Step 3: Generated images are mixed with the original images from the dataset.
- Step 4: Train the discriminator on the mixed dataset to discriminate between synthetic and the original samples.
- Step 5: Generate new random noise vector and create synthetic images, now purposefully label the images as original images to test the discriminator.
- Step 6: Train GAN on the synthetic images labeled as original images.

3.1 Generator and Discriminator Model

The DCGAN class consists of two static methods—Generator and Discriminator—implemented using Python module Keras [5]. Both the models are designed using same height, width, depth, and channel parameters. The layer architecture consists of dense layer, Conv_2d Transpose layer followed by different activation functions and batch normalization. The dense layer is one of the mostly used fully connected network layers in neural architectures and is linked deeply in the model. The neurons in the dense layer communicate with its previous layer for the input. In the backend, the matrix–vector multiplication is performed by this hidden layer. The values used in the matrix are trained and restored with the help of back propagation technique. Operations like rotation, scaling, and translation of output vectors are also performed in the dense layer.

The Conv2dTranpose layer in Keras takes parameters such as: kernel size, strides, and padding. The kernel size is the filter size that means bigger the kernel size, each number in the output layer is a broader interpretation of the input layer and carries more information about the input. The strides show how fast the kernel moves along the rows and columns. If strides are (1, 1), the kernel moves one row/column in one step, whereas strides (2, 2) mean the kernel moves two rows/columns in one step. Hence, this increases the computation speed. For transposed convolution, Keras offers two types of padding—'valid' and 'same'. Padding is done to maintain the size of output layer. In transposed convolution, we expand the input layer. Hence, when 'valid' is chosen the output shape will be larger than the input, and when 'same' is chosen, the output is forced to be same as the input. If the output is smaller than the original output shape, only its middle part is maintained.

Batch normalization (BN) is layer technique which gives ease to each layer to learn independently in the neural network architecture. It resolves the problem of over-fitting of the models, fastens the learning process, makes it efficient, and regularizes the overall process. Sequential modeling adds BN to regulate the input and to assimilate the output from the previous layers. BN layers are set after convolution layers in sequential modeling. In normalization, the main aim is to transform the data in such a way that the mean approaches '0' and standard deviation approaches '1'.

3.2 Modeling Technique and Activation Functions

Sequential modeling technique has been chosen here, which consists of a large variety of inbuilt attributes. For example, each layer in neural network can be represented using attribute 'layer'. Inbuilt functions such as 'add ()' and sequential constructor are available in this modeling technique for adding different activation functions and batch normalization in the neural network. This technique is desirable because layer addition in the network occurs in a step-by step-fashion. The layers in generator model uses ReLU activation functions as AF_0, AF_1, and AF_2 (Fig. 2.). The final activation function AF_3 uses Tanh function because using a bounded function (-1 to 1) in the output layer makes the model learn quickly [8].

Leaky ReLU has been used in this implementation as activation function in discriminator model [9]. In ReLU, the gradient is zero for any input less than zero. This makes the neural network less sparse by keeping only the useful links, and so network is less dense and decreases the computation. But, if the gradient becomes

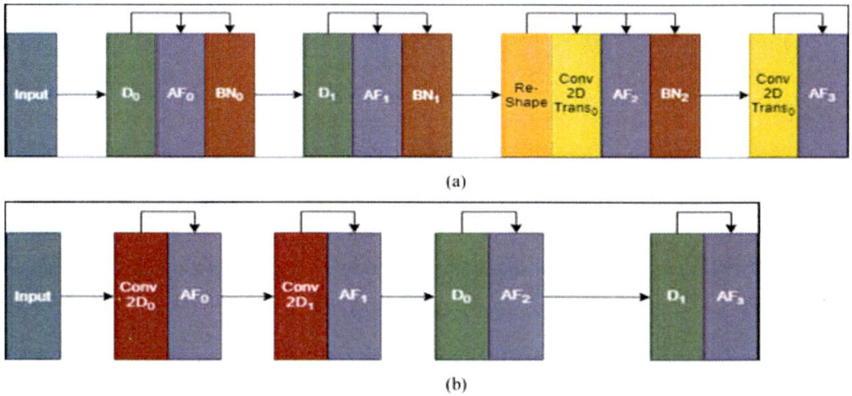

Fig. 2 DCGAN: **a** generator, **b** discriminator model. The symbols D, AF and BN denote dense, activation function, and batch normalization layers, respectively

zero, the corresponding node does not have any effect on the network, and hence the learning stops. A neural network is efficient until it is learning. This is known as dying ReLU problem. To avoid this problem, an alpha parameter, which is the leak, is added, so the gradient will be small but never zero.

$$\frac{\mathrm{d}}{\mathrm{d}x}\,\mathrm{Irelu}(x) = \begin{cases} \alpha \text{ if } x \leq 0 \\ 1 \text{ if } x < 0 \end{cases} \tag{8}$$

Adding a small constant although reduces the sparsity, but the network becomes more rigid for optimization. The nodes which were not active with ReLU are activated by adjusting the weights on the corresponding nodes. In final layer of discriminator model, sigmoid bounded function (to ease the calculation) is used to capture the probability that whether the data obtained is synthetic or original. Figure 2 shows the DCGAN layer architecture for (a) generator (b) discriminator model, respectively.

4 Results and Discussions

4.1 Simulation Scenario

The results in this work are obtained using Fashion-MNIST dataset [10]. The dataset comprises of a test set of 10,000 image samples and a training set of 60,000 image samples. Images in the dataset are grayscale images each of resolution 28 × 28. There are ten classes, and the images are linked with the corresponding labels, for example, 'label 0' depicts 'T-shirt/Top', 'label 1' depicts 'Trouser', and so on. Each image has 784 features (i.e. 28 × 28) where each pixel value is between 0 and 255.The implemented DCGAN model trains the dataset in a batch size of 256 images into

25 epoch outputs. The hardware configuration and the software requirements for the simulation are listed below.

Hardware and Software Requirements:

- Intel Core i5, Ryzen 5or higher Processor, with 8 GB RAM and Integrated Intel HD Graphics 520 Graphics Card
- Anaconda Distribution
- Tensor Flow 0.12.1
- Keras
- OpenCV
- Python 2.7+
- Imutils
- Fashion-MNIST Dataset
- MATLAB R2014a.

The generator model for the DCGAN network has certain layers as shown in Fig. 2, which are implemented using sequential modeling. Table 1a, b, and shows generator discriminator DCGAN model assumptions and initialized values for our work. The model accepts an input vector of 100d and transforms it into 512d using fully-connected (dense) layer.

4.2 Epoch Outputs

Figure 3 shows the first epoch generated from the trained model in which the labels of the mnist dataset are not clearly visible. But, from the fifth epoch onwards, the corrupted pixels get replaced by the synthetic pixels, and the structures are clearly visible. The trained model aims to fill the missing region and the corrupted pixels in very low-resolution images. When model is trained for generating the image, the discriminator and generator loss functions are reduced to minimum so that when the model runs on the given low-resolution image, it tries to reconstruct the image based on its trained model. The visual difference can be observed in the Fig. 3d twenty-fifth epoch that is generated in the twenty-fifth iteration.

4.3 Quality Metrics: PSNR and SSIM

Any image processing algorithm is incomplete without its quality analysis. The objective methods for image quality analysis, in case of full reference method, are built on direct numerical evaluation such as peak signal-to-noise ratio and structural similarity index method. The ratio between the highest possible power of the signal and the corrupting noise power present in the image is known as peak signal-to-noise ratio. SSIM is a structure similarity index method, which calculates the extent

Table 1 Model (generator and discriminator) assumptions

Quantity	Value
Spatial dimension (width × height) for generator	7 × 7
Depth	64
Channel (gray scale)	1
Total number of nodes	7 × 7 × 64 = 3136
Input dimension for the uniformly distributed noise vector	100d
Output dimension for the uniformly distributed noise vector	512d

a

Layers	Parameter
Dense	512
Activation	512
batch_normalization	512
dense_	3136
activation_1	3136
batch_normalization_1	3136
Reshape	7, 7, 64
Conv2d_transpose [2 × 2 strides to increase spatial dimensions using 32 filters]	14, 14, 32
activation_2	14, 14, 32
batch_normalization_2	14, 14, 32
Conv2d_transpose_1 [14 × 14 is transformed into 28 × 28]	28, 28, 1
activation _3	28, 28, 1

b

Layer	Parameter
conv2d	32, 5, 5
leakyReLU	0.2
conv2d	64, 5, 5
leakyReLU_1 dense	0.2 512

c

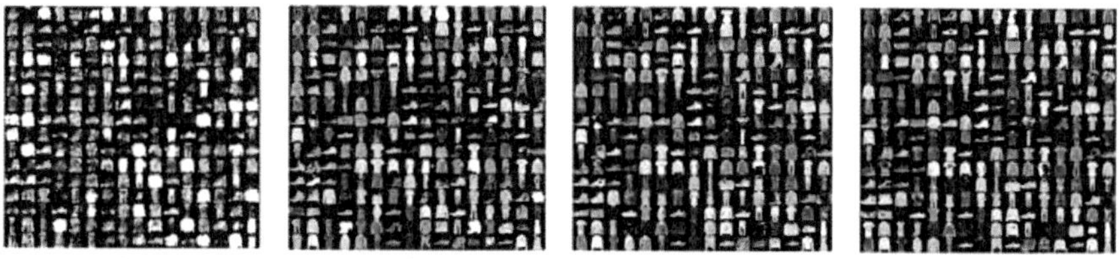

(a) (b) (c) (d)

Fig. 3 **a** First, **b** tenth, **c** fifteenth, **d** twenty-fifth epoch output obtained, respectively

Table 2 PSNR and SSIM calculated for our proposed method using DCGAN

Input image	PSNR (mean)	SSIM
Epoch output 1	13.74	0.419
Epoch output 5	14.65	0.588
Epoch output 10	15.29	0.642
Epoch output 15	16.03	0.681
Epoch output 20	16.67	0.708
Epoch output 25	17.04	0.745

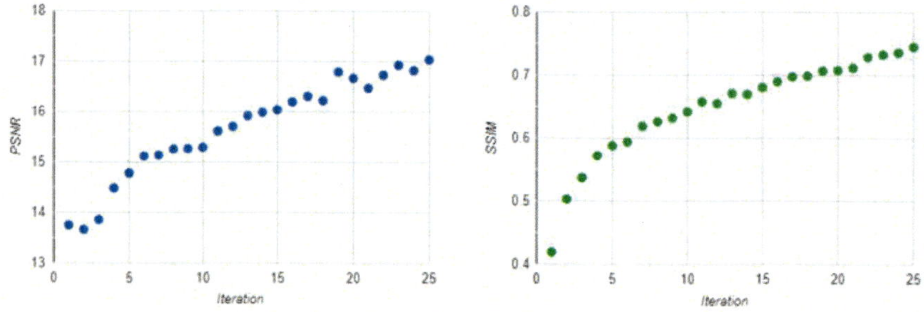

(a) (b)

Fig. 4 **a** PSNR and **b** SSIM calculated for the epoch obtained in every iteration

of similarity between two images. Structure similarity is calculated by calculating image distortion which is generally represented as the amalgamation of luminance, contrast, and loss of correlation. For the trained epochs obtained, the PSNR and SSIM values are calculated as illustrated in [11].

Table 2 depicts the PSNR and SSIM values calculated for some of the epoch outputs. For the implemented algorithm of DCGAN, the PSNR and the SSIM values show a gradual increase with the increasing iterations which means the image quality is being enhanced with the algorithm. Figures 4a and b plot the PSNR and SSIM values for each iteration.

4.4 Comparison with Other Algorithms

In this section, the values obtained for PSNR and SSIM using the DCGAN approach is compared with other existing algorithms. Authors have applied different algorithms on Fashion-MNIST dataset such as: Fourier ptychography, OPCNN-1, and U-Net family of algorithms in the past. Table 3 shows comparable results between DCGAN and other algorithms on Fashion-MNIST datasets. The proposed DCGAN approach shows similar PSNR and better results for SSIM when compared with other algorithms.

Table 3 Comparison of PSNR and SSIM values of different techniques on Fashion-MNIST database

Technique	PSNR	SSIM
Proposed DCGAN implementation	17.04	0.745
OPCNN-1 using convolutional layers [4]	18.42	0.6024
OPCNN-1 using down-sampling layers [4]	17.97	0.6589
FPM for numerical aperture 0.5 [6]	26.88	0.88
FPM for numerical aperture 0.2 [6]	23.88	0.76
FPM for numerical aperture 0.05 [6]	11.96	0.19
U-Net Algo without skip connections in outer encoder-decoder architecture [7]	18.75 (noisy images) 19.62 (de-noised images)	0.4354 (noisy images) 0.6902 (de-noised images)
U-Net Algo with skip connections in outer encoder decoder architecture [7]	18.75 (noisy images) 27.96 (de-noised images)	0.4354 (noisy images) 0.9351 (de-noised images)

The result shows good performance of DCGAN algorithm and confirms the capability of GAN structure in generating samples. Any GAN model learns the distribution much more quickly than CNN model. DCGAN is computationally simple when compared with the other algorithms, for example, in Fourier ptychography, for training and the evaluation of image classification accuracy for different numerical aperture, a greater number of trials are needed, and so, the linear relationship between SSMI, PSNR, and image classification accuracy is calculated based on a greater number of such trials.

The selection of the hidden layers, activation functions, and stability is a challenge. Any deep neural network is successful if it is learning. Once the learning stops, the network gets stagnant, and the decision accuracy decreases. The proposed DCGAN model exhibits great flexibility in choosing and ordering of hidden layers, whereas in OPCNN [4], the hidden layers are ordered in fixed manner. In their architecture, the first layer is convolutional layer, second is a down-sampling layer, and the third is nonlinear activation layer. This restricts the flexibility of the model.

5 Conclusion

In this work, we have applied the concepts of DCGAN on low-resolution fashion-MNIST dataset to enhance its quality. The choice of layers, activation functions in the generator and discriminator network in this DCGAN algorithm has resulted in PSNR and SSIM scores as 17.04 and 0.745, respectively. The overall performance (PSNR AND SSIM considered together) of DCGAN method is better than the OPCNN method of [4], the FPM method for 0.05 NA of [6] and the different U-Net algorithm proposed in [7]. The FPM method of [6] with aperture size 0.5 and 0.2

appears to provide marginally better-quality measures than our technique. However, the method involves more computational complexity because the accuracy during image classification for different aperture values require huge number of trials.

In DCGAN, the network often fluctuates, undermine, unsettle, and never converge. The instability can cause the generator to collapse which can produce limited variety of samples. The generator gradient is lost when the discriminator is very efficient, and this unbalance between D and G can cause overfitting. GAN architecture has shown great results in unsupervised learning process and can be improved in the coming days. The only limitation of this work is that DCGAN model requires huge datasets for making accurate decisions. The intelligence of generator and discriminator model depends upon the dataset. These DCGAN and other derived GAN models are of limited use if the user needs results with less time and less computational complexity.

References

1. Pan Z, Yu W, Yi X, Khan A, Yuan F, Zheng Y (2019) Recent progress on generative adversarial networks (GANs): a survey. IEEE Access 7:36322–36333
2. Creswell A, White T, Dumoulin V, Arulkumaran K, Sengupta B, Bharath AA (2018) Generative adversarial networks: an overview. IEEE Signal Process Mag 35(1):53–65
3. Goodfellow I, Pouget-Abadie J, Mirza M, Xu B, Warde-Farley D, Ozair S, Courville A, Bengio Y (2014) Generative adversarial nets. Adv Neur Inf Process Syst 27
4. Ziyu G, Gao Y, Liu X (2021) Optronic convolutional neural networks of multi-layers with different functions executed in optics for image classification. Opt Express 29:5877–5889
5. Chollet F (2017) Deep learning with Python. Simon and Schuster
6. Zhang H, Zhang Y, Wang L, Hu Z, Zhou W, Tsang PW, Cao D, Poon TC (2021) Study of image classification accuracy with fourier ptychography. Appl Sci 11(10):4500
7. Gandhi ST (2020) Context sensitive image denoising and enhancement using U-nets. Rochester Institute of Technology
8. Radford A, Metz L, Chintala S (2016) Unsupervised representation learning with deep convolutional generative adversarial networks. In: International conference on learning representations
9. Wang Y, Li Y, Song Y, Rong X (2020) The influence of the activation function in a convolution neural network model of facial expression recognition. Appl Sci 10(5):1897
10. Xiao H, Rasul K, Vollgraf R (2017) Fashion-mnist: a novel image dataset for benchmarking machine learning algorithms. arXiv preprint
11. Hore A, Ziou D (2010) Image quality metrics: PSNR vs. SSIM. In: 20th IEEE International conference on pattern recognition, pp 2366–2369

Optical Character Reader

Sanchit Sharma, Rishav Kumar, Muskan Bisht, and Isha Singh

Abstract We present an instance of the available OCR tools and teach the Tesseract tool in English which is transcribed in Latin male or female. OCR accessibility for the blind is a splendid development, where customers can experiment with books, magazines, plugins or other programs. It can also be mixed with sound utilities or sound synthesizers, which read textual content optically recognized by the software. OCR additionally has its own boundaries. This can no longer understand complicated text, mathematical variables and handwriting scripts. The upscale OCR utility is used inside the vicinity needed, but it comes with additional computing fees. This paper targets to offer the OCR software that recognizes text characters, using the device getting to know model.

Keywords OCR · Tesseract · Training · Braille · Thresholding · Grayscale conversion · Processing · Image conversion · Binary image

1 Introduction

All through the beyond 5 a long time, system reading has grown from goals to fact. The introduction of optical characters has grown to be one of the most a hit era programs. Many structures to do OCR exist for numerous packages, despite the fact that the engine nonetheless has now not been capable of compete with human analyzing capabilities [1]. Many valuable files are scanned and stored for backup. Changing scanned text images to machines that can be read with the engine are referred to as the introduction of optical characters (OCR) and this topic has always been an interesting topic in computer science. Braille has been the main reading and writing system for visual disorders since the nineteenth century [2]. The tool

S. Sharma (✉) · R. Kumar · M. Bisht · I. Singh
HMRITM, Delhi, India
e-mail: replysanchit@gmail.com

I. Singh
e-mail: isha.singh@hmritm.ac.in

P. Dutta et al. (eds.), *Emerging Technologies in Data Mining and Information Security*, Lecture Notes in Networks and Systems 490,
https://doi.org/10.1007/978-981-19-4052-1_34

is always needed that can be used to extract the text of the image using a powerful OCR Tesseract Open-source engine and then apply it in a Braille translation by changing the extracted text into Braille letters. This application is thus useful for visual disorders because it can make Braille text from scanned image documents and can be used to translate old valuable books to Braille format.

English is spoken by a critical piece of the populace in North Africa. Being true in Morocco beginning around 2011. Anyway a few examinations with the OCR gadget inquisitive of the language were composed inside the letter set tifinagh or interpreted in Arabic or Latin letters [3]. With the advancements qualified by examination into the notoriety of optical text styles, the extent of studies on example notoriety, on man-made brainpower, and on innovative and portentous PCs, different gear has been intended to accomplish the change of the text content pix with the most noteworthy text-based substance [4]. The gadget committed to OCR is open access or paid by their permit.

Optically processed characters are called man or woman reputation (OCR). OCR is a means used to organize multiple files, PDFs or digital images in Standard Yank Code for Interchange Records (ASCII) or other equivalent enhancement forms that can be edited or searchable using the information [5]. This latest increase in the popularity of models through numerous applications has required, including OCR, file category, information mining, etc. the usage of OCR has a critical function inside the file scanner, individual popularity, language reputation, security, authentication in banks and so forth. OCR is classed into sorts: online man or woman reputation and offline individual reputation gadget [6]. OCR on line OUT Beat OCR Offline as a processed person as written, this avoids the preliminary tiers to become aware Additional offline OCR from print and handwriting OCR. In offline OCR, typically processed by scanning typed/handwritten characters into binary images or grayscale for the creation set of rules. With the development of the OCR scanned documents it is miles extra valuable while the picture report is normal, it will become textual content identified with the aid of the laptop the usual hassle with the OCR lies in character segmentation or symbols that be part of photograph input without delay proportions with OCR accuracy. However there are not any recognition algorithms that compete with the fine of human intelligence, nonetheless proven to be lots quicker this is nevertheless interesting [7].

2 Literature Survey

The manuscripts are available in different sizes and styles. They vary in terms of font, text styles, punctuation, etc. Sentences in the sample text can be frequently linked, segmented to present them as a separate logical entity. Application of textual content recognition as it is very vital kilometers in fields as well as jobs, schools etc. where the strength of the boys has waned now. In addition, handwritten text in various languages or scripts can also be optically recognized. One example is that the Hindi language can be identified as a structural (characteristic) description of male or female. Functions are roughly translated into density, phase and momentum capabilities of the male or female [8]. Other Indian scriptures as well as Kannada and Devanagari are identified through outlines-based networks, affecting the emphasis on the outlines/boundaries of the characters. Thus, an algorithm-based mainly on feature extraction is formulated, which is diagnosed using different characters optically and as input [9]. The sending normalization is a fundamental step for the prediction model and the adaptive normalization ratio (Ran) is one of the effective fields of each field correctly divided into equal zones and the intensity of each zone is recorded. The methodology for creating offline handwritten characters is proposed in [10]. The technique of extracting characteristics of handwritten characters and numbers relied entirely in particular on styles of (a) statistical capabilities from statistical distribution points. (B) structural statistics. The maximum functions generally used for individual representation are: (a) zoning, in which the characters are divided into many areas and capacities extracted from the density in each human contour region or from the woman using the histogram chain code calculation in each quarter, (b) projection and (c) crossing [11]. The individual images are divided into uniform regions which are searched for vertical, horizontal and diagonal segments. A wide variety of segments are provided to the classifier. Provides a method to segment the line of text to be entered for the creation of the Kannada OCR through the use of power minimization problems that use valuable new capabilities. OCR works carried out in Indian cultural writings are proposed in [12]. In most Indians, the textual content line can be divided into 3 areas. The tip sector shows the section above the title line, the middle area is made up of individual fundamental components (and compounds) below the top line, and the bottom quarter is the part below the bottom line. Traditionally, the model introduction process has been divided into completely feature-based models and methods.

The preliminary OCR gadget is more practical when used with a model-based approach, but the advanced machine combines it with a function-based technique to achieve higher results [13]. The feature-based procedures can encompass two types, namely, the spatial domains and the technique of the remodeling domain. The mapping of the emission report determines the effectiveness of the Aran version. In the characteristic extraction segment, the distribution of neighboring travel directions is the most widely used approach. Dimond [14] on the Aran, the image to be normalized fits the plan, with one of the dimensions stuffed according to the ratio of the components. Here, the dimensions of the aircraft are assumed to be rectangular. The L thing ratio is mapped by calculating the width and height of the normalized photo. The component ratio of normalized snapshots can be resized with single images [15]. Directional extraction features can be in the form of string code functions, NCFE functions and gradient functions. In NCFE, the contours and edges of the photograph are mapped to the orientation. Reputation of English textual content using the proposed energy minimization technique [16].

3 Tesseract

Tesseract is an independently well-known open-source optical gadget for various work establishments. Initially created on HP somewhere in the range of 1984 and 1994 [17, 18]. It was altered and sped up in 1995 with more noteworthy accuracy. In late 2005, HP delivered a Tesseract for publicly releasing. Presently advanced and controlled with the assistance of Google. It utilizes a measurable procedure dependent on polygonal assessments and on the computation of the distance between the limits removed [19] tesseract is viewed as one of the most dependable and right OCR machines accessible today. Numerous other OCR programming presently use it as a premise. It is an OCR application, with an incredible assortment of adaptability, a thick code base, and a huge local area associated with individuals intrigued by it [20]. Figure 1 image is taken as input adaptive thresholding is applied then text from the image are detected from which we get the required result.

Tesseract OCR follows regular cylinder handling bit by bit [21]. These means are (•) Adaptive Threshold: changes the photograph into a paired photograph. (•) Analysis of related parts: Used to extricate a blended line of characters. This method can be exceptionally valuable while applying OCR to depictions with white text

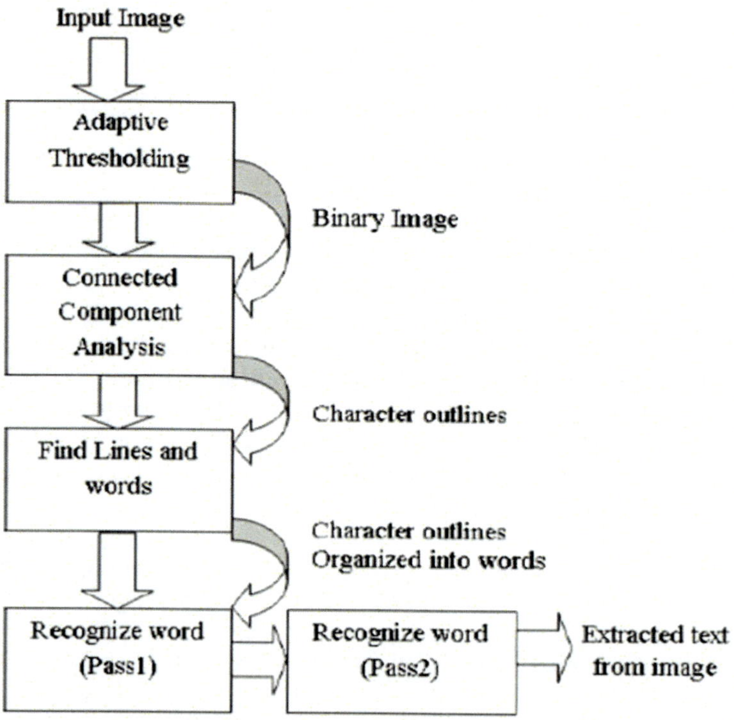

Fig. 1 Architecture of tesseract OCR

and a dark foundation. Tesseract is maybe quick to give this sort of handling at this level, the diagram is gathered on the whole, essentially by settling, being a square. (•) Tufts are characterized in the text line. Furthermore, regions are investigated for some proper tones or relative texts [7]. The text is separated into words utilizing safe zones and ill-defined situations (•) ubiquity accepts the zone as a two-venture strategy: (•) in the first workaround, you attempt to get familiar with each word thus. Each charming sentence passes the versatile classifier as school facts. Therefore, the versatile classifier can see more precise textual content under the web page (•) on the second skip, the adaptive classifier runs through the page to apprehend phrases that are not quite properly identified within the first bypass. The remaining section finished the bushy room and checked the opportunity speculation for X-height to find a small hat text.

4 Methodology

The steps are used to extract the textual content of the scanned image file after which convert text to Braille letters displayed inside the image [22]. (1) The usage of this process someone can convert files and vintage books to spur visually impaired Fig. 2 shows the entire process of image conversion in which it is converted to grayscale then correcting the recognized text and then converting to Braille.

Steps to follow:

1. Preprocessing photograph: on this step the scanned color photograph is converted to a grayscale photo to growth accuracy in the recognition step [23].
2. Step popularity: on this step the text is extracted from the image using the OCR Tesseract machine.
3. Submit textual content processing: on this step the mistake generated within the ultimate step was corrected using the spell checking the fire of JeHo [24].
4. Braille translation: in this step the corrected textual content is transformed to Braille using a set of precise policies. Transformed Braille text can then be

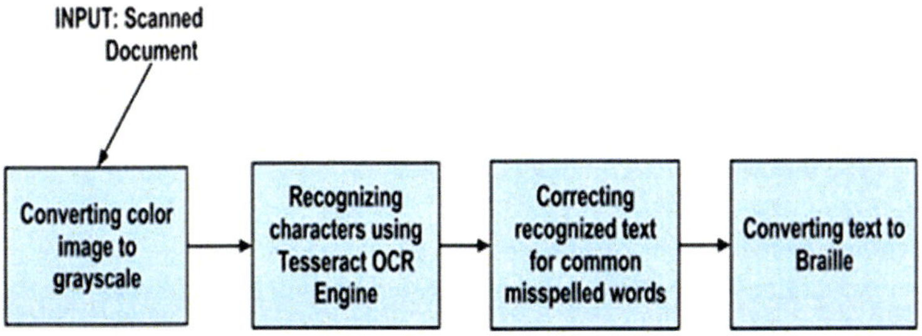

Fig. 2 Steps to be used for the entire conversion

stored as a prepared Braille layout or may be sent to Braille Embosser to print Braille textual content [19]. Every step is explained extra info.

4.1 Picture Processing

To start with color pix are generated after scanning paper documents. Color images must be transformed to a grayscale for more correct reputation inside the 2nd step of the four-degree method [25].

pix with excessive M and N width are represented as a discrete function $f(x, y)$ in the sort of way that

$$f(x, y) = (x_i, y_i), \text{ wherein } 0 <= I <= J <= 1.$$

Right here the pair (X_I, Y_J) is called pixels [26]. The pair (zero.0) is the first pixel and the pair $(N-1, M-1)$ is the closing pixel. Each pixel has its own RGB cost and uses grayscale values calculated.

4.2 Conversion to Grayscale

A Pixel (X_I, Y_J) has the identical RGB price falling inside the gray circle of relatives and the usage of this remark an algorithm is decreased as below [27].

algorithm:

For $I =$ zero to $N-1$ for $J = 0$ to $M-1$ GR $(X_I, Y_J) = 0.299 * R\ (X_I, Y_J) +$ zero.587 $* G\ (X_I, Y_J) +$ zero.114 $* B\ (X_I, Y_J)$

in which r, g, b red, blue and inexperienced value from pixels (x_i, y_j) and $0 <= r, g, b < 256$. Gr (x_i, y_i) is a converted grayscale cost [28].

4.3 Implementing the Algorithm

The use of the algorithm the scanned color photograph is transformed to grayscale as in Fig. 3.

Algorithms exchange shade pictures to grayscale pix and conversions appropriate for applications [29]. Conversion to grayscale images followed with the aid of conversion to binary image.

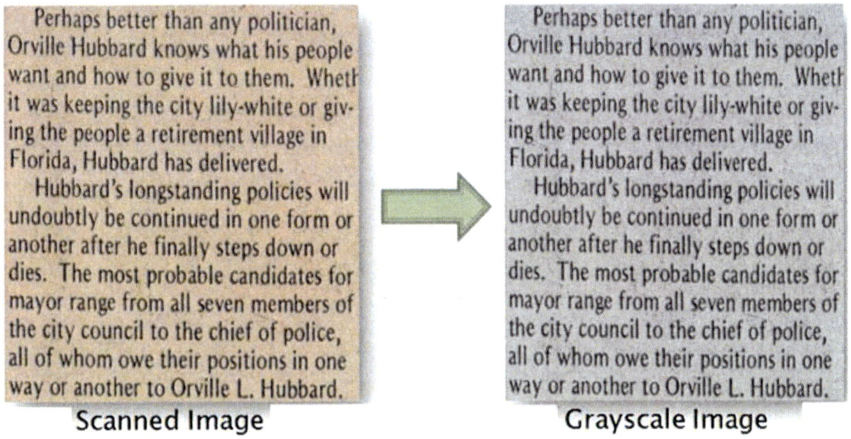

Fig. 3 Conversion of the image

4.4 *The Working Process*

Using Braille representation as given in Fig. 3 snap shots, an set of rules to change the extracted textual content right into a Braille letter made [30].

As given in the waft diagram, the algorithm does the following.

1. Look for compound letters and print a Braille illustration for compound letters if discovered [31].
2. If other if compound letters are not found and if the man or woman is examine is the alphabet, print the Braille alphabet representation that needs to be remembered approximately the representation of the final of the uppercase letter.
3. If the person is not always the alphabet and the character is the variety then prints a Braille illustration for numbers [32].
4. If different if the character is a special character, a print illustration for special characters.
5. Other Print Braille illustration for space.

The usage of this algorithm the text is converted to Braille and now transformed texts may be used in helping visually impaired [33]. Figure 4 Algorithm used to convert the image to Braille languages shown.

4.5 *Results*

Tesseract OCR and the consequences of Python Now that ocr.py has been created, it is time to use Python + Tesseract to do OCR on some sample input snapshots. Run each image through Tesseract Binary. We will then run each image via ocr.py (which

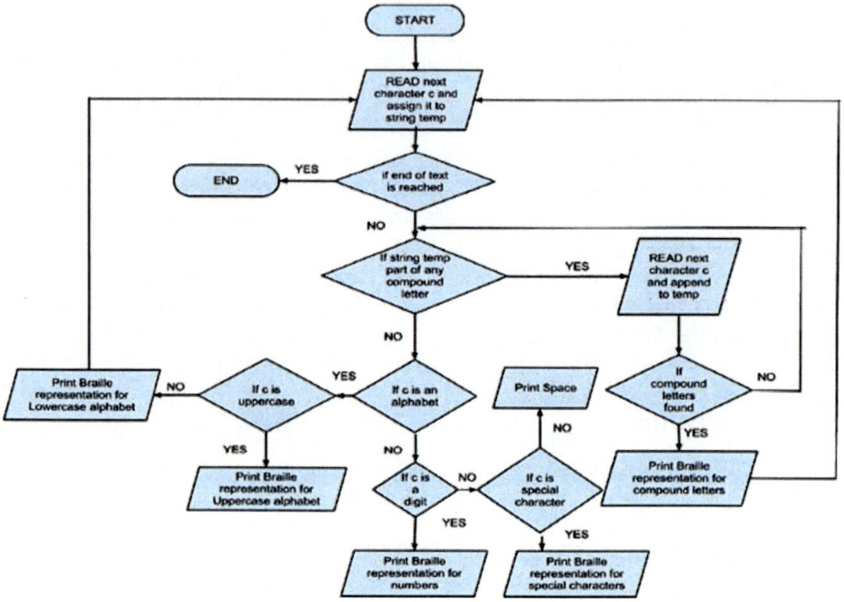

Fig. 4 Flow chart for Braille conversion algorithm

does the preliminary procedure before sending via tesseract) [21]. Finally, we will assess the consequences of these two techniques and file errors. Our first example is a "noisy" photo. This photo incorporates the story's favorite black textual content which is partly white and partly artificially distributed with the round pieces. The lumps act as "problems" for our simple algorithm [34]. Figure 5 Input to the OCR.

As you can see from this screenshot, the image on the edge is very smooth and the old afterlife has been erased. Our script successfully prints the contents of the image to the console [35]. Next, let us look at the tesseract and our preprocessing scripts on the image with the "salt and pepper" noise of the story: Fig. 6 Conversion of noisy image to extract text.

As you could see in this screenshot, the brink photo may be very clean and the history has been deleted [36]. Our script efficaciously prints picture content cloth to the console. Next, allow's check the tesseract and our preprocessing scripts on the photograph with noise "salt and pepper" inside the background: Fig. 7 Image

Fig. 5 Project input

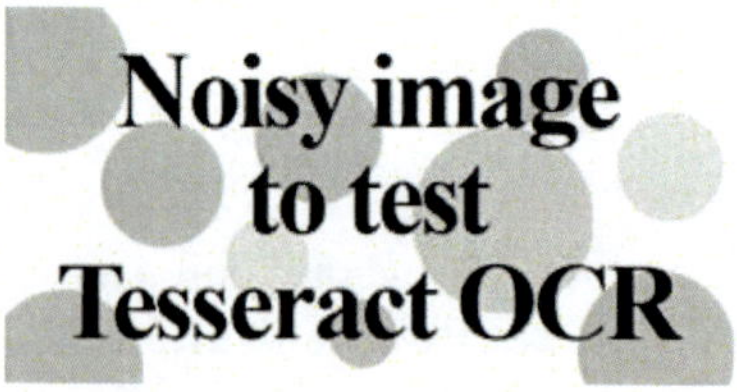

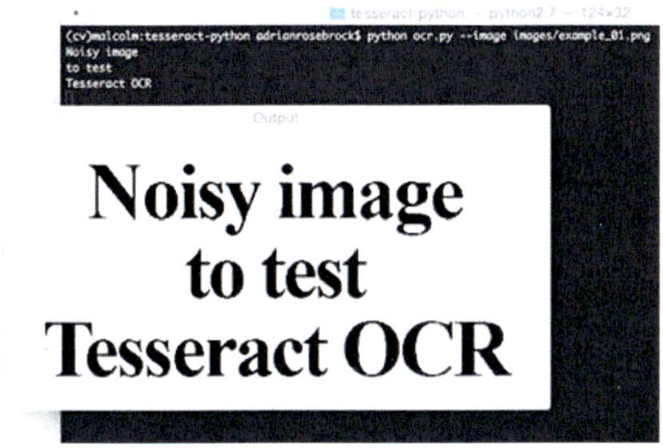

Fig. 6 Applying preprocessing images

Fig. 7 An example input image containing noise. This image will "confuse" our OCR algorithm, leading to incorrect OCR results

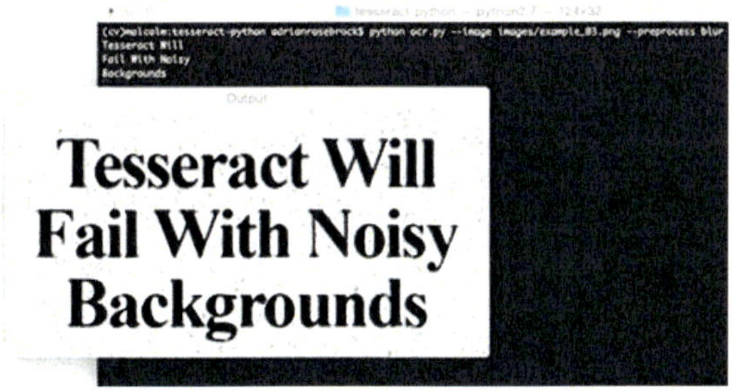

Fig. 8 Output image

showing that will confuse our OCR due to salt and pepper in it [37]. Also Fig. 8 Output of the image from our OCR.

5 Conclusion

We have got delivered the OCR device and its additives. Then, we have got supplied English. Given the achievement that the Tesseract device has been established, we choose to use it to English transcribed into Latin. Language that has not yet been explored in depth in the field of OCR. This look opens up many perspectives while increasing the preprocessing phase by applying various treatments to increase the share of notoriety and enrich the previous corpus. This enrichment consists of adding a chain of the last characters and carrying out the instruction for the brand new characters. For many years ancient languages were not able to decode and the reasons behind those writings were not found but using OCR technology these can be decoded and can reveal the truth about history. OCR can combine both Artificial Intelligence and machine learning which would allow us to analyze and configure data and systems to find and solve any discrepancies in large data. With increase in OCR demand traditional OCR cannot detect multiple files but in this we can extend the functionality of an OCR in which it could handle multiple files by generating multiple OCR simultaneously. OCR can be made with increased functionality in which it could extract text from images in which there is salt and pepper in the image itself. OCR has a lot of functionality with increase in source and libraries offered by tesseract it could be used. To detect text of any type and of any language whether ancient or modern.

References

1. Ahmed Ali AA, Suresha M, Mohsin Ahmed HA (2019) Different handwritten character recognition methods: a review. In: 2019 global conference for advancement in technology (GCAT). https://doi.org/10.1109/gcat47503.2019.8978347
2. Joe KG, Savit M, Chandrasekaran K (2019) Offline character recognition on segmented handwritten Kannada characters. In: 2019 global conference for advancement in technology (GCAT). https://doi.org/10.1109/gcat47503.2019.8978320
3. Siddiqi I, Djeddi C, Raza A, Souici-meslati L (2014) Automatic analysis of handwriting for gender classification. Pattern Anal Appl 18(4):887–899. https://doi.org/10.1007/s10044-014-0371-0
4. Lv Y, Cai F, Lin D, Cao D (2016) Chinese character Captcha recognition based on convolution neural network. In: 2016 IEEE congress on evolutionary computation (CEC). https://doi.org/10.1109/cec.2016.7744412
5. Keshari AK, Sharma R, Nigam MJ (2019) Digitizing physical documents using optical character recognition. Tenth Int Conf Signal Process Syst. https://doi.org/10.1117/12.2516743
6. Arica N, Yarman-Vural FT (2002) Optical character recognition for cursive handwriting. IEEE Trans Pattern Anal Mach Intell 24(6):801–813. https://doi.org/10.1109/tpami.2002.1008386
7. Nagy G (1992) At the frontiers of OCR. Proc IEEE 80(7):1093–1100. https://doi.org/10.1109/5.156472
8. Goel A, Gautam S, Tyagi N, Sharma N, Sagayam M (2021) Securing biometric framework with cryptanalysis. Intell Data Analy Terror Threat Pred 181–208. https://doi.org/10.1002/9781119711629.ch9
9. Impedovo S, Ottaviano L, Occhinegro S (1991) Optical character recognition—a survey. Int J Patt Recogn Artif Intell 05(01n02):1–24. https://doi.org/10.1142/s0218001491000041
10. Strong JP (1973) Binary image operations using non-coherent optical techniques. Pattern Recogn 5(1):51–57. https://doi.org/10.1016/0031-3203(73)90008-3
11. Suen CY (1994) Automatic recognition of handwritten characters. Fundament Handwrit Recognit 70–80. https://doi.org/10.1007/978-3-642-78646-4_4
12. Govindan VK, Shivaprasad AP (1990) Character recognition—a review. Pattern Recogn 23(7):671–683. https://doi.org/10.1016/0031-3203(90)90091-x
13. Goel A, Tyagi N, Gautam S (2019) Comparative analysis of 3-D password using various techniques. ' Comparative analysis of 3-D password using various techniques ',

Int J Emerg Technol Innov Res (www.jetir.org), 6(6):711–718, ISSN: 2349–5162, June 2019, Available: http://www.jetir.org/papers/JETIR1907Q08.pdf
14. Dimond TL (1958) Devices for reading handwritten characters. Papers and discussions presented at the December 9–13, 1957, eastern joint computer conference: computers with deadlines to meet on XX—IRE-ACM-AIEE '57 (Eastern). https://doi.org/10.1145/1457720.1457765
15. Sarkhel R, Das N, Das A, Kundu M, Nasipuri M (2017) A multi-scale deep quad tree based feature extraction method for the recognition of isolated handwritten characters of popular Indic scripts. Pattern Recogn 71:78–93. https://doi.org/10.1016/j.patcog.2017.05.022
16. Tyagi N, Gautam S, Goel A, Mann P (2021) A framework for blockchain technology including features. In: Hassanien AE, Bhattacharyya S, Chakrabati S, Bhattacharya A, Dutta S (eds) Emerging technologies in data mining and information security. Advances in intelligent systems and computing, vol 1286. Springer, Singapore. https://doi.org/10.1007/978-981-15-9927-9_62
17. Priyadharshini S, Alaguvathana P, Muthulakshmi S, Akalya C, Akilaa V, Gayathri V, Lavanya CR (2020) Image text to speech conversion using optical character recognition. Int J Psychosoc Rehabilit 24(5):4199–4205. https://doi.org/10.37200/ijpr/v24i5/pr2020134
18. Mori S, Suen CY, Yamamoto K (1992) Historical review of OCR research and development. Proc IEEE 80(7):1029–1058. https://doi.org/10.1109/5.156468
19. Nagy G, Nartker TA, Rice SV (1999). Document recognition and retrieval VII. https://doi.org/10.1117/12.373511
20. Troxel DE (1976) Feature selection for low error rate OCR. Pattern Recogn 8(2):73–76. https://doi.org/10.1016/0031-3203(76)90035-2
21. Baird HS (n.d.) Digital libraries and document image analysis. In: Proceedings of seventh international conference on document analysis and recognition, 2003. https://doi.org/10.1109/icdar.2003.1227619
22. Shillman RJ, Kuklinski TT, Blesser BA (1976) Psychophysical techniques for investigating the distinctive features of letters. Int J Man Mach Stud 8(2):195–205. https://doi.org/10.1016/s0020-7373(76)80043-0
23. Blesser BA, Kuklinski TT, Shillman RJ (1976) Empirical tests for feature selection based on a psychological theory of character recognition. Pattern Recogn 8(2):77–85. https://doi.org/10.1016/0031-3203(76)90036-4
24. Glassner AS (1990) Planar cubic curves. Graphics G 575–578. https://doi.org/10.1016/b978-0-08-050753-8.50125-x
25. Rousseeuw PJ, Leroy AM (1987) Robust regression and outlier detection. Wiley Series Probab Statist. https://doi.org/10.1002/0471725382
26. Saadouni M, Stroomer H (2020) Tashelhiyt Berber manuscripts in Arabic characters: an update. Études Et Documents Berbères, No 42(2):193–200. https://doi.org/10.3917/edb.042.0193
27. Zramdini A, Ingold R (n.d.) A study of document image degradation effects on font recognition. In: Proceedings of 3rd international conference on document analysis and recognition. https://doi.org/10.1109/icdar.1995.602008
28. Yuan Q, Tan CL (n.d.) Text extraction from gray scale document images using edge information. In: Proceedings of sixth international conference on document analysis and recognition. https://doi.org/10.1109/icdar.2001.953803
29. Dortier JF (2002) Aux origines de la culture. La Culture 103–119. https://doi.org/10.3917/sh.journ.2002.01.0103
30. Federal information processing standards publication: optical character recognition (OCR)—dot matrix character sets for OCR-MA (1987) https://doi.org/10.6028/nbs.fips.129-1987
31. Chaudhuri A, Mandaviya K, Badelia P, Ghosh SK (2016) Soft computing techniques for optical character recognition systems. Opt Charact Recognit Syst Diff Lang Soft Comput 43–83. https://doi.org/10.1007/978-3-319-50252-6_3
32. Dumouchel B, Dercourt J (2009) Assessing the quality and impact of research: practices and initiatives in scholarly communication. Inf Serv Use 28(3–4):193–195. https://doi.org/10.3233/isu-2008-0578

33. Gurung A, Gautam S, Garg T, Bhardwaj Y, Gupta H (2021) Virtual numeric authentication system using contour detection of color-banded fingertips. In: Tavares JMRS, Chakrabarti S, Bhattacharya A, Ghatak S (eds) Emerging technologies in data mining and information security. Lecture notes in networks and systems, vol 164. Springer, Singapore. https://doi.org/10.1007/978-981-15-9774-9_32
34. Singh G, Gautam S, Prachi Verma A, Kaushal T (2021) Analysis of blockchain induced cryptocurrency: regulations and challenges of cryptocurrencies. In: Hassanien AE, Bhattacharyya S, Chakrabati S, Bhattacharya A, Dutta S (eds) Emerging technologies in data mining and information security. Advances in intelligent systems and computing, vol 1286. Springer, Singapore. https://doi.org/10.1007/978-981-15-9927-9_54
35. Sagheer MW, He CL, Nobile N, Suen CY (2009) A new large urdu database for off-line handwriting recognition. Image Analy Process ICIAP 2009 538–546. https://doi.org/10.1007/978-3-642-04146-4_58
36. Toselli AH, Vidal E (2013) Fast HMM-filler approach for key word spotting in handwritten documents. In: 2013 12th International conference on document analysis and recognition. https://doi.org/10.1109/icdar.2013.106
37. Khan RA, Meyer A, Konik H, Bouakaz S (2013) Framework for reliable, real-time facial expression recognition for low resolution images. Pattern Recogn Lett 34(10):1159–1168. https://doi.org/10.1016/j.patrec.2013.03.022

Face Recognition Techniques and Implementation

Inshita Bamba, Yashika, Jahanvi Singh, and Pronika Chawla

Abstract Human face is multidimensional structure that conveys expression. Face recognition is one of the most essential techniques used these days for security purposes. Face recognition is capable of detecting human face. Face recognition has been a great tool used in various real-life scenarios like security systems, social networking platforms and numerous commercial areas. As compared to traditional ML algorithms, deep learning techniques have improved the performance and accuracy of image processing such as contrast adjustment, histogram and image blending. Face recognition is a method of perceiving the face of a person based on distinctive features. In this era of artificial intelligence, face recognition is one of the most popular among other biometric techniques. Other biometric techniques may fail due to some reason but human face recognition will be the same and work properly. Different approaches like statistical approaches, principal component analysis, local binary pattern, hybrid approaches, convolutional neural network and support vector machine are used to perform this task of face recognition. Finally, it is implemented using MATLAB based on eigenvectors.

Keywords Eigenvectors · Face recognition · Image processing · Neural network · Support vector machine

1 Introduction

The face recognition system helps in the identification of an individual through their facial images. Face recognition recent approaches have improved the security checks in various fields. Effective methods used in the extraction of features and classification tasks help to identify key factors for real-world applications. As per the history of face recognition, traditional methods have faced many challenges, i.e., lacked basic image features and artificial intelligence. Facial recognition is a challenging field due to the variations in expressions, poses and lightning. Although much research has

I. Bamba · Yashika (✉) · J. Singh · P. Chawla
Manav Rachna International Institute of Research and Studies, Faridabad, Haryana, India

P. Dutta et al. (eds.), *Emerging Technologies in Data Mining and Information Security*, Lecture Notes in Networks and Systems 490,
https://doi.org/10.1007/978-981-19-4052-1_35

already solved the problem of face recognition still some issues need to be resolved. Differentiations in poses, orientation, color, etc., are various issues that are to be taken into account minutely [1].

Many approaches have been applied to face recognition. Facial recognition has applications in different fields like analyzing characteristics of a person's image distinctively through video capturing. We usually heard of cyber-crimes and credit card frauds as criminals were taking advantage of passwords, PINs, etc., but biometrics technology can be a great tool for complete security solutions. Face recognition gives an authentic identity to an individual.

2 Literature Review

One of the active research areas from the last 40 years is face recognition. It contains several fundaments like image processing, neural network, computer's vision, machine's approach and pattern analysis. The main issue is their grouping. 'The main step in face recognition is to develop the face images from the familiar individuals and after it to compare and classify the upcoming test or algorithm images'. This trouble can be overcome by 'humans', but a fixed amount of storage or memory can be a problem [2]. The basic issues or limitations for machine-based face recognition are as follows:

1. Overlapping due to scarf, mask, etc.
2. Size and dimensions of the image
3. Profile and side angles
4. Aging links and wrinkles
5. Expression changes
6. Pixel distortion
7. Background difference
8. Existence of spectacles or lenses, facial hairs, etc.

In existing face recognition systems (Fig. 1), the main trouble is the involvement of background and difference in facial features due to aging. A complete and successful application needs to cover all these sub-problems [3].

In brief, there are two types of wide categories of face recognition algorithms, i.e., 'image template based, and geometry based'. The main feature of the template-based method is that it correlates face images and one or more test images for the identification of images from the stored database. According to Brunelli and Poggio, optimal strategy of face recognition of 1993 some of the statistical tools for the constructions of an appropriate database are:

Fig. 1 Configuration of the face recognition system [3]

1. Support vector machines (SVM)
2. Kernel methods
3. Principal component analysis (PCA)
4. Neural networks
5. Independent component analysis
6. Linear discriminant analysis (LDA).

On the other hand, there are more approaches to this process like hybrid approaches, i.e., the combination of neural and pattern techniques. For instance, it can include PCA and neural technology. In these approaches', resolutions, pixels difference and overlapping due to specs, scarf, etc., can be a major problem. The hybrid approach of PCA with local feature analysis (LFA) is considered one of the best and successful systems all over the countries. It is also a commercial know in the industry for identifying faces.

3 Existing Methods

The basic classification of all the face recognition methods can be seen in Fig. 2 [4].

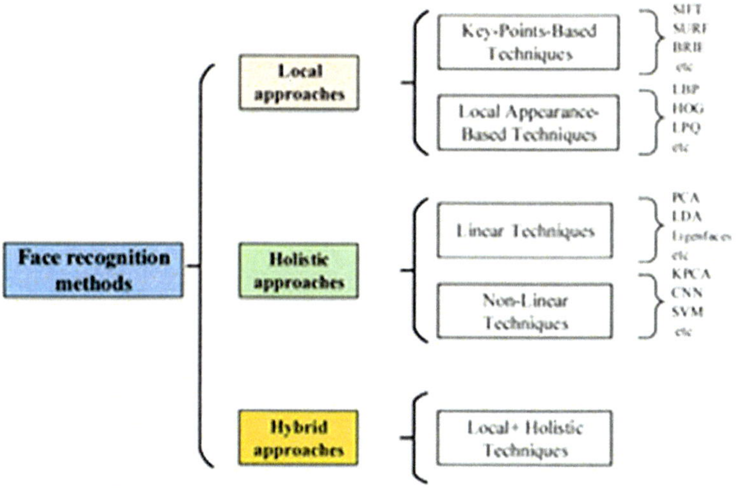

Fig. 2 Classification flowchart

3.1 Local Approaches

Local binary pattern (LBP)

Local binary pattern (LBP) (Fig. 3) is an easy but very efficient texture operator which counts the pixels of a picture by means of thresholding the neighborhood of each pixel and considers the end result as a binary number. It may be visible as a unifying approach to the traditionally divergent statistical and structural models of texture evaluation. It is a popular approach in various applications [5]. Parameters used are radius, neighbors, Grid X and Grid Y. If LBP and histogram are used together, we can represent the face images with a simple data vector.

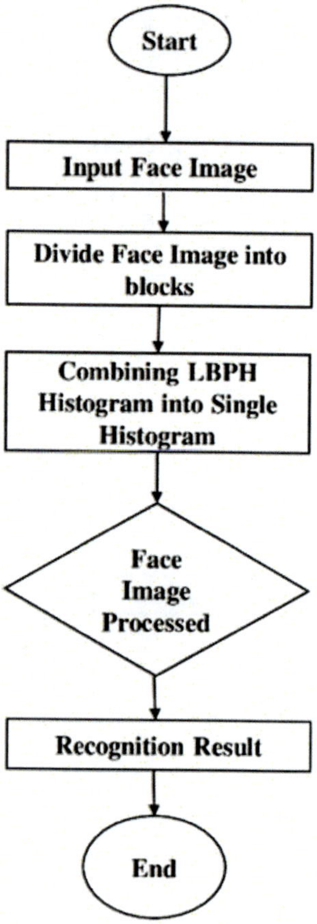

Fig. 3 LBPH algorithm flowchart

3.2 *Statistical Approaches (Holistic approaches)*

Some techniques identify, parameterize, and analyze linear subspaces. Statistical face recognition techniques are also based on nonlinear subspaces, transformation and support vector machine (SVM). Statistical face recognition approaches like PCA, LDA view a 2D face image as a vector in image space memory.

Principal component analysis (PCA)

Principal component analysis (Fig. 4) is used to reduce the image space, by transforming a large set of variables into a smaller one that still contains most of the information in the large set. It includes eigenfaces. Most of the data is reduced into small sets almost without any loss. It is preferred because of its speed, enhanced learning.

Fig. 4 PCA algorithm flowchart

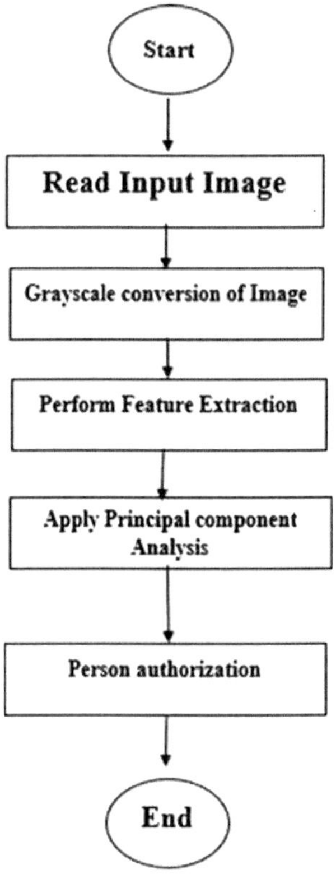

Table 1 Comparison of LBP, PCA and RBF

LBP	PCA	RBF
Local binary pattern (LBP) histogram's algorithm is used to identify the face structure of a human from both front and side face. It is used for characterization of image features	Principal component analysis (PCA) is a nonparametric statistical approach and an unsupervised technique, basically implies to short or cut down the dimension in ML	Radial basis function is a hybrid approach and mostly uses neural classifier to cope up with some of the minute training sets of high dimensions
Parameters used here are—radius, neighbors, Grid X and Grid Y	It is consisted of eigenfaces and the data, here is reduced without any loss	It is used in face recognition to identify various faces
LBP is the most reliable recognition technique	PCA is basically used for preprocessing of data before other analysis	It is used to assign real value to each input from its domain

3.3 Hybrid Approaches (Holistic approaches)

Hybrid approaches are one of the common approaches to face recognition techniques as it uses both statistical pattern and neural networks which help to overcome disadvantages of individual methods.

PCA and RBF (radial basis function)

Er Et Al suggested that a hybrid learning algorithm for the RBF method should be used for data extraction which decreases the dimension of search memory space. Image features are first extracted by principal component analysis (PCA). Backpropagation neural network (BPNN) is used on facial features to classify different face classes.

After an experimental analysis of all approaches, it has been found that the hybrid approach is much more efficient than other facial features extraction approaches individually.

4 Comparison Between LBP, PCA and RBF

Comparison of LBP, PCA and RBF are shown in Table 1.

5 Face Recognition Architecture

There are four main stages in a face recognition system as shown in Fig. 5:

Detection, preprocessing, feature extraction and recognition.

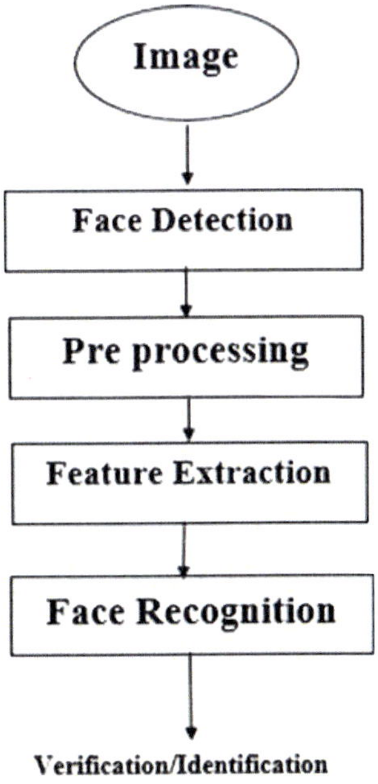

Fig. 5 Stages in face recognition [4]

1. Face Detection:

   This stage mainly detects the captured screen for a face from the database. After detecting the face, the output is then transferred to the next stage for preprocessing.
2. Preprocessing:

   This step makes the captured image ready for feature extraction. It removes the shadow effect, blurriness, lightning effect, etc.
3. Feature Extraction:

   In this step, the face is transformed into a vector with fixed dimensions and points. Features of images are extracted using the feature extraction approach.
4. Face Recognition:

In this step, finally, the image is recognized using different identities of the image. For the recognition, an image database is built. Then, when an image comes for detection, it performs all the steps serial wise first detection then preprocessing next feature extraction and then compares with images in the database.

There are two subdivisions of applications of face recognition, i.e., identification and verification. The term face identification refers to the process of determining a person's identity with a given image without his awareness. Whereas the term face verification refers to the process of guessing the identification, the model should generate true or false according to the guess [4].

6 Neural Networks

Neural network is an efficient classification technique that can be used for both known and unknown datasets. The success rate of the algorithm completely depends on the selection of the features based on the classifier [6].

7 Convolutional Neural Network

Yann n LeCun and Yoshua Bengio proposed the idea of CNN in 1995. CNNs are forward fed networks that extract the properties and features of the raw input image provided. As CNN is unvarying with respect to transformations and distortions made with the image.

A convolutional neural network is given the training to detect and recognize face images, and logistic regression classifier is used to classify those features. Various methods are used for the extraction of features and classification activity. MLP was the most used classifiers, but they were up to the desired accuracy and their performance was depending on the preprocessing stage. Its main drawback was, it was invariant for rotations, scaling and shifting of images. CNN's discovery solved the limitations of MLP. CNNs are networks capable of extracting features and properties from the input images without any preprocessing tasks. An advantage of CNN is that it can recognize features with a great variety of distortions and geometrical transformations [7]. CNN is an effective neural network technique used for image recognition. It is fed with some input data and made of various layers. CNN comprises filters and neurons based on some parameters. Each filter in the CNN takes in the input and applies convolution methodology on it to extract its features.

8 Support Vector Machine

The assist vector system (SVM) is a potential binary line divider. An improbable feature with key forces. This is compared to the retreating divisions and the Naïve Bayes. That is, SVM splits the data across the entire selection limit. The subset of data that supports decision-making is appropriately known as help vectors. Compared to SVMs, probabilistic classifiers extend the model that quality defines detail by thinking of all the information as opposed to just a small subset. Soon, later, the planning stages may require large computer resources.

The banner and line segments, however, are two SVM parameters. Recent developments using the 'kernel trick' have addressed the equality limit at the decision limit. However, the lack of power to distinguish greater facts than training continues to be a place for further studies. Strategies to date consist of creating more than one SVM that tests virtual gadgets among them in broadening the channels, which include one rather than all-in-one (OVA) or all-against-all (AVA) (Bhavsar and Ganatra 2012). The latter is also called one-versus-one (OVO). Fork orders, OVA requires the separation of students in the school so that the same beauty discriminates the final ok-1 training. AVA requires ok (k − 1)/2 classifiers because one size is discriminated against in each category, for all active pairing elements. After constructing a number of binary split desires in both strategies based on tests that provide the most significant discriminatory value [8].

SVM divides all information gadgets into feature space into two categories. Mathematical objects should have functions $\{\times 1 \ldots \times n\}$ and a label; i.e., SVM treats each piece of information as a point in the feature area as if the object belongs to one height or another. In particular, the real object (reflected with the help of its working vector) both belong to a category, where the beauty label is $y_i = 1$, or not to the luxury (meaning it belongs to another category) where the beauty label is $y_i = -1$. Therefore, the mathematical definition in which the p vector of the element is negative and has n the variety of information points [9].

$$\text{Data} = \left\{(x_i, y_i) | x_i \in \Re^p,\ \ y_i \in (-1,\ +1)\right\}_{i=1}$$

9 Experimentation and Simulation

This paper presents a face recognition algorithm implemented using MATLAB based on an eigenvector.

This dataset consists of four images of ten persons each taken from various angles. It is an eigenvector-based approach to do facial recognition [5].

Step 1: Preparing dataset.

The dataset has images of faces of 40 persons. Each of these has ten images of faces with different angles and poses. We have a total of 400 images. For every person, there is a separate folder consisting of ten images. All images are taken in gray scale and have the same resolution, extension and dimension. All the images should have same extension.

Step 2: Loading the dataset

After preparation, we have to load the dataset. For that, we need to use a function to load the dataset.

Firstly, the function name 'load_database()' is announced which consequently gives numeric type of images. The pictures to be returned should be stored in the variable named 'output value'.

After this, two factors named 'loaded' and 'numeric pictures'". These factors have a place with persistent type [10].

Utilizing the 'if' condition, the 'loaded' variable must be checked if the variable is vacant, if empty, we will load the dataset. The persistent factors permanently store the information. We need to get the dataset loaded just a single time. Along these lines, it is important to check if the variable is empty toward the start [11].

We have 40 pictures. Every one of them has $92 \times 112 = 10{,}304$ pixels. That is the reason we should take 10,304 zeros 40 times. Later the pixel values of the pictures will replace these zeros. Then, at that point inside the 'for loop' utilizing strcat, folder names are to be connected (p1, p2, p3, etc.), names of the pictures (1, 2, 3, etc.) and the extension of the pictures (pgm pictures). After that utilizing 'imread' work, the pictures are loaded. It is important to reshape the pictures. After loading 'reshape' work is utilized to change over the pictures into a single column matrix. Then, at that point, 'size' function is likewise utilized which gives the size of the rows and columns of the images.

From that point onward, the pictures are changed over into an 8-bit unsigned number to decrease the memory used. Utilizing the 'uint8' work, we can straightforwardly change over information into an 8-bit unsigned whole number in MATLAB. There is an 'if' condition before the 'for loop'. It will turn out to be false when the 'loaded' variable has a value. At last, this function returns the pictures in an 8-bit unsigned number form [12–14].

Stage 3: Recognizing the face

For acknowledgment of the faces, load the dataset is first and foremost. After that the random function is used to produce random index and with the assistance of the random index, the pictures are stacked which are be perceived later. The remainders of the pictures are additionally stacked into a different variable for additional acknowledgment. [15–18].

After this, the mean of the entirety of the pictures is determined and deducted the mean from them. The eigenvectors had been made on these pictures. After having the eigenvalues, a framework can be made where each column contains the mark of single pictures. Presently by the assistance of eigenvalues and the signature of the picture we can remember them further; the mean is deducted from the picture which we need to recognize. Next, we duplicate it with the eigenvector, finally we can anticipate the picture based on the distinction determined between present signatures of picture with the signature to be recognized.

The exactness of this strategy is around 94% [19–21].

This code will automatically load an image and then will find image of the same person from the dataset.

10 Conclusion and Future Scope

Performance of face detection and recognition is influenced and harmed on the grounds that impediment frequently prompts missed identification. To work on the exactness of face identification and acknowledgment, a visual consideration component direction model is proposed in this paper, which utilizes the visual consideration system to direct the model feature the apparent space of the impeded face. The face recognition issue is streamlined into the undeniable level semantic element discovery issue through the worked on insightful organization, and the area and size of the face are anticipated by the actuation guide to stay away from extra boundary settings. Countless reproduction test results show that our proposed strategy is better than other examination calculations for the precision of impediment face discovery and acknowledgment on the face information base. Furthermore, our proposed technique accomplishes a superior harmony among exactness and speed, which can be utilized in the field of safety observation. Be that as it may, the exhibition of the proposed calculation is touchy to boundaries, and its speculation isn't high. The most effective method to further develop this issue will be more helpful for the model applied to different situations or information.

As per the market demand, the face recognition is the most popular and required need of the society for security and threat issues. So there is a need for dynamic recognition of patterns and numeric values. The static recognition may cause ambiguity in the real-time systems. Dynamic recognition is the future scope of the facial recognition systems.

References

1. Kasar MM, Bhattacharyya D, Kim TH (2016) Face recognition using neural network: a review. Int J Secur Appl 10(3):81–100
2. Bansal A (2016) A literature review on face recognition system. Int J Latest Eng Res Appl 01(03):109–112
3. Kumar P, Singh D (2016) Approach on face recognition & detection techniques. Int J Eng Comput Sci 5(7):17133–17135, ISSN: 2319–7242
4. Gupta P (2018) Deep neural network for human face recognition. Published by MECS Publisher. Selection and/or peer-review under the responsibility of the research association of modern education and computer science
5. Tanwar S, Chawla P, Maadan R, Bhadana P (2020) Authentication of face using Matlab. In: 2020 5th international conference on communication and electronics systems (ICCES), pp 1229–1234. IEEE
6. Barahate SR, Saturwar J (2010) Face recognition using PCA based algorithm and neural network. Association for computing machinery, New York, NY, USA, pp 249–252
7. Facial emotion recognition using convolutional neural networks (FERC) Ninad Mehendale Springer Nature Switzerland AG 2020
8. Singh AP, Manvi SKS, Nimbal P, Shyam GK (2019) face recognition system based on LBPH algorithm. Int J Eng Adv Technol (IJEAT) 8(5S), ISSN: 2249 – 8958
9. Face Recognition Based on Convolutional Neural Network Musab Coşkun Faculty of Electrical and Electronics Engineering, Fırat University, Elazığ, Turkey
10. Kortli Y, Jridi M, Al Falou A, Atri M (2020) Face recognition systems: a survey. MDPI J Sens 20:342
11. ICET2017, Antalya, Turkey 978–1–5386–1949–0/17/$31.00 ©2017 IEEE Understanding of a Convolutional Neural Network
12. Khalajzadeh H, Mansouri M, Teshnehlab M (2014) Face recognition using convolutional neural network and simple logistic classifier. In: Soft computing in industrial applications, pp 197–207. Springer, Cham
13. Bamba I, Singh J, Chawla P, Soni K (2021) Big social data and modelling frameworks. In: the 2021 international conference on artificial intelligence and smart systems (ICAIS), pp 724–727. IEEE

14. Coşkun M, Uçar A, Yildirim Ö, Demir Y (2017) Face recognition based on convolutional neural network. In: 2017 international conference on modern electrical and energy systems (MEES), pp 376–379. IEEE
15. Pranav DS, Punj D, Dubey T, Chawla P (2021) Data mining in cloud computing. In: 2021 5th international conference on computing methodologies and communication (ICCMC), pp 71–78. IEEE
16. Kumar TS (2021) Study of retail applications with virtual and augmented reality technologies quot. J Innov Image Process (JIIP) 3(02):144–156
17. Manoharan JS (2019) A smart image processing algorithm for text recognition, information extraction and vocalization for the visually challenged. J Innov Image Process 1(1):30–38
18. Tyagi SS (2021) Secure data storage in cloud using encryption algorithm. In: 2021 third international conference on intelligent communication technologies and virtual mobile networks (ICICV), pp 136–141. IEEE
19. Jain M, Rajput H, Garg N, Chawla P (2020) Prediction of house pricing using machine learning with Python. In: 2020 International conference on electronics and sustainable communication systems (ICESC), pp 570–574. IEEE
20. Rahman AM, Al Mamun A, Islam A (2017) Programming challenges of chatbot: current and future prospective. In: 2017 IEEE region 10 humanitarian technology conference (R10-HTC), pp 75–78. IEEE
21. Mutreja M, Khandelwal K, Dham H, Chawla P (2021) Perception of IOT: application and challenges. In: 2021 6th international conference on communication and electronics systems (ICCES), pp 597–603. IEEE

Cybersecurity Vis-A-Vis Artificial Intelligence: An Analysis of the International Conventions

Gagandeep Kaur and Prashant Chauhan

Abstract The expansion of digitalization, not only has brought convenience to individuals; however, it has resulted in the vulnerabilities of the data and threats to cybersecurity. Where a new age technology in form of artificial intelligence has intruded into many fields, cyberspace is no exception to it. Artificial intelligence can be a technology to make cyberspace a more secure place, but the same technology can be utilized by the cybercriminal to penetrate cyberspace resulting in the breach of individual privacy. Protection of individual rights has been the sole objective of international organizations and the conventions. However, with the rise of artificial intelligence and its intrusion into the cyberworld, the prevalent international laws do not appear to be competent enough to secure individual rights. Two major questions of research have been addressed in this study, namely (1) what are the issues and challenges posed by artificial intelligence, and (2) whether the laws governing the use of artificial intelligence are self-sufficient to regulate cybersecurity intact or not. Through this paper, an attempt has been made to analyze international conventions governing AI and their relevancy in reference to cybersecurity.

Keywords Artificial intelligence · Cyber crimes · Data protection · Cybersecurity · International law · International conventions

1 Introduction

In every phase of civilizational development, inventions have shaped the course of the future of mankind. The development in the field of computer technology has drastically improvised the face of civilization. Artificial intelligence and machine learning promise to positively impact nearly every aspect of one's life, with the

G. Kaur (✉) · P. Chauhan (✉)
University of Petroleum and Energy Studies, Dehradun, India
e-mail: gkaur@ddn.upes.ac.in

P. Chauhan
e-mail: legum.jus@gmail.com

P. Dutta et al. (eds.), *Emerging Technologies in Data Mining and Information Security*, Lecture Notes in Networks and Systems 490,
https://doi.org/10.1007/978-981-19-4052-1_36

potential to change society in fundamental ways [1]. There is a real fear that smart machines are set to replace many humans as decision-makers as artificial intelligence (AI) has crept into many organizational processes [2].

Intrusion detection and computer security have been strengthened through the integration of machine learning methods. By enhancing data quality and training for IDS using ML techniques, a number of research contributions demonstrate how ML can be used to achieve reliable intrusion detection with an excellent performance [3]. The move to digital technology has resulted in more and more confidential information being exposed, which is a significant threat to our privacy [4]. Because of increasing reliance on AI, not only will attackers have a greater incentive to target those algorithms, but also each successful attack will have more devastating consequences [5].

2 Literature Review

Hackers are increasingly using artificial intelligence, putting sensitive data at risk, and restoring it takes quite some time. AI is far superior to traditional cybersecurity methods. Artificial intelligence holds promise for many organizations, so they need to focus on securing cyberspace based on AI. Alessandro Parisi, in his book, expressed his views that in an obvious sense, an artificially intelligent machine is a creation of human intelligence. Comparing the two is impractical since they operate at different levels. Conflict is bound to happen with the increase of advanced technologies such as artificial intelligence and the use of AI, similar to other fields, in cybersecurity [6].

Whereas Rammanohar Das and Raghav Sandhane in their article expressed their concerns that artificial intelligence must be at an advanced level to deal with the complex technologies of cyberspace, as opposed to a regular or ordinary level. Cyberspace parameters are secured by the neural network. Despite artificial intelligence's use in cybersecurity, it is lacking proper implementation. The issue of data and decision-making needs to be resolved. The issue of artificial intelligence being used to intrude into cyberspace has also been raised, posing a security threat [7]. The authors through their article expressed that artificial intelligence is being used in a lot of fields, including cyberattacks with the use of artificial intelligence. To penetrate cyber systems around the world, cybercriminals are using artificial intelligence. By using artificial intelligence effectively to protect cyberspace, it will only take seconds for it to be destroyed [8]. Whereas authors expressed their opinion in their article, that artificial intelligence is needed to handle cybersecurity issues. Since tech can only be fought with tech. AI is viewed as something that needs to be integrated into current systems with enhanced capabilities [9].

Throwing light on the capability of AI in dealing with cybersecurity, IH Sarker, MH Farhad, and R Nowrozy in their article pointed out that compared with conventional methods of cybersecurity, artificial intelligence-based security systems cope better with cyber threats. In their opinion, artificial intelligence can be a significant factor in managing cyberspace. In a broader manner, artificial intelligence-based systems can handle cybersecurity issues with ease [10]. In a study conducted based on articles from the scholarly database, the authors have concluded that even if cyber-attacks are complex in nature, AI may be very useful for detecting and combating them. According to them, artificial intelligence would reduce the complexity of cybersecurity; they found that cybersecurity will be forever changed by AI, and it will likely play a crucial role in securing the Internet [11]. Joseph Ogaba Oche, on the other hand, in his research article as narrow AI has a limited domain, though general AI is broader in effect, and a slight error can cause the entire system to collapse. They argue that there is no such thing as an entirely secure system. The machine does not seem to understand human values as they are volatile and hard to program and understand. Whether self-learning or self-modifying learning will be able to adapt to human value systems must also be considered [12]. The authors in their research paper stated that with the increase in cyberattacks, organizations of all kinds are creating more secure environments for their stored information. Cyberattacks are increasingly using artificial intelligence, so immediate attention is required. Artificial intelligence must be used to design a better defense mechanism [13].

Attention has been drawn by the authors, that the intruders are exploiting artificial intelligence to bypass security systems is negative. Being human, it is difficult to predict how potential hackers will program their attacks. Artificial intelligence cannot be separated from human influence [14]. Using artificial intelligence to identify potential cyber threats reduces manual effort. AI's effectiveness in cybersecurity is under scrutiny. Machine learning is still in its infant stages [15]. AI's effectiveness in cybersecurity is under scrutiny. The manipulation of technology may result in exposure to its negative aspects. It is agreed that each technology has its pros and cons. But our concern lies in whether international law is sufficiently compatible to regulate effectively the use of this technology. Why international law and not local laws? Due to the fact, the Internet knows no boundaries, restricting it to one geographic area is impractical or impossible.

3 Research Gap and Research Methodology

Based on the review of literature, it has been analyzed that most of the research has been conducted on the multifaceted aspects of artificial intelligence most prominently contemporary threats of AI. However, the novelty of this paper lies in the comparative analysis of the international convention on AI. After exploring the application of these conventions to contemporary challenges, some suggestions have been put forward so as provide a more holistic and comprehensive outcome of this research.

The methodology adopted during this analysis paper is doctrinal, analytical, and comparative. The international laws and global guidelines governing artificial intelligence are critically analyzed with a focus on exploring their application to contemporary challenges. Reliance has been placed upon the research literature by eminent authors in the field of cybersecurity, artificial intelligence, and international law.

4 Cybercrime: A Virtual Enemy of Technological Era

The pandemic in 2021 leads to more workers settling into remote work schedules, making it imperative to safeguard backup files. "The U.S. Department of Justice [DOJ] divides cybercrime into three categories:

1. Crimes in which the computing device is the target—for example, to gain network access;
2. Crimes in which the computer is used as a weapon—for example, to launch a denial-of-service [DoS] attack; and
3. Crimes in which the computer is used as an accessory to a crime—for example, using a computer to store illegally obtained data"[16].

New hardware vulnerabilities were discovered in 2018, and research pointed to the likelihood of powerful cyberattacks by utilizing artificial intelligence [17]. "Cyberattacks on critical infrastructure—rated the fifth top risk in 2020 by our expert network—have become the new normal across sectors such as energy, healthcare, and transportation. Such attacks have even affected entire cities. Public and private sectors alike are at risk of being held, hostage. Organized cybercrime entities are joining forces, and their likelihood of detection and prosecution is estimated to be as low as 0.05% in the United States. Cy-cybercrime-as-a-service is also a growing business model, as the increasing sophistication of tools on the Darknet makes malicious services more affordable and easily accessible for anyone" [18]. "According to an INTERPOL report in one four-month period [January to April] some 907,000 spam messages, 737 incidents related to the malware, and 48,000 malicious URLs—all related to COVID-19—were detected by one of INTERPOL's private sector partners" [19]. "The 2020 Internet Crime Report includes information from 791,790 complaints of suspected Internet crime—an increase of more than 300,000 complaints from 2019—and reported losses exceeding $4.2 billion" [20]. It is clear that malicious intruders are exploiting lacunas to their own benefit [21].

5 Artificial Intelligence as a Tool for Cybercrime

The major objective of cybercriminals is generally monetary, as far as AI is concerned, it is an algorithm-based application having the capability of acting autonomously. If this technology of AI is used to commit a cybercrime that will be a technology-based crime without direct human intervention. Few major types of cybercrimes that may be committed by using AI may include the following:

Cyberespionage: Hacking into a government network or system to obtain confidential information from it. This is often encouraged by profit or ideology. Webcams or CCTV cameras can be used for cyberespionage, as well as emails, texts, and instant messages. In several countries, Pegasus surveillance software targeting iPhone and Android devices were used to observe activists, journalists, and politicians [22].

Online Identity theft: Identity theft occurs when a person accesses and gathers personal information from a computer to gain access to banking accounts and credit cards. Personal health records are another target for identity thieves. "According to the Aite Group, 47% of Americans experienced financial identity theft in 2020. The group's report, U.S. Identity Theft: The Stark Reality, found that losses from identity theft cases cost $502.5 billion in 2019 and increased 42% to $712.4 billion in 2020" [23].

Cyberextortion: Attacks or threats by an attacker include demands for money to protect the victim. Examples include ransomware attacks. Attackers infiltrate a network and encrypt the organization's data, making it unusable until a ransom is paid. Typically, Bitcoin is used for this purpose. "In 2015, it was estimated that Crypto Wall, a specific type of ransomware virus, made off with over $18 million worth of cryptocurrencies" [24].

6 International Regulatory Regime on Cybercrime

As per the discussion under the previous heading where the most common types of cybercrime were discussed. In order to deal with these cybercrimes, the United Nations had taken certain steps in the form of conventions and protocols, with an objective to bring together the member state and stakeholders in dealing with the growing menace of cybercrime. In the following table (Table 1), the authors have tried to bring forth briefly the mechanism of these conventions and protocols in dealing with cybercrime and to what extent the issue of artificial intelligence has been dealt with:

As per the United Nations Conference on Trade and Development, "154 countries [79%] have enacted cybercrime legislation, the pattern varies by region: Europe has the highest adoption rate [93%] and Asia and the Pacific the lowest [55%]" [25].

Having analyzed the conventions and protocol, the authors have concluded that these conventions have been vital in addressing the issue of cybercrime. Nevertheless, the authors have sought to draw attention to the lack of updates of these conventions and protocols in keeping up with technology, including the introduction of artificial intelligence. A recent proposal by the European Union to the European Parliament dates from 21.04.21, proposing regulating the use of artificial intelligence, in order to bring about a more open and non-discriminatory society. A similar initiative needs to be adopted by the international organization in relation to cybercrime and artificial intelligence.

Table 1 Comparative analysis of international convention and protocol

Sr. No	International conventions/protocols	Aim/Objectives	Cybercrime/Cybersecurity issues	Issues related to AI
1	Convention on the Rights of the Child [1989]	Each state parties commit to protect the child from any form of exploitation or abuse that may be harmful to the child	The availability of such kind of material on the dark web makes it an issue of immediate attention	The convention lacks requisite amendment concerning AI
2	United Nations Convention against Transnational Organized Crime [2000]	The main objective is to target organized crime	The provisions are relevant to cybercrime due to its use as a mode of committing cybercrime	The convention is silent as to the use of AI in organized cybercrime
3	Optional Protocol to the Convention on the Rights of the Child [2001]	It acknowledges that there is an increasing availability of content related to child pornography	Child pornography is a big market on the dark web	The issue of utilizing AI in such activities is left unaddressed
4	Convention on Cybercrime [2001]	Promoting international cooperation and adopting relevant legislation are important parts of pursuing a similar criminal policy to protect society against cybercrime	Being a convention specifically related to cybercrime, it has comprehensively dealt with cybercrime	Although this is a specific convention dealing with cybercrime but fails to incorporate AI as a mode of committing cybercrime
5	Additional Protocol to the Convention on Cybercrime Concerning the Criminalisation of Acts of a Racist or Xenophobic Nature Committed Through Computer Systems [2003]	This convention aims to have computer-based propagandistic activities of a racist or xenophobic nature addressed under national and international laws	The world being a global village, these kinds of crime can create disharmony among the races and communities, which can cause disability in countries	This convention fails to recognize AI as a mode in committing crimes prohibited by this convention

(continued)

7 Cybersecurity and Artificial Intelligence

Cybercrime and the laws regulating it might become outdated because of the rate at which technology improves day by day. We are now seeing futuristic technology

Table 1 (continued)

Sr. No	International conventions/protocols	Aim/Objectives	Cybercrime/Cybersecurity issues	Issues related to AI
6	Convention on the Protection of Children against Sexual Exploitation and Sexual Abuse [2007]	Its main objective is to protect children from abuse and exploitation, to fight sexual abuse	The dark web market is a significant conduit for child sexual exploitation	Even this convention fails to recognize AI as a potential threat being a mode of committing cybercrime

taking shape. Artificial intelligence is one technology that has become a global attraction. Automation has heavily infused artificial intelligence into almost all machines today. Artificial intelligence is being emphasized in various sectors. In both cases, technology can be beneficial to humans but can also cause harm. Many such examples are recorded in history. Despite these promising results, futuristic technologies have also generated serious concerns.

8 Conclusion

There is therefore a need to regulate AI not just at the level of its application but also at the level of its research also. The issue of accountability must also be addressed. The authors propose that accountability means that if a program is created that makes use of artificial intelligence, then it is responsible for keeping the technology in a safer environment and that absolute liability must be applied if the program breaches cybersecurity. Presently, artificial intelligence does not fall under any specific rules or conventions that regulate cybersecurity. Dedicated and comprehensive cybersecurity and artificial intelligence convention would be beneficial, according to the authors. Another suggestion is the creation of an international agency that will coordinate with member states on activities relating to cybersecurity and artificial intelligence at the international level. Cyberspace is not limited to a single region, so a global approach is necessary for better-controlling cybercrime. Jurisdictional lines must not impede the prevention of cybercrime.

References

1. Jackson BW (2019) Artificial intelligence and the fog of innovation: a deep-dive on governance and the liability of autonomous systems. Santa Clara High Technol Law J 35(4):35
2. Jarrahi MH (2018) Artificial intelligence and the future of work: Human-AI symbios is in organizational decision making. Business Horizons 61(4):577–586. https://doi.org/10.1016/j.bushor.2018.03.007

3. Guezzaz A, Benkirane S, Azrour M, Khurram S (2021) Tree with enhanced data quality
4. Gheyas IA, Abdallah AE (2016) Detection and prediction of insider threats to cyber security: a systematic literature review and meta-analysis. Big Data Analyt 1(1). https://doi.org/10.1186/s41044-016-0006-0
5. Wolff J (2020) How to improve cyber security for artificial intelligence. Brookings. Retrieved 12 Nov 2021, from https://www.brookings.edu/research/how-to-improve-cybersecurity-for-artificial-intelligence/
6. Parisi A (2019) Hands-on artificial intelligence for cybersecurity: implement smart AI systems for preventing cyberattacks and detecting threats and network anomalies. Packt Publishing.
7. Das R, Sandhane R (2021). Artificial intelligence in cyber security. J Phys: Conf Series. Retrieved 14 Nov 2021, from https://iopscience.iop.org/article/https://doi.org/10.1088/1742-6596/1964/4/042072.
8. Bhatele KR, Shrivastava H, Kumari N (2019) The role of artificial intelligence in cyber security. Countering cyber attacks and preserving the integrity and availability of critical systems 170–192. https://doi.org/10.4018/978-1-5225-8241-0.ch009
9. Panimalar AS, Pai GU, Khan SK (2018) Artificial intelligence techniques for cyber security. Int Res J Eng Technol (IRJET) 5(3)
10. Sarker IH, Furhad MH, Nowrozy R (2021) AI-driven cyber security: an overview, security intelligence modelling and research directions. SN Comput Sci 2(3).https://doi.org/10.1007/s42979-021-00557-0
11. Wiafe I, Koranteng FN, Obeng EN, Assyne N, Wiafe A, Gulliver SR (2020) Artificial intelligence for cyber security: A systematic mapping of literature. IEEE Access, 8, 146598–146612. https://doi.org/10.1109/access.2020.3013145
12. Oche JO (2019) The risk of artificial intelligence in cyber security and the role of humans. Texila Int J Acad Res 1–7. https://doi.org/10.21522/tijar.2014.se.19.01.art001
13. Tao F, Akhtar M, Jiayuan Z (2021) The future of artificial intelligence in cyber security: a comprehensive survey. EAI Endorsed Trans Creat Technol 8(28), 170285. https://doi.org/10.4108/eai.7-7-2021.170285
14. Atiku SB, Aaron AU, Job GK, Shittu F, Yakubu IZ (2020) Survey on the applications of artificial intelligence in cyber security. Int J Sci Technol Res. Retrieved 14 Nov 2021, from https://www.ijstr.org/paper-references.php?ref=IJSTR-0920-42145
15. Kabbas A, Alharthi A, Munshi A (2020) [PDF] Artificial Intelligence applications in cyber security: semantic scholar. [PDF] Artificial intelligence applications in cyber security|semantic Scholar. Retrieved 14 Nov 2021, from https://www.semanticscholar.org/paper/Artificial-Intelligence-Applications-in-Kabbas-Alharthi/92c78be81eb42945f4147198484f4706dc4bfbcf.
16. Brush K, Rosencrance L, Cobb M (2021) What is cyber crime? Definition from search security. Search Security. Retrieved 14 Nov 2021, from https://searchsecurity.techtarget.com/definition/cybercrime
17. The global risks report 2019. World Economic Forum (n.d.) Retrieved 14 Nov 2021, from https://www.weforum.org/reports/the-global-risks-report-2019
18. The global risks report 2020. Global risks report 2020 (n.d.) Retrieved 14 Nov 2021, from https://reports.weforum.org/global-risks-report-2020/
19. INTERPOL report shows alarming rate of cyber attacks during COVID-19. INTERPOL. (n.d.) Retrieved 14 Nov 2021, from https://www.interpol.int/en/News-and-Events/News/2020/INTERPOL-report-shows-alarming-rate-of-cyberattacks-during-COVID-19
20. FBI (2021) IC3 releases 2020 internet crime report. FBI. Retrieved 14 Nov 2021 from https://www.fbi.gov/news/pressrel/press-releases/fbi-releases-the-internet-crime-complaint-center-2020-internet-crime-report-including-covid-19-scam-statistics
21. Covid-19 cyber crime analysis report-Interpol (n.d.) Retrieved 14 Nov 2021, from https://www.interpol.int/content/download/15526/file/COVID-19%20Cybercrime%20Analysis%20Report-%20August%202020.pdf
22. Significant cyber incidents.|Center for strategic and international studies (n.d.) Retrieved 14 Nov 2021, from https://www.csis.org/programs/strategic-technologies-program/significant-cyber-incidents

23. Facts + statistics: identity theft and cyber crime (n.d.) Retrieved 14 Nov 2021 from https://www.iii.org/fact-statistic/facts-statistics-identity-theft-and-cybercrime
24. Baum SC (n.d.) Crypto currency fraud: a look into the frontier of fraud. Digital Commons @ Georgia Southern. Retrieved 14 Nov 2021, from https://digitalcommons.georgiasouthern.edu/honors-theses/375
25. Cyber crime legislation worldwide. UNCTAD. (n.d.). Retrieved 14 Nov 2021, from https://unctad.org/page/cybercrime-legislation-worldwide

Role of Multi-agent Systems in Health Care: A Review

Anubhuti and Harjot Kaur

Abstract A multi-agent system (MAS) is a complex system consisting of multiple intelligent agents which can operate in an autonomous manner, yet can still act in a consistent way due to the information connections amongst them. With the growth in population, demand for the healthcare services has also been increased. Multi-agent systems play a very significant role in the health care services. It has been analysed that very few models have been proposed in the area of women and infant healthcare. In this paper, we have presented a comprehensive review of the research work for the use of multi-agent systems in different aspects of healthcare system and find out the research gaps in the same.

Keywords Multi-agent systems (MAS) · Intelligent agents · Infant care · Foetal monitoring · Health care services

1 Introduction

Healthcare is one of the most swiftly developing sectors. In today's era, the demand for healthcare services is increasing with the growing population. The present healthcare services are unable to cope up with the rising medical demands.

Our research reviews the healthcare sector, with a special focus on:

a. Multi-agent-based Healthcare Systems.
b. Infant and Child Healthcare.
c. Prenatal and postnatal Care of Mothers.

The agent innovation can essentially improve the outline and analysis of problem areas under the accompanying three conditions: (1) The problem space is geologically disseminated; (2) the subsystems of a problem space exist in a dynamic domain; and (3) the subsystems need to associate with each other all the more adaptably. The domain of healthcare systems is well suited to an agent-based approach because

Anubhuti (✉) · H. Kaur
GNDU Regional Campus, Gurdaspur, India
e-mail: anubhuti.cse@gmail.com

P. Dutta et al. (eds.), *Emerging Technologies in Data Mining and Information Security*, Lecture Notes in Networks and Systems 490,
https://doi.org/10.1007/978-981-19-4052-1_37

of its topographically dispersed nature and its need for flexible communication and cooperation [1].

1.1 Definitions

Various definitions related to the presented literature have been covered in this subsection.

Agent: "Schumacher [2] defines Agent as a computer system that is situated in some environment, and that is capable of autonomous action in this environment in order to meet its design objectives."

In other words, Agents are computer systems with two essential capabilities, i.e. firstly they are in any event to some degree fit for self-sufficient activity of choosing for them what they have to do keeping in mind the end goal to fulfil their design objectives and secondly, they are fit for interfacing with different agents—not just by exchanging information, but rather by taking part in analogues of the sort of social movement that we as a whole take part in each day of our lives: participation, coordination, negotiation, and so forth.

Intelligent Agent: "An agent with the following three properties is said to be an intelligent agent [3]:

- Reactivity
- Proactivity
- Social-activity".

Multi-agent System: "A multi-agent system (MAS) is a complex system consisting of multiple intelligent agents which can operate in an autonomous manner, yet can still act in a consistent way due to the information connections amongst them [4]."

Multi-agent System (MAS) according to Wooldridge [3] can be defined as:

> A multi-agent system is one that consists of a number of agents, which interact with one another, typically by exchanging messages through some computer network infrastructure.

For example, a key flight authority system at the fundamental aeroplane terminal of Ruritania all of a sudden fizzles, leaving flights in the region of the air terminal with no airport regulation help. Fortunately, autonomous air-traffic control systems in close-by airports perceive the failure of their associate, and participate to track and manage every influenced flight. The possibly heartbreaking circumstance goes without accident [3], due to the coordination amongst various vehicle monitoring agents in the system.

Women healthcare: Women's Health Care provides comprehensive, quality obstetrical and gynaecological care with the geniality, responsiveness and accessibility of a community-based practise [5].

Women's well-being [6] is committed to encouraging the safeguarding of well-being and avoidance of disease in females, and includes Screening, recognition and administration of conditions that are solely in females and are more serious and are more genuine in females.

Women have unique health issues which includes pregnancy, menopause and conditions of the female organs. Women can assure a healthy pregnancy by getting early and regular prenatal care [7]. Various issues that women can face include: maternal mortality, prenatal care issues, postnatal care issues and support for taking care of infants and children under five.

Maternal Mortality: "Pregnancy-related deaths are defined as death of a woman during or within one year of pregnancy that can be caused by a pregnancy complication, a chain of events initiated by pregnancy, or worsening of an unrelated condition due to the physiologic effects of pregnancy [8, 9]."

Maternal mortality Rate: "The pregnancy-related mortality ratio is the number of pregnancy-related deaths per 100,000 live births [8, 10]."

Under five mortality rate: "It can be defined as Probability of Dying between birth and exactly 5 years of age, expressed per 1000 live births [11]."

Prenatal care: "It aims to provide regular check-ups that allow doctors or midwives to treat and prevent potential health problems throughout the course of the pregnancy and to promote healthy lifestyles that can benefit both mother and child [12, 13]."

Postnatal care (PNC): "The care given to the mother and her newborn baby immediately after the birth and for the first six weeks of life is termed as Postnatal care [14]."

1.2 Why Multi-Agent System?

One of the main issues in medical field is of communication, coordination and cooperation amongst various healthcare workers [15]. There is a need of better coordination component for healthcare delivery frameworks. Due to the enlarging complexities of the real world, the traditional modelling techniques which were used previously have been deficit to deal with the real world situations [16]. The properties of software agents make them favourable in resolving these issues. Following are the reasons which make MAS suitable for the distributed, dispersed and sophisticated healthcare systems:

1. It enhances the ability of the medical workers to manage motile situations, thus ameliorating the quality of medical services they can provide [17].
2. MAS have the ability to adjust to the changes in the system promptly.
3. Due to the properties of intelligent agents, MAS offers an improved choice to settle the complicated issues in healthcare domain [18].

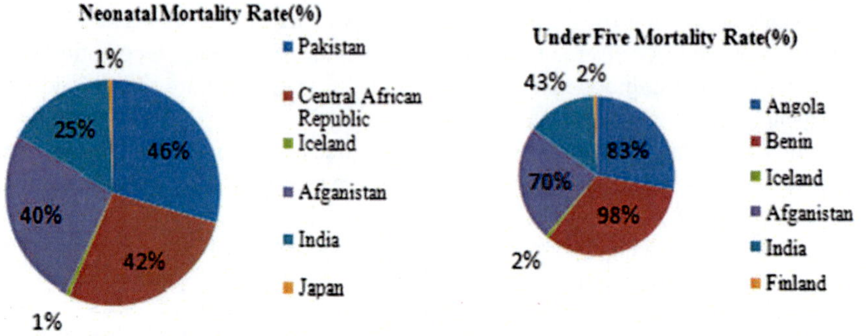

Fig. 1 Graphs showing statistics related to mortality rates

4. Agent-based solutions aid practitioners in better decision making due to its properties like autonomy, intelligence and social ability [19].

1.3 Objective/Motivation of the Research

The presented research tries to review about various works done in the field of healthcare till now, focussing more on the child and infant healthcare systems. It tries to find out the gaps in the existing research in healthcare sector. In short, our research tries to find out the answers to the presented research questions (in Sect. 2.1).

From the graphs presented in Fig. 1, it can be analysed that the mortality rates have been considerably higher in various countries including India.

The paper is structured as follows: Sect. 2 presents the research methodology employed to perform this survey. Section 3 describes the literature review. Section 4 identifies various research gaps in healthcare sector and Sect. 5 closes with conclusion and future recommendations.

2 Research Methodology

This section describes the research methodology for the selection of papers to perform the survey about the work done in the area of development of healthcare systems using multi-agent systems.

2.1 Research Questions

Various research questions were needed to plan a review:

Q1. What work has been done in the field of healthcare till now?
Q2. Any issues or challenges in the field of healthcare?
Q3. What kind of homecare/healthcare systems exist?
Q4. What kind of work already has been done in the field of child healthcare?
Q5. Is there any work related to prenatal and postnatal care of the females? If yes, what are the tools and techniques used?

2.2 Sources of Information

In order to answer the research questions described previously, various databases were searched. Most relevant books, journals and conferences and workshop proceedings were studied. Various databases like ACM Digital Library, IEEE explore, Springer LNCS, Google Scholar, Inderscience, Science Direct, Taylor and Francis and Books, etc. were searched.

2.3 Search Criteria

The articles and journals chosen from amongst the various databases described in the Sect. 2.2 were the ones, which had coverage to varied and relevant topics such as healthcare, childcare, public health, elderly care, maternal health, smart healthcare.

In almost all the searches, keyword "healthcare" was used. The keywords used to obtain the desired results were: "healthcare using MAS", "childcare", "child healthcare" "using" "MAS", "prenatal" and "postnatal" "healthcare", "infant" "healthcare", "home" "healthcare" "systems", "home" "healthcare" "systems" using "MAS", "foetal" "healthcare", "foetal" "monitoring".

2.4 Study Selection Process

The study materials selected for the purpose of presented research follows the following process:

Foremost, the research questions were designed as explained in Sect. 2.1. Besides, various keywords in the form of search criteria (as described in Sect. 2.3) were used to achieve the desired results from the varied databases (as listed in Sect. 2.2). The papers were excluded or included based on:

a. Title of the paper.
b. Abstract and keywords.
c. Content of the paper.

If the resulted papers, had relevant titles or abstract or the complete text that satisfied our research questions, the papers were included else excluded.

3 Literature Review

Multi-agent systems play very important roles in providing the speedy, efficient and accurate healthcare services to the various categories of peoples. This section presents the review of the research work carried out in the area of multi-agents-based healthcare systems. This section is divided into five subsections, i.e. Multi-agent-based healthcare, Child healthcare systems, prenatal and postnatal healthcare of mother, elderly healthcare and home healthcare system in order to understand the work done in these healthcare domains till now. The research work till date in each of these systems has been presented as follows.

3.1 Multi-Agent-Based Healthcare

Tian and Tianfield [20] examined the attributes of health delivery systems, surveys common coordination mechanisms like contract net, auction mechanism and suggests the collaboration of MAS to address the complexities of health delivery systems.

Daknou et al. [21] explained the use of the agent-based system for the healthcare of patients at the emergency department (ED) in a hospital. The proposed system helps the ED staff in decision making, provides scheduling of patients, diagnoses the problem and provides the treatment plan. Three phase reactive scheduling algorithm was applied to the scheduling of patients and tasks.

Cabrera et al. presented an agent-based Decision Support Systems for the operation of healthcare emergency department (ED) [22]. Its objective was to aid ED managers in setting up strategies and managing guidelines effectively and optimally. This was an ongoing research performed by IoM of CAOS in the UABs. The model used MOORE STATE machine-based agents and employed exhausted search technique to obtain optimal results and implemented using Net Logo. In the extended version of the paper, the results are obtained using pipeline scheme.

Hylton et al. in [23], developed an intelligent agent-based system for fixing appointments for the patients and doctors in the hospital. The system takes into consideration the priority of the patients, e.g.: senior citizens, pregnant women were given more priority than others. It was implemented using JADE-LEAP and Android 2.2.

Turcu et al. in [24] explained the role of Internet of Things (IOT), Radio Frequency Identification (RFID) and MAS in the domain of healthcare to improve healthcare systems by enhancing people's access to healthcare services, reducing medical errors, assisting and identifying medical staffs, nurses, etc. Some of the issues of using IOT approach were pointed which included security and privacy, standardization and scalability.

Arsene et al. proposed an expert system using software agents [25] that allowed the automatic exchange of information amongst medical specialists. The system used software agents to represent human body systems (e.g. respiratory, cardiovascular). It comprised three key components: Knowledge Management, Uncertainty Reasoning and Software Agents. Bayesian network has been used for knowledge representation and reasoning.

3.2 Child Healthcare Systems

Mago et al. [26] developed a multi-agent-based system that helped the healthcare practitioners to detect the disease amongst infants/children, select the degree of severity and identify the treatment plan. A mathematical model has been presented to reflect the working of the system. The system helped the rural patients in lowering their travelling costs as it assisted Primary Health Care (PHC) practitioners in diagnosing the childhood diseases. The author adopted JADE for the construction and the execution of the system.

Dezhi et al. [27] presented the design of MAS to minimize malnutrition in children less than 5 years of age in Srilanka and to give nutrition exhortation to residents through the e-govt initiatives. The system consisted of an agent reporting number of malnutrition children in each district. The author used JADE to develop the system.

Zhang et al. In [18] concentrated on MAS for children diabetic healthcare. A multi-agent system IMAS has been designed which provides the real-time monitoring and handling of glucose level in patients. In case of an acute situation, it alerts the patients and care providers by sending an alarm and also conducts a meeting amongst agents.

Saadatian et al. [28] proposed a low-cost mobile-based intelligent system which supervised the biological conditions of the baby including movement, temperature, and heart-beat using reasonable sensors. The model also conveyed the status of the infant to parents of the child and in case of the abnormality or emergency; the alert messages were sent to the healthcare people (clinics).

Goto et al. designed a system that recognized the activities of children such as sleeping, eating, sitting, walking, running, playing and hand motion by using 3-axis acceleration sensor attached to child's upper arm [29]. In case of medical emergency of infants, this model lacked treatment options as well as a consultation with specialist physician. Also, it did not signify anything regarding the diet plan of a child and first-aid instructions.

Wettasinghe et al. [30] developed a knowledge sharing and forecasting system for motherhood and infant care in Sri Lanka by using Neural Networks. It aided the users to monitor the fitness of mothers and children and gratified individual needs that can emanate during various stages of motherhood. The system provides the following facilities to mothers: health analyser, meal and nutrition analyser and special needs analyser. The system provided suggestions to the mother and midwives to pinpoint critical situations and to tackle them efficaciously.

Fonseca et al. in [31] developed an expert diagnosis prototype that tried to predict and detect the aberrant state of a baby and alerted the concerned contacts in order to prevent sudden infant death. It used sensors fitted in the baby cradle which further connected to a mobile device via Wi-Fi or Bluetooth for measuring physiological signals of a baby. The system only notified about occurrence of any abnormal situation or circumstances and did not address anything about diet plan and immunization plan of a baby, which is one of the significant factors of health.

Gupta et al. in [32] developed a system "Aarogya" that made the communication between a paediatrician and the patient compelling as well as solid. The disease of the patient has been diagnosed by analysing the answers given by the users to the set of questions asked to them. The treatment was then provided accordingly. It also helped in scheduling an appointment with the doctors, in case of emergency.

3.3 Prenatal and Postnatal Care of Mothers

Nunes et al. [19] developed MAS-based prenatal care unified system (SUAP) for assisting and observing the prenatal care of pregnant mothers. This project systemizes the Brazilian prenatal care as a web-based application. It utilized agent technology to supervise healthcare records, to behave as a clinical decision support system and to handle high-risk pregnancy cases. The proposed system used the seam framework with JBoss application server where software agents were developed using JADE framework.

Liu et al. in [33] reviewed the need of social support to the parents of high-risk infants and importance of communication amongst caregivers and professionals. A prototype system named Estrellita was developed to improve communication and social support for caregivers. The system in turn reduced language and literacy barriers, balanced work and benefit and provided control and feedback for managing privacy and uncertainty of shared information.

Mahunnah and Tavetar [34] focused on individualized care for a healthy pregnancy in the areas of limited connectivity. The author concentrated on uncontrolled physiological conditions such as blood glucose, blood pressure and weight gain during pregnancy which may lead to various critical healthcare conditions such as diabetes, hypertension and obesity and recommended lifestyle changes based on patient allergies, hobbies and medical history.

Su and Chu in [35] developed an electronic foetal tracking system which supervised the foetus in real-time from any area using Personal Digital Assistant, smart phone, laptop or desktop computer. This research proposed a Mobile Multi-Agent Information Platform (IMAIS) in which multiple agents collected foetal vital signs (such as foetal heart rate and uterine contractions) and automatically notified the physician about abnormal situations in real life.

3.4 Elderly Healthcare

Shakshuki and Reid [36], discussed MAS applications in healthcare domain such as telemedicine, lifestyle monitoring, detection and assistance.

Cioara et al. [37] presented an expert system for the elderly nutrition care. Ontologies had been used to capture nutrition-related knowledge which further comprised four sub-ontologies: nutrition monitoring ontology for collection of required data, nutrition assessment ontology which involved assessment of the collected information, problem identification ontology and the nutrition intervention ontology. The proposed expert system was compared with nutritionist assessment and was proved to be as competent as a nutritionist.

Mendes et al. in [38] presented a multi-agent-based m-health system in order to monitor routine activities (i.e. sleep quality, no. of steps, activity levels, etc.) of elderly. It used a set of Fit bit charge HR wristbands for collection of data and k-means clustering algorithm. It was concluded that almost all elderly people exhibited similar activities and physiological patterns, which can be characterized in quantitative and qualitative manner to monitor their daily conditions and health progress.

3.5 Home Healthcare

Koufi et al. proposed [39] a chronic disease management system. It is patient health record (PHR)-based system, which consisted of 3 subsystems: PHR subsystem, which held the complete record of the patients, Workflow subsystem, for managing workflows related to the medication of patients and providing remedy plans and the security subsystem, which is important for governing access to workflow job and related web services.

Marcon et al. [40] addressed the caretaker's routing problem in home healthcare. A MAS has been proposed for solving the problem in a dynamic and distributed manner based on four decision rules, i.e. balanced route rule (BRR), no wait rule (NRR), nearest patient rule (NPR), shortest route rule (SRR). It has been concluded that the BRR rule is better than the other three at minimizing the caregivers' travel time, with or without disruptions.

4 Research Gaps in Healthcare Sector

1. Most of the surveyed models, In case of medical emergency of infants, lack treatment options as well as a consultation with specialist physician.
2. There is lesser number of models which signify anything regarding the diet plan of a child and first-aid instructions in case of an emergency.

3. The reviewed systems rarely provide any immunization plan for the infants and the mothers during prenatal and postnatal stages. Besides, most of these systems do not provide any method for monitoring of foetus during prenatal stage.

5 Concluding Remarks and Future Research Directions

Healthcare is one of the emerging trends in today's era. With the growth in population, demand for the healthcare services has also been increased. It has been analysed that very few models have been proposed in the area of women and infant healthcare. These areas require urgent attention as it has been explored that the infant mortality rates have been considerably higher in various developing nations including India. Henceforth, it can be concluded that a MAS can be proposed in future for the prenatal and postnatal care of women for the welfare of society. From the survey, following future lines of research can be outlined:

- MAS can be proposed which provides better treatment options for women and infants during prenatal and postnatal stage as well as a consultation with specialist physician.
- Diet is one of the important factors that affect health. So, MAS can be proposed that is capable of providing the proper diet plan for mother and child and first-aid instructions in case of an emergency.
- MAS can be designed for monitoring the foetus during prenatal stage as well as for providing the immunization plan for infants and mothers.

References

1. Chen B, Cheng HH (2010) A review of the applications of agent technology in traffic and transportation systems. IEEE Trans Intell Transp Syst 11(2):485–497. https://doi.org/10.1109/TITS.2010.2048313
2. Schumacher M. Objective coordination in multi-agent system engineering. Lecture notes in computer science, vol 2039. Springer, Berlin, Heidelberg
3. Wooldridge M (2009) An introduction to multiagent systems. John Wiley & Sons. ISBN-10: 0470519460
4. Mahmoud MS (2018) Multiagent systems. In: Advanced control design with application to electromechanical systems, Elsevier, 2018, pp 313–338
5. https://www.whcobgyn.org/
6. Karney P (2000) Women's health: an evolving mosaic. J Gen Intern Med 15(8):600–602. https://doi.org/10.1046/j.1525-1497.2000.00623.x
7. https://www.medicalnewstoday.com/categories/womens_health
8. Hirshberg A, Srinivas SK (2017) Epidemiology of maternal morbidity and mortality. Sem Perinatol 41(6):332–337. https://doi.org/10.1053/j.semperi.2017.07.007
9. Blencowe H, Calvert C, Lawn JE, Cousens S, Campbell OMR (2016) Measuring maternal, fetal and neonatal mortality: challenges and solutions. In: Best practice and research: clinical obstetrics and gynaecology, vol 36. Bailliere Tindall Ltd, pp 14–29, Oct 01, 2016. https://doi.org/10.1016/j.bpobgyn.2016.05.006

10. https://data.unicef.org/topic/maternal-health/maternal-mortality/
11. Jamison DT, Murphy SM, Sandbu ME (2016) Why has under-5 mortality decreased at such different rates in different countries? J Health Econ 48:16–25. https://doi.org/10.1016/j.jhealeco.2016.03.002
12. https://en.wikipedia.org/wiki/Prenatal_care
13. https://www.plannedparenthood.org/learn/pregnancy/prenatal-care
14. http://www.open.edu/openlearncreate/mod/oucontent/view.php?id=335&printable=1
15. Jemal H, Kechaou Z, Ben Ayed M, Alimi AM (2015) A multi agent system for hospital organization. Int J Mach Learn Comput 5(1):51–56. https://doi.org/10.7763/ijmlc.2015.v5.482
16. Moreno A. The application of agent technology to health care
17. Han BM, Song SJ, Lee M, Jang KS, Shin DR. Multi agent system based efficient healthcare service
18. Zhang P, Carlsson B, Johansson SJ (2010) Enhanced collaboration in diabetic healthcare for children using multi-agent systems. Int J Agent-Oriented Softw Eng 4(2):155–173. https://doi.org/10.1504/IJAOSE.2010.032800
19. Nunes I, Choren R, Nunes C, Fabri B, Silva F, Carvalho G, Lucena C (2010) Supporting prenatal care in the public healthcare system in a newly industrialized country. In: Proceedings of 9th international conference on autonomous agents and multi agent systems: industry track, May 2010, pp 1723–1730
20. Tian J, Tianfield H (2009) Health delivery systems—a case for multi-agent systems. In: Conference proceedings—IEEE international conference on systems, man and cybernetics, pp 2718–2722
21. Daknou A, Zgaya H, Hammadi S, Hubert H (2009) Agent-based architecture for the healthcare of patients at the emergency department. IFAC Proceedings, vol 2, no PART 1, pp 175–180. https://doi.org/10.3182/20090610-3-it-4004.00035
22. Cabrera E, Taboada M, Iglesias ML, Epelde F, Luque E (2011) Optimization of healthcare emergency departments by agent-based simulation. Procedia Comput Sci 4:1880–1889. https://doi.org/10.1016/j.procs.2011.04.204
23. Hylton A, Sankaranarayanan S (2012) Application of intelligent agents in hospital appointment scheduling system. Int J Comput Theory Eng 4(4):625–630. https://doi.org/10.7763/ijcte.2012.v4.545
24. Turcu CE, Turcu CO (2013) Internet of Things as key enabler for sustainable healthcare delivery. Procedia—Soc Behav Sci 73:251–256. https://doi.org/10.1016/j.sbspro.2013.02.049
25. Arsene O, Dumitrache I, Mihu I (2015) Expert system for medicine diagnosis using software agents. Expert Syst Appl 42(4):1825–1834. https://doi.org/10.1016/j.eswa.2014.10.026
26. Mago VK, Syamala Devi M (2007) A multi-agent medical system for Indian rural infant and child care. In: Proceedings of 20th international joint conference on Arificial Intelligence (IJCAI2007), Hyderabad, pp 1396–1401
27. Dezhi X, Ganegoda GU (2010) Multi agent system to reduce malnutrition (MASRM) in children. In: 2010 international conference on computational intelligence and software engineering, CiSE 2010, pp 1–4.https://doi.org/10.1109/CISE.2010.5676807
28. Saadatian E et al (2011) Low cost infant monitoring and communication system. In: 2011 IEEE colloquium on humanities, science and engineering, CHUSER 2011, no. Chuser, pp 503–508. https://doi.org/10.1109/CHUSER.2011.6163782
29. Goto J, Kidokoro T, Ogura T, Suzuki S (2013) Activity recognition system for watching over infant children. In: Proceedings—IEEE international workshop robot human interaction communication, pp 473–477. https://doi.org/10.1109/ROMAN.2013.6628549
30. Wettasinghe RI, Perera PB, Aponsu GR, Jayathilake AV, Gamage MP, De Silva KP (2013) Knowledge sharing and prediction system for maternity and infant care in Sri Lanka. In: Proceedings of the 8th international conference on computer science and education, ICCSE 2013, no. Iccse, pp 292–297, 2013. https://doi.org/10.1109/ICCSE.2013.6553927
31. Fonseca ÂM, Horta ET, Sendra S, Rodrigues JJ, Moutinho JA (2014) A sudden infant death prevention system for babies. In: 2014 IEEE 16th international conference on e-health networking, applications and services (healthcom), pp 525–530. https://doi.org/10.1109/HealthCom.2014.7001897

32. Gupta A, Pawar C, Antony B, Ringe S (2016) Aarogya—an intelligent multi-agent paediatric system. Int J Sci Eng Res 7(2):1–5
33. Liu LS, Hirano SH, Tentori M, Cheng KG, George S, Park SY, Hayes GR (2011) Improving communication and social support for caregivers of high-risk infants through mobile technologies. In: Proceedings of the ACM conference on computer supported cooperative work, CSCW, pp 475–484
34. Mahunnah M, Taveter K (2013) A scalable multi-agent architecture in environments with limited connectivity: case study on individualised care for healthy pregnancy. IEEE International Conference on Digital Ecosystems and Technologies, pp 84–89. https://doi.org/10.1109/DEST.2013.6611334
35. Su CJ, Chu TW (2014) A mobile multi-agent information system for ubiquitous fetal monitoring. Int J Environ Res Public Health 11(1):600–625. https://doi.org/10.3390/ijerph110100600
36. Shakshuki E, Reid M (2015) Multi-agent system applications in healthcare: current technology and future roadmap. Procedia Comput Sci 52(1):252–261. https://doi.org/10.1016/j.procs.2015.05.071
37. Cioara T et al (2018) Expert system for nutrition care process of older adults. Futur Gener Comput Syst 80:368–383. https://doi.org/10.1016/j.future.2017.05.037
38. Mendes S, Queiroz J, Leitão P (2017) Data driven multi-agent m-health system to characterize the daily activities of elderly people. In: 12th Iberian conference on information systems and technologies (CISTI), pp 1–6. https://doi.org/10.23919/CISTI.2017.7975725
39. Koufi V, Malamateniou F, Vassilacopoulos G (2010) An agent-based application of personal health record in homecare. In: Proceedings of the 3rd international conference on pervasive technologies related to assistive environments, PETRA 2010, Samos, Greece, June 23–25, 2010. ACM International Conference Proceeding Series. ACM
40. Marcon E, Chaabane S, Sallez Y, Bonte T, Trentesaux D (2017) A multi-agent system based on reactive decision rules for solving the caregiver routing problem in home health care. J Simul Model Pract Theory 74:134–151. ISSN 1569-190X

Iterated Function System in Fingerprint Images

M. Raji and G. Jayalalitha

Abstract Biometric identification is the process of determining a person's identity. Everyday people want to protect their personal belongings and have easy accessibility to it at a fair cost. Biometric-based identification is more secure than any other technique since it binds an identity to a specific person rather than a password or a code that anybody could use. Other security measures, such as smart ID cards and chips are frequently paired. One of the important biometric systems is fingerprint images. Iterated function systems are self-similarity-based method for creating fractals. In this paper, it obtains an iterated function system for fingerprint images which leads a good representation of the given images.

Keywords Biometric system · Fingerprint · Graphs · Fractals

1 Introduction

The word *biometric* is meant about combining life and measures. Because of substantial improvement in the computer field, automated biometric systems have been developed. However, many of these new automated processes are founded on concepts that date back hundreds of years ago.

Biometric system is a method of identifying and verifying people by measuring and analysing their physical and behavioural traits. For security and privacy reasons, authentication is required. The importance benefit of biometrics is identification. It is more comfortable than other security systems such as the use of passwords, pins and patterns. Biometric authentication compares data for a person's attributes to a biometric template for that person. The objective is to obtain a unit of biometric data from this individual. It could be an image of their face, a recording of their voice, or a fingerprint image. This information is compared to biometric information from

M. Raji (✉) · G. Jayalalitha
Department of Mathematics, School of Basic Sciences, Vels Institute of Science, Technology and Advanced Studies, Chennai, India
e-mail: rajialagumurugan@gmail.com

P. Dutta et al. (eds.), *Emerging Technologies in Data Mining and Information Security*, Lecture Notes in Networks and Systems 490,
https://doi.org/10.1007/978-981-19-4052-1_38

several other people stored in database. The concepts of biometric system can be found in [1, 2].

Fingerprints are decided by our genes. The fingerprint patterns have the similarity in which patterns run in families. Every person in the world has a unique fingerprint pattern. Fingerprints are one of the simplest biometric patterns to identify a concern person. The formal use of fingerprints by police departments was the other way. South America, Asia and Europe had all undergone this phenomenon. Edward Henry, Inspector General of Bengal Police devised the first such comprehensive method for indexing fingerprints in India. This approach is known as Henry system and its variants are still used to classify fingerprints today.

Fingerprints have been utilized in forensic applications for over a century. They are patterns of ridges and valleys on the "friction ridge" surfaces of fingers. Even identical twins do not have the same fingerprints because friction ridges are generated in utero during foetal development. The performance of currently existing fingerprint-based recognition systems that use of multiple fingerprints is quite good. The details of fingerprint concepts are in [3, 4].

There are some types of fingerprint patterns. Arches, loops and whorls are three main fingerprint patterns (Fig. 1).

The graph theory relates points and lines of the graphs. A graph is defined as (V, E) with V, a finite non-empty set of vertices and E, a finite set of edges. The graph theoretical concepts are given in [5–7].

Fractals are of irregular geometric shape which can be subdivided and each subdivision gives reduced similar copy of the whole. Geometrical representation of the fingerprint is fractal. Obviously, a fractal looks irregular. An iterated function system is used to specify this behaviour under magnification [8–12].

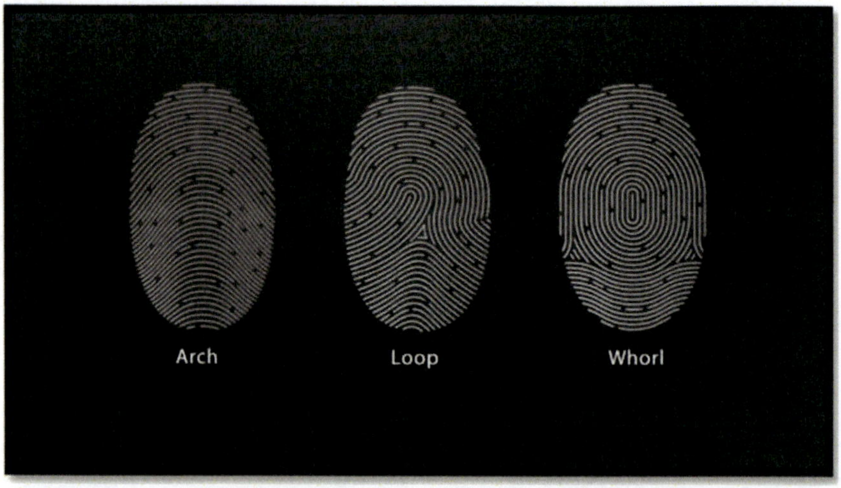

Fig. 1 Main three types of fingerprint images

Assume that S is a metric space. Assume A and B are both subsets of S. d is a metric which satisfies the following properties (i) positive definite, (ii) symmetric and (iii) triangle inequality, $d(A, B) \leq d(A, C) + d(C, B)$ for all $A, B, C \in S$. The metric d is known as the *Hausdorff metric* on S. The definitions of contraction, attractor and an iterated function system are referred in [8–12].

Here, the curves in the fingerprint images, taking a graph G_f can be considered as a set in the plane and each curve (say, G_{f_i} where $i \in (\infty, -\infty)$) is a subset of the line R and G_f cannot be an ordinary set of geometry. If $x \in G_f$, for every > 0, the interval $(x - \varepsilon, x + \varepsilon)$ contains points of G_f in addition to x.

2 Iterated Function System for Fingerprint Images

This section describes that the fingerprint shape produces an image consisting of a specified number of affine transformations which gives the good approximation by attracter to the fingerprint. From the shape of fingerprint, it can be obtained that a metric space is (G_f, d) where G_f is a non-empty set and the metric is function.

$d : G_f \times G_f \to R$ is a real valued function with metric properties.

2.1 Proposition

Let $\{G_{f_1}, G_{f_2}, \ldots, G_{f_m}\}$ *be an IFS and suppose that* $|G_{f_i}(x) - G_{f_i}(y)| \leq c|x - y| \forall x,\ y \in R^n$ and $\forall i$, where $c > 1$.

Let $A \subset R^n$ be any non-empty compact set. Here, B is the attractor for the IFS and d is the Hausdorff metric. Then,

$$d(A, B) \leq d\left(A, \bigcup_{i=1}^{m} G_f(A)\right)\frac{1}{1 - C}.$$

Proof *For the Hausdorff metric*

$$\begin{aligned} d(A, B) &\leq d\left(A, \bigcup_{i=1}^{m} G_f(A)\right) + d\left(\bigcup_{i=1}^{m} G_f(A), B\right) \\ &= d\left(A, \bigcup_{i=1}^{m} G_f(A)\right) + d\left(\bigcup_{i=1}^{m} G_f(A), \bigcup_{i=1}^{m} G_f(B)\right) \\ &\leq d\left(A, \bigcup_{i=1}^{m} G_f(A)\right) + cd(A, B) \end{aligned}$$

3 Conclusion

In this paper, it discusses about iterated function system for fingerprint images. By proposition, it can be concluded that the good approximation of attractor by combining the smaller copies which gives the fingerprint shape. The approximation by attractor of IFS is very useful to encode the images. Future work will be gone on demonstration of encoding a complicated fingerprint images.

References

1. Sharma A et al (2015) Biometric system—a review. Int J Comp Sci Inf Technol 6(5):4616–4619
2. Shrivastava S (2015) Biometric: types and its applications. Int J Sci Res NCKITE-2015, pp 204–207
3. Tamboli D, Patil SB (2015) A survey on fingerprint matching techniques. Int J Eng Res Technol 3(20):1–3
4. Barua K, Bhattacharya S, Mali K (2011) Fingerprint identification. Glob J Comp Sci Technol 11(1):61–64
5. Bondy JA, Murty USR (1982) Graph theory with applications. Elsevier Science Publishing Co. Inc., New York, pp 1–3
6. Deo N (1974) Graph theory with applications to engineering and computer science. PHI Learning Pvt. Ltd., New Jersey, pp 19–21
7. Arumugam S, Issac AT, Somasundaram A (2019) Modern analysis. Yes Dee Publishing Pvt. Ltd, Chennai, pp 87–207
8. Falconer KJ (1990) Fractal geometry, mathematical foundations and applications. Wiley, New York, pp 1–25
9. Edgar GA (2008) Measure, topology and fractal geometry. Springer, New York, pp 117–133
10. Falconer KJ (1997) Techniques in fractal geometry. John Wiley & Sons Ltd., Chichester, pp 1–150
11. Barnsley MF (2006) Super fractals. Cambridge University Press, New York, pp 1–30
12. Triebel H (1997) Fractals and spectra. Birkhauser Verlag, Switzerland, pp 147–158

An Overview of Efficient Regression Testing Prioritization Techniques Based on Genetic Algorithm

R. Adline Freeda and P. Selvi Rajendran

Abstract Testing is crucial in delivering a bug-free application to the client. Retesting the application whether there is a modification in different functionalities of the application when there is a code change is called regression testing. It is discovered to be expensive as it should be introduced after every code change. Among the numerous strategies existing for regression testing, reordering the test cases through prioritization and just the low prioritized test cases are ignored without executing which saves cost and time. The fault detection rate is found to be more for search-based techniques that apply genetic algorithms. Multi-objective search-based prioritization was proved for its better performance when compared to the single objective genetic algorithms. Hence, this paper reviews and analyzes the multi-objective search algorithms, and it is subjected to improvements. The review proves that genetic algorithms when integrated along with an efficient approach yield a better result compared to other prioritization algorithms.

Keywords Regression · Multi-objective · Prioritization

1 Introduction

Nowadays, the role of software testing is crucial in the SDLC of software development. The application is tested for its quality and reliability once it becomes a deliverable application [1]. Regression testing ensures the application's quality by continuously testing it when it is subjected to change, and the application has refrained from bugs which make regression testing an expensive approach [2]. The following are identified as state of art techniques among the various techniques exiting to perform regression testing.

R. A. Freeda (✉) · P. S. Rajendran
Department of Computer Science, Hindustan Institute of Technology & Science (Deemed To Be University), Padur, Chennai, Tamil Nadu, India
e-mail: rp.20603036@student.hindustanuniv.ac.in

P. S. Rajendran
e-mail: selvir@hindustanuniv.ac.in

P. Dutta et al. (eds.), *Emerging Technologies in Data Mining and Information Security*, Lecture Notes in Networks and Systems 490,
https://doi.org/10.1007/978-981-19-4052-1_39

- Test suite reduction (TSR);
- Regression test case selection;
- Test case prioritization.

The size of the test suite is reduced to a greater extent by both test case reduction and selection techniques at the same time retaining its efficiency of fault detection [1]. However, fault detection happens at the earlier stage of regression testing with the help of test case prioritization techniques [3, 4]. The test cases which meet the goals are selected among the pool of test cases [5]. The objective of performance can be easily met with the implementation of the prioritization of test cases. It also takes into account the fault detection rate, which measures how much earlier the faults are detected during the regression testing. For regression testing, the high fault detection rate at the earlier stages as a result of prioritizing the test cases is helpful and given as feedback to the tester about the system under test assist in correcting the faults much earlier which otherwise would be impossible [6, 7]. Test case prioritization methods are classified into many categories as follows. (i) coverage based (ii) requirement based (iii) history-based (iv) fault based (v) search based [8]. Requirement-based prioritization groups the test cases based on the requirements from the customers [9]. The test cases are prioritized based on the history of the software [10]. Search-based prioritization techniques are a part of machine learning algorithms. These techniques can be either single objective or multi-objective [11, 12]. In this paper, multi-objective search-based techniques are reviewed.

2 Related Work

The genetic algorithm, a search-based technique for test case prioritization, had become a state of art algorithm due to its efficiency and increasing popularity. The researchers had reviewed the performance of the genetic algorithm on test case prioritization.

Yoo and Harman [11] conducted a complete review of regression techniques such as test suite minimization, selection, and prioritization. And also, different prioritization techniques and the various metrics were reviewed.

Singh et al. [6] also reviewed the prioritization techniques and identified the research gap.

Elanthiraiyan and Arumugam [13] when changes are made to the SUT, both the original test cases and the modified test cases will be available in the test suite, and the size of the suite also increases. The multi-criteria optimization techniques generate test suites that are found optimal. The author proposes ant colony optimization techniques with maximum fault detection.

Khatibsyarbini et al. [14] classified the TCP techniques into several types and revealed the research gap about the datasets and how the metrics should be applied.

Catal and Mishra [15] provided a study of various TCP algorithms, metrics, and datasets involved. The most frequently used TCP approaches were also reviewed.

2.1 Genetic Algorithm in TCP

Only, a few works exist on genetic algorithm-based test case prioritization in spite of many TCP algorithms were explored and studied extensively.

This work presents the study of two test case prioritization techniques based on a genetic algorithm.

The evolutionary algorithm (EA): Xia et al. [1] developed software must go through a quality check to deliver bug-free and high-quality software. As a result, software testing had become mandatory in addition to the different phases of the software development life cycle. Regression testing confirms the quality and dependability of the software developed, despite changes to the software. The test suite is prone to expansion as when the software developed is subjected to modification, in turn, increases the time and cost for testing. Various optimization criteria such as maximum statement coverage, fault detection, and minimum execution cost are achieved through the solution of minimizing the size of the test suite as explored by many researchers.

The authors had proposed an evolutionary multi-objective optimization algorithm for minimizing the test suite.

As the test suite is a collection of a substantial quantity of test cases, minimization of the test suite occurs in two steps.

Step 1: *K*-Means Clustering Algorithm

The test cases are initially grouped into clusters using the *K*-means clustering algorithm. The authors have done extensive research on various clustering algorithms existing; *K*-means clustering was found efficient by the researchers, and finally, the authors had opted for *K*-means clustering algorithms as it is widely adopted for clustering test cases in regression testing (Table 1).

Step 2

After the test cases are clustered, an evolutionary algorithm which is a multi-objective optimization algorithm is further implemented as the next step. The objectives include maximum fault detection rate, statement coverage, and minimum execution time.

The non-dominated genetic algorithm is applied to the generated cluster individuals. The new clusters formed are evaluated through selection, crossover, and mutation phases.

Table 1 Steps of *K*-means algorithm

1. As *K* value indicates the no. of clusters, identify the no. of clusters
2. Select *k* objects for centroids randomly
3. Repeat the above steps until there is no change in centroid
4. The distance between the objects and centroids of each cluster is calculated assigned to the nearest centroid

Selection: The fittest objects from the cluster are selected in this phase. The selected objects bring out the offspring. The phase is re-iterated until the newly generated cluster reaches the size of the original cluster.

Crossover: This is the most significant phase in the genetic algorithm. It is also called as a reproduction phase. The crossover point is identified randomly between two selected parents, and offspring are generated by exchanging the genes.

Mutation: In the new offspring generation formed, some of the genes may be subjected to mutation to improve the quality of the offspring. A test case is then randomly inserted into the newly generated cluster if one of the same types does not exist already.

The experimental evaluation proves that the proposed method meets the optimization objectives of fault detection and removing redundant test cases.

Advantages: The algorithm outperforms the existing algorithms in terms of identifying missing fault rates and test suite minimization.

Drawbacks: The execution time of the proposed algorithm is slightly high when compared to the other algorithms taken for comparison. Therefore, the testing efficiency is not satisfactory concerning the algorithms taken for comparison.

(i) **Mutant coverage-based multi-objective approach (MC)**: Gupta et al. [2] instead of the traditional approaches that cluster the test cases and then apply the minimization techniques an integrated approach is proposed for test suite minimization which includes fault detection and fault localization. Fault detection is identifying the location where the test case fails for the SUT. And fault localization is a subsequent action to be performed after fault detection where the detected faults are debugged and resolved for their correctness (Table 2).

To achieve test suite minimization, the statement coverage, branch coverage, and diversity-aware mutation adequacy criterion are maximized. The statement coverage criterion is used primarily for fault detection, whereas fault detection and localization were achieved through the implementation of the D-criterion. The first two criteria require information about the test cases' maximum coverage of statements and branches, whereas the D-criterion requires information about mutants that were ignored.

Table 2 Multi-objective optimization steps

1	The test cases are scrutinized for their maximum statement and branch coverage
2	The adequacy criterion/D-criterion is applied to identify the mutants in the test suite
3	The NSGA-II algorithm takes the output of steps 1 and 2 as input from which a subset of test cases generated as output
4	The test cases are evaluated with respect to the objectives as a maximum statement and branch coverage and minimum D-criterion score, and minimized test suite is generated
5	The final minimized test suite is evaluated for fault detection and localization

Maximum D-Criterion/D-score

The D-score is calculated as the no. of individual mutants produced by the test suite to the maximum no. of individual mutants. The maximum score ranges from [0, 1].

Maximizing D-criterion minimizes the test case utilization within the test suite.

Statement Coverage

The maximization of statements covered for the SUT by the test cases reduces the size of the test suite.

Branch Coverage

If a test case under a test suite can cover the maximum no. of branches under a control structure, then the no. of test cases to be executed is minimized; in turn, the reduction in the size of the test suite is achieved.

In the above approach, fault detection and localization are well achieved using the D-criterion mutation partitioning. The efficiency of the proposed method can be increased by maximizing the partitions of mutants. This integrated approach is explored only for minimizing the test suite.

Advantages: With the integrated approach, two possible activities such as reduced test suite and fault detection and localization are accomplished. In comparison to other approaches, the average ratio of test suite minimization, fault detection, and localization is good.

Drawback: The analysis of time consumption and cost factor is not taken into consideration. It will be effective if the time consumption is analyzed as most mutant-based fault localization techniques are time-consuming, and it is primarily used in the integrated approach.

3 Experimental Analysis

An extensive study was conducted to compare the efficiency of the two algorithms. The five C programs from the Siemens suite from SIR were considered for experimentation [16].

The two algorithms are evaluated for their efficiency in context to percentage of test suite reduction and percentage of fault detection from the experimental analysis; the multi-objective algorithm performs better for minimizing the test suite, and the evolutionary algorithms perform better for the percentage of fault detection (Tables 3 and 4). Test suite reduction comparison and fault detection comparison are represented in Figs. 1 and 2, respectively.

Table 3 Siemens suite of programs for experiments

Sl. No.	Program name	L O C	Faulty version	Original suite size	Mutants
1	Printtokens	40	7	4130	582
2	Replace	51	32	5542	666
3	Schedule	29	9	2650	253
4	Tcas	13	148	1608	291
5	Print-token2	78	10	4115	375

Table 4 Experimental analysis of test suite

Sl. No.	Program name	Test suite reduction		Fault detection rate	
		EA (45%)	MC (78%)	EA (97)	MC (95%)
1	Printtokens	46.5	52.3	95.7	94.2
2	Replace	49.8	55.9	96.8	96.2
3	Schedule	48.3	54.6	100	99.3
4	Tcas	46.5	53.3	97.6	98.3
5	Print-token2	43.1	49.5	100	99.1

Fig. 1 Test suite reduction comparison

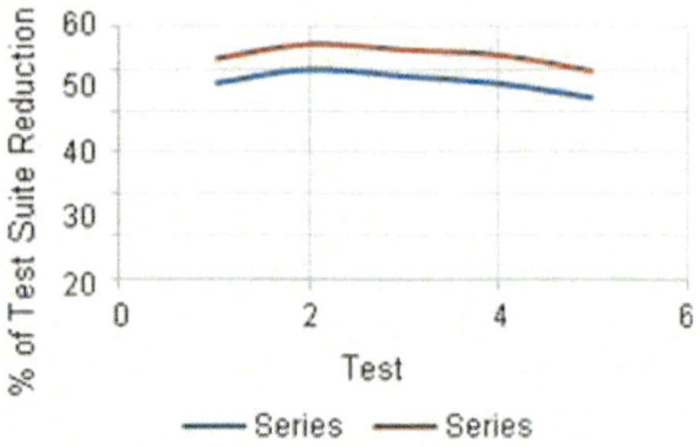

Fig. 2 Fault detection comparison

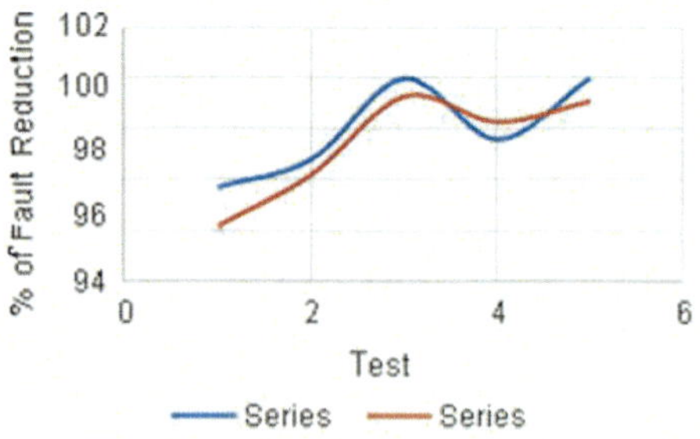

4 Conclusion

Prioritization techniques unlike other techniques had gained increasing popularity in recent years as the test cases are reordered, and only, the high priority test cases are executed instead of removing the test cases from the test suite. In recent researches, the genetic algorithms of the various meta-heuristic algorithms had gained popularity. Therefore, we have reviewed two different approaches that implement a genetic algorithm for prioritization, in turn, minimize the test suite. The advantages and drawbacks of both approaches are also reviewed.

References

1. Xia C, Zhang Y, Hui Z (2021) Test suite reduction via evolutionary clustering. IEEE Access 9:28111–28121
2. Gupta N, Sharma A, Pachariya MK. Multi-objective test suite optimization for detection and localization of software faults, J King Saud Univ—Comp Inf Sci. https://doi.org/10.1016/j.jksuci.2020.01.009
3. Rothermel G, Harrold MJ (1996) Analyzing regression test selection techniques. IEEE Trans Softw Eng 22(8):529–551
4. Selvi Rajendran P, Anithaashri TP Dr. CNN based framework for identifying the Indian currency denomination for physically challenged people. In: IOP conference series: materials science and engineering for the publication
5. Rothermel G, Untch RH, Chu C, Harrold MJ (2001) Prioritizing test cases for regression testing. IEEE Trans Softw Eng 27(10):929–948
6. Singh Y, Kaur A, Suri B, Singhal S (2012) Systematic literature review on regression test prioritization techniques. Informatica 36(4):379–408
7. Selvi Rajendran P, Geetha A Dr (2021) Optimization of hospital bed occupancy in hospital using double deep Q network. In: International conference on intelligent communication technologies and virtual mobile network (ICICV-2021), 4–6 Feb 2021
8. Rothermel G, Untch R, Chu C, Harrold MJ (1999) Test case prioritization an empirical study. In: Proceedings of international conference software maintenance, pp 179–188
9. Huang YC, Peng KL, Huang CY (2012) A history-based cost-cognizant test case prioritization technique in regression testing. J Syst Softw 85(3):626–637
10. Di Nucci D, Panichella A, Zaidman A, De Lucia A (2018) A test case prioritization genetic algorithm guided by the hypervolume indicator. In: IEEE transactions on software engineering, 2018
11. Yoo S, Harman M (2012) Regression testing minimization, selection and prioritization: a survey. Verification Reliab 22(20):v67–120
12. Pappala LK, Selvi Rajendran P Dr (2021) A novel music genre classification using convolution neural networks. In: IEEE conference on communication and electronics system, July 8–10, 2021
13. Elanthiraiyan N, Arumugam C (2014) Parallelized ACO algorithm for regression testing prioritization in Hadoop framework. In: 2014 International conference on advanced communication control and computing technologies (ICACCCT). IEEE
14. Khatibsyarbini M, Isa MA, Jawawi DN, Tumeng R (2018) Test case prioritization approaches in regression testing: a systematic literature review. Inf Softw Technol 93:74–93

15. Catal C, Mishra D (2013) Test case prioritization: a systematic mapping study. Software Qual J 21(3):445–478
16. Rothermel G, Harrold MJ, Ostrin J, Hong C (1998) An empirical study of the effects of minimization on the fault detection capabilities of test suites. In: International conference of software maintenance, Bethesda, Maryland, November 1998, pp 34–43

Activation Functions for Analysis of Skin Lesion and Melanoma Cancer Detection

Damarla Anupama and D. Sumathi

Abstract Skin cancer is the most common occurred cancer on the body which is time consuming and tedious. Detection of this cancer at early stage has a high effect on the disease treatment. Skin cancers are defined as either malignant or benign cells. With the advancements in deep learning algorithms in image processing and computer vision, there was a great evolution in medical image analysis especially for the classification of melanoma from skin cancer diagnosis. A convolution layer with activation function will move to next layer. The activation function plays a prominent role in the deep neural networks. On the proposed transfer learning method, various activation functions are implemented to check the model accuracy and to classify the melanoma. For experimental purpose, we use the datasets collected from various patients toward melanoma detection.

Keywords Skin cancer · Skin lesion images · Deep learning · Melanoma · Activation functions · Classification

1 Introduction

The entire human body is covered with a largest organ called skin. It protects the body against injuries and infections and helps regulate the body temperature. Cancer is a disease that starts in one part or area of the body and spreads to other organs. One of the most common types of cancer in human beings is skin cancer [1]. It is classified as melanoma and non-melanoma. Melanoma is the worst form of skin cancer. Early detection of this disease can affect the result of the illness and improve the chance of surviving [2]. The key to effective treatment of skin cancer is early detection [3]. These approaches are complex, error prone, time consuming and subjective [4]. Due

D. Anupama (✉) · D. Sumathi
VIT-AP University, Vijayawada, Andhra Pradesh, India
e-mail: anupama.20phd7135@vitap.ac.in

D. Sumathi
e-mail: sumathi.d@vitap.ac.in

P. Dutta et al. (eds.), *Emerging Technologies in Data Mining and Information Security*, Lecture Notes in Networks and Systems 490,
https://doi.org/10.1007/978-981-19-4052-1_40

to the excessive exposure to dust, sun rays and other impurities there occurs a lesion on the skin. Not all the lesions lead to cancer. Computer vision applications are extremely implemented with deep neural networks especially the CNN. It is difficult to identify, analyze and interpretation of skin lesion images due to multiple factors [5]. The images include bubbles, hair, some contain blood vessels, etc. Also, the lesion images are also of different shapes, size and color and these factors also limit the efficiency of the build model. Melanoma leads to highest mortality rate and can be cured if detected early [6]. The traditional approaches are implemented by steps of low-level pixel processing methods [7] (Fig. 1).

Dermatologists identify the skin lesions using dermatoscopy, an instrument which visualizes through a lens with a source of light that identifies color, pigmentation, ramification, shape, etc. (Fig. 2). They also identify with some visibility feature rules like ABCDE rule [8], seven-point check list method, Menzies method [9]. Some of the computer-aided detection and pigmented skin lesion techniques help patients in early detecting and traumatizing medical procedures [10]. Generally, there are few steps in diagnosing the melanoma cancer. Those comprised preprocessing, segmentation, feature extraction and classifications of lesions [11]. Few researchers have reviewed on deep learning methods for detecting, recognizing and classifying the melanoma

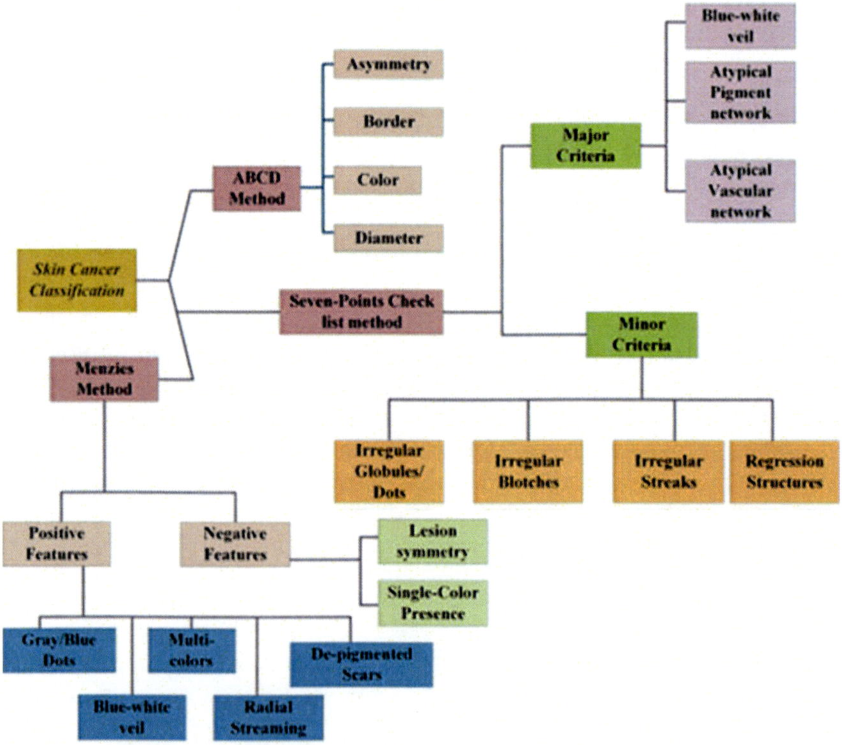

Fig. 1 Skin cancer classification methods

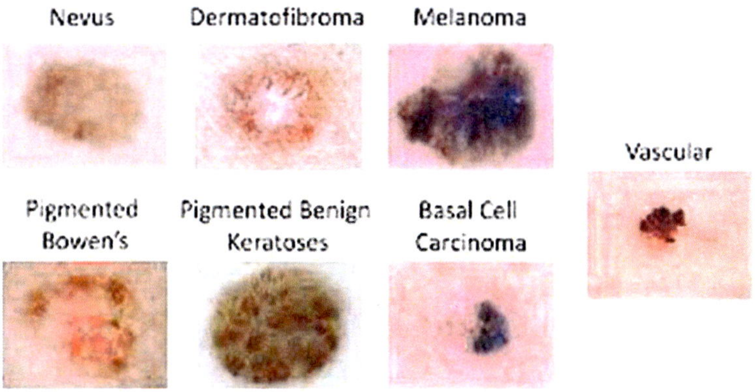

Fig. 2 Types of skin lesions [2]

lesions on medical images [12]. Deep learning models especially CNN have gained fast performance with the availability of large datasets. The activation functions in the hidden layers act as an important parameter for feature extraction which decides a neuron will fire or not [2]. Additionally, few more algorithms also proved to be excellent in image classification [13].

The rest of the paper is organized as follows: Sect. 2 presents steps for diagnosing skin cancer and classification. Section 3 discusses about the structure of deep neural network; Sect. 4 depicts the experimental measures, and finally, Sect. 5 concludes with the result.

2 Background Theory

2.1 Challenges in Detection of skin Lesions

There are multiple difficulties in detecting skin lesions due to the variations in image type and sources. Various challenges include:

- Different shape and with multi-size.
- Artifacts presence and noise.
- Fuzzy irregular boundaries.
- Low contrast.
- Color illumination.

2.2 *Skin Classifications*

2.2.1 ABCDE Rule

This is a rule that is used to examine the skin lesions on the body and identify the classification.

A-Asymmetry property: Skin lesions are accessed for similarity to find the edges, shapes and color.

B-Border property: Edges are assessed to determine if they are well-defined and smooth. The lesions are likely to be melanoma if the edges are fuzzy, uneven and jagged.

C-Color property: Visibility of melanoma shows color contrast between different skin types of shades varying from red, black, tan and brown.

D-Diameter property: If the diameter of the lesion is above 6 mm, then it is a sign of melanoma occurrence.

E-Evolution property: When the skin lesion was started on the human body.

To calculate the final score (total dermoscopy score—TDS), the following formula is used: TDS = $[(A \text{ score} \times 1.3) + (B \text{ score} \times 0.1) + (C \text{ score} \times 0.5) + (D \text{ score} \times 0.5)]$, [14].

2.2.2 Seven-Point Checklist Method

Two types of criteria exist in this method. They are classified into major criteria and minor criteria. Each major criterion is further divided into three types and has a score metric of two and minor is further divided into four and each has a score metric of one.

If (Score metric) > 3 then

Malignant

Else

Benign

Major Criteria

Blue-white veil: Blue bubbles in color and no structure.
Atypical pigment network: Thickness and color are heterogeneous, asymmetric distribution.
Atypical vascular pattern: Dotted vessel or linearly irregular globular.

Minor Criteria

Irregular globules/dots: Dots are irregular size, and shape are less than 0.1 mm.
Irregular blotches: No defined structure and color may vary in black, white and brown.

Irregular streaks: Radial streak type pattern situated at the lesion edges.
Regression structures: Scars are white in color and de-pigmentation.

2.2.3 Menzies Method

These methods are defined based on the positive and negative features. The negative indicates that the lesion is malignant and positive indicates benign.

2.3 Steps to Diagnose Malignant Melanoma

Skin cancer is caused by various changes in our DNA cells that allow growing uncontrollably apart to the exposure to UV rays emitted from the sun. Melanoma is a specific type of skin cancer that occurs in pigment level containing cells known as melanocytes which are found primarily in the skin. Dermatologists use a tool known as TNM classification system to stage different types of cancer.

T-Tumor size (How deep the growth is?).
N-Involvement of lymph node.
M-Metastasis (How much cancer has spread?).

Skin cancer diseases such as melanoma cancer has a highest mortality rate and can, however, be cured easily when detected at early stage [6]. Melanoma increase access to health services and early detection of disease will reduce mortality. A self-investigation of the body from head to toe is needed to identify skin cancers.

2.3.1 Preprocessing

This is the initial step in the process of detection on melanoma. The raw images captured contain noise. These noises should be removed so that it will not affect the performance of the model. There are multiple types of filters such as wiener filter, Gaussian filter, mean filter, adaptive filter and median filter that remove noise from the images.

2.3.2 Segmentation

This is an important phase that performs image recognition. It separates the foreground and background of a given image. It obtains region of interest. Upon separating labels are assigned to the regions and further they are analyzed. Segmentation is basically classified into two types: local segmentation (concerned within a selected region or part of image) and global segmentation (concerned with an entire image)

[15]. The detection of lesions also starts from this phase. Segmentation is classified into different types based on the threshold, region, pixel, model, color and cluster.

2.3.3 Feature Extraction

The segmented skin lesion is used for feature extraction in this section. There are various methods used for feature extraction which include ABCD rule, HOG, GLCM and ABCDE rule. Some extracts based on the textural features, and some are extracted based on the shape and edges. These features are the primary elements that affect the classification system [11]. With these extracted lesion features, it is further classified whether it is a malignant or not [16]. With the evolution of deep learning, extracting the features has great extent [17]. There are various feature extraction techniques: image transformation, template matching, graph analysis, contour methods, project histograms, Fourier technology and gradient approach [7].

2.3.4 Classification

Deep learning-based algorithms have brought effective outcomes compared to the conventional methods. When the model is trained successfully these algorithms has outstanding accuracy and generate excellent results. It also identifies the type of skin tumor. There are different models that distinguish between benign or malignant [18]. There are different classifiers such as SVM, Naïve Bayes, KNN, neural network, etc. (Fig. 1).

3 Proposed Approach

3.1 Dermoscopic Image Datasets

There are various dermoscopic image datasets that are open access for classification purposes. The biggest dataset from International Skin Imaging Collaboration (ISIC) repository contains 25,331 images. These images are collected from various clinical institutions [19]. There are several disease categories: dermatofiboma, nevus, melanoma, vascular, pigmented Bowen's, pigmented benign keratoses and Basal cell carcinoma.

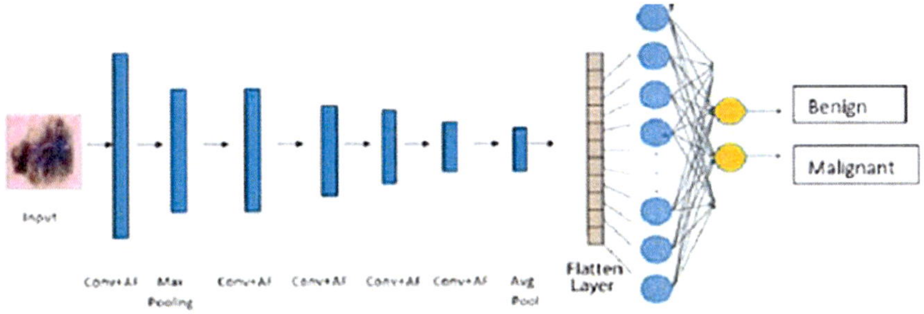

Fig. 3 Proposed convolution method

3.2 *CNN Model*

A CNN models (Fig. 3) in deep learning are used for feature extraction and lesion classification. Using different images, the proposed model has classified the lesions as benign or malignant. The input for the proposed model is an RGB images. The model is constructed as input layer, hidden layers and output layer. The internal structure of the model contains five convolutional layers with activation functions, two fully connected layers and an output layer. The model was completed in Python, making use of the packages NumPy [20], Keras [21] and TensorFlow [22]. The hyperparameters used where loss function was binary cross entropy. The optimization algorithm used were Adam with $\alpha = 0.001$.

Numerous activation functions like ReLU, PReLU, ELU, LeakyReLU, swish, Tanh, etc., have been proposed to improve the model. In this work, the parametric ReLU activation function considered the best accuracy and produced desired values with the proposed model. The dataset considered was a private dataset which contains 135 images of which 70 images are training dataset, 35 images are validation dataset and rest are considered for 30 images.

3.2.1 Input Layer

Suppose $x = x_1, x_2, \ldots, x_N$ denote all the N training samples. X, i.e. $R\ m \times n$ is the ith sample in X with size of $m \times n$.

Convolutional layers: In a convolution layer, each neuron is only linked to the next local layer of input neurons instead of a long-length connection so that the count of parameters to be learned is decreased significantly and the network grows deeper with a fewer number of parameters. Convolution: The outcome of the convolution operation is calculated by convolving an input (I) with several filters as follows:

$$j = I * W_j + b_j \tag{1}$$

$j = 1, 2, \ldots, F$ Where F is said to be the set of filters, x_j is the output for the jth convolution filter, W_j is the weights of the jth filter, and b_j is the jth bias.

3.2.2 Activation Functions [24]

ReLU: Most prominently used element-wise nonlinear activation function which replaces all negative feature values with the following equation. It is the most used activation function in hidden layers.

$$f = \max(0, y) \tag{2}$$

ELU: The ELU hyperparameters control the negative net inputs from moving to saturation point which is usually set to 1.

$$f(x) = \begin{pmatrix} x, & \text{if } x > 0 \\ \alpha \exp(x) - 1, & \text{if } x \leq 0 \end{pmatrix} \tag{3}$$

PReLU: It generalizes the traditional rectified unit with a slope for negative values.

$$f(y) = y \quad \text{if } y \geq 0 \tag{4}$$

$$f(y) = ay \quad \text{if } y < 0 \tag{5}$$

Leaky ReLU: Instead of considering the zero value for negative values, the LReLU activation function will have a small slope for negative values.

$$f(y) = 1 \quad \text{if } y < 0 \tag{6}$$

$$f(y) = \alpha(y) + 1 \quad \text{if } y \geq 0 \tag{7}$$

3.2.3 Fully Connected Layer

The input image from the previous layers is flattened and passed to this layer. It undergoes few more fully connected layers with the mathematical functions. In this stage, the classification process starts.

3.2.4 Output Layer

This is the last layer of the model which classifies where the object in the input image is benign or malignant.

4 Results and Discussions

In the experiment, we randomly sampled on 135 images. The splitting of training and testing dataset were done randomly on the major criteria for performance evaluation of classification model include accuracy [11], sensitivity [18] and specificity [18]. We proposed a model that accepts different activation functions and generate a desirable output to overcome the gradient problem. Across the dataset, it showed that PReLU has outperformed well than the ReLU, ELU and Leaky ReLU with minimal increase in number of parameters. By placing the PReLU, there is a decrease in loss function and significantly increase in the parameters. The comparison of the activation function accuracy and parameters for the proposed model and the results are calculated.

The above results obtained on the proposed model as shown in Figs. 4 and 5.

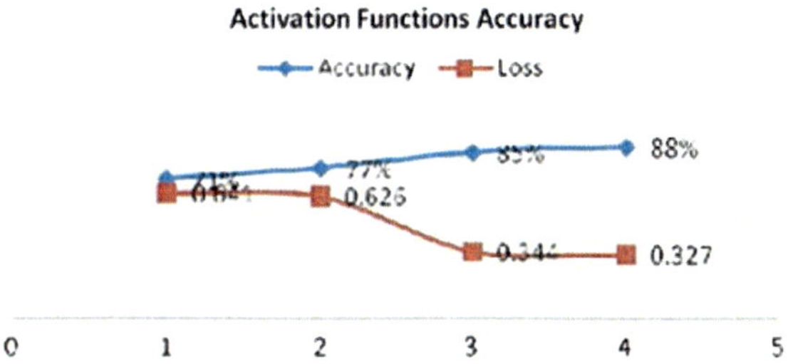

Fig. 4 Activation function accuracy

Activation function	Accuracy	Loss	Parameters
ELU	71	0.641	84,357
Leaky ReLU	77	0.626	86,357
ReLU	85	0.344	86,357
PReLU	88	0.327	96,919

Fig. 5 Activation function performance on dataset

5 Conclusion

A deep learning framework can accurately predict in images and videos classification. In medical imaging, deep learning-based skin disease recognition is promising. The models trained on a single skin dataset may not equally work well on different skin disease datasets [23]. This is because multiple different datasets were collected from a variety of clinical settings and patient cohorts, using different imaging devices and labeling standards. The activation functions also differ and impact on the accuracy. The ReLU function creates dying rule problem. Leaky ReLU creates the vanish gradient problem with multiplying with 0.01. Though the ELU function satisfies the model, but it increases the computational time. Finally, the PReLU, the value of a is a learning parametric which is not fixed and learns dynamically. Therefore, with the minimal addition of parameters will lead to a vast improvement in performance.

References

1. Sung H et al (2021) Global cancer statistics 2020: GLOBOCAN estimates of incidence and mortality worldwide for 36 cancers in 185 countries. Cancer J Clin 68(6) (Feb 2021), caac.21660. ISSN: 0007-9235. https://doi.org/10.3322/caac.21660
2. Namozov A et al (2018) Convolutional neural network algorithm with parameterized activation function for melanoma classification. In: ICTC 2018, 978-1-5386-5041-7/18/2018. IEEE
3. Johanen TH et al (2019) Recent advances in hyperspectral imaging for melanoma detection. WIREs Comput Stat. 2019;e1465, Wiley Periodicals, Inc. https://doi.org/10.1002/wics.1465
4. Yu L, Chen H, Dou Q, Qin J, Heng P-A (2016) Automated melanoma recognition in dermoscopy images via very deep residual networks. IEEE Trans Med Imaging 36(4):994–1004
5. Adegun A, Viriri S (2019) An enhanced deep learning framework for skin lesions segmentation. International conference on computational collective intelligence. Springer, Cham, pp 414–425
6. Al-Masni MA, Al-antari MA, Choi M-T, Han S-M, Kim T-S (2018) Skin lesion segmentation in dermoscopy images via deep full resolution convolutional networks. Compute Methods Programs Biomed 162:221–231
7. Adegun A et al (2020) Deep learning techniques for skin lesion analysis and melanoma cancer detection: a survey of state of the art. Springer Nature B.V., pp 811–841
8. Ozkan IA, Koklu M (2017) Skin lesion classification using machine learning algorithms. Int J Intell Syst Appl Eng 5(4):285–289
9. Munir K et al (2019) Cancer diagnosis using deep learning: a bibliographic review. Cancers 11:1235. https://doi.org/10.3390/cancers11091235
10. Almaraz-Damian J-A, Ponomaryov V, Sadovnychiy S, Castillejos-Fernandez H (2020) Melanoma and nevus skin lesion classification using handcraft and deep learning feature fusion via mutual information measures. Entropy 22(4):484
11. Saeed JN et al (2021) Skin lesion classification based on deep convolutional neural networks architectures. J Appl Sci Technol Trends 2(1):41–51
12. Okur E, Turkan M (2018) A survey on automated melanoma detection. Eng Appl Artif Intell 73:50–67
13. He K, Zhang X, Ren S, Sun J (2016) Deep residual learning for image recognition. In: Proceedings of the IEEE conference on computer vision and pattern recognition, pp 770–778
14. EL Abbadi NK, Faisal Z (2017) Detection and analysis of skin cancer from skin lesions. Int J Appl Eng Res 12(19):9046–9052. ISSN 0973-4562, Research India Publications. http://www.ripublication.com

15. Kaur D et al (2014) Various image segmentation techniques: a review. IJCSMC 3(5):809–814
16. Filali Y, Abdelouahed S, Aarab A (2019) An improved segmentation approach for skin lesion classification. Statis Optimiz Inf Comput 7(2):456–467
17. Khan MA, Javed MY, Sharif M, Saba T, Rehman A (2019) Multimodel deep neural network based features extraction and optimal selection approach for skin lesion classification. In: 2019 International conference on computer and information sciences (ICCIS), IEEE, pp 1–7
18. Vidya M et al (2020) Skin cancer detection using machine learning techniques. 978-1-7281-6828-9/20. IEEE
19. Tschandl P, Rosendahl C, Kittler H (2018) The HAM10000 dataset, a large collection of multi-source dermatoscopic images of common pigmented skin lesions. Sci Data 5: 180161. https://doi.org/10.1038/sdata.2018.161
20. Harris CR et al (2020) Array programming with NumPy. Nature 585(7825):357–362
21. Chollet F (2015) Keras. Available: https://keras.io
22. Abadi M et al (2016) Tensorflow: a system for large-scale machine learning. In: 12th {USENIX} symposium on operating systems design and implementation ({OSDI} 16), 2016, pp 265–283
23. Litjens G, Kooi T, Bejnordi BE, Setio AA, Ciompi F, Ghafoorian M, Van Der Laak JA, Van Ginneken B, Sánchez CI (2017) A survey on deep learning in medical image analysis. Med Image Anal 42:60–88
24. Inturrisi J, Khoo SY, Kouzani A, Pagliarella R.Piecewise linear units improve deep neural networks

Automated Real-Time Face Detection and Generated Mail System for Border Security

Khushboo Tripathi, Juhi Singh, and Rajesh Kumar Tyagi

Abstract This paper focuses on face recognition and automatic mail generated system for observations by using a solitary camera. The application is created for detection and capturing the face, which can be applicable in surveillance system at border gateway. The camera is fixed for human face identification. A framework that perceives human information is handled in a second and an email alert is produced for security reasons. The key regions for examination of face detection have taken as consideration in scenario. After applying in real-time state Viola–Jones algorithm is used for face detection from different angles. The better pixels based image is used for sending emails. The performance has been checked and analyzed after repeating iterations of experiments through simulation.

Keywords Recognition system · Face detection algorithms · Viola–Jones algorithm

1 Introduction

The facial recognition framework is a technology that can recognize or verify a person from a video outline taken from a high-resolution image or video source. Analyzing face information features effectively and efficiently is a difficult undertaking that takes a lot of time and effort. A facial recognition framework can function in a number different ways, but it usually works by comparing selected facial highlights of a specific photo to faces in the database. It is likewise portrayed as a biometric application dependent on man-made consciousness that can interestingly distinguish an individual by breaking down examples dependent on the surface and state of the individual's face. The basic role of the observation framework here is to identify and follow people utilizing a solitary camera. The camera is fixed where it's needed. At the point when a human substance is distinguished, a following line conforms to the individual to follow the item. A framework that perceives human information

K. Tripathi (✉) · J. Singh · R. K. Tyagi
Department of Computer Science and Engineering, Amity University, Gurugram, Haryana, India
e-mail: khushbootripathi.cse@gmail.com; ktripathi@ggn.amity.edu

P. Dutta et al. (eds.), *Emerging Technologies in Data Mining and Information Security*, Lecture Notes in Networks and Systems 490,
https://doi.org/10.1007/978-981-19-4052-1_41

is prepared in a second and an email alert is produced for security reasons. The fundamental goal is to build up a security framework progressively. Continuous face discovery presents the framework with a progression of casings for distinguishing faces, so the space-time channel (discover contrasts between resulting outlines) can be utilized to recognize territories of the casing that have changed.

2 Related Background

The improved LBP method was used to explain an improved face recognition algorithm and its implementation in an attendance management system by Bah and Ming [1]. Lighting, sharpness, noise, resolution, size, and position were used to improve the LBP algorithm's face recognition accuracy. Images that expose more information about image features, allow for more precise feature extraction and comparison. Dobhal et al. [2] created an automatic process to take attendance by using face detection using LBPH algorithm, viola–Jones face detection. The author has discussed about existing systems which is currently lacking proper time management and doesn't work in real time. It doesn't monitor the students which involves problems like what to do if the students leave the class [1, 2].

Face recognition in a crowded environment was presented by Akhila and Hemanth [3]. Motion detection processes and a canny edge detection algorithm were used in this study. The devised system was highly efficient when dealing with large crowds, which is why it may be employed in military applications, trains, and other applications [3]. Zhu et al. [4] presented a high-fidelity pose and expression normalization approach based on a 3D morphable model that can build a natural face, image in frontal stance and neutral expression, and recover canonical view expression free photos, all using a 3D morphable model. The authors used the Basel face model, 3D morphable model, 3D millimeter fitting, 3D meshing and normalization, labeled faces in the wild (LFW), and the CMU multi-pieda tab in this paper. High-fidelity pose and normalization (HPEN) was used to bridge the gap in the unseen reason based on face symmetry [4]. Huang et al. [5] research proposes a method to solve the problem of face recognition with image sets and does facilitate more robust classification. The methodology used was discriminant analysis on Riemannian manifold of Gaussian distribution algorithm (DARG). For future work, more probabilistic kernel forgot scenes and more conventional learning methods will be extended to Romanian manifold of Gaussian distribution [5]. Reddy and Raju [6] mention about local ternary pattern, local binary pattern, no processes normalization Weber faces normalization, wavelength diagnosing, and gradient faces [6]. Authors used different normalization techniques used to eliminate the different lighting changes, local texture patterns used for robust features extractions, nearest neighborhood classifier used to measure feature. Experimental results are showing that local ternary pattern achieved better recognition rate than that of local binary patterns. Future scope can extend to fusion of multiple normalization technique for pattern recognition rate. In another paper, Ryu and Kim [7] describe efficient face tracking and recognition algorithm and system presented for DVR [7]. The methodology used was DCT—HMM (hidden Markov model) method. It focuses on detecting enough face area with eyes. This shows frequency coefficient around eye is considered to play an important role for overall performance so it is very important to implement database that focuses on it.

The following are the facial recognition algorithms such as:

(a) **Kanade–Lucas–Tomasi (KLT)**: The KLT calculation was presented by Lucas and Kanade, and their work was stretched out by Tomasi and Kanade. The calculation is utilized to discover dissipated component focuses that are sufficiently finished to follow the necessary focuses with a proper norm. The Kanade–Lucas–Tomasi (KLT) calculation is utilized to consistently follow human faces in video outlines [8]. This technique is accomplished via scanning for parameters that decrease the divergence estimations between the trademark focuses related with the first interpretation model.

(b) **Principle Component Analysis (PCA)**: PCA is a method for simplifying the difficulty of selecting an eigenvalue and associated eigenvector representation to obtain a consistent representation. This can be accomplished by reducing the representation's dimension space. The dimension space must be decreased in order to provide rapid and reliable object detection. Furthermore, PCA

(c) **Linear Discriminant Analysis (FLD)**: It also saves the original information from the data. The eigenface-based approach [9] uses the PCA basis. Fisher's linear discriminant analysis (LDA) is a type of linear discriminant analysis (FLD). The FLD approach is used to minimize the dimension space. The FLD method makes use of within-class information to reduce variation within each class and increase class separation [10].

(d) **Viola–Jones Algorithm**: The Viola–Jones computation is a widely used tool for identifying objects. The main benefit of this computation is that it takes only a few minutes to prepare but only a few seconds to identify. This calculation does not use duplication and uses Haar's base trademark channel. By initially establishing the essential picture, the productivity of the Viola–Jones calculation can be increased indefinitely. The discovery takes place within the location window. A base and most extreme window size is chosen, and a sliding advance size is chosen for each size.

At that point the discovery window travels through the picture as follows:

- Determine the smallest window size and the sliding step that corresponds to it.
- Slide the window vertically and horizontally in the same process, depending on the window size. A set of N face recognition filters is applied at each phase. The face is detected in the current widow if one of the filters returns a positive result.
- When the window has reached its maximum size, stop the operation. If not, raise the window size and corresponding sliding step to the next desired size and go to step 2.

Viola and Jones [11] presented this calculation. This computation was made mostly due to the issue of face position.

3 Proposed Methodology

A. Scenario Description

In our research work, which is divided into three main sections are: The first section was focused on mainly capturing the face; the second section was detection of face of human being, and third was sending the detected face on an email by applying Viola–Jones algorithm. In the first section, a digital static live camera was used at the entrance to capture images of people entering a room or a building, and advanced image processing techniques such as contrast adjustment, noise reduction using a bilateral filter, image histogram equalization, angel and pixel filtrations were applied to the captured images to improve their quality, and then the Viola–Jones algorithm was applied to the captured images to detect individual faces, which subsequently led to the development of the Viola–Jones algorithm. The experimental results for this part are shown in Figs. 1, 2, 3, and 4. Twelve samples of mobile human faces have been taken for experiment randomly. Using our approach in day time and in light on scenario, we were able to capture, detect, and send the final image to the primarily made Gmail account.

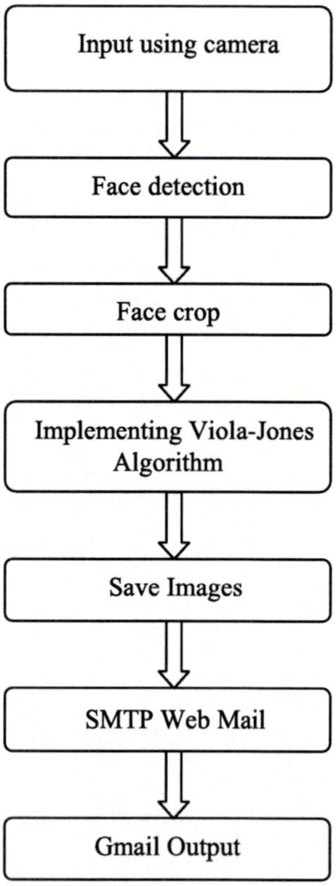

Fig. 1 Steps of implementation

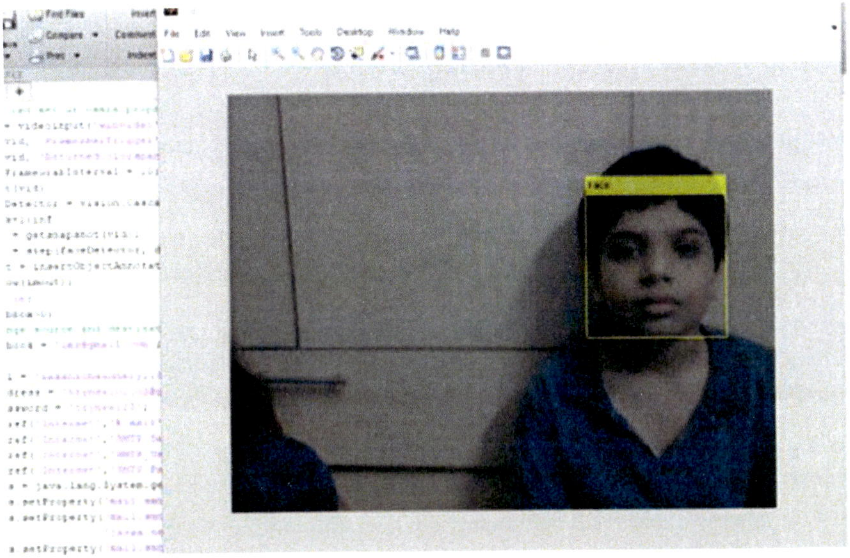

Fig. 2 Image detection

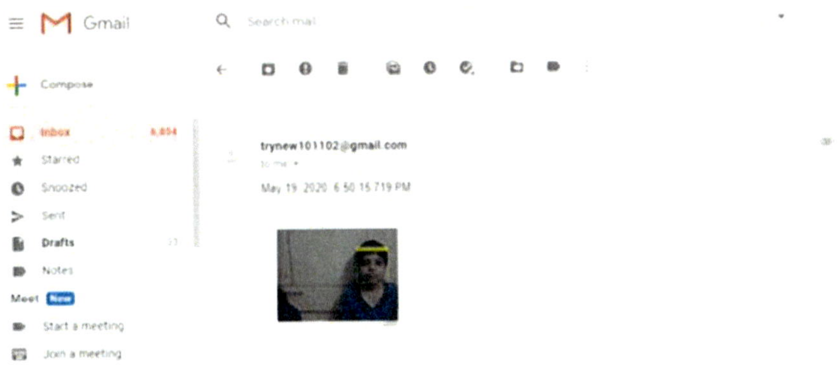

Fig. 3 Gmail output

Fig. 4 Number of frames per second

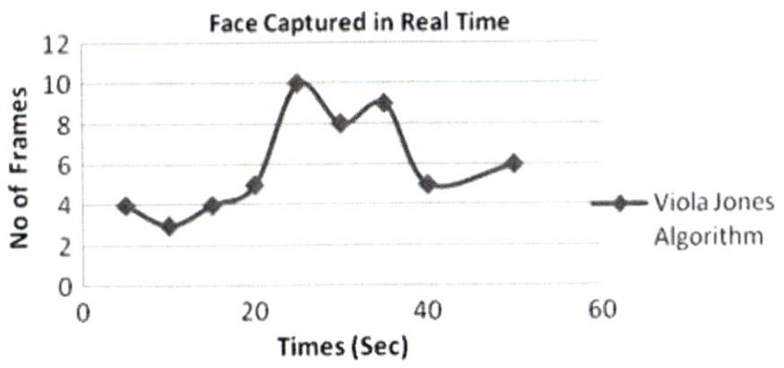

B. **Experiment**

Using above scenario for the experiment MATLAB vs. 13 B tools was used for simulation as in Fig. 1. The following steps were used for the experiment as below:

(a) Read a video frame and run the face detector. Camera is continuous checking and giving output images.
(b) Facial detection takes camera/video sequences as input and locates face areas within these images. Face areas are separated from non-face backdrop regions in this way. Within the detection process, facial feature extraction locates essential features (eyes, mouth, nose, and eye-brows).
(c) The video/image format is used to capture real-time input. The video (a series of images) is turned into frames before being processed further. Each cropped facial image was reduced to a size of 20×20 pixels.
(d) Applying Viola–Jones algorithm.
(e) After cropping face and automatic image was saved in folder.
(f) Automatic attach crop photo in SMTP WEB MAIL will be achieved.
(g) Automatic images will be sending in Gmail account.

4 Results and Analysis

After implementation of above steps as in Fig. 1, the following results have been obtained:

(a) Image detection—Image detection is the process of capturing and identifying the human face by a static camera. After capturing the face and running the program with Viola–Jones algorithm and cropping steps of implementation of different samples with different deviations, camera has detected the face of a person, who went across the camera in real time as in Fig. 2. This is one real-time capturing and detection of human face after applying cropping and selection of pixels' strategies.

(b) Detection of face in real time through Gmail output—After image detection in real-time scenario, the system saves the picture and sends it to the mail ID through SMTP. So an automatic system is generated for the storing image immediately in Gmail account after implementation of detection algorithm as given in Fig. 3.

(c) Number of frames per second—Fig. 4 graph is about the number of frames captured in seconds after applying Viola–Jones algorithm for face detection.

5 Conclusion

Despite the fact that the topic of face recognition has been explored for the past two decades, the majority of the research has been done on still photos. There have been very few online face recognition systems designed to examine the topic of face recognition in a real-time scenario with predefined limits. Various strategies for detecting a face have been proposed in the literature; nevertheless, for real-time applications, they are computationally expensive. The purpose of an automated video surveillance system is to observe people in a busy setting in real time and describe their actions. To maintain security, safety, and site management, persons must be detected and tracked. The implementation scenario for detecting and tracking human faces with a single camera was successful. The camera was installed in the proper location. This was an attempt to secure border crossings. If a human entity is recognized on the border lines, the object is tracked, and the observed image is promptly communicated to the base station via the email system. The email notice is generated for security reasons.

6 Future Work

The proposed research and work serve as inspiration for a real-time human identification and tracking system. Any recognition algorithm or system relies heavily on feature extraction. Using preprocessing and alternative feature extraction, a significant change in recognition rate has been seen. There is still more work to be done in this field, particularly in trimming an image before running it through a recognition engine. It would be fascinating to investigate novel preprocessing strategies that would result in the best identification rates when using different face detection algorithms. Despite the fact that individual discovery and verification frameworks are already accessible, more research is expected to solve the challenges of real-life circumstances. Despite the fact that there are many surveillance cameras around us, it is quite improbable that they will be constantly monitored. The problems in this field are improving face detection accuracy, overall face recognition accuracy, and lowering the amount of false positives and false negatives.

References

1. Bah SM, Ming F (2020) An improved face recognition algorithm and its application in attendance management system. Array 5:100014
2. Dobhal S, Singh AK, Kumar A (2020) Automated attendance system using multiple face detection and recognition. Int Res J Eng Technol 7(4):920–923
3. Goud GA, Hemanth B (2015) Face recognition using edge detection of layer. Int J Scien Res Sci Eng Technol 1(2):378–380
4. Zhu X, Lei Z, Yan J, Yi D, Li SZ (2015) High-fidelity pose and expression normalization for face recognition in the wild. In: Proceedings of the IEEE conference on computer vision and pattern recognition, pp 787–796
5. Huang Z, Wang R, Shan S, Chen X (2015) Projection metric learning on Grassmann manifold with application to video based face recognition. In: Proceedings of the IEEE conference on computer vision and pattern recognition, pp 140–149

6. Reddy BA, Raju TA (2015) Extraction of hidden features using preprocessing techniques and texture analysis for face recognition. Int J Adv Res Electron Commun Eng 4(8):2241–2247
7. Ryu JS, Kim ET (2006) Development of face tracking and recognition algorithm for DVR (Digital Video Recorder). Int J Comp Sci Netw Secur 6(3A):17–24
8. Mstafa RJ, Elleithy KM (2016) A video steganography algorithm based on Kanade-Lucas-Tomasi tracking algorithm and error correcting codes. Multimedia Tools Appl 75(17):10311–10333
9. Chen S, Zhu Y (2004) Subpattern-based principal component analysis. Patt Recogn 37(5):1081–1083
10. Kumar N, Andreou AG (1996) A generalization of linear discriminant analysis in maximum likelihood framework
11. Viola PA, Jones MJ (2007) Detecting pedestrians using patterns of motion and appearance in videos. U.S. Patent 7,212,651

A Review of Time, Frequency and Hybrid Domain Features in Pattern Recognition Techniques

Pooja Kataria, Tripti Sharma, and Yogendra Narayan

Abstract The treatment and rehabilitation of individuals with motor disabilities are one of the key application areas. Therefore, in biomedical engineering, the detection of EMG signals using efficient and advanced methodologies is becoming a critical requirement. Clinical diagnosis and biomedical applications are the key reasons for EMG signal analysis popularity. Modern signal processing techniques, which can provide a time–frequency representation, are one of the possible solutions for automated EMG analysis. Furthermore, EMG-based prosthetic control supports varying levels of freedom operations, enabling amputees to operate the device intuitively. This paper highlights a detailed comparison among the data acquisition methods, features extracted like time-domain, frequency–domain and time–frequency domain by deploying different feature extraction techniques, distinct machine learning and deep learning classification methods and their respective accuracy for decoding the various limb movements done by the normal and disabled subjects.

Keywords EMG · DWT · KNN · SVM · ANN

1 Introduction

World Health Organization (WHO) in 1970s stated stroke as a loss of neurological activity and in terms of medical name given is cerebrovascular accident (CVA) mentioned for stroke, that remained for more than 24 h or can lead to death within span of one day [1]. The main cause of occurrence of stroke is reduction or interruption of blood supply to brain, thus blocking the oxygen to the tissues of brain. This

P. Kataria (✉) · T. Sharma · Y. Narayan
Chandigarh University, Mohali, India
e-mail: pooja.e8835@cumail.in

T. Sharma
e-mail: triptisharma.ece@cumail.in

Y. Narayan
e-mail: yogendranarayan.cse@cumail.in

P. Dutta et al. (eds.), *Emerging Technologies in Data Mining and Information Security*, Lecture Notes in Networks and Systems 490,
https://doi.org/10.1007/978-981-19-4052-1_42

can lead to death of brain cells in few minutes. So, stroke is considered as medical emergency and quick attention and treatment is very crucial. We can categorize the cerebrovascular accident (CVA) into mainly two parts, first one is that which is due to blockage is called ischemic and another one which is due to damage of blood vessels is known as hemorrhagic stroke. In both cases, part of brain did not get sufficient oxygen supply and blood which further leads to death of brain cells [2]. Majority of strokes are ischemic, rest are hemorrhagic. Ischemic stroke condition has been given the name of atherosclerosis.

Stroke treatment mainly depends upon the how long symptoms have been occurred and to know about its severity and the main reason for its occurrence. If early treatment with medication is provided, it can reduce the chances of brain damage and further treatment targets on how to resist further stroke. Some medications which can help to dissolve the clot of blood must be taken within early hours of symptoms [3, 4]. After that period, which is usually 2–3 h, further treatment process can include complete rest, intravenous liquids and further medicines which can make thinner of blood so that chances of further stroke can be avoided. Various therapies such as physical, occupational and speech therapy specialist generally go through the patient to cope up with any loss due to stroke.

2 Robotic Exoskeletons in Rehabilitation

The main purpose of stroke rehabilitation is to help the patient to relearn the various skills that have been damaged or diminished [5], whenever portion of the brain is affected by the stroke, so that the patient can return to their normal life as they were living before the stroke. Neurorehabilitation gives great advantage for the patients those have been suffered from injuries related to neuromuscular [6]. Neurons properties can be changed along with muscular properties by this process which exploits the plasticity of neuromuscular system of human, which includes how the connectivity pattern is and about their function. Sensory motor therapy is one of the major and very beneficial therapy which helps the patients for physical movements of upper or lower extremity generally helped by the robot or human so that patient again captures the art of movement as it was doing earlier. By rehabilitation robotics, we mean combination of industrial robotics and medical rehabilitation leads to covering multiple areas such as biomedical engineering, mechanical engineering, artificial intelligence, electrical engineering, actuator and sensor technology. Human function may be physical or cognitive which is considered for medical rehabilitation is tried to restore into its normal behavior may be partially. Robotic systems used for rehabilitation purposes won't take into account with the fact that their goal is to replace weak or unavailable limbs of human with mechanical function. Functional re-impose is to be considered as an important aspect for rehabilitation robotics. Main goal of rehabilitation robotics is to help the patient or user so that it can perform with maximum efficiency or freedom.

3 EMG-Based Control Scheme of Exoskeletons

If we talk about rehabilitation in terms of robotics, there are multiple developments phases that has been occurred, the main factor is how to gain control just similar to human like as exoskeletons comes under wearable robots [7]. From safety point of view, patients should not be hurt as exoskeletons have to be worn by patient. Various information sources can be used to identify or address to generate control signal for orthosis and active prosthesis [8]. The major categories are central nervous system signals, biochemical signals, electromyography (EMG) signals and peripheral nervous system signals. Among them, EMG signals suited the most for controlling active orthosis. Biochemical signals come under the category of low-frequency and low-amplitude electric signals which can be measured from biological being such as humans. From various known bioelectric signals, the best that can be sensed over the surface of skin is EMG signal [9–12]. Its generation is due to muscles fiber activity that is electrical in nature, which arises due to contraction of muscles or relaxation of muscles. As the muscle undergoes movement, it will reflect the particular pattern of multiple muscle fiber that is why in order to find out the movement, multiple-channel EMG recording can be utilized.

sEMG signals are often used to control most hand prosthesis. In Fig. 1, steps for using EMG signal for controlling the external devices as a prosthetic aid is depicted. A substantial percentage of muscles persists in the stub of arm after the amputation of a hand. As a result, the sEMG can be used to regulate the prosthetic hand. However, using sEMG signals to recognize hand movements create a multitude of challenges. An appropriate blend of feature extraction methodologies and dimension reduction strategies should be used to overcome these problems and improve performance of the classifier. To enhance classification efficiency, an optimal classifier should be employed [13, 14]. The biosignals have multidimensionality. As a result, choosing a suitable attributes extraction, dimensionality minimization, and machine learning methods for prosthetic hand control is challenging [15–18].

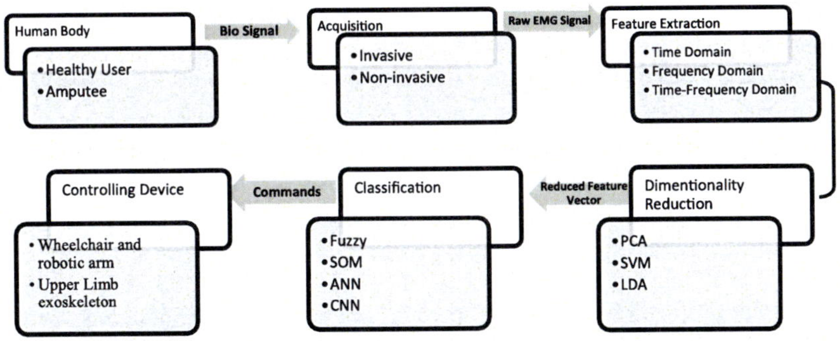

Fig. 1 Steps for EMG-based control system

4 Literature Review

Sarasola-Sanz and Irastorza-Landa in [19], for stratifying surface electromyography (sEMG) signals for recognition of physical intended motions used flexible analytic wavelet transform (FAWT). sEMG signal is decomposed into eight sub-band features, namely entropy, mean absolute value, variance, MMAV Type1, WL, SSI, Tsallis entropy and integrated EMG (IEMG) using FAWT. Features that were extricated are fed into an extreme learning machine classifier with sigmoid activation function. When comparison was made among all the sub-bands, the seventh sub-band clinched the best performance with the precision of 99.36%, sensitivity of 99.36% and specificity of 99.93%. Zhuojun et al. in [20] suggested window sample entropy and window kurtosis as a new feature extrication method in order successfully identify intended motion for intelligent bionic limb. Data is recorded for five hand motion tasks. Two new features and two classical features: Root mean square (RMS) and integrated electromyography (IEMG) were fed as the feature vector into the wavelet neural network (WNN) to perceive the muscle force. It is suggested that force should be ventured by neural network which is trained by contralateral data due to the truth that amputates' residuary limb does not offer complete educate facts for motion recognition. Both ipsilateral and contralateral experimental results demonstrated the practicability of the suggested features extrication methods. Motion classification precision with normalized mean square 0.58 was achieved through ipsilateral experiment. In the contralateral experiment, suggested method will benefit unilateral transradial amputees to retrain the intelligent bionic limb by their own sEMG.

Shafivvulla in [21], this paper suggested a method to recognize hand gestures using muscle activity separated from EMG using artificial neural network. EMG identification is limited to three arm muscles temporal activities to retain a constraint-free user environment. A unique signature for each particular posture was constructed with the attainment of these parameters. A success rate of 97.5% was achieved in recognizing six gestures. Peternel et al. [1] assessed identification of the paralyzed limb's motion intention using intense EMG recording and pattern discrimination techniques. EMG data was measured from 12 hemiparetic stroke subjects when they were asked to perform 20 distinct possible arm motion, hand motion and finger motions along with the paralyzed upper arm, for extracting sEMG signals comprising of 89 channels.

Ghannadi et al. [3] developed a virtual reality system to observe and promote cortical reorganization using motor unreal coaching in VR environment. They give training to nine healthy users. As per the capabilities of the user, virtual avatar dynamically adjusts its difficulty level in a motor imagery training task. About 85% of the experimental performance is achieved by detecting the SMR and to perform training in our virtual reality environment. But precision control of the arms is much lower. Author by repetitive motion of the task, improved their virtual arms control. Gao et al. in [22] suggested a technique for the persons affected from stroke to perform two important hand motions, i.e., hand opening and hand closing functional task. For this, EMG signals from their hemiplegic side were recorded and then trained the impaired hand by exploitation associate degree frame robotic hand. This system is based on the activity of their muscle signals as it recognizes the stroke person's motion intention. Bakardjian et al. in [23] developed a system for hand functions task to which training is provided with a purpose of the stroke rehabilitation by the exoskeleton device. It detects two important hand motions, i.e., hand opening or closing from the stroke affected person or healthy person by recording from the hemiplegic side, the electromyography (EMG) signals. For training the system, embedded controller and a robotic hand module were used.

Bin et al. [24] suggested a method related to event desynchronization and synchronization and state control for 2D wheelchair control. The 2D wheelchair can be designed to move in all the fundamental directions such as left or right or to move straight or stop. There are two important characteristics associated with sEMG signal, namely ERD/ERS that help in precise stratification of intended motion of the user. The result proved a high precision rate between a range of 87.5 and 100% when the wheelchair was handled while continuously moving in a two-dimensional plane according to the imagination of motion intension by the subject. For result validation, five healthy subjects were given the training for two forums, i.e., motor execution and motor imagery. In every forum, a 20 min calibration was done and for duration of 30 min, two sets of games were played.

Harwing et al. in [15] suggested an exoskeleton orthotic device for a human arm by indulging the neuromuscular signal (EMG) as the main controlling signal for the exoskeleton device. For experimental execution of an elbow joint regulated by the subject, a BMI was set at the neuromuscular level. A hill-based prototype was fed with features extricated from the sEMG signal and the joint kinematics in order to precisely stratify the motion performed on the elbow joint. Bermúdez et al. [25] suggested a method in which a synchronous training was given to the volunteer where each volunteer was asked to perform 160 trials. Two DOF electrical hand prostheses were controlled effectively by the suggested steady-state visual evoked potentials (SSVEPs). Classification precision between 44 and 88% was clinched during training to the four healthy participants; the participants clinched a precision of 75.5–217.5 s to reprint a sequence of motions, thereby controlling the prosthetic hand asynchronously. The quantity of fake negative (FN) choices varied from zero to ten (the maximal viable choices had been 34).

León et al. [26] suggested an efficient technique for controlling the single-DOF powered exoskeleton devices which are used in the activity of physiotherapy and restoration of the patient forelimb. Suggested technique recorded EMG signals from single muscles as well as dueler muscle pairs, to enhance the user's spontaneous control over the exoskeleton device. Freed et al. in [27], to discriminate from the muscle activation pattern, the deliberate physical tasks of stroke survivors suggested a method for electromyography (EMG) pattern classification technique. This study established the feasibility of the EMG pattern stratification method to discriminate the intention hand motion of stroke survivors. Kakoty and Hazarika in [28] used wavelet packet decomposition method for motion identification. With the help of wavelet packet decomposition, sEMG signal was extricated, and then, it is fed to BP neural network for the classification of motion during offline measurement. So, after a thorough literature survey, a tabular comparison (Table 1) is also drawn on the articles on the basis of certain identified parameters like data acquisition, feature extraction, classifiers and the accuracy achieved.

Table 1 Parametric comparison

References	Subject	Hand motions	Feature extraction	Classifiers	Accuracy
[29]	Ten subjects	Six upper-limb motions	DWT	Linear discriminant analysis (LDA)	Good classification results came when new traits were extricated from the second-level restored sEMG signal traits like MAV, ZC and WAMP traits by using Db7 as the mother wavelet and the MYOP trait extricated from first-level restored sEMG signal (D1) with the Db8 as the mother wavelet clinched the best motion stratification
[30]	Ten normal subjects	Six hand movements	DWT	Decision tree algorithms	An accuracy of 96.67% was reported by the amalgamation of DWT and a classification method named random forest for multiple DOF

(continued)

Table 1 (continued)

References	Subject	Hand motions	Feature extraction	Classifiers	Accuracy
[31]	Eight subjects	Six upper-limb motions including no movement	Time-domain features	LDA and ANN classifiers	An accuracy of 92% was clinched by using a new set of time-domain traits extricated from sEMG signal, giving a jump of 6.49% in accuracy compared to the readily used time-domain traits
[32]	Five normally limbed subjects	Four types of forearm movements	DWT	Gustafson-Kessel clustering classifier	Classification accuracy of 97% was achieved with GK classifier to discriminate four arm movements
[33]	Four subjects	Seven motions patterns	WPT	SVM classifier	WPT, NWFE and SVM have good real-time performance and classification accuracy
[34]	18 subjects	Eight dynamic motions	Time-domain features	LDA, QDA, KNN DT, Naive Bayes (NB) and Mahalanobis distance (MD)	An improvement of 2–8% was reported, when the stratification of movement was done using the transformed signal for 6 different classifiers than the classification done with original signals

(continued)

Table 1 (continued)

References	Subject	Hand motions	Feature extraction	Classifiers	Accuracy
[35]	12 healthy volunteer	Six hand movements	WPT	PSO-ISVM	The average movement stratification rate peaks to 90.66% and time required to train the classifier can be sawed-off to 0.042 s
[36]	20 healthy subjects	11 classes of motion	Time-domain, frequency domain and time–frequency domain	LDA, K-NN, DT, MLE, MLP and SVM	Peak stratification rate of 95% was reported by the proposed traits, namely logarithmic root mean square (LRMS) and normalized logarithmic energy (NLE)
[19]	Four subjects	Ten physical actions	Flexible analytic wavelet transform (FAWT)	Extreme learning machine (ELM)	Seventh sub-band clinched the optimum stratification result, where an precision of 99.36%, sentience of 99.36% and particularity of 99.93% was reported
[37]	Eight healthy subjects	Six am movement	DWT	SVM and ANN	Standard deviation measurements in Butterworth filter proved better

(continued)

Table 1 (continued)

References	Subject	Hand motions	Feature extraction	Classifiers	Accuracy
[38]	Ten healthy volunteers	Six hand movement	Time-domain features	LDA, KNN, medium tree (MT) and quadratic discriminant analysis (QDA) classifiers	Attained 6% gain in classification accuracy by opting first-order differentiation of EMG signal
[39]	Four healthy person	Three different hand motions	Discrete wavelet transform (DWT)	ANN classifier	Better results obtained for sEMG classification using ANN
[20]	Seven right-handed healthy volunteers	Five hand movement	Time-domain features	Wavelet neural network	0.58 ± 0.05 pattern classification precision with normalized mean square
[21]	20 subjects	Six arm gestures	Time-domain features	ANN	System can identified the six gesture with 97.5% precision

5 Discussions and Conclusion

EMG-based control scheme for gesture detection in upper-limb amputees has drawn a significant attention from researchers in the field of rehabilitation robotics over the years. Prosthetic systems with several degrees of freedom can be constructed using this control method, giving wearers a greater level of agility and responsiveness. The nature of sEMG signals is typically complex. Various researches have been conducted in order to analyze its complexity, but they have been ineffective due to large variances in muscle activity of some muscles than others. Also none of the pattern recognition system and feature extraction technique is 100% accurate for movement detection. As a matter of concern, all the testing has been done on healthy or able users, a very few study is done on stroke affected person. Based on a comprehensive review of several research articles, it has been found that in contrast to frequency domain and time–frequency features like autocorrelation parameters, spectral metrics, (STFT) short-time Fourier transform, (WT) wavelet transform and (WPT) wavelet packet transform, the most informative features are grounded on time statistics such as mean value, variance of EMG signals, (ZC) zero crossing, (SSC) slope sign changes, which entail less computational resources. The time-domain features have been commonly utilized due to its relative ease of execution and high efficiency. Wavelet analysis gathers insights in time–frequency domain which is best suited for sEMG attributes extrication. Also fractal Fourier transformations could also improve non-stationary signal frequency and temporal domain resolution, making them a viable feature extraction technique. Finally, as an end note, when compared to other models, a deep learning model can be utilized for automatic EMG segmentation and categorization, leading to higher recognition rate.

References

1. Peternel L, Fang C, Tsagarakis N, Ajoudani A (2019) A selective muscle fatigue management approach to ergonomic human robot co-manipulation. Robotics and Computer Integrated Manufacturing 58:69–79
2. Nougarou MA, Campeau-Lecours A (2019) Pattern recognition based on HD-sEMG spatial features extractionfor an efficient proportional control of a robotic arm. Biomed Signal Process Control 53:101–550
3. Ghannadi B, Razavian RS, McPhee J (2019) Upper extremity rehabilitation robots: a survey. In: Handbook of Biomechatronics
4. Kwee HM (2015) Rehabilitation robotics—softening the hardware. IEEE Engineering in Medicine and Biology
5. Simonetti D, Zollo L (2016) Multimodal adaptive interfaces for 3D robot-mediated upper limb neuro-rehabilitation: an overview of bio-cooperative systems. Robotics and Autonomous Systems 85:62–72
6. Wei H, Bu Y, Zhu Z (2020) Robotic arm controlling based on a spiking neural circuit and synaptic plasticity. Biomed Signal Process Control 55:101–640
7. Birbaumer N, Kubler A, Ghanayim N, Hinterberger T, Perelmouter J, Kaiser J, Iversen I, Kotchoubey B, Neumann N, Flor H (2012) The thought translation device (TTD) for completely paralyzed patients. IEEE Trans Rehabil Eng
8. Finkea A, Lenhardt A, Ritter H (2014) The MindGame: a P300-based brain-computer interface game. Elsevier
9. Narayan Y et al (2018) Surface EMG signal classification using ensemble algorithm, PCA and DWT for robot control. In: International conference on advanced informatics for computing research. Springer, Singapore
10. Krusienski DJ, Sellers EW, Cabestaing F, Bayoudh S, McFarland DJ, Vaughan TM, Wolpaw JR (2016) A comparison of classification techniques for the P300 Speller. J Neural Eng
11. Muller-Putz GR, Scherer R, Brauneis C, Pfurtscheller G (2015) Steady-state visual evoked potential (SSVEP)-based communication: impact of harmonic frequency components. J Neural Eng
12. Ahlawat V, Narayan Y, Kumar D (2021) DWT-based hand movement identification of EMG signals using SVM. In: Proceedings of international conference on communication and artificial intelligence. Springer, Singapore
13. Narayan Y (2021) Comparative analysis of SVM and Naive Bayes classifier for the SEMG signal classification. Materials Today: Proceedings 37:3241–3245
14. Narayan Y (2021) Direct comparison of SVM and LR classifier for SEMG signal classification using TFD features. Materials Today: Proceedings 45:3543–3546
15. Harwing WS, Rahman T, Foulds RA (2015) A review of design issues in rehabilitation robotics with reference to North America research. IEEE Transactions on Rehabilitation Engineering
16. Narayan Y, Mathew L, Chatterji S (2017) sEMG signal classification using discrete wavelet transform and decision tree classifier. Int J Control Theory Appl 10(6):511–517

17. Narayan Y (2021) SEMG signal classification using KNN classifier with FD and TFD features. Materials Today: Proceedings 37:3219–3225
18. Ahlawat V, Thakur R, Narayan Y (2018) Support vector machine based classification improvement for EMG signals using principal component analysis. J Eng Appl Sci 13:6341–6345
19. Sarasola-Sanz A, Irastorza-Landa N (2017) A hybrid brain-machine interface based on EEG and EMG activity for the motor rehabilitation of stroke patients. In: 2017 international conference on rehabilitation robotics (ICORR), pp 17–20
20. Zhuojun X, Yantao T, Yang L (2015) sEMG pattern recognition of muscle force of upper arm for intelligent bionic limb control. J Bionic Eng 12:316–323
21. Shafivulla M (2016) sEMG based human machine interaction for robotic wheelchair using ANN. In: International conference on computational modeling and simulation, pp 949–953
22. Gao X, Xu D, Cheng M, Gao S (2013) A BCI-based environmental controller for the motion-disabled. IEEE Trans Neural Sys Rehab Eng
23. Bakardjian H, Tanaka T, Cichocki A (2010) Optimization of SSVEP brain responses with application to eight-command brain–computer interface. Neurosci Lett 469:34–38
24. Bin G, Gao X, Yan Z, Hong B, Gao S (2019) An online multi-channel SSVEP-based brain–computer interface using a canonical correlation analysis method. Journal of Neural Engineering
25. Bermúdez i Badia S, Morgade AG, Samaha H, Verschure PFMJ (2013) Using a hybrid brain computer interface and virtual reality system to monitor and promote cortical reorganization through motor activity and motor imagery training. IEEE Transactions on Neural Systems and Rehabilitation Engineering 21(2)
26. León M, Gutiérrez JM, Leija L, Muñoz R (2011) EMG pattern recognition using support vector machines classifier for myoelectric control purposes. In: Proceedings of the IEEE International Conference on Health Care Exchanges (PAHCE), Rio De Janeiro, Brazil, March 28–April 1, pp 175–178
27. Freed A, Chan ADC, Lemaire ED, Parush A (2011) Wearable EMG analysis for rehabilitation (WEAR): surface electromyography in clinical gait analysis. In: Proceedings of the IEEE international conference on medical measurements and applications proceedings (MeMeA), pp 601–604
28. Kakoty NM, Hazarika SM (2011) Recognition of grasp types through principal components of DWT based emg features. In: Proceedings of the IEEE international conference on rehabilitation robotics rehab week Zurich, ETH Zurich Science City, Switzerland, June 29–July 1, 2011, pp 1–6
29. Luo S, Xia H, Gao Y, Jin JS (2018) EEG based brain computer interface for controlling a robot arm movement through thought. Elsevier Masson SAS, pp 1–7
30. Mathew L, Chatterji S (2018) SEMG signal classification with novel feature extraction using different machine learning approaches. IFS 35:5099–5109
31. Ghaemia A, Rashedia E, Pourrahimib AM, Kamandara M, Rahdaric F (2017) Automatic channel selection in EEG signals for classification of left or right hand movement in brain computer interfaces using improved binary gravitation search algorithm. Biomed Signal Process Control 33:109–118
32. Elamvazuthia I, Zulkiflia Z, Alia Z, Balajid M, Chandrasekarane M (2015) Development of electromyography signal signature for forearm muscle. IEEE International Symposium on Robotics and Intelligent Sensors 76:229–234
33. Raza H, Prasad G (2017) EEG-EMG base hybrid brain computer interface for triggering hand exoskeleton for neuro-rehabilitation, pp 377–378
34. Chaudhary S, Taran S, Bajaj V, Siuly S (2020) A flexible analytic wavelet transform based approach for motor-imagery tasks classification in BCI applications. Computer Methods and Programs in Biomedicine 15–45
35. Sfax, Ecole Nationale d'Ing´enieurs de Sfax (ENIS), Computer and Embedded System (2019) A survey on different human-machine interactions used for controlling an electric wheelchair. In: 23rd international conference on knowledge-based and intelligent information and engineering systems, pp 398–407

36. Liu H, Tao J, Lyu P (2019) Human-robot cooperative control based on sEMG for the upper limb exoskeleton robot. Robotics and Autonomous Systems 125:103350
37. Chowdhury A, Raza H, Meena YK (2018) An EEG-EMG correlation-based brain-computer interface for hand orthosis supported neuro-rehabilitation. Journal of Neuroscience Methods
38. Song Y, Du Y, Wu X, Chen X, Xie P (2014) A synchronous and multi-domain feature extraction method of EEG and sEMG in power-assist rehabilitation robot. In: IEEE International Conference on Robotics and Automation (ICRA)
39. Mane SM, Kambli RA, Kazi FS, Singh NM (2015) Hand motion recognition from single channel surface EMG using wavelet & artificial neural network. In: 4th international conference on advances in computing, communication and control, pp 58–65

Supervised Learning Techniques for Sentiment Analysis

Nonita Sharma, Monika Mangla, and Sachi Nandan Mohanty

Abstract Data mining implies the application of techniques of obtaining useful knowledge from a huge data. Another term for data mining is knowledge discovery from data. For the same, various data mining technologies are available such as statistics (lay the foundation of data mining), artificial intelligence (applying human thoughts like processing of data), and machine learning (union of statistics and artificial intelligence). In this research work, authors employ natural language processing in order to perform sentiment analysis using various feature extraction techniques of NLP. Sentiment analysis is especially important to gain users' feedback and opinion about products. In this paper, authors perform sentiment analysis of twitter data. Each data point (tweets in considered case) will be classified as "positive tweet" or "negative tweet". For this classification, six different techniques, i.e., information gain, Gini index (GI), Naive Bayes, K-nearest neighbor, random forest, and gradient boost are used. In the end, classification through all these techniques are analyzed and a comparative analysis is made based upon accuracy, precision, recall, and F1-score. Experimental results suggest that random forest aces the current analysis by yielding an accuracy of 97%.

Keywords Sentiment analysis · Classification · Data mining · Naïve Bayes · Random forest · Gradient boost

---

N. Sharma
Department of Information Technology, Indira Gandhi Delhi Technical University for Women, Delhi, India
e-mail: nonitasharma@igdtuw.ac.in

M. Mangla (✉)
Department of Information Technology, Dwarkadas J Sanghvi College of Engineering, Mumbai, India
e-mail: monika.mangla@djsce.ac.in

S. N. Mohanty
School of Computer Science & Engineering, VIT-AP University, Amaravati, India
e-mail: sachinandan09@gmail.com

P. Dutta et al. (eds.), *Emerging Technologies in Data Mining and Information Security*, Lecture Notes in Networks and Systems 490,
https://doi.org/10.1007/978-981-19-4052-1_43

1 Introduction

Sentiment analysis (SA) is an important tool to expand business for big business giants like Flipkart, amazon, etc. The advantage of SA cannot be underestimated even for small companies to understand and analyze about their product performance through user reviews and comment [1]. As user reviews and comments data is gigantic, it cannot be processed manually. Hence, leading tech-giants rely on sentiment analysis for processing huge user data. Through SA, user comments are classified as positive, negative, and neutral comments. The underlying principle of SA is that if the number of positive comments exceeds negative comments, then product is successful otherwise not. So, sentiment analysis is highly used in real life [2].

Classification is mainly used for extracting hidden information to find meaningful data for manual analysis and making data visualization easier [3]. It is applied in various fields as it makes data visualization more effective. Consequently, authors in this manuscript devise a model for sentiment analysis that processes Twitter dataset and analyzes it based on different classification algorithms such as random forest, Naïve Bayes, decision tree, random forest, K-nearest neighbor (KNN), and support vector machine (SVM) to process the voluminous database. The model also employs logistic regression which gives the information according to the probabilistic approach and assumes linearity among dependent and independent variables. Naïve Bayes is derived from Bayes' theorem [4] and works on the assumption that features are independent. KNN is a lazy learning method and returns the majority classification in k closest points. However, it is an easy and quick method, but its main challenge is to decide the value of k [5]. Decision tree is a greedy algorithm and works by selecting the best available feature. There is no need of feature scaling, and it usually causes overfitting. Random forest makes many decision trees and select the answer with maximum frequency [6]. It is less prone to overfitting, outputs the important features but the computations can sometimes become complex. SVM uses a plane to separate the features. The main advantage of SVM is prevention of overfitting but it is not a good choice when numerous values are involved [7].

The manuscript is divided into the following sections. Section 2 presents various classification algorithms employed as materials and methods for experimentation purpose. Section 3 discusses the methodology and steps performed in experimentation. Section 4 presents the results and discussion. Section 5 presents the conclusion and future scope of the research work.

2 Materials and Methods

The methods followed for classification of tweets are described as under:

2.1 Decision Tree Based on Information Gain

The information gain can be quantified by subtraction of weighted entropies of each branch from the original entropy. By using these metrics to train a decision tree, the best split is determined by optimizing information gain. It examines all potential consequences of a decision and follows each direction to a conclusion [8]. Further, it generates a detailed overview of the implications along each branch and determines decision nodes that need further investigation. However, it is unstable as a small change in the data may lead to a significant (exponential) change in the configuration of the optimal decision tree.

2.2 Decision Tree Using Gini Index

The Gini index (GI) is a metric that determines the probability of incorrect detection of a randomly selected variable. It is calculated using (1)

$$GI = \sum_{j} P_j^2 \tag{1}$$

It is straightforward to understand as it follows a constant method whereas creating any call-in real life. It is very helpful to make decisions [9]. There is less demand for data cleansing compared to other algorithms. However, the decision tree contains legion layers that make it advanced. It may have an associated overfitting issue, which could be resolved by exploiting the random forest formula. For loads of class labels, the method quality of the selection tree might increase.

2.3 Naïve Bayes

Naïve Bayes classifier is derived from Bayes' theorem which assumes that features are independent [10]. Main goal of Bayesian classification is to determine the likelihood of an event/attribute provided certain observed characteristics, *P*(*L*|features). We can represent Bayes' theorem using (2)

$$P(c|x) = \frac{P(x|C)p(C)}{p(x)} \quad (2)$$

Naïve Bayes makes the computations easier and for big datasets it demonstrates better speed and accuracy. However, in certain instances, when there is a dependence between variables, it does not provide consistent results.

2.4 K-nearest Neighbor (KNN)

An instance-based learning method (lazy learner) which means that it requires minimal to none training phase. It returns the majority classification in k closest points [11]. Its implementation is trouble free, and there is no need to worry about information loss. The disadvantage of KNN is huge requirement of memory. Another disadvantage is that it may be easily fooled by irrelevant attributes.

2.5 Random Forest

In this algorithm, combination of various trees is responsible for result. Consider a scenario where there is a dataset containing several fruit images. As a result, the random forest classifier is given this dataset. During the training process, each decision tree generates a prediction result, and when a new data point appears, the random forest classifier forecasts the final decision-based result's weight [12]. It is vigorous against outliers and performs well on large datasets. In terms of accuracy, it is better than alternative classification algorithms. However, random forests are considered to be biased while handling categorical variables. Moreover, it is not appropriate for linear methods with a lot of sparse features.

2.6 Gradient Boost

Gradient Boosting have mainly three parameters that are to be considered, i.e., depth, number of trees and learning rate, and each tree build is normally shallow [13]. The user defines the size of each tree, resulting in a set of stumps that can be interpreted as an additive model, making it more interpretable than bagged trees/random forest. Categorical features are to be hold easily. However, if the number of trees is too high, it can overfit, unlike bagging and random forests.

3 Experimentation

In this, authors illustrate the performance of experiments. Authors extract the features using bag-of-words (bow) and term frequency inverse document frequency (TFIDF). Thereafter, using machine learning models, it is predicted whether the tweet is positive and negative. The various steps performed are given in the pseudocode given below:

Input: Tweets Dataset

1. Import Dataset
2. Preprocessing the Data

a. Removing user-handles and Normalization
b. Removing Urls, numbers and Hashtags
c. Tokenization
d. Removing stopwords
e. Stemming and lemmatization

3. Feature Extraction using BoW and TFIDF
4. Split the dataset into Training and validation set
5. Performance Evaluation

The various steps are explained in detail below:

3.1 Data Collection

The dataset employed in this research work is a sentiment with 140 datasets. It has 1.6 million tweets taken from API of twitter out of which authors have taken 20,000 tweets. It contains the following 6 columns:

- tweet's sign (0 = 'negative', 4 = 'positive')
- tweet's identification which is unique (2086)
- tweet's time and date (for example Sat May 16 23:58:44 UTC 2009)
- value is 'NO-QUERY', if no query is there
- tweet's user (Joy Wolf)
- tweet's text (he is good).

Important attributes for our project are target and text. Text works as input and target works as output.

3.2 Data Preprocessing

Following are few steps that are performed during data preprocessing:

- User handles are not important while analyzing the data. So it is better to remove them from data. Data normalization is also very important. Normalization consists of two parts—one is converting the data into lower case and other is removing punctuations from data.
- Urls, numbers, and hashtags also do not contribute into sentiment analysis, so these are also removed from data.
- Tokenization means splitting the data into tokens. Here, tokens mean words. Data can also be split into sentences.
- Removing stop words, the words which are too common and does not add to sentiment analysis.
- Stemming and lemmatization are used to convert the words into dictionary order

3.3 Feature Extraction

It is important to extract the features from the cleaned and preprocessed text data. As discussed earlier, for extracting the features from data, two common approaches are used—bag-of-words and TFIDF [14]. Bag-of-words is a simple method in which text data is considered as collection of words where there is no importance of grammar and order of words. But the only thing which matters is the frequency of words. However, in case of TFIDF, with the increase of frequency of word in document, TFIDF also increases. It is evaluated using (3)

$$ifidf_{(i,j)} = tf_{(i,j)} \times \log \frac{N}{df_i} \tag{3}$$

where i is the term in document j. $tf_{(i,j)}$ is the number of occurrences of i in j. df_i is the number of documents containing i. N is the total number of documents.

3.4 Splitting the Dataset

By using the features from bag-of-words for training set, dataset is split into training and validation sets.

4 Results and Discussion

After the experimental evaluation, various classification models are applied for evaluation purpose. Classifiers are compared on the basis of accuracy, precision, recall, F1-score, and support as evaluated by the following equations. There are two classes of tweets: positive represented by class 0 and negative represented by class 4. Further, authors present the metrics obtained through various classification algorithms.

4.1 Naïve Bayes

Initially, Naïve Bayes is applied to features obtained from BoW and TFIDF. It achieves an accuracy of 67% and 68% for BoW and TFIDF, respectively. Comparison of accuracy on BoW and TFIDF is done in Fig. 1. The comparison of precision, recall, F1-score for both classes 0 and 4 is given in Table 1.

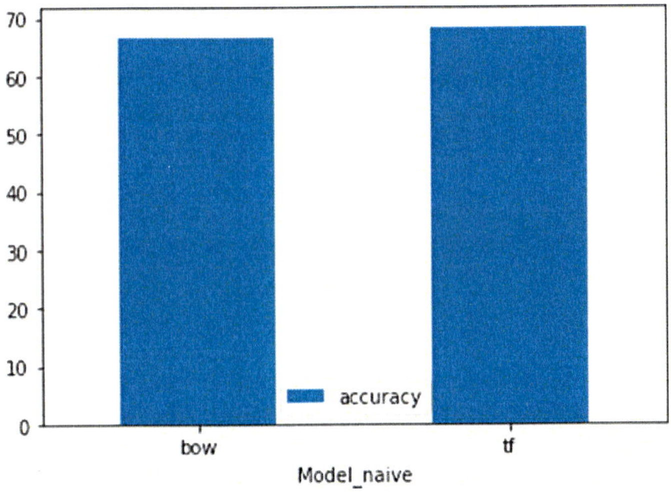

Fig. 1 Accuracy comparison of Naïve Bayes on BoW and TFIDF

Table 1 Performance comparison of Naïve Bayes

Classifier	Class	Precision	Recall	F1-score	Support
Naïve Bayes (BoW)	0	0.73	0.58	0.64	3005
Naïve Bayes (BoW)	4	0.62	0.77	0.69	2736
Naïve Bayes (TFIDF)	0	0.73	0.63	0.67	3005
Naive Bayes (TFIDF)	4	0.65	0.74	0.69	2736

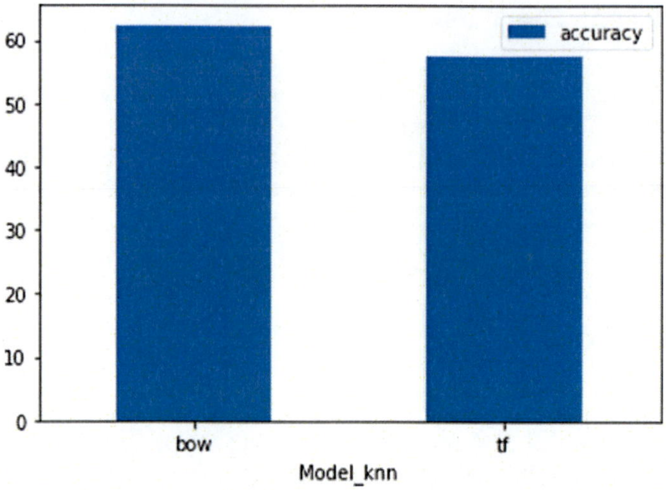

Fig. 2 Accuracy comparison of KNN on BoW and TFIDF

Table 2 Performance comparison of KNN

Classifier	Class	Precision	Recall	F1-Score	Support
KNN(BoW)	0	0.62	0.73	0.67	3005
KNN(BoW)	4	0.63	0.51	0.56	2736
KNN(TFIDF)	0	0.58	0.69	0.63	3005
KNN(TFIDF)	4	0.57	0.45	0.50	2736

4.2 KNN

KNN achieves the accuracy of 62% and 57% for BoW and TFIDF, respectively. Comparison of accuracy on BoW and TFIDF is done in Fig. 2. The comparison of precision, recall, F1-score for both classes 0 and 4 is given in Table 2.

4.3 Random Forest

Random forest achieves the accuracy of 97% for BoW and 62% for TFIDF. Comparison of accuracy on BoW and TFIDF is demonstrated Fig. 3. The comparison of precision, recall, F1-score for both classes 0 and 4 is given in Table 3.

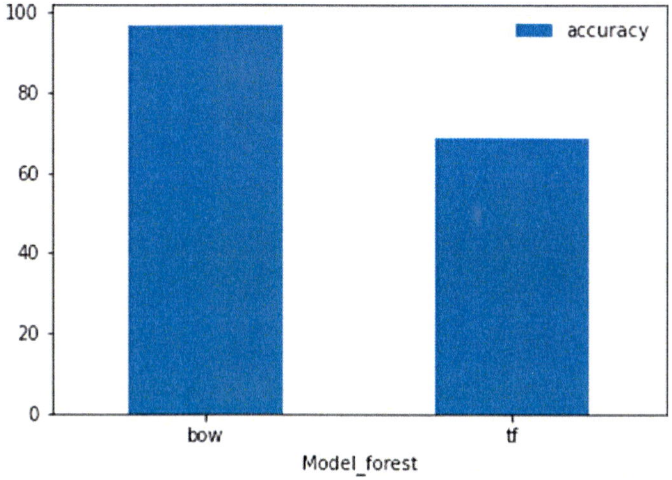

Fig. 3 Accuracy comparison of random forest on BoW and TFIDF

Table 3 Performance comparison of random forest

Classifier	Class	Precision	Recall	F1-score	Support
RF(BoW)	0	0.99	0.95	0.97	3005
RF(BoW)	4	0.95	0.99	0.97	2736
RF(TFIDF)	0	0.79	0.55	0.65	3005
RF(TFIDF)	4	0.63	0.84	0.72	2736

4.4 Decision Tree

Decision tree is implemented using information gain and GI. Decision tree using information gain achieves the accuracy of 73% for BoW and 55% for TFIDF. Decision tree using information gain achieves the accuracy of 73% for BoW and 67% for TFIDF.

Decision trees obtained from information gain and GI are given in Fig. 4 and Fig. 5, respectively. The comparison of precision, recall, F1-score for both classes 0 and 4 is given in Table 4.

4.5 Gradient Boost

Gradient boost (GB) achieves the accuracy of 69% for BoW and 68% for TFIDF. Comparison of accuracy on BoW and TFIDF is done in Fig. 6. The comparison of precision, recall, F1-score for both classes 0 and 4 is given in Table 5.

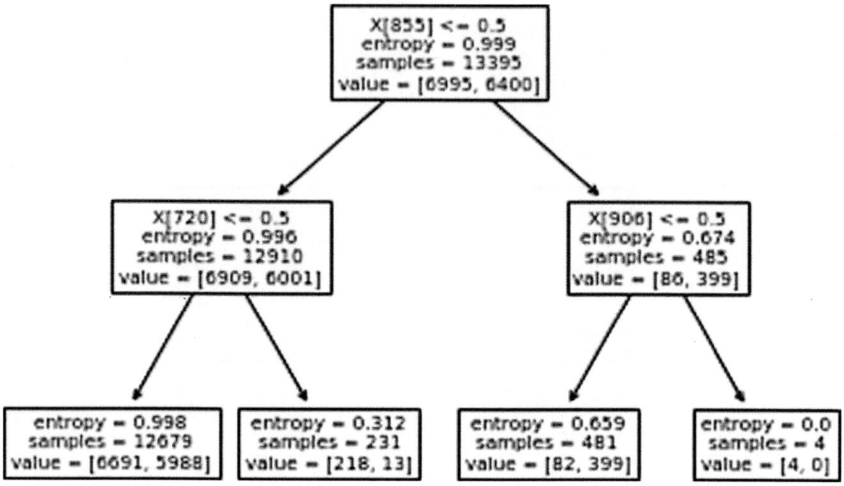

Fig. 4 Decision tree obtained from information gain

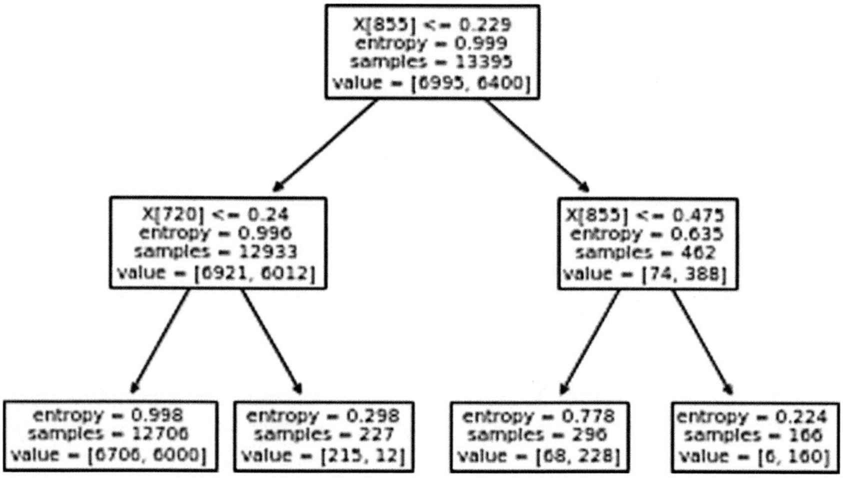

Fig. 5 Decision tree obtained from Gini index

4.6 Comparative Analysis of Classifier

A comparative analysis of all the classifier is demonstrated in Fig. 7. The comparison of accuracy is drawn on both BoW and TFID. It can be seen that random forest performs well in both the scenarios. This is a majority voting ensemble algorithm based on decision tree and exhibit best accuracy for our dataset. However, decision trees based on information gain and GI displayed the worst performance.

Table 4 Performance comparison of decision tree

Classifier	Class	Precision	Recall	F1-score	Support
IG(BoW)	0	0.53	1.00	0.69	3005
IG(BoW)	4	0.94	0.02	0.04	2736
IG(TFIDF)	0	0.54	0.99	0.69	3005
IG(TFIDF)	4	0.81	0.06	0.12	2736
GI(BoW)	0	0.52	1.00	0.69	3005
GI(BoW)	4	0.94	0.02	0.04	2736
GI(TFIDF)	0	0.54	0.99	0.69	3005
GI(TFIDF)	4	0.81	0.06	0.12	2736

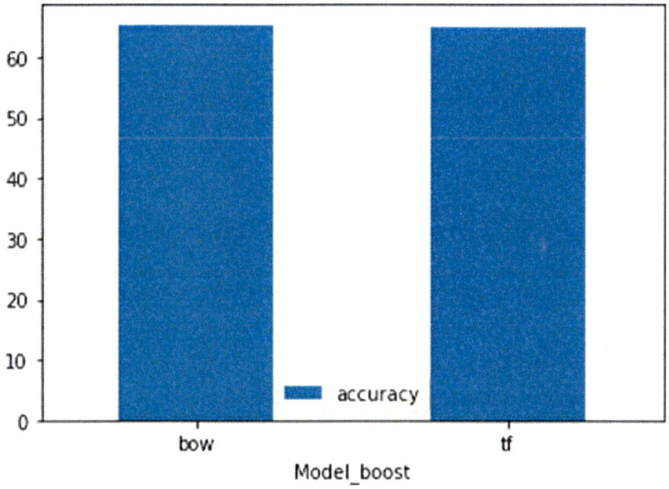

Fig. 6 Accuracy comparison of gradient boost on BoW and TFIDF

Table 5 Performance comparison of gradient boost

Classifier	Class	Precision	Recall	F1-score	Support
GB(BoW)	0	0.62	0.88	0.73	3005
GB(BoW)	4	0.75	0.41	0.53	2736
GB(TFIDF)	0	0.62	0.89	0.73	3005
GB(TFIDF)	4	0.76	0.39	0.52	2736

5 Conclusion and Future Scope

It can be concluded that random forest methods have the highest accuracy in comparison with other classifier methods for sentiment analysis on tweets, because it is an

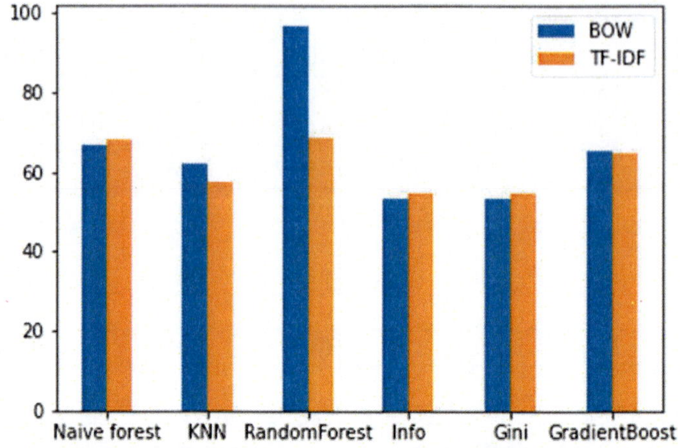

Fig. 7 Comparative analysis of all classifiers for sentiment analysis

ensemble technique, therefore, rather than depending on one classifier it computes the result by considering the results of multiple classifiers. Thus, a sentiment analysis model driven by the random forest method. The current work can be further extended in the direction of exploring new techniques giving better accuracy metrics than our current model. One such technique that can be applied is to use feature elimination, which may yield better accuracy. Availability of a better dataset than our current dataset may also contribute toward enhancing efficiency of the current model.

References

1. Sharma N, Sikka G (2020) Multimodal sentiment analysis of social media data: a review. In: The international conference on recent innovations in computing, pp 545–561. March 2020, Springer, Singapore
2. Singh B, Kumar P, Sharma N, Sharma KP (2020) Sales forecast for amazon sales with time series modeling. In: 2020 first international conference on power, control and computing technologies (ICPC2T), pp 38–43, January 2020, IEEE
3. Mangla M, Sharma N, Mohanty SN (2021) A sequential ensemble model for software fault prediction. Innovations in Systems and Software Engineering 1–8
4. Aggarwal CC, Zhai C (2012) A survey of text classification algorithms. Mining text data. Springer, Boston, pp 163–222
5. Deshwal A, Sharma SK (2016) Twitter sentiment analysis using various classification algorithms. In: 2016 5th international conference on reliability, infocom technologies and optimization (Trends and Future Directions) (ICRITO), pp 251–257, September 2016. IEEE
6. Xia R, Zong C, Li S (2011) Ensemble of feature sets and classification algorithms for sentiment classification. Inf Sci 181(6):1138–1152
7. Sharma N, Mangla M, Mohanty SN, Satpaty S (2021) A stochastic neighbor embedding approach for cancer prediction. In: 2021 international conference on emerging smart computing and informatics (ESCI), pp 599–603. March, 2021. IEEE

8. Go A, Bhayani R, Huang L (2009) Twitter sentiment classification using distant supervision. CS224N project report, Stanford, 1(12):2009
9. Pandarachalil R, Sendhilkumar S, Mahalakshmi GS (2015) Twitter sentiment analysis for large-scale data: an unsupervised approach. Cogn Comput 7(2):254–262
10. Ortega R, Fonseca A, Montoyo A (2013) SSA-UO: unsupervised Twitter sentiment analysis. In: Second joint conference on lexical and computational semantics (*SEM), vol 2, pp 501–507, June 2013
11. Dinsoreanu M, Bacu A (2014) Unsupervised twitter sentiment classification. In: KMIS, pp 220–227, October 2014
12. Chauhan P, Sharma N, Sikka G (2021) The emergence of social media data and sentiment analysis in election prediction. J Ambient Intell Humaniz Comput 12(2):2601–2627
13. Tang D, Wei F, Yang N, Zhou M, Liu T, Qin B (2014) Learning sentiment-specific word embedding for twitter sentiment classification. In: Proceedings of the 52nd annual meeting of the association for computational linguistics, vol 1: Long Papers, pp 1555–1565
14. da Silva NFF, Coletta LF, Hruschka ER, Hruschka ER (2016) Using unsupervised information to improve semi-supervised tweet sentiment classification. Inf Sci 355:348–365

A Future Perspectives on Fingerprint Liveness Detection Requirements and Understanding Real Threats in Open-Set Approaches to Presentation Attack Detection

Riya Chaudhary and Akhilesh Verma

Abstract In the last few years, researchers are reporting a gap in the requirements of fingerprint biometric systems and their implementation which is evident from increased personal identity theft. However, authentication using fingerprint biometrics is in wide use in a variety of domain applications. The fair uses of biometric systems are questionable even now in conventional uses, and frequent incidences are available which report security and privacy risk along with other problems. At the time of designing of biometric system, security and privacy are the main keys. For example, unauthorized person access in a biometric system is possible with spoof samples generated from a latent fingerprint image. In this paper, we discuss problems included in the biometric system, i.e., performance, security, bias, interpretability, and privacy. These problems help to access the real threat through a critical review of LivDet 2017, 2019, and 2021 competition. It is observed that the accuracy of the system is increased in the past decades but the error rate is still not zero, false-negative error detection is a challenging task, and it is difficult to handle in generalized cases. The various performance measures are showing that current solutions need a re-look to operate in an open-set environment. The literature work presented here shows current trends and practices that evolved during the last five years. This work proposes a simple architecture for fingerprint presentation attack detection using naturalness image quality evaluator (NIQE) model and perception-based image quality evaluator (PIQE) no-reference image quality score. To handle the generalized cases, this proposal propounds a person-specific liveness model for each individual which is independent of spoof sample training requirements.

Keywords Open-set approach · Close-set approach · Image quality evaluator

---

R. Chaudhary · A. Verma (✉)
Ajay Kumar Garg Engineering College, Ghaziabad, UP, India
e-mail: vermaakhilesh@akgec.ac.in

R. Chaudhary
e-mail: riya2010008m@akgec.ac.in

P. Dutta et al. (eds.), *Emerging Technologies in Data Mining and Information Security*, Lecture Notes in Networks and Systems 490,
https://doi.org/10.1007/978-981-19-4052-1_44

1 Introduction

Today, the use of biometric system has increased from 2 to 8 folds since 2010. Most of us carry smartphone for access to our bank account, email in an organization due ease of use. The biometrics system identifies a person automatically by using various biometric traits such as fingerprint, face, and iris for identification and verification. Even though dissemination of biometric systems is visible throughout the world, it is also reported that there is a feeling of mistrust among the users and authorities, for example—hacking of biometric databases, high price for high-security level, and lack of transparency towards privacy. Due to these concern, this article tries to articulate important findings in the last few years reported in various research journals. This work attempts to highlight the number of unanswered questions and doubts regarding various sub-modules of the biometric-based recognition system that are still a part of the research question in the research community. The element of skepticism comes into the picture while a user [scientific community + public] encounters it. These are (i) recognition performance, (ii) security issues (spoof, adversarial, template reconstruction, and leakage attack), (iii) bias and fairness (iv) interpretability of decisions, and (v) privacy issues with a centralized system [13].

Fingerprint-based identification systems do not ensure authentication due to spoof attacks. It is necessary to identify the authenticity of an individual while using fingerprints in biometric for identification. Authentic users or individuals must be present and give live samples while using the biometric system. It is reported that the biometric system is vulnerable to malicious attacks. One of the malicious attacks is a presentation attack where a copy of a biometric sample is present on a biometric sensor to get access as they make identification without error. It is a necessity to remove this attack because it affects acceptance on the biometric system. A presentation attack is also known as a spoof attack. The success rate of these attacks is over 70%, so it is very important to identify these attacks [17].

The researcher highlighted the method of generating spoof samples or duplicate samples; it includes copying original specimens from the user directly or from a latent sample on a surface as a negative impression. In the literature, capturing a negative impression on a mold is also named as cooperative and non-cooperative methods. The basic molding and casting method creates cooperative spoofing. We also know it as consensual spoofing, in this spoofing method live fingers create the finger's plastic mold, then fill it with material such as gelatin, play-doh, silicone, wood glue to create a spoof fingerprint. Here, personal participation is necessary. In the non-cooperative method, spoofs are created without personal participation. These are divided into four parts—latent fingerprint, fingerprint reactivation, cadaver, and synthesis [16]. Nowadays, the screen spoof method is widely used for spoof fabrication because these are easy to use and take less time and effort [6]. State-of-the-art also mentions that the varieties in the spoof sample are endless, so a model trained earlier for the old sample fails to work on a new sample. To overcome endless varieties in spoof samples, this work tries to eliminate the need for spoof samples during training.

2 Liveness Detection Approaches and Their State-of-the-Arts

There are two techniques used for liveness detection: Hardware-based and software-based techniques [16]. In the hardware-based approach, different devices are added to the sensor to detect the characteristics of individuals such as blood pressure, distortion, and odor. Because of the additional sensors, its cost is very high. Hardware-based require additional sensor due to which its cost is high. This approach is only used for the known samples it can't use for unknown samples. These extra sensor detect the characteristics of vitality, such as temperature [22], blood pressure [3], odor [2], and so on. In [23], poor generalization is a major challenge in fingerprint spoof detection. The material used for spoof fabrication is not seen in the training phase cause the generalization problem, which arises the open-set problem discussed above in detail. With the help of the above explanations, it is clear that in the open-set approach, all classes are not known during the training.

In a software-based approach, researcher distinguish live or fake fingerprint images with no additional device. It is not expensive because the extra device is not needed and works on two types of features; static features like sweat pores, valley features, perspiration, and ridges, etc., and dynamic features like skin color change due to pressure, skin elastic properties etc. It is easy to overcome presentation attack problem with software-based approach. The software-based approach deals with known as well as unknown attacks and is cheaper than the hardware-based approach. They do not require any extra sensor device. The software-based approach utilizes one of the following approaches: (i) anatomical features (e.g., sweat pores [8] and perspiration [22]), (ii) texture-based features in this method textural features are used to recognize real or spoof fingerprints some techniques are used, i.e., local phase quantization (LPQ) [10], binarized statistical image features (BSIF) [9], and Weber local descriptor [28]. Menotti et al. [18] and Nogueira et al. [20] proposed a convolutional neural network (CNN) which gives better results to detect the live or spoof fingerprints.

Another way to classify liveness detection methods is based on close-set and open-set approaches. Both approaches are listed below in detail.

2.1 Closed-Set Approach a Brief Overview

In the closed-set approach, complete knowledge of spoof samples is a prerequisite, also we give only available material/sample knowledge during the training and testing phases. For cross material, it does not provide excellent results. Closed set does not show good reliability when tested for novel spoof materials. It requires a large training dataset of live and fake fingerprints on which the model is examined. With the help of known classes, we get both training and testing samples. It works on a limited dataset so this model is weak in detecting unknown samples. Poor generalization and over-fitting are the key challenges in fingerprint spoof detection using the close-set approach. Many researchers work on it using various approaches, but they cannot show the interoperability in their solutions where closed-set strategies are used. Closed set is a problem of recognition in which the training and testing datasets contain the same type of spoof samples.

The closed-set approach to solving a problem is not reliable as compared to the open set. The accuracy rate is very low of the closed-set problem in terms of the true detection rate (TDR) parameter. In the work [25], the authors used a patch-based voting approach for fingerprint liveness detection. The authors used the convolution neural network model for differentiating live and spoof fingerprint images in a patch-based system. This approach comprises 8000 live and 8000 fake fingerprint images for the training set, 8000 live and 8000 fake fingerprint images for the testing set. This model gives the best performance as compared to other previous models.

In a work by [15], the researchers proposed a uniform local binary pattern (ULBP) technique. The ULBP technique is a software-based liveness detection technique used for discriminating live and fake images. This approach comprises 4471 live and 3979 fake fingerprint images for the training set, 4403 live and 4000 fake fingerprint images for the testing set. The ULBP model achieved 21.20% ACE.

In the paper [1], the researchers introduced artificial intelligence (AI) to learn a model for spoof fingerprint detection. It is an incremental learning model which takes the features of live and fake fingerprint images. This approach comprises 2000 live and 4000 fake fingerprint images for the training set, 2000 live and 2000 fake fingerprint images for the testing set from LivDet 2011, 2013, and 2015 and which achieved an accuracy of 49.57%. In the paper [4], the authors used Slim-Res CNN technique. It is a lightweight yet powerful network structure that comprises stack series of improved residual blocks used for the fingerprint presentation attack detection. This approach comprises 4000 live and 4000 fake fingerprint images for the training set, 4000 live and 4000 fake fingerprint images from LivDet 2013 and LivDet 2015 dataset and achieved an overall accuracy of 96.82%. In a similar work [14], the authors proposed the densely connected convolutional network (DenseNet) technique for fingerprint liveness detection. This work compares the performance using average classification error (ACE) value and variance. This approach comprises 38,910 fingerprint images or training sets and 40,340 fingerprint images for the testing sets. The accuracy achieved by DenseNet is 98.22% (Table 1).

2.2 *Open-Set Approach a Brief Overview*

In the open-set approach, incomplete or no knowledge of spoof material is provided during the training and testing phases [27]. For cross material, it provides excellent results. Open-set methods show good reliability when tested for novel spoof material. This approach is effective even with fewer datasets of live and spoof fingerprints on which model has been examined. Training and testing samples come from both known and unknown classes. It comprises multiple known classes and many

Table 1 Performance comparison of various close-set approach in terms of average classification error (ACE)

Approach	Datasets	Sensors							
		Biometrica	Italdata	Crossmatch	Digital persona	swipe	Sagmen	GreenBit	**ACE**
CNN patch-based voting strategy [25]	LivDet 2011	4.0	6.3	NA	4.5	NA	3.7	NA	4.625
	LivDet 2013	0.4	0.5	5.4	NA	1.3	NA	NA	1.9
Uniform local binary pattern [15]	LivDet 2013	10.68	13.7	46.09	NA	14.35	NA	NA	21.20
AI learn model to implement spoof detection material [1]	LivDet 2011 LivDet 2013 LivDet 2015	NA	NA	NA	NA	NA	NA	NA	49.57% Accuracy
Slim-ResCNN [4]	LivDet 2013	0.47	5.21	NA	NA	NA	NA	NA	2.84
	LivDet 2015	2.78	NA	3.03	4.48	NA	NA	2.14	3.10
DenseNet, Genetic Net [4]	LivDet 2009	8.91	2.67	2.98	NA	NA	NA	NA	4.85
	LivDet 2011	7.70	12.25	5.95	NA	5.12	NA	NA	7.76
	LivDet 2013	2.70	2.00	9.18	NA	NA	10.47	NA	6.16
	LivDet 2015	6.15	NA	4.71	7.20	NA	NA	2.32	5.09

unknown classes, where unknown classes are submitted during the testing. Open set contains the different spoof samples during the training and testing phases. It reduces the chances of dataset bias because of less knowledge of data. A highly unbalanced training dataset or incomplete class knowledge in the open-set approach sometimes causes overspecialization. The open-set approach is further divided into two categories: one-class and two-class open-set problems. In a one-class problem, only live samples are addressed in the training set and it rejects the fake fingerprint. In two-class problems, the researcher used both live and spoof samples for the training set. The work done under these categories discussed in [12].

The open class problem is more authentic for variety in spoof material as the true detection rate (TDR) is very high. The researchers [24] deployed the static software-based technique used ridge and valley features of fingerprint. These local quality features used for detecting the liveness in fingerprints. This approach comprises 4000 live and 4000 fake fingerprint images for the training set, 4473 live and 4000 fake fingerprint images for the testing set. The module used the publicly available dataset LivDet 2019 and achieved 5.3% average classification error. In another work by [29], the authors used the fusion of fingerprint matching and liveness detection module. Logistic regression is applied to the feature vectors to differentiate live and spoof samples. This approach consists of 2000 live and 2000 fake fingerprint images for the training set, 2000 live and 2000 fake fingerprint images for testing sets using LivDet 2013 and LivDet 2015 and achieved overall accuracy of 96.88%. The authors of [7] used a DCNN-based approach that used minutiae points for differentiating live and spoof samples. This approach consists of 10,510 live and 10,500 fake fingerprint images for the training set, 10,473 live and 11,948 fake fingerprint images for the testing set. This approach achieved reduction in error rates for cross material (78%), cross sensor (17%) and cross dataset (38%). In the work [26], the researchers proposed discriminative restricted Boltzmann machines (DRBM) to recognize the live or spoof images. It deals with the complex texture of fingerprints. This approach consists 2000 live and 2000 fake fingerprint images used for the training set, 2000 live and 2000 fake fingerprint images used for the testing set. The model used publically available datasets LivDet 2013 and LivDet 2015. The model achieved ACE of 3.6%. In the paper [23], the researchers implement Weibull-calibrated SVM technique which is used to provide the security of open-set recognition for novel set and spoof material. This approach consists of 2000 live and 2000 fake fingerprint images for the training set, 2000 live and 2000 fake fingerprint images for the testing set. For detecting spoofs in fingerprint, they achieve 44% accuracy (Table 2).

3 Performance Evaluation and Fingerprint Datasets

The performance of fingerprint liveness or spoof detection methods is evaluated under the following scenario in the PAD system (Table 3).

Table 2 Performance comparison of open-set approach report on LivDet 2011,LivDet 2013, LivDet 2015, and LivDet 2019 with respect to average classification error (ACE), accuracy, and error rate parameter

Approach	Dataset	ACE	Accuracy	Error rate
Sharma and Dey [24] Local quality patches	LivDet 2015	9.45	NA	NA
(YONGLIANG [29] A score-level fusion of fingerprint matching	LivDet 2019	NA	96.88%	NA
Inutiae-based local patches [7]	LivDet 2011 LivDet 2013 LivDet 2015	2.6 0.5 1.4	Cross material—3.5% Cross dataset—2.59% Cross sensor—16.6%	NA
Jung et al. [26] DRBM/DBM model	LivDet 2015 LivDet 2013	NA	Cross datasets:18.9%	NA
Ajita [23] Weibull-calibrated SVM	LivDet 2011	NA	44%	NA

1. Cross-dataset evaluation: Fingerprint presentation attack detection model is trained or tested on different databases. For example, the PAD model was trained on LivDet 2009 and tested from different datasets LivDet 2011 or 2015.
2. Cross-sensor evaluation: Fingerprint PAD model is trained and tested on different sensors. For example, a model trained on a dataset taken from Biometrica sensor and tested on data taken from Italdata sensor.
3. Cross-material evaluation: Fingerprint PAD model trained and tested on datasets made by various materials. For example, the PAD model was trained on a dataset taken from silicone and tested on a dataset taken from gelatin or play-doh.

To implement and evaluate the above cases, research community uses well-known publicly available fingerprints for anti-spoofing. Some of these datasets are—NIST, ATVS-FFp, and LivDet. The most widely used database is from the liveness detection competition known as LivDet. The competition had various forms like LivDet 2009, LivDet 2011, LivDet 2015, LivDet 2017, LivDet 2019, and LivDet 2021. These databases contain a large amount of live and spoof fingerprint samples used for fingerprint spoof detection. The glimpse of LivDet 2009–2015 is shown [11] which mention its various attributes like sensors, model, ACE score, etc. The LivDet competition 2017 onward is recently been incorporated for the research community, and in Table 3, LivDet 2021, 2019, and 2017 show performance comparison of LivDet dataset 2021 [5], 2019 [21], and 2017 [19]. LivDet 2021 is based on cross-dataset evaluation, LivDet 2019 is based on cross-sensor evaluation, while LivDet 2015 is based on the cross-material evaluation.

Table 3 Detail performance comparison of LivDet dataset 2021, 2019, and 2017

Datasets	Algorithms	Accuracy			Avg. Accuracy (%)
LivDet 2017		GreenBit	Digital persona	Orcathuns	97.06
	PDFV	93.58	93.31	NA	
	LCPD	89.87	88.84	86.87	
	Hanulj	97.06	97.06	92.04	
	PB LivDet 2	92.86	90.43	92.60	
	ModuLAB	94.25	90.40	90.21	
	ZYL2	96.26	94.73	93.17	
	SSLFD	93.58	94.33	93.14	
LivDet 2019	JungCNN	98.31	88.56	97.99	99.73
	JLWs	98.85	94.00	97.80	
	Unina	95.54	87.62	92.32	
	LivDet2019-CNN	95.67	90.84	92.77	
	Unina	92.65	53.77	96.82	
	FSb	99.73	83.64	97.50	
	JLWa	99.20	88.81	97.45	
	JLWs	99.20	88.81	97.45	
	JungCNN	99.06	81.23	99.13	
LivDet 2021	LivDet21DanC1	91.16/98.64	99.09/88.59	NA	96.27
	LivDet21CanC1	89.83/91.55	98.90/79.9		
	Fingerprint anti-spoof	84.53/74.81	98.81/74.81		
	JLWLivDetD	96.13/80.57	98.81/66.35		
	JLWLivDetW	96.13/80.57	98.81/66.35		
	hallyMMC	85.77/89.94	96.19/89.94		
	Damavand	60.80/67.73	68.35/52.69		

4 Challenges Identified in Literature

The aforesaid discussion concludes a challenging problem with the biometric systems. Authentication using biometrics has less acceptance and no wide inclusiveness, as an effective presentation attack detection sub-system is missing. We see this shortcoming in end-user behavior, as we can see a reluctance to use biometrics for authentication—in financial transactions and access control for critical establishment. To accept any biometric system, a presentation attack detection system must work well when attacks are unknown. In addition, the biometric authentication market will grow only when the present model addresses the above research issues identified in the open-set approach to FPAD. Some challenges which are faced at the time of the literature survey are listed below:

1. The machine learning (ML) model suffers from the irrelevant use of handcrafted features, and it reduces the performance of presentation attack detection (PAD)
2. The fingerprint detection module may not work effectively in case of unknown material
3. The anti-spoofing techniques are not robust to other conditions—like varieties in sensors.
4. Also, the average classification error is still not zero and error control means are not generalized enough.

5 Proposed Futuristic Perspectives on Fingerprint Liveness Detection

In this work, the proposed architecture anticipates liveness by applying the fusion rule (Fig. 1). It carried a score-level fusion of judgments out on the decision made by the classifier. Here, high-quality live samples are utilized for the training liveness model, and for the testing, spoof samples are employed. This work indicates that fingerprint liveness information of an individual can be represented by understanding the variation in quality values in different live samples of the same person and that is innate. Therefore, each live specimen has 'liveness' information. In this plan, a no-reference (NR) image quality measure (IQM) is exploited for discriminating live and spoof samples.

This model uses IQM-based feature vector representation for training different one-class classifiers to predict adversarial input. Further, for improving accuracy, a classifier can be trained on feature vectors from augmented inputs for efficient outlier detection with a significant gain in accuracy. So, by using the fusion rule, a score-level fusion predicts negative cases (or outliers) of liveness. Three image quality assessment metrics—PIQE, NIQE, and BRISQUE are proposed to be applied on PIQE models activity mask, artifacts mask, and noise mask.

6 Conclusion and Future Scope

In summary, this work unfolds a novel approach for a near-ideal spoof detection system within the current scope of the problem, solving the FPAD problems with an open-set approach. In near future, applications will heavily rely on trustworthy biometric system, so the proposed architecture makes a way forward toward the standard FPAD systems design and a resolution on development issues. A proposal is put forward to model the liveness detection using the patches of artifacts in a natural scene, considering it as an urgent need for future application. The possible applications of this research direction envision a wide variety of applications, a new set of uses that may soon become a reality in the presence of a trustworthy FPAD system.

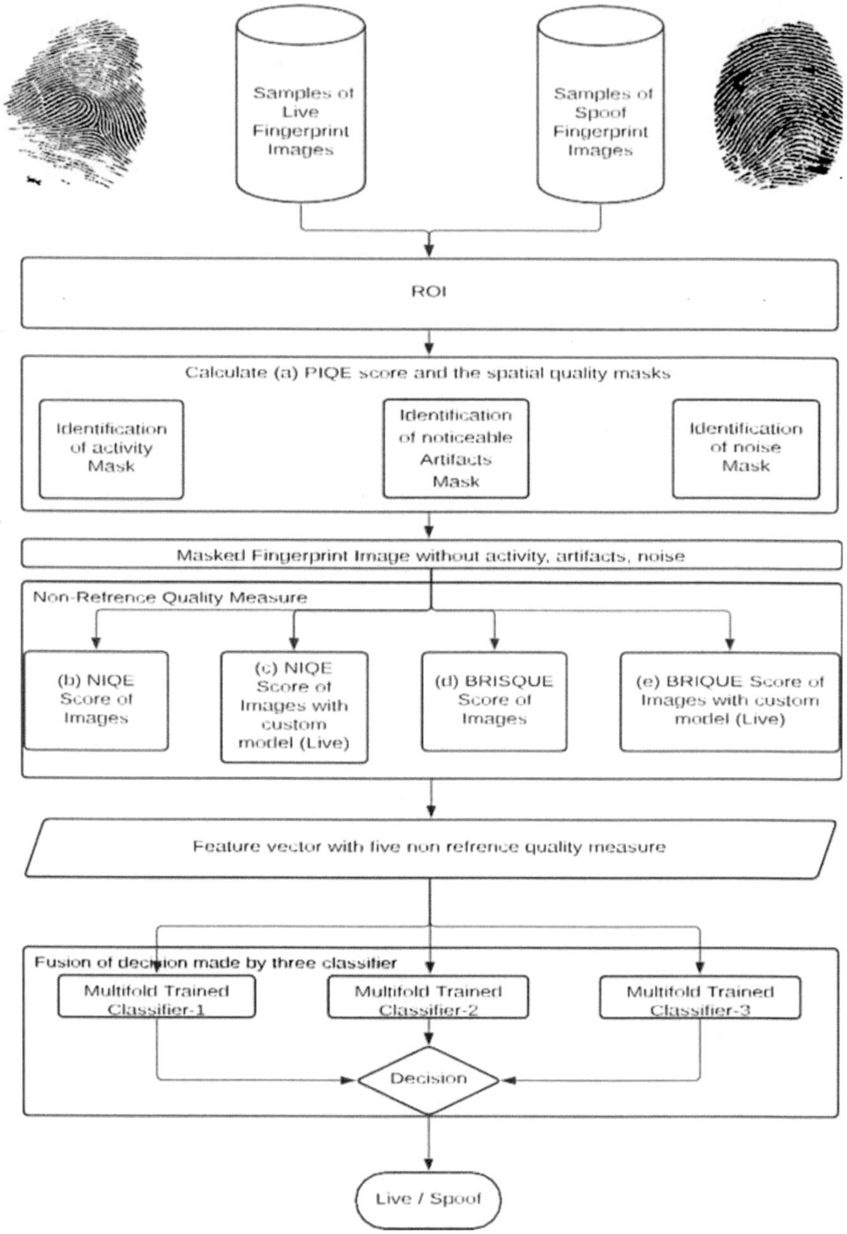

Fig. 1 Proposed futuristic perspectives of person-specific PAD model

References

1. Agarwal S, Rattani A, Chowdary CR (2020) AILearn: an adaptive incremental learning model for spoof fingerprint detection
2. Baldisserra D, Franco A, Maio D, Maltoni D (2006) Fake fingerprint detection by odor analysis. Lecture notes in computer science (including sub-series lecture notes in artificial intelligence and lecture notes in bioinformatics), vol 3832, LNCS, pp 265–272
3. Busch C, Sousedik C (2014) Presentation attack detection methods for fingerprint recognition systems: a survey. IET Biometrics 3(4):219–233
4. Cao DI, Zhu K, Li Z (2019) Slim-ResCNN: a deep residual convolutional neural network for fingerprint liveness detection. IEEE Access 7:91476–91487
5. Casula R, Micheletto M, Orru G et al (2021) Livdet 2021 fingerprint liveness detection competition—Into the unknown. In: 2021 IEEE international joint conference on biometrics, IJCB 2021
6. Casula R, Orru‘ G, Angioni D, Feng X, Marcialis GL, Roli F (2020) Are spoofs from latent fingerprints a real threat for the best state-of-art liveness detectors? In: Proceedings—international conference on pattern recognition, pp 3412–3418
7. Chugh T, Cao K (2017) Jain AK (2018) Fingerprint spoof detection using minutiae-based local patches. IEEE International Joint Conference on Biometrics, IJCB 2018:581–589
8. Espinoza M, Champod C (2011) Using the number of pores on fingerprint images to detect spoofing attacks. In: 2011 international conference on hand-based biometrics, ICHB 2011—proceedings, pp 1–5
9. Ghiani L, Hadid A, Marcialis GL, Roli F (2013) Fingerprint liveness detection using binarized statistical image features. In: IEEE 6th international conference on biometrics: theory, applications and systems, BTAS 2013
10. Ghiani L, Marcialis GL, Roli F (2012) Fingerprint liveness detection by local phase quantization. In: Proceedings—international conference on pattern recognition, no. Icpr, pp 537–540
11. Ghiani L, Yambay DA, Mura V, Marcialis GL, Roli F, Schuckers SA (2017) Review of the fingerprint liveness detection (LivDet) competition series: 2009 to 2015. Image and Vision Computing 58:110–128
12. Gupta A, Verma A (2021) Fingerprint presentation attack detection approaches in open-set and closed-set scenario. J Phys: Conf Ser 1964(4):042050
13. Jain AK, Deb D, Engelsma JJ (2021) Biometrics: trust, but verify. Journal of Latex Class Files 14(8):1–20
14. Jian W, Zhou Y, Liu H (2021) Densely connected convolutional network optimized by genetic algorithm for fingerprint liveness detection. IEEE Access 9:2229–2243
15. Jiang Y, Liu X (2018) Uniform local binary pattern for fingerprint liveness detection in the Gaussian pyramid. Journal of Electrical and Computer Engineering 2018:1–9
16. Kulkarni SS (2015) Survey on fingerprint spoofing, detection techniques and databases Hemprasad Y Patil. International Journal of Computer Applications, no. Ncac, 975–8887
17. Matsumoto T, Matsumoto H (2002) Impact of artificial gummy fingers on fingerprint systems. In: Proceedings of optical security and counterfeit deterrence techniques, vol 4677, no. May, pp 275–289
18. Menotti D, Chiachia G, Pinto A et al (2015) Deep representations for iris, face, and fingerprint spoofing detection. IEEE Trans Inf Forensics Secur 10(4):864–879
19. Mura V, Orru G, Casula R et al (2018) LivDet 2017 fingerprint liveness detection competition 2017. In: Proceedings—2018 international conference on biometrics, ICB 2018, pp 297–302
20. Nogueira RF, De Alencar LR, Campos MacHado R (2016) Fingerprint liveness detection using convolutional neural networks. IEEE Trans Inf Forensics Secur 11(6):1206–1213
21. Orru G, Casula R, Tuveri P et al (2019) LivDet in action—fingerprint liveness detection competition 2019. In: 2019 international conference on biometrics, ICB 2019
22. Putte T, Keuning J (2000) Biometrical fingerprint recognition: don't get your fingers burned. Smart Card Research and Advanced Applications. Springer, Boston, pp 289–303

23. Rattani A, Scheirer WJ, Ross A (2015) Novel fabrication materials. IEEE Trans Inf Forensics Secur 10(11):2447–2460
24. Sharma RP, Dey S (2019) Fingerprint liveness detection using local quality features. Visual Computer 35(10):1393–1410
25. Toosi A, Cumani S, Bottino A (2017) CNN patch-based voting for finger-print liveness detection. In: IJCCI 2017—proceedings of the 9th international joint conference on computational intelligence, pp 158–165
26. Uliyan DM, Sadeghi S, Jalab HA (2020) Anti-spoofing method for fingerprint recognition using patch based deep learning machine. Engineering Science and Technology, an International Journal 23(2):264–273
27. Verma A, Gupta VK, Goel S, Yadav AK, Yadav D (2021) Modeling fingerprint presentation attack detection through transient liveness factor-a person specific approach. Traitement du Signal 38(2):299–307
28. Xia Z, Yuan C, Lv R, Sun X, Xiong NN, Shi YQ (2020) A novel weber local binary descriptor for fingerprint liveness detection. IEEE Transactions on Systems, Man, and Cybernetics: Systems 50(4):1526–1536
29. Zhang Y, Gao C, Pan S, Li Z, Xu Y, Qiu H (2020) A score-level fusion of fingerprint matching with fingerprint liveness detection. IEEE Access 8:183391–183400

Review of Toolkit to Build Automatic Speech Recognition Models

G. P. Raghudathesh, C. B. Chandrakala, and B. Dinesh Rao

Abstract Speech is one of the most significant types of communication between human beings. It is beginning to be a preferred means for communication between machines and humans. The mechanism of transforming human speech into its equivalent textual format is known as speech recognition. Various toolkits are being used to automate the process of speech-to-text conversion, and this process is referred to as automatic speech recognition (ASR). The usage of the ASR system is becoming prevalent with the implementation of human–machine interaction. Numerous speech-based assistive systems are available today used in several different areas. This paper provides insight into the ASR domain and toolkit used in ASR system—HTK, CMU Sphinx, Kaldi, and Julius with their comparative analysis in terms of installation, ease of use, and accuracy assessment.

Keywords Automatic speech recognition · ASR toolkit · Acoustic models · Language models · Lexicon

1 Introduction

Speech is one of the primary forms of human–human communication. With the rise in the human–machine interaction, increased research is going in the areas of automatic speech recognition domain. Developing an interactive system using an easy-to-use interface is a major technological challenge. The development of the

G. P. Raghudathesh · B. D. Rao
Manipal School of Information Sciences, MAHE, Manipal 576104, India
e-mail: raghudathesh.gp@manipal.edu

B. D. Rao
e-mail: dinesh.rao@manipal.edu

C. B. Chandrakala (✉)
Department of Information and Communication Technology, Manipal Institute of Technology, MAHE, Manipal 576104, India
e-mail: chandrakala.cb@manipal.edu

P. Dutta et al. (eds.), *Emerging Technologies in Data Mining and Information Security*, Lecture Notes in Networks and Systems 490,
https://doi.org/10.1007/978-981-19-4052-1_45

ASR system is the need of the time for enabling the population to seamlessly integrate into the digital eco-system via speech medium. Today, ASR-based systems are being used for various exciting tasks in different application domains [1]. Some of the application domains include health care—medical transcription, assistive system for visually impaired; interactive voice response systems (IVRS)—railway, gas bookings, weather forecasting, agriculture commodity price inquiry systems, ATM; real-time voice writing—press conference, court reporting; data entry systems—entering personal credentials to a portal; command and control systems—military, robotics, office suite with voice support, speech-based systems for mobile applications to name a few.

The work on speech recognition dates back to 1879 when Thomas Edison invented the first dictation machine. Increased research interest began in the early 1950s with the development of Audrey—a digit recognition system developed by Bell Labs [2]. From machine learning, prospective speech recognition can be viewed as finding the best sequence of words according to the acoustic, the pronunciation lexicon, and the language model. Researchers are thriving to improve the accuracy of speech recognition system for years using various signal processing and machine learning techniques. Recognizing and interpreting speech is a knowledge-intensive task that must consider several aspects of speech. Automatic speech recognition is one of the challenging research domain because with the research is going on from several decades but accuracy of ASR system is not comparable to the human capability, and also due to its dependency on large number of parameters like acoustic environment in which the recording is taken, i.e. noisy or clean, variations in the speakers, type of recognition, size of vocabulary, are few to name. This paper throws a light on the ASR domain and availability of tools to develop the models for the automatic recognition of the speech in the textual format.

2 Automatic Speech Recognition (ASR) System

Automatic speech recognition is a technique of deriving the transcription (text format) of a spoke speech. Sometimes ASR systems are referred to as speech-to-text conversion entities. An ASR system comprises of a microphone to take the speech recording, sometimes a speech preprocessing software—MATLAB for speech enhancement, Toolkit like HTK, Kaldi, CMU Sphinx, Julius, etc., to build the models to recognize the spoken speech, a word processing software.

2.1 ASR System Overview

The ASR systems produce the most likelihood sequences for the given input speech data. The schematic representation of the ASR system depicting its basic building block is illustrated in Fig. 1. The architecture of the ASR system consists

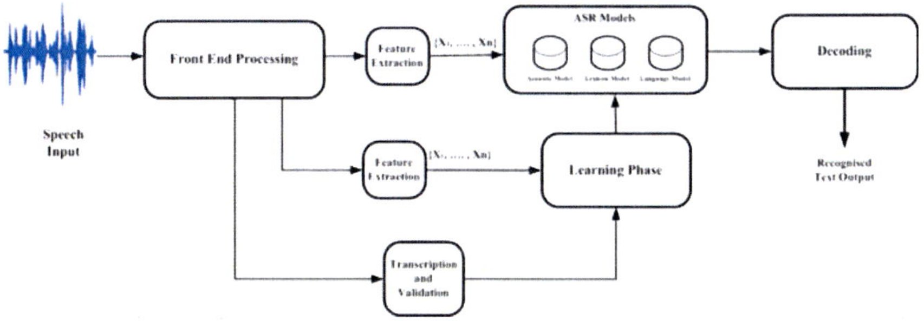

Fig. 1 ASR system

of the following components: front end processing, transcription and validation, feature extraction, learning phase, ASR models (language, lexicon/pronunciation, and acoustic), and decoding modules.

Front end processing is the first stage in ASR model creation. This step determines the parameters for collecting and improving the quality of voice data, such as sampling rate, encoding format, file format, and speech enhancement algorithm.

Transcription is the process of converting audio files to text. Depending on the data, transcription can be done at the word or phoneme level. A single word might have different phoneme sequences depending on the pronunciation style. Table 1 shows a word and phoneme-level transcriptions. When a transcriber misses a section of speech or noise in the speech file, the validation tools are utilized to repair it.

Feature extraction involves translating the voice recordings into a parametric representation with good discriminating between phonemes that should be perceptually relevant and invariant. Feature extraction has a big impact on ASR quality. Table 2 lists multiple ASR feature extraction approaches against various parameters [3]. Each phoneme in any language is represented by a collection of feature vectors. During the learning phase, the set of feature vectors are mapped to phonemes using statistical modelling approaches as hidden Markov model (HMM), Gaussian mixture model (GMM), and deep neural network (DNN).

Once the feature vectors are extracted various models—lexical (pronunciation dictionary), language, and acoustic model are built.

A language's lexicon is also a dictionary. The lexicon for the given voice data corpus is shown in Table 1. The vocabulary is divided into two parts: word-level transcription and phoneme-level transcription, or alternate pronunciations of each spoken

Table 1 Word and phoneme-level transcriptions

Word level	Phoneme level
Manipal	ma ni pa l
Manipal	ma ni pa la
Heart	ha ar t
Heart	ha ar tu

Table 2 Feature extraction techniques

	Discrete wavelet transform (DWT)	Line spectral frequencies (LSF)	Linear prediction cepstral coefficient (LPCC)	Perceptual linear prediction (PLP)	Linear prediction coefficient (LPC)	Mel frequency cepstral coefficient (MFCC)
Reliability	Medium	Medium	Medium	Medium	High	High
Noise resilience	Medium	High	High	Medium	High	Medium
Computational speed	High	Medium	Medium	Medium	High	High
Nature of Modelling	–	Human vocal tract	Human vocal tract	Human auditory system	Human vocal tract	Human auditory system
Coefficient	Wavelets	Spectral	Cepstral	Cepstral and autocorrelation	Autocorrelation coefficient	Cepstral
Filter shape	–	Linear	Linear	Trapezoidal	Linear	Triangular
Filter type	Low pass and high pass	Linear prediction	Linear prediction	Bark	Linear prediction	Mel

word. Phoneme-based lexicon modelling—mono-phone, bi-phone, tri-phone, etc., are used to build lexicon model.

The language model identifies regularities in the language spoken and is used by the recognizer to systematize the word sequence as part of the language. These language models will be created by using language resources such as phonemes, dictionary, silence, and non-silence phones. N-gram modelling techniques are used to build language model in speech recognition. Tri-gram modelling is most commonly used an optimum balance between robust estimation and complexity of processing.

The properties of voice data, their equivalent transcriptions, and modelling techniques used to construct statistical representations for each phoneme in a language are used to build an acoustic model. HMM, GMM, SGMM, DNN, etc. Models are statistical representations.

A computer module compares a user's sound to the acoustic model. When a match is detected, the decoder determines the phoneme. It keeps matching phonemes until the user stops speaking. Then, searches the language model for a matching phoneme set. If a match is found, the spoken word's text (phoneme sequence) is returned.

2.2 *Mathematical Representation for Speech Recognition*

From machine learning prospective, speech recognition is seen as finding the best sequence of words according to the acoustic, pronunciation lexicon, and language model. Figure 2 illustrates the sequence of steps to be followed to

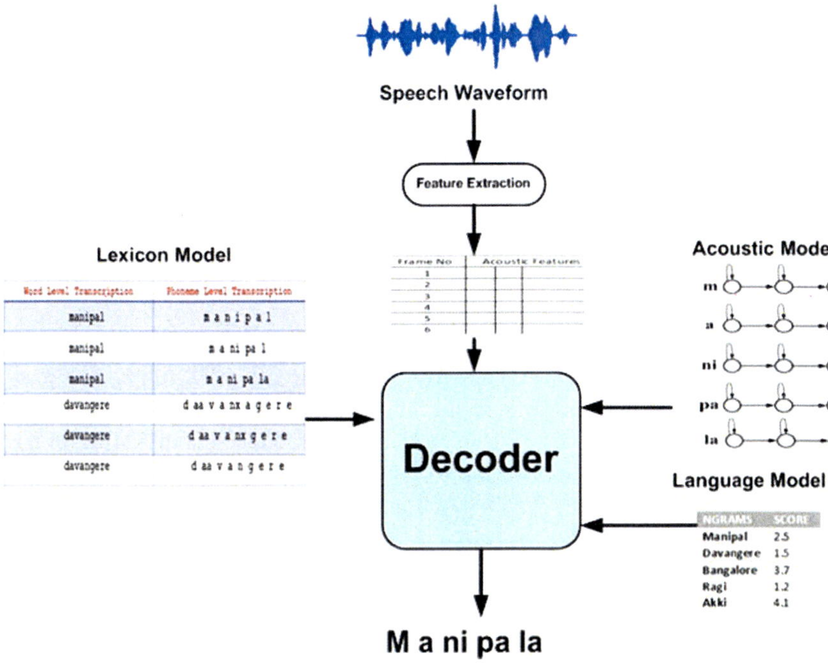

Fig. 2 Steps to be followed to create models for ASR systems

create models for ASR systems. Given the word sequence represented as $W = [w_1, w_2, \ldots, w_3]$, sequence of acoustic feature vector sequence for uttered speech is $Y = [y_1, y_2, \ldots, y_3]$, size of the dictionary is represented as N then fundamental equation of statistical speech recognition is given in (1)

$$W^* = arg \overset{\max}{W} P(W|Y) \tag{1}$$

2.3 *Performance of the ASR System*

The performance of the ASR system is evaluated using word error rate (WER) calculated using the Eq. (2) as the result of division of number of word uttered wrongly to the total number of word uttered in the original speech, where I is the count of word inserted, D is the count of word deleted, *S* is the count of the word substituted, and *T* is the number of word spoken. It is desirable to have lower WER for the ASR system to perform well.

$$WER = (I + D + S)/T \tag{2}$$

3 Speech Recognition Toolkits

This section provides insight on different toolkits that are being used in speech recognition like—HTK, CMU Sphinx, Julius, and Kaldi with some of the related work. Table 3. illustrates the comparison of these toolkits.

3.1 *Hidden Markov Toolkit (HTK)*

HTK is a toolkit for building hidden Markov models. HMMs can be used to model any time series, and the core of HTK is similarly general purpose. HTK offers multiplatform support both on Windows and Linux. HTK is primarily designed for building HMM-based speech processing tools in particular speech recognizers. Much of the functionality of HTK is built into the library modules and is developed in C language. HTK tools are designed to run with a traditional command line command line style interface. HTK has four steps of processing—data preparation, learning/training, decoding/recognition, and analysis.

Kumar et al. [3] proposed a connected-word Hindi speech recognition system using HTK. Authors used a data corpus of 102 words, and a total of 17 speakers having both female and male are used. Acoustic models are built on the word levels by estimating the HMM parameters, the achieved WER was 12.99%.

Thalengala and Shama [4] developed Kannada isolated word recognition system using the HTK. Two types of dictionaries have been developed for the recognition of Kannada isolated speech data. The experiments on Kannada speech data were conducted by different Gaussian mixtures in every state. The accuracy of speech recognition for mono-phone and tri-phone models is 60.2 and 74.35%.

Aggarwal et al. [5] developed an text independent speaker recognition system for isolated English words speaking Hindi natives for a NATO standard vocabulary of 23 words having 100 speakers having both male and female speakers using HTK achieved accuracy of 75%.

Table 3 Feature comparison of ASR toolkits

Toolkit	Programming language	Operating system support	Modelling techniques	Open source
HTK	C language	Cross-platform	HMM	YES
CMU Sphinx	Java	Cross-platform	HMM	YES
Julius	C language	Cross-platform	HMM	YES
Kaldi	C + + , scripts	Cross-platform	HMM, GMM, SGMM, DNN	YES

3.2 CMU Sphinx

Sphinx-4 hosted in Carnegie Mellon University (CMU) repository [6] was a collaborative work of different laboratories and universities. Compared to previous Sphinx versions, the Sphinx-4 has more advantages like modularity, reliability, flexibility, and algorithmic aspects. It endorses various dialects and various inquiring strategies. The complete Sphinx-4 kit was built using JAVA and supports multithreading ideas.

Wang and Zhang [7] develop a system for rapidly identifying the Mandarin continuous digits using Sphinx toolkit. Authors used CASIA speech dataset has continuous speech spoken digit sequence where each recording has variable length of spoken digit string by 55 male speakers speaking 4400 sentences. Authors employed MFCC feature extraction technique with a vector length of 51 dimension with HMM modelling technique to achieve a WER of 2.80%.

Abushariah et al. [8] use Sphinx toolkit to build and realization of continuous speech-to-text recognition systems for the natural Arabic language. The authors researched the performance of Sphinx models and built a comprehensive Arabic ASR entity. In contrast to the standard Arabic ASR, the newly developed Arabic speech recognition uses HTK methods and Sphinx in the development of LM's and AM's. MFCC techniques were employed to extract feature vectors from the speech signal. The system uses 500 senses with 16 GMM's and 5 state HMM's for 7 h of verified transcribed speech data. An hour of data was used for decoding/validation, achieved recognition accuracy of 92.67%.

Bassan and Kadyan [9] developed a continuous speech ASR system using Sphinx toolkit for speaker-independent and dependent environments. The ASR system was trained with 422 sentence spoken by 15 speaker having 6 male and 9 female speakers using HMM modelling technique achieved a WER of 11.04% for speaker-independent and 6.15% for speaker-dependent datasets.

3.3 Julius

For speech-related developers and researchers, Julius is free, high performance, large vocabulary continuous vocabulary speech recognition software written in C language [10]. It performs multi-model decoding with a single processor, a recognition that uses other LMs and AMs simultaneously. During run time, it supports the "hot plugging" arbitrary modules. Support multiple operation system works with a minimum of 32 Mb of memory requirement—precise, hi-speed, and real-time recognition based on 2-pass strategy. Julius supports HMM for acoustic modelling and language models, such as the statistical N-gram model and rule-based grammars. It includes LM's based on—grammar, isolated words, and N-gram. Input acts as on-the-fly recognition for microphone and network. Input rejection is based on GMM. For decoding, Julius supports N-best output, word-graph, and confusion network output. On word, phoneme and state levels, there is forced alignment.

Kokubo et al. [11] have proposed a modified version of Gaussian mixture selection (GMS) to port Julius toolkit onto the microcontroller to be used with mobile environment. Based on simulation results, computational cost has reduced by 20% compared to conventional GMS. Modified version of Julius was evaluated on T-engine hardware platform.

Sharma et al. [12] proposed a study of live continuous speech recognition systems' based on Julius and HTK toolkits for Kannada language. Authors used a speech corpus having 16 speakers uttering 115 Kannada phrases having a total of 7728 voice samples using HMM-based acoustic modelling technique achieved a WER of 85% for live-female and 75% for live-male speakers.

3.4 Kaldi

Kaldi is a toolkit written in C++ for speech recognition (SR) for the development of the ASR system together provided with shell scripts for various tasks like preparation of dataset, master script for running complete work, etc. Kaldi is open-source offers multiplatform support both on Windows and Linux. Kaldi has better features compared to other toolkits. Kaldi supports both conventional modelling techniques like HMM, GMM, and SGMM together with neural network capabilities like deep neural network support (DNN). The main advantages of Kaldi over others are cleanly structured code, flexible, better weighted finite-state transducers (WFST) integration, math support, and the license terms of Kaldi are open source. Because Kaldi uses finite-state transducers (FST)-based structure, any LM that can be represented as an FST can be used [13]. PyTorch-Kaldi is another open-source toolkit having detailed documentation designed to work on a standalone system or on high-performance computing (HPC) clusters having interfaces for user-defined acoustic models together with numerous pre-implemented neural networks which can be personalized using configuration files [14].

Ali et al. [15] developed a Arabic speech recognition system having large lexicon and data highly sparse with GMM and SGMM modelling techniques were employed. For Kaldi system learning and decoding, the authors used 200 h recordings comprising of 36 phonemes achieved a WER of 15.81% for broadcast report (BR), 32.21% for broadcast conversation (BC), 26.95% of total speech recordings using Kaldi recipe together with Arabic-language resources.

Shahnawazuddin et al. [16] describe spoken request mechanism developed by IITG. The models were created using Kaldi. The ASR setup comprises of IVRS call flow, IMD, and agriculture market network databases and models develop for the setup. Previous IITG speech dialogue setup leads to higher WER as various noises were present in the collected speech information. To enhance recognition accuracy, the authors developed a noise elimination mechanism, zero frequency filtered signal (ZFFS), and introduced before the MFCC extraction step. The legacy system was built employing GMM HMM-based modelling mechanism. The authors then developed the ASR models using the modelling methodology based on SGMM-DNN resulting in a reduction of 6.24% WER.

Yadava and Jayanna [17] addressed the creation of ASR models using Kaldi toolkit. The language models (LMs) and acoustic models (AMs) were created for the Kannada speech archive. The Kannada speech database was developed by collecting it in an uncontrolled environment from Karnataka state farmers. The dictionary and phoneme set for the Kannada language is also described by the authors. The experimental results show that the WERs achieved are 11.40%, 11.10%, 9.34%, and 11.25%, respectively, for districts, mandis, commodities, and general speech information. A speech enhancement algorithm was recommended to reduce noise in degraded speech information by combining spectral subtraction with voice activity detection (SS-VAD) and a zero-crossing minimum mean square error spectrum power estimator (MMSE-SPZC) [18, 19]. Three types of estimators were studied and implemented in detail. The best estimator (MMSE-SPZC) among three was combined with the SS-VAD algorithm to reduce musical and babble noises in corrupted speech records. The experiments were conducted on both Kannada and TIMIT speech databases. The experimental results suggest that the recommended combined approach gave improved speech quality and reduction in noise compared to individual methods.

4 Performance Comparison

ASR models were built and compared for HTK, CMU Sphinx, and Kaldi on Hindi dataset having 150 grammatically rich Hindi sentences uttered from 7 male speakers, thus a total of 1050 recordings were obtained. HTK took the longest time to set up, prepare, run, and optimize, while Sphinx took less and Kaldi took the least. The Sphinx toolbox has a subset of training tool compared to Kaldi, resulting in a lower level of accuracy, but also enables training and speech recognition as soon as the system is installed. In comparison with recognizers, Kaldi's exceptional performance is a watershed moment for open-source voice recognition technology. The system has out-of-the-box support for nearly all state-of-the-art techniques—LDA, TDNN, LSTM, DNN, etc. Even if the user is not an expert, the offered recipes and scripts enable rapid implementation of all of these strategies. The HTK toolset is the toughest. To get the system working, a time-consuming and error-prone training pipeline must be constructed. Creating strategies beyond the tutorials takes far more skill and work than other systems. Table 4 gives a comparisons of ASR model built using tri-phone modelling.

Table 4 Performance comparison of ASR toolkits

	Setting time	Ease of use	WER (tri-phone) (%)
HTK	Highest	Lowest	15
CMU Sphinx	Medium	Medium	13
KALDI	Least	Highest	8

5 Conclusions

In this paper, we have presented insight on automatic speech recognition system (ASR), its classification, accessing the performance of ASR systems based on word error rate (WER), and toolkit used to build ASR system. It can be seen from various works, the performance of the ASR system (WER) depends of the size of the corpus, modelling techniques used to build the various models—lexicon, acoustic, language, and conditions in which the recording has been taken, techniques/algorithms used during speech decoding. The creation of automated systems for the understanding and recognition of spoken languages as a human being is a complex task. The aim of research on automated speech recognition is to address the different issues of speech recognition. ASR system built for languages like Arabic, English, and French has attained a high recognition accuracy. For Indian languages, significant work is being carried out in recent years, enabling population to seamlessly integrate into the digital eco-system.

References

1. Caranica A, Cucu H, Burileanu C, Portet F, Vacher M (2017) Speech recognition results for voice-controlled assistive applications. In: 2017 9th international conference on speech technology and human-computer dialogue, SpeD 2017. https://doi.org/10.1109/SPED.2017.7990438
2. Juang BH, Rabiner LR (2004) Automatic speech recognition—a brief history of the technology development. Elsevier Encycl Lang Linguist 50(2):637–655
3. Kumar K, Aggarwal RK, Jain A (2012) A Hindi speech recognition system for connected words using HTK. Int J Comput Syst Eng 1(1):25. https://doi.org/10.1504/ijcsyse.2012.044740
4. Thalengala A, Shama K (2016) Study of sub-word acoustical models for Kannada isolated word recognition system. Int J Speech Technol 19(4):817–826. https://doi.org/10.1007/s10772-016-9374-0
5. Agrawal SS, Bansal S, Pandey D (2013) A hidden Markov model based speaker identification system using mobile phone database of North Atlantic Treaty Organization words. J Acoust Soc Am 133(5):3247–3247. https://doi.org/10.1121/1.4805213
6. Lamere P et al (2003) Design of the CMU Sphinx-4 decoder. In: EUROSPEECH 2003—8th European conference on speech communication and technology, pp 1181–1184
7. Wang Y, Zhang X (2010) Realization of Mandarin continuous digits speech recognition system using Sphinx. In: 3CA 2010—2010 international symposium on computer, communication, control and automation. https://doi.org/10.1109/3CA.2010.5533801
8. Abushariah MAM, Ainon RN, Zainuddin R, Elshafei M, Khalifa OO (2010) Natural speaker-independent Arabic speech recognition system based on hidden Markov models using Sphinx tools. In: International Conference on Computer and Communication Engineering, ICCCE'10. https://doi.org/10.1109/ICCCE.2010.5556829
9. Bassan N, Kadyan V (2019) An experimental study of continuous automatic speech recognition system using MFCC with reference to Punjabi language. In: Advances in Intelligent Systems and Computing, pp 267–275. Springer Singapore
10. Lee A, Kawahara T (2009) Recent development of open-source speech recognition engine Julius. In: APSIPA ASC 2009—Asia-Pacific signal and information processing association 2009 annual summit and conference

11. Kokubo H, Hataoka N, Lee A, Kawahara T, Shikano K (2007) Real-time continuous speech recognition system on SH-4A microprocessor. In: Proceedings of 2007 IEEE 9th international workshop on multimedia signal processing, MMSP 2007. https://doi.org/10.1109/MMSP.2007.4412812
12. Sharma RS, Paladugu SH, Jeeva Priya K, Gupta D (2019) Speech recognition in Kannada using HTK and julius: a comparative study. In: Proceedings of the 2019 IEEE international conference on communication and signal processing, ICCSP 2019. https://doi.org/10.1109/ICCSP.2019.8698039
13. Povey D et al (2011) The Kaldi speech recognition toolkit. In: Proceedings of ASRU. https://doi.org/10.1017/CBO9781107415324.004
14. Ravanelli M, Parcollet T, Bengio Y (2019) The Pytorch-kaldi speech recognition toolkit. In: Proceedings of ICASSP, IEEE international conference on acoustics, speech and signal processing. https://doi.org/10.1109/ICASSP.2019.8683713
15. Ali A, Zhang Y, Cardinal P, Dahak N, Vogel S, Glass J (2014) A complete KALDI recipe for building Arabic speech recognition systems. In: Proceedings of 2014 IEEE workshop on spoken language technology, SLT 2014, pp 525–529. https://doi.org/10.1109/SLT.2014.7078629
16. Shahnawazuddin S, Thotappa D, Dey A, Imani S, Prasanna SRM, Sinha R (2017) Improvements in IITG Assamese spoken query system: background noise suppression and alternate acoustic modeling. J Signal Process Syst 88(1):91–102. https://doi.org/10.1007/s11265-016-1133-6
17. Yadava TG, Jayanna HS (2017) Creating language and acoustic models using Kaldi to build an automatic speech recognition system for Kannada language. In: Proceedings of RTEICT 2017—2nd IEEE international conference on recent trends in electronics, information and communication technology, vol 2018-Janua, pp 161–165. https://doi.org/10.1109/RTEICT.2017.8256578
18. Yadava TG, Jayanna HS (2019) Speech enhancement by combining spectral subtraction and minimum mean square error-spectrum power estimator based on zero crossing. Int J Speech Technol 22(3):639–648
19. Yadava TG, Jayanna HS (2018) Creation and comparison of language and acoustic models using Kaldi for noisy and enhanced speech data. Int J Intell Syst Appl 10(3):22–32. https://doi.org/10.5815/ijisa.2018.03.03

Loyalty Score Generation for Customers Using Sentimental Analysis of Reviews in e-commerce

N. Vandana Raj and Jatinderkumar R. Saini

Abstract Presently, the existing systems use the star ratings of products as the quality factor, but ignore the tangible feedback. It is problematic since the star ratings may differ from the intensity of written feedback given by the user. This motivated authors to come up with the system that will use both ratings and reviews together. In this paper, the author has proposed a model in which sentimental score will be generated for the reviews given by the user by using sentimental analysis and shall derive the correlation between the ratings and sentimental score generated by analyzing the customer reviews. By doing the same for all the reviews given by the user, the model will assign a loyalty score automatically to the customer. This loyalty score can be used to build customer profile, and also the sentimental score generated can be used to boost the products when the user applies a sort by ratings filter as the user will get to see not only the products that are having good ratings, but also the products that got better reviews along with it.

Keywords Customer loyalty score · Sentimental analysis · Ratings · Review analysis

1 Introduction and Literature Review

Customer opinion is vital to consumer brand success. We have to step beyond the positives and negatives to find the real tone of what has been said by the customer. The revenues will be solely dependent on the user experience, so it is very imperative for us to maintain the brand value. People opt to buy from unicorns like Amazon, Flipkart, and Myntra because of the trust they have built with customers over years. The Internet will become the source of market revenue [1]. Once the e-commerce store is set up and the trust is gained, Sales go up and there will be a huge inflow

N. V. Raj · J. R. Saini (✉)
Symbiosis Institute of Computer Studies and Research, Pune, India
e-mail: saini_expert@yahoo.com

Symbiosis International (Deemed University), Pune, India

P. Dutta et al. (eds.), *Emerging Technologies in Data Mining and Information Security*, Lecture Notes in Networks and Systems 490,
https://doi.org/10.1007/978-981-19-4052-1_46

in traffic, and likewise just one bad review on a popular product can significantly jeopardize the sales of that product. So, analysis is much needed for the reviews that enter the site. Sentimental analysis will help in analyzing the reviews. Sentimental analysis (SA) can be done in many ways, one of them is by building lexicons. The system will take in the sentiment embeddings and does SA on feedback made by the customer and generate polarity scores of the comments made. This entire score is taken product wise. Similar words will have the same polarity score; this will be identified using word similarity. By this, we will be able to segregate the positive and negative scores in two ends of the spectrum. Then, the polarity scores from the feedback of the particular product are added to obtain an overall score for the product, which is then displayed as a graph to the administrator based upon which he can manage the inventory in order to improve the overall quality of the website [1].

Same user can have multiple accounts and then comment from them to bring down the ratings of a few products, in order to tackle that media access control address (MAC address) can be used and then identify if the user is the same or different. By this, if a user is found to be a fraud, further steps to abstain could be taken [2]. But in this paper, the author has observed that still users can use multiple devices and access the accounts, and in this case, same user still can have multiple MAC addresses. Lexicons are frequently used for polarity and intensity classification in sentimental analysis. But there is an issue with the accuracy of representations of emotional categories as a lexicon uses words to analyze the emotion, but not the concept and opinion. We consider the set of emotional categories in WorldNet Affect to be excessively broad. Second, there is an issue of labeling ambiguity. Hence, the author has proposed SentiSense [3], an effective lexicon that attaches emotional categories to WordNet Synsets. One of its main advantages is the availability of different sets of tools and algorithms that allow users to easily expand the coverage of the lexicon both manually and automatically, in order to cover the emotional vocabulary of each specific application domain. "Enhancement bag-of-words" technique can be used to check the sentiment polarity, dividing them into positive, negative and neutral and to score automatically by using the words weight method instead of term frequency [4].

Not only the search results, but recommendations also play a vital role in attracting the customers to a website. Hence, we can use sentimental analysis in improving the recommendations in e-commerce site. Admin will have access to the reviews and various types of reviews. Reviews can be ratings, reviews, text reviews and smiley reviews. All reviews are pushed into the database for future evaluation. Ratings, reviews and emoticons are the input data for this analysis. Rating, reviews and emoticons are the parameters to measure the quality (as with a critic rating a product recently, he/she has bought), quantity or some combination of both. Ratings, reviews and emoticons are analyzed and segregated as positive and negative, and the products will be searched with the help of review-based filtering. MAC-based filtering approach can be used to get rid of fake reviews. This method was tested against real-time user data collected from an online website, by using a list of the products liked by each user as an input to the system [5].

Word ambiguity and detection of sarcasm in the comments made are one of the major challenges in sentimental analysis. A mutual bootstrapping algorithm is proposed to automatically construct the determination of the sentiments and polarity-ambiguous words, utilizing relationships among sentiments, polarity words and analysis of the words being used. Then, the sentiment of polarity-ambiguous words in context can be decoded by looking at the entire spectrum built by analyzing the polarity-ambiguous words. At the sentence level, experiments have shown that this method is effective [6]. Numerical sarcasm is a special case of sarcasm where analysis arises between textual and numerical content. Previous approaches were unable to capture numerical sarcasm because of them being sensitive only to normal sarcasm. In the future, all the past approaches for sarcasm detection can benefit by separating the normal sarcasm from numerical sarcasm and improve performance of sentimental analysis [7]. Many researchers have proposed ways to check the reviews given manually and then had a system in place by using e-bay website as an example. They analyze the feedback given through sentimental analysis and get a score for a buyer and as well as the seller, by this, we can get to know the credibility of both and the loyalty of the customers. The mechanism was built where the sellers having such high scores are given more visibility in the site so that they can sell the products through e-bay [8].

Fraud detection is one of the important aspects to handle to build customer trust. One of the parameters which we have to consider is the profit factor while detecting fraud. As the fraud detection improves the consumer will buy more goods and products will be sold online. The proposed paper will consider quality and user reviews indicating fraud detection of consumers toward the vendor. Mistakes in the fraud detection will cause loss of the vendor and owners of the organization. Fraud detection will be easy to lose but difficult to gain [9, 10]. Collated trust profiles are calculated for sellers using CommTrust. It includes complete reputation scores and weights and overall trust scores by collating the dimension scores of reputations of the sellers and comments made. The first system which calculates micro level of multidimensional trust profiles automatically by mining feedback comments is by CommTrust. Later, we use the terms reputation score and trust score interchangeably [11]. There are many data mining techniques [12] which will look at the transactions made and help in identifying the potential risks and frauds [13]. Microsoft has developed a fraud management system (FMS) which will use ML models to construct real-time archiving, and helps in building dynamic risk tables and knowledge graphs [14].

Customer satisfaction and customer value are one of the important factors that will decide the loyalty of the customer toward the site [15]. Apart from them, service quality will also be a critical factor to consider [16]. In order to retain the customers and gain e-loyalty, the above-mentioned factors need to be taken care of and also, one should determine how to give value added services and can use CRM to improvise the core capability [17]. The other factors that affect the loyalty of the customer are guarantee, fulfillment, security, reliability [18], quality of the website, user-friendliness of the site [19], products available, service provided, brand established by the company, emotional experience [20], perceived privacy and awareness about the e-commerce [21]. In building the customer trust, one should take care that the accurate information about the product should be available on the website, product quality will always have the strongest and also long-lasting effect on product experience, thus customer loyalty [22]. Product returns or product replacements activities will talk about the service provided by the company, whereas apparel brand will have effect on brand experience. Shopping journey of the user on the website will reflect the quality and user-friendliness of the website [19, 20].

There are too many e-commerce sites in the market leading to cut throat competitions, and many sites are trying to come up with innovative ways to retain the old customer base and also to attract new potential customers. Some of the factors that will help in attaining customer stickiness to the website are: price points, capturing the needs and interests of the customer and showcasing them on the website which will lead to have personal experience [19], good content, user friendly designs on the website as one should not confuse the customer by overdoing it, performance of the website, customer support, site utility [21].

In the literature review done, it is noticed that there is no such research work to the best of my knowledge based on using reviews and ratings together and then building customer loyalty profile on it. This was the motivation behind the work, and authors have suggested the model which intakes the customer reviews given on different products and awards customer loyalty points based on it.

2 Methodology

Figure 1 depicts the proposed methodology. The cleaned data is taken as input for the model. In ratings related analysis, the author has verified cases such as those in which a single user has given all the ratings resulting in bulk reviews. Once reviews are found to be genuine, review related SA is done. By which sentimental score for the reviews given by the user is recorded, the correlation between the sentimental score of the review made by the user for a particular product and sentimental score generated for the overall reviews for that product is noted, similarly correlation is noted for all the reviews that a user has made across all the product and their corresponding correlation noted. If correlation is found to be positive, points are awarded accordingly to the user. If the correlation is found to be negative, then the points will be relatively less. If a user has reviewed it negatively, a factor like shipping is also considered to see if shipping is the issue, in that case points are not reduced as it is on the service provider and not on the customer.

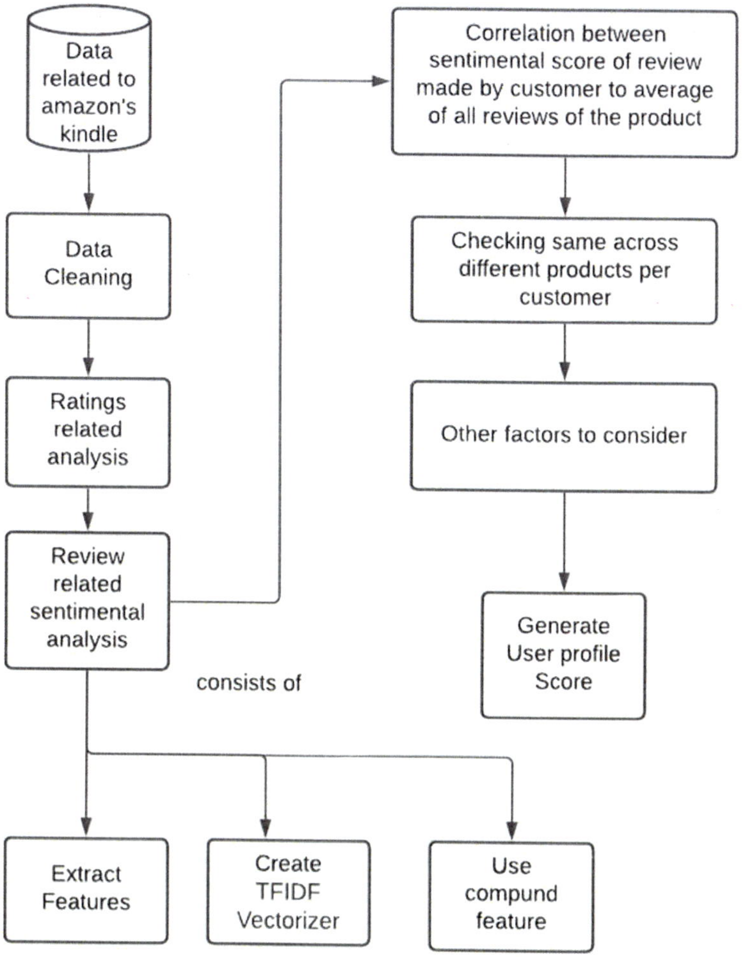

Fig. 1 Proposed methodology

2.1 *Dataset and Experimental Setup*

This section describes the dataset considered during the study. The data is taken from Kaggle website [23]. The number of entries in the dataset is 34,660 entries with 18 columns.

Dataset belongs to Amazon having below mentioned attributes:

- ID: ID of the product
- Name: Name of the product
- Brand: Brand of the product
- Categories: Category of the product
- Keys, source URLs: Product URL related information

- Manufacturer: Merchant related information
- Date: Date on which the product was added to the site
- dateAdded: Product made seen for the customers
- dateSeen: Customer added product into the cart
- didPurchase: Date related information regarding when the customer bought the product
- doRecommend: Boolean field, having true if the product is used for recommendations
- reviews.id: Unique IDs for the reviews
- Rating: Rating information related to the product
- Text: Reviews made
- Title: Title of the review
- userCity, userProvince, username: User related details.

Among above-mentioned data, all the attributes got object as data type, i.e., categorical data except Id, reviews.id, and rating attributes.

2.2 Experiment Steps

2.2.1 Data Cleaning and Preprocessing

As part of data cleaning, the author removed all the stop-words and special characters from the dataset by using the NLTK library.

In data preprocessing, the author has removed a few fields from the dataset as they were not considered for the model like didPurchase, ID, userCity, userProvince and then verified to see if there are any null values in the data. There were few rows whose reviews were null as an entry. Such entries are removed from the dataset.

2.2.2 Ratings Related Analysis

There is chance that the same set of users could have given ratings to all the products; in this case, there will be huge bias and the model built will not efficiently cover all the scenarios; hence, authors made sure that there is no such case where single user itself has given ratings/reviews for the majority of the data we are considering for the study.

From Table 1, it can be inferred that only 0.55% of the users are bulk users, but only 9% of the ratings in the dataset have been given by just 0.55% users. This is a bit ambiguous and cannot conclude that the data is biased, so histogram is used to delve deeper into the bulk ratings.

From Fig. 2, it is noted that there are no discrepancies between ratings distribution for the bulk users in case of positive and negative reviews. Hence, it can be concluded that there is no bias in the ratings made. From Fig. 3, the overall distribution of the

Table 1 Statistical inferences related to the ratings

Total ratings—Sum of all the ratings per person	34,658
Total users—Total number of unique usernames that the data is having	26,789
Users giving bulk ratings (more than 10)—Taken the sum of users who have given ratings and reviews more than 10 times	146
Populations of bulk ratings which is given by the below formula: bulk_rating*100/sum(rating_per_person) Where rating_per_person is given by unique names in the username attribute	9.12
Populations of bulk users are calculated with the formula: sum(rating_per_person > 10)*100/len(rating_per_person)	0.55

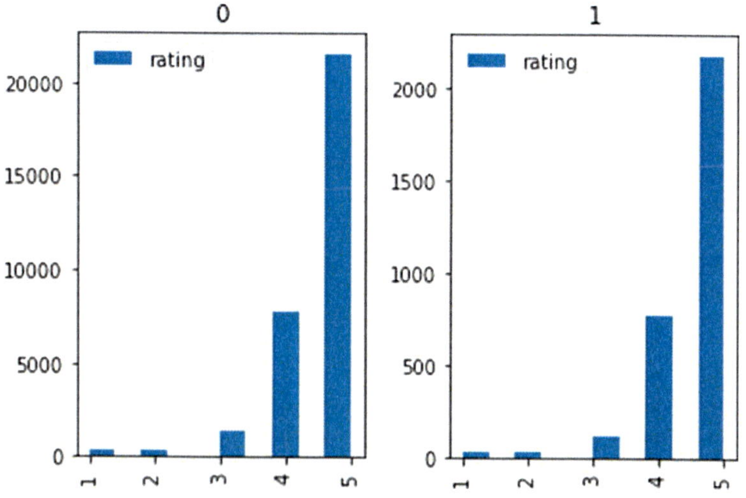

Fig. 2 Ratings distribution of positive and negative reviews of the ratings given by bulk user

ratings is shown and it is observed that there are totally 23,775 entries with 5-star ratings, 8541 products 4-star ratings, 1499 entries got 3-star ratings, 402 products got 2 stars and 410 entries got 1-star ratings.

2.2.3 Review Related Analysis

Net promoter score will always help in understanding the customer satisfaction with given website. For the proposed algorithm, for the review related analysis, we find the NPS score for the dataset and proceed with the dataset only if the score is greater than 50%. This condition will help in adding value to the sections/categories/products which are already having customers' attention. Ratings with 1, 2 and 3 are known as detractors. Four can be regarded as the passive ones, and five will be the promoters. The formula used to calculate NPS is percentage of

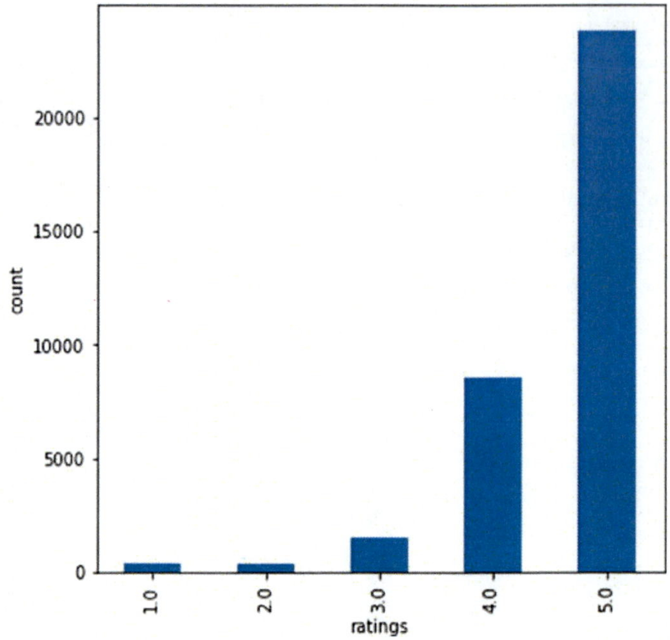

Fig. 3 Overall distribution of the ratings given by bulk user using histogram

((Promoters—Detractors)/Total_ratings). The value for the dataset we have taken is 61.99%.

As part of data cleaning in the reviews, the author has removed punctuations, stop-words and converted all the reviews into lower case. As a part of feature extraction, TfidfVectorizer method from sklearn is used and the feature vectors are built from the text column matrix. Both stemmed (3693 features) and lemmatized (4547 features) analyzers are used, and lemmatized have 18% more than stemmed. In this model, the author will be proceeding with a stemmed analyzer because the information density is more in the compressed columns.

Once the feature extraction is completed, the sum of all the TF-IDF scores is built and sorted in ascending and descending manner as shown in Table 2. This

Table 2 Top five and bottom five TF-IDF scores for the features extracted

Serial number	Keywords	Bottom 5 TF-IDF score	Serial number	Keywords	Top 5 TF-IDF score
1	Fool	217.397630	1	Gripe	0.076083
2	Summer	209.501804	2	Grandkid	0.076083
3	Matt	198.910788	3	Old	0.076083
4	Lighter	181.218969	4	Cow	0.076083
5	Purs	147.851506	5	Felt	0.076083

will help us in understanding the spread of the feature vectors built. Keywords like gripe, grandkid, old, cow and felt are the top five keywords with score 0.076, and the keywords like fool is having the highest TF-IDF score of 217.398.

Sentiment intensity analyzer from NLTK library is used to deduce the polarity scores of the reviews given by the customers. A separate method is written to append the compound polarity score generated for every review into the data frame. Although pos and neg of the sentiment Intensity analyzer talks about the positive and negative emotions in the sentence, compound is taken into consideration as it has got the overall final effect. The matrix shown in Tables 3 and 4 is the outcome of sentiment intensity analyzer.

Figure 4 depicts the flow chart for calculating the loyalty score. In the previous step, the compound score of the reviews is calculated which will reflect the sentiments of the user, and now, that is used as the review score. The ratings columns were replaced with the manually configured scores to find the correlations.

Tables 3 and 4 show the reviews and its corresponding score generated. It is noticed that the reviews whose ratings is greater than or equal to 3, got an average

Table 3 Reviews for whose ratings are greater than or equal to 3

Text	Rating	Sentimental score
If, like I, you own an earlier generation Kindle that isn't a Paperwhite, this latest model is WELL WORTH THE UPGRADE. Love my new Kindle	5.0	0.9035
It was so easy to set up the first time and add books. Easy to read clear screen and big words. Super easy to set up and use	5.0	0.9508
WORKS PERFECTLY, NO GLICHES. GREAT SIZE FOR TRAVEL AND IS GOOD IN BRIGHT OR DIM LIGHT. PERFECT FOR MY NEEDS	4.0	0.9485
My first attempt at kindle. It is nice. Not a very quick processor but gets the job done. Great device for a beginner	4.0	0.8201

Table 4 Reviews whose ratings are 1 and 2

Text	Rating	Sentimental score
Not for the average computer person. Very confusing instructions. I spent several hours trying to install the "overdrive" so I could get e-books from the library. Ran the battery down	1.0	– 0.7630
I HATE this machine. First, I specifically asked the salesperson if I could connect to the internet. He assured me I could. It does not. Second complaint is that you can't just press the book and delete it	1.0	– 0.8703
Used it for 2 days and the screen froze 5 times—the kindle just won't switch on	1.0	0.1002
Asked a Bestbuy employee if I could connect to my library. He told me yes so I bought the Kindle. When I got it home, found out that downloading books from my library was not an option	1.0	-0.1867

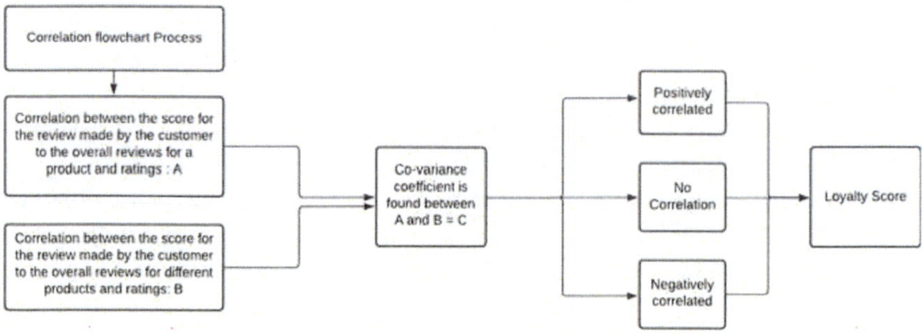

Fig. 4 Correlation between the ratings and the reviews given by the user

sentimental score of 0.7 and the reviews whose ratings are less than or equal to 2 got an average score of 0.3.

Since there are entries for the users who have given more than 10 reviews, author has considered that data for the model, and the product considered for calculating is Amazon Kindle paperwhite. As per the flowchart, first the correlation coefficient is found between the compound sentiment score of the reviews made by the particular user, overall compound sentimental scores of the reviews and ratings belonging to Amazon Kindle (A). As part of the next step, the correlation coefficient is calculated between the overall compound sentimental score made by the user for all products, he has reviewed to the overall compound sentimental scores for those respective products along with the ratings (B). Table 5 shows the ratings and manually configured sentimental score for the ratings. By doing this, the model can be dynamically configured as per the needs.

Further, based on the covariance matrix generated when both matrices are taken into consideration, the average value of the covariance matrix is calculated, i.e., C = average (covariance (A, B)). The value will be segregated into three categories as shown in Fig. 4. Loyalty score will then be rewarded to the user by multiplying the value obtained previously (C) by 100.

Table 5 Ratings and score distribution

Ratings provided	Manually configured score for that particular rating
5.0	1.0
4.0	0.75
3.0	0.5
2.0	0.25
1.0	0.0

3 Results

As mentioned in the previous section, different set of users have been identified from the dataset and analyzed their reviews and ratings for Amazon Kindle product and also to their overall ratings and reviews provided for the other products as well.

From Table 6, the author has taken different users as an example to show how the proposed model is working. In case of User T, we see that both A and B are positively correlated and variance is positive. Hence, the score will be multiplied with 100. User U got positive correlation for both metrics, but here it can be noticed that the B got high correlation. Hence, the score for U is more than T. In the case of V, the author noticed that although A is positively correlated, B is negatively correlated and variance is also negative. Hence, from the algorithm, whenever the covariance is negative, the score is taken as zero. User W, although A is negatively correlated, B is positively correlated and covariance is also positive, which is good and score is generated. In the last case, both are negatively correlated and the variance is also negatively related. Hence, the score is zero.

From the results, the author can interpret that whenever the correlation between A and B is positive, and C is positive, the score generated is more. This will help in identifying the customers who are more loyal to the site. In the next case, it is highlighted that A got more correlation, but B got less or negative correlation. The score generated in this case is less or zero. If B is negative or zero, the customer is always reviewing negatively although many other customers are commenting positively about all the products he has reviewed. Because of this, the user will be rewarded zero loyalty points as it looks like a suspicious behavior.

Now, the next case is taken where A is negative, but still B is positive and covariance derived is positive. In this case, it is believed that the customer might have faced some problem; hence, the review is negative as his other reviews still look genuine. Thus, the model will automatically still reward him with few loyalty points. In the last case, the reviews provided by the author are always negatively correlated. This

Table 6 Sample results for few selected users

User	Correlation between the score for the review made by the customer to the overall reviews for a product and ratings: A	Correlation between the score for the review made by the customer to the overall reviews for different products and ratings: B	Covariance is found between A and B = C	Loyalty Score = C*100
T	0.4046	0.6201	0.1771	17.71
U	0.3385	0.6675	0.2187	21.87
V	0.2182	– 0.0332	– 0.5338	0.00
W	– 0.3993	0.4626	0.1443	14.43
X	– 0.0248	– 0.2797	– 0.9790	0.00

shows that the customer is always commenting negatively for all the products he bought. But the overall reviews and ratings of those products are positive. From this, the author can conclude that the reviews provided by the customer are inaccurate. Hence, the loyalty score is rewarded as zero in this case. One can promote products having good reviews when a customer selects a sort by rating filter by implementing proposed methodology. e-Loyalty will definitely increase as an implication of this development.

Most of the models proposed via research are having manual processes involved in handling the reviews made by the customers on e-commerce sites. But in this work, the author has proposed a way to completely automate the process and build customer loyalty scores based on the same. Since there has been no such work done previously to the best of author's knowledge, model comparison is not done in this work. But the author has covered all the possible scenarios and the results are promising.

4 Conclusion

In today's digital era, the easier the things are accessible to us at our fingertips, the more we are exposed to frauds and security breaches. Thus, using sentimental analysis to avoid such things is very important. As we segregate the sentiments associated with the reviews into positive, negative or neutral and derive the correlation between the reviews made and reviews provided for the product, one will be able to judge the loyalty of the customer and the seller who carry out their transaction on this e-commerce platform. Loyalty scores can also be used for customer profiling, customer segregation, and by this, we can provide curated vouchers/coupons to such customers as a token of their loyalty. Doing this will attract more customers as well.

The major challenges in this design are maintenance of the data and capturing the genuine comments made by the user. Also, since there is no such system in existence as of now, there is a need to build a prototype system as proposed in the work and assess the model's real-time behavior. There are chances that merchants selling on the site themselves can act as customers and add more positive reviews for the products sold by them. This problem can be taken as the future work to be done on this. We need to come up with an algorithm which can capture such fake reviews too. Also, in the algorithm, we need to handle sarcasm and word ambiguity challenges as future scope of work.

References

1. Murugavalli S, Bagirathan U, Saiprassanth R, Arvindkumar S (2017) Feedback analysis using sentiment analysis for E-commerce. Feedback 2(3):84–90
2. Revathy R (2020) A hybrid approach for product reviews using sentiment analysis. Adalya J 9(2):340–343

3. Cao Y, Zhang P, Xiong A (2015) Sentiment analysis based on expanded aspect and polarity-ambiguous word lexicon. Journal of Advanced Computer Science and Applications 6(2)
4. Albornoz JC, Plaza L, Gerv P (2010) SentiSense: an easily scalable concept-based affective lexicon for sentiment analysis
5. Chaturvedia I, Cambria E, Welschb RE, Herrerac F (2005) Distinguishing between facts and opinions for sentiment analysis: survey and challenges. Information Fusion 44:65–77
6. Firake VR, Patil YS (2015) Survey on CommTrust: multi-dimensional trust using mining e-commerce feedback comments. Proc Int J Innovative Res Comput Commun Eng 3(3)
7. Barbosa L, Feng J (2010) Robust sentiment detection on twitter from biased and noisy data
8. Kumar L, Somani A, Bhattacharyya P (2010) Having 2 hours to write a paper is fun! Detecting Sarcasm in Numerical Portions of Text
9. Osimo D, Mureddu F (2010) Research challenge on opinion mining and sentiment analysis
10. Sakaki T, Okazak M, Matsuo Y (2010) Earthquake shakes twitter users: real-time event detection by social sensors
11. Yang N, Liu S , Li M, Zhou M, Yu N (2013) Word alignment modeling with context dependent deep neural network
12. Nanduri J, Jia Y, Oka A, Beaver J, Liu Y-W (2020) Microsoft uses machine learning and optimization to reduce E-commerce fraud. INFORMS Journal on Applied Analytics 50(1):64–79
13. Eid MI (2011) Determinants of e-commerce customer satisfaction, trust and loyalty in Saudi Arabia
14. Huang A (2017) A risk detection system of e-commerce: researches based on soft information extracted by affective computing web texts. Electron Commer Res 18(1):143–157
15. Mikolov T, Sutskever I, Chen K, Corrado G, Dean J (2013) Distributed representations of words and phrases and their compositionality
16. Sun H, Li S, Wu H, Lu M (2010) The customer loyalty research based on B2C ECommerce sites
17. Hu Y (2009) Study on the impacts of service quality and customer satisfaction on customer loyalty in B2C E-commerce
18. Mikola R (2019) The impact of trust factors on customer loyalty in B2C e-commerce
19. Nemzow M (2021) Ecommerce "stickiness" for customer retention
20. Al Adwan A, Zamil AMA, Areiqat AY (2021) Factors affecting online shopping behavior of consumer understanding factors leading to customer's loyalty
21. Yin W, Xu B (2021) Effect of online shopping experience on customer loyalty in apparel business-to-consumer ecommerce. Textile Research Journal 91(23–24):2882–2895
22. Aslam W, Hussain A, Farhat K, Arif I (2020) Underlying factors influencing consumers' trust and loyalty in E-commerce. Business Perspectives and Research 8(2):186–204
23. Data is taken from kaggle.com. Link

A Novel Approach for Iris Recognition Model Using Statistical Feature Techniques

Sonali S. Gaikwad and Jyotsna S. Gaikwad

Abstract Numerous researchers have proposed iris recognition systems that use various techniques for extracting features for accurate and dependable biometric authentication. This study proposes and implements a statistical feature extraction approach based on the correlation between neighboring pixels. Image processing and enhancement methods are utilized to identify the iris. Statistical characteristics have been used to assess a system's performance. Experiments on the influence of a wide range of statistical characteristics have also been carried out. The studies' findings, which were based on a unique collection of statistical properties of iris scans, reveal a considerable improvement. A receiver operating characteristic curve is used to do the performance analysis.

Keywords Iris recognition system · Biometric · Statistical feature extraction

1 Introduction

Automated security of information and authentication of persons has invariably been an intriguing research subject. Biometric authentication methods are based on facial, finger, voice, and/or iris features [1–5]. In high-security areas, the iris recognition technology is often employed. Various algorithms for feature extraction have been proposed by a number of researchers. However, a small amount of work [6, 7] has been reported utilizing statistical techniques directly on pixel values to extract features. The term "preprocessing" refers to the process of converting an image of the eye into a format from which the required features may be extracted and utilized to identify an individual. Iris localization, iris normalization, and image enhancement are the three steps in the image processing process. Iris localization entails locating

S. S. Gaikwad
Shri Shivaji Science and Arts College, Chikhli, Buldhana, India

J. S. Gaikwad (✉)
Deogiri College, Aurangabad, India
e-mail: jyotsna10g@gmail.com

P. Dutta et al. (eds.), *Emerging Technologies in Data Mining and Information Security*, Lecture Notes in Networks and Systems 490,
https://doi.org/10.1007/978-981-19-4052-1_47

the inner and outer boundaries of the iris, as well as locating and removing any eyelashes of eyelids that may have obscured the iris region. Iris normalization is a process that converts an iris image from Cartesian coordinates to polar coordinates. A rectangular iris image with angular and radial resolutions is a normalized iris image. While capturing the image of an eye, normalization helps in removing the dimensional anomalies that arise due to variation in illumination, camera distance, angle, and so on. Now, the obtained normalized image is enhanced to adjust for the low contrast, poor light source, and light source position. Various researchers have proposed and implemented a variety of preprocessing algorithms [4, 8–11].

2 Preprocessing

The first step of a biometric system is image preprocessing. Preprocessing is used to separate the iris area from an eye picture. Noise in the iris area caused by reflection, light, and occlusion owing to eyelids or eyelashes is reduced in this stage. Several techniques for three steps of preprocessing have been developed by various researchers: iris localization, iris normalization, and iris picture enhancement.

2.1 Conversion of Iris Image

The method of converting a 2D picture to a 3D image is based on the grayscale and luminescence settings. A 2D image has just two dimensions: height and width, but a 3D image has a third dimension called depth in addition to height and breadth. In comparison with 2D images, 3D images provide more information and provide a better real-time world experience.

$$\text{grayscale}(2D) = \frac{R + G + B}{3} \tag{1}$$

Calculate the average of the three colors. Because the three separate colors have three different wavelengths and each contributes to the image format.

2.2 Intensity Transformation Function

When dealing with grayscale images, we sometimes want to adjust the intensity values. For example, we could want to flip the black and white intensities, or you might want to make the shadows darker and the lights brighter. Intensity transformations are used to increase the contrast between certain intensity values so that you can pick items out of an image. Changing the intensity is usually done by intensity transformation functions. The intensity transformation function is as follows:

$$G(j, k) = T[F(j, k)] \tag{2}$$

where $F(j, k)$ is the input image, $G(j, k)$ is the output image, and T is operator defined over a neighborhood.

2.3 Photographic Negative

The photographic negative is probably the easiest of the intensity transformations to describe. Assume we are working with grayscale double arrays with black at 0 and white at 1. The idea is that 0 s become 1 s, 1 s become 0 s, and all other gradients are reversed as well. In terms of intensity, this indicates that real black becomes true white and vice versa. If the image has an intensity level in the range $[0, L - 1]$, the intensity transformation is as follows:

$$s = L - 1 - r \tag{3}$$

where s and r represent the intensities.

2.4 Gamma Transformation

You may curve the grayscale components using gamma transformations to brighten the intensity (when gamma is less than one) or darken the intensity (when gamma is greater than one). The transformation function represents 1×1 neighborhood operation as point processing.

$$s = T(r) \tag{4}$$

where s and r represent the intensities of the image.

2.5 Logarithmic Transformations

To brighten the intensities of an image, logarithmic transformations, such as the gamma transformation, can always be applied. It is most commonly used to boost low-intensity values' detail or contrast. It is particularly good at bringing out detail in extended forms.

$$g = c * \log(1 + (f)) \tag{5}$$

Here, g is output image, c is the constant value of log function, and f is the input image.

2.6 Contrast-Stretching Transformations

Contrast-stretching transformations increase the contrast between the darks and the lights. Everything was kept at relatively similar levels, and the histogram was simply stretched to cover the image's intensity domain. Stretch the intensity around a certain level, with just a few levels of gray around the level of interest. The original value r is mapped to output value s utilizing the function.

$$s = (r - c)\left(\frac{b - a}{d - c}\right) \tag{6}$$

where a and d are lower limits of the input image and b and are upper limits of the input image, respectively.

2.7 Histogram Equalization

Histogram equalization is a computer-assisted image processing technique that improves image contrast. It does this by efficiently spreading out the most common intensity values, i.e., stretching the image's intensity range. When its usable data is represented by close contrast values, this method usually increases the global contrast of images. This allows areas with a lower local contrast to get a higher one. Following formula shows the contrast limited adaptive histogram equalization (CLAHE).

$$g = [g_{\max} - g_{\min}] * p(f) + g_{\min} \tag{7}$$

where

$g_{\max}$ Maximum pixel value
$g_{\min}$ Minimum pixel value
g Computed pixel value
$p(f)$ Cumulative probability distribution (CPD)

For exponential distribution gray level can be adapted as

$$g = g_{\min} - \left(\frac{1}{\alpha}\right) * \ln[1 - p(f)] \tag{8}$$

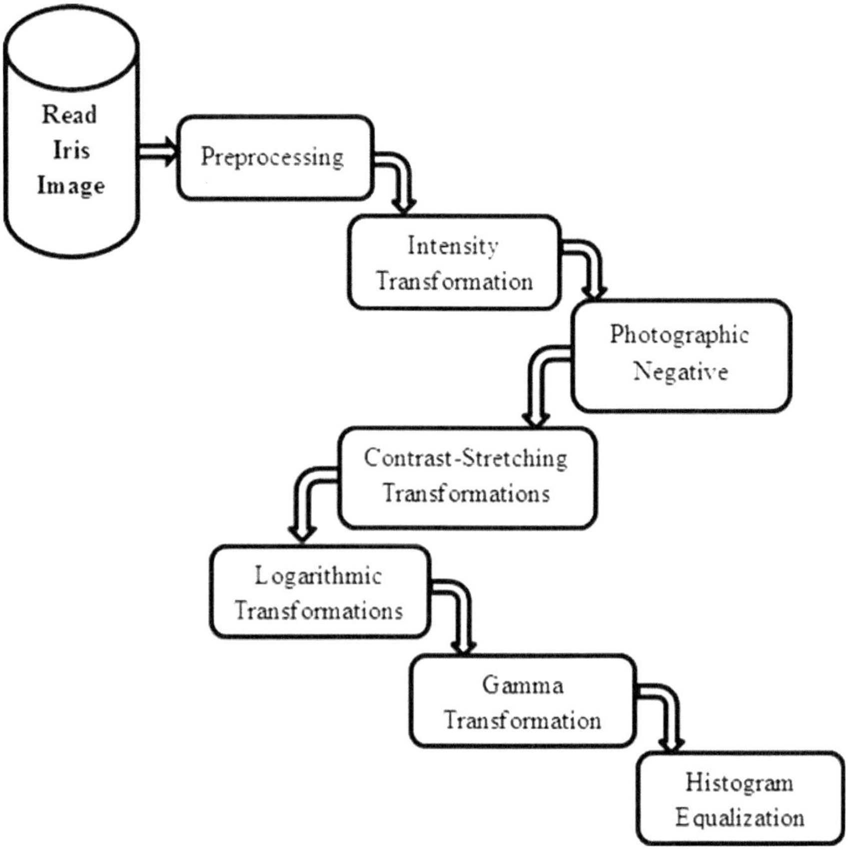

Fig. 1 Iris image enhancement

where $\alpha =$ Clip parameter.

Figure 1 shows the preprocessing operation on iris image for enhancement.

3 Methodology

After the iris picture has been preprocessed, feature extraction takes place. Feature extraction is a kind of dimensionality reduction in which a large number of pixels of an image are efficiently represented in such a way that interesting parts of the image are captured effectively. The initial step in feature extraction is to define edges that may be found in the image and then attempt to combine them into objects and shapes that are assumed to be in the image. The most basic definition of an edge is that it represents a change in the luminescence levels of neighboring pixels. Following the segmentation of iris images, feature extraction is performed by computing the area, length, thickness, diameter, and mean diameter.

Area

$$\text{Area} = \pi \times r^2 \tag{9}$$

Diameter

$$\text{Diameter} = \sqrt{\text{Area}/\pi} \tag{10}$$

Length

$$\text{Length} = \frac{\text{Area}}{2} \tag{11}$$

Thickness

$$\text{Thickness} = \frac{\text{Area}}{\text{Length}} \tag{12}$$

Mean Diameter

$$\text{Mean} = \frac{\sum X}{n} \tag{13}$$

Figure 2 shows the workflow of statistical feature extraction of iris.

3.1 Conversion of Iris Image:

The receiver operating a characteristic curve is an excellent method for determining a classification model's performance. For the probabilities of the classifier predictions, the true positive rate (TR) is pitted against the false positive rate (FR). The area under the plot is then calculated. Using varied probability thresholds, ROC curves summarize the trade-off between the true positive (TP) and false positive (FP) rates for a predictive model. When the observations are balanced between each class, ROC curves are appropriate, while precision-recall curves are appropriate for imbalanced datasets.

4 Result

The statistical features of iris are computed in order to identify people based on their iris. Area, length, thickness, diameter, and mean diameter of extracted iris are the

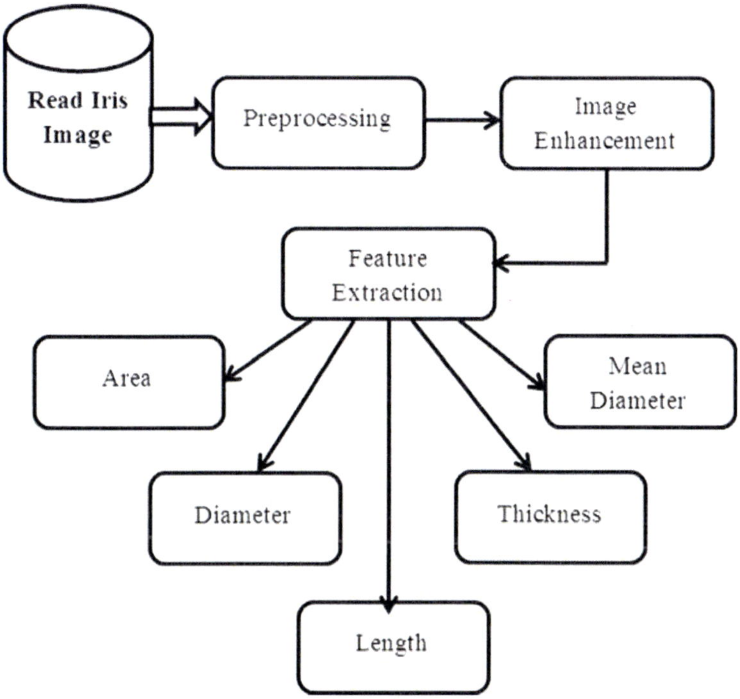

Fig. 2 Workflow of statistical feature extraction of iris

features. We can identify the person based on these statistical features. Tables 1, 2, and 3 show the output of statistical features (Figs. 3, 4 and 5).

Table 1 Statistical features of IIT Delhi iris image database

Sr. No	Area	Dia	Len	Thick	Mean Dia
1	3208	180	1604	2.00	20.00
2	5291	232	2646	1.95	19.98
3	7223	271	3612	1.93	19.97
4	2605	162	1303	1.95	19.94
5	6599	259	3300	1.97	19.98
6	4515	214	2258	1.98	19.94
7	1858	137	929	1.99	20.00
8	5489	236	2745	1.95	19.92
9	5822	243	2911	1.97	20.00
10	2291	152	1146	1.99	19.92

Table 2 Statistical features of UBIRIS iris image database

Sr. No	Area	Dia	Len	Thick	Mean Dia
1	3208	180	1604	2.00	20.00
2	5291	232	2646	1.95	19.98
3	7223	271	3612	1.93	19.97
4	2605	162	1303	1.95	19.94
5	6599	259	3300	1.97	19.98
6	4515	214	2258	1.98	19.94
7	1858	137	929	1.99	20.00
8	5489	236	2745	1.95	19.92
9	5822	243	2911	1.97	20.00
10	2291	152	1146	1.99	19.92

Table 3 Statistical features of UPOL iris image database

Sr. No	Area	Dia	Len	Thick	Mean Dia
1	135,189	1170	67,595	1.92	19.93
2	133,057	1161	66,529	1.94	19.23
3	135,026	1170	67,513	2.00	20.01
4	132,266	1158	66,133	2.10	20.05
5	132,712	1160	66,356	2.12	20.09
6	132,330	1158	66,165	2.23	20.15
7	131,260	1153	65,630	2.33	20.14
8	130,645	1151	65,323	1.97	19.97
9	131,699	1155	65,850	1.98	19.92
10	13,046	1150	65,231	1.97	19.94

5 Conclusion

Humans cannot fail to recall or lose their a physical characteristics in the manner that they can lose passwords or identity cards, so biometric techniques, which recognize individual's dependant on physical o. Among these biometric approaches, iris is now regarded as one of the most solid biometrics owing to its arbitrary variety of remarkable surface. Additionally, iris ends up being all around sheltered from the outside climate behind the cornea, moderately small to secure and stable everywhere in the individual's existence. Use open-source iris databases (IIT Delhi, UBIRIS, and UL iris database) to implement this algorithm. In this work, statistical feature extraction has been proposed and implemented. It has been shown that statistical features may be calculated using the area, diameter, length, thickness, and mean diameter of an iris image. System performance is satisfactory in both directions. Experimental results obtained using statistical feature extraction technique are encouraging. It has

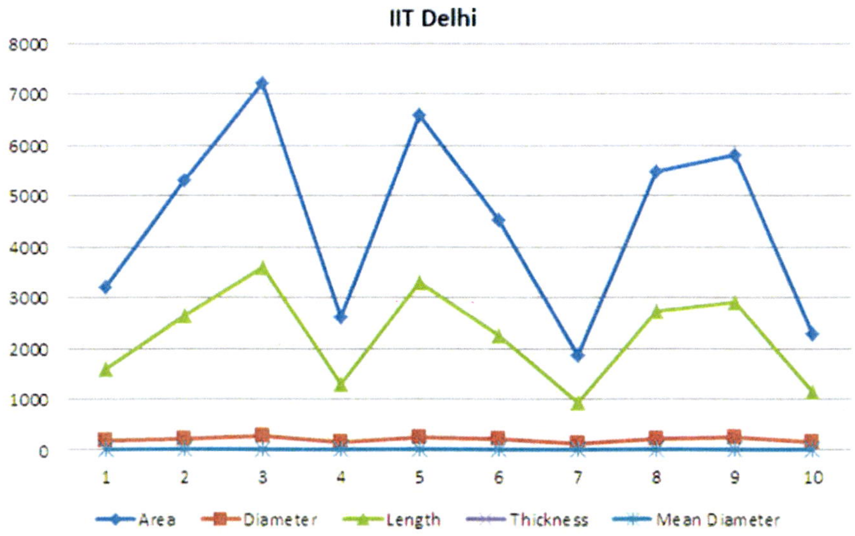

Fig. 3 Statistical features of IIT Delhi iris image database

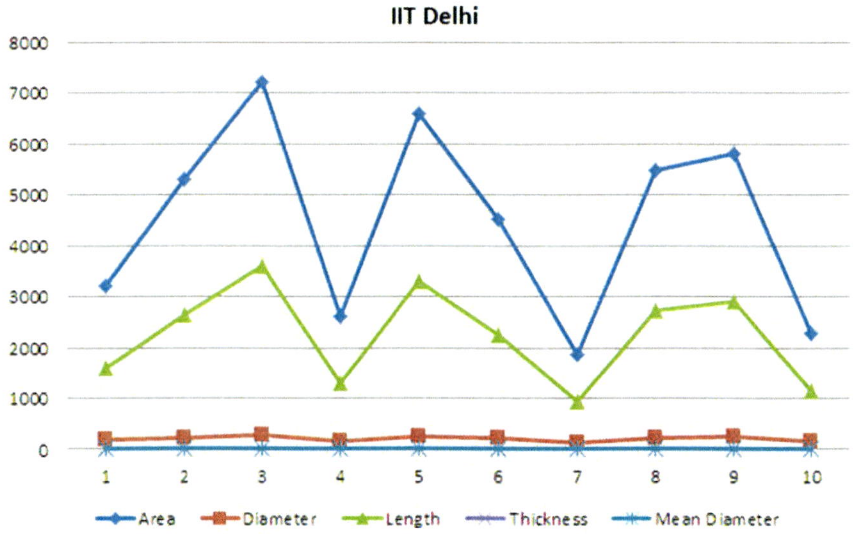

Fig. 4 Statistical features of UBIRIS iris image database

been shown that system performance improves as the number of statistical features increases.

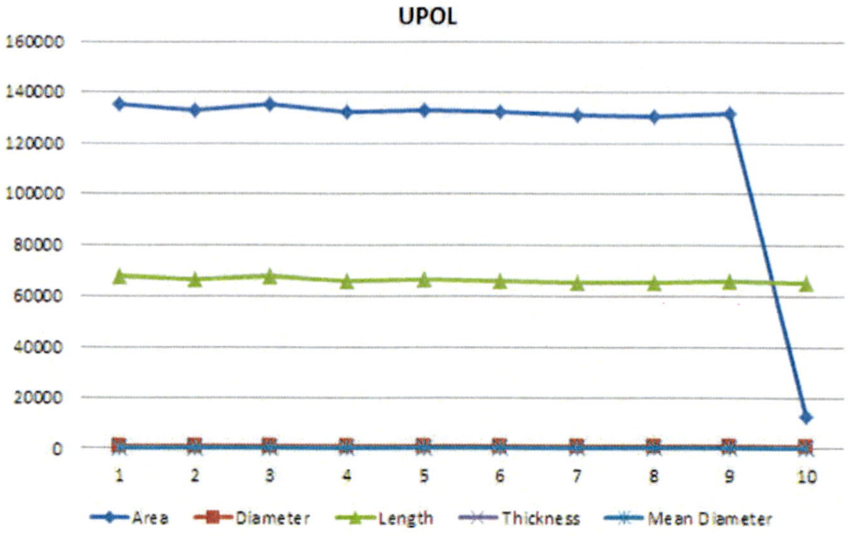

Fig. 5 Statistical features of UPOL iris image database

References

1. Giot R, Hemery B, Rosenberger C (2010) Low cost and usable multimodal biometric system based on keystroke dynamics and 2-D face recognition. In: Proceedings of twentieth IEEE international conference on pattern recognition, 23–26 August 2010, pp 1128–1131
2. Cao K, Eryun L, Jain AK (2014) Segmentation and enhancement of latent fingerprints: a coarse to fine ridge structure dictionary. IEEE Trans Pattern Anal Mach Intell 36(9):1847–1859
3. Senoussaoui M, Kenny P, Stafylakis T, Dumouchel P (2014) A study of the cosine distance-based mean shift for telephone speech diarization. IEEE Trans Audio, Speech Language Process 22(1):217–227
4. Daugman J (1993) High confidence visual recognition of persons by a test of statistical independence. IEEE Trans Pattern Anal Mach Intell 15:1148–1161
5. Daugman J (2004) How iris recognition works? IEEE Trans Circuits Syst Video Technol 14(1):21–30
6. Gook KJ, Hee GY, Hee YJ (2006) Iris recognition using cumulative sum based change analyses. In: International symposium on intelligent signal processing and communication system, pp 275–278
7. Sint KKS (2009) Iris recognition system using statistical features for biometric identification. In: Proceedings of international conference on electronic computer technology, pp 554–556
8. Bansal A, Agarwal R, Sharma RK (2010) Trends in iris recognition algorithm. In: Proceedings of IEEE fourth Asia international conference on mathematical/analytical modeling and computer simulation, pp 337–340
9. He Z, Tan T, Sun Z, Qiu X (2009) Toward accurate and fast iris segmentation for iris biometrics. IEEE Trans Pattern Anal Mach Intell 1670–1684
10. Kumar A, Passi A (2010) Comparison and combination of iris matchers for reliable personal authentication. Pattern Recognit 43(3):1016–1026
11. Su L, Li Q, Yuan X (2011) Study on algorithm of eyelash occlusions detection based on endpoint identification. In: Proceedings of third international workshop on intelligent systems and applications (ISA), pp 1–4

Sentiment Analysis of User Groups in an Working Environment Using CNN for Streaming Data Analysis

G. T. Tarun Kishore, A. Manoj, G. Sidharth, T. R. Abijeeth Vasra, A. Sheik Abdullah, and S. Selvakumar

Abstract The domain of sentiment analysis is mainly concerned with observing the nature of text with positive, negative, and neutral effects in a given environment. It is also called the process of opinion mining in which the emotion behind the idea, service, or product completely signifies the nature of the environment as observed. The insights from this unorganized textual data can be evaluated along with the methods available in machine learning. This research work completely focuses on the mechanism of observing the project teams with a significant analysis in monitoring the percentage of happiness involved in executing a specified project. In addition to the collection of views from the different groups of employers, we have observed their nature through a Webcam-enabled platform to best determine the work nature of team members. We have used CNN with the available streaming data and captured the nature of workers in a dignified environment. A set of statistical measures has been evaluated to best validate the proposed method which extracts sentiments for the observed data. Future work can be progressed with the extraction of organizational data rather than focusing on working teams in a given environment.

Keywords Data analytics · Machine learning · Sentiment analysis · Neural network · Statistical analysis

G. T. T. Kishore (✉) · A. Manoj · G. Sidharth · T. R. A. Vasra
Department of Information Technology, Thiagarajar College of Engineering, Tamil Nadu, Madurai 625 015, India
e-mail: tarunksu30@gmail.com

G. Sidharth
e-mail: sidharth@student.tce.edu

T. R. A. Vasra
e-mail: abijeeth@student.tce.edu

S. Selvakumar
Department of CSE, Visvesvaraya College of Engineering and Technology, Hyderabad, India

A. S. Abdullah (✉)
School of Computer Science and Engineering, Vellore Institute of Technology, Chennai, India
e-mail: sheikabdullah.a@vit.ac.in

P. Dutta et al. (eds.), *Emerging Technologies in Data Mining and Information Security*, Lecture Notes in Networks and Systems 490,
https://doi.org/10.1007/978-981-19-4052-1_48

1 Introduction

The target behind the successful execution of a project lies in the involvement of the working teams in a given environment. The work process, nature of the people, and their innovations lie at the support provided by the industrial and organizational practices. Collecting the user responses and making their inferences at the higher level need some form of analysis which in turn provides greater insights for the user to observe. The aspect of analyzing the unstructured and semi-structural data can be made using sentiment analysis in order to observe the effects focusing on positive, negative, and neutral responses [1].

The measure of happiness in the execution of a project provides an environment with complete progress toward the project. The process of extracting opinions lies at different stages of analysis. The polarity of opinions is completely based on the nature of the subjective information or the attitudinal content which is collected from the user groups. At certain stages, the honest feelings and the relevance of a particular user can be easily determined with the classification schemes available in machine learning [2].

In this research work, we have deployed convolutional neural network (CNN) to classify the observed set of sentiments of a particular user group in a dignified environment. The main focus to use CNN is that it automatically detects the best set of features without any intervention by programmers. Also, the computational cost and its efficiency of CNN are found to be good when compared to other such algorithms.

2 Literature Review

The work by the authors [3] provided a significant and robust model for gesture recognition systems in accordance with the interpretation of human gestures. In this work, real-time data are used for analysis to model out the sequential frames to achieve a good rate of classification. The experimental results have been made using 3D-CNN along with the LSTM model to extract the set of features of shorter duration. From this combined model evaluation process, a set of pretrained features has been used at intermediate levels of frame execution and incorporation. The future work can be focused on further improvisation of the model space with regard to time and space complexity.

The incorporation of morphological filters and approximation strategies is well used in different image processing units. The work by the authors [4] provided experimentation for gesture recognition using CNN. They have used different forms of filters for the best determination of gesture recognition in the given environment. The evaluated metrics provided valid results with the significant identification of gestures for the observed data.

The work by the authors [5] provided a vision-based gesture recognition system using the recurrent network in order to determine the hand gestures and their significant variations. In this work, the authors have analyzed different forms of video segments with a multi-frame classification process. With the execution of multiple frames, there arises computational complexity during the execution process. In order to overcome this problem, the authors proposed novel tiled image and binary patterns with segment-based evaluation process using deep CNN.

The work by the authors [6] provided a decision support model for the assessment of risk factors using step-wise regression upon statistical evaluation. The model is used for the best determination of risk factors that specifically focus on the determination of the occurrence of a disease. The experiments showed that the proposed model provided an improvement in accuracy of about 89.72%.

Meanwhile, this research work focuses on the applicability of CNN along with the sentiment extraction of opinions for a set of users in a dignified environment. The evaluation has been made in such a way that the model variants are evaluated with a significant difference in performance metrics. The implementation includes qualitative questions to allow valuable interpretations and efficient discussion within the team.

3 Proposed Methodology

The private dataset has the following contents:

- Train data folder
- Validation data folder
- train.csv
- test.csv

  The train and validation folder consists of gestures that have been continuously monitored and recorded by the Webcam mounted on the laptop. Each gesture corresponds to a specific command:
- Thumbs-up (Fig. 1)
- Thumbs-down
- Left swipe
- Right swipe
- Stop.

Every video captured will be split into a sequence of 30 frames. The training data consist of 663 sequences of video data categorized into one of the abovementioned five classes. Similarly, the validation data consist of 100 sequences of video data categorized into one of the abovementioned five classes. All the images in a particular video subfolder have the same dimensions, but different videos may have different dimensions. Specifically, videos have two types of dimensions—either 360 × 360 or 120 × 160 which is dependent on the type of the Webcam that has been used to record the video.

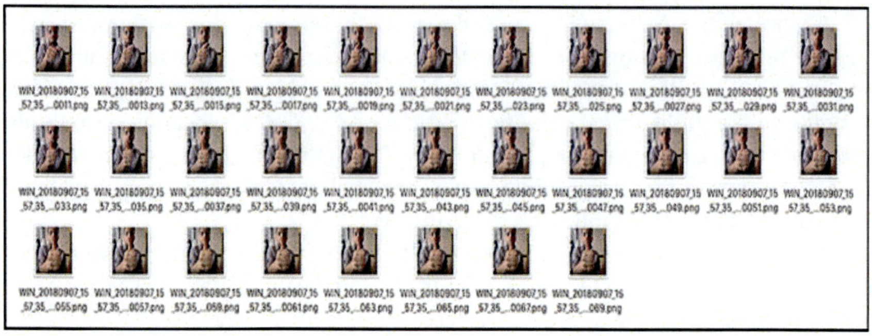

Fig. 1 Video frames in one thumbs-up folder

The CNN algorithm consists of about convolutional and the sub-layers which are fully connected in a network of action. The input is given in the form of $m \times m \times r$ image in which the value of m and n corresponds to the height and width of the image used for analysis. The value of r corresponds to the number of channels, and here, we have set the value of $r = 3$. The filters at a significant level are used in accordance with the input image in order to generate the k-feature maps. The size of the map is assigned to be $m - n + 1$ with sub-sampling across maximum pooling over the contiguous regions. Here, the maximum pooling is set as 2 for the images along with the size considered for evaluation. Also, an additive bias and a sigmoid nonlinear function are applied during the execution of the convolution layer and its processing [7].

In order to make the dataset available for analysis, we have adopted the following preprocessing techniques (Fig. 2):

- Images resize
- Image cropping
- Normalization.

As we already know, in most deep learning projects, we have to feed data to the model in batches as the data size will be huge and to utilize the full potential of the CPU and the GPU. This is generally done using the concept of generators.

Creating data generators is probably the most important part of building a training pipeline. Although libraries such as Keras provide built-in generator functionalities, they are often restricted in scope, and we have to write our own generators from scratch. In this project, we will implement our own custom generator; our generator will feed batches of videos, not images [8].

For example, assume we have 23 samples, and we pick batch size as 10. In this case, there will be two complete batches of ten each.

- Batch 1: 10
- Batch 2: 10
- Batch 3: 3.

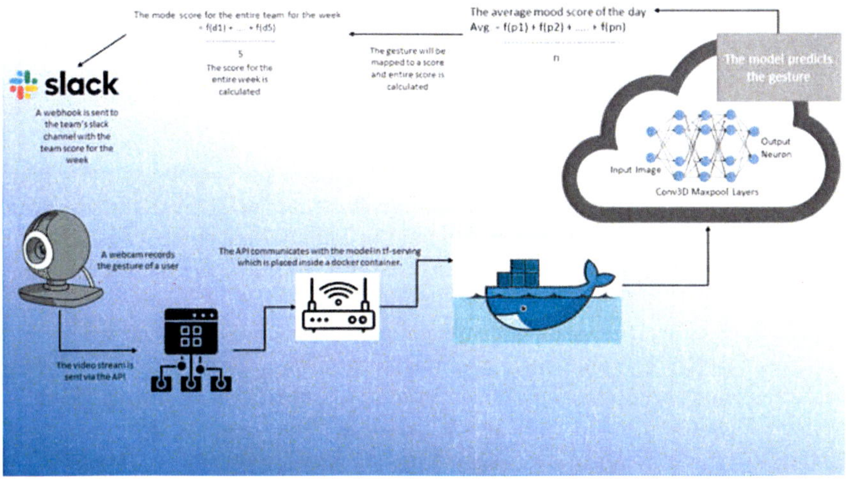

Fig. 2 Proposed methodology

The final run will be for the remaining batch that was not part of the full batch. Full batches are covered as part of for loop, and the remaining batches are covered. In a case, where in the batch size is 30, but, we have only 23 samples, then there will be only one single batch with 23 samples.

A Web page will be designed and hosted on a Docker container which puts forth questions which can be used to determine the mood of the user. The user will respond accordingly using one of the three gestures thumbs-up, thumbs-down, and neutral. A timer will start, and the Web cam on the user's laptop opens. It will record a 5 s video of the gesture that the user performs [9].

This video will be sent as an input via an API to the model which is resting on the Docker container with tf-serving. The video here will be broken down into frames. 30 frames which consist the clear gesture data will be sent to the model which will contain the pretrained weights. The model will predict the probability of what the gesture could be. The maximum predicted score by all the neurons will be chosen as the correct prediction, and it will be mapped to its corresponding label.

The label will then be mapped to a corresponding mood value. In this way, the mood of the entire team will be calculated for a particular day. Then, the average score of the entire team for a particular week will be calculated. This score will be the ultimate team mood value and will be sent through a Web hook to the team's slack channel [10].

4 Experimental Results and Discussion

During the initial phases of experimentation, the necessary library packages are imported, and the respective TensorFlow package is chosen. The training and the validation path is set across the training and the testing frames with the respective gestures and its parametric values. The frames are fed in such a way that the height, width, number of channels, and number of frames are initialized in accordance with the input as fed at the initial level. Continuously, batch-wise input is fed along with the labels. Similar augmentation methods like affine transformation and horizontal flipping are used to observe the best results from the input data. A generator function is used which collects the path corresponding to the source file, batch size as the input parameters. The entire channel is transformed into 4D shape for processing each frame along with modeling the architecture. The architecture is described in Fig. 3.

Once the model weights are collected in a.h5 file, a signature for the model is created. SignatureDefs aim to provide generic support to identify inputs and outputs of a function and can be specified when building a SavedModel [11, 12]. Once the

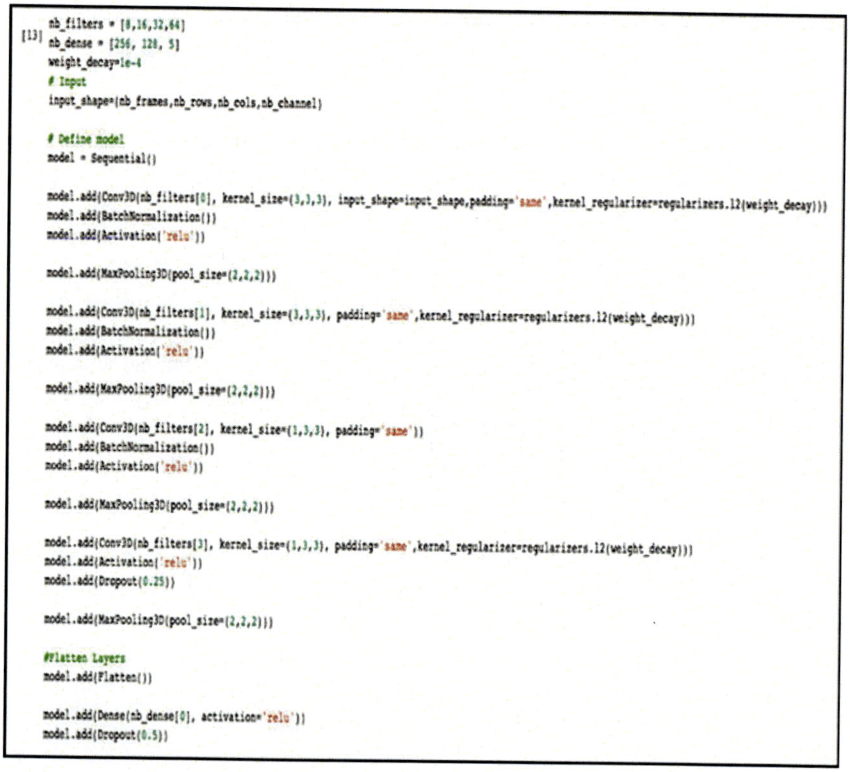

```
[13] nb_filters = [8,16,32,64]
nb_dense = [256, 128, 5]
weight_decay=1e-4
# Input
input_shape=(nb_frames,nb_rows,nb_cols,nb_channel)

# Define model
model = Sequential()

model.add(Conv3D(nb_filters[0], kernel_size=(3,3,3), input_shape=input_shape,padding='same',kernel_regularizer=regularizers.l2(weight_decay)))
model.add(BatchNormalization())
model.add(Activation('relu'))

model.add(MaxPooling3D(pool_size=(2,2,2)))

model.add(Conv3D(nb_filters[1], kernel_size=(3,3,3), padding='same',kernel_regularizer=regularizers.l2(weight_decay)))
model.add(BatchNormalization())
model.add(Activation('relu'))

model.add(MaxPooling3D(pool_size=(2,2,2)))

model.add(Conv3D(nb_filters[2], kernel_size=(1,3,3), padding='same'))
model.add(BatchNormalization())
model.add(Activation('relu'))

model.add(MaxPooling3D(pool_size=(2,2,2)))

model.add(Conv3D(nb_filters[3], kernel_size=(1,3,3), padding='same',kernel_regularizer=regularizers.l2(weight_decay)))
model.add(Activation('relu'))
model.add(Dropout(0.25))

model.add(MaxPooling3D(pool_size=(2,2,2)))

#Flatten Layers
model.add(Flatten())

model.add(Dense(nb_dense[0], activation='relu'))
model.add(Dropout(0.5))
```

Fig. 3 Model architecture

```
Epoch 18/20
67/67 [==============================] - ETA: 0s - loss: 0.2158 - categorical_accuracy: 0.9204
Epoch 00018: val_loss did not improve from 0.35137
67/67 [==============================] - 31s 455ms/step - loss: 0.2158 - categorical_accuracy: 0.9204 - val_loss: 0.4958 - val_categorical_accuracy: 0.8300 -
Epoch 19/20
67/67 [==============================] - ETA: 0s - loss: 0.1426 - categorical_accuracy: 0.9685
Epoch 00019: val_loss did not improve from 0.35137

Epoch 00019: ReduceLROnPlateau reducing learning rate to 3.051757957450718e-08.
67/67 [==============================] - 31s 461ms/step - loss: 0.1426 - categorical_accuracy: 0.9685 - val_loss: 0.5152 - val_categorical_accuracy: 0.8000 -
Epoch 20/20
67/67 [==============================] - ETA: 0s - loss: 0.2528 - categorical_accuracy: 0.9270
Epoch 00020: val_loss did not improve from 0.35137
67/67 [==============================] - 30s 449ms/step - loss: 0.2528 - categorical_accuracy: 0.9270 - val_loss: 0.3662 - val_categorical_accuracy: 0.8400 -
<tensorflow.python.keras.callbacks.History at 0x7f57d2043b70>
```

Fig. 4 Output model predicting the gestures

Table 1 Performance analysis of algorithms with different subsets

Model with Variants	Train (epoch 10)	Val (epoch 10)	Train (epoch 20)	Val (epoch 20
3c	0.6169	0.7300	0.6401	0.5700
3e	0.7197	0.5500	0.8126	0.6200
4c	0.5266	0.4500	0.6039	0.6500
4d	0.7751	0.6500	0.91358	0.7667
4e	0.8740	0.7600	0.9088	0.8100
4f	0.9337	0.8300	0.9270	0.8400

signature is created, the model will be saved using the save_model syntax which will save the model in the format of .pb.A folder will be created with variables, assets folders, and a.pb file which will be used to deploy in the TensorFlow serving. Start the Docker in the system. Pull the TensorFlow-serving image into a Docker container using the Docker-pull command. Then, run this model inside the Docker container on the TensorFlow-serving image using the following command:

docker run -p 8501:8501 –mounttype = bind, source = /absolute/path/to/the model/model_name, target = /models/model_name.

-e MODEL_NAME = model_name -t tensorflow/serving (Fig. 4).

Performance analysis of algorithms with different subsets is represented in Table 1.

5 Conclusion

The feedback from employees at regular intervals plays a significant role in determining the features supported/provided by the organization. In this research work, we have evaluated the process in three different stages with the target in determining the gestures using CNN. Here, TensorFlow with a Docker-based evaluation paradigm is used along with the production environment. The system is justified with end-to-end testing in local machines, and the interaction is made through an application program interface. Once the model classifies the gesture, an evaluation function is used to calculate the average mood of all the team employees for the entire week. This mood report will then be sent to the team's slack channel. Then, the model will be integrated with the CI/CD framework. As an outcome, the team mood and their representation are evaluated through sentiments, i.e., opinion extraction process in order to best determine the nature of employees in an environment. As future work, we look forward to getting the feedback of the model's working and retraining the model with additional dataset and hybrid approach to get high accuracies and avoid cases of failure of the model in certain cases.

References

1. Akash K, Sheik Abdullah A (2020) A new model of zero energy air cooler: an cost and energy efficient device in exploit. Electric power and renewable energy conference (EPREC-2020) organized by the Department of Electrical Engineering, National Institute of Technology (NIT) Jamshedpur, India during May 29th–30th, 2020. Lecture Notes in Electrical Engineering, Springer
2. Araque O, Zhu G, Iglesias CA (2019) A semantic similarity-based perspective of affect lexicons for sentiment analysis. Knowledge-Based System 165:346–59. Available from: https://doi.org/10.1016/j.knosys.2018.12.005
3. Mullick K, Namboodiri AM (2017) Learning deep and compact models for gesture recognition. In: 2017 IEEE international conference on image processing (ICIP), IEEE. Sept 2017. Available from: http://dx.doi.org/https://doi.org/10.1109/icip.2017.8297033
4. Hatibaruah D, Talukdar AK, Kumar Sarma K (2020) A static hand gesture based sign language recognition system using convolutional neural networks. In: 2020 IEEE 17th India council international conference (INDICON) [Internet], IEEE. 10 Dec 2020; Available from: http://dx.doi.org/https://doi.org/10.1109/indicon49873.2020.9342405
5. John V, Boyali A, Mita S, Imanishi M, Sanma N (2016) Deep learning-based fast hand gesture recognition using representative frames. In: 2016 international conference on digital image computing: techniques and applications (DICTA) [Internet]. IEEE. Nov 2016. Available from: http://dx.doi.org/https://doi.org/10.1109/dicta.2016.7797030
6. Harsheni SK, Souganthika S, GokulKarthik K, Sheik Abdullah A, Selvakumar S (2019) Analysis of the risk factors of heart disease using step-wise regression with statistical evaluation. Emerging techniques in computing and expert technology, pp 712–718. Springer. https://doi.org/10.1007/978-3-030-32150-5_70, Online ISBN 978-3-030-32150-5
7. Sheik Abdullah A, Selvakumar S, Parkavi R, Suganya R, Venkatesh M (2019) An introduction to survival analytics, types, and its applications. Biomechanics, Intech Open Publishers, UK. https://doi.org/10.5772/intechopen.80953
8. Abdullah AS, Selvakumar S, Karthikeyan P, Venkatesh M (2017) Comparing the efficacy of decision tree and its variants using medical data. Indian Journal of Science and Technology 10(18). https://doi.org/10.17485/ijst/2017/v10i18/111768
9. Hung C, Chen S-J (2016) Word sense disambiguation based sentiment lexicons for sentiment classification. Knowledge-Based Systems 110:224–232. Available from: https://doi.org/10.1016/j.knosys.2016.07.030
10. Patel A, Tiwari AK (2019) Sentiment analysis by using recurrent neural network. SSRN Electronic Journal. https://doi.org/10.2139/ssrn.3349572
11. Naz S, Parveen S (2021) Twitter sentiment analysis using convolutional neural network. SSRN Electronic Journal Available from: https://doi.org/10.2139/ssrn.3852906
12. Abdullah AS, Ramya C, Priyadharsini V, Reshma C, Selvakumar S (2017) A survey on evolutionary techniques for feature selection. In: 2017 conference on emerging devices and smart systems (ICEDSS), IEEE. Available from: https://doi.org/10.1109/icedss.2017.8073659

A Short Review on Automatic Detection of Glaucoma Using Fundus Image

Neha Varma, Sunita Yadav, and Jay Kant Pratap Singh Yadav

Abstract Among the eye diseases, glaucoma is considered the second leading disease all around the world that develops intraocular pressure and damages optic nerve head (ONH) within the human eye, which causes vision loss. Early stage diagnosis of glaucoma can provide various benefits using fundus images. This study introduces various types of glaucoma, their associated risk factors, and a short review of various methods used to identify glaucoma automatically. This review discusses various image processing and deep learning-based techniques in detail. In this review, in the image processing-based techniques section, a high emphasis is given to segmentation techniques used in handcrafted features extraction. However, features are extracted automatically in deep learning techniques, so high emphasis is given to various deep learning models and classifiers in the deep learning-based techniques section. This study briefly described the details about the publicly available datasets. The study also discussed the analysis of various medical features such as CDR, ISNT, NRR, RNFL, GRI, and performance metrics, namely sensitivity, specificity, and accuracy, used for glaucoma detection.

Keywords Automatic diagnosis · CDR · Fundus image · Optic disk · Optic cup · Chronic glaucoma · Acute angle-closure glaucoma · ISNT · NRR · RNFL

1 Introduction

Glaucoma is the second leading visual-impaired eye disease that causes damage to the optic nerve and may lead to vision loss. Due to intraocular pressure increases, the risk of glaucoma increases or when an excessive amount of fluid is developed in the eye, resulting in the blockage of outflow channels of the eye. The significant symptoms

N. Varma · S. Yadav · J. K. P. S. Yadav (✉)
Department of Computer Science and Engineering, Ajay Kumar Garg Engineering College, Ghaziabad, Uttar Pradesh, India
e-mail: yadavjaykant@akgec.ac.in

S. Yadav
e-mail: yadavsunita@akgec.ac.in

P. Dutta et al. (eds.), *Emerging Technologies in Data Mining and Information Security*, Lecture Notes in Networks and Systems 490,
https://doi.org/10.1007/978-981-19-4052-1_49

Table 1 Comparison between chronic glaucoma and acute angle-closure glaucoma

Chronic glaucoma	Acute angle-closure glaucoma
Another name of open-angle glaucoma is primary or chronic glaucoma	Another name of angle-closure glaucoma is closed-angle glaucoma or acute glaucoma
It develops very slowly	It develops very fast

of glaucoma are blurred vision, redness in the eye, head pain, and high intraocular pressure. The risk factors associated with glaucoma are high blood pressure, thin cornea eye, family history of glaucoma, diabetes, severe myopia, and age. Glaucoma can be broadly classified into chronic glaucoma and acute angle-closure glaucoma, described in detail in Table 1.

Now a day, glaucoma is a common disease and known for affecting a significant part of the population worldwide, with more than 64 million cases globally reported in 2013, while the number will be increased up to 111.8 million by 2040 [1, 2]. It is also reported in the literature that chronic glaucoma affects around 90% of total patients that are affected by glaucoma worldwide. In India, this figure is 11.2 million and contributes by 5.9% to the country's total blindness [3].

1.1 Fundus Images

Fundus image gives detailed information about the internal structure of the human eye and is popularly used in the diagnosis of eyes. It includes both types of images, namely normal and abnormal images. This section elaborated and discussed some essential fundus image features like an optic disk, optic cup, and blood vessels used to diagnose glaucoma disease at the early stage. Optic disk (OD) is a circular area located at the back position inside the human eye that transfers the signals where the optic nerve connects to the retina, allowing us to see beautiful objects. The optic cup lies within the central region of the optic disk. Generally, the color of the optic cup is white and more petite in size [4]. Retinal blood vessels transfer the blood to the inner neuron within the human eye [4]. It contains essential information that helps to diagnose glaucoma disease with deep learning models. Figure 1a shows the glaucoma fundus image. As the intraocular pressure increases, the value of both CDR and RDR will be affected. Both CDR and RDR are the key features used for diagnosing glaucoma. When the intraocular pressure increases in the human eye, it reduces the flow of blood passage of the optic nerve, thus resulting in glaucoma disease. Figure 1b shows the normal fundus image.

The remaining paper is organized as follows. Publicly available datasets for glaucoma detection are discussed in Sect. 2. Related works are presented in Sect. 3. Section 4 presents a brief overview of the proposed methodology. Section 5 includes

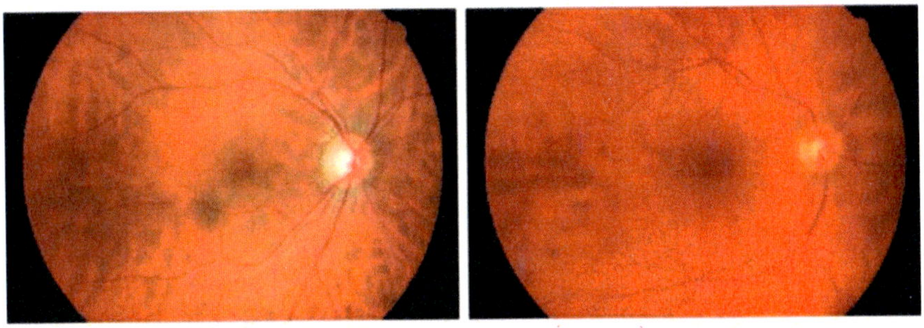

(a) (b)

Fig. 1 Sample of fundus image. **a** Glaucoma image. **b** Healthy image [4]

the results done by researchers in their past work. Section 6 includes some analysis part done in the past based on the performance metric parameters for glaucoma diagnosis. Finally, in Sect. 7, the conclusion is presented.

2 Publicly Available Dataset

This section presents an overview of various publicly available datasets, for example, REFUGE, ORIGA, HRF, DRISHTI-GS, and Sichoi-86 HRF, PRV glaucoma, and DRIONS-DB. The Retinal Fundus Glaucoma Challenge (REFUGE) [5] dataset contains 1200 total fundus images that perform mainly three jobs: glaucoma classification, optic disk segmentation, and fovea localization that consists of 300 and glaucoma of 101, and each image size is 2124 × 2056 pixels. This dataset has been created to evaluate the ratio of the optic disk to the cup. Online Retinal Fundus Image Dataset for Glaucoma Analysis (ORIGA) [5] dataset is used to optimize the image processing technologies and help to build new tools for diagnosing glaucoma. There is a total of 650 fundus images, of which 168 images are glaucomatous and 482 are normal images. The HRF [6] dataset contains a total of 45 images, from which 15 images are considered glaucomatous, and the remaining images are normal, and each image pixel size is 2336 × 2056 pixels. This dataset is used to diagnose glaucoma disease automatically. High-resolution fundus (HRF) dataset used to detect and diagnose glaucoma. In Sichoi-86 HRF [6] dataset, there were 401 fundus images. In the DRISHTI-GS dataset, there are 110 fundus images of size 2896 × 1944 pixels and used for segmentation purposes. PRV glaucoma [6] dataset contains 659 fundus images of size 2880 × 2160 pixels, from which 424 images are glaucomas and 235 images are normal. DRIONS-DB [6] is also used for segmentation purposes. Among 110 fundus images, 60 were normal images, and 50 exhibited the sign of glaucoma, and the size of each image is 600 × 400 pixels. All of these publicly available datasets are briefly summarized in Table 2.

Table 2 Summary of various publicly available dataset names with image types and their dimensions in pixels

Dataset name	Images	Dimension
REFUGE [5]	1200 Images	2124 × 2056 pixels
ORIGA [5]	650 Images (168 glaucomas, 482 normal)	420 × 420 pixels
HRF [6]	45 Images (15 glaucomas, 15 normal)	2336 × 2056 pixels
DRISHTI-GS [5]	101 Images (50 glaucomas, 51 normal)	2896 × 1944 pixels
Sichoi-86 HRF [6]	401 Images (300 normal, 101 glaucomas)	–
PRV Glaucoma [6]	659 Images (424 glaucomas, 235 normal)	2880 × 2160 pixels
DRIONS-DB [6]	110 Images (60 glaucomas, 50 normal)	600 × 400 pixels

3 Related Work

In the medical field, various techniques for glaucoma diagnosis and classification like image processing, machine learning, and deep learning were utilized by the researchers in their work. This section presents a literature survey of glaucoma diagnosis using fundus images are.

(A) **Methods using image processing techniques for glaucoma diagnosis**

Mukherjee et al. [7] introduced a system that includes the feature analysis part of the focal notching by using the neuroretinal rim and cup-to-disk ratio values. They used the thresholding approach by reducing the vessel and segment OD and OC that made segmentation easier. Then, they computed the value of NRR by using the ISNT rule. They also calculate the CDR value for glaucoma detection. They used a machine learning approach named SVM for decision-making that classified the dataset into two categories: glaucoma and normal. They achieved an accuracy of 87.128% for glaucoma detection. The advantage of using the thresholding approach is that it gives better results for cup extraction, and the drawback of the SVM algorithm is that it takes more time while using large datasets. Jain et al. [8] suggested a method for glaucoma identification using fundus images. They used tensor-based empirical wavelet transform based on the single decomposition of the Fourier spectrum. Correntropy features were examined; they then used a t-test approach to extract them. They also used the SVM classifier to classify the fundus images into normal and glaucoma images. The proposed method gained an accuracy of 98.7% for glaucoma detection. The advantage of the t-test is that it provides the best result on the green channel, and EWT does not include any predefined function. Carrillo et al. [9] designed a system for glaucoma detection using fundus images. The drawback of this proposed system is that it uses a small dataset. They used the red channel for disk segmentation and the green channel for vessel segmentation. For cup segmentation, they used the ISNT rule. After that, they calculated the CDR value for each image. They achieved 88.5% of accuracy for glaucoma detection using fundus images. They achieved 95% accuracy for the disk segmentation, which reduces the noise problem. A study conducted by Veena et al. [5] worked on many segmentation and classification techniques for glaucoma detection. They also include the challenge that arises when too much optic disk and cup boundary segmentation is performed because many blood vessels are present within the retinal images. To avoid this problem, disk segmentation is performed on the red channel. They used the KNN approach, but the drawback of the proposed system was the cost of computation. Mohamed et al. [10] designed a system that segments the optic cup and disk area. After applying the segmentation process, they also extract the features like mean, variance, kurtosis, and skewness. For the classification purpose, they used an SVM classifier in combination with linear function kernel and RBF. The proposed system achieved accuracy, sensitivity, and specificity of 98.63%, 97.60%, and 92.3%, respectively. Rehman et al. [11] have proposed a system for glaucoma identification using SVM with a superpixel technique. They used image preprocessing steps, namely removal of noise, image enhancement, and cropping OD-edge enhancement. The proposed system gained accuracy, sensitivity, and specificity as 99.30%, 99.40%, and 96.90%, respectively.

From the review work, we have seen that the performance of image processing-based methods depends on manual feature extraction. Hence, these methods suffered from certain limitations: Manual feature extraction is a time-taken step and needs expert advice from ophthalmologists. In the presence of large blood vessels, sometimes, segmentation of cup and disk becomes a difficult task. So, we need to design a system using deep learning models for automatic feature extractions that can overcome all of the above-given limitations.

(B) **Methods using deep learning approach for glaucoma diagnosis**

Thakur et al. [12] suggested a technique based on the OHTS retinal image. They presented an alternative approach for glaucoma detection. The approach uses contrast enhancement and the Gaussian filtering method to enhance the retinal image features. After that, they used CNN, Mobile Net V2 model to train the model. The proposed system achieved 94% of accuracy. The drawback of the OHTS dataset was collected from a restricted clinic. Karthiyayini and Shenbagavadivu [13] performed the experiment using 400 retinal images. They performed grayscale conversion into a binary image, crop ROI, and resized the image and separation of RGB channels. After that, these images were segmented by using RGB channels. Then perform feature extraction to extract the optic cup and disk features. They proposed an algorithm as EARMAM for the prediction of disease. The proposed EARMAM method achieved 89% of accuracy. Saxena et al. [14] designed a system using the CNN model. For proper extraction of ROI, the system makes faster that helps for glaucoma detection easier. They used SCES and ORIGA datasets for glaucoma detection. The success rate was 82.2 and 88.2% for the SCES and ORIGA datasets. The advantage of CNN is that it reduces the extra work and enhances the system's performance. Masot et al. [15] proposed a system for glaucoma detection using machine learning and segmentation techniques using fundus images to detect optic disk and cup independently, then combine them and extract the features. They utilized U-net architecture to segment the optic disk and cup from fundus images. The success rate was 84% for cup segmentation using the RIM-ONE dataset and 89% using the DRISHTI-GS dataset. The success rate was 92% for disk segmentation using the RIM-ONE dataset and 93% using the DRISHTI-GS dataset. Liu et al. [16] introduced an approach by using the CNN model for automatic classification and detection of glaucoma to screen a large-scale database of images. Fu et al. [17] developed an automatic system for glaucoma detection by using M-Net to calculate the CDR ratio, and it contains U-shape CNN and DENet. The advantage of the proposed system is that segmentation is not needed. They obtained an accuracy of 89.9% on M-Net using the SCES dataset, whereas they achieved an accuracy of 91.8% on the DENet using the SCES dataset. Christopher et al. [18] proposed a system by using ResNet50 and transfer learning approach for glaucoma detection, and they achieved accuracy, sensitivity, and specificity as 97%, 93%, and 92%, respectively. They preprocessed the fundus images to segment the optic disk for glaucoma detection. Li et al. [19] designed a system using the ResNet101 technique for glaucoma detection. The proposed system gained accuracy, sensitivity, and specificity as 94.1%, 95.7%, and 92.9%, respectively.

From the review, we stated some drawbacks of the existing system methods applied to detect glaucoma: The SVM algorithm gives higher accuracy than ResNet, but SVM does not work for large datasets, whereas CNN takes less time on a large dataset. CNN model obtained high accuracy on ORIGA dataset as compared with SCES dataset. The CNN model shows better performance for classifying the images similar to the dataset. However, if the images were rotated, CNN has difficulty performing the task. This problem can be overcome by adding multiple copies of images, and this process is known as the augmentation of data.

4 Proposed Methodology

This section is used to propose a method for automatic glaucoma detection and classification. Figure 2 shows the general architecture diagram of the proposed glaucoma detection and classification method using fundus images. The steps are described as follows:

(a) Collect the fundus images as input that are publicly available on the Internet like HRF, ORIGA, REFUGE, DRISHTI-GS, PRV Glaucoma, and DRIONS-DB.
(b) Preprocess the input fundus image before they are used to train the model for better segmentation of retinal images.

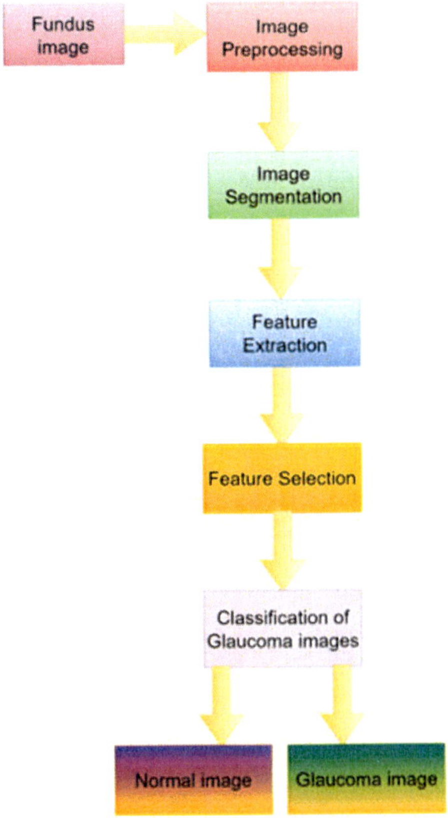

Fig. 2 Generic architecture diagram for glaucoma detection and classification

(c) Segmentation of preprocessing images for feature extraction.

In image preprocessing, image enhancement is an essential step that is used to enhance the images by adjusting the intensity values, improving the contrast of the image, and sharpening the edge. Another commonly used image preprocessing method that is used to diagnose glaucoma diseases is, namely, RGB channel separation that is used for the enhancement of features. Histogram equalization is also an important preprocessing step used for adjusting the intensity values of the images. Noise removal is used for removing the noise that occurs in differentiating the object from the background image. Image resizing is performed by resizing the image.

In image segmentation, fundus images are partitioned into multiple segments to simplify analysis. There are various techniques available to date used for segmentation purposes. Out of them, thresholding is the easiest method that is used to segment the grayscale images into binary images. Another approach is region-based segmentation that is used to detect regions directly. In literature, edge detection is another important segmentation method that is used to find the edges of objects in the images. In order to determine the endpoint of an edge, Hough transform is a popularly used method. (d) For extracting the fundus images features, feature extraction methods are applied for selecting the best feature when we have a large dataset of images. Correct feature extractions of fundus images are to be analyzed for glaucoma diagnosis. Fundus image contains the information about the core part of the retina, namely blood vessels, optic cup, and disk are the main features of the retinal images. CNN model utilized the input image for feature extraction. (e) The feature selection method chooses the relevant features for improving the classifier performance. (f) Image classification is used to classify the images into two categories: normal images and glaucoma images. Some popular techniques for glaucoma classification are SVM, Naive Bayes, and ANN.

5 Results

This section briefly summarizes the existing literature on image processing, machine learning, and deep learning algorithms for glaucoma detection on retinal images shown in Table 3

As seen above from Table 3, the image processing-based method that works on handcrafted features like notch filter, CDR, NRR degrades the system's performance and requires consultation from an expert eye specialist. This disadvantage can be overcome by developing an automatic glaucoma diagnosis system using machine learning and deep learning method. SVM algorithm gives better performance in terms of accuracy as compared to other classifiers. Hence, the combination of ensemble learning with an SVM algorithm gives higher accuracy that is 99.3%.

Table 3 Main studies using image processing, machine learning, and deep learning techniques

Methods	Year	Architecture	Results in terms of accuracy, sensitivity, specificity
Raghavendra et al. [20]	2018	CNN	98.13%, 98%, 98.3%
Christopher et al. [18]	2018	ResNet50	97%, 93%, 92%
Mukherjee. et al. [7]	2019	Notch filtering	87.128%
Jain et al. [8]	2019	Wavelet transform, SVM	98.7% of accuracy
Mohamed et al. [10]	2019	SVM	98.63%, 97.60%, 92.30%
Rehman et al. [11]	2019	Ensemble learning, SVM	99.30%, 99.40%, 96.90%
Thakur et al. [12]	2020	CNN, Mobile Net V2	94% of accuracy
Li et al. [19]	2020	ResNet101	94.1%, 95.7%, 92.9%

6 Analysis

This section presents the analysis for the glaucoma detection based on some literature work in which we have considered some factors, namely analysis based on medical features, analysis based on performance metrics.

6.1 Analysis Based on Medical Features

(a) **CDR:** Cup-to-disk ratio (CDR) is used in ophthalmology as it gives the relation between OD and cup area for the progression of the glaucoma disease. In Eq. 1, CDR is defined as a cup-to-disk ratio, and in normal patients, the CDR value is less than 0.5 is defined in Eq. 2.

$$\text{CDR} = \frac{\text{area cup}}{\text{area disk}} \tag{1}$$

$$\text{CDR}_{\text{Normal}} \leq 0.5 \tag{2}$$

(b) **ISNT:** It is a method that is used to calculate the width of the neuroretinal rim and optic disk. The inferior region contains the maximum width for the normal human eye. ISNT rule is stated as below in Eq. 3.

$$\text{Inferior} > \text{Superior} > \text{Nasal} > \text{Temporal} \tag{3}$$

(c) **NRR:** NRR is stated as neuroretinal rim ratio that is used for classification of fundus images into two categories as glaucomatous and normal.
(d) **RNFL:** RNFL is known as retinal nerve fiber layer. It helps in detecting glaucoma disease at the early stage.
(e) **GRI:** GRI (Glaucoma risk index) is used to the classified eye as normal or abnormal by using various principal components.

Table 4 shows the summary of medical features used by researchers in the past work based on feature extraction analysis such as CDR, ISNT, NRR, RNFL, and GRI.

Table 4 Summary of medical features based on literature paper

Medical features	Works
ISNT	[7, 13]
CDR	[7, 9, 21]
RNFL	[22]
GRI	[5]
NRR	[7, 13]

Table 5 Summary of sensitivity, specificity, and accuracy for glaucoma detection based on literature work

Metrics	Sensitivity	Specificity	Accuracy	AUC
70–80%	[23]	[24]	[24]	–
80–90%	[24, 25, 21]	[23]	[7, 13, 26]	–
Above 90%	[26, 26–29]	[26, 25–30]	[9, 12, 14, 26, 27, 25–29, 11]	[20]

6.2 *Analysis Based on Performance Metrics*

Accuracy: Accuracy is defined as how much we are close to the actual value. Accuracy (Acc.) is defined as below in Eq. 4.

$$\text{Acc.} = \frac{T.P. + T.N.}{T.P. + F.P. + F.N.} \tag{4}$$

Sensitivity: True-positive rate defines the term sensitivity. Sensitivity is defined as below in Eq. 5.

$$\text{Sensitivity(or Recall)} = \frac{T.P.}{T.P. + F.N.} \tag{5}$$

Specificity: True-negative rate defines the term specificity. Specificity is defined as below in Eq. 6.

$$\text{Specificity} = \frac{T.N.}{T.N. + F.P.} \tag{6}$$

where TP represents the true positive, $T.N.$ represents the true negative, FP represents the false positive, and FN represents the false negative. In this section, Table 5 shows the analysis of literature work for glaucoma detection.

7 Conclusion

This review paper provides a brief introduction about glaucoma, types, risk factors, and symptoms of glaucoma methods used by ophthalmologists. We used fundus images for glaucoma diagnose as these images provide detailed information. We provide a brief review of some datasets that are publicly available on the Internet for glaucoma detection. We also discussed different image processing and deep learning techniques that diagnosis glaucoma diseases. At last, we discussed performance parameters for glaucoma diagnosis such as CDR, ISNT, NRR, RNFL, and

GRI. Performance analysis based on sensitivity, specificity, and accuracy for glaucoma detection and classification was summarized. These techniques will help in the medical field for automatic glaucoma detection that may be helpful for not losing our vision.

References

1. Wong WL et al (2014) Global prevalence of age-related macular degeneration and disease burden projection for 2020 and 2040: a systematic review and meta-analysis. Lancet Global Health 106–116
2. Tan JH et al (2018) Age-related macular degeneration detection using deep convolutional neural network. Futur Gener Comput Syst 87:127–135
3. Nikam S, Patil CY (2017) Glaucoma detection from fundus images using MATLAB GUI. In: 3rd IEEE conference on advances in computing, communication and amp automation
4. Sekhar S et al (2008) An automated localization of optic disk and fovea in retinal fundus images. Proceedings of Signal Processing Conference 80:24205–24220
5. Veena HN et al (2020). A Review on the optic disc and optic cup segmentation and classification approaches over retinal fundus images for detection of glaucoma, vol 2, Issue 9. Springer, Switzerland
6. Abbas Q (2017) Glaucoma-deep: detection of glaucoma eye disease on retinal fundus Images using deep learning. (IJACSA) International Journal of Advanced Computer Science and Applications 8(6)
7. Mukherjee R et al (2019) Predictive diagnosis of glaucoma based on analysis of focal notching along the neuro-retinal rim using machine learning. Pattern Recognit Image Anal 29(3):523–532
8. Jain S et al (2019) Detection of glaucoma using two dimensional tensor empirical wavelets transforms. Springer, Switzerland
9. Carrillo J et al (2019) Glaucoma detection using fundus images of the eye, symposium on image, signal processing and artificial vision, pp 1–4
10. Mohamed NA et al (2019) An automated glaucoma screening system using cup-to-disc ratio via simple linear iterative clustering super pixel approach. Biomed Signal Process Control 53:101454
11. Rehman ZU et al (2019) Multi-parametric optic disc segmentation using super pixel based feature classification. Expert System with Application 120:461–473
12. Thakur A et al (2020) Predicting glaucoma before onset using deep learning. American Academy of Ophthalmology 3(4):262–268
13. Karthiyayini R, Shenbagavadivu N (2020) Retinal image analysis for ocular disease prediction using rule mining algorithms. Interdiscip Sci 13(3):451–462
14. Saxena A et al (2020) A glaucoma detection using convolutional neural network. In: 2020 international conference on electronics and sustainable communication systems (ICESC), pp 815–820
15. Masot JC et al (2020) Dual machine-learning system to aid glaucoma diagnosis using disc and cup feature extraction, IEEE Access, pp 1–9
16. Liu H et al (2019) Development and validation of a deep learning system to detect glaucomatous optic neuropathy using fundus photography. Jama Ophthalmology 137(12):1353–1360
17. Fu H et al (2019) Glaucoma detection based on deep learning network in fundus image. Interdiscip Sci 13(3):451–462
18. Christopher M et al (2018) Performance of deep learning architectures and transfer learning for detecting glaucomatous optic neuropathy in fundus photographs. Sci Rep 8(1):1–13

19. Li F et al (2020) Deep learning-based automated detection of glaucomatous optic neuropathy on color fundus photographs. Graefe's Archive for Clinical and Experimental Ophthalmology 258(4):851–867
20. Raghavendra U et al (2018) Deep convolution neural network for accurate diagnosis of glaucoma using digital fundus images. Inf Sci 441:41–49
21. Bajwa MN et al (2019) Two-stage framework for optic disc localization and glaucoma classification in retinal fundus images using deep learning. BMC Medical Informatics Decision Making 19(1)
22. Geetha A et al (2020) Image processing techniques for diagnosis of glaucoma from retinal image: brief review. J Clin Diagn Res 14(2):1–9
23. Agrawal V et al (2018) Enhanced optic disk and cup segmentation with glaucoma screening from fundus images using position encoded CNNs
24. Prasad DK et al (2018) Improved automatic detection of glaucoma using cup-to-disk ratio and hybrid classifiers. ICTACT Journal on Image and Video 9(2)
25. Orlando JI et al (2020) REFUGE challenge: a unified framework for evaluating automated methods for glaucoma assessment from fundus photographs. Medical Image Analysis 59
26. Kim J et al (2019) Optic disc and cup segmentation for glaucoma characterization using deep learning. In: IEEE 32nd international symposium on computer-based medical systems (CBMS)
27. Fujita H et al (2018) Deep convolution neural network for accurate diagnosis of glaucoma using digital fundus images. Inform Science 441:41–49
28. Septiarini A et al (2018) Automatic glaucoma detection method applying a statistical approach to fundus images. Healthc Inform Res 24:53–60
29. Hagiwara Y et al (2018) Computer-aided diagnosis of glaucoma using fundus images: a review. Computer Methods Program Biomed 165:1–12
30. Raimundo M (2019) Sensitivity of psychological, electrophysiological and structural tests for detection and progression monitoring in ocular hypertension and glaucoma. Revista Sociedade Portuguesa De Oftalmologia 42(1)

Information Retrieval

Machine Learning Methods to Identify Aggressive Behavior in Social Media

Varsha Pawar and Deepa V. Jose

Abstract With the more usage of Internet and online social media, platforms creep with lot of cybercrimes. Texts in the online platforms and chat rooms are aggressive. In few instances, people target and humiliate them with the text. It affects victim mental health. Therefore, there is a need of detecting the abuse words in the text. In this paper, a study of machine learning methods is done to identify the aggressive behavior. Accuracy can be improved by incorporating additional features.

Keywords Cyber-aggressive · Machine learning · Cyber-bullying

1 Introduction

Social media became a platform that offers abundant information about the human activities and their relationships. Due to the enormous usage of social media networks, people are propagating illegal activities in them. In the recent years, cybercrime and its forms has adverse consequences. Cybercrime is an act performed illegally by using digital media. Cybercrime creeps in many forms like cyber-stalking, cyber-bullying, and profile hacking and online scams.

Cyber-bullying is an act performed in social media to agitate others. Victim gets humiliated with these unwanted activities Bully targets the victim's personal information shared in social media and pose indelicate comments. They will pose electronic text. Victims undergo mental illness with the bullying act and may sometimes have a risk of suicidal attempt. Twitter is one of the social media platforms where

V. Pawar (✉)
Assistant Professor, Department of Computer Applications, CMR Institute of Technology, Bengaluru, India
e-mail: varshapawar1917@gmail.com

Research Scholar, Department of Computer Science, Christ Deemed to be University, Bengaluru, India

D. V. Jose
Christ Deemed to be University, Bangalore, India
e-mail: deepa.v.jose@christuniveristy.in

P. Dutta et al. (eds.), *Emerging Technologies in Data Mining and Information Security*, Lecture Notes in Networks and Systems 490,
https://doi.org/10.1007/978-981-19-4052-1_50

most of the cyber-bullying incidents occurred [1]. With the advancements of online services, people will spend most of the time irrespective of geographical constraints [2]. Cyber-bullying is increasing as the bully's information is confidential, and they have a high scope of bullying with fake identities.

The different forms of cyber-bullying are as follows:

(i) Chat Rooms: Posting a message by dissing or threaten others to be viewed by all in the chat rooms.
(ii) Harassment: When a bully sends offensive material to hurt the victims.
(iii) Impersonation: While sending messages, one will behave on behalf of others.
(iv) Denigration: Sending a text message which are gossips or rumors.
(v) Outing: Sending confidential information of others.
(vi) Trickery: Accessing the secret information by performing some tricks.
(vii) Exclusion: Messages to forcefully exclude someone.
(viii) Cyber-stalking: Intentionally harassing or threatening someone.
(ix) Blogging: Taking public forum as a platform to abuse others.

Detecting cybercrime and its forms is crucial and is very important. Manual detection of these activities takes too long duration and is tedious. Hence, there is a need of automatic processes to detect cybercrime and their communication.

Detection of cyber-bullying in text is considered as a classification problem. The existing models found the cyber-bullying behavior by classifying the messages as bullying and non-bullying text. This paper motive is to study and compare the performance of various machine learning algorithms.

In this paper, Sect. 2 deals with the literature review. Section 3 performs the analysis of various machine learning algorithms Sect. 4 provides the result for identifying the best model. The dataset used for illustrating derives from Twitter tweets.

2 Literature Review

Singh et al. [1] performed SNA techniques by analyzing 'momochallenge' tweets on Twitter. Data collection is done through NodeXL. Word and word pairs are taken for performing sentiment analysis. Using SNA and Clauset–Newman–Moore algorithm, they are able to find the people who are involved in communication. Chavan et al. [3] proposed a method with classification techniques like logistic regression and SVM. They found that SVM increases its accuracy by 4%. They have used chi-square method for feature selection. Al-garadi et al. [2] used various features like user, activity, network, and content for feature analysis. Various classifiers like NBA, Random Forest, SVM, and KNN are used. Their results found that random forest using SMOTE was the best technique for classifying tweets. Mangaonkar et al. [4] worked with three collaboration techniques as heterogeneous, homogenous, and selective collaboration. Along with these technique, following methods are used—OR, AND, Random 2 OR parallelism. They found that best recall results are with OR parallelism. Novianto et al. [5] found that best accuracy is given by SVM method by poly-kernel. They have applied on different classes for n-grams from 1 to 5. Jain et al. [6] provided a statistical model for measuring the percentage of cyber-bullying act. Manova, Spearman correlation and multiple logistic regression tests are performed. Arora et al. [7] detected parameter-based threats using classifiers on WEKA tool. They have implemented random forest algorithm on data with parameters like synonyms, hashtags, age, keywords, gender, and location. Anomalies are predicted by considering multiple attributes using this model.

Andleeb et al. [8] implemented SVM and Bernoulli NB with extracted data comprising different features like textual, behavioral, and demographic features. The results found that SVM is having high accuracy than Bernoulli NB.

Van Hee et al. [9] performed automatic detection on Dutch and English corpus with binary classifiers and ten-fold cross-validation. A keyword-based system is also performed. The keyword-based system has low performance comparatively. Pawar and Raje [10] proposed a multi-lingual model to detect bullying behavior on a dataset of tweets in Indian languages on Marathi and Hindi. Logistic regression performs outstandingly on both languages corpus. Rafiq et al. [11] proposed a model with two steps, i.e., dynamic priority scheduler and incremental classifier. Logistic regression is used in the incremental model. In this model, the scalability issues are addressed. Sintaha and Mostakim [12] implemented Naïve Bayes and SVM classifiers. Their results show that SVM performs more accurately than Naïve Bayes. They also found that Naïve Bayes works well on corpus data rather than tweets. Ting et al. [13] proposed an approach with opinion mining and also included keyword matching technique in that approach. Their results can be improved by applying different weights for various features. Shekhar et al. [14] used Soundex algorithm to extract unigram features. Using pronunciation features identifies the misspelled text and censored words. Silva et al. [15] proposed a model which automatically identifies the bullying act. Bullying rank is computed based on the insulting content and the levels of vulnerability. Nandhini et al. [16] proposed an architecture where genetic algorithm is used. The fuzzy rule set is used in the evaluation of the chromosomes. Vimala et al. [17] implemented various classifiers NB, random forest, and J48. J48 provides high accuracy compared with other models.

3 Machine Learning Methodologies

3.1 Preprocessing Techniques

Preprocessing has to be done on the dataset by performing various listed techniques tokenization for breaking the text into small chunks. Those small chunks of data need to be stemmed to retrieve the actual root of the word. Stop words, unwanted characters, and symbols need to be swept out. Change of case will be applied. Spell checker is used to correct the spelling mistakes to get the correct word.

3.2 Feature Extraction

Social media data contain different features of data. Network features include ratio of followers of a message and the verification status. Pro-active users and their activities should be considered. User features include the personal details of user like age and gender. Content features include the messages which they are sharing. All these features should be considered for detecting the cyber-bullying act.

The feature sets are extracted by using TF-IDF. The importance of words is determined with the most occurrences of words. Chi-square method is used for accessing the best feature.

Classification: A pool of machine learning algorithms is used to classify the text as bullying and non-bulling.

A. Bagging Classifier:

Bagging classifier fits base classifier with the random subsets of the dataset and joins their each predictions of the subset.

B. SGD Classifier:

SGD classifier updates the weight matrix by computing the gradient on subsets of training data.

C. Logistic Regression:

This classifier outcomes the probability of the bullying text in the given dataset.

D. Decision Tree:

Decision tree classifies based on root which is test on a variable and leaves are labels passed to it, and the outcome of test is organized as branches.

E. Linear SVC:

Linear SVC is more flexible in choosing the loss functions and is better choice to work with large number of samples.

F. Random Forest Classifier:

It is an ensemble tree-based algorithm. It is a set of decision trees from randomly selected sub-set of dataset.

G. AdaBoost Classifier:

It helps in boosting the performance of decision trees. It improves the accuracy.

H. Multinomial NB:

It works better for the word counts in a text and predicts the tag of the text.

I. XGB Classifier:

It classifies by applying parallel boost on the set of data.

J. KNN:

It works on Euclidean distances as distance metric without prior knowledge.

4 Results

A dataset has 28,000 tweets out of which 9000 tweets are non-offensive. The evaluation parameters used are accuracy, precision, recall, F1-score (Fig. 1). A bagging classifier provides the best accuracy [18] (Fig. 2).

The accuracy is not too high as traditional features are applied and can also be increased by adding the additional features.

	Algorithm	Accuracy: Test	Precision: Test	Recall: Test	F1 Score: Test
0	BaggingClassifier	0.929026	0.964735	0.925850	0.944893
1	SGDClassifier	0.927909	0.964013	0.924830	0.944015
2	LogisticRegression	0.926344	0.964089	0.922279	0.942721
3	DecisionTreeClassifier	0.923662	0.953253	0.929422	0.941187
4	LinearSVC	0.916732	0.946599	0.925510	0.935936
5	RandomForestClassifier	0.914496	0.945015	0.923639	0.934205
6	AdaBoostClassifier	0.907567	0.972508	0.884354	0.926338
7	MultinomialNB	0.893372	0.901663	0.940306	0.920579
8	XGBClassifier	0.890578	0.986307	0.845238	0.910340
9	KNeighborsClassifier	0.857606	0.895161	0.887245	0.891186

Fig. 1 Evaluation parameters of algorithms

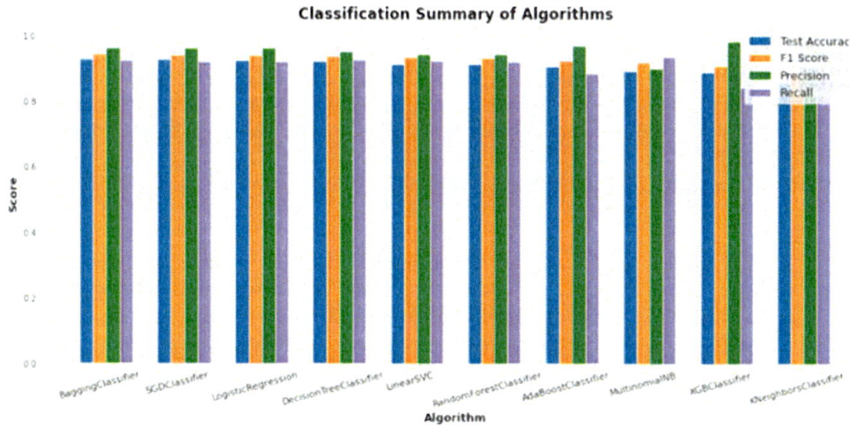

Fig. 2 Classification results

5 Conclusion

In this paper, a set of machine learning algorithms are performed in the detection of cyber-bullying. Most of the research should be targeting in the identification of bullying text in other languages other than English and the native language text written in English. The mechanisms are needed to detect and prevent the bullying behavior in online platforms instantly.

References

1. Singh S, Thapar V, Bagga S (2020) Exploring the hidden patterns of cyberbullying on social media. Procedia Computer Science 167:1636–1647. https://doi.org/10.1016/j.procs.2020.03.374
2. Al-garadi MA, Varathan KD, Ravana SD (2016) Cybercrime detection in online communications: the experimental case of cyberbullying detection in the Twitter network. Computers in Human Behavior 63:433–443. https://doi.org/10.1016/j.chb.2016.05.051
3. Chavan VS, Shylaja SS (2015) Machine learning approach for detection of cyber-aggressive comments by peers on social media network. In: 2015 international conference on advances in computing, communications and informatics (ICACCI). https://doi.org/10.1109/icacci.2015.7275970
4. Mangaonkar A, Hayrapetian A, Raje R (2015) Collaborative detection of cyberbullying behavior in twitter data. In: 2015 IEEE international conference on electro/information technology (EIT). https://doi.org/10.1109/eit.2015.7293405
5. Noviantho SMI, Ashianti L (2017) Cyberbullying classification using text mining. In: 2017 1st international conference on informatics and computational sciences (ICICoS). https://doi.org/10.1109/icicos.2017.8276369
6. Jain O, Gupta M, Satam S, Panda S (2020) Has the COVID-19 pandemic affected the susceptibility to cyberbullying in India? Computers in Human Behavior Reports. https://doi.org/10.1016/j.chbr.2020.100029
7. Arora T, Sharma M, Khatri SK (2019) Detection of cyber crime on social media using random forest algorithm. In: 2019 2nd international conference on power energy, environment and intelligent control (PEEIC). https://doi.org/10.1109/peeic47157.2019.8976474
8. Andleeb S, Ahmed R, Ahmed Z, Kanwal M (2019) Identification and classification of cybercrimes using text mining technique. In: 2019 international conference on Frontiers of information technology (FIT). https://doi.org/10.1109/fit47737.2019.00050
9. Van Hee C et al (2018) Automatic detection of cyberbullying in social media text. PLoS ONE 13(10):e0203794
10. Pawar R, Raje RR (2019) Multilingual cyberbullying detection system. In: 2019 IEEE international conference on electro information technology (EIT). https://doi.org/10.1109/eit.2019.8833846
11. Rafiq RI, Hosseinmardi H, Han R, Lv Q, Mishra S (2018) Scalable and timely detection of cyberbullying in online social networks. In: Proceedings of the 33rd annual ACM symposium on applied computing. https://doi.org/10.1145/3167132.3167317
12. Sintaha M, Mostakim M (2018) An empirical study and analysis of the machine learning algorithms used in detecting cyberbullying in social media. In: 2018 21st international conference of computer and information technology (ICCIT). https://doi.org/10.1109/iccitechn.2018.8631958
13. Ting I-H, Liou WS, Liberona D, Wang S-L, Bermudez GMT (2017) Towards the detection of cyberbullying based on social network mining techniques. In: 2017 international conference on

behavioral, economic, socio-cultural computing (BESC). https://doi.org/10.1109/besc.2017.8256403
14. Shekhar A, Venkatesan M (2018) A bag-of- phonetic-codes modelfor cyber-bullying detection in twitter. In: 2018 international conference on current trends towards converging technologies (ICCTCT). https://doi.org/10.1109/icctct.2018.8550938
15. Silva YN, Rich C, Hall D (2016) BullyBlocker: towards the identification of cyberbullying in social networking sites. In: 2016 IEEE/ACM international conference on advances in social networks analysis and mining (ASONAM). https://doi.org/10.1109/asonam.2016.7752420
16. Nandhini BS, Sheeba JI (2015) Online social network bullying detection using intelligence techniques. Procedia Computer Science 45:485–492. https://doi.org/10.1016/j.procs.2015.03.085
17. Balakrishnan V, Khan S, Arabnia HR (2020) Improving cyberbullying detection using twitter users' psychological features and machine learning. Comput Secur 90:101710. https://doi.org/10.1016/j.cose.2019.101710
18. https://github.com/dhavalpotdar/cyberbullying-detection/tree/master/data

A Comprehensive Study on Robots in Health and Social Care

Adil Khadidos

Abstract The world have experienced a severe human-health crisis as a result of the emergence of a novel coronavirus (COVID-19), which was declared a global pandemic by WHO. As close human-to-human contact can spread the COVID-19 causing virus, keeping social distance is now an absolute necessity as a preventative measure. At a time of global pandemic, there is a huge need to treat patients with little patient–doctor interaction by using robots. Robots can be characterized as machines that can execute a wide range of tasks with greater autonomy and degree of freedom (DoF) than humans, making it difficult to identify them from other machines. A wide range of equipment, sensors, and information and communication technology (ICT) are now part of the healthcare system, which has become increasingly complicated. Protecting front-line personnel from virus exposure is the primary goal of using robots in health care. The aim of this study is to emphasize the evolving importance of robotics applications in health care and related fields. This paper examines in depth the design and operation of a wide range of healthcare robots in use around the world.

Keywords COVID-19 · Robotics · Medical robots · Artificial intelligence · Health care

1 Introduction

Global emergency was announced on January 30, 2020, owing to the outbreak of novel coronavirus COVID-19 by World Health Organization (WHO). There were 266 cases of coronavirus in Wuhan, China, in December 2019 year. More than one million people were confirmed to have been infected with the new coronavirus on April 2, 2020, with 52,000 deaths reported worldwide, in less than a month [1]. This COVID-19 pandemic had a negative impact on almost every country in the world,

A. Khadidos (✉)
Department of Information Technology, Faculty of Computing and Information Technology, King Abdulaziz University, Jeddah, Saudi Arabia
e-mail: akhadidos@kau.edu.sa

P. Dutta et al. (eds.), *Emerging Technologies in Data Mining and Information Security*, Lecture Notes in Networks and Systems 490,
https://doi.org/10.1007/978-981-19-4052-1_51

causing widespread concern about healthcare facilities and economic crises. So far, the number of people infected with COVID-19 has been reported to be up to 23 million worldwide. The virus can spread through close human interactions, posing a risk to front-line workers who are treating COVID-19-infected patients [2]. This situation has prompted many researchers to consider the development of robotic solutions to assist healthcare personnel in serving patients effectively while avoiding infection [3].

This study precisely highlighted various contribution of robotics in health care. Section 2, which describes how various robots are classified and operated, and Sect. 3, which covers all types of healthcare robots used to combat the pandemic; Sect. 4, which combines these results and future reviews of what can be done to improve current robots' efficiency and reliability.

2 The Classification of Medical Robots

The use of robots to alleviate the problems caused by COVID-19 [4], presents no better alternative than the application of robotics in the healthcare industry, which offers infinite advantages on its own. The types of medical robots listed in Table 1 will be discussed briefly in this section, as well as how they work and what medical responsibilities they may accomplish. We will also look at how they have aided in the fight against various viruses and diseases.

2.1 Disinfecting/Spraying Robots

Portable robots are becoming increasingly popular for cleaning and disinfecting objects around the world. In COVID-19-like pandemic conditions, cleaning and cleanliness are critical for safe indoor/outdoor habitats [5].

Door knobs are extremely sensitive points of touch that can be infected by these viruses. As a result, a cleaning activity that is automated not only increases safety but also improves efficiency. A human support robot (HSR) is proposed in this category as a way to automate the cleaning process using artificial intelligence (AI) [6]. Among other things, this includes washing door handles and the HSR, as well as the general cleaning of the facilities. Machine learning is used to organize the visual space and provide instructions to the robot in the identification phase. The robotic operating system provides useful control over the spraying and cleaning process. When the discovery module has gathered information, it is used by the control module to construct an operating space for the robot and determine the best condition for controllers to operate in. Figure 1 depicts the cleaning robots' sequence of operations [7].

Below is the list of disinfecting robots used amidst the outbreak of the COVID-19:

- **UVD-bot:** A germicidal robot that employs ultraviolet light and is self-driving was developed in Denmark (UV C-254nm) By disrupting the DNA base pairing, the

Table 1 Preview of Indian robots and their utilities

Name of the robot	Cleaning	Hospitality	Treatment
Sona2.5		✓	
Zafi and zafi medic robot		✓	
Karmi-bot		✓	✓
Sayabot		✓	✓
UV-bot	✓		
CO-bot		✓	
NIGA-bot			✓
Nightingale-19		✓	
Milagrow i-map9	✓		
Rail-bot		✓	✓
Mitra		✓	
Maitri			✓
Wegree robot	✓	✓	
Starship robot		✓	
Zorobot's Cruzr robot			✓
Pepper robot		✓	

UVC radiation utilized by this robot is effective against the coronavirus. Figure 2a shows how the robot is utilized to sanitize hospital areas, preventing direct contact between patients and contaminated areas [8]. An entire room may be disinfected by the robot in less than ten minutes, and it is completely self-sufficient and incredibly effective. Moreover, its operation is simple enough that it can be operated by regular cleaning staff [9]. These UVD bots are also being used for the disinfection of documents and papers for some organizations because of their proven disinfection capabilities.

- **iMap9 (Milagrow i-map9):** Automated cleaning of the floor is possible, as seen in Fig. 2b. Following the advice of the ICMR, NaOCl (Sodium Hypochlorite) solution is used to remove COVID-19-carrying spores from the surface. Particulate matter smaller than 0.3 micrometers is eliminated 99.97% of the time by its HEPA filter.

2.2 *Robotic Hospitality*

Due to the pandemic crisis, which significantly increased the death rates among healthcare personnel, the role of receptionist and nursing robots has risen to new heights [10]. The proposed system comprises of three parts they are receptionist, nurse robot system, and a medical servant where the receptionist robot and the nurse

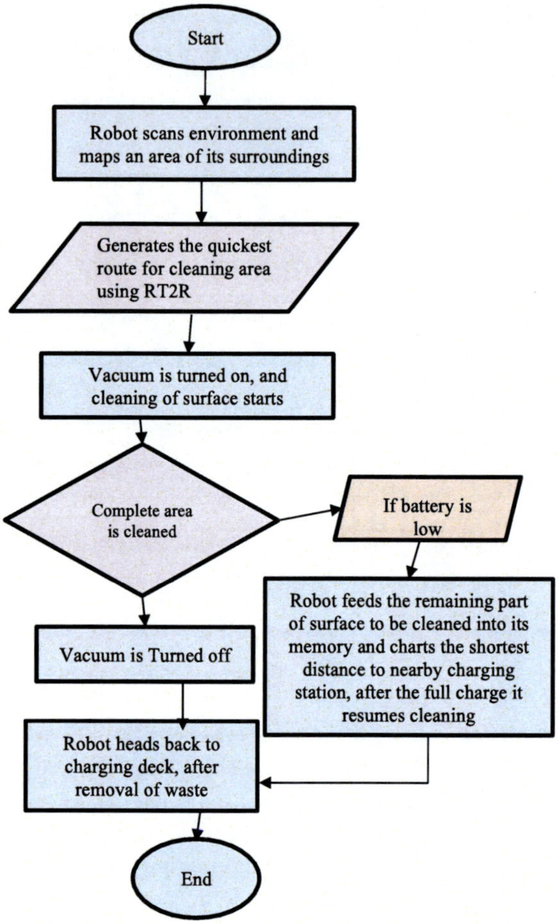

Fig. 1 Working flowchart of a cleaning robot

assistant bot are addressed to assist nurses [11]. Patients' medical data are obtained and stored on the medical server by healthcare robots [12, 13]. The robots deliver summaries of this data to human caretakers via a Web interface. This could have a huge impact on the ability of the hospital staff to get in touch with those who have been impacted. One of a nurse robot's primary duties is to provide patients with medication and nourishment [14, 15]. There is a distinct difference between a receptionist robot and a human receptionist. While there were several robots used to distribute and monitor medication during this pandemic, human nurses were kept out of it.

Below is the list of a few robots which are deployed to indulge in tasks as of a medical server or nurse:

- **Sona-2.5:** It is employed to control the distribution of medications and food to hospitalized patients and to monitor their body temperatures [16], as shown in

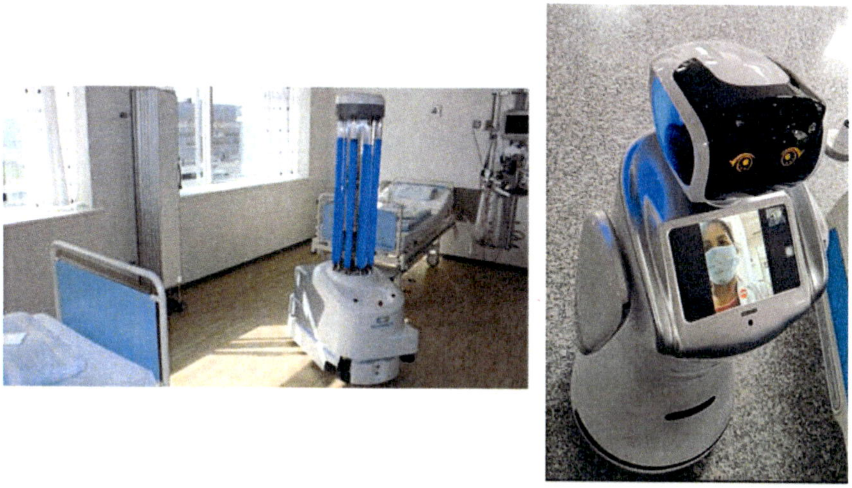

(a) (b)

Fig. 2 **a** UV-Bot used for disinfecting the hospital premises. **b** Structure of Milagrow robot

Fig. 3a. There is a vision camera for face detection, and it can carry a weight of 15 kg on this robot's shoulders.

- **Zafi and Zafi medic robots:** These robots perform the tasks to deliver food and medicines to the affected as shown in Fig. 3b. While Zafi has a payload of eight kilograms and can go as a clinical partner for contact-less consultation, Zafi Medic is an off-base robot that can deliver supplies of as much as 20 kg, and can be controlled from a range of one kilometer with live view support. The robots are enclosed by cling film before they are dispatched to the wards. Therefore, the film will protect the machines from contamination and can be easily replaced after the required work is done [17].
- **KARMI-Bot:** When a vacant ward is found, it can analyze and map the area, and then carry out functions such as feeding patients according to their schedules and monitoring their wards via videoconferencing with doctors, as illustrated in Fig. 3. 3c. The robot can also have added assets like self-charging capacity and a load strength of 25 kg [4].
- **CO-Bot:** Corona Combat robot is the acronym for it. Serving food and water to COVID-19 patients as depicted in Fig. 3d and returning empty trays and plates is its major goal. It can serve multiple people at once and can carry up to 20 kg of weight [18].
- **Wegree Robot:** A Polish business created this robot, Fig. 3e, to help healthcare personnel avoid contact with people who may be carriers of the COVID-19 virus. When visitors come into contact with the robot, it invites them to wash their hands, use a non-contact thermometer attached to the robot to obtain temperature readings, and wear a protective face mask [19].

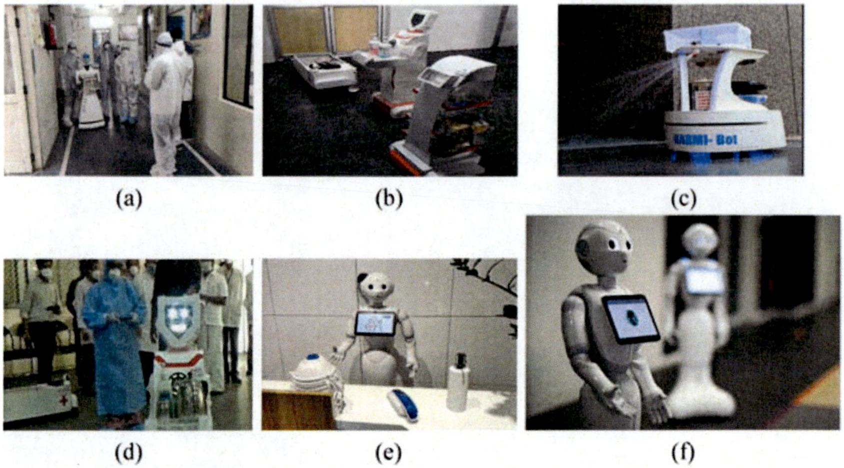

Fig. 3 **a** Sona-2.5 robot used in a hospital, for delivering medicines and food to the quarantined patients. **b** Zafi and Zafi medic robots used in Trichy hospital for delivering basic commodities to the quarantined. **c** Karmi-bot used for delivery of medicines and food. **d** CO-bot used for delivery, disinfection. **e** Wegree robot in hospital. **f** Pepper robots in display

- **Pepper Robot:** Softbank Robotics developed the Pepper Robot, a 28 kg humanoid robot with a battery life of up to 12 h (Fig. 3f) [20]. In addition to 15 languages, this robot is able to speak and communicate with patients. Aside from detecting whether or not guests are wearing masks, Cloud Pepper can also form relationships with individuals and understand their emotions by using facial recognition and natural language processing. Doctors can avoid unnecessary contact with patients for minor issues because to this robot's ability to assist doctors in communicating with patients remotely [21].

2.3 *Telepresence System*

The robot's forearm is equipped with a capacitive touch screen. Figure 4 depicts the audio/video conference system based on Web-real-time communication (WebRTC) used for patient-provider communication Fig. 4 [22, 23]. The robot is fitted with speech recognition so that it can recognize the patient's voice and communicate with them. The patient's emotional state is also monitored using a deep neural network [24–26].

In such a situation, a mobile robot with appropriate sensors in a tiny form factor would be the best option for rescue. The little robot is anticipated to roam the premises on its own and collect data on safe and hazardous environmental conditions. Healthcare staff can, of course, use the system to locate and assist victims [27]. In order

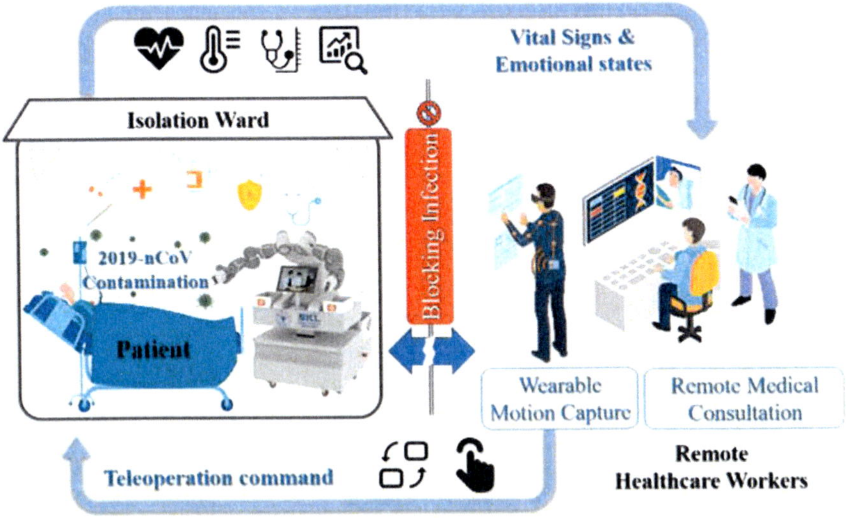

Fig. 4 Working of telepresence [24, 26]

to perform 3D mapping, a self-enabled robot is utilized. SLAM technology (generic environment and mapping) can be used to frame this [27]. In a three-dimensional environment, the robot is designed to move using six degrees of freedom methods. However, the limitations imposed by odometry information from wheel encoder sensors add another layer of complexity. Thus, because of the potentially hazardous nature of the location, the entire system is utterly untrustworthy. If robot movement is restricted, tilting laser range finders or other motion sensors can be employed to verify that 3D mapping is achievable while navigating between robots [28].

2.4 *Surgical Robot*

Due to its multiple advantages, including as high mechanical accuracy, endurance, and the ability to work in dangerous settings, engineers and researchers are continuing to work on autonomy in surgery [29]. While some surgical treatments are significantly easier to perform than others, such as autonomous cardiac ablation of the pumping heart, others are extremely complex, such as requiring the employment of a robot to generate exact lesions in the heart [30, 31]. Since surgical robots can be used to undertake difficult procedures on COVID-19 infected patients and relieve the overburden on medical staff, they are a significant asset during a pandemic [32].

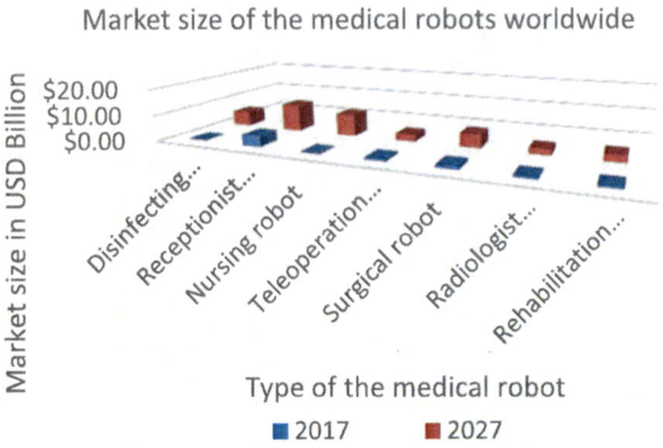

Fig. 5 Market size of the various medical robots used worldwide. Adapted from [33]

3 Scenario of Medical Robots Worldwide

In this study, we have looked into the medical robots at risk owing to recent healthcare crises, like the Ebola epidemic and the COVID-19 pandemic. As observed in Fig. 5 following the pandemic, it is projected that the market size of medical robots will skyrocket in several categories such as disinfection, nursing, tele-operation, radiologist, and rehabilitation robots, which had a market size of less than $1 billion in 2017. The average growth in the market size of these 5 medical robots is expected to be averagely increasing 3.8 times by 2027. Whereas, the other medical robots like receptionist and surgical robots will have a substantial growth [33].

4 Results and Conclusion

This research gives a thorough examination of the many types of robots used in the therapeutic sector to perform tasks in human-hazard-related industries.

Following the pandemic, the new normal world of health care appears to be more reliant on robots to prevent casualties caused by insufficient humanitarian limitations. The outbreak of novel coronavirus (COVID-19) has led to a wide range of changes in the clinical working patterns in practically all nations across the world, and it has been announced that medical robots are in more demand than ever before because of the numerous features where they are superior to humans both in function and also in their ability to restrict the spread of this virus. These weaknesses mean that robotics will continue to be used more frequently and at a faster rate, and this tendency will continue indefinitely. In order to obtain financial and medical stability, all countries must increase their interest and use of robotic innovations. In the post-pandemic context, it is expected that the use of medical robots will significantly outpace the expansion of other robots.

4.1 Future Scope

As shown in the previous section, healthcare facilities around the world are in severe need of improvement, and as a result, several research are being conducted to identify ways to improve existing robots while also reducing their cost and increasing their reliability.

The settings of healthcare robots that aid children and the elderly should be basic and straightforward to use, as well as interesting and amusing [34]. The robot's ergonomic design, as well as the software program, should be evaluated in order to make it more cost-effective and reliable to use [35]. It is observed that the improvement in the two important medical robots, namely surgical robots, rehabilitation robots, was identified as the main setbacks because of a few limitations like high cost and complexities. The need for these robots is paramount and therefore has to be deployed effectively in near future. As a result, for the implementation of these robotic devices, a broad modularization method is required. Few of the ways by which the needful can be done is by adapting minimization in the costs of making which involves the usage of cheaper computer systems and sensors [36]. Therefore, the outcome of the study confers that amidst the pandemic, the healthcare sector will undergo huge technological advancements accountably worthwhile to combat these uncertain situations and also elevate the quality working of health-care sector.

References

1. Fanelli D, Piazza F (2020) Analysis and forecast of COVID-19 spreading in China, Italy and France. Chaos Solitons Fractals 134:109761
2. Gostic K, Gomez AC, Mummah RO, Kucharski AJ, Lloyd-Smith JO (2020) Estimated effectiveness of symptom and risk screening to prevent the spread of COVID-19. Elife 9:e55570
3. Kannan S, Ali PSS, Sheeza A, Hemalatha K (2020) COVID-19 (novel coronavirus 2019)-recent trends. Eur Rev Med Pharmacol Sci 24(4):2006–2011
4. Aymerich-Franch L, Ferrer I (2020) The implementation of social robots during the covid-19 pandemic. *arXiv preprint* arXiv:2007.03941
5. Klingbeil E, Saxena A, Ng AY (2010) Learning to open new doors. In: 2010 IEEE/RSJ international conference on intelligent robots and systems. IEEE, 2010, pp 2751–2757
6. Ramalingam B, Yin J, Rajesh Elara M, Tamilselvam YK, Mohan Rayguru M, Muthugala M, Félix G6mez B (2020) A human support robot for the cleaning and maintenance of door handles using a deep-learning framework. Sensors 20(12):3543
7. De Carvalho RN, Vidal H, Vieira P, Ribeiro M (1997) Complete coverage path planning and guidance for cleaning robots. In: ISIE'97 proceeding of the IEEE international symposium on industrial electronics, vol 2. IEEE, pp 677–682
8. Vijayalakshmi M, Baljoshi B, Lavanya G, Master G, Sushil G (2020) Smart vacuum robot. In: ICT for competitive strategies. CRC Press, pp 81–90
9. Ackerman E (2020) Autonomous robots are helping kill coronavirus in hospitals. IEEE Spectr 11
10. Macalam TM, Locsin R (2020) Humanoid nurse robots and compassion: dialogical conversation with Rozzano Locsin. J Health Caring Sci 2(1):71–77

11. Ahn HS, Lee MH, MacDonald BA (2015) Healthcare robot systems for a hospital environment: carebot and receptionbot. In: 24th IEEE international symposium on robot and human interactive communication (RO-MAN). IEEE 2015, pp 571–576
12. Giorgi I, Watson C, Pratt C, Masala GL (2021) Designing robot verbal and nonverbal interactions in socially assistive domain for quality ageing in place. In: Human centred intelligent systems. Springer, pp 255–265
13. Li H, John-John C, Tan YK (2011) Towards an effective design of social robots. Int J Soc Rob 3(4):333–335
14. Mukherjee UK, Sinha KK (2020) Robot-assisted surgical care delivery at a hospital: policies for maximizing clinical outcome benefits and minimizing costs. J Oper Manage 66(1–2):227–256
15. Edelman LS, McConnell ES, Kennerly SM, Alderden J, Horn SD, Yap TL (2020) Mitigating the effects of a pandemic: facilitating improved nursing home care delivery through technology. JMIR Aging 3(1):e20110
16. Nahla N (2020) Medical robots to the rescue in the battle against coronavirus. April 7, 2020. https://www.thehindu.com/sci-tech/technology/gadgets/how-medical-robots-are-helping-doctors-in-the-fight-against-coronavirus/article31271989.ece
17. Kumar A, Sharma GK (2020) Artificial intelligence technologies combating against COVID-19. Dev Sanskriti Interdisc Int J 16:56–60
18. Malik AA, Masood T, Kousar R (2020) Repurposing factories with robotics in the face of COVID-19. Sci Rob 5(43)
19. Kaminski J (2020) Informatics in the time of COVID-19. Can J Nurs Inform 15(1)
20. Podpora M, Gardecki A, Beniak R, Klin B, Vicario JL, Kawala-Sterniuk A (2020) Human interaction smart subsystem? Extending speech-based human-robot interaction systems with an implementation of external smart sensors. Sensors 20(8):2376
21. Kyrarini M, Lygerakis F, Rajavenkatanarayanan A, Sevastopoulos C, Nambiappan HR, Chaitanya KK, Babu AR, Mathew J, Makedon F (2021) A survey of robots in healthcare. Technologies 9(1):8
22. Uday Girish M, Harsha Vardhan G, Sudheer A (2020) Riggu: a semi-humanoid robot platform for speech and image recognition. In: Intelligent systems, technologies and applications. Springer, pp 31–41
23. Wan S, Gu Z, Ni Q (2020) Cognitive computing and wireless communications on the edge for healthcare service robots. Comput Commun 149:99–106
24. Hai NDX, Nam LHT, Thinh NT (2019) Remote healthcare for the elderly, patients by telepresence robot. In: 2019 international conference on system science and engineering (ICSSE). IEEE, 2019, pp 506–510
25. Mariappan M, Nadarajan M, Porle R, Parimon N, Khong W (2016) Towards real-time visual biometric authentication using human face for healthcare telepresence mobile robots. ResearchGate 8(11)
26. Rincon F, Vibbert M, Childs V, Fry R, Caliguri D, Urtecho J, Rosenwasser R, Jallo J (2012) Implementation of a model of robotic tele-presence (RTP) in the neuro-ICU: effect on critical care nursing team satisfaction. Neurocrit Care 17(1):97–101
27. Panzirsch M, Weber B, Rubio L, Coloma S, Ferre M, Artigas J (2017) Tele-healthcare with humanoid robots: a user study on the evaluation of force feedback effects. In: IEEE world haptics conference (WHC). IEEE, 2017, pp 245–250
28. Khouri III GA, Blanton AT (2020) Method and apparatus for improving subject treatment and navigation related to a medical transport telepresence system, Feb 20 2020, US Patent App. 16/102,808
29. Lipow K (2004) Surgical robot and robotic controller, US Patent App. 10/738,359
30. Yip M, Das N (2019) Robot autonomy for surgery. In: The encyclopedia of medical robotics: volume 1 minimally invasive surgical robotics. World Scientific, pp 281–313
31. Hemli JM, Patel NC (2020) Robotic cardiac surgery. Surg Clin 100(2):219–236
32. Shah SK, Felinski MM, Wilson TD, Bajwa KS, Wilson EB, Next-generation surgical robots. Digit Surg 401

33. Raje S, Reddy N, Jerbi H, Randhawa P, Tsaramirsis G, Shrivas NV, Pavlopoulou A, Stojmenović M, Piromalis D (2021) Applications of healthcare robots in combating the COVID-19 pandemic. Appl Bionics Biomech 2021
34. Broadbent E, Stafford R, MacDonald B (2009) Acceptance of healthcare robots for the older population: review and future directions. Int J Soc Rob 1(4):319
35. Prasad A (2013) Robotic surgery in India. J Young Med Res 1(1):e4
36. Haider H (2020) Barriers to the adoption of artificial intelligence in healthcare in India. Brighton: Institute of Development Studies (UK). Available online: https://opendocs.ids.ac.uk/opendocs/handle/20.500.12413/15272

Integrated Health Care Delivery and Telemedicine: Existing Legal Impediments in India

Meera Mathew

Abstract The technological innovation in the healthcare sector has contributed to the growth of telemedicine in India. "Health services" fall under State responsibility as per the Indian Constitution by virtue of Schedule 7—although policy and planning framework are under the scope of Central government. Telemedicine cannot not work as an autonomous service, rather, ought to be subjected to different regulations having complex ethical, medico-legal manifestations. As far as India is concerned, Ministry of Health and Family Welfare of India (MoHFW) is the body responsible for initiating the policy of digitization of healthcare. However the point is—how far digital health services going appropriately in India. Based on NDHB's comprehensive architectural framework of "Federated National Health Information System" in January 2020 and as the pandemic strategy Medical Council of India and the NITI Aayog released new guidelines on telemedicine with respect to registered medical practitioners, this research needed to be checked. Thus, the examination was done in these aspects. Guidelines were revisited to see how the hospitals in Delhi and NOIDA function based on the records submitted in medical consultation given to patients using telemedicine. It is felt that telemedicine being a nebulous concept in India, it needs to be analyzed in the light of prospective opportunities it would offer. There is a need for collaborative approaches on digital health, revision in the prevailing legal and ethical frameworks, the clinical practices corresponding to standing medical guidelines. Also, it is found that there exist no uniform telemedicine practices balancing the privacy norms, medico-legal responsibility and regulatory standards. To arrive at conclusion, the best practices prevailed in other countries are examined and adopted. It is felt that the policies existing in telemedicine need to be bifurcated as digital consultation, digital photography, remote patient monitoring (RPM) separately.

Keywords Digital healthcare · Telemedicine · Medico-legal standards · Consent

M. Mathew (✉)
Associate Professor, School of Law, Christ (Deemed to be University) Delhi N.C.R, Delhi, India
e-mail: meera.mathew@christuniversity.in

P. Dutta et al. (eds.), *Emerging Technologies in Data Mining and Information Security*, Lecture Notes in Networks and Systems 490,
https://doi.org/10.1007/978-981-19-4052-1_52

1 Introduction

As healthcare institutions get magnified and assimilated, health service is less subdued by topography, remoteness, number of patients or even by institutional limitations. With the thriving technology, digital health care or e-governance in healthcare is a bourgeoning sector. The visible progresses in this area include accessibility of health information via web sites, online customer support, computerized and automated analysis, collaborative health advancement programs and electronic mail exchanges with medical service providers. These electronically integrated health care system—though often termed as e-health [1] represents technological innovations merging telecommunications, audio/graphic technologies and computers in numerous ways that vary from the provision of medical information to diagnosis and treatment [2]. Globally, the healthcare as a segment is having a paradigm shift in the mode and applications in which health services are provided. This paper would critically examine the existing Indian (national and state) collaborative approaches on digital health, the prevailing legal and ethical frameworks, the clinical practices corresponding to standing medical guidelines and analyze if there need any reforms or standards. It would further analyze how other countries positively have taken forwarded the telemedicine practices balancing the privacy norms, medico-legal responsibility and regulatory standards, which India can adopt as best practices [3]. Health service delivery is fluctuating from "consumer-supplier relationships to collaborations, and this change is imminent balancing public interests". Although the Indian laws governing the healthcare industry has yet to be applied to e-health, amid this period of COVID-19 pandemic, it was necessary to promote health in a digital manner when the lock-down happened. One major step taken was legally giving significance to telemedicine [4].

2 Materials and Methods

This section deals with the kinds of documents, literature, materials assessed and the kind of methodology adopted to understand inherent issues.

2.1 Conceptualizing the Meaning of Telemedicine

Telemedicine implies the application of health care services via technology to patients. When remoteness and distance are issues for the proper delivery of service information technology is utilized so that any patient who endeavors to consult a doctor can get the diagnosis of disease done and treated. Telemedicine implies intersection between medicine and technology wherein telecommunication technology is used to deliver medical services [5]. There are commonly two classes of interactive platforms available in telemedicine [6].

(a) store-and-forward technology—It allows for an "image to be scanned" and accordingly to be stored, later forwarded anywhere in the world reducing any distance barriers
(b) interactive video conferencing—It is "virtual interaction" for practitioners to deal with patients where by one-on-one consultation is possible.

By virtue of telemedicine, those in needs of medical consultation, regardless of location can connect with health care professionals and thus access to health care is possible. There lies a facility to make a rapid medical involvement and maintenance for the patient on site, instead of conveying the patient to alternative setting. Terrestrial remoteness no longer would remain separation from therapeutic care [7]. By way of technology; medical practitioners are enabled to deliver "audio, visual and other data sharing communications" to facilitate the patients reducing their healthcare costs thus making the access to primary and specialty care possible. Medical sector receives an expanding market base and the countries need to neither establish large health centers nor train medical practitioners, but merely cultivate the expertise and rely on technology. As technology endures to enable quicker and further effectual communication, the logistical hurdles that once hindered the diffusion of telemedicine are fading. Nevertheless, the legal system has yet to catch up. Telemedicine encroaches into this conventional legal framework in three ways. Firstly, in terms of regulation in medical practice, secondly in terms of dispute resolution where negligence case is put forth and thirdly lack of legal compliance to protect patient-information confidentiality.

The downside of telemedicine is the lack of a standardized arrangement to the interaction. In India, Medical Council of India (MCI) used to exist under the Indian Medical Council (IMC) Act, 1956. In 2019, the National Medical Commission Act, 2019 (NMC Act) came into effect and substituted the National Medical Commission (NMC). Other than NMC, there also exist the Indian Nursing Council, the Dental Council of India, the Rehabilitation Council of India and the Pharmacy Council of India. Under the aegis of Ministry of Health and Family Welfare (MoHFW), NMC Act set out regulatory and advisory role of NMC and Medical Advisory Council. Meaning thereby, through this Act, the Ministry would not only frame strategies for regulating medical institutions and medical professionals but also maintains minimum ideals of medical education. Further, in India, the medical devices are governed by the Drugs and Cosmetics Act, 1940 along with Medical Device Rules, 2017 (the Rules) which covers a wide variety of drugs, therapeutic substances, diagnostics and medical devices. There is Central Drug Standards Control Organization (CDSCO) as regulatory body for pharmaceuticals and medical devices. These regulatory frameworks are consistent with pertinent technical recommendations from WHO.

2.2 MoHFW Guidelines and Impact

The infamous 2018 Mumbai High Court judgment in *Deepa Sanjeev Pawaskar* v *State of Maharashtra* had put forwarded the perils due to the telemedicine practice where in the medical practitioner was found guilty under medical negligence under Sect. 304 A of IPC-causing death due to negligence [8]. Under the conventional set up, medical negligence can be established by proving the elements of: (a) a duty by medical practitioner to perform along with certain standards; (b) a breach of this standard of care; (c) an injury; and (d) causation between the breach of care and the patient's injury. But when the judgment stated "prescription without physical diagnosis and henceforth resulting into death of the patient amounts to criminal negligence on the part of the doctors" [9] the question is in a telemedicine mode, what are the checklists to safeguard the essence of a reasonable degree of care and skill? Although, MoHFW issued the Telemedicine Practice Guidelines affixed as Appendix 5 to the Indian Medical Council (Professional Conduct, Etiquette and Ethics) Regulations, 2002 in tandem with NITI Aayog's "The National Institution for Transforming India" there still remain lot of loopholes [10]. These are mere guidelines and are lacking enforceability. This led for examination of the effectiveness telemedicine policies in India. Accordingly, a survey was done with the phone calls made and enquiries done to fifteen hospitals within Delhi and fifteen hospital within Uttar Pradesh (NOIDA) areas comprising different districts. It thus was observed that under telemedicine, where means being technology, what precautions are to be taken by doctors and patients if errors take place due to a breakdown in communication because of technical glitches, etc. are ambiguous to service providers. Hence juxtaposed against conventional structure is telemedicine's facility to enable the inaccessible patient diagnosis in India. Primarily, many hospitals did not use their services as telehealth or telemedicine—rather they simply acknowledged their service as online/digital/tele-consultation. Secondly there was no coherence found in answers given. Many staff members were not providing the exclusive telehealth service numbers rather the common phone number that they diverted to the doctors as needed. The means of data collection/utilization by way of recording of conversation, taking informed consent, etc. were in fact was left blank. Some hospitals just responded as not applicable and could not be commented. This again pin pointed at the legitimacy of services provided.

3 Results

This section deals with the observations made on the basis of what had been tested as issues therein.

3.1 Pandemic and Misinformation Within Forwarding of Prescriptions

As per the guidelines, it stated that "an enrolled in the State Medical Register or the Indian Medical Register can only practice". But there is no mention of any authority who cross checks the entry of names of such medical practitioners in telemedicine service. Conversely, how can a medical practitioner rely upon the patient's identity and his or her medical history when enters a virtual office via video-conference [11]. Licensing being significant to enforce continuing medical standards, unfit practitioners' licenses can be withdrawn. However, the current incoherent system prevailing in India on telemedicine prevents the distribution of evidence amid licensing authorities. There is no check list without proper regulation on if the medical practitioner who is into telemedicine is licensed at all or if under suspension, or has lost his license to practice. Let us take the example of a case reported recently about a prescription that had been shared by many users in social media in the name of one "Dr Raj Kamal Agarwal, a Senior Consultant, who works in the Department of Anaesthesiology in Sir Ganga Ram Hospital, New Delhi" as—"In fighting coronavirus as per Indian Council of Medical Research (ICMR) guidelines". In that forged prescription, it was recommended that one should take "HCQ, 400 mg" once a week along with the vitamin C to gain immunity to fight the pandemic. The matter came to the notice of Gangaram hospital officials and they condemned it also filed a case against such mal-information. On due verification, it was found that the ICMR had never issued such guideline for HCQ tablets [12]. If during pandemic such prescriptions can be made rounds as forwards, what would happen to those audio clips or video clips that had been used differently via telemedical services? This again poses the safeguards on prevailing current regulation. Also, is the concern pertaining to "informed consent" under a vital prerequisite under medico-legal jurisprudence while treating a patient?

3.2 Professional Standards and Informed Consent

The responses received from the hospitals again are vague on the "essential communication between patients and medical practitioners in the course of medical treatment" which is fundamentally embodied in the legal dogma of "informed consent". Corresponding to the professional standards, there exists "the practice of direct diagnosis where medical practitioner could converse personally with the patient, providing the required revelations. The patient's consent should go hand in hand with the trust he or she has toward the medical practitioners at the time of disclosure dialog and there lies confidentiality" [13, 14]. In the telemedicine background, the utmost accountability concerns relate predominantly to the establishment of the physician–patient association, the applicable standard of care and the essential informed consent. With no clear explanation provided in the "Telemedicine Practice Guidelines of MoHFW" and zero case precedents, whether a standard of precaution that reflects in face to face consultation is suitable within the telemedicine setting, particularly given the use of this advanced technology remain uncertain. With the telemedicine services considerably alter the conventional face to face consultation, would it entail a transformation with regard to proving the standard of care-needs precision. The other issue that needs to resolve is regarding "vicarious liability" principle under the common law. The civil law recognizes liability upon the employer if employee performed anything arising in course of his/her employment resulting the omissions. So, if telemedicine happens with the good will of a practitioner being associated with a hospital, should hospital be brought in is still a contentious question [15, 16].

4 Discussion

Telemedicine being an innovative model concerning processing of medicinal data, it comprises various legal frameworks dealing with medical regulation, data protection, data sharing information, communication technologies and further aspects of scientific research. Hence it involves few aspects of cyber law especially under the Information Technology (Reasonable Security Practices and Procedures and Sensitive Personal Data or Information) Rules, 2011, the Information Technology (Intermediaries Guidelines) Rules, 2011 and the Telecom Unsolicited Commercial Communications Regulations, 2007 and Telecom Commercial Communication Customer Preference Regulations, 2010. Whether the collected data from patients would be transferred to any third parties, the mode of retaining such data, would there be any detrimental effect on data later, etc. are not mentioned in the MoHFW governmental guidelines. These administrative aspects pertaining to telemedicine services are very important because these entail elements such as terms and conditions relating to transmission, keeping the recordings and footage of sessions and patient data, the upkeep and enhancing of software, the capacity building and infrastructure and creating training programs for health-workers and technicians. Besides that it is pertinent to comprehend the role of different stakeholders involved in this service. In addition to the patient and the doctors, there involve telemedicine technicians, paramedical service providers, tele-service providers, pharma industry also the Insurance companies. Hence, the issue pertains to protection of patient privacy with the digitization of medical records need to be settled. In telemedicine scenario, there lie new complications on providers to guarantee adequate privacy protections of audio and video communications. The underpinning of data privacy in health care is a balance between utility and security. MoHFW guidelines do not deal with "who should ultimately be responsible when a security breach occurs"? Telemedicine implicates a relentless interchange of information among the patient and the service provider. The personal information of patient—be it—medical history or the physical disorders or physiological conditions only if safeguarded as sensitive personal data, the said patient will have an opportunity for legal recourse. With the Personal Data Protection Bill, 2019 pending before parliament, providing safeguard to patient's data is another challenge India needs to overcome. The potential use of cryptography and password protected attachments, etc. have been proposed by various jurists in other countries as a solution to information privacy [17, 18].

5 Conclusion

Because telemedicine is a relatively new field and is indispensable in this pandemic time, patients need to get a proper understanding of its merits and demerits and hence the outlining of its framework needs paramount significance. Hence in India in the absence of current personal data protection legislation, guidelines for telemedicine with respect to clinical, technical and operational aspects need to be outlined which then need to be approved by National Medical Commission (NMC). Also, NMC need to make a framework before any patient agreeing to a telemedicine consult such as—"what are the limitations of telemedicine service, what all dos and do nots are to be charted, what happens if one or more of the avenues of communication/examination are lost? Where do the patient's records being kept?". While, telemedicine is still in the pilot phase, it is important that it should be firstly applied to a specific disease level so as to test the solutions and assess the minor problems so that it can be used for a wide range of diseases to ensure that it is cost effective. For example, using fundus camera for both diabetic retinopathy and for diagnosis based on picture of many illnesses of teeth, skin and similar aspects. Such approach can share expense between several specialties and can improve the cost effectiveness of the technology [17, 19]. Also, focus on challenges like reliable power supply and suitable hardware. India is a country with huge fallout in the required number of doctors owing to which the existing doctors are overloaded with work pressure. To tackle this situation, telemedicine is an effective solution. But it needs explicit strategies so as to make telemedicine workable. Emphasis on explicit diseases can significantly bound the bandwidth needed for judgemental ways of problem solving and decision making of doctors thereby having negligible influence on the demanding routine of doctors. Telemedicine can thus overcome the widespread shortage of paramedical personnel and training the staff thus making it important to extend the usage of telemedicine across villages.

References

1. High-Level Expert Group on AI Assessment (2020) List for Trustworthy artificial intelligence (ALTAI) for self-assessment. European Commission. Also, WHO on E-health in WHA58.28 resolution passed (on Fifty-eighth World Health Assembly, Geneva, Switzerland, 2005). WHO on E-health in "WHA58.28" resolution passed (on Fifty-eighth World Health Assembly) in Geneva, Switzerland, 2005). It is mentioned therein E-health is a broader term understood as juncture of therapeutic informatics, public health and trade, referring to health services and information distributed or improved through the Internet and associated technologies

2. Della Mea V (2001) What is e-health: the death of telemedicine? J Med Internet Res 3(2):e22. https://doi.org/10.2196/jmir.3.2.e22
3. See WHO Eleventh General Programme of Work (2006–2015) In its report, provides a global health agenda for WHO's Member States, wherein it focuses on four elements—availability, accessibility, acceptability and quality—as it calls as AAAQ
4. Viegas S, Dunn K (1998) Telemedicine: practicing in the information age. Lippincott, Raven Publishers, Philadelphia, PA. It is stated "There exists a difference between telehealth and telemedicine. Telemedicine, when has a clinician aspect telehealth is any use of information technology for health purposes". Though both involve application of electronic information and communication technologies for healthcare, telemedicine specifically implies "long-distance patient care"
5. Daley HA (2000) Telemedicine: The Invisible Legal Barriers to the Health Care of the Future Annals of. Health Law 9(1):73–106
6. Daar JF, Koerner S (1997) Telemedicine: legal and practical implications. Whittier Law Review Fall 19(1):328
7. Hollander JE, Carr BG (2020) Virtually perfect? Telemedicine for Covid-19. N Engl J Med 382:1679–1681. https://doi.org/10.1056/NEJMp2003539
8. Deepa Sanjeev Pawaskar vs. State of Maharashtra, 2018 SCC OnLine Bom 1841, order dated 25 July 2018
9. Jain T, Mehrotra A (2020) Comparison of direct-to-consumer telemedicine visits with primary care visits. JAMA Netw Open 3(12):e2028392–e2028392
10. Telemedicine Practice Guidelines by The Ministry of Health and Family Welfare on March 25, 2020. Available at https://www.mohfw.gov.in/pdf/Telemedicine.pdf. Last seen on 15 Jan 2022
11. Andriola M (2019) Telemedicine and legal disruption. Health Law and Policy Brief 13(2). Available at: https://digitalcommons.wcl.american.edu/hlp/vol13/iss2/2
12. Kapoor A, Pandurangi U, Arora V, Gupta A, Jaswal A, Nabar A et al (2020) Cardiovascular risks of hydroxychloroquine in treatment and prophylaxis of COVID-19 patients: a scientific statement from the Indian Heart Rhythm Society. Indian Pacing Electrophysiol J 20(3):117–120. https://doi.org/10.1016/j.ipej.2020.04.003. Also see, https://www.altnews.in/a-fake-prescription-coronavirus-has-been-circulating-in-the-name-of-ganga-ram-hospital-delhi/. Last seen on 7 Nov 2021
13. Daar J (1995) Informed consent: defining limits through therapeutic parameters, 16 Whittier L Rev 187
14. Ose D, Kunz A, Pohlmann S, Hofmann H, Qreini M, Krisam J et al (2019) A personal electronic health record: study protocol of a feasibility study on implementation in a real-world health care setting. JMIR Res Protoc 6(3):133
15. Lee TH (2010) Turning doctors into leaders. Harv Bus Rev 88(4):50–58
16. Lustgarten SD, Colbow AJ (2021) Ethical concerns for telemental health therapy amidst governmental surveillance. Am Psychol 72(2):159–170
17. Stanberry B (2006) Legal and ethical aspects of telemedicine. Journal of Telemedicine and Telecare 12(4):175. Also see, the Health Insurance Act, 2004 in France
18. Benjamens S, Dhunnoo P, Meskó B (2020) The state of artificial intelligence-based FDA-approved medical devices and algorithms: an online database. NPJ Digit Med 3:118. https://doi.org/10.1038/s41746-020-00324-0
19. Pennington K (2015) Legal and ethical concerns of digital media and technology in healthcare. Maryland medicine: the Journal of Maryland State Medical Society 16(3):29–30

Wheat Head Detection from Outdoor Wheat Field Images Using YOLOv5

Samadur Khan and Ayatullah Faruk Mollah

Abstract Automatic wheat head detection is an important problem having applications in wheat production estimation, wheat breeding, crop management, etc. Since the introduction of popular object detection model You Only Look Once (YOLO) in 2015, a number of advancements have come into picture. In this paper, YOLOv5, the latest of the YOLO family of models, is deployed for automatic wheat head detection from outdoor wheat field images. Experimenting on a publicly available wheat head dataset, mean average precision of 92.5% is achieved, which reflects its capability in learning and prediction wheat heads. The present method also outperforms some other methods on the same dataset. YOLOv5, being open source, may receive more commits in future paving the possibility of further improvement in wheat head detection performance.

Keywords Deep learning · Wheat head detection · GWHD dataset · YOLOv5

1 Introduction

Wheat is one of the most harvested crops in India and abroad. In the current scenario, 108.75 million tons of wheat are produced by India which is the second largest producer after China [1]. Automatic wheat head detection may facilitate not only wheat head identification but also wheat production estimation, wheat breeders, crop management, etc. However, it is very challenging due to (a) overlap of dense wheat plants, (b) degradation of acquired images, and (c) difference in appearance in terms of color, maturity, genotype, and head orientation.

Until recently, methods involving image processing technology and shallow learning were mainly applied for detection of wheat heads. Zhu et al. [2] presented

S. Khan (✉) · A. F. Mollah
Department of Computer Science and Engineering, Aliah University, Kolkata 700160, India
e-mail: samadur7726@gmail.com

A. F. Mollah
e-mail: afmollah@aliah.ac.in

P. Dutta et al. (eds.), *Emerging Technologies in Data Mining and Information Security*, Lecture Notes in Networks and Systems 490,
https://doi.org/10.1007/978-981-19-4052-1_53

a two-step wheat head detection method. Binarization approach is followed to identify wheat heads from image background by Dammer et al. [3]. Bi et al. [4] applied geometric properties to detect wheat heads. On the other hand, deep learning emphasizes on the structure of detection models [5] and turns out to be effective in detection problems.

Wheat head detection may be thought of as object detection as well. Therefore, deep learning-based object detection frameworks can be applied in detecting wheat heads from images or video frames. In this paper, a popular and fast object detection model known as YOLOv5 is custom-trained on the training set of wheat head images, and wheat heads are detected from the test set of images by the trained model. It is noticed that the developed model is able to identify wheat heads from images even if they are significantly overlapped. Experiments on the Global Wheat Detection dataset [6] reflect that the developed model detects wheat heads accurately and quickly.

2 Related Work and Motivation

In this section, some popular deep learning methods used in object detection are discussed, and then, the progress in wheat head detection is reported. The algorithms and approaches proposed in this field may be broadly divided into two categories: (a) two-stage detection and (b) one-stage detection. Two-stage detectors reported in literature include R-CNN [7], Fast R-CNN [8], Faster R-CNN [9], and FCOS [10]. One-stage detectors reported in literature are mainly the YOLO family models [11–15]. Such detectors are usually fast and high performing.

As evident from above discussion, one-stage detection is more powerful in terms of speed and accuracy. There are only a few related works on Global Wheat Head Detection on public datasets such as Global Wheat Head Dataset (GWHD). Like, it was trained on YOLOv3, YOLOv4, Faster R-CNN, etc. Although they all performed moderately well, there were some limitations. Their accuracies were not satisfactory for real-life scenario. In this paper, we apply a later version of YOLO family of models, i.e., YOLOv5 [15] to get it custom-trained for detection of wheat heads from outdoor wheat field images.

3 Methodology

YOLOv5 [15], the latest version of YOLO family of models, supports custom training. It was introduced by Mr. Glenn Jocher (Founder and CEO of Utralytics), and it is one of the best available models for object detection at this moment. It is the first YOLO model to be implemented in PyTorch entirely. In comparison to its previous version, i.e., YOLOv4, it has slightly less performance in COCO dataset, but it is superior in respect of ease of use, exportability, memory requirements, speed,

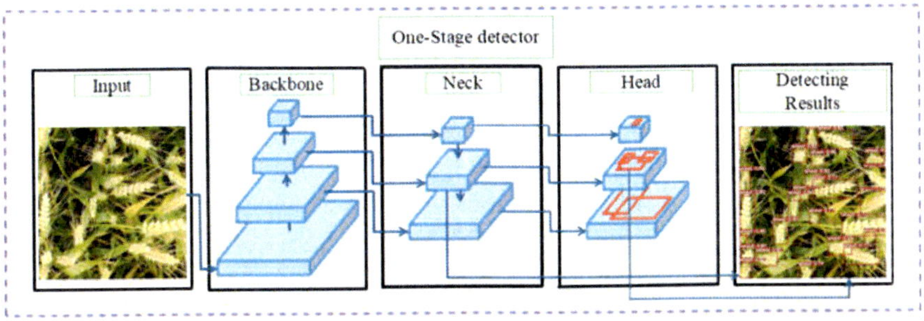

Fig. 1 YOLOv5 architecture for wheat head detection (The backbone is used for feature extraction; the neck is used for feature fusion, and the head is used for detection)

good mean average precision, and market size. YOLOv5 architecture is shown in Fig. 1.

3.1 Data Preprocessing

Now, for the training purpose, the whole data should be prepared as per requirement of YOLO. Data labeling, splitting of dataset, and creation of the YAML file take place in this stage. YOLO expects one text file per image having information of wheat heads as demonstrated in Fig. 2.

After labeling the images, a YAML file is created containing (a) the path for training and validation images, (b) number of classes present, and (c) the names corresponding to those classes.

3.2 Training of Model

In order to train this model with a custom dataset, certain steps are followed. Firstly, the dataset needs to be arranged in accordance with the YOLO format by installing dependencies, setting up the data, and the YAML files. After configuring all these setups, the training process may begin with the following specified parameters.

img: define input image size
batch: determine batch size
epochs: define the number of training epochs.
data: set the path to the YAML file
cfg: specify our model configuration
weights: specify a custom path to weights
name: result names
nosave: only save the final checkpoint

Fig. 2 YOLO formatted bounding box of an object (Wheat head)

cache: cache images for faster training.

3.3 Running Inference

Using the final trained weights obtained after completion of training, we can use the model for inference. To predict the wheat heads from a test image, we pass (i) the input image, (ii) path of the trained model, and (iii) the confidence threshold.

4 Experiments and Results

The Kaggle dataset of global wheat detection [6] is used in this work. It consists of 3373 outdoor wheat field images recorded from many locations around the world. It contains images of with bounding boxes for each wheat head to be identified. Ground truth information, i.e., the bounding rectangles enclosing wheat heads are also available. These images have been divided in 7:2:1 ratio for training, validation, and prediction, respectively.

Fig. 3 Comparison of ground truth rectangle and detected rectangle

4.1 Experimental Setup

The training is carried out using Google Colab platform for its faster GPU. In this project, maximum 150 epochs are taken for better performance. The libraries used are pandas, matplotlib, numpy, opencv-python, tensor-board, torch, tqdm, etc. Finally, testing of the developed model is performed using the test dataset. Then, we compare the obtained bounding boxes and the ground truth annotations to measure quantitative performance.

4.2 Evaluation Metrics

Intersection over Union (IoU): IoU is used to determine whether a predicted output is true positive (TP) or false positive (FP) [16, 17] as shown in Fig. 3. Ground truth annotations are used to determine true negative (TN) or false negative (FN) in case the model fails to predict a positive wheat head. Using these measures, precision may be obtained as $P = \frac{TP}{TP+FP}$, and recall may be obtained as $R = \frac{TP}{TP+FN}$.

Mean Average Precision (mAP): mAP is a widely used evaluation metric in wheat head detection performance measurement. Two variants of mAP, i.e., mAP_{50} and mAP_{95} are used in this work. In mAP_{50}, confidence threshold is 50%, and in mAP_{95}, confidence threshold is 50–95%.

4.3 Detection Performance

The results of the developed model are presented here. It may be noticed from Table 1 and Fig. 4 that the model detected the wheat head(s) from the images reasonably accurately.

Table 1 Performance of our model for different epochs

Epochs	Precision (%)	Recall (%)	mAP_{50} (%)	mAP_{95} (%)	Training time
10	73.1	62.7	67.6	28.5	5 min 51 s
50	90.8	84.1	90.6	46	18 min 20 s
100	91.4	86.2	91.7	48	36 min 59 s
150	91.7	86.9	92.5	48.8	54 min 55 s

Fig. 4 Detected wheat heads from two samples images of the test set

Here, basically, the statistical graphs are shown about the precision and recall of the model with increasing epochs. From Fig. 5, it may be noticed that the model is performing well as the precision and recall values increase over time up to a certain range during the training process.

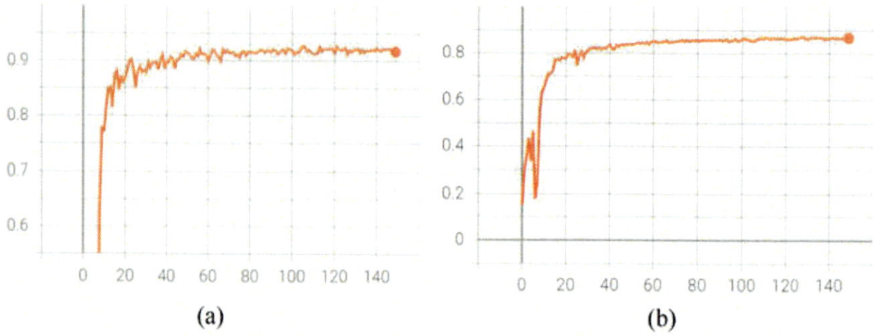

Fig. 5 Graphical representation of varying. **a** Precision and **b** recall with increasing epochs

Table 2 Performance comparison of our model with other methods

Method	Datasets	mAP_{50}	mAP_{95}
YOLOv3 [13]	GWHD	90.5%	47.7%
YOLOv4 [14]	GWHD	91.4%	**51.2%**
Faster R-CNN [9]	GWHD	76.6%	49.1%
Our method (YOLOv5)	GWHD	**92.5%**	48.8%

Bold indicates the highest measure in the respective parameter

4.4 Discussion

Performance of the present wheat head detection cum localization model is compared with some of the reported methods on the same wheat head dataset, i.e., the GWHD dataset [6] in Table 2. It may be noted that the said comparison is reported in terms of standard metrics, viz., mAP_{50} and mAP_{95} adopted for quantifying wheat head detection performance. In case of mAP with threshold 50, the present method performs very well with 92.5% precision. But, at mAP with threshold 95, our method performs slightly less than YOLOv4.

Moreover, as reported in literature, YOLOv5 is much faster than YOLOv4, and it is a one-stage object detection model. Thus, YOLOv5 appears to be a prospective trainable model for wheat head detection problems as well.

5 Conclusion

In this paper, YOLOv5, the latest model of YOLO family, is deployed on a publicly available wheat head dataset, and the accuracy obtained with the trained model is 92.5%. Compared to some other methods that reported results on the same dataset, the present method is found to detect wheat heads effectively, and it marginally outperforms other methods as well. Moreover, the built-in PyTorch framework of YOLOv5 is user-friendly, and it may receive more contributions in near future as it is an open-source model, which in turn may further improve wheat head detection performance.

References

1. Geography and You. Available online: https://geographyandyou.com/wheat-crop/. Accessed on 18 Oct 2021
2. Zhu Y, Cao Z, Lu H, Li Y, Xiao Y (2016) In-field automatic observation of wheat heading stage using computer vision. Biosyst Eng 143:28–42
3. Dammer K, Möller B, Rodemann B, Heppner D (2011) Detection of head blight (Fusarium ssp.) in winter wheat by color and multispectral image analyses. Crop Prot 30:420–428

4. Bi K, Jiang P, Li L, Shi B, Wang C (2010) Non-destructive measurement of wheat spike characteristics based on morphological image processing. Trans Chin Soc Agric Eng 26:212–216
5. LeCun Y, Bengio Y, Hinton G (2015) Deep learning. Nature 521:436–444
6. Global Wheat Head Detection (GWHD) Dataset (2021) Available in Online: https://www.kaggle.com/c/global-wheat-detection/data. Accessed on 20 Oct 2021
7. Girshick R, Donahue J, Darrell T, Malik J (2014) Rich feature hierarchies for accurate detection and semantic segmentation. In: Proceedings of IEEE conference on computer vision and pattern recognition, Columbus, OH, USA, pp 580–587
8. Girshick R (2015) Fast R-CNN. In: Proceedings of the 2015 IEEE international conference on computer vision (ICCV), Santiago, Chile, pp 1440–1448
9. Ren S, He K, Girshick R, Sun J (2017) Faster R-CNN: towards real-time object detection with region proposal networks. IEEE Trans Pattern Anal Mach Intell 39:1137–1149
10. Tian Z, Shen C, Chen H, He T (2019) FCOS: fully convolutional one-stage object detection. In: Proceedings of IEEE/CVF international conference on computer vision (ICCV), Seoul, Korea, pp 9626–9635
11. Redmon J, Divvala S, Girshick R, Farhadi A (2016) You only look once: unified, real-time object detection. In: Proceedings of IEEE conference on computer vision and pattern recognition (CVPR), Las Vegas, NV, USA, pp 779–788
12. Redmon J, Farhadi A (2017) YOLO9000: better, faster, stronger. In: Proceedings of the 2017 IEEE conference on computer vision and pattern recognition (CVPR), Honolulu, HI, USA, pp 6517–6525
13. Redmon J, Farhadi A (2020) YOLOv3: an incremental improvement. arXiv 2018, arXiv:1804.02767. Available online: http://arxiv.org/abs/1804.02767. Accessed on 29 Dec 2020
14. Bochkovskiy A, Wang C-Y, Liao H-YM (2020) YOLOv4: optimal speed and accuracy of object detection. arXiv 2020, arXiv:2004.10934
15. Ultralytics.YOLOv5.2020. Available online: https://github.com/ultralytics/yolov5#readme
16. Khan T, Sarkar R, Mollah AF (2021) Deep learning approaches to scene text detection: a comprehensive review. Artif Intell Rev 54:3239–3298
17. Saha S, Chakraborty N, Kundu S, Paul S, Mollah AF, Basu S, Sarkar R (2020) Multi-lingual scene text detection and language identification. Pattern Recogn Lett 138:16–22

Prediction of COVID-19 Disease by ARIMA Model and Tuning Hyperparameter Through GridSearchCV

Aisha Alsobhi

Abstract The COVID-19 pandemic has significantly impacted the mental, physiological, and financial well-being of people around the globe. It has threatened lives and livelihoods and triggered supply chain disruptions and economic crises. In every country, there are risks and long-term implications. Planners and decision-makers could benefit from a forecasting model that anticipates the spread of this virus, thereby providing insight for a more targeted approach, advanced preparation, and drive better proactive collaboration. The signs and symptoms of a disease like COVID-19 are hard to define and predict, particularly during times of pandemic. Several epidemiological studies have been successful in identifying predictors, using artificial intelligence (AI). This paper explores various methodologies for tuning the hyperparameters of the auto-regressive integrated moving average (ARIMA) model, using GridSearchCV, to predict and analyze the occurrence of COVID-19 in populations. In time series analysis, hyperparameters are crucial and the GridSearchCV methodology results in greater predictive accuracy. The parameters proposed for the analysis of daily confirmed cases, recovered cases, and deceased cases in India were ARIMA (4, 1, 5), ARIMA (5, 1, 1), and ARIMA (5, 1, 1), respectively. The performance of the model with different configurations was evaluated using three measurements: root mean square error (RMSE), R2 score, and mean absolute error (MAE). These results were compared with a state-of-the-art method to assess model selection, fitting, and forecasting accuracy. The results indicated accuracy and continuous growth in the number of confirmed and deceased cases, while a decreasing trend was graphed for recovered cases. In addition, the proposed ARIMA using a GridSearchCV model predicted more accurately than existing approaches.

Keywords Machine learning · Prediction and analysis · ARIMA · Optimization · GridSearchCV · SARS-CoV2 · Hyperparameter tuning

---

A. Alsobhi (✉)
Department of Information Systems, Faculty of Computing and Information Technology, King Abdulaziz University, Jeddah, Saudi Arabia
e-mail: ayalsobhi@kau.edu.sa

P. Dutta et al. (eds.), *Emerging Technologies in Data Mining and Information Security*, Lecture Notes in Networks and Systems 490,
https://doi.org/10.1007/978-981-19-4052-1_54

1 Introduction

A new version of severe acute respiratory syndrome coronavirus (SARS-CoV2) disease was identified in 2019 (COVID-19) and caused a global pandemic that is still a major and urgent threat to health around the world. It was estimated in October 2020 that the COVID-19 outbreak, allegedly originating in the Chinese region of Hubei in early December 2019, had infected 39,500,000 people, though actual numbers are likely to be significantly higher. Currently, we calculate the world death rate from COVID-19 to be 1.1 million. Health systems and medical networks are challenged daily by the spread of this disease, the increased demand for hospital beds, and the associated shortages of medical supplies [1]. The ability to make fast clinical choices and use healthcare resources efficiently are critical, particularly during a pandemic, but developing countries had difficulty obtaining the most reliable COVID-19 diagnostic test which uses reverse transcriptase polymerase chain reaction (RT-PCR). This scarcity results in increased infection rates and delays in important preventive measures [2].

During the early months of the COVID-19 epidemic in Israel, the Israeli Ministry of Health mandated that all COVID-19 diagnostic laboratory tests be carried out in accordance with its standards. The Ministry used a nasopharyngeal swab test, and the results were made public. It is acknowledged several variables may have affected/altered the reports, including whether a patient had COVID-19 at the start of the trial, exposure to infected individuals, certain geographic locations, and the potential for complications if a subject became sick. Except for a tiny number of healthcare personnel who were tested as part of a survey, all subjects had justifications for testing [3]. Therefore, referral bias did not appear to be an issue, in contrast to prior research. In addition, the Ministry used RT-PCR assays to confirm all negative and positive COVID-19 results for their dataset.

Currently, machine learning (ML) is being used to assist with understanding and identifying who is most affected by COVID-19. It is facilitating the production of medications and vaccines, the identification of patients, and the research of drug behaviors from similar viral infections. ML is also helping with locating the origin of this virus and projecting the next global pandemic outbreak. No cure for COVID-19 has yet been discovered, so social distance is our only solution to stop its spread. It is an underlying factor for this study that a solid mathematical foundation must be developed to track and make decisions about this pandemic's behavior [3].

There are not enough COVID-19 analysis kits in hospitals to handle the growing number of cases. This fact, along with ignorance and fear, means there are not enough people being tested. As a result, an automated prediction system can be a powerful tool to support the data from analysis kits and to extrapolate results to deliver sufficient treatment during the initial virus stage, while halting disease transmission between people. In the fight against COVID-19, ML can be used to perform intense data management and accurate disease spread prediction. This technology can aid in diagnosis and prediction of the presence of COVID-19. It can help track down COVID-19 instances, create alerts and dashboards for things like social distancing, make diagnoses, and administer suitable treatments.

2 Literature Review

In [4], four time series models were presented for predicting the count of infected, death, and recovery cases. The models, are accessible in the prediction package in R, were applied to the publicly available dataset of COVID-19 from Italy and the United States. In comparison with the Holt and trigonometric exponential smoothing state-space model with Box-Cox transformation (TBATS) models, ARIMA and Spline models showed small prediction errors and shorter prediction ranges. In comparison with the other models, ARIMA produced more consistent results. The akaike information criteria (AIC) values of the TBATS and Spline models were identical and were frequently less than the AIC of the ARIMA model, indicating that the TBATS and Spline models matched the data better. These results indicate the ARIMA model could be optimized to produce more accurate results. Using data between February 21 and April 15, 2020, an ARIMA model was created from parameters p, d, and q. The COVID-19 epidemiological pattern was tested in the three most disease impacted nations in Europe: Spain, Italy, and France. Orders p, d, and q were analyzed for best performance, and the lowest MAPE values for the three countries [5] were selected as the basis for use with the India dataset. The ARIMA model [6] was used to forecast the cases of COVID-19 in an epidemiological COVID-19 dataset from Johns Hopkins taken from January 20 to February 10, 2020. This ARIMA model comprises an auto regression (AR) model, moving average (MA) model, and seasonal ARIMA (SARIMA). To stabilize the time series log transformation, differences were preferred. The ARIMA model was evaluated by autocorrelation function (ACF) and partial autocorrelation (PACF) correlograms. The results showed that ARIMA (1, 0, 4) performed better, while ARIMA (1, 0, 3) excelled at determining the incidence of COVID-19. Seasonality had no effect on the incidence or prevalence of COVID-19, according to ACF and PACF. The number of confirmed cases continued to rise, but the frequency dropped somewhat. Using the SVM model [7], patients were classified by the seriousness of their COVID-19 symptoms. In [8], the SVM was used for binary class labeling on 137 records including blood and urine test data, and then combined with patients who presented with severe illness and patients who presented with minor symptoms. The findings demonstrated a link of 0.815 between severe COVID-19 and 32 other variables. Notably, age and gender were the most significant determinants of the seriousness of a case. Severe instances were found in patients aged 65 and older. Males were more likely than females to experience life-threatening COVID-19 symptoms. In a comparison of urine and blood test results, there were greater disparities between serious and moderate cases that in urine test results. The severity of COVID-19 patient disease was determined by E. Team [9] using an LR model. A dataset including clinical and demographic COVID-19 case data for 115 patients in a non-serious state and 68 patients in a serious state was used [10]. To distinguish between mild and severe instances, four significant criteria were selected: age, high-sensitivity C-reactive protein level, lymphocyte count, and d-dimer level (a marker of inflammation). The results showed the prediction was accurate, with an area under the receiver operating characteristic values of 0.881, 0.839 sensitivity, and 0.794 specificity. In [11], 3927 COVID-19 patient samples were analyzed to predict mortality risk using XGBoost software. The study employed demographic and clinical data from 33 hospitals to reflect the patient population. The model's accuracy was 0.85, and its AUC was 0.90. Furthermore, [12] used 1969 COVID-19-positive patients to create LR-based mortality predictions. Age and O2 were determined to be important factors with AUCs of 0.89, 0.82 sensitivity, and 0.81 specificity. Since ML algorithms have been proposed in other studies [13] to predict COVID-19 positive instances, deep learning network models were used in this study for classification, decision-making, and regression tasks. The categorization models were used to detect when an infection had spread to humans. This study made use of widely disparate scientific datasets, including infectious and non-infectious COVID-19 cases. When evaluated with Naive algorithms, there was a high degree of accuracy (95%) as a categorization model. High-dimensional datasets with more complexity were error-prone. For assessing COVID-19 cases [14], three ML models were constructed: polynomial regression, decision tree regressor, and random forest method. The polynomial regression model was found to have the highest accuracy (91%). This level is insufficient for use in medical data analysis. ML models such as the susceptible models [15] used exposed infectious recovered (SEIR) and regression models to study the change in the propagation of the COVID-19 disease. The SEIR model's root mean adjusted for a log error rate of 1.52, whereas the regression model rate was 1.75. The AdaBoost approach was used by Shankar et al. [16] to improve the random forest model which forecasted how serious the positive instances would be based on the data. The model's accuracy and $F1$ score were 94% and 0.86, respectively, indicating that accuracy and $F1$ needed to be improved. Few studies have attempted to predict the spread of disease, but those that have attempted it have relied on rudimentary statistical methods or modest survey. The use of AI in the form of advanced ML concepts in such investigations is still being developed.

3 Methodology

This study follows a step-by-step process to look at infection transmission, as well as illness progression and progression prediction. Prior to applying ML algorithms to a dataset, the correct dataset with relevant parameters was selected and pre-processed in the appropriate format. Adjusting the percentage of testing and training data was the main step to prepare the classifier. This was accomplished through a process of trial and error and followed by training and testing different categorization algorithms. All ML algorithms used in this study were compared using the performance indicators outlined in Sect. 4. Figure 1 shows the steps taken to implement the operation.

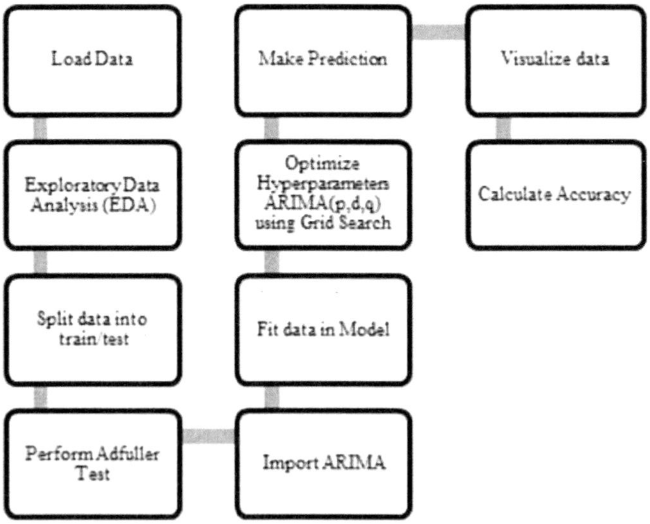

Fig. 1 Proposed workflow diagram

3.1 Dataset Description

In the evaluation, the India dataset was collected from Kaggle 2020. This is the world's largest dataset repository. The information gathered pertained to all COVID-19 cases reported in India between February 1, 2020 and March 9, 2021. The dataset was in CSV format and contained multiple attributes related to COVID-19 including date, timestamp, daily confirmed cases, total cases, daily recovered, total recovered, daily deceased, and total deceased in India as shown in Table 1. A training set and a test set were created from the dataset.

Table 1 Dataset attributes information

Date	Timestamp	Daily confirmed	Number of cases	Daily recovered	Total recovered	Daily deceased	Total deceased
2021-03-04	1614816000	16,824	11,173,495	13,788	10,837,845	113	156,993
2021-03-05	1614902400	18,324	11,191,819	14,186	10,852,031	109	157,102
2021-03-06	1614988800	18,724	11,210,543	14,379	10,866,410	100	157,202
2021-03-07	1615075200	18,650	11,229,193	14,303	10,880,713	97	157,299
2021-03-08	1615161600	15,353	11,244,546	16,606	10,897,319	76	157,375

3.2 ARIMA Model

The ARIMA model was introduced by Box and Jenkin in [17]. It uses the discrepancies between values of a time series to forecast future values. ARIMA consists of three other models in which every element is a parameter [18]: autoregressive (AR) model, moving average (MA) model, and seasonal ARIMA (SARIMA) [19]. The time series must be stationery and trend-free, so the augmented Dickey-Fuller (ADF) was used to test the dataset. It provided the ability to log transformation and differences. The tuples of the ARIMA model included (p, q, r) (P, Q, R) S.

- p: order of AR
- d: rate of difference in trend
- q: order of MA
- P: seasonal AR lag value
- D: rate of seasonal difference
- Q: seasonal MA value
- S: height of cyclical pattern.

3.3 ARIMA Model Optimization Using GridSearchCV

Grid search is a strategy for determining the best hyperparameters for a model. Finding hyperparameters in training data, unlike parameters, is considered impossible. As a result, a model was developed for each combination of hyperparameters to ensure the best hyperparameters were employed. To acquire a good approximation of the predictive capacity of a model, cross-validation was used on each new dataset. Two datasets were presented: an independent dataset and a training dataset. We partitioned each dataset into two sets [20]. Figure 2 shows the proposed workflow for optimizing the ARIMA model using GridSearchCV.

The comparison was made using the previous collected data (April 22 to May 21, 2020). Showing the forecast for a 30-day time period using ARIMA, the metrics are displayed in Table 2.

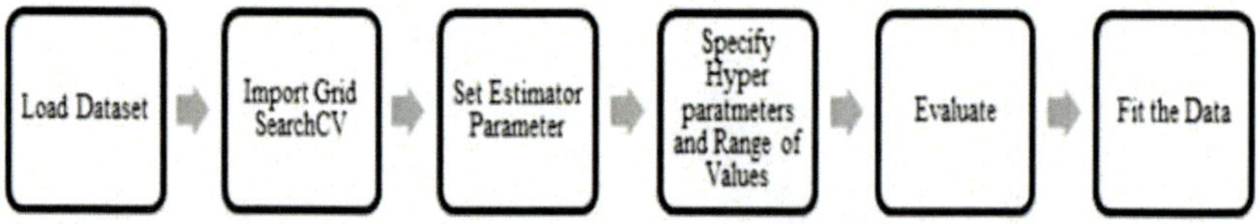

Fig. 2 shows the proposed workflow for optimizing the ARIMA model using GridSearchCV

Table 2 Evaluation metrics for ARIMA and optimized ARIMA model for deceased, confirmed, and recovered cases

	Deceased cases		Confirmed cases		Recovered cases	
	ARIMA	Optimized ARIMA	ARIMA	Optimized ARIMA	ARIMA	Optimized ARIMA
R2 score	−11.406655	0.92	0.255996	0.5612	0.310856	0.87
MAE	617.6	15	2967	1016	961.1666667	1990
RMSE	388,366.6667	12	9,814,486.067	1260	1,024,253.9	1011

Table 3 Evaluation metrics showing comparison with various state-of-the-art models

Performance metrics	Linear regression	Polynomial regression	Prophet	ARIMA (without grid search)	ARIMA (with grid search)
R square	0.01	0.31	0.46	0.255996	0.5612
RMSE	284,809.4	149,117.8	568.58	3132.808	1260

4 Results and Discussion

Results obtained through the optimal ARIMA model showed a significant improvement in performance, tuning the hyperparameters through GridSearchCV. The parameters p, d, q could be optimally obtained. The outcome of the prediction and estimation was determined by the 'event' description and data gathering method used. We found that future comparisons or viewpoints, case definition, and data collection must be preserved in real time. In general, the fitted values and anticipated values obtained using the following approaches (MAE, R2, RMSE) accurately matched the real incidence of COVID-19 diseases. Although more information would be needed to make a more precise projection, it appeared the virus's spread was slowing dramatically. In addition, the frequency had fallen slightly, despite the number of confirmed cases climbing. If virus mutations do not develop, the number of cases will reach a high.

Table 3 shows the comparison of various state-of-the-art models with the proposed model. The comparison demonstrates better performance in terms of minimum loss. The ARIMA model with a grid search approach produced better results as compared with ARIMA without grid search approach.

5 Conclusion

The COVID-19 pandemic has proven to be a deadly sickness transmitted by infected people, much like any other widely spread illness. As the number of patients increases rapidly during a pandemic, the healthcare sector struggles to identify effective treat-

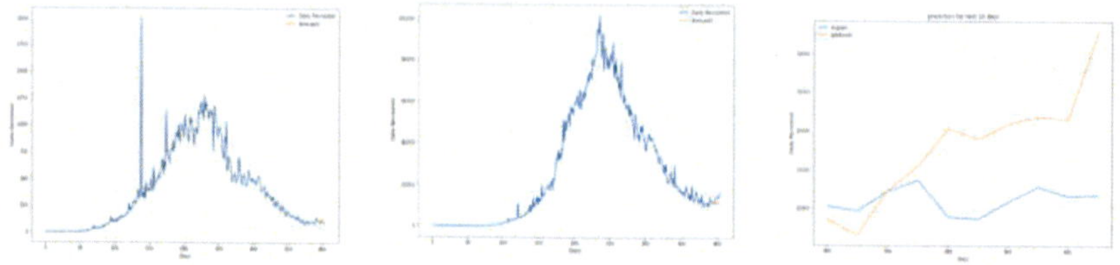

(a) Deceased Cases (b) Recovered Cases (c) Daily Recovered Cases

Fig. 3 Prediction of cases using proposed ARIMA with GridSearchCV model

ments. This study predicts confirmed cases, recovered cases, and deceased cases of COVID-19 for 20 days ahead, until March 29, 2021, using an optimized ARIMA statistical model with GridSearchCV as shown in Fig. 3. Despite fast viral mutation and the structure of the dataset dependency on time and date, the research attempted to limit data variability by using just the dataset. Predicted results were more accurate with actual data after using ARIMA with GridSearchCV model. Our forecast indicated that the count of confirmed and deceased cases increased while recovered cases were predicted to decrease. On the basis of our prediction, strategy administrators of health care could make necessary decisions for the timing of providing healthcare aids to the public as well as transporting equipment to hospitals. The prediction could help policymakers frame their pandemic policies. The result has also focused and emphasized the significance of social distancing and the execution of various protective measures. The future work will yield more substantial results utilizing a variety of additional ML algorithms for producing estimates to help physicians and other medical professionals, and government agencies plan ahead for possible pandemic diseases. Deep Learning models can also be used to anticipate COVID instances.

References

1. Muhammad L, Algehyne EA, Usman SS, Ahmad A, Chakraborty C, Mohammed IA (2021) Supervised machine learning models for prediction of covid-19 infection using epidemiology dataset. SN Comput Sci 2(1):1–13
2. Monica G, Bharathi Devi M (2020) Using machine learning approach to predict covid-19 progress. Int J Mod Trends Sci Technol 6(1):58–62
3. Kumar PR, Sarkar A, Mohanty SN, Kumar PP (2020) Segmentation of white blood cells using image segmentation algorithms. In: 2020 5th international conference on computing, communication and security (ICCCS). IEEE, pp 1–4
4. Cascella M, Rajnik M, Aleem A, Dulebohn S, Di Napoli R (2021) Features, evaluation, and treatment of coronavirus (covid-19). StatPearls
5. Abu Al-Qumboz MN, Abu-Naser SS (2019) Spinach expert system: diseases and symptoms. Int J Acad Inf Syst Res (IJAISR) 3(3):16–22
6. Pasayat AK, Pati SN, Maharana A (2020) Predicting the covid-19 positive cases in India with concern to lockdown by using mathematical and machine learning based models. medRxiv

7. Petropoulos F, Makridakis S (2020) Forecasting the novel coronavirus covid-19. PLoS One 15(3):e0231236
8. Grasselli G, Pesenti A, Cecconi M (2020) Critical care utilization for the covid-19 outbreak in Lombardy, Italy: early experience and forecast during an emergency response. JAMA 323(16):1545–1546, e0231236
9. E. Team (2020) The epidemiological characteristics of an outbreak of 2019 novel coronavirus diseases (covid-19)-China. China CDC Wkly 2(8):113
10. Makridakis S, Spiliotis E, Assimakopoulos V (2018) Statistical and machine learning forecasting methods: concerns and ways forward. PLoS One 13(3):e0194889
11. Maxwell JC (1873) A treatise on electricity and magnetism, vol 1. Clarendon Press
12. Jacobs I (1963) Fine particles, thin films and exchange anisotropy. Magnetism 271–350
13. Yorozu T, Hirano M, Oka K, Tagawa Y (1987) Electron spectroscopy studies on magneto-optical media and plastic substrate interface. IEEE Transl J Magn Jpn 2(8):740–741, e0194889
14. CDC COVID-19 Response Team, Bialek S, Boundy E, Bowen V, Chow N et al (2020) Severe outcomes among patients with coronavirus disease 2019 (covid-19)-United States. Morbidity and mortality weekly report, Feb 12–Mar 16, 2020, vol 69, no 12, p 343
15. Dong E, Du H, Gardner L (2020) An interactive web-based dashboard to track covid-19 in real time. Lancet Infect Dis 20(5):533–534, e0194889
16. Shankar K, Mohanty SN, Yadav K, Gopalakrishnan T, Elmisery AM (2021) Automated covid-19 diagnosis and classification using convolutional neural network with fusion based feature extraction model. Cogn Neurodyn 1–14
17. Box, George EP (1976) Science and statistics. J Am Stat Assoc 71(356):791–799
18. Hernandez-Matamoros A, Fujita H, Hayashi T, Perez-Meana H (2020) Forecasting of COVID19 per regions using ARIMA models and polynomial functions. Appl Soft Comput 96:106610
19. Fattah J, Ezzine L, Aman Z, El Moussami H, Lachhab A (2018) Forecasting of demand using ARIMA model. Int J Eng Bus Manage 10:1847979018808673
20. Collins C, Landivar LC, Ruppanner L, Scarborough WJ (2021) COVID19 and the gender gap in work hours. Gender, Work & Organization 28:101–112

A Spatio-demographic Analysis Over Twitter Data Using Artificial Neural Networks

Tawfiq Hasanin

Abstract The demographic and population modeling methods have been under investigation trends since the 1980s. Extrapolation, prediction, and theoretical computational analysis of exogenous variables are approaches to the forecasting of population processes. Such methods can be exploited to predict individual birth preferences or experts' views at the population level. Predicting demographic changes have been problematic while its precision usually depends on the case or pattern; numerous methods have been explored; however, so far there is no clear guidelines where the proper approach ought to be. Like certain fields of industry and policy, planning is focused on projections for the future composition of the population, the potential creation of population sizes and institutions which are significant. In order to recognize potential social security issues as one determinant of overall macroeconomic growth, countries that have reduced mortality and low fertility, the case with some of the Asian nations, desperately require accurate demographic estimates. This introduction provides a stochastic cohort model that uses stochastic fertility, migration, and mortality modeling approaches to forecast the population by gender and literacy. This work focus on the population and literacy ratio of India as this nation holds the second largest population in the world. Our approach is based on artificial neural network algorithm that can forecast the population literacy ratio and gender differences based on living states populations using social networks data. We concentrated primarily on quantifying future planning challenges as previous research appeared to neglect potential risks. Our model is then used to forecast/predict gender-wise population for each major state/city. The findings offer clear perspectives on the projected gender demographic composition, and our model holds high precision results.

Keywords Demographics · Machine learning · Neural network · Population forecasting · Social media

T. Hasanin (✉)
Department of Information Systems, Faculty of Computing and Information Technology, King Abdulaziz University, Jeddah, Saudi Arabia
e-mail: thasanin@kau.edu.sa

P. Dutta et al. (eds.), *Emerging Technologies in Data Mining and Information Security*, Lecture Notes in Networks and Systems 490,
https://doi.org/10.1007/978-981-19-4052-1_55

1 Introduction

The scale of population data requires an interpretation of the flow of knowledge across dynamic networks which can be hard to examine through traditional methods. For one, the data are usually inexpensive in contrast to traditional methods of data collection [1]. Such data have been used by the experts of ethnic and public health institutional networks of different areas [2]. Population growth can be studied through population forecasts by shifting infertility [3, 4]. Such factors may also be influenced by other external impacts. Population prediction models use recent birth, mortality, and migration flow data to generate accurate forecasts of the demographic makeup of the future. The combined consequences of a natural transition (birth and death) and net migration are likely to be our foundation. In the twenty-first century, population transition is a critical issue [4–6]. For governments, corporations and culture in general, the increasing scale and image of people are of considerable significance. To policymakers and decision-makers, sustained cycles of new housing growth, aging demographics, increasing diversity, and average household sizes are important considerations. Our demographic prediction uses simulation approaches to determine possible demographic outcomes. These are a product of all population patterns, and all state, provincial, and local decision-making should be focused on political initiatives. Both macro- or micro-graphs and for infinite projections are used for our population supply models.

Our work focuses on India as a case study on the ground that it has the second largest country population in the world where 1.38 billion humans live in India,[1] right after China which has 1.402 billion population and before the USA which has the population of 329.5 million. Moreover, the availability of public data plays a major factor to select India as a case study in this experimental work.

The remaining of the paper is organized as follows: Section 2 provides our literature review on the related works. Sections 3 and 4 present the mythology and results, respectively. Finally, Sect. 5 draws a conclusion from this work.

2 Literature Review

Demographic analysis is one of the most important fields for research over the years [4, 7]. For this field, many of the recent technological advances can be manipulated to strengthen and to improve the spatial demographic analysis. Some of the recent technologies used with huge capabilities to evaluate demographic data are machine learning (ML), deep learning (DL), and Internet of Behaviour (IOB).

The work in [8] provided a systematic study of analysis of risk factors for the loss in the mobility of elderly people living in the community. In their literature review, they identified comparative correlations of underlying risk factors and their

[1] https://www.worldometers.info/—Worldometer provides a world statistics available in a simulated and time manner.

corresponding functional role and carried out a comprehensive analysis of clinical studies. All publications mentioned in this review have been completed by a Lexicon-based emotion analysis which has a cycle that contains three stages. i. e, the overall feelings, the classification of emotions, and the compilation of results.

The work in [9] highlighted the crossroads between gender, location, age, and ethnicity between Twitter users, particularly from U.S. The Power-Law [9, 10] behavior approach and the Intertweet Interval (ITI) [9] scheme are two methods used to measure the amount of user interaction on Twitter.

The researchers in [11] demonstrated the social issues between those who use the available position services for themselves and others who do not. The authors explained Twitter users' demographic attributes. Social media users have the ability to enable their location services, and a previous study by the same research team in [12] shows that approximately only 0.85% are geotagged, i. e, using longitude and latitude of the users when every tweet was posted, and thus, the exact location was recorded. That led to collecting two separate datasets in their work, one was based on geo-service facilitation, and another was based on tweet geo-tagging service. The behavior of Twitter users has been analyzed based on their age, status (economic and social), gender, and language.

The demographic change from high to low fertility was noticed by Refs. [13–15] which explained Europe's population history and gave a valuable guide to the relationship between innovations and population growth. In the developing nations, the population change was proposed to follow a far different direction from one taken in the USA and Europe. The work in [16] offers a statistical description of various demographic shifts. It describes the various phases of demographic transition, changes in population patterns of global populations.

Through closely analyzing the evidence provided by various secondary data, the works in [17] map demographic patterns in Kerala. They also investigated the multiple conflicting theories suggested to clarify the drop in birth and death rates. The generational change in the Kerala economy in the 1990s was clarified by the same team in [18]. The analysis reveals the changes in fertility based on secondary data sources.

The work in [19] proposed a model that predicts the occurrence of a criminal incidents based on feature-based approach and textual analysis in the USA to prevent crimes. Their work concludes that using Twitter data automation outperforms other models while benefits from cases where only unstructured data are available.

3 Methodology

In this section, we introduce our model for population forecasting using a neural network (NN), which is robust model for spatio-demographic analysis [20, 21]. An NN is a collection of algorithms that aim to identify the underlying connections of the data using a mechanism that imitates the workings of the human brain [22, 23]. For this respect, neural networks apply to biological or artificial neuron structures.

Neural networks may be modified to alter data, ensuring that the network delivers the optimal performance without the need to rewrite output parameters. The idea of NNs, rooted in artificial intelligence, is gaining rapid prominence in trading systems growth. A neural network behaves like the neural network of the human brain. A neuron is a mathematical entity in the neural network that gathers and classifies information through a particular architecture [22]. The network is closely linked to statistical approaches like curve correction and regression analysis.

There are variety of options for the neuron to make a decision using the activation function [24]. Some of the most common activation functions are listed below:

- **Rectified Linear Units (ReLU)** $\text{Relu}(x) = \max(0, x)$: With this function, the output cannot be less than zero (or negative). The output is x if x is greater than zero. The output is zero if x is less or equal to zero. In short, the function set the output between 0 and z for the full plot.
- **Tanh**: $f(x) = \tanh(x)$: To simplify it, the function finds the hyperbolic tangent of z and return it.
- **Sigmoid** $f(x) = \frac{1}{1+e^{-1*x}}$: This function is commonly used in feedforward NN due to its derivative being simple compatibility and its nonlinear behavior [25]. However, the gradients may fade when the input is distant from zero.

Figure 1 shows our NN structure. The left side holds the input layer such as total population, total literacy, the total population of males and females, and the total literacy of males and females. The two hidden layers are created with ten nodes and one threshold node. After that, we got our forecast output layer with one node.

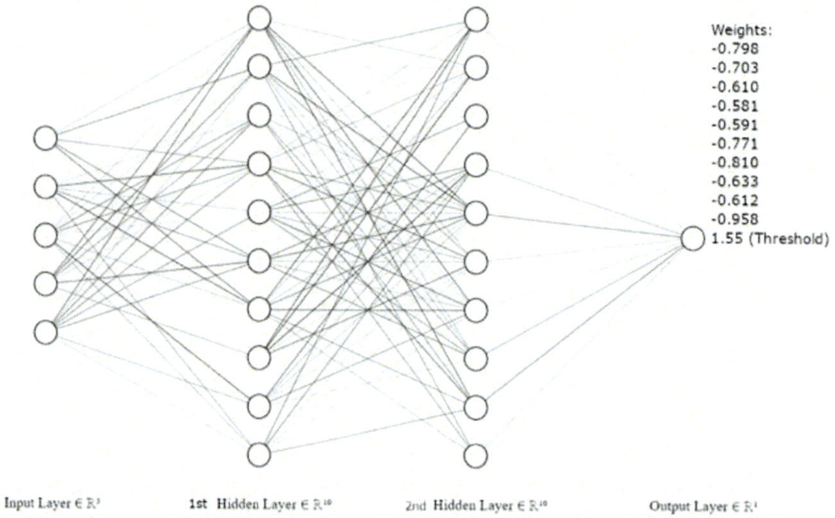

Fig. 1 The applied neural network structure is fully connected neural network with five inputs, two hidden layers, ten nodes, and one output node

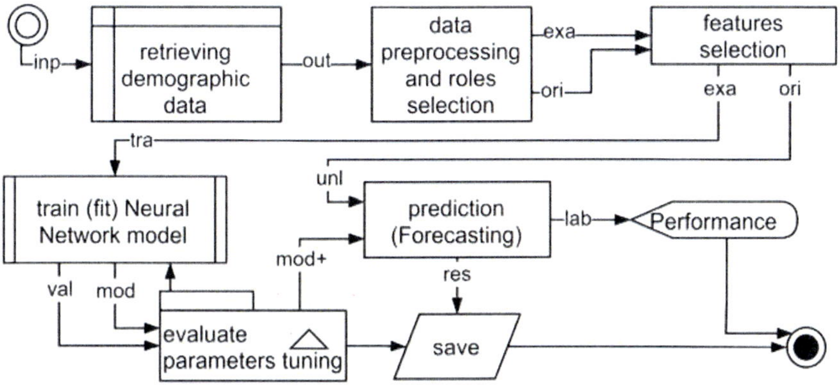

Fig. 2 Our proposed forecasting module using neural networks consisting of six major steps to reprocess the demographic data and forecast the population

We developed our model using the desired NN and processed over Twitter dataset through windowing module (WM), and it passes through sub-process. The sub-process filtered through generating the macro-module, after that the dataset passes through the k folds cross-validation module. The output model then will be applied on the testing dataset and record the performance. Finally, the forecast model can be put into production to predict users demographic based on their living states.

Our proposed method can be summarized into the following six steps which also can be recapitulate from Fig. 2. Note that *inp, out, exa, ori, tra, val, lab, unl, mod, per, and res* are shorts for input data, output data, example set, original set, training set, validation set, labeled data, unlabeled data, model, performance, and result set, respectably. The proposed steps are as follows:

1. Import the dataset of the entire population state-wise.
2. Select role by defining the aimed attribute to predict and classify.
3. Select the attribute as a subset of the dataset in which the state name is taken.
4. Specify the neural network module parameters with 2000 training cycle, 0.01 learning rate, and 0.9 momentum with normalization is to be performed.
5. Forecasting/predicting using the applied model for the states population.
6. Check the performance of our final model and output of the dataset. The sub-process filtered through generating the macro-module, after that the dataset passes through the performances.

This work's experimental case study was performed using the RapidMiner [2] platform to prepossess the data, engineer and prepare the feature space, and perform the NN algorithm to forecast the population.

[2] https://rapidminer.com/—RapidMiner is a data science software platform developed by the company of the same name that provides an integrated environment for data preparation, machine learning, deep learning, text mining, and predictive analytics.

4 Experimental Results

Every case study, were repeated five times to avoid randomness that happens during the ML algorithm and/or data preparations and prepossessing. For each state in the country, the results of the prediction were recorded and averaged among the five repetition.

Figure 3 shows the state-wise population forecasting without separating the gender among the community. It is observed that Maharashtra has highest population and Uttar Pradesh has second highest population which is represented by the actual number of people on the ground. Note that the solid line plot indicates original population and the diamond symbol indicates the predicted population using our NN method. It is noticeable that the higher the actual state population is, the bigger the

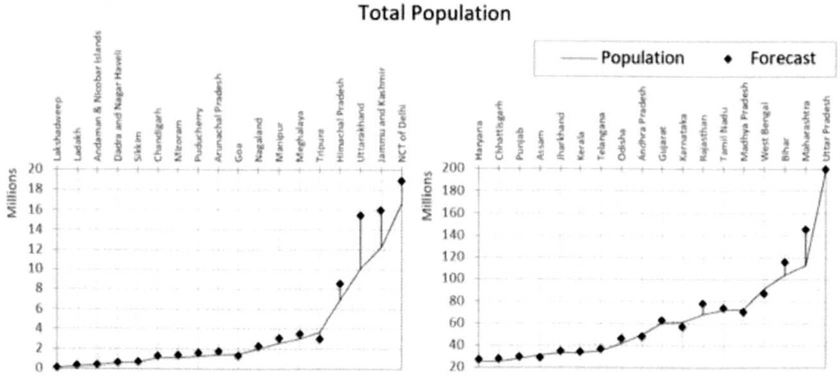

Fig. 3 Projected the forecasting total state-wise population for both genders with comparison to the latest actual population

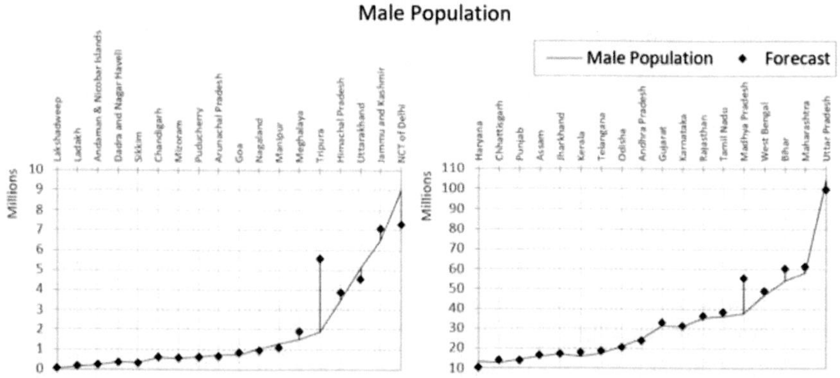

Fig. 4 Projected the forecasting male gender state-wise population with comparison to the recent male population

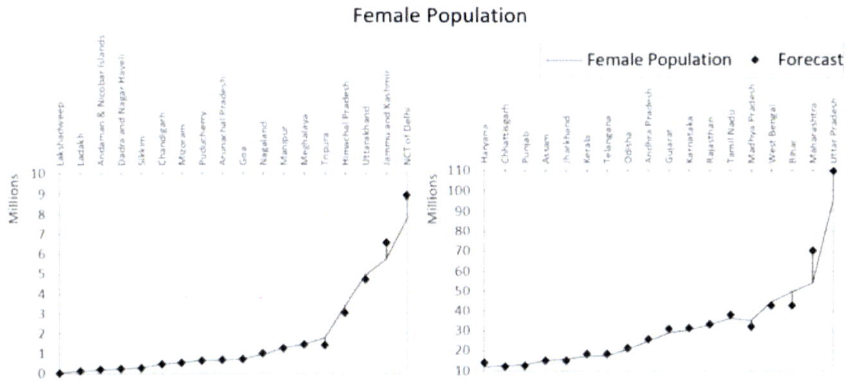

Fig. 5 Projected the forecasting female gender state-wise population with comparison to the recent female population

difference is between the prediction and the actual numbers. Figure 4 shows state-wise male actual and predicted population with our forecasting NN. On the other hand, Fig. 5 shows state-wise female population forecasting. We can also observe that plots in the graph are similar to the ones from the male forecasting. Both Figs. 4 and 5 show that in some cases, the model failed to predict/forecast the population accurately when the population do not have enough samples in some states. However, the models yield high-performance results when the highest population states were classified. Moreover, Figs. 4 and 5 assert that male prediction yield better results than female forecasting in the case study under investigation for most of the states. After repetition of the experimental works, we found that parameters such as the activation function play a great role in the entire process throughout good population forecasting results. In general, our method performs very well in comparison with other techniques. Note that not all results were displayed in this section due to the limited space.

5 Conclusion

In this paper, we proposed a method to forecast the state-wise population of India, on account of being the second largest world nation in terms of population. This can help to draw a decent case study to be applied on other nations. The focus of this work was mainly on the gender population that is male and female. The experimental work extracted their behaviors from Twitter social media platform based on geographic location. We have considered four categories of population datasets, that is total population, male population, female population, and highest population state. Then, we proposed a method to forecasting the predicted population using neural network technique. The results show that the highest male population was predicted to be in

Madhya Pradesh and the second highest male population was in Tripura. We also found that the highest female population was in the Maharashtra state and the second highest female population in Uttar Pradesh. Our findings indicate that the more data available in the training population dataset there are, the better the prediction in the forecasting process for both male and female cases while the male yields better prediction results than female population. On the other hand, the total gender population yields different performance values than separate genders forecasting for several states. Our results also indicate that building the NNs models by applying the ReLU activation function yields better performance than the other two activation function performed in this study. Our proposed methodology can be generalized and applied on other nations with large populations.

References

1. De Leeuw ED (2005) To mix or not to mix data collection modes in surveys. J Official Stat 21(5):233–255
2. Smith Jervelund S, Vinther-Jensen K, Ryom K, Villadsen SF, Hempler NF (2021) Recommendations for ethnic equity in health: a Delphi study from Denmark. Scand J Public Health. https://doi.org/10.1177/14034948211040965
3. Thoma ME, McLain AC, Louis JF, King RB, Trumble AC, Sundaram R, Louis GMB (2013) Prevalence of infertility in the United States as estimated by the current duration approach and a traditional constructed approach. Fertil Steril 99(5):1324–1331
4. Weeks JR (2004) The role of spatial analysis in demographic research. In: Spatially integrated social science, pp 381–399
5. Lutz W, Butz WP, Samir KE (2014) World population and human capital in the twenty-first century. OUP Oxford
6. Short S (2008) Transition and challenge: China's population at the beginning of the 21st century
7. Billari FC (2001) Sequence analysis in demographic research. Can Stud Popul 439–458
8. Stuck AE, Walthert JM, Nikolaus T, Büla CJ, Hohmann C, Beck JC (1999) Risk factors for functional status decline in community-living elderly people: a systematic literature review. Soc Sci Med 48(4):445–469
9. Murthy D, Gross A, Pensavalle A (2016) Urban social media demographics: an exploration of Twitter use in major American cities. J Comput Mediated Commun 21(1):33–49
10. Clauset A, Shalizi CR, Newman ME (2009) Power-law distributions in empirical data. SIAM Rev 51(4):661–703
11. Sloan L, Morgan J (2015) Who tweets with their location? Understanding the relationship between demographic characteristics and the use of geoservices and geotagging on Twitter. PLoS One 10(11):e0142209
12. Sloan L, Morgan J, Housley W, Williams M, Edwards A, Burnap P, Rana O (2013) Knowing the tweeters: deriving sociologically relevant demographics from Twitter. Sociol Res Online 18(3):74–84, e0142209
13. Mislove A, Lehmann S, Ahn YY, Onnela J-P, Rosenquist J (2011) Understanding the demographics of Twitter users. In: Proceedings of the international AAAI conference on web and social media, vol 5, no 1
14. Mislove A, Lehmann S, Ahn YY, Onnela J, Rosenquist J (2021) The pulse of the nation. In: Atlas of forecasts: modeling and mapping desirable futures, p 131
15. Rich W (1973) Smaller families through social and economic progress. Technical report
16. Jhingan ML (2011) The economics of development and planning. Vrinda Publications

17. Rajan SI, Sarma PS, Mishra U (2003) Demography of Indian aging, 2001–2051. J Aging Soc Policy 15(2–3):11–30, e0142209
18. Rajan IS, Aliyar S (2004) Demographic change in Kerala in the 1990s and beyond. In: Kerala's economic development: performance and problems in the post liberalisation period, pp 61–81
19. Wang X, Brown DE, Gerber MS (2012) Spatio-temporal modeling of criminal incidents using geographic, demographic, and twitter-derived information. In: 2012 IEEE international conference on intelligence and security informatics. IEEE, pp 36–41
20. Ebrahimi M, ShafieiBavani E, Wong R, Chen F (2018) A unified neural network model for geolocating Twitter users. In: Proceedings of the 22nd conference on computational natural language learning, pp 42–53
21. Pereira-Kohatsu JC, Quijano-Sánchez L, Liberatore F, Camacho-Collados M (2019) Detecting and monitoring hate speech in twitter. Sensors 19(21):4654
22. Abdi H (1994) A neural network primer. J Biol Syst 2(03):247–281, e0142209
23. Landi A, Piaggi P, Pioggia G (2009) Backpropagation-based non linear PCA for biomedical applications. In: 2009 ninth international conference on intelligent systems design and applications. IEEE, pp 635–640
24. Sharma S, Sharma S, Athaiya A (2017) Activation functions in neural networks. Towards Data Sci 6(12):310–316
25. Han J, Moraga C (1995) The influence of the sigmoid function parameters on the speed of backpropagation learning. In: International workshop on artificial neural networks. Springer, Berlin, pp 195–201

Credit Risk Analysis Using EDA

Prakriti Arora, Siddharth Gautam, Anushka Kalra, Ashish Negi, and Nitin Tyagi

Abstract Organizations or banks are providing funds to support people monetarily, keeping assets in return till the amount is repaid to the company with interest encounter loss at many instances when the borrower or the client fails to repay the loan appearing to be a defaulter. Also, when the firm disapproves the loan of an applicant who is likely to repay the sum, the loss is again withstood by the firm. Therefore, to avoid this loss, this research is performed deeply analyzing the factors using exploratory data analysis, affecting the trend of defaulters as well as non-defaulters, helping the firm recognize the defaulters, and disapproving their request to borrow. The exploratory data analysis is performed by visually performing univariate, bivariate, and multivariate analysis on almost all the aspects of the two credit history datasets. The patterns and learnings were noted based on the visual as well as statistical analysis to determine creditworthiness of a client.

Keywords Credit risk analysis · Loan EDA case study · Exploratory data analysis · Univariate analysis · Bivariate analysis · IQR · EDA

1 Introduction

In the financial field, credit risk analysis has always been a topic of major concern. The banks and the agencies that sanction loans find it difficult to approve loans for people who lack credit history. They are usually at a high risk of being used smartly by the customers as they get a plus point if they are defaulters [1]. Loans are a huge amount of monetary funds borrowed for a fixed tenure and expected to be repaid

P. Arora · A. Negi · N. Tyagi (✉)
HMRITM, Delhi, India
e-mail: nitintyagi5631@gmail.com

S. Gautam
NSUT, Delhi, India

A. Kalra
Mahavir Swami Institute of Technology, Sonipat, India

P. Dutta et al. (eds.), *Emerging Technologies in Data Mining and Information Security*, Lecture Notes in Networks and Systems 490,
https://doi.org/10.1007/978-981-19-4052-1_56

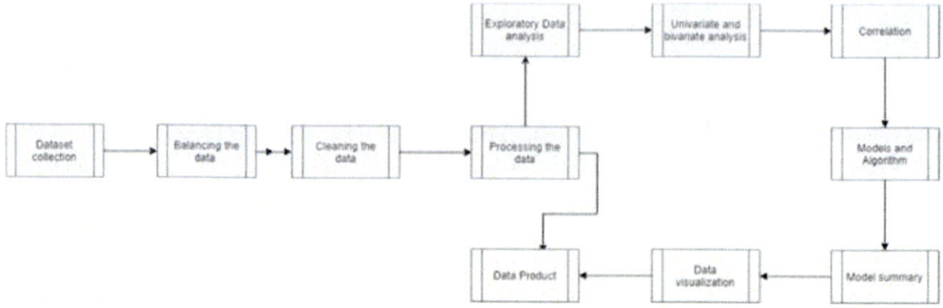

Fig. 1 Process of analysis of the credibility of an individual

to the bank usually with a surplus of interest rate. It would enable us to analyze the data with the help of both statistics and graphical visualizations with the help of univariate as well as multivariate analyses which are further branches of EDA [2]. These analyses enable the bank to analyze the loan application and decide the acceptance or rejection of any application keeping in mind the repayment of loan and the financial profit of the bank. So, when a client requests a loan from a bank, he is likely to get a response of either the application being approved, cancelled, or rejected based on the requirements that need to be fulfilled by the client [3].

The visual representation and inferences from the current, previous as well as the combination of the two datasets enables agencies to judge whether the person would be eligible and capable enough to repay the amount in monthly installments which makes credit risk problem easier to solve. The structured bar graphs, histograms, scatter plots help in distinguishing the two datasets effectively but at the same time makes the determination of the relationship between variables evident enough for effective study [4]. The data are filtered from all the outliers, and any imbalances present in the dataset are eradicated completely before the analysis for a better accuracy. Figure 1 depicts the process of how the analysis of the credibility of the person is analyzed, and the application is rejected or accepted.

2 Literature Review

Many studies in the financial and banking sectors have been undertaken using data mining. This section summarizes some of the approaches highlighted along with their key findings. A. J. Hamid and T. M. Ahmed [5] employed the "Tree model," "Random Forest," and "SVM model," and integrated the above models into an ensemble model. In article [6], an illustration has been examined so that the banking industry can accept or deny credit requests from their clients. Real-coded genetic methods are the primary algorithm employed. The tree algorithm is determined to have an 81.25% precision. The authors in Ref. [7] use data mining to inspect the metadata. The data mining process gives excellent perception in credit prediction systems, as it quickly identifies clients who are able to revert back the loan in a reasonable amount of time. The "J48 algorithm," "Bayes net," and "Naive Bayes" algorithms are employed. When these methods were applied to the metadata, it was discovered that the "J48 algorithm" has a high precision of 78.378%. Default probability (PD) is the important stage for clients who come for a loan in a bank; an enhanced risk analysis congregating method is utilized in research [8] utilizing R-language to discover the inferior loan clients. Data mining methods give a framework for locating PD in a metadata collection. When there are missing values in data collection, the KNN method in the R-programming language is used to conduct multiple accusation computations.

3 Exploratory Data Analysis

EDA is preferred by data scientists to inspect and scrutinize data and note the major traits with visualization techniques. The purpose of using EDA is to shed light on a concept of preparing a rough outline of the data and infer the basic crux of the analysis and get maximum out of analysis. There are numerous tools utilized to perform analytical judgment such as gathering and dimension trimming approach, univariate and bivariate envision of every aspect in dataset which provides access to study correlation between varying entities and construct a summary [9]. Though EDA is used in various sectors such as businesses and service, but it can be used to a greater extent in the banking sector for maintaining loans, in managing transaction history of the person, and account of customers. It includes data mining and enables maintenance of a systematic loan prediction model and helps to analyze financial risks involved [10]. The set of information would go through a procedure of sieving information and standardizing required columns, recognizing values, and representing the filtered entities in graphical form. This enables the loan providing companies to approve loan after comparisons such as purpose of loan and the earnings and the term and the working years of a person [11]. EDA is primarily branched into two separate ways where every technique involves variables and can be represented graphically as well as in a non-graphical way. The types are classified as:

3.1 Univariate Analysis

Uni stands for one and variant means something that keeps on changing which means only one dependable variable. The major purpose of univariate analysis is to gather a brief of data and give it a structured analysis; each variable is treated separately and does not handle the relations between variables. Univariate analysis can be graphical or non-graphical where non-graphical is the ground level analysis and does not supply a clear picture of the dataset [12]. On the other hand, univariate graphical analysis provides a clearer picture and includes stem and leaf plotting, representation through histograms, depiction of data using the scatter plot and bar graphs [13].

3.2 Bivariate Analysis

Bi alludes to two and variate stands for a varying quantity, which means an analysis with two variables. Bivariate analyses and patterns permit us to assess the relation of every varying quantity in the data and throw light on the focused value with which we are concerned [14]. This technique is concerned with effects and consequences of the relations between varying variables.

4 Research Work

The money lending companies experience financial loss in two ways, by providing and not providing credit to their clients in a way that if the client is likely to repay loan, then not approving the loan will out-turn to be a loss, while if the client is likely to default, then approving would bring in loss again [15]. The research work, with the help of credit risk analysis using the current and previous application data of the clients, aims to fathom and analyze the problem mentioned above to inspect for the creditworthiness of a client. In this exploratory data analysis, we are going to dig deep into the details to analyze each angle of the previous data as well as current data [16, 17].

4.1 Data Reading and Inspection

The exploratory data analysis will be conducted on the two humongous real-world datasets, namely application_data consisting of around 307,511 rows and 122 columns of current applications of the applicants and previous_application consisting of about 1,670,214 rows and 37 columns of the credit history of the applicants [18]. In this step, current as well as previous application data was read using and inspected the shape, datatype of the features, and values of the features [19].

4.2 Data Preprocessing

For the data preprocessing step, we inspected and imputed null values and removed unnecessary columns from both the datasets. In the current dataset, 41 columns with more than 50% null values were removed, and 4 columns with approximately 13% null values were imputed with mode technique. While in the previous application dataset, we removed 4 columns with more than 50% null data [20]. Also, columns not beneficial for the further analysis were removed, and the current application dataset was reduced to 38 columns from 122 and previous application dataset from 37 to 33 columns. Rescripted the columns with negative data with the help of abs() function [21].

Further, to transform the data, we identified outliers, analyzed, and removed them with the box plot and IQR technique. IQR, interquartile range, is the central 50 points of the dataset or the measure of where the bulk of data lies. Quarters are referred to as quartiles, and they are labeled Q1, Q2, Q3, and Q4 in increasing order. The fifty percent of sample tips that are over Q1 and under Q4 is included in the interquartile range [22]. To comprehend the center and disperse the sample, combine the IQR with an estimate of middle tendency, such as the median. Numerically, it is calculated as IQR = Q3 − Q1. Lastly, binning of two continuous variables (AGE_GROUP, AMT_CATEGORY) was performed to make them categorical for easy analysis [23].

4.3 Data Visualization

Throughout our analysis, visualization was performed to have a clear view since the conduct is all about exploratory data analysis [24]. We visualized features having less than 15% null values with the help of boxplot to determine which technique is the best for the imputation ample number of visualizations were made performing univariate, bivariate, and multivariate analysis on the current, previous, and combined application dataset [25]. The current data were highly imbalanced, hence were divided for analysis on the categorical column, target with value 0 and 1 symbolizing non-defaulters and defaulters, respectively [26]. Univariate and bivariate analysis for 27 categorical and 9 numerical variables of current data was plotted as a histogram and visualized, interpreting insights from the visualization, one of which is show in Figs. 2 and 3.

Then, correlation for numerical columns was represented using a heat map, as shown in Fig. 4, followed by conclusions [27]. The concept of correlation depicts the relationship or interdependency between two variables and a degree by which a couple of variables are linearly associated [28].

Similarly, EDA for previous application data was performed separating the categorical as well as numerical columns. Finally, after the separate analysis of the two

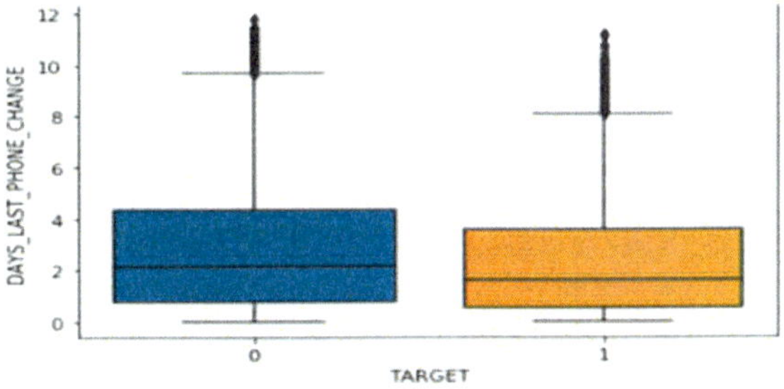

Fig. 2 Univariate analysis of AMT_GOODS_PRICE with target = 0 and 1, respectively

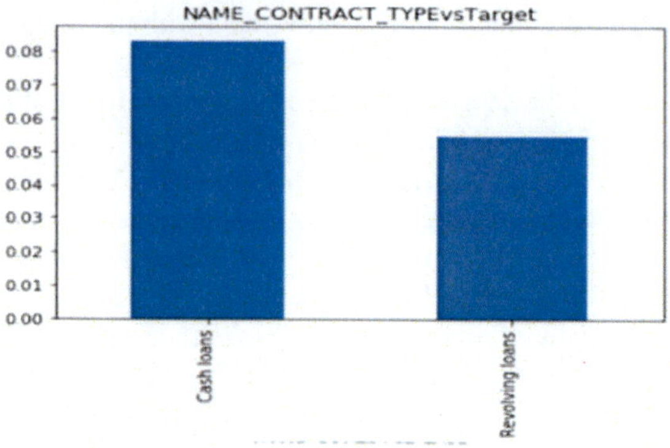

Fig. 3 Bivariate analysis between contract types versus target

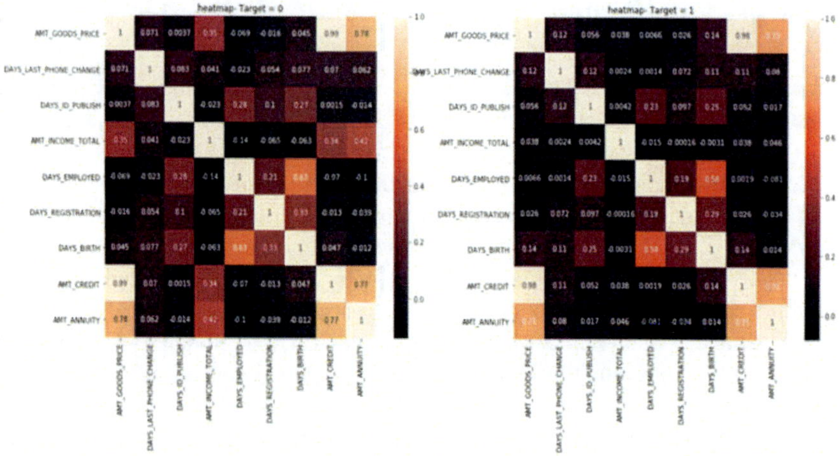

Fig. 4 Correlation based on target = 0 and target = 1 represented using heat map

datasets, they were merged on the SK_ID_CURR column and analyzed for values of the columns NAME_CONTRACT_STATUS exhibiting the status of previous applications of the clients were approved, cancelled, refused, or reused [29]. Then, for each status, categorical as well as numerical columns were separated and analyzed by performing univariate and bivariate analysis. Lastly, exploratory data analysis as well as the aim was accomplished by drawing and gathering all the insights and conclusions from the research.

4.4 Methodology

The project began with importing libraries and reading the two datasets, consequently understanding the structure of the dataset [30]. Then, working on the data quality check, examining for null values, and outliers in the dataset. Eventually, after preprocessing, the current data were reduced from (307,511, 122) to (307,511, 83) while previous data from (1,670,214, 37) to (1,670,214, 33). Outliers were detected and were removed with the help of IQR as mentioned. The analysis was initiated with univariate analysis on categorical and continuous variables of the current dataset and correlation analysis for the numerical columns, as shown in Fig. 5. We went ahead with the bivariate and multivariate analysis for the categorical and numerical columns separately; hence, a lot of insights were noted. Following the analysis of the current dataset, the previous application dataset was analyzed in the same manner with ample of insights. Once individual analysis of the two datasets was complete; the datasets were then merged on the column SK_ID_CURR, the application id, and further, univariate, bivariate, and multivariate analysis were conducted, concluding insights [31].

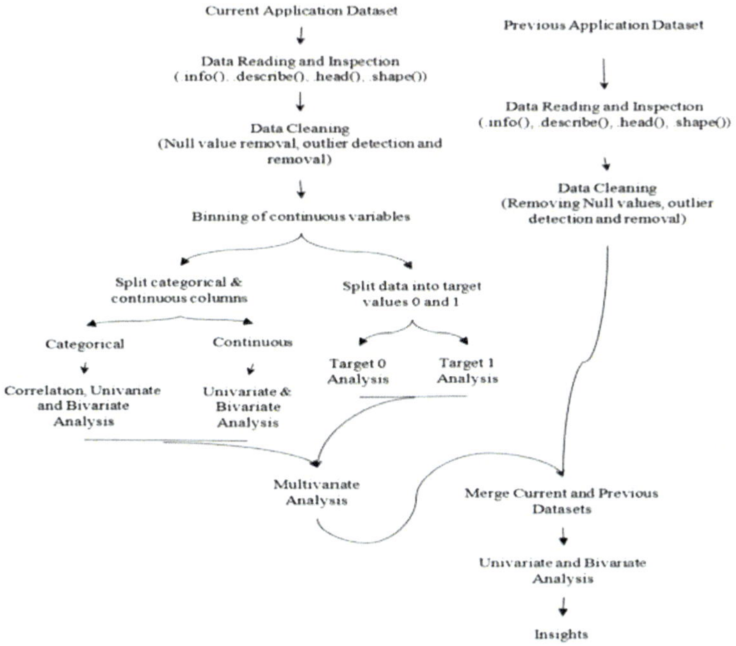

Fig. 5 Workflow of the project

4.5 Insights

The following Table 1 depicts the inferences concluded from the EDA performed, from each dataset and various features:

5 Conclusion and Future Scope

For clients with insufficient or non-existent credit history, it is quite strenuous for the loan providing companies to recognize defaulters. Hence, the chief intention of this EDA is to examine and classify the trends and patterns of the loan applicants based on the current as well as previous applications. Based on the inspection and analysis, plenteous visualizations using the Matplotlib and Seaborn libraries were made followed by the patterns observed. Once the data were well preprocessed, exploratory data analysis including univariate, bivariate, and multivariate analysis on the current application, previous application then on the combined of the datasets was successfully completed.

Moreover, the chief motive behind using EDA was to highlight elementary organization of quite a large number of features and records using extensive visualization methods. With the analysis and visualization of almost all the features and correlation between the features were analyzed. Consequently, an ample number of insights were noted, and conclusions were made based on the patterns inferred. Thus, based on the inferences, actions can be taken in future to avoid approving the loan for defaulters. At the niche level, the major goal of the research was to determine whether the applicant is a defaulter or a non-defaulter. Further, the project can be carried on to a remarkable level owing to the predictive algorithms of machine learning focusing on the insight which is concluded from our research.

Table 1 Insights of the analysis

Dataset	Features	Inference
Current application dataset	Code_Gender	Majorly, it is more likely for a male to default rather than other genders
	Age_Group	Defaulters have a higher probability of being in the age group of 20 s or 30 s
	NAME_INCOME_TYPE	Up to a large extent, the defaulter is likely to be employed as a working client or on maternity leave; else, they are unemployed, while it is less likely for pensioners to default [32]
	OCCUPATION_TYPE	Broadly laborers constitute a high proportion of defaulters
	REGION_RATING_CLIENT_W_CITY	Clients with a rating of 3 are more likely to default
	DAYS_LAST_PHONE_CHANGE	Defaulters seem to change phone numbers prior to application submission [33]
	DAYS_ID_PUBLISH	Defaulters are expected to change IDs more frequently than non-defaulters
	DAYS_REGISTRATION	Defaulter client usually changes registration on a date near the submission of application [34]
Previous application dataset	Approved status (NAME_CONTRACT_STATUS) versus family status	Married clients are expected to get loans approved
	Approved status (NAME_CONTRACT_STATUS) versus income type	Student customers are least expected to get the credit approved, while working customers are more probable
	Approved status (NAME_CONTRACT_STATUS) versus age	Clients in the age group of 30–50 are more likely to get credit approved in contrast to the ones in 20s and 60s [35]
	Cash loans	Cash loans are expected to be reapplied or cancelled

References

1. Huang S-H, Tu W-P, Yeh H-H, Chi MC (2013) An EDA course module for the topic of reliability using automotive electronics as applications. In: 2013 3rd Interdisciplinary engineering design education conference. https://doi.org/10.1109/iedec.2013.6526768
2. Malik A, Gautam S, Khatoon N, Sharma N, Kaushik I, Kumar S (2020) Analysis of black-hole attack with its mitigation techniques in Ad-hoc network. In: Deep learning strategies for security enhancement in wireless sensor networks advances in information security, privacy, and ethics, pp 211–232. https://doi.org/10.4018/978-1-7998-5068-7.ch011
3. Schiantarelli F, Stacchini M, Strahan P (2016) Bank quality, judicial efficiency and borrower runs: Loan repayment delays in Italy. https://doi.org/10.3386/w22034
4. Pulakkazhy (2013) Data mining in banking and its Applications-A Review. J Comput Sci 9(10):1252–1259. https://doi.org/10.3844/jcssp.2013.1252.1259
5. Rustagi A, Manchanda C, Sharma N (2020) IoE: a boon & threat to the mankind. In: 2020 IEEE 9th international conference on communication systems and network technologies (CSNT). https://doi.org/10.1109/csnt48778.2020.9115748
6. Arif M, Khatak A, Hussain M (2015) A framework for Data Warehouse using data mining and knowledge discovery for a network of hospitals in Pakistan. Int J Bio-Sci Bio-Technol 7(3):217–222. https://doi.org/10.14257/ijbsbt.2015.7.3.23
7. Tiwari A, Sharma N, Kaushik I, Tiwari R (2019) Privacy issues & security techniques in big data. In: 2019 International conference on computing, communication, and intelligent systems (ICCCIS). https://doi.org/10.1109/icccis48478.2019.8974511
8. Predicting student's academic performance using data mining techniques (2020) Int J Eng Adv Technol 9(3)"215–219. https://doi.org/10.35940/ijeat.b4521.029320
9. Zurada J, Zurada M (2002) How secure are good loans: validating loan-granting decisions and predicting default rates on consumer loans. Rev Bus Inf Syst (RBIS) 6(3):65–84. https://doi.org/10.19030/rbis.v6i3.4563
10. Zurada J (2002) Data mining techniques in predicting default rates on customer loans. Databases Inf Syst II:285–296. https://doi.org/10.1007/978-94-015-9978-8_22
11. Xin L, Guozi S, Huakang L (2017) Overdue prediction of bank loans based on Deep Neural Network. In: International symposium on computer science and artificial intelligence (ISCSAI). https://doi.org/10.26480/iscsai.01.2017.26.28
12. Comparative analysis and study of data mining techniques used for IOT based Smart Healthcare System (2020) Int J Emerg Trends Eng Res 8(9):6131–6138. https://doi.org/10.30534/ijeter/2020/198892020
13. Data mining techniques for analysing employment data (2019) Int J Eng Adv Technol 9(2):555–556. https://doi.org/10.35940/ijeat.b3311.129219
14. Purohit SU, Mahadevan V, Kulkarni AN (2012) Credit evaluation model of loan proposals for Indian Banks. Int J Model Optim 529–534.https://doi.org/10.7763/ijmo.2012.v2.176
15. Yuanyuan L (2008) Research on personal credit evaluation system of commercial banks. In: First International Workshop on Knowledge Discovery and Data Mining (WKDD 2008). https://doi.org/10.1109/wkdd.2008.147
16. Jafar Hamid A, Ahmed TM (2016) Developing prediction model of loan risk in banks using data mining. Mach Learn Appl Int J 3(1):1–9. https://doi.org/10.5121/mlaij.2016.3101
17. Li W, Liao J (2011) An empirical study on credit scoring model for credit card by using Data Mining Technology. In: 2011 Seventh international conference on computational intelligence and security. https://doi.org/10.1109/cis.2011.283
18. Jayasree (2013) A review on data mining in banking sector. Am J Appl Sci 10(10):1160–1165.https://doi.org/10.3844/ajassp.2013.1160.1165
19. Kamatchi K, Siva Balan A (2013) Multiphase text mining predictor for market analysis. In: 2013 International conference on current trends in engineering and technology (ICCTET). https://doi.org/10.1109/icctet.2013.6675990

20. Samanta D, Dutta S, Galety MG, Pramanik S (2022) A novel approach for web mining taxonomy for high-performance computing. In: Tavares JMRS, Dutta P, Dutta S, Samanta D (eds) Cyber intelligence and information retrieval. Lecture Notes in Networks and Systems, vol 291. Springer, Singapore. https://doi.org/10.1007/978-981-16-4284-5_37
21. Grover M, Sharma N, Bhushan B, Kaushik I, Khamparia A (2020) Malware threat analysis of IoT devices using deep learning neural network methodologies. In: Security and Trust Issues in Internet of Things, pp 123–143.https://doi.org/10.1201/9781003121664-6
22. Goel A, Tyagi N, Gautam S (2019) Comparative analysis of 3-D password using various techniques (June 16, 2019). ' Comparative analysis of 3-D password using various techniques &#39. Int J Emerg Technol Innov Res, 6(6):711–718 (www.jetir.org), ISSN:2349-5162. http://www.jetir.org/papers/JETIR1907Q08.pdf
23. Goyal S, Sharma N, Kaushik I, Bhushan B (2021) Industrial revolution: blockchain as a wave for industry 4.0 and iiot. In: Advances in computing communications and informatics, pp 108–130. https://doi.org/10.2174/9781681088624121010008
24. Kaushik I, Sharma N (2020) Black hole attack and its security measure in wireless sensors networks. In: Advances in intelligent systems and computing handbook of wireless sensor networks: issues and challenges in current scenarios, pp 401–416.https://doi.org/10.1007/978-3-030-40305-8_20
25. An exploratory analysis of corporate governance using supervised data mining learning (2019) Int J Recent Technol Eng 8(3):3546–3557. https://doi.org/10.35940/ijrte.c5279.098319
26. Kaieski N, Oliveira LP, Villamil MB (2016) Vis-health: Exploratory analysis and visualization of dengue cases in Brazil. In: 2016 49th Hawaii international conference on system sciences (HICSS). https://doi.org/10.1109/hicss.2016.385
27. Tyagi N, Gautam S, Goel A, Mann P (2021) A framework for blockchain technology including features. In: Hassanien AE, Bhattacharyya S, Chakrabati S, Bhattacharya A, Dutta S (eds) Emerging technologies in data mining and information security. Advances in Intelligent Systems and Computing, vol 1286. Springer, Singapore. https://doi.org/10.1007/978-981-15-9927-9_62
28. Gautam S, Malik A, Singh N, Kumar S (2019) Recent advances and countermeasures against various attacks in IoT environment. In: 2019 2nd international conference on signal processing and communication (ICSPC), pp 315–319. https://doi.org/10.1109/ICSPC46172.2019.8976527
29. Singh G, Gautam S, Prachi VA, Kaushal T (2021) Analysis of blockchain induced cryptocurrency: regulations and challenges of cryptocurrencies. In: Hassanien AE, Bhattacharyya S, Chakrabati S, Bhattacharya A, Dutta S (eds) Emerging technologies in data mining and information security. Advances in intelligent systems and computing, vol 1286. Springer, Singapore. https://doi.org/10.1007/978-981-15-9927-9_54
30. Sharma N, Kaushik I, Bhushan B, Gautam S, Khamparia A (2020) Applicability of WSN and biometric models in the field of healthcare. In: Deep learning strategies for security enhancement in wireless sensor networks advances in information security, privacy, and ethics, pp 304–329.https://doi.org/10.4018/978-1-7998-5068-7.ch016
31. Sharma N, Kaushik I, Rathi R, Kumar S (2020) Evaluation of accidental death records using hybrid genetic algorithm. SSRN Electron J. https://doi.org/10.2139/ssrn.3563084
32. Gurung A, Gautam S, Garg T, Bhardwaj Y, Gupta H (2021) Virtual numeric authentication system using contour detection of color-banded fingertips. In: Tavares JMRS, Chakrabarti S, Bhattacharya A, Ghatak S (eds) Emerging technologies in data mining and information security. Lecture notes in networks and systems, vol 164. Springer, Singapore. https://doi.org/10.1007/978-981-15-9774-9_32

33. Rustagi A, Manchanda C, Sharma N, Kaushik I (2020) Depression anatomy using combinational deep neural network. In: Advances in intelligent systems and computing international conference on innovative computing and communications, pp 19–33.https://doi.org/10.1007/978-981-15-5148-2_3
34. Goel A, Gautam S, Tyagi N, Sharma N, Sagayam M (2021) Securing biometric framework with cryptanalysis. In: Intelligent data analytics for terror threat prediction, pp 181–208.https://doi.org/10.1002/9781119711629.ch9
35. Kathuria RS, Gautam S, Singh A, Khatri S, Yadav N (2019) Real time sentiment analysis on twitter data using deep learning (Keras). In: 2019 international conference on computing, communication, and intelligent systems (ICCCIS), pp 69–73. https://doi.org/10.1109/ICCCIS48478.2019.8974557

A COVID-19 Infection Rate Detection Technique Using Bayes Probability

Arnab Mondal, Ankush Mallick, Sayan Das, Arpan Mondal, and Sanjay Chakraborty

Abstract The main objective of this paper is to detect the infection rate of the SARS-Cov-2 virus among patients who are suffering from COVID with different symptoms. In this work, some data inputs from the intended patients (like contact with any COVID infected person and any COVID patient within 1 km.) are collected in the form of a questionnaire and then applied Naïve Bayes probabilistic technique to evaluate the probability of how much that patient is affected in this deadly virus. Following this process, we collect sample data of 80 patients and apply the proposed analysis process using the C programming language. This approach also shows the comparison for different test cases with respect to the feedbacks of actual patient data analysis.

Keywords COVID-19 · Naïve Bayes · Health care · Machine learning · SARS-Cov-2 · Infection detection

1 Introduction

According to World Health Organization (WHO), the COVID-19 pandemic is caused by severe acute respiratory syndrome coronavirus 2 (SARS-Cov-2) virus infection which is a novel strain of coronavirus. In the current scenario, it is one of the most dangerous and critical diseases; over 200 countries are infected with this disease. It is a type of RNA virus; it can spread from the cough, sneeze, speak, breathe of the infected person [1]. By following the rules of WHO, we have tried to detect if any person is COVID infected or not and the rate of his/her infection using his/her traveling history, contact with an infected person, previous record of any other kind of diseases etc., in this paper. This paper mainly focuses on a popular supervised machine learning algorithm (Naïve-Bayes) for detecting the COVID-19 infection rate by collecting active patient questionnaires. In future work, we are already planning to build a Webpage and an android application based on our proposed approach.

A. Mondal · A. Mallick · S. Das · A. Mondal · S. Chakraborty (✉)
Computer Science and Engineering Department, JIS University, Kolkata, India
e-mail: schakraborty770@gmail.com

P. Dutta et al. (eds.), *Emerging Technologies in Data Mining and Information Security*, Lecture Notes in Networks and Systems 490,
https://doi.org/10.1007/978-981-19-4052-1_57

In this pandemic time, we have to try to stay at home, but peoples who have some symptoms or any doubt of being affected by this virus, then they can visit our Webpage or download our mobile application and can check how much possibility they have of effecting in this disease at their home and proceed for the treatment.

The rest of the paper is organized as follows: Sect. 2 describes the related works and their impact to boost up this proposed idea; Sect. 3 explains the proposed work with some suitable illustrations, and Sect. 4 describes the detailed result analysis, and at last, Sect. 5 deals with the conclusion of this paper.

2 Literature Review

There is a number of research works have been conducted in the last 2 years on COVID-19 detection using AI and machine learning. The simple slope test and the Johnson-Neyman technique were used to test the interaction of fear and collectivism on preventive intention toward COVID-19 [2]. One study aims to assess the usability of a novel platform used in military hospitals in Korea to gather data and deploy patient selection solutions for COVID-19 [3]. AI and ML can be useful for medicine development, designing efficient diagnosis strategies, and producing predictions of disease spread, and Internet of Things (IoT) can also help with applications such as automated drug delivery, responding to patient queries, and tracking the causes of disease spread [4]. One work conducted a secondary analysis of a cross-sectional, nationally representative survey of Internet users to evaluate whether there were racial/ethnic disparities in self-reported telehealth use early in the pandemic [5]. The analysis is widened by investigating artificial intelligence (AI) approaches for the diagnosis, anticipating infection, and mortality rate by tracing contacts and targeted drug designing [6]. In this study [7], ACoS system was introduced for preliminary diagnosis of nCOVID-19 infected patients so that proper precautionary measures (like isolation and RT-PCR test) can be taken to prevent the further outbreak of the infection. In this article [8], various machine learning models are built to predict the PPIs between the virus and human proteins that are very useful to the biological world. At the same time, an ensemble voting classifier is used which gives more accuracy compared to other techniques. In this study [9], a new COVID-19 diagnosis strategy called feature correlated Naïve Bayes (FCNB) has been introduced. From this study, it has been seen that the decision tree appears to be the best model [10]. In this paper, it classified textual clinical reports into four classes by using classical and ensemble machine learning algorithms. These term frequency/inverse document frequency (TF/IDF) features were supplied to traditional and ensemble machine learning classifiers [11]. A priori power analysis using G*Power (version 3.1.9.2) is used to analyze data in paper [12]. The primary thing is they measured the concrete behaviors in response to COVID-19 pandemic as well as the psychological responses. At the same time, they explore whether fear and the self-perceived likelihood of contracting the virus were associated with risk-mitigating behaviors [12]. In this paper, we have seen that they have used the Naive Bayes [13] algorithm as well as the J48 decision tree to build the algorithm. They collected 1082 records, and after analysis, they established those old patients are at high risk of developing MERS-CoV complications [14]. In a paper, we can learn how ML is used to identify high-risk patients before irreversible lesions occur. At the same time, it has found a way to collect these from hospitals LDH, lymphocytes, and hs-CRP. The significance of the work is two folds. It provides a probabilistic model to find the risk of death in COVID [15]. We have known about distance-biased Naive Bayes (DBNB) technique that is used to diagnose COVID-19 patients using advanced particle swarm optimization (APSO) selection technique [16].

3 Proposed Work

3.1 Proposed Algorithm

Let x is event, and c is the class.

Step 1: Start.

Step 2: Initialize COUNT < - 0.

Step 3: IF user's age > 13 then COUNT++.

Step 4: IF they travel across India then COUNT++.

Step 5: IF they contact with COVID patients then COUNT++.

Step 6: IF there have any COVID patients within 1 km then COUNT++.

Step 7: IF they are suffering from cancer, lung's disease or diabetes then COUNT++.

Step 8: IF user's temperature is ≥38 °C then COUNT++.

Step 9: IF user is suffering from any of one like Dry Cough, Shortness of Breath, Headache, Aches and pain, sore throat, Fatigue, Diarrhea then COUNT++.

Step 10: IF user has BP or sugar problem then COUNT < - COUNT +1.

*Step 11: Initialize COUNT < - COUNT*12.5.*

Step 12: IF COUT ≤ 30 then user is in green zone.

Step 13: ELSE IF(COUT > 30 and COUT < 60) then user is in orange zone.

Step 14: ELSE user is in red zone (High chance).

Step 15: From 80 user's answers has taken for the previous following 8 questions.

For any test case two values are calculated P(yes/x) and P(No/x) from the

Frequency and Likelihood Table (Discussed below).

Step 16: IF P(yes/x) > P(No/x).

Then, the user has high chance to be infected in COVID-19.

Step 17: ELSE the user has low chance to be infected in COVID-19.

Step 18: Stop.

3.2 Solution with Illustration

The solution is represented by the frequency and likelihood tables based on different attributes collecting from various patients as questionnaire. Some important tables are summarized below as demo (Tables 1, 2, 3, and 4).

From the above tables, seven people have suffered from cancer, diabetes, or lung's diseases. Out of 7, six people have a high chance (Programming output: >60%) of suffering from COVID-19, and 1 person has a low chance (Programming output: <60%) of suffering from COVID-19. Out of 80, 73 people have not suffered from cancer, diabetes, or lung's diseases. Out of 73, 8 people have a high chance (Programming output: >60%) of suffering from COVID-19, and other 65 people have a low chance (Programming output: <60%) of suffering from COVID-19.

Table 1 Age

Frequency table		Probability of infection in COVID-19		
		Yes	No	
Age > 13	True	14	66	
	False	0	0	
Likelihood table		Probability of infection in COVID-19		
		Yes	No	
AGE > 13	True	14/14	66/66	80/80
	False	0	0	0/80
		14/80	66/80	

Table 2 Suffering from cancer, diabetes, or lungs diseases

Frequency table		Probability of infection in COVID-19	
		Yes	No
Cancer, diabetes	True	6	1
	False	8	65

Likelihood table		Probability of infection in COVID-19		
		Yes	No	
Cancer, diabetes	True	6/14	1/66	7/80
	False	8/14	65/66	73/80
		14/80	66/80	

Table 3 Body temperature

Frequency table		Probability of infection in COVID-19	
		Yes	No
Body temperature (≥38 °C)	True	14	35
	False	0	31

Likelihood table		Probability of infection in COVID-19		
		Yes	No	
Body temperature (≥38 °C)	True	14/14	35/66	49/80
	False	0	31/66	31/80
		14/80	66/80	

Body Temperature (in degree Celsius):

In the below table (Table 3), it is given that 49 people have their body temperature (≥38). Out of 49, 14 people have a high chance (Programming output: >60%) of suffering from COVID-19, and 35 persons have a low chance (Programming utput: <60%) of suffering from COVID-19. Rest 31 people have their body temperature under 38 °C. None of them have a high chance (Programming output: >60%) of suffering from COVID-19.

4 Result Analysis and Comparison

We have taken 80 random sample patients inputs through online Google from questionnaires. We have asked eight questions through the Google Form and similarly inserted those questions in our program source code. After giving the inputs, we have

Table 4 Diseases like (dry cough, shortness of breath, headache, aches, sore throat, fatigue, diarrhea)

Frequency table			Probability of infection in COVID-19	
			Yes	No
Different types of diseases	True	Dry cough	9	1
		Shortness of breath	2	1
		Headache	1	8
		Aches and pain	1	3
		Sore throat	1	3
		Fatigue	0	0
		Diarrhea	0	0
		False	0	50

Likelihood table		Probability of infection in COVID-19		
		Yes	No	
Different types of diseases	True			
	Dry cough	9/14	1/66	10/80
	Shortness of breath	2/14	1/66	3/80
	Headache	1/14	8/66	9/80
	Aches and pain	1/14	3/66	4/80
	Sore throat	1/14	3/66	4/80
	Fatigue	0	0	0
	Diarrhea	0	0	
	False	0	50/66	50/80
		14/80	66/80	

applied Naïve Bayes probabilistic model to obtain the output and classify infected patients from the non-infected. However, we have defined that <30%—in green zone, ≥30% and <60%—in orange zone, ≥60%—in red zone. Therefore, we have divided three zones for the users. At the same time, we have considered that ≥60% results have highly chance to be infected in the COVID-19 disease. In this experiment, 'accuracy' is mainly used for performance evaluation metric. The needed configuration for the experiment is described below,

- Intel core i3 3rd-generation processor
- 1 TB HDD
- 4 GB RAM
- Windows 10 operating system.

From the Table 1, we can see there have 14 cases; those people have high chance to be infected in COVID. Those data are marked with blue highlighter. However, the rest

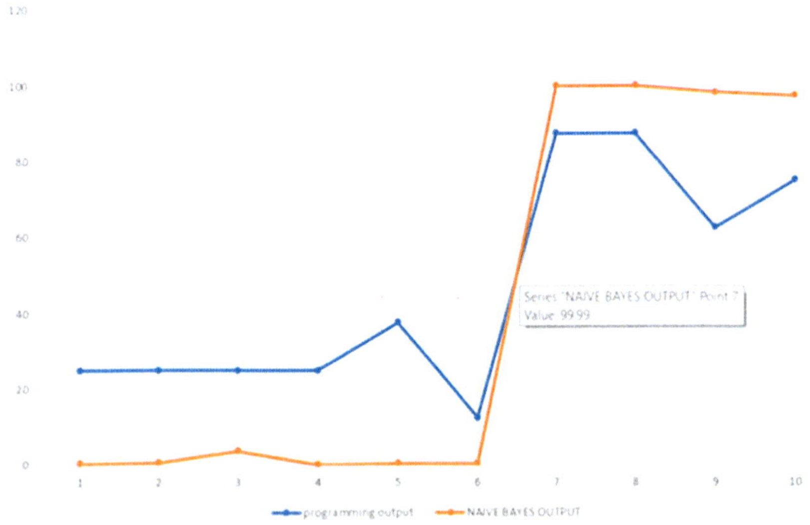

Fig. 1 Actual output versus target output

of the data have no high chances to be infected in COVID-19. Now, we will take ten data values for accuracy and comparison purpose. We are taking randomly first six test cases and last four test cases from the data value Table 1. However, from the Fig. 1, we have seen that the blue line represents actual output, and orange line represents Naïve Bayes output. Then, we can conclude that the two lines are intersected (optimum point) near the point 50 of *y*-axis. So, there have 50–50 (WIN–WIN) chances for every user to be infected in COVID-19 with respect to programming model as well as proposed model. At the same time, we can say that Naïve Bayes output is more stable than programming output because there are more fluctuation in the orange line rather than the blue line (Fig. 1). Accuracy calculations for ten random sample test cases are shown in Table 6. Infected vs non-infected output for 80 sample patients' data is shown in Table 5.

Test Case—1:

***P* (yes/*x*)** = *p* (Age-True/yes) × *p* (Travel Across India-False/yes) × *p* (Contact with COVID patient-False/yes) × *p* (Any COVID patient within 1 km-False/yes) × *p* (Have any Diseases-False/yes) × *p* (Temperature-True/yes) × *p* (Suffering from any disorder-False/yes) × *p* (BP of sugar-False/yes) × *p* (yes).

$= 14/14 \times 5/14 \times 3/14 \times 2/14 \times 8/14 \times 14/14 \times 7/14 \times 14/80.$

$= 0.00055.$

***P* (No/*x*)** = *p* (Age-True/No) × *p* (Travel Across India-False/No) × *p* (Contact with COVID patient-False/No) × *p* (Any COVID patient within 1 km-False/No) × *p* (Have any Diseases-False/No) × *p* (Temperature-True/No) × *p* (Suffering from any disorder-False/No) × *p* (BP of sugar-False/No) × *p* (No).

Table 5 Infected versus non-infected output for 80 sample patients data

	PROGRAGMMING OUTPUT	PROBABILITY OF INFECTION IN COVID 19(>60% -> YES)
2	25% (IN GREEN ZONE)	NO
3	25% (IN GREEN ZONE)	NO
4	25% (IN GREEN ZONE)	NO
5	37.5% (IN ORANGE ZONE)	NO
6	12.5% (IN GREEN ZONE)	NO
7	12.5% (IN GREEN ZONE)	NO
8	25% (IN GREEN ZONE)	NO
9	12.5% (IN GREEN ZONE)	NO
10	12.5% (IN GREEN ZONE)	NO
11	50% (IN ORANGE ZONE)	NO
12	25% (IN GREEN ZONE)	NO
13	12.5% (IN GREEN ZONE)	NO
14	37.5% (IN ORANGE ZONE)	NO
15	37.5% (IN ORANGE ZONE)	NO
16	37.5% (IN ORANGE ZONE)	NO
17	25% (IN GREEN ZONE)	NO
18	37.5% (IN ORANGE ZONE)	NO
19	25% (IN GREEN ZONE)	NO
20	37.5% (IN ORANGE ZONE)	NO
21	50% (IN ORANGE ZONE)	NO
22	37.5% (IN ORANGE ZONE)	NO
23	25% (IN GREEN ZONE)	NO
24	37.5% (IN ORANGE ZONE)	NO
25	37.5% (IN ORANGE ZONE)	NO
26	25% (IN GREEN ZONE)	NO
28	62.5% (IN RED ZONE)	YES
29	37.5% (IN ORANGE ZONE)	NO
30	37.5% (IN ORANGE ZONE)	NO
31	12.5% (IN GREEN ZONE)	NO
32	25% (IN GREEN ZONE)	NO
33	25% (IN GREEN ZONE)	NO
34	37.5% (IN ORANGE ZONE)	NO
35	37.5% (IN ORANGE ZONE)	NO
36	12.5% (IN GREEN ZONE)	NO
37	37.5% (IN ORANGE ZONE)	NO
38	37.5% (IN ORANGE ZONE)	NO
39	87.5% (IN RED ZONE)	YES
40	25% (IN GREEN ZONE)	NO
41	75% (IN RED ZONE)	YES
42	37.5% (IN ORANGE ZONE)	NO
43	50% (IN ORANGE ZONE)	NO
44	50% (IN ORANGE ZONE)	NO
45	25% (IN GREEN ZONE)	NO
46	25% (IN GREEN ZONE)	NO
47	37.5% (IN ORANGE ZONE)	NO
48	37.5% (IN ORANGE ZONE)	NO
49	25% (IN GREEN ZONE)	NO
50	12.5% (IN GREEN ZONE)	NO
51	25% (IN GREEN ZONE)	NO
52	12.5% (IN GREEN ZONE)	NO
53	25% (IN GREEN ZONE)	NO
54	25% (IN GREEN ZONE)	NO
55	50% (IN ORANGE ZONE)	NO
56	25% (IN GREEN ZONE)	NO
57	25% (IN GREEN ZONE)	NO
58	37.5% (IN ORANGE ZONE)	NO
59	50% (IN ORANGE ZONE)	NO
60	50% (IN ORANGE ZONE)	NO
61	12.5% (IN GREEN ZONE)	NO
62	75% (IN RED ZONE)	YES
63	25% (IN GREEN ZONE)	NO
64	25% (IN GREEN ZONE)	NO
65	25% (IN GREEN ZONE)	NO
66	50% (IN ORANGE ZONE)	NO
67	75% (IN RED ZONE)	YES
68	87.5% (IN RED ZONE)	YES
69	87.5% (IN RED ZONE)	YES
70	62.5% (IN RED ZONE)	YES
71	50% (IN ORANGE ZONE)	NO
72	75% (IN RED ZONE)	YES
73	25% (IN GREEN ZONE)	NO
74	25% (IN GREEN ZONE)	NO
75	100% (IN RED ZONE)	YES
76	87.5% (IN RED ZONE)	YES
77	87.5% (IN RED ZONE)	YES
78	62.5% (IN RED ZONE)	YES
79	75% (IN RED ZONE)	YES
80	37.5% (IN ORANGE ZONE)	NO

Table 6 Accuracy calculations for ten random sample test cases

Test cases	Actual output (execution) (%)	Target output (%)	Accuracy (%)
TEST CASE—1	25	0.31	98.72
TEST CASE—2	25	0.62	97.52
TEST CASE—3	25	3.67	85.32
TEST CASE—4	25	0.11	99.7
TEST CASE—5	37.5	0.33	97.36
TEST CASE—6	12.5	0.34	97.28
TEST CASE—76	87.5	99.99	93.75
TEST CASE—77	87.5	99.99	93.75
TEST CASE—78	62.5	98.13	80.32
TEST CASE—79	75	97.29	86.12

$= 66/66 \times 60/66 \times 61/66 \times 42/66 \times 65/66 \times 35/66 \times 50/66 \times 64/66 \times 66/80.$
$= 0.16924.$

- Probability of being infected in COVID
  $= 0.00055/(0.00055 + 0.16924) \times 100\%$
  $= 0.32\%$
- Programming output is $= 25\%$.
  Accuracy $= (25 - 0.32)/25 \times 100\%$
  $= 98.72\%$.

5 Conclusion

In this research work, a model is proposed to detect the rate of infection with the help of natural symptoms of this disease and a popular probabilistic technique. In the future, automated and rapid detection of suspected infections will become more effective. In this paper, the various necessary symptoms have been collected from a set of sample patients. It helps to know whether they have a probability of being infected or not. Those who already have a cold, fever, and cough kind of symptoms can get the idea of their chance of being infected by COVID through our approach and can take some necessary precautions to isolate themselves from others.

However, we are planning to build a mobile application in android that will help to detect the rate of COVID-19 for a patient at any place. In addition, the diagnosis accuracy can be increased by improving the symptom management algorithm and adapting it to the pandemic.

References

1. Chakraborty S, Dey L (2021) The implementation of AI and AI-empowered imaging systems to fight against COVID-19—a review. In: Smart healthcare system design: security and privacy aspects. Wiley, p 301
2. Huang F, Ding H, Liu Z, Wu P, Zhu M, Li A, Zhu T (2020) How fear and collectivism influence public's preventive intention towards COVID-19 infection: a study based on big data from the social media. BMC Public Health 20(1):1–9
3. Heo J, Park JA, Han D, Kim HJ, Ahn D, Ha B, … Park YR (2020) COVID-19 outcome prediction and monitoring solution for military hospitals in South Korea: Development and evaluation of an application. J Med Internet Res 22(11):e22131
4. Abir SM, Islam SN, Anwar A, Mahmood AN, Oo AMT (2020) Building resilience against COVID-19 pandemic using artificial intelligence, machine learning, and IoT: A survey of recent progress. IoT 1(2):506–528
5. Campos-Castillo C, Anthony D (2021) Racial and ethnic differences in self-reported telehealth use during the COVID-19 pandemic: a secondary analysis of a US survey of internet users from late March. J Am Med Inform Assoc 28(1):119–125
6. Rasheed J, Jamil A, Hameed AA, Aftab U, Aftab J, Shah SA, Draheim D (2020) A survey on artificial intelligence approaches in supporting frontline workers and decision makers for COVID-19 pandemic. Chaos Solitons Fractals 110337
7. Chandra TB, Verma K, Singh BK, Jain D, Netam SS (2021) Coronavirus disease (COVID-19) detection in chest X-ray images using majority voting based classifier ensemble. Expert Syst Appl 165:113909
8. Dey L, Chakraborty S, Mukhopadhyay A (2020) Machine learning techniques for sequence-based prediction of viral–host interactions between SARS-CoV-2 and human proteins. Biomed J 43(5):438–450
9. Mansour NA, Saleh AI, Badawy M, Ali HA (2021) Accurate detection of COVID-19 patients based on Feature Correlated Naïve Bayes (FCNB) classification strategy. J Ambient Intell Hum Comput 1–33
10. Muhammad LJ, Islam MM, Usman SS, Ayon SI (2020) Predictive data mining models for novel coronavirus (COVID-19) infected patients' recovery. SN Comput Sci 1(4):1–7
11. Khanday AMUD, Rabani ST, Khan QR, Rouf N, Din MMU (2020) Machine learning based approaches for detecting COVID-19 using clinical text data. Int J Inf Technol 12(3):731–739
12. Harper CA, Satchell LP, Fido D, Latzman RD (2020) Functional fear predicts public health compliance in the COVID-19 pandemic. Int J Mental Health Addict 1–14
13. Dey L, Chakraborty S, Biswas A, Bose B, Tiwari S (2016) Sentiment analysis of review datasets using naive bayes and k-nn classifier. arXiv preprint arXiv:1610.09982
14. Al-Turaiki I, Alshahrani M, Almutairi T (2016) Building predictive models for MERS-CoV infections using data mining techniques. J Infect Public Health 9(6):744–748
15. Yan L, Zhang H, Goncalves J, Xiao Y, Wang M, Guo Y, … Yuan Y (2020) A machine learning-based model for survival prediction in patients with severe COVID-19 infection
16. Shaban WM, Rabie AH, Saleh AI, Abo-Elsoud MA (2021) Accurate Detection of COVID-19 patients based on distance biased Naïve Bayes (DBNB) classification strategy. Pattern Recog 108110

A Python-Based Virtual AI Assistant

Tirthajyoti Nag, Jayasree Ghosh, Manona Mukherjee, Subhadip Basak, and Sanjay Chakraborty

Abstract Artificial intelligence is a core specialization of making smart machines, especially computer programs. It relates to the familiar task of using computers to understand human intelligence. AI is broadly classified to study computations that permit perception, reason, and action. This paper gives an overview idea of a Python-based personal assistant for Windows-based systems. This paper got its motivation from popular AI assistants like Windows's Cortana, Siri of iOS, Amazon's Alexa, and Google Assistant of Google. It is designed in a good user-friendly interface to execute various types of tasks by just saying certain set of instructions. As an AI personal assistant, this paper focuses mainly upon the assistance of the users with their daily activities like asking questions on Google, staying updated with the latest news, video watching in YouTube, sending emails, live weather forecasts, simple calculations, Wikipedia-powered AI, utilization of Google Maps, keep tracks on time, operating system-based tasks like shutting down the PC and restarting the PC, plays songs, tells pyjokes, makes notes, and last but not the least, greets the user when called by name.

Keywords AI assistant · Python · Personal assistant · Artificial intelligence · Smart assistance

1 Introduction

Today, the advancement of AI systems can create a human–machine which can interact through voice commands, over communication, even in facial expressions and gestures, etc., which are gaining popularity [1, 2]. According to Moustafa Elshafei [3], virtual personal assistant (VPA) is the next generation of carrier services for mobile and smart device users. Recently, most popular voice assistants are Apple's Siri, Amazon's Alexa, Microsoft's Cortana, and Google Assistant of Google. This paper consists of some brief description of the construction of AI voice assistants.

T. Nag · J. Ghosh · M. Mukherjee · S. Basak · S. Chakraborty (✉)
Computer Science and Engineering Department, JIS University, Kolkata, India
e-mail: schakraborty770@gmail.com

P. Dutta et al. (eds.), *Emerging Technologies in Data Mining and Information Security*, Lecture Notes in Networks and Systems 490,
https://doi.org/10.1007/978-981-19-4052-1_58

The ultimate objective of this particular work is to establish a specific voice assistant that can carry out the works of human and the daily task that a human need to perform in a day-to-day life. It has some features like assuming the weather around and getting the climatic conditions at any desired location. It can open chrome, or do basic searches, do simple calculations, greets you when called, use Google Maps, does some basic operating system tasks, and many more. We have decided to choose this topic to help the people who are old, imperceptible, even physically challenged, and also handicapped children. To confirm that the interconnection with a machine is not a trouble for people, anymore. The blind people who are deceptive, even they can also interact with it using their voice commands only. A different aspect of speech recognition has been implemented here along with the voice commands [4] "AI Voice Assistant" is used in Alexa, Siri, Cortana, Google Assistant, but these applications mostly work in Internet connections. The current voice assistant system that basically exists on Windows OS is Cortana which is a completely online-based system and requires high-speed fast Internet and a regular Microsoft account for login. Talking about the other existing voice assistant system, it is the "OK-Google" voice assistant which is browser dependent. Our proposed AI voice assistant has the potential to work using very slow Internet connectivity. It is named as Python-based AI Assistant, as it takes the user input in form of voice or text and then processes it, and then returns the output in a manner like it performs the tasks that have been provided.

The remaining part of the paperwork is arranged as follows. In Sect. 2, it describes the proposed paperwork and the workflow of the proposed system. Section 3 discusses the salient features and applications of the proposed system, Sect. 4 describes the result analysis of our work, and at last, Sect. 5 deals with the conclusion of this paper.

2 Proposed Work

The system (Fig. 1) is formatted by recognizing the voice commands provided by a user via a microphone. Our work comes up with various tasks like searching queries on Google, sending emails, watching videos on YouTube, live weather forecasts, simple calculations, utilization of google maps, keeping tracks on time, operating system-based tasks like shutting down and restarting the desktop [5], playing songs, making notes, and last but not the least, greet the user when called by name. When the assistant captures the voice commands given by the user, it will convert the voice into text according to the underlying flowchart [6]. Audio command is an eloquent and natural way for people to interact with applications by either enhancing or replacing the utility of mice, keyboards, controllers, and gestures. A different aspect of speech recognition has been implemented here for people with functional disabilities or other kinds of handicaps. To make their daily chores facile, we have decided to build a voice-commanding AI control assistant, which could be helpful. In our framework, as specified previously, the input process goes on till the end, until the user on one side ends his speech after the recognizer waits for an approximate time (nearly about

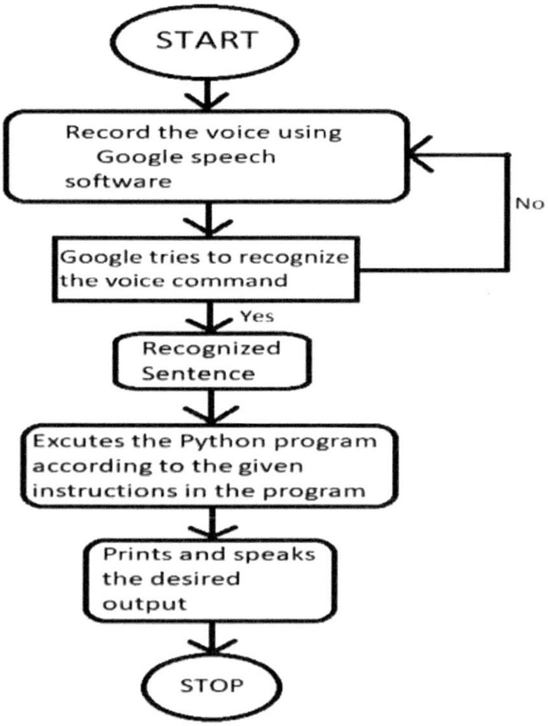

Fig. 1 Flowchart of our proposed working system

5 s) before it switches and converts the subsequent input to text. Then, the speech recognition module completes the recognizing process and generates a search thread.

In this system, as specified, the input process goes on till the end which is quite effective as if the user on one side has ended his speech input, but the recognizer waits for ample time before it switches and converts the subsequent input to text. Then, the speech recognition module completes the recognizing process and generates a query. Initially, this system remains in idle mode. As it receives any wake-up call, it begins execution. The generated query is identified whether it is a questionnaire or a task to be performed, and on the basis of that, the following is carried out. After the question is being answered or the task is being performed, the system waits for another command. This loop continues unless it receives the command to *stop listening*. At that moment, it goes back to sleep mode [7].

3 Salient Features and Applications

3.1 Speech Recognition with Google Web Speech API

The speech recognition module is just like a wrapper for many popular speech APIs, and it is extremely flexible. The Google Web Speech API is one of them. It supports a default API key that is hard-coded into the speech recognition library. Speech recognition helps to retrieve the audio input from the user really easy. Instead of facing the problems of building scripts for getting access to microphones and processing the inputted audio files from scratch, speech recognition will have you up and running in just a few minutes [8].

3.2 Web Scraping Using Beautiful Soup Module

The search results are directed back to the user by scraping the contents of the Web pages using the Beautiful Soup module of Python. The search results are placed in an order such that it helps in accessing the Web page quickly according to the user's desire and thus saving time, as opposed to the user reading out the whole search result that he wishes to select [9, 10].

3.3 Searches from the Web

Searching is one of the common things that everybody does in their lives. Our project has also been provided with some search engines which is an essential part of a netizen life [11]. It supports multiple search engines like Google, Yahoo, and Microsoft Bing.

3.4 Accessing Videos from YouTube

Videos are one of the most important source of entertainment and one of the most prioritized tasks of virtual assistants. Besides entertainment, they are also important for educational purposes as mostly all research activities and teaching-based videos presently are done using YouTube. These help in the creation of more practical learning process as well as out of the four walls of the classroom.

3.5 Weather Report

Getting the live weather condition of a location is a vital task of most virtual assistants. Our assistant fulfills this task with the help of Python programming [12].

3.6 Sending Emails

Incorporating the mailing features eases the job of mailing, which otherwise would consume more time by logging in to the dedicated email address [13]. Here, the users don't need to change tabs to do one of the most important tasks of our daily life, i.e., mailing. The users can easily send emails to the concerned receiver by just using voice commands.

3.7 Making a Note

As we all know, every single word is important, and to track what is, we try taking notes [14]. However, we are stuck at knowing about what is going to come in handy, and this results in taking down too many notes. This program can take down notes for the user.

3.8 Simple Calculations

Calculating is one of the vital things that everybody does and wants it to be quick and accurate. For a developer, working on Windows is also same. So, in our program also, we have imprinted such an essential part of a netizen's life by enabling our voice assistant, which we believe is going to be very useful.

3.9 Operating System-Based Tasks

It is the simplest feature of shutting down and restarting our device. Our proposed AI system does that basic thing.

3.10 Greets User When Called

Just like other AI voice assistants, our voice assistant has also been provided with a name "*Olivia.*" It greets the user when called. It interacts with the users using its voice commands and assists them to proceed with their work.

3.11 Tell Me Your Features

This feature has newly been added to our program. By just asking "tell me your features," it displays all its features, so any new user can understand which kind of assistance it can come up with.

4 Result Analysis

All the experiments have been conducted on the Python platform. The machine hardware configuration is mentioned below:

- Intel core i5 10th generation processor
- 1 TB HDD
- 16 GB RAM
- Windows 10 operating system.

Figure 2 show how the assistant opens different apps on desktop.

Figure 3 shows how the assistant opens different Web sites on the default Web browser.

Fig. 2 Proposed assistant opens different apps on desktop

Fig. 3 Proposed assistant opens different web sites on the default web browser

Figure 4 shows how the assistant checks the time or searches any location.

Figure 5 shows how the assistant does weather forecasting.

Figure 6 shows how the assistant keeps us update on COVID in different countries or all around the globe.

Figure 7 shows how the assistant searches any questions we ask.

Figure 8 shows how the assistant sends email.

Figure 9 shows how the assistant write a note.

Figure 10 shows how the assistant does basic operating system tasks like locking the device, hibernating the device, shutting down the device, and restarting the device.

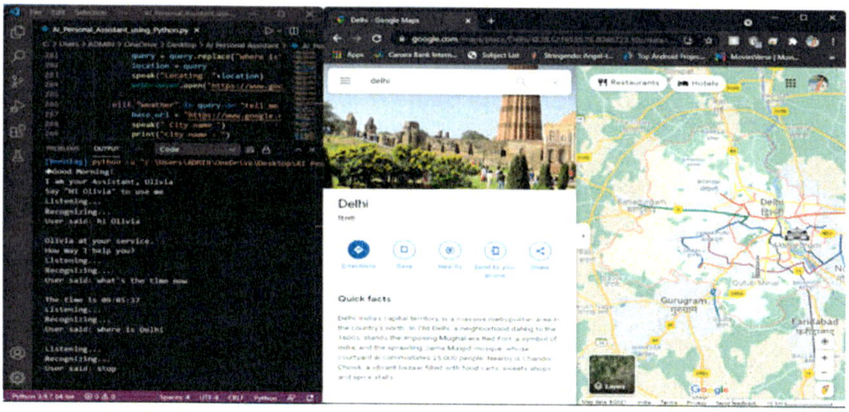

Fig. 4 Proposed assistant checks the time or searches any location

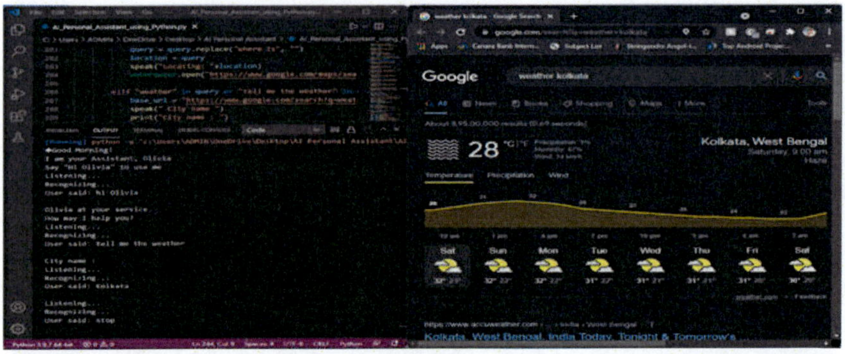

Fig. 5 Proposed assistant does weather forecasting

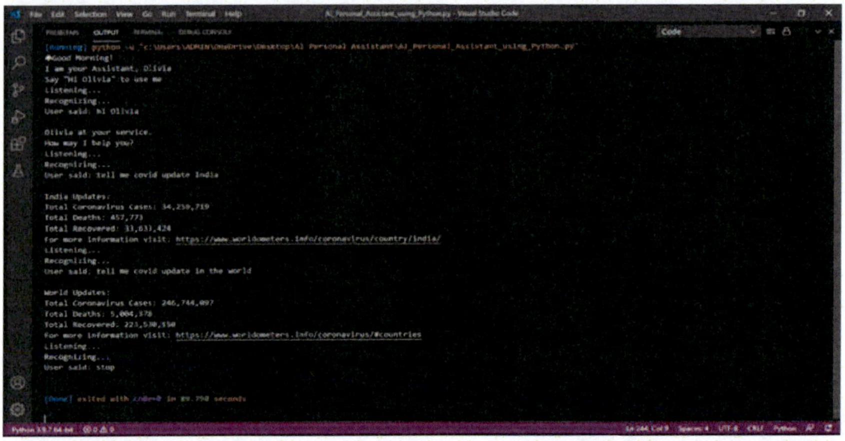

Fig. 6 Proposed assistant keeps us update on COVID in different countries

5 Conclusion

By using this voice assistant, that we have created, has computerized various facilities by only applying a single-line command. It makes most of the tasks of the user easier, like Web searches, weather forecast details, calculations, making notes, and sending emails. The functionality of the current system is limited to working in online mode only. Those virtual AI assistants that are available nowadays in the market are very swift and also very responsive. We still have a long way to go. There are many avenues left for future work. Many reliability portions on our current system are needed to be improved a lot later [15].

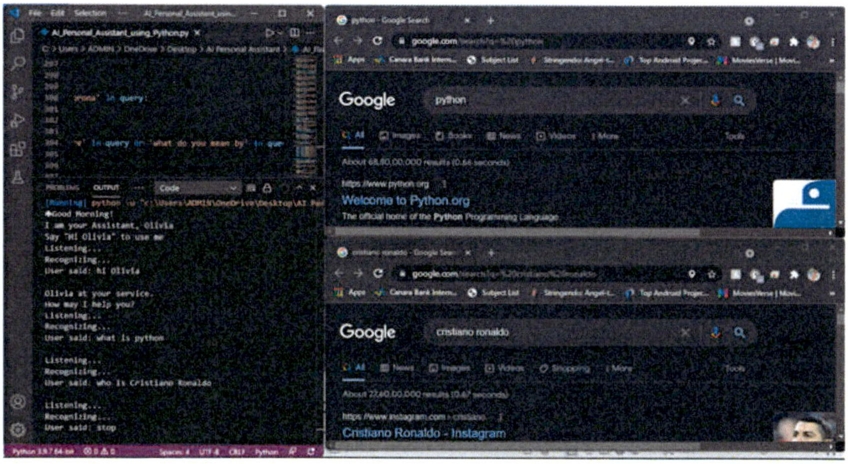

Fig. 7 Proposed assistant works as a search engine

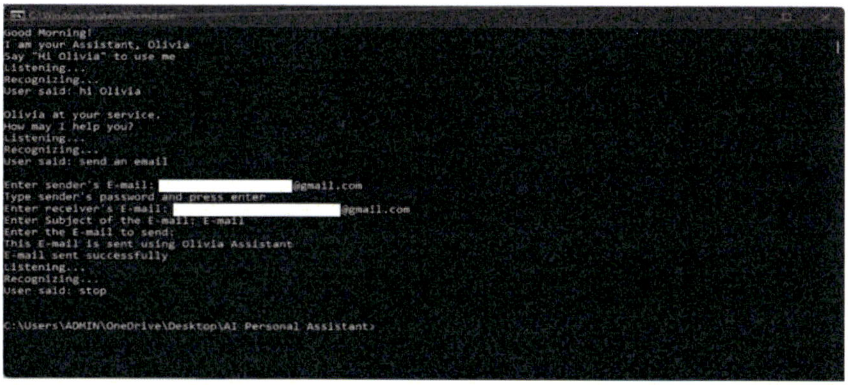

Fig. 8 Proposed assistant shows how to access emails

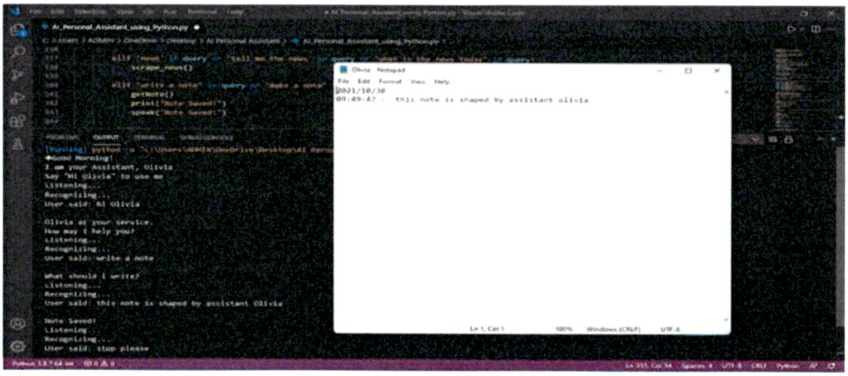

Fig. 9 Proposed assistant calls a node editor for editing

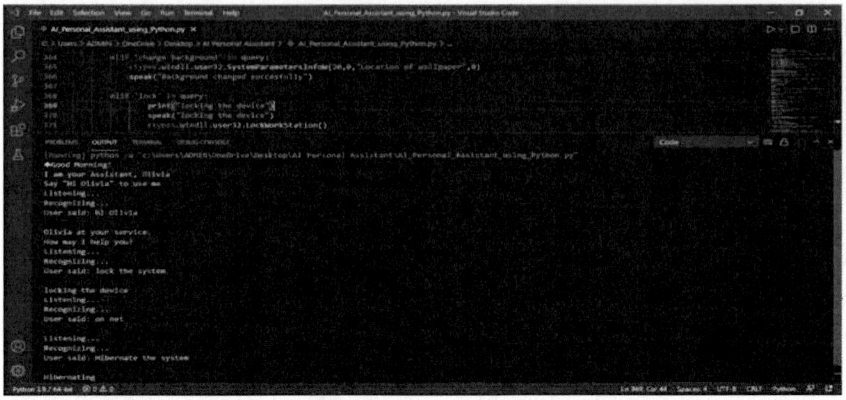

Fig. 10 Proposed assistant does basic operating system tasks

References

1. Sakharkar A, Tondawalkar S, Thombare P, Sonawane R (2021) Python based AI assistant for computer. Int Res J Eng Technol (IRJET)
2. Divya K, Rahul P, Raja A (2020) Jarvis AI using python (No. 4818). EasyChair
3. Elshafei M, Virtual personal assistant (VPA) for mobile users. Mitel Networks (2000–2002)
4. Geetha V, Gomathy CK, Vardhan KMS, Kumar NP (2021) The voice enabled personal assistant for Pc using python. Int J Eng Adv Technol (IJEAT) 10(4). ISSN: 2249-8958
5. Yadav A, Singh A, Sharma A, Sindhu A, Rastogi U (2020) Desktop voice assistant for visually impaired. Int J Recent Technol Eng (IJRTE) 9(2). ISSN: 2277-3878
6. Pandey A, Vashist V, Tiwari P, Sikka S, Makkar P (2020) Smart voice based virtual personal assistants with artificial intelligence. In: Artificial & computational intelligence
7. Sayyed A, Shaikh A, Sancheti A, Sangamnere S (2021) Desktop assistant AI using python. Int J Adv Res Sci Commun Technol (IJARSCT) 6(2)
8. Sakharkar A, Tondawalkar S, Thombare P, Sonawane R (2021) Python based AI assistant for computer. In: Conference on advances in communication and control systems (CAC2S)
9. Le V, Bahadur T, Shah J, Ara R (2020) A web-based AI assistant Application using Python and JavaScript. School of Professional Studies 63
10. Priyam R, Thakur R, Thakur VK (2013) Artificial intelligence applications for speech recognition. In: Conference on advances in communication and control systems (CAC2S)
11. Kulhalli KV, Sirbi K, Patankar MAJ (2017) Personal assistant with voice recognition intelligence. Int J Eng Res Technol (IJERT)
12. Shende D, Umahiya R, Raghorte M, Bhisikar A, Bhange A (2019) AI based voice assistant using python. J Emerg Technol Innov Res 6(2):506–509
13. Vardhan GH, Charan GH (2014) Artificial intelligence & its applications for speech recognition. Int J Sci Res (IJSR). ISSN 2319-7064 (Online)
14. Rabiner L (1993) Fundamentals of speech recognition. PTR Prentice Hall. ISBN 0-13-015157-2
15. Pérez A, Lopez-Otero P, Parapar J (2020) Designing an open source virtual assistant. In: Multidisciplinary digital publishing institute proceedings, vol 54, issue 1, pp 30

A User Independent Recommendation System for Web Series

Aditya Vikram Singhania, Anuran Bhattacharya, Priyanka Banerjee, Ritajit Majumdar, and Debasmita Bhoumik

Abstract The number of people streaming content over the Internet has boomed in the recent years. However, streaming platforms have to promote the right kind of titles to the right people in order to keep them interested in using their platform and that is why a recommendation engine is important. Different people choose to like Web series in different ways. Some like to watch multiple shows of their favorite actor, while some prefer some genres over others. Therefore, not only do the recommendations have to be accurate in the sense of similarity, but that similarity also has to be localized to within parameters, set by the user. In this paper, we have proposed a recommendation system that can recommend Web series to a user without any extensive data. Our recommender can recommend TV shows, both using all possible and localized parameters, tailoring the recommendations to exactly what the user wants, without imposing the bias of other users on them.

Keywords Recommendation-engine · Correlation · Jaccard-index · Web series recommender

1 Introduction

Since their inception, both Amazon Prime Video and Netflix have become huge giants in media streaming over the years. During the Coronavirus pandemic in 2020, they both recorded over 200 million users each [1, 2]. But to maintain such a huge user-base, only creating and maintaining good content is not enough. Streaming services have to promote the right kind of titles to the right people in order to keep them interested in using their platform. And that's why recommendation engines are important. Different people choose to like TV shows and movies in different ways.

A. V. Singhania (✉) · A. Bhattacharya · P. Banerjee
Department of Computer Science, The Bhawanipur Education Society College, Kolkata, India
e-mail: mailingadityavs@gmail.com

R. Majumdar · D. Bhoumik
Advanced Computing and Microelectronics Unit, Indian Statistical Institute, Kolkata, India

P. Dutta et al. (eds.), *Emerging Technologies in Data Mining and Information Security*, Lecture Notes in Networks and Systems 490,
https://doi.org/10.1007/978-981-19-4052-1_59

Some like to watch multiple shows of their favorite actor(s), while some prefer some genres over others. This means that, not only do the recommendations have to be accurate in the sense of similarity, but that similarity also has to be localized to within parameters, set by the user.

Streaming apps have found an easy solution to this problem, where they list TV shows segregated by their genres, and users can choose which genre they want to watch. And while it is a great solution that works for them, users want much more than genres. Theoretically speaking, it would be great to watch a show with a story-line similar to a show we like; or to watch multiple shows of an actor we like.

In [3], the authors used a support vector machine (SVM)-based model [4] to classify TV shows based on like/dislike input given by users. While they had a baseline accuracy of about 75%, the major drawback is that they only considered a two-class case classifier to predict TV shows liked by the users, thus resulting it to perform recommendations accurately only upon a very few sets of TV show titles owing to differences in what each user likes.

A dataset collected using about 340,000 devices with over 19,500 unique items was used by the authors in [5] on which they applied the linear discrimination analysis (LDA) [6] technique. They split the dataset into a 75:25 training-testing ratio and reached a final accuracy of about 80.76%. The drawback of their system is that their results are solely on the basis of top ratings and the total number of views, not tailoring the recommendations to suit individual users.

As mentioned above, the previous works in the field either go for a *one size fits all* approach for user preferences, or disregard user preferences completely. Nevertheless, they fail to analyze the content of a title and other important aspects that make people interested in watching a TV show. This is where our recommender differs from the ones used in-house by streaming apps and the previous works. Without extensive data such as user watch history and user preferences, our recommender can recommend TV shows both using all possible parameters, and also using localized parameters first, such as cast, genres, or story-line; tailoring the recommendations to exactly what the user wants, without imposing the bias of other users on them. Using the description of the titles and comparing similarity between them using Jaccard's index [7] is a novel idea, which has not been implemented before. This enables our recommender to confidently recommend titles which have a story-line very similar to each other.

The rest of the paper is organized as follows: In Sect. 2, we discuss the proposed techniques used for (i) creation of dataset, (ii) selecting the most significant attributes, and (iii) preparing the final dataset of recommendation based on those. Sections 3 and 4, respectively, present the details of developing a Web series recommendation system and how well it performs against famous shows which have a well-known plot-line and huge fan base. We conclude in Sect. 5.

2 Background

There are quite a few datasets available for movie prediction on the Internet. However, Web series and online TV shows are relatively newer concepts, which gained an upward trajectory only in recent years, resulting in fewer available datasets to work on. While the majority of our dataset is made using the Internet movie database (IMDb) dataset [8], we added an important column, *description*, using the movie database's (TMDb) *application programming interface (API)* [9]. The *description* attribute is what gives us the story-line of the tv shows, which we use in Sect. 3 to make the recommendations.

The final dataset contained ≈365 million unique pairs of titles, and to optimize the various filtering and sorting operations in the method proposed by us, we computed a correlation matrix to reduce the dataset on the basis of the most significant attributes among the entire dataset. This allows us to make recommendations quickly while still accomplishing what we set out to do.

A correlation matrix [10] is a table created to obtain the statistical measure of the strength of relationship between the variables. It is used to sum up the data, as an input into a more advanced analysis, and as a diagnostic for advanced computations [11]. There are mainly three methods for computing a correlation matrix:

- Pearson parametric correlation test [12]
- Spearman rank-based correlation analysis [13]
- Kendall rank-based correlation analysis [14].

We have used the Spearman rank-based correlation analysis technique which takes two attributes into consideration and results to 1 if the attributes are similar. A correlation matrix shown in Fig. 1 is computed, and the final dataset is made for the recommendation purpose.

Based on the correlation matrix shown in Fig. 1, we chose the genre as our most significant attribute and reduced the dataset to ≈100 using the method explained in Sect. 3.1, and for each pair of publication in our dataset, we computed the number of common genres and the number of common cast members between them. The similarity between the descriptions of each pair is computed using Jaccard's similarity index. Jaccard's index is a measurement used to get the similarity between two objects, such as two text documents or two sets of elements. Consider the description of publication A be defined as Set A and publication B be defined as Set B. Thus, the Jaccard similarity can be computed as

$$J(A, B) = \frac{|A \cap B|}{|A \cup B|} \tag{1}$$

Here, $|A \cap B|$ is the number of members shared between both sets, and $|A \cup B|$ is the total number of members in both sets (common and distinct). The Jaccard similarity will be 0 if the two sets do not share any values and 1 if the two sets are identical.

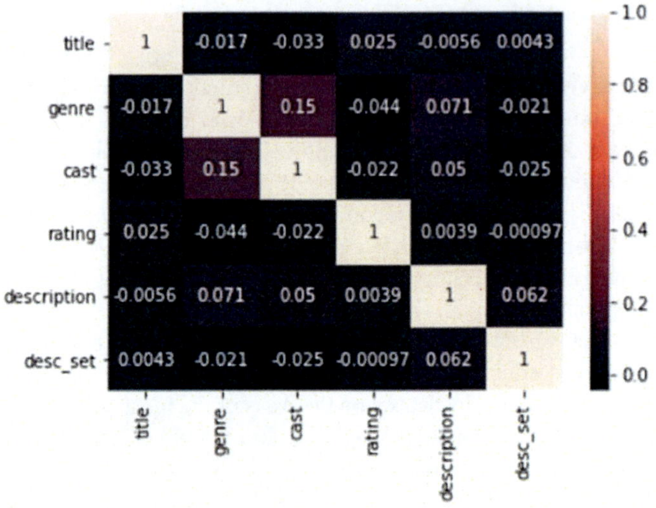

Fig. 1 Correlation matrix using Spearman's rank correlation

A final table is created which consists of the Jaccard's index values, numbers of common genres, and cast members between two publications and is used to make recommendations to the users. This table can be used to recommend titles in any order, such as genre first, or similar story-line, or even common actors.

3 Methodology

3.1 Processing

The recommendation engine uses three parameters to recommend Web series related to the input, which are

- Similarity between the description of two publications
- Number of common genres between two publications
- Number of common cast members between two publications.

To calculate these values, any dataset with the following details can be used: (i) list of titles, (ii) description for each title, (iii) at least 1 genre for each title, (iv) a list of main/all cast members for each title.

The parameters were designed in this way because the most common ways a publication can be similar to another one are if their descriptions are similar, if they have the same or mostly same cast members, or if they have multiple common genres.

Our dataset contains $N \approx 27{,}000$ rows and the total number of unique pairs of publications (T), for which we would need to calculate the above mentioned characteristics would be in the order of $O(N^2)$, which lead to an empirical value of 364.5 million pairs. As mentioned earlier, we used a correlation matrix and chose the most significant attribute in our dataset, *genre*, to reduce our dataset size, to reduce the amount of time taken by our recommender during the filter and sorting steps. We only kept rows with titles having at least 1 genre in common with at least 1 other title reducing our number of rows to ≈100 million.

Now, for each pair of publications in our list, we calculated the values listed above, creating a new table with the following columns:

- *Title 1*—Title of the first publication
- *Title 2*—Title of the second publication
- *JI*—Jaccard's index between the description of both publications
- *CG*—Number of common genres between the two publications
- *CC*—Number of common cast members between the two publications.

This final table is all that is needed to make the recommendations for any show, genre, or actor there is in the entire dataset.

By default, the program takes input the name of a show, and then outputs top 5 show based on the Jaccard's index, number of common genres, and number of common cast members, in that order. But it can be modified to put those three parameters in any order by passing additional parameters.

3.2 Dataset Used

We used IMDb's datasets [8], namely title.basics.tsv.gz, title.principals.tsv.gz, along with TMDb's API [9] to get descriptions for all relevant titles.

The names and a brief description of each of the columns used are as follows:

- *tconst*—Unique identifier provided by IMDb for each publication.
- *primaryTitle* (further renamed to *title*)—The primary title for the publication, translated to English, if originally in a different language.
- *genres*—A list of all genres a publication belongs to.
- *cast*—A list of primary cast members for that publication, along with directors and writers, represented by a special unique id belonging to each individual.
- *description*—A short description/synopsis of the publication's story.

The two datasets were combined using the *tconst* provided by IMDb, and all rows with missing data were subsequently dropped. The combined dataset was used for the calculations in Sect. 3.1. The entire workflow is also given in Fig. 2 (Tables 1 and 2).

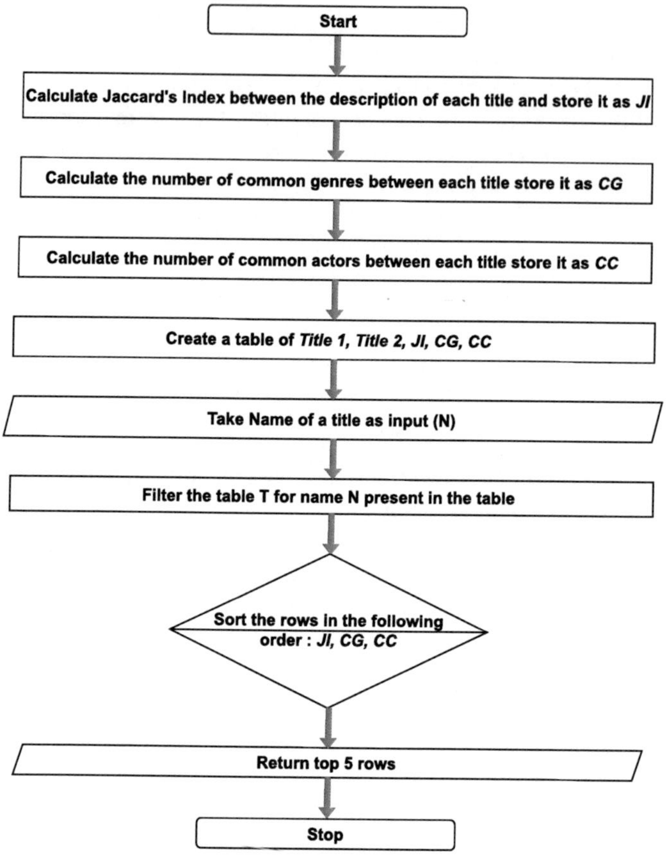

Fig. 2 Workflow of the proposed work

Table 1 Recommendations for title *Game of Thrones* for *same genre(s)*

Name	Genre(s)	JI	Cast
Kamen Rider Kiva	Action, adventure, drama	0.820	Rina Koike, Kôji Seto, Kôhei Takeda, Yû Takahashi
Colony	Action, adventure, drama	0.805	Jacob Buster, Josh Holloway, Sarah Wayne Callies, Peter Jacobson
Ren	Action, adventure, drama	0.777	Coral Adlam, Sophie Skelton, Duran Fulton Brown, Nick Cornwall
2091	Action, adventure, drama	0.757	Jason Day, Manolo Cardona, Angie Cepeda, Christopher Von Uckermann
Extinct	Action, adventure, drama	0.757	Anna Vocino, Chad Michael Collins, Victoria Atkin, Yorke Fryer

Table 2 Recommendations for title *Dexter* for *different genres*

Name	Genre(s)	JI	Cast
Blackout	Game-show	0.801	Phyllis Diller, Bob Goen, Johnny Gilbert, Pat Harrington Jr.
Police Diary 2.0	Drama	0.800	nm10353316, Anjana Jayaprakash, John Kokken
The Mill	Drama	0.799	Hugh Simon, Kerrie Hayes, Ciaran Griffiths, Barbara Marten
Claws	Comedy, Crime, Drama	0.794	Harold Perrineau, Niecy Nash, Carrie Preston, Judy Reyes
Black Sun	Crime, Drama, Thriller	0.791	Aleksandar Stojkovic, Andrija Kuzmanovic, Marija Bergam, Dragan Bjelogrlic

4 Results

Tables 3 and 4 help us to calculate distances between different genres, when our recommender is being used to switch from one genre to another. Starting from a known show of a known genre, we take one unique recommendation from the recommender, and use it as an input back to the recommender, until we are recommended a tv show of the desired genre. The column *Steps* in these tables starts at 0 listing the tv show that we started with and is incremented by 1 for each iteration of input and output. In Table 3, we start with the tv show *Game of Thrones* which has the known genre *Drama*. At Step 2 of the process, we are being recommended the tv show *Being Erica*, belonging to the *Comedy* genre, while *Fortitude* of the *Horror* genre is recommended at Step 6. Thus, we can conclude that *Comedy* is closer to *Drama* than *Horror*. Similarly, in Table 4, we begin with the tv show *High-stakes Poker* of the *Game-Show* genre and are recommended a tv show of the genre *Family* in step 1, while the tv show *Four Star Playhouse* of the *Crime* genre is recommended to us at Step 7. Thus, we are able to understand that *Game-Show* and *Family* are closer genres than *Game-Show* and *Crime*.

To the best of our knowledge, no other recommendation system provides or suggests such a method to change genres while still recommending similar tv shows, and thus we are unable to compare our results to any previous works.

5 Conclusion and Future Work

In this paper, we have proposed a recommendation system that can provide tailored recommendations of Web series based on titles previously liked by the user. Our proposed technique provides recommendations based on both a general overview as well as localized parameters of the title's metadata. The application of this approach

Table 3 Number of steps to reach from genre *Drama* to genre *Horror*

Name	Genre(s)	Cast	JI	Steps
Game of Thrones	Drama			0
Rikoshet	Action, Crime, Drama	Ingrid Olerinskaya Aleksandr Ustyugov Ekaterina Volkova Sergey Shnyryov	0.823188	1
Being Erica	Adventure, Comedy, Drama	Paula Brancati Erin Karpluk Reagan Pasternak Michael Riley	0.853165	2
The Case-Book of Sherlock Holmes	Crime, Drama, Mystery	Peter Vaughan Jeremy Brett Edward Hardwicke Rosalie Williams	0.849315	3
MI-5	Action, Crime, Drama	Richard Armitage Peter Firth Hugh Simon Nicola Walker	0.836842	4
Colony	Action, Adventure, Drama	Jacob Buster Josh Holloway Sarah Wayne Callies Peter Jacobson	0.819372	5
Fortitude	Drama, Horror, Mystery	Darren Boyd Richard Dormer Björn Hlynur Haraldsson Sienna Guillory	0.869333	6

Table 4 Number of steps to reach from genre *Game-Show* to genre *Drama*

Name	Genre(s)	Cast	JI	Steps
High Stakes Poker	Game-Show			0
You Bet Your Life	Comedy, Family, Game-Show	Reg Lewis Groucho Marx George Fenneman Melinda Marx	0.827148	1
Hot Dog	Family	Woody Allen Jonathan Winters Jo Anne Worley Tom Smothers	0.838266	2
The Cosby Show	Comedy, Family, Romance	Geoffrey Owens Bill Cosby Phylicia Rashad Keshia Knight Pulliam	0.823344	3
Q5	Comedy	Bob Todd Spike Milligan John Bluthal Alan Clare	0.828996	4
The Brittas Empire	Comedy	Judy Flynn Chris Barrie Harriet Thorpe Michael Burns	0.831087	5
His Lordship Entertains	Comedy	Barbara Loynes Ronnie Barker Josephine Tewson David Jason	0.835327	6
Four Star Playhouse	Comedy, Crime, Drama	Robert Bice David Niven Dick Powell Charles Boyer	0.830102	7

on the IMDb dataset shows that our recommendation system is able to recommend Web series which have similar characteristics when compared to a title provided by the user. Future scope of this research is the inclusion of a given specific region, either by user input, or by using a database which has that attribute available, and recommending titles which are based in the same region or language. Also, it is to be noted that at present, the program recommends titles solely on the basis of persistent data, but in future, user ratings of titles, which are volatile and ever changing, may also be taken into account. User preferences and watch history of users are also important parameters that should be looked into in future.

References

1. Netflix keeps growing in pandemic, tops 200 mn subscribers (2021). https://tinyurl.com/netflix200
2. Amazon prime hits 200 million users globally, says CEO Jeff Bezos (2021). https://tinyurl.com/amazon200m
3. Xu JA, Araki K (2006) A SVM-based personal recommendation system for TV programs. In: 2006 12th international multi-media modelling conference. IEEE, p 4
4. Cortes C, Vapnik V (1995) Support-vector networks. Mach Learn 20(3):273–297
5. Aharon M, Hillel E, Kagian A, Lempel R, Makabee H, Nissim R (2015) Watch-it-next: a contextual TV recommendation system. In: Joint European conference on machine learning and knowledge discovery in databases. Springer, pp 180–195
6. Fisher RA (1936) The use of multiple measurements in taxonomic problems. Ann Eugen 7(2):179–188
7. Jaccard P (1901) Distribution de la flore alpine dans le bassin des dranses et dans quelques régions voisines. Bull Soc Vaudoise Sci Nat 37:241–272
8. Imdb datasets (2021). https://www.imdb.com/interfaces
9. The movie database API (2021). https://developers.themoviedb.org
10. Godfrey KR (1980) Correlation methods. Automatica 16(5):527–534
11. Benesty J, Chen J, Huang Y, Cohen I (2009) Pearson correlation coefficient. In: Noise reduction in speech processing. Springer, pp 1–4
12. Norman G (2010) Likert scales, levels of measurement and the "laws" of statistics. Adv Health Sci Educ 15(5):625–632
13. Alvo M, Cabilio P (1993) Rank correlations and the analysis of rank-based experimental designs. In: Probability models and statistical analyses for ranking data. Springer, pp 140–156
14. Hervé A (2007) The Kendall rank correlation coefficient. In: Encyclopedia of measurement and statistics. Sage, Thousand Oaks, CA, pp 508–510

More Patients or More Deaths: Investigating the Impact of COVID-19 on Important Economic Indicators

Debanjan Banerjee and Arijit Ghosal

Abstract The impact of COVID-19 over various sectors has continued to interest researchers. This work will investigate the impact of COVID-19 for economic uncertainty and gold price across various countries. This investigative research is done with the help of panel data concept programming packages associated with the R programming language, and therefore, we can call it longitudinal research. The current research deals with two important features related to COVID. These two are monthly COVID patient average and monthly COVID death average, and data is collected from countries such as India, China, Russia, USA, UK, Japan and Germany and apply panel data regression techniques to understand whether these two COVID-related factors significantly impact upon economic uncertainty and gold prices. The current work also observes that the monthly confirmed COVID patient average has more impact across the spectrum of countries on the economic policy uncertainty and the gold price in these countries compared to the other feature, i.e. monthly average COVID death tolls. The same conclusions have been made in the graphical abstract as depicted below.

Keywords COVID-19 · Country wise influence · Gold price · Geopolitical · Economic uncertainty

1 Graphic Abstract

The abstract of the current research has been very well depicted in Fig. 1 as observed by the authors of the current paper. This section, with the help of the below diagram, actually depicts the influence that the two respective COVID-related features which are monthly confirmed COVID patient average and monthly COVID death average

D. Banerjee (✉)
Sarva Siksha Mission, Kolkata, India
e-mail: debanjanbanerjee2009@gmail.com

A. Ghosal
St. Thomas' College of Engineering and Technology, Kolkata, India

P. Dutta et al. (eds.), *Emerging Technologies in Data Mining and Information Security*, Lecture Notes in Networks and Systems 490,
https://doi.org/10.1007/978-981-19-4052-1_60

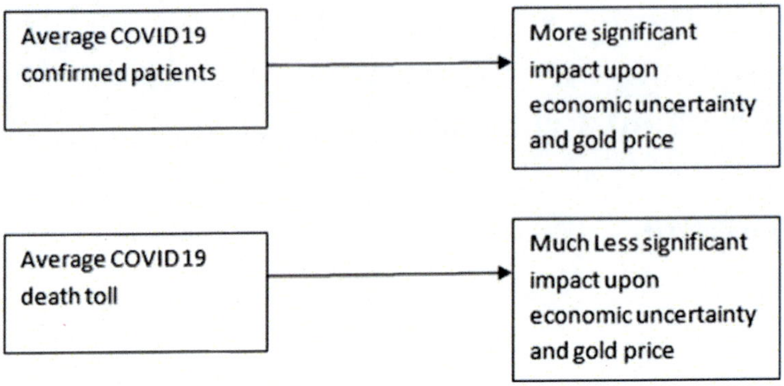

Fig. 1 Graphical abstract

do have over the dependent variables such as the economic policy uncertainty (EPU) and the gold prices.

2 Introduction

The impact of COVID pandemic is a significant matter of interest for the research community. Thus, the current investigation utilizes two COVID-related features such as monthly average COVID confirmed patient counts and monthly average COVID confirmed death tolls across countries such as India, China, Russia, USA, UK, Japan and Germany and observe whether across countries the two above-mentioned COVID-related features, i.e. monthly average COVID confirmed patient counts and monthly average COVID confirmed death toll significantly impact the economic policy uncertainty index (EPU) and gold prices. The current work also aims at investigating whether the two COVID-related factors have similar significant impact over the economic policy uncertainty (EPU) index and gold prices. The economic policy uncertainty (EPU) indexes across the countries for all the countries have been taken from the free online resource (https://www.policyuncertainty.com/), and the gold prices have been taken online stock market indexes.

3 Related Work

The researcher community has been investigating whether and how much the COVID-related lockdowns did impact the economies across the world. Anh and Gan [1] observe the impact of COVID lockdowns on the Vietnamese before and after the COVID lockdowns, and they observe it by using various panel data models. The researchers did observe that the financial sector was the hardest hit due to the COVID lockdowns, whereas the stock markets did perform well during the COVID lockdowns after initially suffering during the early phases of the worldwide lockdowns. Ashraf [2] observes that the news of growth in COVID-19 confirmed cases makes stock markets react with significant negative returns, whereas the news of growth in COVID-19 deaths does not have similar significant impact on stock market negative returns. Cao et al. [3] investigate the impact of COVID spread variance over various stock market indexes during the first six months of 2020, and they observed that a stock market index's performance should be able to improve as international efforts to combat COVID are put into focus while utilizing various panel data concepts.

Crossley [4] et al. observe by using high-quality panel data of probability samples in the UK that the lowest income and minority groups are the hardest hit by the advent of countrywide lockdowns, and they have to depend upon borrowing as well as private wealth transfers to mitigate their lot. Hassan and Gavilanes [5] emphasize that the impact of the global spread of the COVID virus does have a significant short-termed negative impact on the returns of the stock market indices given that these two researchers had utilized panel least squares vector auto-regressive models on stock market indices from China, South Korea, Japan, South Korea, etc. Liu et al. [6] forecast initially strong but then gradually receding impacts on fourteen Asia-Pacific countries by using a dynamic panel data model for the same. Mohiuddin [7] et al. investigated the impact of COVID pandemic before and after the event by focusing panel VAR data models on crude oil, exchange rate and gold price on KSE100 indexes. Rabhi [8] et al. related make an empirical study on the impact of pandemics like COVID on the emerging Asian economies' stock market indices by using the ARDL panel data approach, and they observe that the reported daily COVID-19 growth cases do significantly impact the stock markets as well as other factors such as commodity prices, US stock market and USDollar.

Rajarathinam and TamilSelvan [9] made an empirical study that reveals that the COVID pandemic lockdowns did make significant income reductions on the battered communities in all the districts of Tamil Nadu during the 2020 COVID lockdown while using the panel data regression techniques by collating and aggregating data from various newspaper articles across Tamil Nadu and India. Salisu et al. [10] constructed a COVID global fear index to optimally predict the global commodity price returns. Singh [11] et al. studied the impact of COVID pandemic on the stock market of G20 countries while using the panel data regression techniques on the G20 countries.

Sun et al. [12] utilized panel data regression techniques across the spectrum of stocks on the Chinese stock market, and they observed that in the initial 5–6 weeks, the COVID-related negativity sentiment is the highest amongst the stocks; however, they gradually recover from these sentiments after the same given timeline and some sectors like agriculture, digitization and pharmacy even see a boost in their performances post-COVID. Topcu and Gulal [13] observed that the emerging markets' stock indices tend to gradually teper off the negative impacts of the pandemic once the 5–6 weeks of crisis moment are gone and given the size of government spending or government stimulus they tend to slowly recover given the duration of the time.

4 Proposed Method

The present work investigates the influence of the COVID-19-related indicators, i.e. monthly confirmed COVID patient average and monthly COVID death average on economic policy uncertainty (EPU) index and gold prices.

- The present work assumes that the monthly average of COVID confirmed patients does significantly impact the economic policy uncertainty and gold prices. This particular feature can be computed by first subtracting the total number of confirmed COVID patients at the beginning of a month from the total number of confirmed COVID patients at the end of a month and then dividing the derived subtracted value with the total number of days in that given month.
- The present work also assumes that the monthly average COVID death toll does significantly impact the economic policy uncertainty and gold prices. This particular feature can be computed by first subtracting the total number of confirmed COVID deaths at the beginning of a month from the total number of confirmed COVID death toll at the end of a month and then dividing the derived subtracted value with the total number of days in that given month.
- The present work assumes that the two COVID-related features in question, which are monthly average COVID confirmed patients and monthly COVID death average do significantly influence the economic policy uncertainty (EPU) index and gold prices across countries and across time frames. Thus, a balanced panel data set that comprises economic policy uncertainty and gold prices across a diverse spectrum of countries such as India, China, Russia, USA, UK, Japan and Germany from January 2020 till August 2021 to investigate whether the above-mentioned COVID-related features do significantly impact the EPU and gold prices.
- The present work makes a null hypothesis that there is no influence caused by the COVID-related features over the EPU and the gold price, and the alternative hypothesis can only be accepted once the F-statistic p-value regression statistic from panel data regression methods is always less than or equal to 0.01.

5 Experimental Results

This work performs the panel data regression techniques with the help of the R programming language libraries *plm* and *foreign*. The *plm* is utilized since all the two main independent and COVID-related features do have continuous type, and therefore, using longitudinal or panel data-based regression technique in this case will be very useful. The target variable that has been used in these regression techniques is the economic policy uncertainty and gold price for all respective countries. Table 1 depicts all the results obtained by experimenting with the above-mentioned techniques.

Table 1 Correlationship between COVID factors economic uncertainty factors

Features	EPU F-statistics p-value	Gold price F-statistics p-value
COVID monthly avg. confirmed patient count	0.0127	0.01061
COVID monthly avg. confirmed death count	0.04233	0.04423

The work performs regression based upon the following criterion.

- It can be derived from Table 1 that COVID-related features, i.e. monthly avg. COVID patient count, do influence the EPU and the gold prices.
- The conclusion that can be made from Table 1 is that monthly avg. COVID patient count does more significantly impact the EPU and the gold price across countries and time spectrum compared to the other COVID feature like monthly avg. COVID death count since the F-statistic p-value is lesser than 0.01 for the former compared to what can be observed for the latter.

6 Discussion

The current work concludes from the observations as depicted in Table 1 that the following features do not influence the target variables since

- COVID average monthly patient count has F-statistic p-value of 0.0127 and 0.01061 for target values as EPU and gold price, respectively.
- COVID average monthly death count has F-statistic p-value of 0.04233 and 0.04423 for target values as EPU and gold price, respectively.
- Thus, we can safely deduce from the above regression statistics that COVID average monthly patient count has more significant influence upon the EPU and gold price compared to COVID average monthly death count.

7 Conclusion

Observing the regression statistics, we can safely deduce that the COVID monthly average confirmed case count is a far more significant influencer of important macroeconomic indicators such as economic policy uncertainty index as well as the gold price across a wide spectrum of countries. The explanation for the above observed phenomenon could be since the death rate per thousand patients during the COVID pandemic is much lesser compared to the pandemics of previous era such as the

Spanish Flu, Black Death and Ebola virus. Also the death rate across continents varies widely since countries like USA and UK have much higher death rate per thousand patients compared to countries like China, Japan and India; thus, COVID monthly average death rate may not be the biggest influencer of gold prices as well as economic policy uncertainty.

However since higher monthly average COVID confirmed patient counts indicate more stricter and longer lockdowns and the more the people are locked down, the more there will be impact in terms of economic policy uncertainty as well as gold prices; therefore, the monthly average COVID confirmed patient counts more significantly influence macroeconomic indicators compared to monthly average COVID confirmed death tolls across a wide spectrum of countries across the globe.

8 Future Work

The future scope of work would include that more COVID-19 indicators such as R-factor monthly growth and fluctuating dead count per thousand patient should be considered alongside more micro-economic factors such as inflation, GDP, stock market indices, policy rates and economic stimulus as declared by the governments. There is also a need to investigate the impact of COVID-19 factors across various geographical regions as well as regional economic blocks.

References

1. Anh DLT, Gan C (2020) The impact of the COVID-19 lockdown on stock market performance: evidence from Vietnam. J Econ Stud
2. Ashraf BN (2020) Stock markets' reaction to COVID-19: cases or fatalities? Res Int Bus Financ 54:101249
3. Cao KH, Li Q, Liu Y, Woo CK (2021) COVID-19's adverse effects on a stock market index. Appl Econ Lett 28(14):1157–1161
4. Crossley TF, Fisher P, Low H (2021) The heterogeneous and regressive consequences of COVID-19: evidence from high quality panel data. J Publ Econ 193:104334
5. Hassan S, Gavilanes JMR (2021) First to react is the last to forgive: evidence from the stock market impact of COVID 19. J Risk Financ Manag 14(1):26
6. Liu L, Moon HR, Schorfheide F (2021) Panel forecasts of country-level COVID-19 infections. J Econ 220(1):2–22
7. Mohiuddin I, Alvi J, Inam SG, Rehan M (2021) Impact of crude oil, exchange rate and gold price on KSE100 index: before and during COVID-19 pandemic by using VAR model. J Contemp Issues Bus Gov 27(06)
8. Rabhi A (2020) Stock market vulnerability to the COVID-19 pandemic: evidence from emerging Asian stock market. J Adv Stud Financ (JASF) 11(22):126–131
9. Rajarathinam A, TamilSelvan P (2021) Estimating trends in COVID-19 infected cases based on panel data regression modelling. J Appl Sci 21(1):31–41
10. Salisu AA, Akanni L, Raheem I (2020) The COVID-19 global fear index and the predictability of commodity price returns. J Behav Exp Financ 27:100383

11. Singh B, Dhall R, Narang S, Rawat S (2020) The outbreak of COVID-19 and stock market responses: an event study and panel data analysis for G-20 countries. Glob Bus Rev 0972150920957274
12. Sun Y, Wu M, Zeng X, Peng Z (2021) The impact of COVID-19 on the Chinese stock market: sentimental or substantial? Financ Res Lett 38:101838
13. Topcu M, Gulal OS (2020) The impact of COVID-19 on emerging stock markets. Financ Res Lett 36:101691

Systems Biology Paradigm for Exploring the Relation Between Obesity and Ovarian Cancer with a Focus on Their Genome-Scale Metabolic Models

Priyanka Narad, Romasha Gupta, Sabyasachi Mohanty, Ritika Sharma, Nagma Abbasi, and Abhishek Sengupta

Abstract Obesity is an intricate condition that has been related to the precarious development of lethal gynecologic cancers including ovarian cancer. Ovarian cancer offers a significant number of metabolic dysregulation profiling opportunities, which aids drug discovery and development. Despite the reality that extensive studies have been undertaken on the area, the connection between body weight and ovarian cancer remains uncertain. Genome-scale metabolic models (GEMs) are a useful tool in this situation because they represent the metabolic framework of cells and provide a practical platform for simulating and measuring metabolic fluxes in human beings using constraint-based mathematical methodologies. The model obtained can be utilized for understanding cell metabolism. The dataset GSE130350 was utilized to evaluate and create a model to comprehend their reactions and metabolites. The flux score was calculated by the Recon3D-based metabolic model using MATLAB's COBRA toolbox and Python's COBRApy, and data analysis was done. The resulting model consists of 10,610 reactions and 5837 metabolites with a flux score of 1.72. Absolute quantitative gene expression data have been defined to predict genome-scale flux patterns. RNA sequencing methods like single-cell RNA-seq (scRNA-seq) method and bulk RNA-seq method were used for understanding gene expression. scRNA-seq was designed to exhibit distinct cell behavior in relation to its environs; bulk RNA-seq, on the contrary, was never meant to characterize a cancer formation precisely.

P. Narad · R. Gupta · S. Mohanty · R. Sharma · A. Sengupta (✉)
Amity Institute of Biotechnology, Amity University Uttar Pradesh, Noida, Uttar Pradesh 201301, India
e-mail: asengupta@amity.edu

P. Narad
e-mail: pnarad@amity.edu

N. Abbasi
NextGen Life Sciences Pvt. Ltd., New Delhi 110092, India
e-mail: nagma@nextgenlife.com

P. Dutta et al. (eds.), *Emerging Technologies in Data Mining and Information Security*, Lecture Notes in Networks and Systems 490,
https://doi.org/10.1007/978-981-19-4052-1_61

Keywords Genome-scale metabolic model · Obesity · Ovarian cancer · Systems biology

1 Introduction

Obesity, a compulsive disorder developed by an extreme growth of adipose tissue, is expanding around the world. It is a clinical issue that increases the risk of certain cancers including ovarian cancer. Generally, obesity results from a blend of inherited factors, along with the environment and individual diet, and choice of exercise. Fat mass and obesity-associated (FTO) genes associated with body mass index (BMI) higher risk of obesity were distinguished in numerous populations, making FTO the principal locus unequivocally connected with adiposity. Even though many secretory and metabolic pathways have been investigated to learn more about the influence of obesity on ovarian cancer, the link between obesity and ovarian cancer is not fully understood, and oncologic therapeutic programs rarely involve weight and lifestyle management. Both the growth of omics sciences and the steady advancement of related technologies encourage the exploration of new strategies based on the reconciliation of knowledge from various sources, to gain a better insight into the interaction and relationships among various genetic molecules.

In this scenario, genome-scale metabolic models (GEMs) have emerged as a useful tool [1] to express and simulate an event based on the meaning of particular arrangements of objects and constraints, particularly a system. GEM provides a useful framework for constraint-based modeling (CBM) strategies aimed at increasing metabolic fluxes in living organisms [2]. The increasing availability of high throughput data is encouraging the development of novel methodologies in which structural data represented by GEMs serve as a scaffold for the coordination of omics data to construct condition-specific metabolic states. To develop condition-specific networks to govern metabolic fluxes, the workflow outlined in our study incorporates gene expression data acquired from GFP (control) and FTO (treatment) subjects into a published rebuilt GEM of humans, Recon3D.

2 Materials and Methods

Manually curated transcriptomics data were derived from the Gene Expression Omnibus for control and healthy patients (GEO). A substantial percentage of bulk and single RNA sequencing samples was collected, and evidence for metabolic model construction and differentially expressed genes was given. During the identification of metabolic genes, metabolic anomalies in ovarian cancer cells were revealed by comparing the expression of metabolism-related genes in control and healthy patient samples. Data on metabolism-related genes were extracted using the Recon3D model. Optimization of a computational approach for gene network analysis that can locate

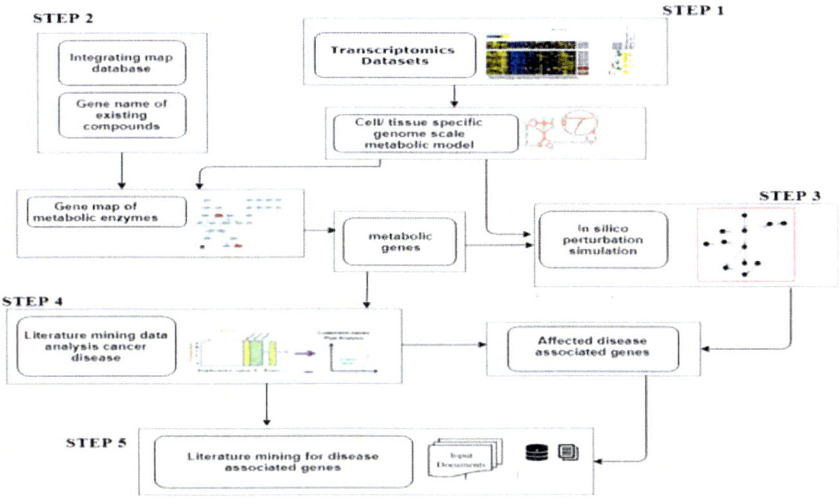

Fig. 1 Figure demonstrates the approach we utilized for metabolic model construction using the transcriptomic datasets

and quantify disease-associated genes in silico while doing in silico perturbation simulation. We examined differential gene expression and differentially expressed metabolic genes (set-top 35% of expressed genes) in the cells of ovarian cancer between control and healthy samples using literature mining data analysis. Different metabolic genes and pathways were found by enrichment analysis. By integrating more datasets, we were able to validate our findings with literature and aid in the discovery of new data (Fig. 1).

3 High-Throughput Data Integration

From the GEO [4] database, we collected bulk RNA-seq data [3] of cancer cells from the ovary. A total of six samples were chosen that were associated with ovarian cancer cells. The data were analyzed with the help of "limma," "Deseq," "edgeR" packages in the Bioconductor of R-programming language. To locate the differentially expressed genes (DEGs) between control and treatment patients, limma package is utilized. We aimed to characterize gene expression levels; therefore, we mixed samples from several cell types. Gene expression data were collected from the dataset GSE130350 present in the gene expression omnibus portal. This dataset contains three biological replicates for GFP and three biological replicates for FTO-GFP, which were downloaded. The fat mass and obesity-associated protein (FTO) were recognized as that of an oncogenic factor in ovarian cancer and reported significant differential expressed genes related to body weight. From all the six samples, expressed values attached to the same group were calculated using mean and standard deviation (SD). The outcome files, one for the control (GFP) as well as one for the treatment (FTO), containing mean and standard deviation for each gene IDs, were utilized as input for the simulation of the model and validation. The false discovery rate was applied for significant enrichment. Next, we extracted single-cell RNA-seq [5] data GSE158937 present in the GEO portal. In this dataset, a total of three samples are there. Analysis of data was accomplished utilizing "Seurat," "dplyr," "cowplot" packages in R. Seurat R package is used to explore single-cell RNA-seq data and to perform QC analysis. We used two datasets to improve the data's resilience, being tested on a similar dataset.

4 Metabolic Modeling and Differential Expression Analysis

To create the metabolic model of GFP and FTO, we used the COBRA toolbox's GIMME [6] technique. The general human model Recon3D [7] was utilized as an input template for GIMME, which was downloaded in the compressed system biology markup language (SBML) format and reconstructed and curated, as well as gene expressions relying upon combined multi-omics data. The reconstruction included the relationship between genes and their reactions, as well as response directionality. Gene–gene reactions in the early Recon3D template employed transcript IDs. Utilizing our data, which comprises gene names, we associated the transcript IDs with the gene names. Aside from that, new reactions were added to the biomass objective function, and a few reactions were removed. Transcript IDs obtained from Recon3D were updated to Entrez gene IDs, and RNA-seq data contain information about genes, so the data collected for genes were mapped. The reconstructed model involved a total of 26,854 genes. The dead-end reactions were also removed, reducing the size of the models. Extracellular, nucleus, cytoplasm, endoplasmic reticulum, Golgi apparatus, mitochondria, and lysosomes are among the compartments where reactions are circulated in the model. The models were explored utilizing leak testing, gene deletion, and iterative methods to accomplish important attributes. Simulating the models along with all the closed exchange reactions and checking for non-zero flux in each reaction, an assessment was done on the leaks. The mass imbalance was rectified if models created metabolites. Gene deletion analysis was used to see if the model could detect gene essentiality. The models were validated using flux balance analysis (FBA). There are 5837 metabolites as well as 10,610 reactions in the model, with an FBA value of 1.722. GFP was the term given to the models (control).

5 Model Validation

The algorithm we refer to as targets is used in FBA approaches to avoid the need for a prior-specific objective function. The algorithm operates the optimization process by

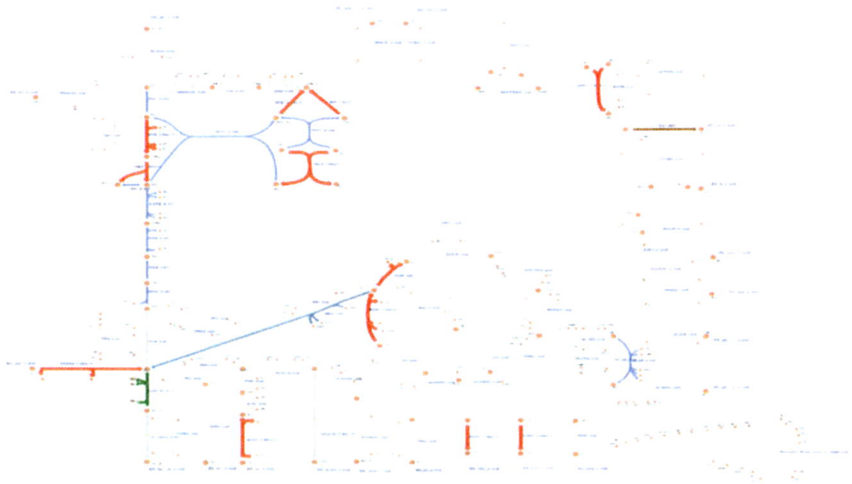

Fig. 2 Escher flux model of active metabolic pathways in obesity and ovarian cancer

enhancing the association between the steady-state arrangements of the flux solutions and comparing gene expression data of the condition under investigation using the available expression data. The algorithm makes use of the constraint-based reconstruction and analysis (COBRA) toolbox, and it was also utilized to import the "Recon3D" GEM in SBML format into the environment. The control and treatment samples' mean and standard deviation values of gene expression were transferred independently. To validate our method, we used the transcriptomics dataset described by the "Recon3D" authors and compared our results to theirs, which were obtained using a standard FBA method. As a result, we integrated the flux balance analysis approach in all the models to derive the flux distribution networks under treatment conditions and to assess the non-zero fluxes across the previously established pathways. Escher Web [8] applications were used to build flux maps. The flow distribution metabolic pathways from genome-scale metabolic models are visualized using Escher (Fig. 2). The links were entered into an Escher map and examined using topological features. Glycolysis, TCA cycle, glucose to lactate conversion, and model fluxes are depicted in this Escher map. The colors represent the reactive flux in the model that was created. Gray arrows show the zero flux. The non-zero fluxes are represented by the blue, red, and green colors.

6 Metabolic Differentially Expressed Genes in Cancer Disease

To get gene expression data from the GEO database, bulk RNA-seq and single-cell data were employed. The relevant raw files were retrieved from GEO portal datasets

GSE158937 [9] and GSE130350 [10]. The procedure is enhanced further by classifying the most recent datasets as the major dataset of interest and the primary cell types as blood or PBMC. For single-cell RNA-seq, the data are in the matrix, barcode, and feature file format. We used a two-fold cutoff for substantial differentially expressed genes. For datasets where two-fold resulted in a low number to zero differentially expressed genes, a cutoff of 1.5 overlaps was used. We utilized DAVID V6.8 [11] and the STRING database, as well as Gene Ontology, KEGG, and REACTOME pathways, to analyze biological processes enrichment. For different testing adjustments, a *P*-value of significance level 0.05 along with a false discovery rate limit of 5% was utilized.

7 Results

The methodology involved a number of steps: (1) using a transcriptomics dataset for constructing a metabolic model, (2) recognition of metabolic genes, (3) in silico perturbation focuses on preexisting information to establish whichever reactions are affected, (4) finding genes that have been differentially expressed in cancer patients and incorporating them with flux ratios, (5) and verifying with literature and predicting novel targets. The utilization of two approaches of RNA-seq that are routinely employed to research onco-immunology is the single-cell RNA-seq approach and bulk RNA-seq approach in statistical studies of gene expression. Bulk RNA-seq and scRNA-seq data analysis combine expressed genes list. Also, using Cobrapy plus Cobra toolbox, the model was built, and the base template was constructed using Recon3D model. The majority of bulk RNA-seq expression esteems are negativebinomial or log-normal distributions and a zero-expanded distribution from scRNA-seq. As a result, scRNA-seq was designed to investigate a specific cell's or cell type's behavior in response to its microenvironment, whereas bulk RNA-seq wasn't really designed to characterize a cancer formation in detail. RNA-seq analysis of bulk seq data contains three biological replicates for GFP and three biological replicates for FTO-GFP, ovarian cancer cells. Patients with high-grade serous ovarian cancer had their single-cell RNA-seq data evaluated as well (Figs. 3, 4, 5, and 6).

The voom plot shows how the coefficient of variation of the count relies upon the count size. Black spots with a LOWESS pattern are used to represent gene-wise means and fluctuations in RNA-seq data. Plots are arranged by expanding levels of biological variation in datasets. Voom trends for GFP (control) and FTO (treatment) genes for six different samples (Fig. 3a). Volcano plot commonly plots some proportion of impact on *x*-axis (normally the fold change) in addition to the statistical importance on *y*-axis (ordinarily the—log10 of the *p*-value). Highly dysregulated genes are located more to the left and right sides of the plot, whereas massive substantial alterations appear to be higher on the plot (Fig. 3b). A MA plot with a substantial proportion of information focuses falling over the one edge on the *y*-axis would show a higher amount of genes being upregulated, whereas more than 1 would show

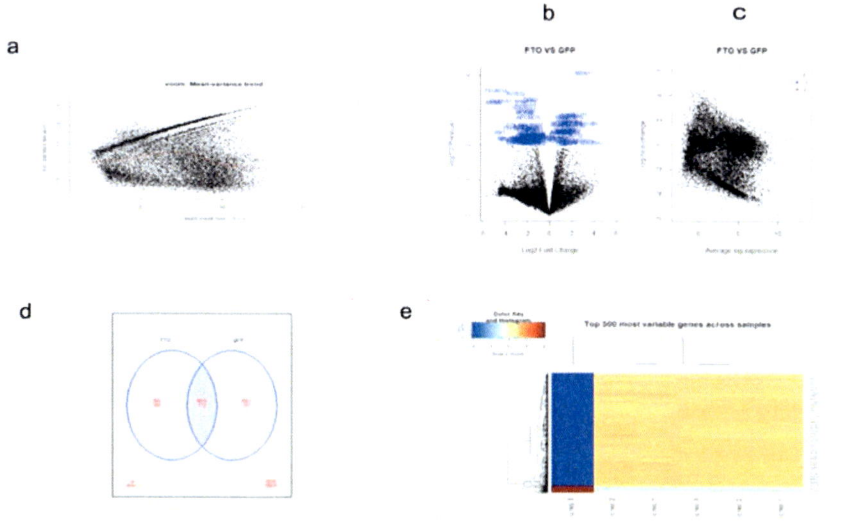

Fig. 3 **a** Voom plot depicts the mean variance relationship. **b** Volcano plot considering differential expression data. **c** MA plot of differentially expressed genes. **d** Venn diagram between differentially expressed genes in FTO versus GFP. **e** Heatmap of differential expressed genes

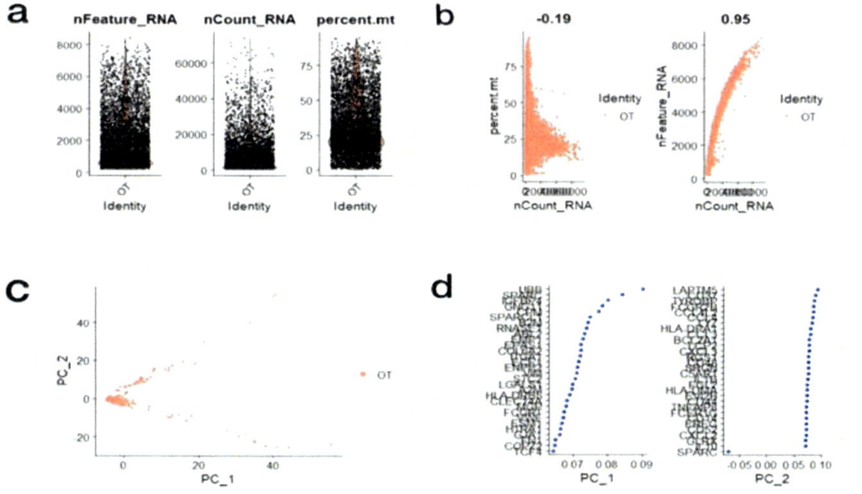

Fig. 4 **a** Plot shows QC matrices. **b** The plot depicts the gene–gene relationship which can also be used as well for object-related calculations. **c** The dimplot visualizes both cells and genes. **d** The plot depicts principal component analysis results

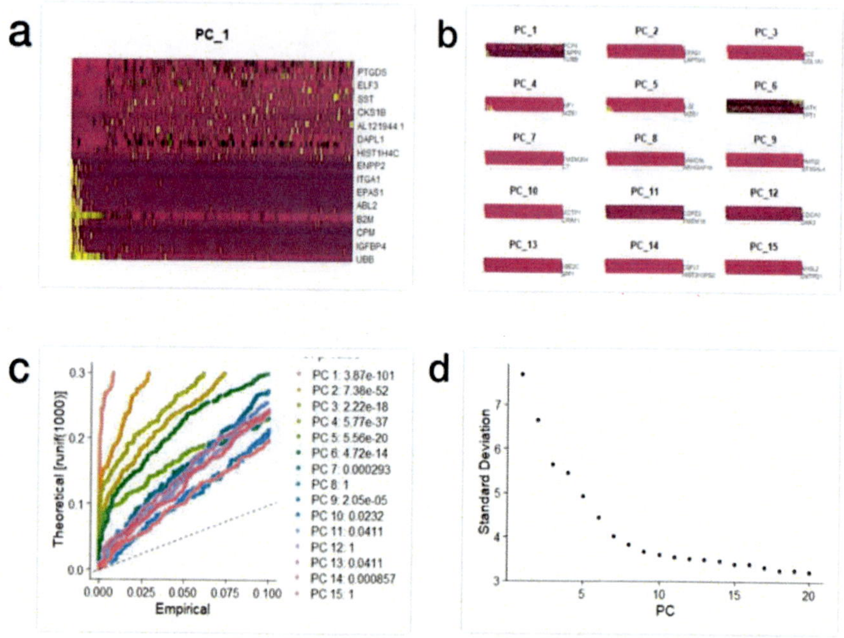

Fig. 5 **a** Heatmap permits easy exploration of primary heterogeneity sources in the dataset. **b** This heatmap shows multiple repetitions until the final expression is reached. **c** The significant principal component will reveal a large enrichment of genes with low *p*-values (solid curve just above the dashed line segment). PCs 1–15 appear to be crucial in this scenario. **d** Elbow plot falls around PC 8 showing the change in expression. Here, it shows the ranking of the principal component data based on its variance percentage

increased amounts of gene downregulation (Fig. 3c). Venn diagram is showing differentially expressed genes in FTO and GFP. The overlap of differentially expressed genes can be seen here. In this, we see that we have 6049 upregulated genes and 16,624 downregulated genes in FTO versus GFP patients in ovary cells. Commonly, differentially expressed are 8074 and 112 (Fig. 3d). Differentially expressed gene (DEG) analysis is used to locate differences in gene expression between samples. Nonetheless, genes that add to cell-to-cell contrasts are not discoverable with RNA-seq because RNA-seq samples are obtained from a combination of cells. Single-cell RNA-seq (scRNA-seq) permits the identification monitoring gene expression in every cell. High variable genes (HVGs) revelation allows for the discovery of genes that strongly contribute to cell-to-cell variation using scRNA-seq. Deseq is used for statistical analysis. The blue scale indicates high expression, and the red scale indicates low expression. Top 500 most variable genes across different samples of obese and lean patients. Heatmap of the distance matrix between every single individual GEM, where each line toward the finish of the dendrogram compares to one individual GEM (Fig. 3e).

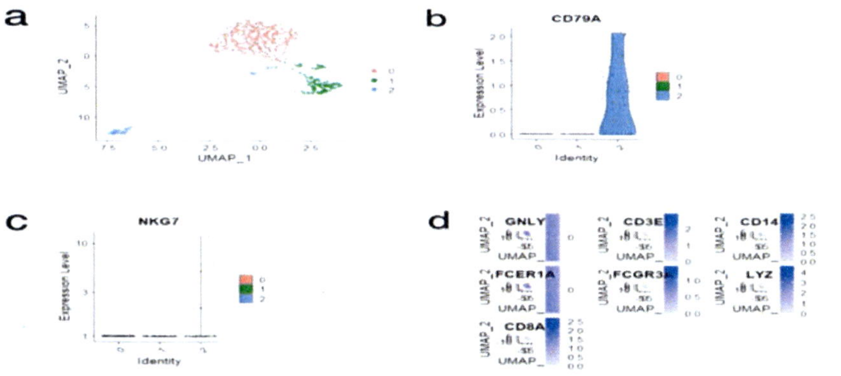

Fig. 6 **a** Plot shows colocalization on the dimension reduction plots. UMAP or tSNE is the two main algorithms that can be employed. This is because the UMAP expects to group cells in low-dimensional space that have similar neighboring areas within high-dimensional region. **b, c** The VlnPlot shows expression probability distributions across clusters. **d** The feature plot depicts visualization of the extent of the expression of differentially expressed genes on a UMAP or PCA plot

Quality check matrices are displayed, and cells are filtered based on gene and molecular counts and relationships, with cells with an obvious outlier exception group of genes identified as probable multiplets being removed. Mitochondrial genome mapping percentage reads cells are filtered that have greater than five percent, mitochondrial count (Fig. 4a). Gene–gene relationship is shown but is also used for object-related calculations. This is a rather uncommon subpopulation of cells having a high mitochondrial proportion. Principal component analysis is indeed a dimensionality-reduction approach that decreases the dimensionality of large datasets by condensing a set of variables into a smaller group that preserves the majority of the data in the larger set (Fig. 4b). Justification of the principal component analysis through dimplot visualization is depicted in Fig. 4c. The results of principal component analysis on gene expression are depicted in the given plot (Fig. 4d). From this graph, we get different expression of genes that are accountable for obesity and ovarian cancer-ubiquitin B (UBB), insulin-like growth factor-binding protein 4 (IGFBP4), secreted protein acidic and cysteine rich (SPARC), G protein subunit gamma 11 (GNG11), carboxypeptidase M (CPM), SPARC-like protein 1 (SPARCL1), B2M, epithelial membrane protein 1 (Emp1), endothelial PAS domain protein 1 (EPAS1), collagen VI (COL6A2), Integrin α1 (ITGA1), ECE-1, ectonucleotide pyrophosphatase (ENPP2), VIM, STC2, galectin 1 (LGALS1), HLADRB5, CLEC14A, matrix Gla protein (MGP), von Willebrand factor (vWF), endothelial cell-specific molecule-1 (ESM-1), HTRA, gap junction protein alpha 1 (GJA1), fibronectin 1 (FN1), coat complex subunit zeta 2 (COPZ2), TCF2.

8 Discussion

Obesity-related adipose tissue dysfunction gets triggered by a collection of signaling pathways that impair mechanisms related to the storage of energy and thus participate in promoting tumor cell growth. Leptin [12] had large metabolic effects on various bodily tissues adding to its influence on satiety. The leptin receptor (LEP-R) aids in the monitoring of leptin overexpression in cancer, allowing leptin to act on the transcription activators Janus kinase and signal transducer (JAK/STAT) and PI3K/Akt/mTOR. The procedure described above reshapes internal metabolic processes as in previous research. Finally, it affects a few key metabolic pathways involved in cell synthesis and proliferation. Restriction of leptin to LEP-R, for example, stimulates JAK, resulting in STAT3 phosphorylation as well as dimerization. Dimerization of STAT3 makes it travel toward the core in which it plays the role of a record factor that upregulates a variety of glycolysis-related characteristics, providing indications that this mechanism is HIF-1α-dependent. Cooperation between STAT3 and HIF-1α is anticipated to build a complete response to hypoxic environments. The recruitment of insusceptible cells to adipose tissue is also accelerated by obesity-related metabolic dysregulation. Expression of monocyte chemoattractant protein 1 significantly rises in the adipose tissue of obese patients, causing an inflow of monocytes to be separated into macrophages. The ensuing spike of cytokine release, particularly TNF, appears to be a cause of insulin resistance in adipose tissue due to obesity by accelerating IR substrate-1 serine phosphorylation and, as a result, obstructing downstream insulin signaling. Insulin blockage of adipocytes induced by cytokines generates compensatory hyperinsulinemia, which promotes carcinogenesis via the previously described components. Adipocytes with insulin resistance also have persistent basal lipolysis. This occurs in the influx of free unsaturated fats (FFAs), which may provide additional substrate to cancer development cells via exogenous free unsaturated fats uptake. The adipocyte-saturated free unsaturated fats also act as ligands for toll-like receptor 4, which activates the nuclear factor-kappa B (NF-B) signaling pathway within macrophages to keep the proliferative movement and cytokine production rolling. Remarkably, cancer cells could have a paracrine lipolytic effect upon nearby adipose tissues, supplying vitality to cancer cells and fostering cell invasion and metastasis. Moreover, cytokines do have a number of legitimate effects on the cancer cells of the ovary. Interleukin 6 (IL-6) adheres to the interleukin 6 receptors and activates the Janus kinase and signal transducer and activator of transcription pathways, in the same way as leptin does. Organization of IL-6 is shown to accelerate glycolysis in vitro via STAT3-subordinate activating hexokinase 2 as well as PFKFB3 amid STAT3-mediating glycolysis. By transmitting impulses downstream to its receptors, NF-κB is triggered by IL-6, IL-1β, TNFα, and FFAs [13]. By acting as a transcription factor over many genes with pleiotropic tumorigenesis, NF-κB is a significant modulator of inflammatory signals. NF-κB also rebuilds metabolic processes to promote cellular proliferation, including the activation of oxygen-consuming glycolysis with angiogenesis [14]. Such effects are primarily achieved to a limited extent via the NF-κB pathway's transcriptional and posttranslational activation of HIF-1α.

Upregulation of transcription of several cytokines, for instance, IL-6 and TNF, is typical of JAK/STAT3 and NF-κB signaling. While these cytokines might stimulate those pathways which promote their production, this propagates proinflammatory signals. Ultimately, insusceptible cells inside the adipose tissue ambiance release responsive oxygen species in an oxidative explosion to damage foreign or harmful cells. Such radicals are thought to contribute toward DNA damage and mutagenesis of surrounding tissues. This indicates how obesity-related inflammation inside the adipose tissue potentially plays a role in tumorigenesis.

9 Conclusion

We propose a precise method for understanding complicated diseases that rely on the reconciliation of gene expression knowledge to generate genome-scale metabolic models in this research. We investigated the correlation between obesity and ovarian cancer using expressed genes to develop and analyze models as part of a contextual analysis. We observed a lot of boundaries and noise when examining the single-cell RNA-seq data and the bulk RNA-seq data, and we had to alter the dataset as a result. We had to manually evaluate and filter out the genes because there are no surefire techniques to assess any biological information; thus, we used R as a key tool for analysis. The dataset (GSE130350) utilized to evaluate and create a model has 10,610 reactions and 5837 metabolites. A flux score of 1.72 was generated from the model. The varied plots allowed us to gain diverse insights into the data, allowing us to better understand and construct models. We used a combination of MATLAB's COBRA toolbox and Python's COBRApy to create a model and analyze the data. Along these lines, a combined effect could determine the cancer prognosis associated with obesity that is substantially poorer. Furthermore, we discovered significant changes in the glucose and glutamine influencing responses during the analysis, which will be included in the future model.

We are also working with the wet lab, which will help us validate the results and locate fresh data, as well as complete more examinations and approvals, such as extra examinations by integrating more datasets and healthy control tests, as well as using various illness models. Regardless, we believe that this study addresses a significant verification of ideas when it comes to fresh frameworks and science integrative approaches that will help us better understand complicated natural biological systems.

Acknowledgements We would really like to thank Dr. Ashok K. Chauhan, the Founder President of Amity University Uttar Pradesh, for allowing us with the opportunity to conduct our research. We would also like to acknowledge the Centre for Computational Biology and Bioinformatics at the Amity Institute of Biotechnology for providing the resources for this study.

References

1. Baart GJE, Martens DE (2011) Genome-scale metabolic models: reconstruction and analysis. In: Methods in molecular biology, pp 107–126.https://doi.org/10.1007/978-1-61779-346-2_7
2. Varma A, Palsson BO (1994) Metabolic flux balancing: basic concepts, scientific and practical use. Bio/Technology 12(10):994–998. https://doi.org/10.1038/nbt1094-994
3. Barrett T, Wilhite SE, Ledoux P, Evangelista C, Kim IF, Tomashevsky M, … Soboleva A (2012) NCBI GEO: archive for functional genomics data sets—update. Nucleic Acids Res 41(D1):D991–D995.https://doi.org/10.1093/nar/gks1193
4. Kuksin M, Morel D, Aglave M, Danlos FX, Marabelle A, Zinovyev A, … Verlingue L (2021) Applications of single-cell and bulk RNA sequencing in onco-immunology. Eur J Cancer 149:193–210. https://doi.org/10.1016/j.ejca.2021.03.005
5. Chen G, Ning B, Shi T (2019) Single-cell RNA-Seq technologies and related computational data analysis. Frontiers in Genetics, 10. Published. https://doi.org/10.3389/fgene.2019.00317
6. Becker SA, Palsson BO (2008) Context-specific metabolic networks are consistent with experiments. PLoS Comput Biol 4(5):e1000082. https://doi.org/10.1371/journal.pcbi.1000082
7. Brunk E, Sahoo S, Zielinski DC, Altunkaya A, Dräger A, Mih N, … Palsson BO (2018) Recon3D enables a three-dimensional view of gene variation in human metabolism. Nat Biotechnol 36(3):272–281.https://doi.org/10.1038/nbt.4072
8. Rowe E, Palsson BO, King ZA (2018) Escher-FBA: a web application for interactive flux balance analysis. BMC Syst Biol 12(1). https://doi.org/10.1186/s12918-018-0607-5
9. Izar B, Tirosh I, Stover EH, Wakiro I, Cuoco MS, Alter I, … Regev A (2020) A single-cell landscape of high-grade serous ovarian cancer. Nat Med 26(8):1271–1279.https://doi.org/10.1038/s41591-020-0926-0
10. Huang H, Wang Y, Kandpal M, Zhao G, Cardenas H, Ji Y, … Matei D (2020) FTO-Dependent N6-Methyladenosine modifications inhibit ovarian cancer stem cell self-renewal by blocking cAMP signaling. Cancer Res 80(16):3200–3214. https://doi.org/10.1158/0008-5472.can-19-4044
11. Huang DW, Sherman BT, Lempicki RA (2008) Systematic and integrative analysis of large gene lists using DAVID bioinformatics resources. Nat Protoc 4(1):44–57. https://doi.org/10.1038/nprot.2008.211
12. Ramos-Lobo AM, Donato J (2017) The role of leptin in health and disease. Temperature 4(3):258–291. https://doi.org/10.1080/23328940.2017.1327003
13. Ben-Neriah Y, Karin M (2011) Inflammation meets cancer, with NF-κB as the matchmaker. Nat Immunol 12(8):715–723. https://doi.org/10.1038/ni.2060
14. Tabruyn SP, Griffioen AW (2008) NF-κB: a new player in angiostatic therapy. Angiogenesis 11(1):101–106. https://doi.org/10.1007/s10456-008-9094-4

Analysis of China's New Media, Regulations, and Violation of Human Rights

Yashi Shrivastava, Rishabh Malhorta, and Gagandeep Kaur

Abstract The famous line given by George Orwell "Until they became conscious they will never rebel, and until after they have rebelled they cannot become conscious." There is no country that doesn't follow certain regulations on their e-platforms. These regulations are justified only till it is helping to maintain peace and tranquility of the nation. These regulations should not be such as to trample on the rights of individual to have freedom of speech and expression. In this paper, author has analyzed the China's new regulations dealing with social media and cybercrime and to what extent these regulations are stringent in order to control the media and violation of fundamental right of freedom of speech and expression which is against the basic human right, briefly discussed PIPL regulation.

Keywords China · Violation of human right · Freedom of speech and expression · Personal information protection law · Censorship

1 Introduction

The strict policy of Chinese censorship has been there from a long period of time. **"The Great Firewall of China"** is the policy adopted by China to control the social media like Face book and Google Scholar. This censorship is basically because a riot was broke out in July 2019 in western province of Xinjiang [1].

The work of censorship is mostly delegated to the private companies, and the government doesn't take responsibility publicly for this censorship [2]. The Chinese authorities often ban the letters or statements which are controversial. This ban often

Y. Shrivastava · R. Malhorta (✉) · G. Kaur (✉)
University of Petroleum and Energy Studies, Dehradun, India
e-mail: malhotrarishabh15@stu.upes.ac.in

G. Kaur
e-mail: gkaur@ddn.upes.ac.in

Y. Shrivastava
e-mail: 500096563@stu.upes.ac.in

P. Dutta et al. (eds.), *Emerging Technologies in Data Mining and Information Security*, Lecture Notes in Networks and Systems 490,
https://doi.org/10.1007/978-981-19-4052-1_62

lead people into microblogging (short and frequent blogs to escape the censorship) [2].

There have been instances in China in the past where people have come together and took the help of social media which has forced Chinese government to take action against the corrupt officials that where harming the society. This phenomenon is called as "human flesh search engine" where the power of social media combined with cloud sourcing [2]. There are netizens known as "fifty cent army." The government pay them for their posts that are pro-government [2]. Hence, in this paper, the author will discuss in detail of the concept of control of Chinese government over the society and recent legal perspective for the same.

2 Control Over Society

The control by Chinese authorities goes way back. The era of Ming Dynasty was the darkest period where the scholars were controlled and were forced to work under the imperial house [3]. In modern era, this control is still prevailing, the Uighur Muslims that are citizens of China face such repression by individual to individual surveillance. This is basically to keep a check on anyone who is showing any active faith to Islam [4]. The launch of the Golden Shield project in 1997 changed the entire scenario of Internet. There are regulatory bodies that control the online content, news media, television broadcast, etc. The regulation on administration of publication regulates body that oversees all the online publications across the country. These regulatory bodies restrict the action of online press on the basis of constitution [5].

3 Latest Chinese regulations—An Overview

The recent report by the freedom house's freedom on the net 2021 held China as the worst abuser of freedom of Internet. The first comprehensive legislation of China on personal information protection lays down the scope of the legislation. The new law on personal data protection for the first time at national level has broadened the grounds on which the personal data have to be processed [6]. The law is introduced to control on that how companies will collect, store, and use the personal data. The law of personal information protection data will be working together with cybersecurity law and the data security law to establish a broader regulatory framework for governing cybersecurity and data privacy protection in China. This law is not just going to affect the domestic companies but even the international companies [7]. The first comprehensive legislation of China on personal information protection lays down the scope of the legislation. It clarifies the legal bases on which the personal data will be processed. It lays down the responsibilities and duties of the processors. The laws lay the stringent requirements on data localization, safeguarding the interest of China in the case of cross-border transfer of personal information [7]. The law has extra territorial application. It covers the processing of personal data within the China and the data located outside [8]. The personal information protection law it is given that the company even if they located outside the territory of China will be bound by the rules and regulations of Chinese law [9]. The concept of personal information processor was introduced which control all the individual processors [10]. The law on personal information protection law sets out detailed requirements for notifications and consent. The processor needs to take consent from individual in certain personal information [11]. The law of personal information on personal data provides the processors obligations in handling the personal data [11]. Under Article 58 of the personal information on protection law, the processor that provides an Internet platform needs to build a further robust data compliance program and needs to establish a body to supervise such implementations [11]. The PIPL law provides with obligations for the special processing activities. It provides with joint processing and entrusted processing [12]. The PIPL also imposes tortuous liability on the processor if they in-fringe the rights and interest of personal information. The burden of proof lies on the processor in a civil action to facilitate damage [13]. If such infringement affects large number of individuals, then the processor may face civil or criminal charges that are brought by the consumer groups [5]. These guidelines have given full authorities in hand of Chinese government to exploit the freedom of speech and expression. The terms such as public safety and national security leaves a big gap as it is not clear as to what situations will be taken as to be against the national security [14].

4 Conclusion

These new rules and regulations brought by China are a clear violation of human right. In the name of data protection, China just wants to keep surveillance on the data of individual person. These laws are against the internationally recognized right of speech and expression.

References

1. Bamman D, O'Connor B, Smith N (2012) Censorship and deletion practices in Chinese social media. First Monday
2. Fadino D (2020, September 14) Hong Kong and China: two systems of social media, one country. Retrieved from https://www.digitalamerica.org/hong-kong-and-china-two-systems-of-social-media-one-country-daniel-fandino/.
3. Mishra V, Polcumpally AT (2021) Understanding Chinese media censorship: From Ming to Jinping
4. Roth K (2020) World report 2020. Retrieved from https://www.hrw.org/sites/default/files/world_report_download/hrw_world_report_2020_0.pdf
5. Foot R (2021) R2P sidelined: the international response to China's repression of Muslim minorities in Xinjiang. Global Responsib Protect 13(1):29–32

6. Cook S (2021, October 12) On Internet freedom china and Taiwan are world apart. The Diplomat. Retrieved from https://thediplomat.com/2021/10/on-internet-freedom-china-and-taiwan-are-worlds-apart/
7. Guo B, Kelly C, Tang P, Li B (2021, September 22) China personal information Protection law will become effective soon. Retrieved from https://www.whitecase.com/publications/alert/china-personal-information-protection-law-will-become-effective-soon
8. Personal Information Protection Law 2021 a. 3 (CN)
9. Personal Information Protection Law 2021 a. 53 (CN)
10. Personal Information Protection Law 2021 a. 73 (CN)
11. Personal Information Protection Law 2021 (CN)
12. Personal Information Protection Law 2021 a. 20 (CN)
13. Personal Information Protection Law 2021 a. 66 (CN)
14. Ruan L, Knockel J, Crete-Nishihata M (2021) Information control by public punishment: the logic of signalling repression in China. China Inf 35(2):133–157

GDPR Oriented Vendors Contracts in Relation to Data Transfer: Analysis of Standard Clauses 2010 and 2021

Gagandeep Kaur, Rishabh Malhorta, and Vinod Kumar Shukla

Abstract Nowadays, data protection and its laws have been in an unstoppable debating. Every country have established their own data protection law in respect to European Union Global Data Protection Regulation (GDPR). Each and every organizations of any sector have to comply with some global standards whilst dealing with some other organizations which have been termed as vendors and to ensure them that they have took appropriate measures for protecting the information of consumers or employees. Standard contractual clause has been established by European Union (EU) in which two major terms have been discussed under which liabilities have been defined. These two major terms, i.e. Data Processors and Data Controllers discussed in whole contract. In 2021, new Standard Contractual Clauses (SCC) has been published which has defined the similar liabilities of Data Processors and Data Controllers unlike of 2020 SCC, the liabilities are different. In this paper the author have analysis both 2020 and 2021 Standard Contractual clauses and their implications.

Keywords Standard Contractual Clauses · GDPR · European Union · Data Processors · Data Controllers

1 Introduction

Every organization collects numerous personal information of consumers (name, email, address, phone number and so on). Sometimes one organization hires or contract with other organizations to collect personal information of consumers and

G. Kaur (✉) · R. Malhorta (✉)
University of Petroleum and Energy Studies, Dehradun, India
e-mail: gkaur@ddn.upes.ac.in

R. Malhorta
e-mail: malhotrarishabh15@stu.upes.ac.in

V. K. Shukla
Department of Engineering and Architecture, Amity University, Dubai, UAE

P. Dutta et al. (eds.), *Emerging Technologies in Data Mining and Information Security*, Lecture Notes in Networks and Systems 490,
https://doi.org/10.1007/978-981-19-4052-1_63

provided to them for processing the same. The entity involved in collecting the personal information considered as "Data Controller" and the entity involved in processing the personal information considered as "Data Processor" [1]. The information are being collected of consumers around the globe, hence in order to transfer the data from one country to third country there are certain guidelines described in Global Data Protection Regulation (GDPR). This GDPR applies to only organizations which are based out in European Economic Area (EEA) or the information transferred from EEA to third countries. As per GDPR Article 44 provides the common principles related to transfers of information from controller to processor (EEA to third/non-EEA countries or third/non-EEA countries to EEA) [2]. Additionally, Sub-Article 1 of Article 46 (GDPR) explains that controller/processor can transfer personal information to non-EEA country, if the controller/processor have applied suitable safeguards for protecting the information with respect to Article 46 (3) in regard to appropriate safeguards provided by controller must be authorized by competent supervisory authority and also be detailed in SC clauses between the controller and the processor or the recipient of the personal information/data in the non-EEA country or international organization [3].

Here, the concept of Standard Contractual Clauses (SCC) comes into play as provided by GDPR and mandatory to have contract for the same between the controller and processor where needs to be define the liabilities of controller as well as processor, jurisdictional clause, criteria of collecting as well as transferring of information with regard to list of appropriate safeguards. These SC clauses apply to every organization that is based out in EU and transferring to non-EU countries. But there are certain countries which do not fall under the ambit of EU, i.e. non-EU countries and does not apply Standard Contractual Clauses. Recently, the European Commission on 4 June 2021 has issued 2 new sets of Standard Contractual Clauses with the aim to replace the older one (2010 SCC). New 2021 clauses reflect certain changes which is implemented with the EU's latest privacy law, i.e. General Data Protection Regulation (GDPR) of 2018 [4].

New SC clauses (2021) come in one document but divided into 4 modules and provided a detailed view on the liability of controller and processor with respective to each modules. In this paper, author briefly described the purpose of new 2021 clause and their timeline, the major differences between 2010 and 2021 clauses.

2 Standard Contractual Clauses—An Overview

2.1 **Meaning**: In general, Standard Contractual Clauses as the term suggest providing contractual terms and conditions between the two parties ensuring certain rights and obligations of each party and in case of non-compliance how it can be resolved. But in case of personal information transfer from European Union (EEA) countries to non-EEA countries, it defined as standard sets of terms and conditions laid down in contract or agreement which both the controller (sender) and the processor (receiver) of the personal data sign up to and make sure that the rights and freedoms of the data subject (individual) are considered and upheld [5].

2.2 **Purpose**: Standard Contractual Clauses (SCCs) are considered very important in the domain of personal information/data protection by donating towards a consistent method which concerns cross-border processing might affects the free-flow of personal information/data in the EU/European Economic Area (EEA) permitting for the steady implementation of specific provisions of the GDPR's. In other words, aim of SCCs to provide protection to personal data whilst departure from the EEA and to those countries who don't have a sufficiency decision and unable to afford equal level of security to personal information/data [6]. For example, an organization gathers a consumer (Data Subject) personal information and handovers it to other organizations established in a country where provisions of GDPR does not apply. In that case, the consumer (data subject) has a risk of losing the protection granted by GDPR over that information/data which includes the ability to exercise their data subject rights. By the help of SCC, these two organizations can make the cross-border transfer subject in a legally-binding contract/agreement constituting the clauses ensuring that the third or non-EEA country recipient will protect the personal data. Hence, this example gives a clear picture the need of Standard Contractual Clauses (SCC) whilst transferring the personal information of any consumer [7, 8].

2.3 **Relation with GDPR**: The main aim of General Data Protection Regulation (GDPR) are to safeguard the fundamental rights and freedom of individuals and in specific their rights in protecting of personal information; free movement or free-flow of personal information within the ambit of Union—neither restricted nor prohibited unless-n-until for reasons associated with the safeguarding of processing personal information of natural persons; enlisting down the rules in relation to the protection of individuals—processing of personal data and rules relating to the free-flow of personal information [9]. Whereas SCCs are a contractual obligations for aiding people residing in EEA to preserve their rights and control over their personal information even when it leaves the EEA. Moreover, Article 45 and 46 (1) and (3) of the GDPR provide some principle to be followed whilst transferring the personal information; appropriate safeguards need to be applied at both the ends of transfer, regard to safeguards there should be a contract between the controller and processor mentioning the details of safeguards applied which needs to authorized by competent supervisory authority. Hence, Article 46 (3) of GDPR clearly states the relation of GDPR with the Standard Contractual Clause (SCC) [10].

3 Difference Between SCC 2010 and SCC 2021

In June 2021, the new standard contractual clauses were released by European Commission which has replaced the old 2010 standard contractual clauses. The reason behind the upgradation in SCC was that the 2021 SCC has widened the purpose and scope whilst transferring of personal information whereas in the 2010 SCC only considered for proposing satisfactory safeguards in respect to the protection of the privacy, fundamental rights and freedoms of natural persons and also consider the exercise of the equivalent rights as essential by Article 26 (2) of Directive 95/46/EC. Moreover, the 2010 SCC came in 2 separate documents which are (i) cross-border transfer of personal data from Controller to Controller (C2C); (ii) cross-border transfer of personal data from Controller to Processor (C2P). However, in 2021 SCC, it comes in one document divided into 4 modules to account for four cross-border scenarios. There are numerous changes been made in new 2021 SCCs as compared to 2010 SCCs, some of the major differences have listed in Table 1 [11–14].

4 Conclusion

The Standard Contractual Clauses plays an important role in cross-border transferring of personal information of the consumers from EEA to non-EEA countries. This clause ensures the privileges and freedoms of the individual to be protected with appropriate safeguard. SCC is mandatory as per GDPR Article 46(3) which provides the rights, obligations and much more related to personal information of consumers. The old 2010 SCCs does not give the detail view of the clauses what the processor or controller needs to do and how but after the establishment of new 2021 SCCs by European Commission widen the purpose and scope of SCC. Moreover, several clauses such as "Docking Clause" inserted, also, the contractual document come into in one document under which it divided into 4 modular approach, i.e. Controller to Processor (C2P), Processor to Controller (P2C), Processor to Processor (P2P) and Controller to Controller (C2C). However, the new 2021 SCCs provides a benefit to the organization for forming new contracts but the issue is that to convert old contracts with respect to new contractual clauses will be a difficult task. The important thing which new 2021 lacking is that Technical and Organizational Measures which are listed out in Annexure II only applicable to Module 1, 2 and 3 not to Module 4 which needs to be inserted.

Table 1 Difference between 2010 SCC and 2021 SCC

	SCC 2021	SCC 2010
Modular approach	SCC 2021 Guidelines com-bine several general clauses with a modular approach to offer a range of transfer situations. The following trans-missions to non-EEA countries are enclosed: Module I: C2C Module II: C2P Module III: P2P Module IV: P2C	Processor to Processor (P2P) and Processor to Controller (P2C) part were not previously prophesied in the 2010 SCCs
Docking clause	The 2021 SCCs encompass a docking clause which would permit additional data exporters or importers to be included As per the 2021 clauses, during the course of the contract even to those who are not a party to these clauses by completing the Appendix and signing Annex I.A. (Clause 7)	The 2010 clauses deal with bipartite contracts/agreements—with-out any exceptional clauses for addition of parties to join directly
Liability	In New SCC, liability was defined module wise, where C2C and P2C liability are same and data importer may not get invoked by the conduct of a processor or sub-processor to evade their own liability Similarly, C2P and P2P liabilities are same and the data importer may not invoke the conduct of a sub-processor to avoid its own liability	In 2010 SCC, data importer and data exporter liabilities were different. As per Clause 6, The data importer may not rely on a breach by a sub-processor of its obligations in order to avoid its own liabilities Clause 7 in old SCC list out the grounds on the basis of which the case can be referred to mediation or courts
Jurisdiction	Clause 18 in new SCC provides the choice of forum and jurisdiction	Clause 7 in old SCC list out the grounds on the basis of which the case can be referred to mediation or courts
Extended geographical reach	It comprise no express limitation when comes to the location of the data exporter The 2021 SCCs generally be applied by a data exporter subject to GDPR—regardless of location	The data exporter must be established in the EU, making the tool absent for a data sender recognized outside the EU but still (for Instance) subject to the GDPR under the authority of Article 3(2) GDPR

(continued)

Table 1 (continued)

	SCC 2021	SCC 2010
Transfer impact assessments	The parties (controller and processor) should carry out a "Transfer Impact Assessment (TIA)", which also be made accessible to the capable supervisory authority, up-on request Clause 14 of the new 2021 SCCs appears to be closer to standard risk-based approach of the GDPR, sanctioning for definite experience-based elements to be part of the Assessment	Not foreseen in OLD SCC
Technical and organizational measures	In 2021 clauses, Annex II has been attached in the last which deals with Technical And Organizational Measures and laid down the possible measures list to be included for ensuring security of data	Not foreseen in OLD 2010 SCC
ANNEX I	In New SCC, Annex I consists of: A. List of Parties B. Description of Transfer C. Competent Supervisory Authority	In old SCC, Annex have Standard Contractual Clauses (Processors) general format

References

1. Ivanova Y (2020) Data controller, processor, or joint controller: towards reaching GDPR compliance in a data and technology-driven world. In: Personal data protection and legal developments in the European Union. IGI Global, pp 61–84
2. General Data Protection Regulation 2018 (EU)
3. Mantovani M (2020) Contractual obligations as a tool for international transfers of personal data under the GDPR. Available at SSRN 3522426
4. Petroli M (2021, August 9) New standard contractual clauses under the GDPR. Retrieved from https://www.natlawreview.com/article/new-standard-contractual-clauses-under-gdpr
5. Vella S (2019, July 26) Why use the standard contractual clauses? Retrieved from https://www.lexology.com/library/detail.aspx?g=5978ddbe-e156-4c5a-a70f-fdc457cbeb67
6. What are Standard Contractual Clauses? Retrieved from https://www.privacytrust.com/guidance/standard-contractual-clauses.html
7. Bateman R (2021, January 18) Using standard contractual clauses. Retrieved from https://www.termsfeed.com/blog/using-standard-contractual-clauses/#Understanding_Standard_Contractual_Clauses
8. Cory N, Dick E, Castro D (2020) The role and value of standard contractual clauses in EU-US digital trade. Information Technology and Innovation Foundation.
9. Article 1, General Data Protection Regulation 2018 (EU)
10. Whitney N (2021, April 8) GDPR: standard contractual clauses vs binding corporate rules. Retrieved from https://www.grcilaw.com/blog/international-data-transfers-model-contract-clauses-vs-binding-corporate-rules

11. Boardman R (June 2021) Replacement standard contractual clauses (SCCs): European commission. Retrieved from https://www.twobirds.com/en/news/articles/2021/uk/replacement-standard-contractual-clauses
12. Standard Contractual Clauses 2021 (EU)
13. Standard Contractual Clauses 2010 (EU)
14. McGeachy S (2021, June 4) International data transfers: European commission publishes final version of new standard contractual clauses. Retrieved from https://www.burnesspaull.com/insights-and-events/news/international-data-transfers-european-commission-publishes-final-version-of-new-standard-contractual-clauses

Enabling Secure and Transparent Crowd Funding Approach Powered by Blockchain

Anurag Mishra, Harsh Khatter, Gopal Gupta, Aatif Jamshed, and Akhilesh Kumar Srivastava

Abstract Now a day, crowdfunding is the most prominent area which the people are using to generate and create the funds. This is not only popular in developed countries, but also popular in developing countries. In India, the scope of the crowdfunding is increasing exponentially. Blockchain is one of the best way to do the fundraising and Crowdfunding. Many MSE and organizations are practicing the crowdfunding to raise funds for their infrastructures and the projects. The large number of people can easy contribute their funds electronically. Traditionally, the task was done in form of bank loans and partnerships, now the digital world has given so many options to raise the funds. Entrepreneur can easily generate the fund through the crowdfunding platforms. The proposed system addressed the various issues and their solutions to use the crowdfunding platforms. The article shows an approach and methodology which is secure and transparent to raise the fund using blockchain architecture.

Keywords Blockchain · Crowd · Crowdfunding · Entrepreneurship · Fund raising · Smart contracts · Social entrepreneurship · Transparency

A. Mishra (✉) · G. Gupta · A. Jamshed · A. K. Srivastava
ABES Engineering College, Ghaziabad, India
e-mail: anurag.mishra@abes.ac.in

G. Gupta
e-mail: gopalgupta@abes.ac.in

A. Jamshed
e-mail: aatif.jamshed@abes.ac.in

A. K. Srivastava
e-mail: akhilesh.srivastava@abes.ac.in

H. Khatter
KIET Group of Institutions, Delhi NCR, Ghaziabad, India
e-mail: harsh.khatter@kiet.edu

P. Dutta et al. (eds.), *Emerging Technologies in Data Mining and Information Security*, Lecture Notes in Networks and Systems 490,
https://doi.org/10.1007/978-981-19-4052-1_64

1 Introduction

While talking about the entrepreneur, the major issue is of the success of the business. Many initiators back off due to the seed capital to invest and how they will move forward with just an idea to develop [1, 2]. Though, the government has taken various initiatives to promote the skill development and the entrepreneur and start-ups, still the people are struggling in developing countries. As discussed by Refs. [3, 4], the seed capital plays a major role to help a start-up, a successful company. Reference [5] mentioned that the seed capital can be from any source, might be the investors from market, Friends and family financial help, the business angels which have trust on the proposed plans and products, or the venture capital funds. Now a days, the fundraising and crowdfunding is novel method which is added to the list. Figure 1 shows the major contributors of seed capital.

In India, the extent of the crowdfunding is expanding dramatically. Blockchain is one of the most incredible method for doing the raising money and crowdfunding. Numerous MSE and associations are rehearsing the crowdfunding to raise assets for their foundations and the tasks. The enormous number of individuals can simple contribute their assets electronically. Customarily, the undertaking was done in type of bank advances and organizations, presently the computerized world has given such countless choices to raise the assets. Business person can without much of a stretch produce the asset through the crowdfunding stages.

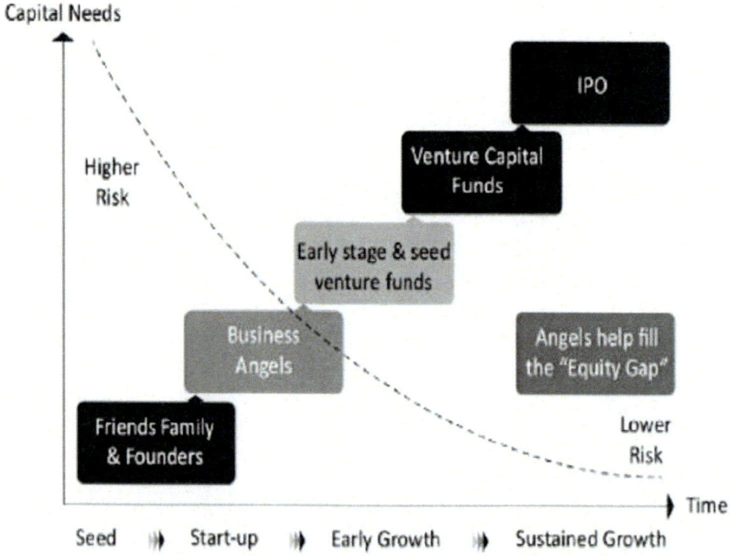

Fig. 1 Capital needs for an emerging business [6]

1.1 Role of Blockchain

Blockchain is one which have driven the digital world into another dimension. There are many incredible methods which are available using blockchain to raise the funds. Crowdfunding is the one which is easily possible through the blockchain. Some of the major applications of blockchain are mentioned in Fig. 2 [7]. The applications can be grouped as:

a. Business and industry
   (a) Energy sector
   (b) Supply chain
b. Data Management
   (a) Human Resource
   (b) Data Distribution

Fig. 2 Applications of Blockchain Architecture

c. Privacy and Security
   (a) Anonymization
   (b) Secure Storage
d. Education
   (a) Certification Management
   (b) Reputation
e. Health Sector
f. Financial Sector
   (a) Cryptocurrency
   (b) Prediction Market Place
g. Governance
   (a) E-Voting
   (b) Notary and Law
   (c) Identity Management
   (d) Public Administration

The scope of the blockchain usage is endless. In the context of the crowdfunding, the blockchain plays a crucial role.

The blockchain is an unchanging record that tracks each exchange. It utilizes shared organization in which all the framework are going about as a customer just as server and every one of the records are put away on each hub. Every one of the exchanges are put away in the blockchain with the assistance of a Smart Contract. Crowdfunding helps in financing a task thought where financial backers can put away cash on a venture. The issue with current crowdfunding frameworks is that they are charging a tremendous amount of cash as an expense, the exchange of cash is not straightforward, and there are additionally instances of trick. We can keep away from these issues and foster a dependable application by fusing blockchain into our application where each exchange is being put away on the blockchain. Every one of the procedure on the blockchain are constrained by a Smart Contract which helps in making this application safer and straightforward [8–11].

2 Existing Systems

Crowdfunding is one of the latest practice in which the digital partners has involved with their policies and terms. Now easily, an individual or an organization can raise funds as the seed capital for their ideas, start-ups and projects. The best thing about the model is the small investor can also contribute in the proposed business idea. All the contributions are through the digital payment options. All check points and conditions are checked and verified to proceed further. Several existing systems are there in market which are helping entrepreneur and start-ups, like Kickstarter.

2.1 Kickstarter

It is New York based company. Kickstarter is one of the online crowdfunding platform where an individual or an organization can place their ideas about music, development of a product, games, comics, technology, etc. The company's stated mission is to "help bring creative projects to life". The people around the world look into the ideas and contribute to a project and helps in raising funds in this reward based crowdfunding system.

2.1.1 Some Features of Kickstarter

- The user has to first register on the website, select the category of project, verify his identity using a government-issued id card, verify his account details, etc.
- The user can launch a campaign and provide details about the campaign like what he wants to develop in this campaign, how much fund he needs to develop the product, the time duration of the project, etc.
- The contributors can read the description of the project and if they like the idea they can contribute to the campaign.
- Once the funds cross the required amount, the funded amount is transferred to the campaign owner's account.
- If the funding does not reach the required amount then the project will be declared as unsuccessful and the backers are not charged any money.

2.2 Related Previous Work

From the last one decade, the boom in the crowdfunding market has been seen. The new companies are coming and convincing people to be a part of this. As of late, crowdfunding has turned into a significant option for creating assets for business visionaries and presently with the assistance of crowdfunding, the sponsor are not restricted to a specific group or space, presently any individual can turn into a supporter by subsidizing a thought over the web.

In the Table 1, fewer recent works and their contributions are shown.

Haas et al. [6] has discussed the ecosystem for the crowdlending services using blockchain architecture [26, 31, 32]. This is the base model for the proposed architecture. The model is shown in Fig. 3.

Kranz et al. [23] discussed the token-based ecosystem for the blockchain and the crowdfunding method. Most of the papers are using the same ecosystem for the crowdfunding using blockchain as shown in the Fig. 4.

The objective of the above study is to get the facts available for the researchers to know in depth about the concept and working areas in crowdfunding and the fundraising using blockchain.

Table 1 Key points of previous research work

S. No	Title	Highlights of study
1	Crowdfunding: Tapping the right crowd [12]	Authors highlighted the issues of the investors as well as the new comers. The decision for the seed capital and the funds are to be done at early stage Authors clearly mentioned that the crowding investors focus on the plans and the managerial decisions
2	Persuasion in crowdfunding: An elaboration likelihood model of crowdfunding performance [13]	Authors discussed the performance model of crowdfunding. The article is for influencing the first time investor and what exactly the first time investor looks in the company to invest Authors used the logistics regression analysis to check and see the patterns of success in crowdfunding
3	Unchaining Social Businesses—Blockchain as the Basic Technology of a Crowdlending Platform [14]	Discussed the issues and the challenges related to the crowdfunding and what are the solutions to the issues Authors also focused on the various types of cryptocurrency and how these are used in market
4	The Applications of Blockchain Technology in Crowdfunding Contract [15]	Authors discussed the role of crowdfunding for business start-ups. In paper, the framework is shown and focused on the security and the effectiveness of it in the system
5	The Application of Blockchain Technology in Crowdfunding: Toward Financial Inclusion via Technology [16]	Authors discussed the scenario and the crowdfunding initiatives in Malaysia. The new ways of crowdfunding using financial technology are represented in the paper The paper highlighted the Islamic crowdfunding methods and compare it with various existing models
6	Crowdfunding the Insurance of a Cyber-Product using Blockchain [17]	Authors discussed the scenarios of cyber risks, and how cyber risks are putting impact on crowdfunding Authors suggested a third party concept to ensure the prevention of cyber risks and the insurance policies
7	Disruption of financial intermediation by FinTech: a review on crowdfunding and blockchain [18]	The review has been done by the authors of around 400 papers and concluded the gaps and the issues in crowdfunding Authors also discussed about the coins system and cryptocurrency concept in terms of crowdfunding using blockchain

(continued)

Table 1 (continued)

S. No	Title	Highlights of study
8	Don't Slip on the Initial Coin Offering (ICO)—A Taxonomy for a Blockchain-enabled Form of Crowdfunding [19]	Authors tried to explain the concept of initial coin offerings (ICO). The crowdfunding has various features and the parameters to work on. The concept of blockchain was well explained in paper and proposed ICO archetypes for smooth conduct. Authors also show cased the differences between proposed approach and the traditional approaches
9	Equity Crowdfunding Based on the Blockchain?—A Delphi Study [20]	Authors structured their paper in terms of various points related to blockchain and crowdfunding. The key points discussed in the paper are Purpose, Findings, Design and Layout, research limitations, the practical practices and the Originality of the works discussed in the paper
10	Crowdsourcing and Crowdfunding Platform using Blockchain and Collective Intelligence [21]	Authors talked about the scope of crowdsourcing and crowdfunding in India. The crowdfunding and its platform are using blockchain and the collective Intelligence method to make it successful
11	A systematic literature review of blockchain-based applications: Current status, classification and open issues [1]	Authors highlighted blockchain-based applications in the paper. A systematic review of various papers has been discussed in the paper on multiple domains
12	Yes, We Can! Blockchain-based Crowdfunding and Crowdworking [22]	Authors discussed about the challenges and the issues in the crowdfunding platforms. How the crowdfunding is using the blockchain-based technologies Authors also talked about the smart contracts and the blockchain inconsistencies
13	Blockchain Token Sale [23]	Authors highlights the working of tokens in crowdfunding and initial coin offering methods. The blockchain is the major contributor to work with crowdfunding and cryptocurrencies
14	Designing for Crowdfunding Co-creation [24]	Authors discussed about the enterprise crowdfunding. Authors highlighted the foundations and applications of the crowdfunding. In paper the research findings are also discussed
15	The Investment Model of Crowdfunding for MSME (Micro, Small and Medium Enterprises) in India [25]	This paper is related to the MSME and the start-ups in India. How the seed amount is collected in India if the start-up is initiated

(continued)

Table 1 (continued)

S. No	Title	Highlights of study
16	Secure and Transparent Crowdfunding using Blockchain [26]	Authors discussed about the Ethereum model and the method to generate the seed capital through the crowdfunding and how the smart contracts help in the seed funding In the paper, the Indian scenario is discussed and shown with the help of an experimental model
17	The new European crowdfunding regulation: facilitating cross-border services [27]	Authors discussed the scenario of crowdfunding in the European countries and how the investors are investing using crowdfunding The cross boarder services are also discussed in the paper
18	A qualitative investigation into crowdfunding framework as a source of financing for waqf land development [28]	Authors shown the framework as the source of financing in Indonesia and Malaysia. Authors highlighted the purpose of work, research findings and the challenges to work on The crowdfunding waqf model has been discussed in the paper
19	Helper Networks and Crowdfunding: Mobilizing Social Entrepreneurship [29]	Authors main motivation toward the paper was the role of individual and the organizations in crowdfunding. The social impact, the networking and the profiling of the start-ups are very much required to attract the investors. The term social entrepreneurship was discussed in that paper
20	Influence of perceived risks on the decision to invest in equity crowdfunding: a study of Malaysian investors [30]	In this paper, authors highlighted the frequency analysis of the age group and the people category to approach for the crowdfunding. In addition, the authors shown a design and model to explain the risks and challenges associated with the crowdfunding

3 Proposed System

The survey in the previous section highlights the various recent works. Many approaches are seen and studied to conclude the previous section and proposed a new model to work on. Proposed model is derived to give better, streamlined and convenience services to the end users. Figure 5 shows the layout of the website or an application to communicate over the network.

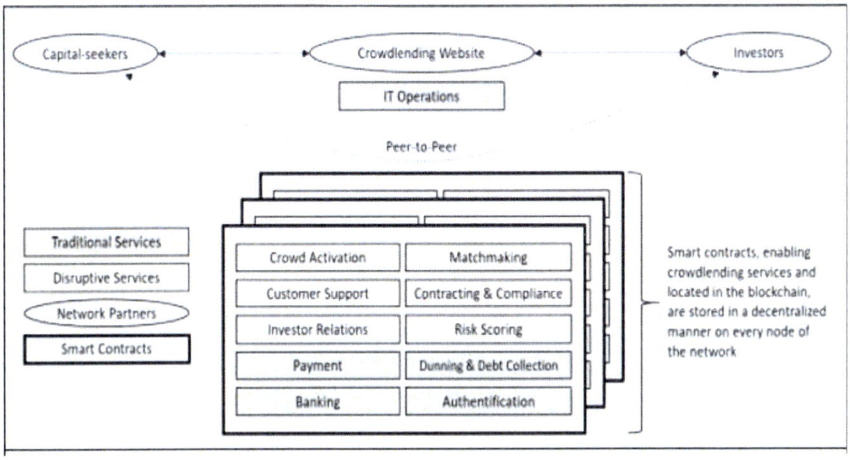

Fig. 3 Ecosystem and services of crowdlending using blockchain [32]

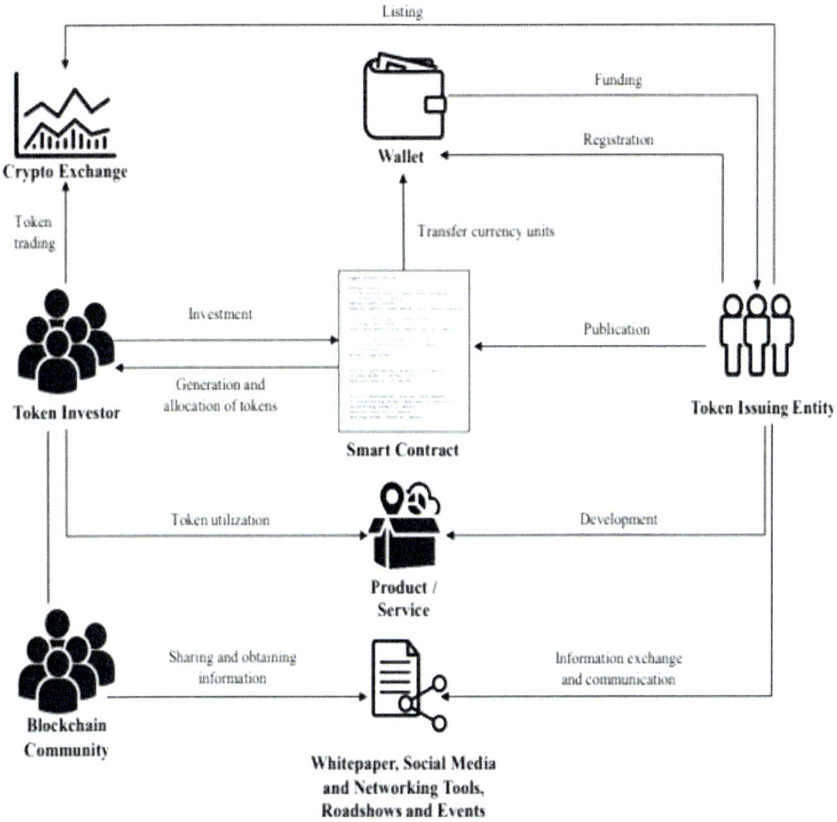

Fig. 4 Token-based crowdfunding system [19]

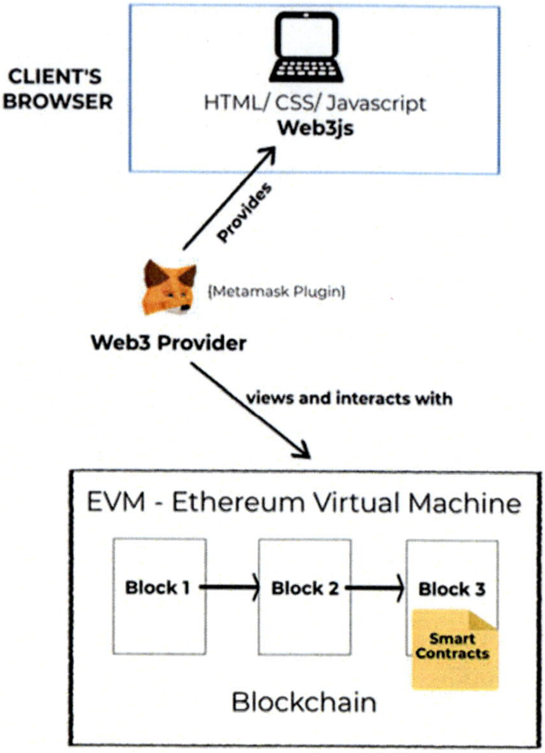

Fig. 5 Architecture of a blockchain application

3.1 Architecture Diagram

The application architecture consists of following parts:

Client's Browser

This is a web browser on the client's system which is used to access the application on Internet.

Web3.js

This allows us to read and write data on blockchain, transfer ether from one account to other and many more functions.

3.2 Proposed Model

The model is derived to see all the constraints mentioned in the review section and how the things can be done to make the system better than the existing systems.

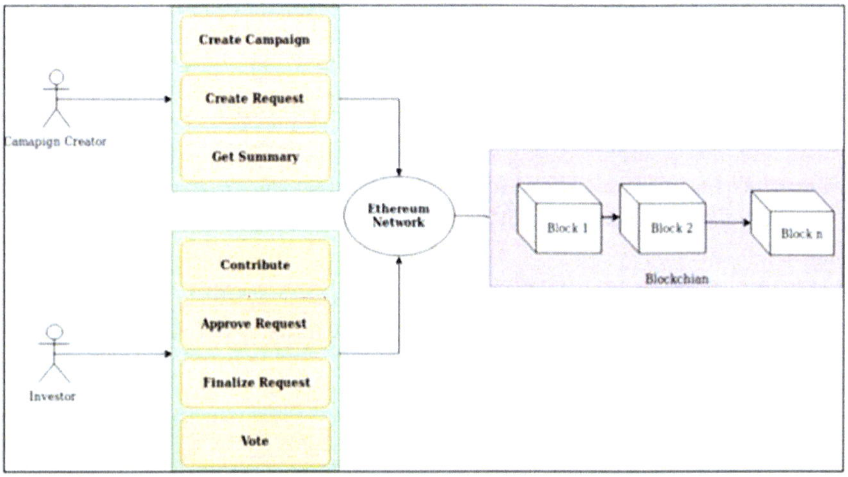

Fig. 6 Proposed model [23]

Figure 6 shows the use case diagram for our crowdfunding support. The user can access this app using a web browser. The user can be a campaign creator or an investor, accordingly different function are available on interface application which can be selected by the user according to his needs.

3.3 *Work Flow of the System*

The work flow of the proposed system is shown in the Fig. 7. How the contributors are joining the campaign and contributing to the start-ups seed fund as a crowdfunding.

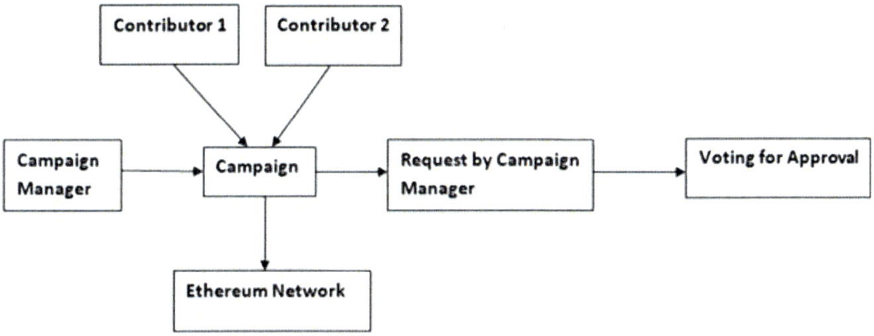

Fig. 7 Work flow of the system

4 Conclusion

In this paper, more than twenty works have been discussed and the hybrid model is derived to apply with ensuing algorithms for the better management and system for crowdfunding using blockchain. The concept of Ethereum is used as a base, which can further be being associated with the cryptocurrencies and the other coming digital modes. The paper highlights the challenges, the research findings and the proposed architecture to work on.

References

1. Casino F, Dasaklis TK, Patsakis C (2019) A systematic literature review of blockchain-based applications: current status, classification and open issues. Telematics Inform 36:55–81. ISSN 0736-5853. https://doi.org/10.1016/j.tele.2018.11.006
2. Kim MJ, Hall CM (2021) Do value-attitude-behavior and personality affect sustainability crowdfunding initiatives? J Environ Manage 280:111827. https://doi.org/10.1016/j.jenvman.2020.111827
3. Saniei S, Kent ML (2021) Social license to operate in crowdfunding campaigns. Public Relat Rev 47(2):102008. https://doi.org/10.1016/j.pubrev.2020.102008
4. Ren J, Raghupathi V, Raghupathi W (2021) Exploring the factors that determine the success of litigation crowdfunding: implications for social justice. Technol Forecast Soc Change 169:120813. https://doi.org/10.1016/j.techfore.2021.120813
5. Ferretti R, Venturelli V, Pedrazzoli A (2021) Do multiple competing offerings on a crowdfunding platform influence investment behavior? J Behav Exp Financ 30:100506. https://doi.org/10.1016/j.jbef.2021.100506
6. Lee Y, Coyle JR, Chen ANK (2021) Improving intention to back projects with effectivedesigns of progress presentation in crowdfunding campaign sites. Decis Support Syst 113573.https://doi.org/10.1016/j.dss.2021.113573
7. Wachs J, Vedres B (2021) Does crowdfunding really foster innovation? Evidence from the board game industry. Technol Forecast Soc Change 168(January). https://doi.org/10.1016/j.techfore.2021.120747
8. Yasar B (2021) The new investment landscape: equity crowdfunding. Central Bank Rev 21(1):1–16. https://doi.org/10.1016/j.cbrev.2021.01.001
9. Hurst B, Johnston KA (2021) The social imperative in public relations: Utilities of social impact, social license and engagement. Public Relat Rev 47(2):102039. https://doi.org/10.1016/j.pubrev.2021.102039
10. Farhoud M, Shah S, Stenholm P, Kibler E, Renko M, Terjesen S (2021) Social enterprise crowdfunding in an acute crisis. J Bus Ventur Insights, 15(November 2020):e00211. https://doi.org/10.1016/j.jbvi.2020.e00211
11. Wesemann H, Wincent J (2021) A whole new world: Counterintuitive crowdfunding insights for female founders. J Bus Ventur Insights 15(February):e00235. https://doi.org/10.1016/j.jbvi.2021.e00235
12. Belleflamme P, Lambert T, Schwienbacher A (2014) Crowdfunding: tapping the right crowd (July 9, 2013). J Bus Venturing 29(5):585–609, CORE Discussion Paper No. 2011/32, Available at SSRN: https://ssrn.com/abstract=1578175 or https://doi.org/10.2139/ssrn.1578175
13. Allison TH, Davis BC, Webb JW, Short JC (2017) Persuasion in crowdfunding: An elaboration likelihood model of crowdfunding performance. J Bus Ventur 32(6):707–725. https://doi.org/10.1016/j.jbusvent.2017.09.002

14. Schweizer A, Schlatt V, Urbach N, Fridgen G (2017) Unchaining social businesses—blockchain as the basic technology of a crowdlending platform. In: Proceedings of the 38th international conference on information systems (ICIS), Seoul, South Korea, December 2017, pp 1–21
15. Zhao H, Coffie CPK (2018) The applications of blockchain technology in crowdfunding contract (January 11, 2018). Available at SSRN: https://ssrn.com/abstract=3133176 or https://doi.org/10.2139/ssrn.3133176
16. Muneeza A, Arshad NA, Arifin AT (2018) The application of blockchain technology in crowdfunding: towards financial inclusion via technology. Int J Manag Appl Res 5(2):82–98. https://doi.org/10.18646/2056.52.18-007
17. Vakilinia I, Badsha S, Sengupta S (2018) Crowdfunding the insurance of a cyber-product using blockchain. In: 2018 9th IEEE annual ubiquitous computing, electronics & mobile communication conference (UEMCON), pp 964–970. https://doi.org/10.1109/UEMCON.2018.8796515
18. Cai CW (2018) Disruption of financial intermediation by FinTech: a review on crowdfunding and blockchain. Account Financ. https://doi.org/10.1111/acfi.12405
19. Fridgen G, Regner F, Schweizer A, Urbach N (2018) Don't slip on the ico—a taxonomy for a blockchain-enabled form of crowdfunding. Research Papers, 83. https://aisel.aisnet.org/ecis2018_rp/83
20. Heieck F, Ermakova T, Fabian B, Lessmann S (2018) Equity crowdfunding based on the blockchain?—A Delphi study (June 18, 2018). Available at SSRN: https://ssrn.com/abstract=3198083 or https://doi.org/10.2139/ssrn.3198083
21. Waheeda Dhokley E, Gupta S, Pawar G, Shaikh A (2019) Crowdsourcing and crowdfunding platform using blockchain and collective intelligence
22. Golrang H, Safari E (2021) Applying gamification design to a donation-based crowd-funding platform for improving user engagement. Entertainment Comput 38:100425. https://doi.org/10.1016/j.entcom.2021.100425
23. Kranz J, Nagel E, Yoo Y (2019) Blockchain token sale. Bus Inf Syst Eng 61:745–753. https://doi.org/10.1007/s12599-019-00598-z
24. Lipusch N, Dellermann D, Bretschneider U et al (2020) Designing for crowdfunding co-creation. Bus Inf Syst Eng 62:483–499. https://doi.org/10.1007/s12599-019-00628-w
25. Srivastava R (2016) The investment model of crowdfunding for MSME (micro, small and medium enterprises) in India. In: Méric J, Maque I, Brabet J (eds) International perspectives on crowdfunding. Emerald Group Publishing Limited, Bingley, pp 169–184. https://doi.org/10.1108/978-1-78560-315-020151010
26. Khatter H, Chauhan H, Trivedi I, Agarwal J (2021) Secure and transparent crowdfunding using blockchain, In: 2021 international conference on recent trends on electronics, information, communication & technology (RTEICT), pp 76–80. https://doi.org/10.1109/RTEICT52294.2021.9573956
27. Brand M (2021) The new European crowdfunding regulation: facilitating cross-border services. J Invest Compliance 22(3):248–252. https://doi.org/10.1108/JOIC-04-2021-0021
28. Hapsari MI, Bin Mohd Thas Thaker MA, Mohammed MO, Duasa J (2021) A qualitative investigation into crowdfunding framework as a source of financing for waqf land development. J Islamic Account Bus Res vol ahead-of-print, No ahead-of-print. https://doi.org/10.1108/JIABR-10-2020-0311
29. Mitra P, Kickul JR, Robb C (2021) Helper networks and crowdfunding: mobilizing social entrepreneurship. In: Wasieleski DM, Weber J (eds) Social entrepreneurship (Business and Society 360, vol 5), Emerald Publishing Limited, Bingley, pp 233–251. https://doi.org/10.1108/S2514-175920220000005011

30. Wasiuzzaman S, Chong LL, Ong HB (2021) Influence of perceived risks on the decision to invest in equity crowdfunding: a study of Malaysian investors. J Entrepreneurship Emerg Econ vol ahead-of-print, No ahead-of-print. https://doi.org/10.1108/JEEE-11-2020-0431
31. Simons A, Kaiser LF, vom Brocke J (2019) Enterprise crowdfunding: foundations, applications, and research findings. Bus Inf Syst Eng 61:113–121. https://doi.org/10.1007/s12599-018-0568-7
32. Haas P, Blohm I, Peters C, Leimeister JM (2015) Modularization of crowdfunding services: designing disruptive innovations in the banking industry. In: Leidner D, Ross J (eds) Proceedings of the 36th international conference on information systems, Fort Worth, TX, USA, pp 1–12

Cloud-Based COVID-19 Analysis and Future Pandemics Prediction Using IoE

Shivangi Sharma and Ajay Kumar Singh

Abstract Each zone could be very well investigated and ability packages, social and financial affects, and limitations to mass rehabilitation are discussed in greater detail. It describes in element the demanding situations and opportunities for IoE framework solutions inside the post-COVID-19 generation. The COVID-19 epidemic has furnished a far-wanted check of IoE-stimulated frameworks and answers. IoE solutions together with a ways flung fitness monitoring and phone tracking have supplied the guide of authorities to successfully manage to unfold the corona contamination. The principle contribution of this compare paper is the in-intensity analysis and segmentation of vicinity-unique IoE technology, which have the potential to be remarkable applications inside the new present day. So far, privacy worries and the protection of IoE applications are thoroughly analyzed with rising standards, and the code for multitasking practices is likewise being mentioned.

Keywords Internet of Everything · COVID-19 · Pandemic · IoE solutions · IoE applications

1 Introduction

IoE [1] is a set of linked gadgets, human beings, items, and facts-sharing offerings to advantage the identical motive in a variety of environments and packages, as defined which allows a tool labeling. IoT [2] currently plays a critical feature in lots of fields. Health care is one of the important sectors that has benefited from IoE development within the health area; IoE has dramatically modified the lives of young and old alike, as it may continuously show their fitness. Those gadgets are widely used with decreased fees and fast diagnostics. The function of the IoE environment has modified

S. Sharma (✉) · A. K. Singh (✉)
Computer Science and Engineering Department, Meerut Institute of Engineering and Technology, Meerut 250005, India
e-mail: shivangisharma1231@gmail.com

A. K. Singh
e-mail: ajay41274@gmail.com

P. Dutta et al. (eds.), *Emerging Technologies in Data Mining and Information Security*, Lecture Notes in Networks and Systems 490,
https://doi.org/10.1007/978-981-19-4052-1_65

dramatically because of the COVID-19 [3] pandemic. A kind of uses is right away related to lowering the spread of the virus, e.g., contact monitoring, temperature finding out, and plenty of others, while others need to simplify the general epidemic, e.g., operates at domestic, reading at home, staying at domestic, and so forth. A plan is wanted to keep corporations strolling human beings healthy. IoE era with RFID, close to area verbal exchange (NFC), Wi-Fi [4], Bluetooth low electricity (BLE), and GPS [5] permits offer higher answers to previously stated times. Further to patient desires, IoE lets in to resolve tool problems quick in wireless-care humans, in fee of pharmacy controls, and to hold tune of tagged device, which includes nebulizers and oxygen tanks.

Our research makes a special of the location of IoE all through COVID-19, IoE-based wireless solutions for COVID-19 in numerous industries, challenges, and approaches to lessen IoE threats. This observes is based totally mostly on the whole evaluation made by the authors. As a result, it became discovered that IoE supplied multi-motive and multi-enterprise answers that health wireless care, transportation, endeavor, education, and communications all through the COVID-19 pandemic.

2 Related Works

The COVID-19 pandemic has issued diverse predictions. It is far perception that IoE will play a main characteristic inside the new famous. This article introduces, analyzes, and discusses IoE-stimulated solutions at some degree within the pandemic in diverse fields, annoying situations, and opportunities beyond the pandemic. IoE gadgets that can be used to restrict infectious sellers collectively with COVID-19 and to beautify healthcare shipping. Some signs may be smoothed out with the usage of IoT gadgets. If signs and symptoms and signs and symptoms of the virus are detected, the tool may additionally moreover notify the patron and the closest fitness department. IoT also can serve as a studies community that identifies over crowded areas or regions with an excessive quantity of instances and an excessive risk of infection. This may improve the health branch's operational interventions to locate and rescue people in emergencies (e.g., an affected character is not able to contact the fitness department in a timely way because of signs and signs) and percentage information with diverse departments to rush up recovery and maintain the network safe.

With using era which includes 4G [6], 5G, and the cloud, it is far plenty less difficult to expose sufferers remotely, specifically for people who find out it tough to get proper of entry to or access health facilities. Initial trying out of group of workers, disaster survivors, or patients is the primary line of protection in opposition to the virus. This doesn't verify the analysis but can be used to decide if someone with the sickness may also have signs of the virus, and further, tests are performed to determine the very last diagnosis.

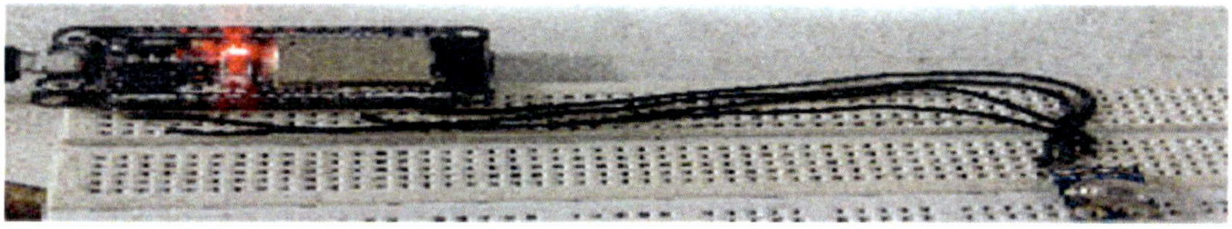

Fig. 1 Connection of device ESP-32 and MLX90614

3 Proposed Methodology

The studies goals under are set out:

(a) The top-level view of the socioeconomic effect, technological adulthood, and timeline of huge distribution.
(b) Identity of key IoE technologies, which has contributed to the control of COVID-19 for a whole evaluation of the literature.
(c) Assessment and dialogue of coverage, privacy, and protection traumatic situations, and sooner or later the identification and evaluation Ranger, 2020 of first-class practices and code of IoE generation strategies to obtain the above studies goals, the subsequent research methodology is accompanied in this look at. Searches of diverse sticks would possibly no longer have determined the applicable research, and evaluations had to solution the research questions of our take a look at. Key-phrase mixtures encompass "COVID-19," Coronavirus, IoT, IoE, IIoT, IoMT, and distinctive applicable key terms. Launch situations were now not applied. Figure 1 shows connection of device ESP-32 and MLX90614.

4 Results

Table 1 shows sample temperature detecting data.

Table 1 Temperature detecting data

Ambient	26.55*C	Object	27.97*C
Ambient	81.16*F	Object	82.02*F
Ambient	27.31*C	Object	27.99*C
Ambient	81.16*F	Object	82.38*F
Ambient	27.27*C	Object	28.25*C
Ambient	81.09*F	Object	82.85*F
Ambient	27.25*C	Object	28.17*C
Ambient	81.05*F	Object	82.71*F

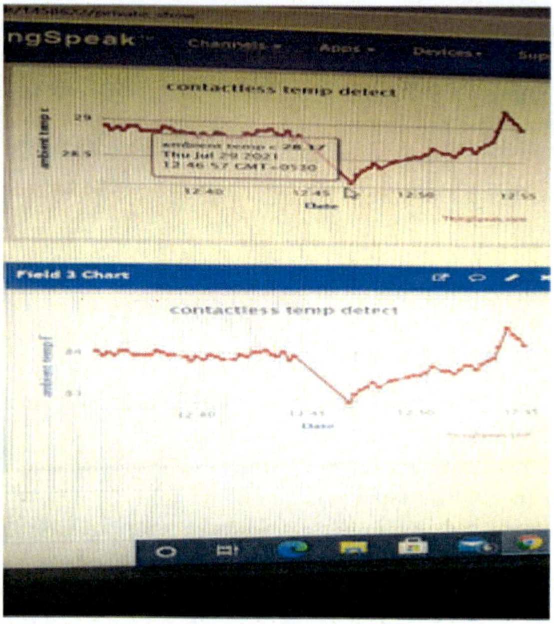

Fig. 2 Graph related to temperature detecting data

Steps To Get Output In Cloud (Fig. 2):

Step 1: First, we connect the ESP8266 Wi-Fi with contactless sensor MLX90614 in breadboard. As shown in Fig. 1.

Step 2: After connection, go to ThingSpeak.com. As shown in Fig. 3

Step 3: Then, click on ThingSpeak one dialogue box opens, and after that click on the Get-Started For Free project.

Step 4: Then, make your official or college ID and password in this ThingSpeak clicking on "create one" for making your own "Math-work Account."

Fig. 3 ThingSpeak

Step 5: After creating Math-work Account there, one dialogue box is open for creating your own channel.

Step 6: In this channel, we give the name to our project and give the field name after clicking on the create button our unique channel ID is made. As shown in Fig. 4.

Step 7: Now, we do coding in Arduino Uno software related to the connection of ESP8266 Wi-Fi and contactless sensor MLX90614

Step 8: After that we go to my channel where we create our unique channel ID and click on API key button and copy "Write Code" from it and paste this code in your coding.

Step 9: In ThingSpeak, our code is compiled in private view.

Step 10: In private view, sensor values generated in a graph form. As shown in Figs. 5, 6, and 7.

Future of IoE

As proven in above Fig. 8, the Internet of Everything (IoE) is constructed at the top of the net of things (IoT), it is far vital to have a close to appearance on IoT which could have three paradigms—net orientated (middleware), things orientated (sensors), and semantic oriented (knowledge). Steady with RFID institution, the IoT can be described wireless as "international network of interconnected devices which are uniquely addressable based totally on the same old protocols of communication." The actuating and sensing devices are interconnected with every exclusive and percentage the information across heterogeneous systems via the framework that consequences in develop the progressive packages in diverse sectors. That is carried out with the useful resource of the information analytic, cloud computing, ubiquitous sensing that acts as a unifying framework clever connectivity with context-aware computation; the usage of modern network is an essential a part of IoT [7].

In present day trends, there may be speedy boom of Wi-Fi and 4G-LTE [6] wireless net get right of entry to which suggests the evolution inside the direction of the verbal exchange networks and ubiquitous information [8, 9]. But, IoT is going past the conventional cellular computing surroundings and evolve into connecting normal present things and embed intelligence into these days' surroundings. IoT

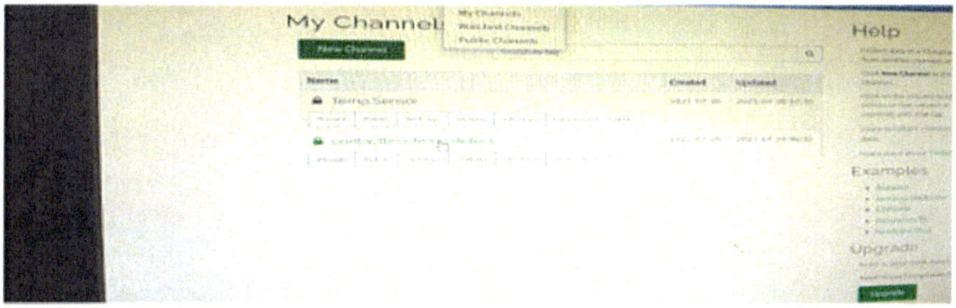

Fig. 4 Channel name in ThingSpeak

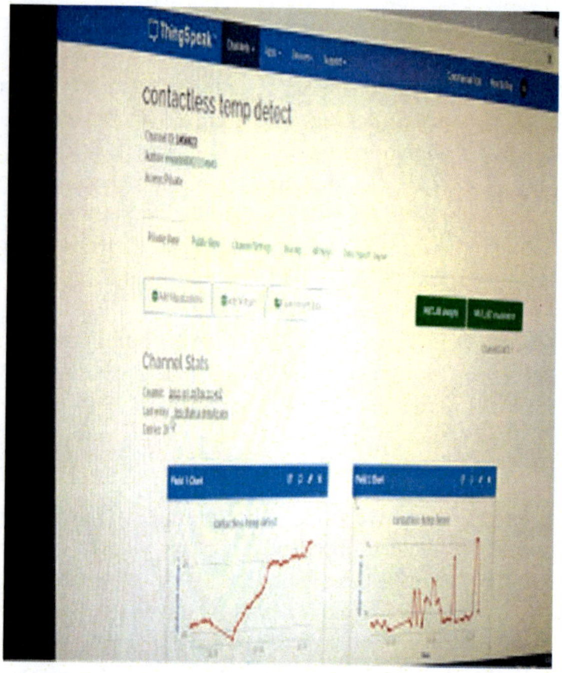

Fig. 5 Graph detect from temperature sensor

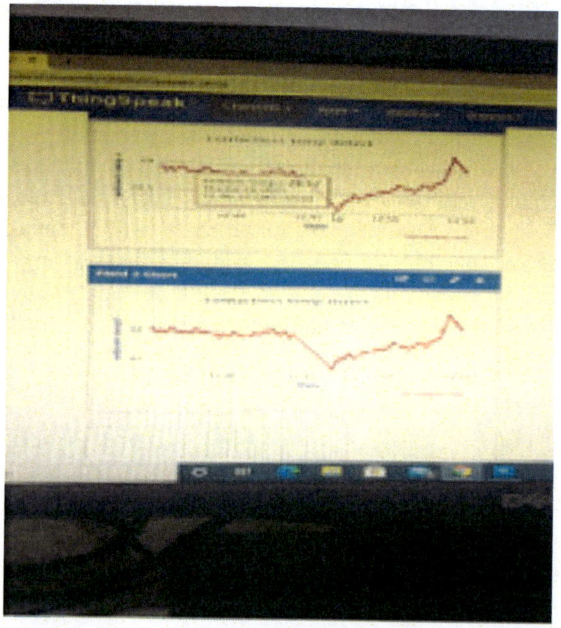

Fig. 6 Graph 2

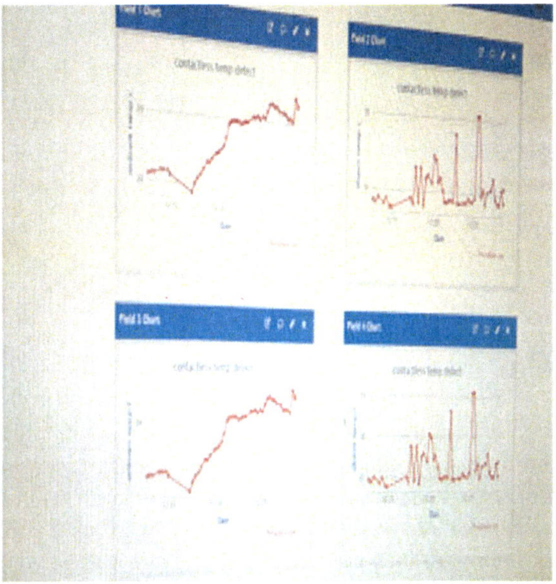

Fig. 7 Graph 3

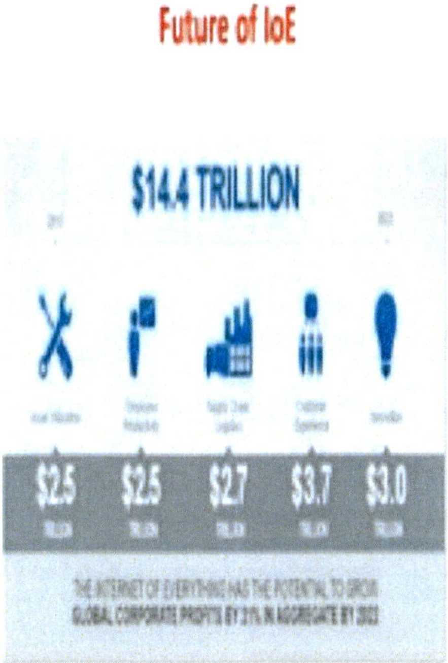

Fig. 8 Future prediction of IoE

Fig. 9 IoE

attention on these factors share the understanding of the state of affairs of its clients and their domestic equipment, pervasive communication networks, and software program architectures to deliver and method the contextual statistics to in which it is miles relevant and the analytic system to IoT that aims for clever conduct and self-sustaining. IoE generation has the capability to convert the world. These are pushed by manner of technical improvement inside the vicinity of interconnectivity, massive information, Wi-Fi intelligence, and semantic operability. The private groups play the essential roles to steer the manner of IoE with a view to trade our societies. The IoE is expected to bring about unheard of societal alternate, starting up new opportunities, and posing new worrying conditions.

Figure 9 illustrates a specific IoE programs and usage of IoE in particular sectors. The following sub-sections discuss IoE solutions and frameworks that help specific enterprise sectors to decrease the impact and created with the aid of the COVID-19 pandemic. Figure 10 demonstrates IoE applications based on key observations.

5 Conclusions

In this research paper, we have got a framework of COVID-19 that lets in growing IoE packages to be evaluated in a variety of application regions. The IoE provided progressed fitness services and enabled advanced clinical selections at a few things are inside in the pandemic.

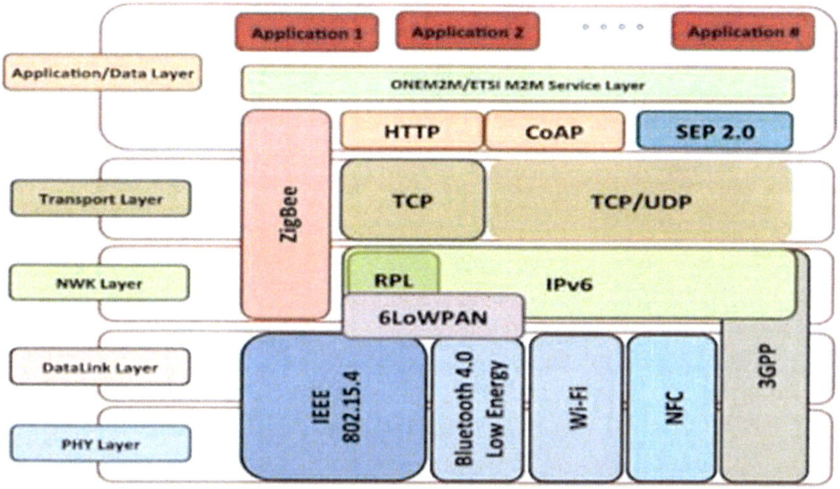

Fig. 10 IoE application

References

1. Singh RP, Javaid M, Haleem A, Suman R (2020) Internet of Everything's (IoE) applications to fight against COVID-19 pandemic. Diabetes Metab Syndr Clin Res Rev 14:521–524
2. Javaid M, Khan IH (2021) Internet of Things (IoT) enabled healthcare helps to take the challenges of COVID-19 Pandemic. J. Oral Biol Craniofacial Res 11:209–214
3. Hakak S, Khan WZ, Imran M, Choo KKR, Shoaib M (2020) "Have you been a victim of COVID-19-related cyber incidents?" Survey, taxonomy, and mitigation strategies. IEEE Access 8:124134–124144
4. Martindale J (2021) What is Wi-Fi, March 28, 2021
5. Parkinson B, Spilker J, Elkaim G (2003) Global positioning system (GPS). First published: 15 April 2003
6. Dahlman E, Parkvall S, Sköld J (2011) 4G LTE/LTE-Advanced for mobile broadband (15th ed). Elsevier/Academic Press, Amsterdam (Chicago)
7. Schulz P et al (2017) Latency critical IoT applications in 5G: perspective on the design of radio interface and network architecture. IEEE Commun Mag 55(2):70–78
8. Ahmed M, Litchfield AT (2018) Taxonomy for identification of security issues in cloud computing environments. J Comput Inf Syst 58(1):79–88
9. De Bernardini L (2021) Why IIoT Demand Continues to Grow. CEO, Autoware May 3rd, 2021

Cognitive Study for Online Education in COVID Using Python

Ankita Sharma

Abstract This paper highlights the impact on education due to the corona virus/COVID 2019 crisis across the world. Over, 100 million students in India will be phasing out this issue from 2019 onward. The impact of this pandemic is not only on education but even on all sectors of the world. The enforced lockdown is having a bad effect on the students' lives. Due to the pandemic, students as well as educational institutions have to attain technology in the education sector. This paper is going to use the online survey method to gather the responses of some 20 schools and 10 college students from different states in India by using self-developed questionnaires to analyze the effect of online-based learning on students. We compare the different surveys of other countries to find out what major steps help in the education sector of those countries. Using Python discovered that 90% of students are currently dealing with health issues, and 70% are not receiving financial assistance. Only, 30% of students are found proficient in digital skills, and many of them feel a lack of teacher–student interaction. We found some negative and positive impacts of COVID-19 on students. Also, we are suggesting some solutions for the educational sector to opt for ICT at a low cost.

Keywords COVID-19 · Education classroom · ICT ecosystem · Online learning · Digital skills · Python

1 Introduction

The deadly COVID-19 pandemic has not stopped at national borders but has spread worldwide at a very fast speed and has affected all the states and cities in each country. This pandemic was first identified in December 2019 in Wuhan, China. On Jan 27, 2020, a 20-year-old lady from Kerala suffered from COVID-19, the first case in India. In the first case, the Indian government imposed a countrywide lockdown, shutting down all industries, businesses, and educational institutions and taking all

A. Sharma (✉)
Chandigarh University, Chandigarh, India
e-mail: ankita.ceh@gmail.com

P. Dutta et al. (eds.), *Emerging Technologies in Data Mining and Information Security*, Lecture Notes in Networks and Systems 490,
https://doi.org/10.1007/978-981-19-4052-1_66

precautions to keep the public safe. This was a good step taken by the government because saving lives is much more important than other things. This sudden lockdown stopped all academic and corporate activities, and this may have led to a great loss of job opportunities, economic growth, human capital, and increased unemployment in the long run. According to the report by UNESCO, 1.2 billion children and young people are facing unprecedented disruption in education throughout the world, and more than 200 million students in higher education institutions are left out of their institutions. According to the MHRD report, 32 core students have started to suffer from the loss of studies, so the only way left is the online mode of learning. As a result of the pandemic, we transitioned our education from offline to entirely online. So, at least, we resolve the one issue that is to continuously provide education to our students; after opting for online-based education again, we have so many issues and problems.

In the early days of the pandemic 2020, it was very difficult to implement online education in India. As we all know, India is home to a diverse population, and not every Indian parent can afford to provide digital education (such as mobile phones, broadband connections, and other digital education-related items) for their children. In 2020, we experienced a very difficult and tough time and witnessed many cases in India where 60% of poor children left school because of online-based education. Many teachers are also phasing the issue of moving offline classes to online and making them more interactive, productive, and fruitful, but this is not possible for teachers who are not from the background of computers. So, the challenges arise in training our teachers as well as providing the infrastructure for students. This paper highlights all of the issues phasing in the education system during the pandemic period. In Sect. 2, we will discuss some relevant literature surveys. In Sect. 3, we discuss the methodology of this survey. In Sect. 4, give the answers to research questions and analyze the students' work on online-based learning. In Sect. 5, we are going to highlight the challenges and issues raised by students and teachers while doing the survey, and lastly, suggest some solutions from literature to improve the ICT system of education

2 Literature Review

In literature review, as the topic is new, very few researchers are working in this field. Sahoo et al [14], this paper does a survey of 318 students, which tells us the e-readiness perception of students' teachers toward online learning in the COVID-19 pandemic, but this paper doesn't give any solutions toward online classes. Pokhrel [12] in this article, you will find some literature review on the teaching and learning impact of COVID-19. Subedi [17], this paper focuses only on nursing students and teachers in Nepal. Through this paper, we identify the impact of e-learning in every country. Bayham [3], this paper outlines the impact of COVID-19 on US schools and also focuses on the healthcare sector phasing out the issue in this pandemic. This article focuses on the challenges reported by the countries due to pandemic phasing in schools and universities. One-half a billion students and young people suffer from this issue. Basilaia [2] highlights some good points in e-learning for students who are physically challenged. They can study with full freedom and participate fully in the virtual environment. Firmansyah [8] focuses on the educational transformation from offline to online and finds out how helpful it is. Research highlights some points such as: comparison of e-learning and classroom learning, impact on the health issues of teachers and students, the limited exposure of teachers to online learning, low income of parents, bandwidth issues in the rural areas of India, and academic issues. This paper focuses on evaluating the impact of e-learning in India on every age group of students' perceptions by conducting an online survey. Through the literature survey and online survey, we find out what tools are out there and compare them with some parameters. In such a critical situation, we have some research questions in the following papers:

- Which countries do the survey on COVID-19 impact on education?
- Whether the students are comfortable with e-learning?
- Which pedagogies used by Indian schools for e-learning?
- What are the challenges and solution for the future online learning?

3 Methodology

This survey was carried out to identify student mental health and acceptability of e-learning as a subject. COVID-19 is a drastic problem phasing all over the world. This pandemic not only speed up the health sector but also speed up every other sector of the organization. The major issue raised in India on a very high scale is the education sector because out of 100 billion students, 50 billion students are living in rural areas, and their facilities are the major concern. But, our education ministry found a solution that lets us start classes through an online platform, and education will never stop. E-learning has been adopted by many schools and universities. A well-structured and closed-ended and open-ended self-developed question was prepared. The tool is divided into eight sections such as basic details, income per capita of the family, e-learning resources, accessibility, digital literacy, health status of students, effectiveness of e-learning, and pedagogies used by schools and universities. The study focused on students between the ages of 14 and 23 who attend engineering colleges and senior secondary schools. The sample size is 700 respondents, of which 300 is engineering students and 400 is school students. All the self-designed questioners will be filled in by the students at the time of data collection. The online survey was carried out during the period of the lockdown period, i.e., from April 10 to May 10. Data were collected from two states of India. Himachal Pradesh cities like Hamirpur, Kangra, and Punjab cities like Mohali, Jalandhar, and Chandigarh. The main respondents are from the engineering and school population. From a rural locality, 37% of students contribute, and from an urban locality, 27% of students contribute. In the sample, 28% of student teachers had prior experience with online learning, while 90% had learned online learning during the pandemic period. After compiling dataset, we used python to generate the graph and identify the results.

4 Results and Discussions

The study revealed the current situation of the education system in India. In this section, we find out the results of the four research questions that are mentioned in Sect. 2 one by one. First, we find a similar survey which helps us to proceed with our studies in an effective manner. After that, we made the self-designed questions for our online survey, which helped us to find some major results. Later, by using graphs, we show the results of the survey.

4.1 Which Countries Do the Survey on Covid-19 Impact on Education?

In this question, we are going to study different countries research papers and articles and identify that what type of problem they are phasing and for that problem what solution they give. In Table 1, we can easily understand the impact on education in other countries also. Those are looking for the solutions Table 1 focus for such researchers from the above table, we identify that every country is taking their best countermeasures to overcome the impact of COVID-19 on education. Out of other

Table 1 Comparative analysis of COVID-19 impact on education

S. No	Author	Country	Year	Impact	Solution
1	Upoalkpajor [19]	Ghana	2020	Due to covid, schools want resources to reconstruct the loss in education	Shifted to digital mode
2	Ferrel [7]	USA	2020	Every education is phasing problem mainly focus on medical students career	NA
3	Hung [9]	USA	2021	92.4% students of medical effecting and 12.6% students of schools effecting	NA
4	Daniel [6]	USA	2020	Impact on teachers, students, parents	Digital format
5	Alemán [1]	Venezuela	2020	Effecting medical education in country	Digital learning
6	Ratten [13]	USA	2021	Effected the entrepreneurship education	NA
7	Sangster [16]	China	2020	Study on the accounting education in 45 different countries around the world	Redesign

(continued)

Table 1 (continued)

S. No	Author	Country	Year	Impact	Solution
8	Carrillo [4]	USA	2020	134 empirical studies of teacher education of literature on online teaching learning were analyzed	Online
9	Chen [5]	China	2020	Review on seven major online platforms	Use artificial intelligence
10	Majeed [11]	Pakistan	2020	Impact on the medical education of students	Digital platforms
11	Tarkar [18]	India	2021	Advantage and disadvantage of online platforms and how Indian accepts the new educations	NA
12	Sahu [15]	Augustine	2020	Impact on education and mental health of students and academic Learning staff	Digital
13	Xue [20]	China	2020	How china education response to COVID-19	Separate groups of students different reasons
14	Jena [10]	India	2020	Impact of pandemic on education in India	Digital learning

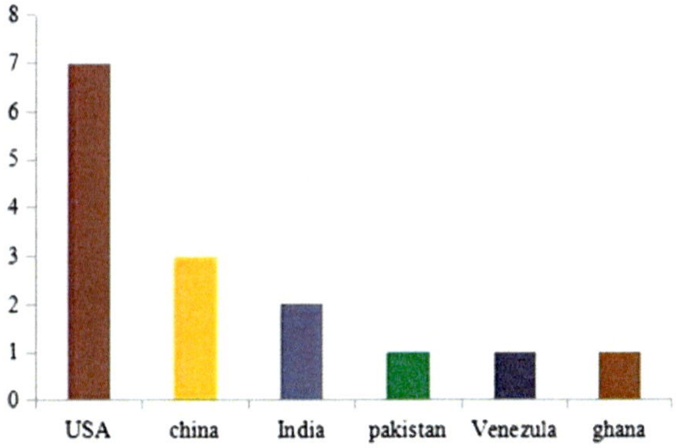

Fig. 1 Countries publishing survey paper on the impact of COVID-19 education

countries, the highest survey on education was done by the USA. After USA, we need to understand how china overcomes from this issue from the mentioned paper in Table 1 [20]. For more clarity, we represent our table in graphical form in Fig. 1.

4.2 *Whether the Students Are Comfortable with E-learning?*

This part is divided into three categories such as: proficiency in using online learning, effect on health of students through online learning. Proficiency in using online learning students' use of the online mode of learning could be understood from some questionnaires. 60% of students felt it was easy to use the online software suggested by teachers for the e-learning process, while 30% of students felt some difficulties in using online software. 90% of students prepares and submits the assignments online.

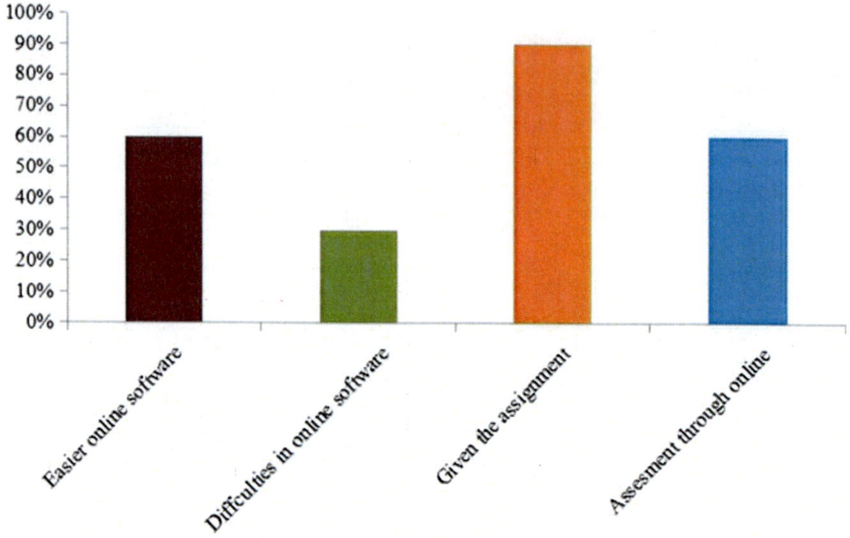

Fig. 2 Graphical representation of proficiency of e-learning

60% of students give the test through an online platform. In Fig. 2, we understand the level of students' digital knowledge.

Health Issues of Students in the e-learning Process: Students phasing health issues during their online live classes. Most of the students said that sitting the whole day in front of a digital device is difficult. According to our results, 80% of the students said their eye sight increased due to continuously watching the digital device. 43% of students reported having a headache issue after their classes. 36% of students are experiencing strains and body pain. 67% of students are phasing out due to mental stress. The graphical view of the health issue is represented in Fig. 3.

4.3 Which Pedagogies Used by Indian Schools for E-learning?

Platform used during the pandemic for e-learning that software must be used on the platform suggested by the schools and concerned subject teachers. 80% of students said they download all the apps on their digital devices, whereas the Zoom app is the first preference chosen by 60% of students for their online sessions, while 43% of students are using Google Meet for their classes. The Microsoft Team is used by 25% of the students only. 56% of high school students use Zoom and Google Meet, whereas universities and colleges adopt their own platforms, such as my class and blackboard. Figure 4 represents the graphic view of this platform used by students.

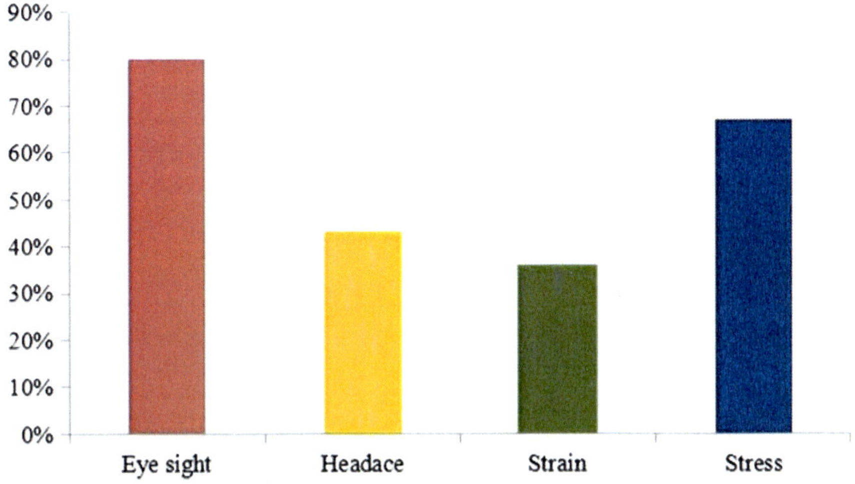

Fig. 3 Health issues during e-learning

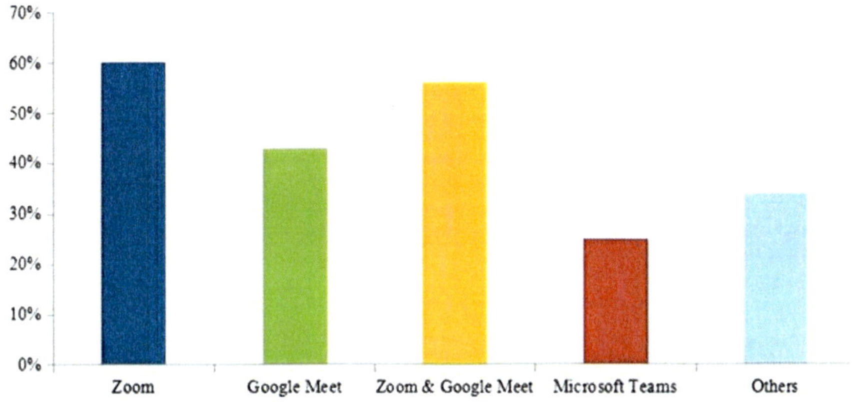

Fig. 4 Platform used for e-learning

Pedagogies Used During E-learning The study revealed that pedagogy varies with teachers, such as some are comfortable with notes, others with blackboards, whiteboards, presentations, and video lectures. According to the student point in Fig. 5 of view, 70% of students understands through the blackboard, whereas 12% of students are comfortable with all pedagogies. 10% felt a lack of confidence. 14% of teachers used presentations and videos to make their classes more interactive. 8% of students felt it was easy to understand with notes. 90% agree that online learning is the better choice in a lockdown.

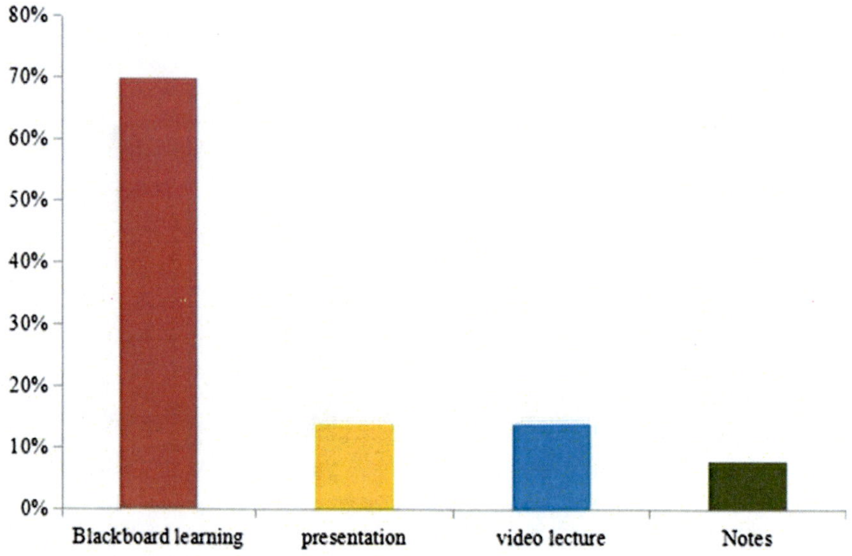

Fig. 5 Graphical representation of pedagogies used for e-learning

4.4 *What Are the Challenges and Solution for the Future Online Learning?*

After the comparative analysis and survey, we found so many challenges in online learning. We categorized the challenges into five parts. This helps the researcher to understand the gaps in digital education

- Infrastructure Issue: 90% of students had digital devices; 70% were having good Internet issues, but 24 h power supply was with only 36% of students; 30% were having good Internet facilities, and 10% were having study space at home.
- Financial Issues: 70% of students in rural and urban areas has a crisis of money. 30% of people lost their jobs and more.
- Digital Skills: 50% of students were proficient in digital skills; 12% of students knew the online study materials, and 13% had adequate knowledge of Internet security, while 70% of students were compatible with digital platforms. Some students did not know how to download the different software online. There may be some lack of awareness between students and teachers.
- Teacher Issues: In this era, teachers are also phasing out so many issues, such as lack of proper communication between teachers and students. Due to this, students have lost their interest in studies. The teacher has no way of knowing whether or not a student understands what is being taught. Time management is also an issue cited by many of the teachers. Students do some unethical activities during the online class.

Solutions

- Knowledge management for e-learning: This e-learning environment not only helps the learner to make sense of content but also provides the proper communication between both learner and tutor.
- Creating learning content and pedagogy: reduce the problem of identifying student behaviors by their expressions. Researchers started working in the area of education to understand how online classes become more interactive and how to increase the quality of online pedagogy. Most of the software provides feedback after the end of classes. Through that feedback, instructors understand that how much content need to redesign and reviewed.

5 Conclusion

The transformation of the education sector from traditional to online mode in a pandemic situation is challenging. Due to the urgent adoption of the online mode of digital learning, teachers and students are left unprepared mentally, financially, socially, and technically. In 2020, there will be many students and teachers who face difficulties because of their unfamiliarity with technology. The study reveals how many countries consider this issue to be important, as well as how many issues students face when phasing in online learning. In our education system of India, we prepare our teachers and students to improve the issues and handle any challenge in future. We need to provide training to our tutors so that they can improve the pedagogical approaches for online delivery of lessons. India will take so many initiatives after 2020. But, we need to refine and recreate more interactive models such as students' comfort zones, parental control, and many more. At the end, we identify some issues phasing by students and highlight solutions that are given in the literature. We need to start learning and motivating tutors and students to move toward the new world of digitalization in education

References

1. Alemán I, Vera E, Patinõ-Torres MJ (2020) Covid-19 and medical education: challenges and opportunities in Venezuela. Educ Med
2. Basilaia G, Kvavadze D (2020) Transition to online education in schools during a sars-cov-2 coronavirus (covid-19) pandemic in Georgia. Pedagogical Res 5(4)
3. Bayham J, Fenichel EP (2020) Impact of school closures for covid-19 on the US health-care workforce and net mortality: a modelling study. Lancet Public Health 5(5):e271–e278
4. Carrillo C, Flores MA (2020) Covid-19 and teacher education: a literature review of online teaching and learning practices. Eur J Teach Educ 43(4):466–487
5. Chen T, Peng L, Jing B, Wu C, Yang J, Cong G (2020) The impact of the covid-19 pandemic on user experience with online education platforms in China. Sustainability 12(18):7329
6. Daniel J (2020) Education and the covid-19 pandemic. Prospects 49(1):91–96
7. Ferrel MN, Ryan JJ (2020) The impact of covid-19 on medical education. Cureus 12(3)

8. Firmansyah R, Putri DM, Galih M, Wicaksono S, Putri SF, Widianto AA, Palil MR (2021) Educational transformation: an evaluation of online learning due to covid-19. Int J Emerg Technol Learn 16(7)
9. Hung M, Licari FW, Hon ES, Lauren E, Su S, Birmingham WC, Wadsworth LL, Lassetter JH, Graff TC, Harman W et al (2021) In an era of uncertainty: impact of covid-19 on dental education. J Dent Educ 85(2):148–156
10. Jena PK (2020) Impact of pandemic covid-19 on education in India. Int J Curr Res (IJCR) 12 (2020)
11. Majeed MM, Durrani MS, Bashir MB, Ahmed M (2020) Covid-19 and dental education in Pakistan. J Coll Phys Surg Pak 30:S115–S117
12. Pokhrel S, Chhetri R (2021) A literature review on impact of covid-19 pandemic on teaching and learning. Higher Educ Future 8(1):133–141
13. Ratten V, Jones P (2021) Covid-19 and entrepreneurship education: implications for advancing research and practice. Int J Manag Educ 19(1):100,432
14. Sahoo S et al (2020) E-readiness and perception of student teachers' towards online learning in the midst of covid-19 pandemic. Available at SSRN 3666914
15. Sahu P (2020) Closure of universities due to coronavirus disease 2019 (covid-19): impact on education and mental health of students and academic staff. Cureus 12(4)
16. Sangster A, Stoner G, Flood B (2020) Insights into accounting education in a covid-19 world. Acc Educ 29(5):431–562
17. Subedi S, Nayaju S, Subedi S, Shah SK, Shah JM (2020) Impact of e-learning during covid-19 pandemic among nursing students and teachers of Nepal. Int J Sci Healthc Res 5(3):68–76
18. Tarkar P (2020) Impact of covid-19 pandemic on education system. Int J Adv Sci Technol 29(9):3812–3814
19. Upoalkpajor JLN, Upoalkpajor CB (2020) The impact of covid-19 on education in Ghana. Asian J Educ Soc Stud 23–33
20. Xue E, Li J, Li T, Shang W (2021) China's education response to covid-19: a perspective of policy analysis. Educ Philos Theory 53(9):881–893

COVID-19 Disease Classification Model Using Deep Dense Convolutional Neural Networks

Anjani Kumar Singha, Nitish Pathak, Neelam Sharma, Pradeep Kumar Tiwari, and J. P. C. Joel

Abstract Preventing the transmission of COVID-19 necessitates diagnosis and identification. Researchers have developed algorithms to detect the presence of COVID-19 in X-ray and CT scans and images. These methodologies produce skewed data and incorrect disease detection. So, in the case of COVID-19 forecasting utilizing CT scans in an IoT setting, the current study paper established an oppositional-based deep dense convolutional neural network (DDCNN) and chimp optimization algorithm. The framework proposed is divided into two stages: preprocessing and estimation. Previously, a CT scan pictures generated from anticipated COVID-19 are acquired utilizing IoT devices from an open-source system. After that, the photos are preprocessed with a Gaussian function. A Gaussian filter can be used to remove undesirable noise from CT scan pictures that have been obtained. The preprocessed photos are then transmitted to the prediction process. DDCNN is applied to the images preprocessed in this step. The recommended classifier is designed to be as efficient as possible using the oppositional-based chimp optimization algorithm (OCOA). This approach is used to choose the best classifier parameters under consideration. Furthermore, the suggested method is applied to forecast COVID-19 and categorizes the findings as COVID-19 or non-COVID-19. The proposed technique was used in Python, and results were assessed using statistical analysis. CNN-EPO and CNN-FA were compared to the new method. The results proved that the proposed model was optimal.

A. K. Singha
Aligarh Muslim University, Aligarh, India

N. Pathak
Parshuram Institute of Technology (BPIT), GGSIPU, New Delhi, India

N. Sharma
Maharaja Agrasen Institute of Technology (MAIT), GGSIPU, New Delhi, India

P. K. Tiwari (✉)
Manipal University Jaipur, Jaipur, India
e-mail: pradeeptiwari.mca@gmail.com

J. P. C. Joel
Senac Faculty of Ceará, Fortaleza, CE, Brazil

P. Dutta et al. (eds.), *Emerging Technologies in Data Mining and Information Security*, Lecture Notes in Networks and Systems 490,
https://doi.org/10.1007/978-981-19-4052-1_67

Keywords Chimp optimization algorithm · Covid-19 · Deep learning · CT images · Deep dense convolutional neural network · Emperor Penguin Optimization (CNN-EPO) · Convolutional neural network-firefly algorithm (CNN-FA)

1 Introduction

For such previous two years, coronavirus disease (COVID-19) has already had a considerable impact on worldwide economic growth [1]. COVID-19 has been designated as either a global health crisis by the World Health Organization (WHO).From March 11th, 2020, there would be an outbreak. Many of the countries that seem to be able to keep COVID from spreading were France, Germany, New Zealand, Vietnam, South Korea, Taiwan, and others have all done well. Many countries, including Honduras, the United States, among India, have struggled to keep the virus for such an extended period of time India was placed under the toughest curfew in its history, beginning in the year 2000. Coronavirus will be contained from April 25th to July 31st, 2020. Therefore, the Indian government began unlock the country in stages beginning on June 1, 2020, and ending on August 31, 2020, because it is impossible for such an improved economy like India to manage the financial damages experienced as just a result of the a prolonged lockdown [2]. Unlock-1 occurred from July 1 to July 31, 2020, with the unlock-2 running from August 1 to August 31, 2020, and the last unlock running from September 1 to September 31, 2020 [3]. So, this same Indian government established guidelines that applied both for individuals and organizations about how to resume requires commercial activities, transport, businesses, religious and various social, COVID-19 has so far eluded researchers' efforts to discover a treatment [4]. Difficulty swallowing, cough, sneezing, diarrhea, loss of taste, and smell dyspnea are some of the symptoms of the condition. Early diagnosis of the disease can aid in the timely isolation and monitoring of patients. Patients with pneumonia, a side consequence of COVID-19 disease, could be diagnosed using machine-based technologies that analyze X-rays/CT scanning pictures of something like the lungs. The machine-based process is useful as either a viable COVID-19 detection instrument option. That is also particularly true in underdeveloped nations, when vast numbers of people were exposed to the virus, and that there is no way to determine whether a person is COVID-19 or non-COVID-19 [5]. Several approaches for detecting COVID-19 using image data have indeed been presented. However, for long-tailed distributed, such techniques failed to obtain the optimum outcomes [6]. The COVID-19 dataset might include triplets-labeled occurrences which can be used to distinguish between afflicted as well as unaffected people [7]. Similarly, some databases contain CT scans that aren't labeled, making it difficult to analyze the findings. Spreadsheets make it more challenging for current algorithms to detect COVID-19 in CT scans. These flaws have prompted researchers to look for new deep learning, and several approaches for detecting COVID-19 using image data have indeed been presented. However, for long-tailed distributed, such techniques failed in obtain the optimum outcomes [6]. The COVID-19 dataset might include triplets-labeled occurrences which can be used to distinguish between afflicted as well as unaffected people [7]. Similarly, some databases contain CT scans that aren't labeled, making it difficult to analyze the findings [8]. Spreadsheets make it more challenging for current algorithms to detect COVID-19 in CT scans [9], and these flaws have prompted researchers to look for new machine learning and deep learning methodologies [10]. So order to identify COVID-19 from datasets, deep learning techniques such as artificial neural network (ANN) [11], support vector machine (SVM) [12], and Bayesian network, are utilized [13]. Such machine learning algorithms, on either side, really aren't ideal for dealing with large image databases in deep neural network (DNN) [14]. As a result, deep learning is now the primary method for detecting COVID-19 in image in deep belief neural network (DBNN) [15]. For COVID-19 identification, deep learning can handle large databases are among the deep learning approaches which can be used to indicate the existence of COVID-19 in images. Several contributions make usage deep learning methods. Among which, CNN is now the most ideal technique for computer vision applications, mainly as geographic area recognition and image [16]. However, as during diagnostic of COVID-19 using pictures, CNN might well be influenced with indeterminate architecture. Several optimization strategies, like as the particle swarm optimization (PSO) [17], ant lion optimization (ALO), and genetic algorithm (GA) [18], are used to improve the performance of CNN structures [19–23]. An optimization approach is used to choose best model parameters [24]. Progress can be trapped by any optimization process [25]. To facilitate for best recognition of COVID-19 from CT scan images in an Internet of Things environment, novel deep learning and optimization techniques are needed [26].

2 The Proposed OCOA-DDCNN Model

So, under common designation 'COVID-19', a total of 201 novel coronaviruses have indeed been gathered, posing a serious health hazard around the world. Those who seem to be able to sleep normally may be impacted since they are subjected to COVID-19, a potentially fatal disease. Early identification as well as diagnosis of those that are infected with COVID-19 is crucial in order to confine them, providing informal for serious medical assistance dependent upon individual outlook, as well as keep the diseases from spreading farther. Deep learning techniques and machine learning are being studied in the recent history for identifying X-ray pictures. The OCOA-DDCNN method's necessary for the process is depicted in Fig. 1. The OCOA-DDCNN model is created again for estimation of COVID-19 in the present study. The proposed methodology's primary objective would be to accurately forecast that existence of COVID-19. Our proposed method is divided into two steps: preprocessing and prediction. COVID-19 CT scan pictures were initially gathered through an open-source platform. After that, the photos were transmitted to a preprocessing stage, where an image enhancement is applied. The preprocessed photos are then

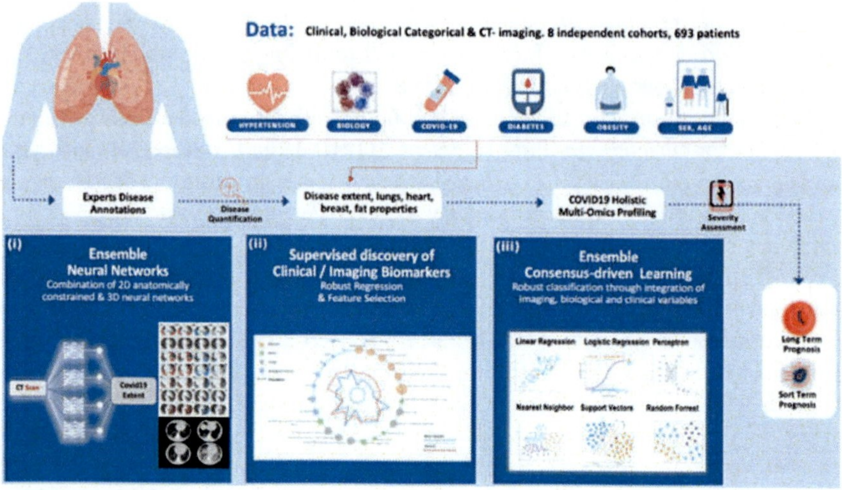

Fig. 1 Proposed methodology's architectural design

transmitted to a prediction step. DDCNN is used in the classification stage to process images. Our suggested classifiers are best constructed with OCOA in mind. This method is used to determine that optimal combination for the classification algorithm, such as kernels of size. Furthermore, the suggested method is used to predict COVID-19 as well as categorize it into two categories: COVID-19 (abnormal) and non-COVID-19 (normal).

2.1 *Stage of Preprocessing*

Its interference as in diagnosis based gets reduced through preprocessing, as well as the dynamic range is improved. The CT scan images differ with one patient to the next. Depending on such characteristic, recognition might lack in produce the optimal outcomes at times. During the feature extraction process, our current proposal employs the Gaussian function to eliminate undesired characteristics or noises from either the image. In frames [27], an image enhancement is used to calculate the dispersion of pixels intensity. Pixel intensity is indeed a mixture of brightness probabilities as well as a stochastic distribution inside the screen at the moment 't'. The following is a formula for a Gaussian model.

$$S(f) = \frac{1}{\sqrt{2\pi\sigma^2}} e^{-\frac{t^2}{2\sigma^2}} \quad (1)$$

Here, s denotes standard deviation, and f denotes interface distance. The operation of standard deviation begins a two-dimensional convolution which is used to obscure photographs. These photos would then be cleaned up to remove any undesirable information or noises. Noise removal, complexity, as softening overall borders in images are all benefits the Gaussian filter. After that, this noise-removed image is passed toward the suggested classifier, which predicts that existence with COVID-19 then categorizes all subjects as corona positive or negative.

2.2 *Stages of Classification*

Our suggested model is used to classify the two COVID-19 classes employing CT images in just this classification stage. Its best hyperparameters of both the OCOA and DDCNN techniques are chosen using an oppositional-based optimization technique. The OCOA and DDCNN techniques be described details in the following.

2.3 *Model of a DDCNN*

DDCNN is a feed-forward neural network with integration as among neurons, distinctive within deep learning approaches. In COVID-19 detection, it has overlapping with different neurons. This chimp optimization technique is used to optimize its DDCNN structure. The model is made up of a complicated structure comprising layers that is used to detect COVID-19 in CT scan pictures. When that comes to extracting a characteristic out of an image of video frames, DDCNN is extremely sensitive and robust. Figure 2 depicts the DDCNN's basic construction.

DDCNN has been designed as well as employed in this study that forecasts that existence with COVID-19. It DDCNN is made up of two convolutional layers, an output nodes, a flattening layer, and a fully connected layer, as well as the input layer. The input layer is the first of those layers but is regarded the most important.

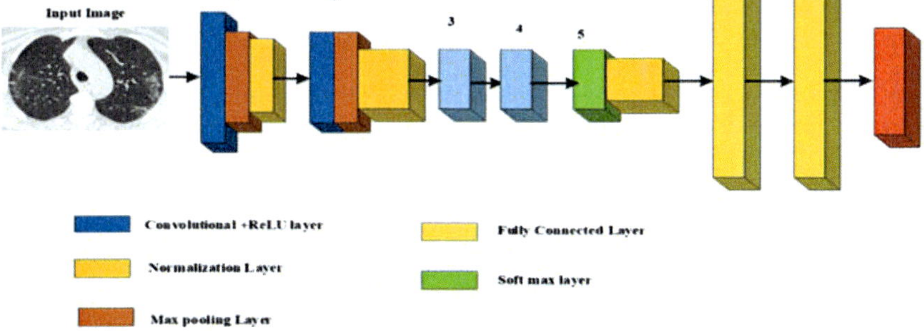

Fig. 2 DDCNN model's architecture

Its second layer is indeed the convolutional layer, which is made up of rectified linear units (ReLU). A convolution layer, a batch normalization (BN) layer and filters. This extracts the features by moving along the time axis. By moving along the horizontal axis, you can see time domain features [28]. Its fourth layer, flatten, corresponds to the C3 vector. Then after, there is a fully connected layer. Neuron is defined in DDCNN as that of the following equation.

$$M(N, J, I) \tag{2}$$

Here, I stands for significant position maps, J for such number of image maps, with N again for number of layers. Here, following is a mathematical model again for relationship across input and output:

$$Y_J^N(I) = E(A) = E\big(Y_J^N(I)\big) \tag{3}$$

The activation function is indicated by E in Eq. (3) and can be written as follows,

$$E(A) = \frac{1}{(1 + e^a)} \tag{4}$$

The output of the second convolution layer is shown below.

$$Y_J^2 = E(A) = E\left(\sum_{I=1}^{M} X_I^1 * W_J^2 + B_J^2(I)\right) \text{ where} J = 1, 2, 3 \ldots M \tag{5}$$

Here, Y_J^2 stands again for $C3$ output layer, B_J^2 again for biases, as well as W_J^2 for the filter. For, the filtering sliding inside a horizontally combine only with feature space. When transmitting the activation layer input, the vectors could be regularized using the BN layer. Its output of layer 4 is then mixed with just a vector.

A forward propagation computational stream of the DDCNN system is determined using the above equations. An error backpropagation method was used to calculate the biases and weights of DDCNN network. At terms of the predicted value, this network can be trained with just a labeled training database. Gradient descent is used to update both the weight and bias of the DDCNN network, as shown above.

$$W^J = W0^J - \eta \frac{\partial E}{\partial W^J} \tag{6}$$

Its local minimum rate is determined as well as corrected during in the DDCNN training phase. There is no pooling layer in this DDCNN structure. The pooling layer gets absent from the DDCNN network, which shortens a network configuration by avoiding the misplaced ideal features.

2.4 Chimp Optimization Algorithm

COA is being used in the suggested methodology because of the benefits of greater convergence, decreased computation, as well as minimal complexity. A mathematical modeling is given below.

(i) **Inspiration**: In most cases, chimp societies are fission–fusion societies. It is one of the communities for which the functionality of the combining of society can also be time variant. Furthermore, each individual in just this society has a certain task as well as capability which may vary throughout time. This decision was made primarily on these grounds. The algorithm develops the goal of separate notions. As a result, each chimp group is unique different tries to locate the search process with its specified unique properties to do out a specific action. Attackers, chasers, barrier, and driving chimps were some of the four categories of chimps found. Their actions of both the chimps are indeed adjusted as during hunt model based on such categories so order to establish more effective hunting operations. These actuators in chimp's program gather the prey not having to just go hunting. Barrier are located at the base of branches and serve as checks for monitoring as well as locking the prey. Catchers quickly seize their prey. Finally, the assailants discover the prey's exit route into lower canopy. Its attackers must be able to identify the target quickly and forecast where it will go. Furthermore, after a successful hunt, all assailants must collect the enormous chunk of meat [29].

(ii) **Prey Chasing and Driving**: This prey could be hunted in COA during the exploitation and exploration stages. As addition to monitoring its prey, a mathematical architecture of pursuit is defined here.

$$D = \left|C.X^{\text{Pray}}(T) - M1.X^{\text{Chimp}(T)}\right| \tag{7}$$

$$X^{\text{Chimp}(T+1)} = X^{\text{Pray}}(T) - BD \tag{8}$$

where X^{Pray} and X^{Chimp} are indeed the chimp and prey location matrices, T is indeed the current iteration number, and B, $M1$, and C are really the effective factors. Its COA's location matrices are calculated by the equation below.

$$B = 2.F.R^1 - a \tag{9}$$

$$C = 2.R^2 \tag{10}$$

$$M1 = \text{Chaotic Value} \tag{11}$$

Here, $R1$ and $R2$ are two randomized parameters that vary within [0, 1], as well as E is indeed the coefficient that fell nonlinearly from 3.4 to 0 over iteration. $M1$ stands for the chaotic parameter, which is calculated using various chaotic maps. As a result, this matrix reflects chimps' sexual motivation in hunting behavior [30].

3 Evaluation of Performance

The section assessed and defended the suggested methodology's efficiency. Statistical metrics such as *F*-measure, specificity, precision, accuracy, recall, and sensitivity were used to verify our suggested methodology. Current techniques like CNN-EPO and CNN-FA were matched to a suggested method. Our suggested approach was evaluated on the 2482 pictures in the COVID-19 CT scan image dataset. These CT pictures are divided into two folders in the database: COVID-19 and non-COVID-19. There seem to be 1325 COVID-19 affirmative image 1321 non-COVID-19 image in the database. A total of 1865 photos are being used for test, with remainder being used to teach our suggested classification. Its suggested product's construction factors are defined in Table 1. Within that part, all statistics measures' performance measures are reported. Figure 3 shows the sequence with sample CT scans. COVID-19 was predicted using CT scan images using the suggested DCNN model. When an individual gets infected with the virus, COVID-19 is positive and detectable, while it is non-COVID-19 if the individual are healthy and afflicted with those other illnesses. These following conditions are used to calculate the confusion matrix:

I. COVID-19 was found positive (TP).
II. COVID-19 was not found, resulting in a true negative result (TN).
III. COVID-19 was not present, which is known as a false positive (FP).

Table 1 Parameters for implementing the recommended methodology

S. No	Method	Description	Item
1	DCNN proposal	The smallest batch size	450
2		Epochs ad infinitum	20
3		Initial rate of learning	0.10
4	COA	Size of population	223
		E	Linearly decreased from 3 to 0
		B	(−2, 2)
		Map with a lot of chaos	{1, 2}
		Criterion for stopping	450

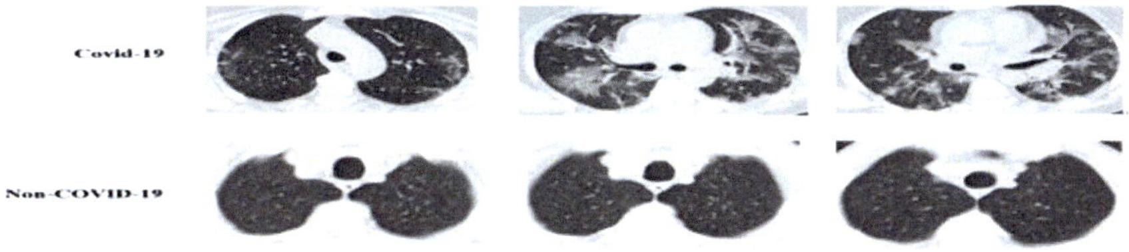

Fig. 3 COVID-19 and non-COVID-19 sample CT images

IV. COVID-19 was not discovered, which is referred to as a false negative (FN).

The proposed methodology is being tested utilizing performance indicators depending just on development of confusion matrix elements.

3.1 Dataset Description

Our proposed method has been tested just on CT scan images that were obtained. Images of COVID-19 patients' CT scans got obtained first from literature [31]. There are 2482 CT images in the this collection, including 1300 X-ray image or 1200 CT image of positive individuals that are not infected with SARS-CoV-2. The data were obtained in live time through patients who are admitted to medical facilities in Sao Paulo, Brazil. One goal of this study is just to promote a unique artificial intelligence method that uses a CT scan image to determine if a person is COVID-19 positive or negative [32–35]. The proposed methodology's performance was validated using the CT scan pictures gathered. Figure 3 shows images of COVID-19 positive and non-COVID-19 instances. Table 2 shows how well the proposed approach got validated utilizing comparison study. Our proposed technique got tested to CNN-FA and models and during comparison. Overall results of a comparison study again for parameter, accuracy, are shown in Fig. 4, and precision analysis is shown in Fig. 5. In epoch 10, the proposed method obtained 0.90 as accuracy (Fig. 4). Likewise, the CNN-FA and CNN-EPO methods achieved 0.85 and 0.83 accuracy, accordingly. In a comparison study, the proposed technique had the highest accuracy in every COVID-19 detection epoch.

Table 2 Analysis of the proposed methodology in comparison

S. No	Method	Image type	Specificity	Recall	Accuracy	F_Messure
1 CNN [15]	Separable	X-ray image	98.65	98.15	98.16	98.10

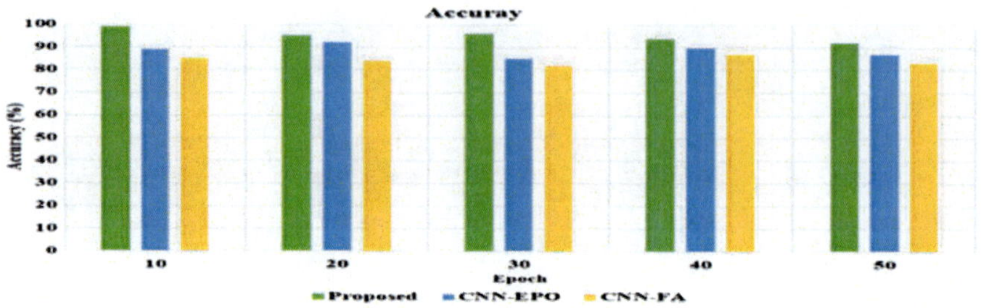

Fig. 4 Accuracy analysis

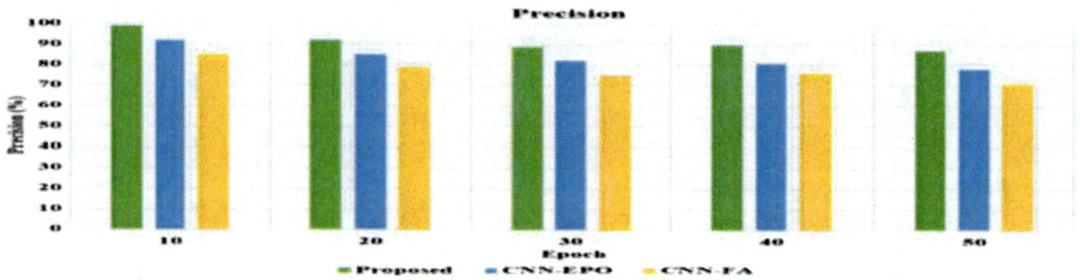

Fig. 5 Precision analysis

4 Conclusion

There in current study, classification model and a COVID-19 prediction associated with the input CT scan images in such an IoT paradigm was provided. Our researchers created an OCOA-based DDCNN framework that accurately predicts and categorizes COVID-19 using images. An OCOA algorithm was used to pick the optimum DCNN hyperparameters. These CT scans first were gathered through sources and afterward forwarded to which was before stage. Noise is eliminated using a Gaussian filter during the preprocessing phase. The images were then categorized to see if COVID-19 was present in the CT scans. There are two parts to the DCNN classifier: training phase. Overall performance of the proposed COVID-19 prediction got examined using MATLAB. For its performance of the project, only, just few metrics was chosen F-measure, accuracy, precision, recall and sensitivity, are some of the proposed methods. The suggested comparative assessment was used to verify the method, which included known methods such as CNN-FA. CNN-EPO, and CNN-EPO both were utilized. The efficiency of the proposed solution was proven by the comparison study. As a result, the proposed model has been validated, and its dominance has been established. Future studies can use large databases and additional classifiers to apply the proposed approach.

References

1. Khan MA, Kadry S, Zhang YD, Akram T, Sharif M et al (2021) Prediction of COVID-19-pneumonia based on selected deep features and one class kernel extreme learning machine. Comput Electr Eng 90:106960
2. Nayak SR, Nayak DR, Sinha U, Arora V, Pachori RB (2021) Application of deep learning techniques for detection of COVID-19 cases using chest X-ray images: a comprehensive study. Biomed Signal Process Control 64:102365
3. Wang Z, Li L, Song M, Yan J, Shi J et al (2021) Evaluating the traditional Chinese medicine (TCM) officially recommended in China for COVID-19 using ontology-based side-effect prediction framework (OSPF) and deep learning. J Ethnopharmacol 272:113957
4. Bhattacharya S, Maddikunta PKR, Pham QV, Gadekallu TR, Chowdhary CL et al (2021) Deep learning and medical image processing for coronavirus (COVID-19) pandemic: a survey. Sustain Cities Soc 65:102589
5. Abbasimehr H, Paki R (2021) Prediction of COVID-19 confirmed cases combining deep learning methods and Bayesian optimization. Chaos Solitons Fractals 142:110511
6. Shiri I, Akhavanallaf A, Sanaat A, Salimi Y, Askari D et al (2021) Ultra-low-dose chest CT imaging of COVID-19 patients using a deep residual neural network. Eur Radiol 31(3):1420–1431
7. Shorten C, Khoshgoftaar TM, Furht B (2021) Deep learning applications for COVID-19. J Big Data 8(1):1–54
8. Sultana J, Singha AK, Siddiqui ST, Pathak N, Sriram AK, Nagalaxmi G, COVID-19 pandemic prediction and forecasting using machine learning classifiers. Intell Autom Soft Comput (Accepted)
9. Singha AK, Kumar A, Kushwaha PK (2018) Patient Cohort Approaches to data science using biomedical field. EPH Int J Sci Eng 1(1):457–462. ISSN: 2454-2016
10. ArunKumar KE, Kalaga DV, Kumar CMS, Kawaji M, Brenza TM (2021) Forecasting of COVID-19 using deep layer recurrent neural networks (RNNs) with gated recurrent units (GRUs) and long short-term memory (LSTM) cells. Chaos Solitons Fractals 146:110861
11. Shatnawi N, Qdais HA (2021) Assessing and predicting air quality in northern Jordan during the lockdown due to the COVID-19 virus pandemic using artificial neural network. Air Qual Atmos Health 14(5):1–10
12. Singha AK, Kumar A, Kushwaha PK (2018) Recognition of human layered structure using Gradient decent model. EPH Int J Sci Eng 1(1):450–456. ISSN: 2454-2016
13. Singha AK, Kumar A, Kushwaha PK (2018) Speed predication of wind using artificial neural network. EPH Int J Sci Eng 1(1):463–469. ISSN: 2454-2016
14. Siddiqui ST, Ahmad MO, Khamruddin M, Gupta AK, Singha AK (2022) Blockchain and IoT for educational certificates generation and verification. In: 2022 2nd International Conference on Computing and Information Technology (ICCIT). IEEE, pp 298–303
15. Singha AK, Zubair S (2022) Machine learning for hypothesis space and inductive bias: a review. In: AIJR Abstracts, p 70
16. Singha AK, Kumar A, Kushwaha PK (2018) Classification of brain tumors using deep Encoder along with regression techniques. EPH Int J Sci Eng 1(1):444–449. ISSN: 2454–2016
17. Singha AK, Pathak SN, Sharma N, Gandhar A, Urooj S, Zubair S, Sultana J, Nagalaxmi G (2022) An experimental approach to diagnose covid-19 using optimized CNN. Intell Autom Soft Comput 34(2):1066–80
18. Sharma S, Jain A (2021) An algorithm to identify the positive COVID-19 cases using genetic algorithm (GABFCov 19). J Interdisc Math 24(1):109–124
19. Gifani P, Shalbaf A, Vafaeezadeh M (2021) Automated detection of COVID-19 using ensemble of transfer learning with deep convolutional neural network based on CT scans. Int J Comput Assist Radiol Surg 16(1):115–123
20. Singha AK, Singla A, Pandey RK (2016) Study and analysis on biometrics and face recognition methods. EPH-Int J Sci Eng 2(6):37–41. ISSN: 2454-2016

21. Jain R, Gupta M, Taneja S, Hemanth DJ (2021) Deep learning based detection and analysis of COVID-19 on chest X-ray images. Appl Intell 51(3):1690–1700
22. Azemin MZC, Hassan R, Tamrin MIM, Ali MAM (2020) COVID-19 deep learning prediction model using publicly available radiologist-adjudicated chest X-ray images as training data: preliminary findings. Int J Biomed Imaging 2020:1–7
23. Narin A, Kaya C, Pamuk Z (2020) Automatic detection of coronavirus disease (covid-19) using x-ray images and deep convolutional neural networks. Image Video Process 1–31
24. Zubair S, Singha AK (2020) Parameter optimization in convolutional neural networks using gradient descent. In: Microservices in big data analytics. Springer, Singapore, pp 87–94
25. Singha AK, Zubair S (2020) Enhancing the efficiency of the stochastic method by using non-smooth and non-convex optimization. J Univ Shanghai Sci Technol 22(10). ISSN: 1007-6735
26. Zubair S, Singha AK (2021) Network in sequential form: combine tree structure components into recurrent neural network. In: IOP conference series: materials science and engineering, vol 1017, no 1. IOP Publishing
27. Chen Z, Zhou Z, Adnan S (2021) Joint low-rank prior and difference of Gaussian filter for magnetic resonance image denoising. Med Biol Eng Compu 59:607–620
28. Heidari M, Mirniaharikandehei S, Khuzani AZ, Danala G, Qiu Y et al (2020) Improving the performance of CNN to predict the likelihood of COVID-19 using chest X-ray images with preprocessing algorithms. Int J Med Informatics 144:104284
29. Singha AK, Zubair S et al. An efficient integrated optimize method based on adaptive meta optimizer. Intell Autom Soft Comput (Accepted)
30. Khishe M, Mosavi MR (2020) Chimp optimization algorithm. Expert Syst Appl 149:113338
31. https://www.kaggle.com/plameneduardo/sarscov2-ctscan-dataset
32. Singha AK, Zubair S et al. Design of ANN based non-linear network using interconnection of parallel processor. Comput Syst Sci Eng (Accepted)
33. Siddiqui ST, Singha AK, Ahmad MO, Khamruddin M, Ahmad R (2022) IoT devices for detecting and machine learning for predicting COVID-19 outbreak. In: Recent trends in communication and intelligent systems. Springer, Singapore, pp 107–114
34. Pathak N, Siddiqui ST, Singha AK, Mohamed HG, Urooj S, Patil AR, Smart quarantine environment privacy through IoT gadgets using blockchain. Intell Autom Soft Comput (Accepted)
35. Singha AK, Zubair S et al. Performance enhancement of adaptive neural networks based on learning rate. Comput Mater Continua (CMC J) (Accepted)

Forecasting COVID-19 Confirmed Cases in China Using an Optimization Method

Anjani Kumar Singha, Nitish Pathak, Neelam Sharma, Pradeep Kumar Tiwari, and J. P. C. Joel

Abstract COVID-19, a brand-new coronavirus, was found in Wuhan, China, in December 2019 and has since spread to 24 additional nations as well as numerous locations in China. The number of confirmed cases continues to rise every day, reaching 34,598 on February 8, 2021. We present our findings a new method was used in this investigation, predictive framework, for such number of reported COVID-19 cases in the China. During the next 10 days, predicated on recently known cases in China. The suggested upgraded adaptable neuro-fuzzy powerful instrument (ANFIS) with an updated floral modeling is used in this model. The salp swarm algorithm (SSA) was used to implement the pollination algorithm (FPA). Generally, SSA is used to enhance FPA in order to minimize its shortcomings. The fundamental theme of the essay FPASSA-ANFIS seems to be a proposed paradigm of improving ANFIS effectiveness through determining FPASSA which was used to determine the ANFIS specifications. The world is also used to analyze the FPASSA-ANFIS model. Statistical figures from the World Health Organization (WHO) on the COVID-19 pandemic for forecast the cases reported these following are indeed the cases for the next 10 days. Most specifically, the FPASSA-ANFIS model in comparison to such a number of other models outperformed them in terms of computing time, root mean squared error (RMSE), and mean absolute percentage (MAP). Researchers also put the suggested model to the tests utilizing two distinct datasets of week pandemic

A. K. Singha
Aligarh Muslim University, Aligarh, India

N. Pathak
Bhagwan Parshuram Institute of Technology (BPIT), GGSIPU, New Delhi, India

N. Sharma
Maharaja Agrasen Institute of Technology (MAIT), GGSIPU, New Delhi, India

P. K. Tiwari (✉)
Manipal University Jaipur, Jaipur, India
e-mail: pradeeptiwari.mca@gmail.com

J. P. C. Joel
Senac Faculty of Ceará, Fortaleza, CE, Brazil

P. Dutta et al. (eds.), *Emerging Technologies in Data Mining and Information Security*, Lecture Notes in Networks and Systems 490,
https://doi.org/10.1007/978-981-19-4052-1_68

confirmed cases from two or more countries: the USA and China. These results also indicated incredible performance.

Keywords Flower pollination algorithm (FPA) · COVID-19 · Salp swarm algorithm (SSA) · Forecasting · Adaptive neuro-fuzzy inference system (ANFIS) · Forecasting

1 Introduction

Coronaviruses, a broad class of viruses, constitute serious human pathogens that cause neurologic disorders, hepatic, respiratory and gastrointestinal. They're dispersed throughout. Humans, livestock, bats, birds, mice, and other wild creatures were among the species affected [1–3]. Major outbreak of such past two years MERS-CoV and SARS-CoV coronaviruses adopted the protocol in 2003 and 2012 [4]. The World Health Organization (WHO) will hold its annual meeting in December 2019. Several instances of respiratory sickness being reported toward the World Health Organization (WHO) by China related to a set of people whom have attended a Wuhan fish market [5]. Wuhan is now dealing also with spreading of such a new coronavirus known as COVID-19.

Although COVID-19 far more consists of two bat-derived coronavirus infections, our investigators determined that something that probably came in bats [6]. Unfortunately, its origin of the COVID-19 is still unknown, and further inquiry seems required. So because majority of cases have been revealed in Wuhan, this same city might have been placed on lockdown at 10:00 a.m. on January 23, 2021, without others permitted out or in. As a result, only those mass transit, including public buses, metro, airports, and trains, had been shut down to stop the COVID-19 on the rise. In addition, a few Hubei cities provinces have been placed under lockdown. Yet, there have been numerous confirmed occurrences in many other Chinese cities as well as in roughly 24 other countries. However, it is less than Wuhan city worldwide (till February 3, 2021). Its Centers for Disease Control and Prevention (CDC) will close on January 30, 2021. COVID-19 was confirmed to be transmitted from person to person by the CDC apparently. A virus could be transferred through air, near personal contact, touching surfaces, or items; according to the CDC, it includes viral particles and face pollution are infrequent. One of the really important issues with COVID-19 would be its incubation period that can last for 14 days, as per [7], which during time this could transmit toward others. Furthermore, a Chinese team reports in [8] that now the typical incubation duration is 3 days, ranging 1 to 25 days, with an average age of 48.0 years. As a response, the number of cases has been confirmed steadily rising, particularly in China. The propagation of this virus is extremely dangerous, necessitating increasingly stringent laws and strategies, that are currently in place in so many Chinese cities, particularly in Hubei province. As a result, forecasting confirmed cases inside the future days is crucial in order to put in place the appropriate protection procedures.

Siddiqui et al. 9] established a mathematical approach to predict the precise figure of cases of COVID-19 that had not been reported during first half of January 2021. Researchers came to the conclusion there were 469 reported incidents between January 1 and January 15, 2021. They also predicted that by the 17th of January 2021, the number of instances would have climbed by a factor of 21. Using datasets from 565 Japanese individuals whose are transported from Wuhan from 29 to 31 January 2021, Pathak et al. [10] established one estimating methodology again for COVID-19 incidence rate in Wuhan, China. Researchers come to the conclusion that expected rate is 10%, with a death rate of 0.4–0.7%. The majority of Japanese citizens relocated from Wuhan, on either hand, is sufficient to assess infection and fatality rates. Tang et al. [11] established a quantitative methodology for calculating COVID-19 transmission risk. They came to the conclusion that its initial sequence number is 6.47. Researchers also anticipated how many confirmed cases there would be in seven days (23 to 29 January 2021). It also predicted that now the top will be achieved in two weeks (from 23 January 2021). Thompson et al. [12] utilized data from 47 individuals to calculate COVID-19 sustained sentient transfer. Its transmission is 0.4, according to the source, from symptoms Half of the data evaluated is related to hospitalization, the transfer is just 0.012. The authors provided a methodology for estimating the COVID-19 carries a high risk of death in [13]. Their estimates the percentages for two possibilities are 6.1% and 9.4%, respectively. Its reproducing numbers both for circumstances were calculated to be 2.1 and 3.2, accordingly. The COVID-19 could potentially spark a pandemic, according to the estimates. As an example, DeFelice et al. [14] established a two—compartment framework to anticipate spillover risk of transmission cases. Its suggested model has been tested using WNY outbreak data from New York Long Island, from 2002 to 2014. Authors tested four methods for forecasting time series that are auto-regressive integrated moving averages (ARIMA), radial basis function (RBF), time-delay neural networks (TDNNs), and the average (ARIMA). The results of the comparisons revealed that the MLP surpassed the other models. In [15], regarding seasonality forecasting, the authors have proposed an ensembles modification of Kalman filter-based model in Influenza epidemics. They tested the suggested model using data from New York's influenza seasons for the past six years (2004–2009), I lived in York City. Singha et al. [16] also presented week virus chance to generate on the vulnerable, ensembles adjustments Kalman filter, with influenza-like sickness. Furthermore, the scientists used a method of Bayesian inference to predict Ebola outbreaks in numerous Guinea, African nations, Liberia in [17]. Massad et al. [18] created a quantitative analytical model as well as forecasting the SARS epidemic's transmission. Researchers came to the conclusion that such reproductive numbers in communities, Toronto and Hong Kong, were 2.2 and 2.7, correspondingly. Singha et al. [19] a flu model has been proposed.

In addition, Nah et al. [20] suggested a possibility methodology for predicting MERS transmission. In time series data, the Adaptive Neuro-Fuzzy Inference System (ANFIS) [21] is commonly used. It performed well in a variety of current applications, including prediction and forecasting. It gives you more options for detecting nonlinearity in time series and combining them both fuzzy sets and artificial neural networks (ANNs) have qualities. It is used in a variety of forecasting applications;

for instance, in, an ANFIS but also empirical mode decomposition stock price forecasting model was developed. Singha et al. [22] presented the time series forecasting model TAIEX, which combines Ordered weighted averaging (OWA) and ANFIS. Another ANFIS-based forecasting of time series approach for electricity prices was given in [23]. Svalina et al. [24] proposed a five-day prediction models built on ANFIS for near market prices in a financial markets. Services also allow electricity load forecasting model was presented by Ekici and Aksoy [25–28]. More specifically, ANFIS is used to forecast electricity loads [29].

2 Materials and Procedures

2.1 *System of Adaptive Neuro-Fuzzy Inference (ANFIS)*

These ANFIS fundamentals were highlighted in the following section; fuzzy logic and neural networks are combined in the ANFIS model [21]. It creates a map between input and output using IF–THEN concepts. The ANFIS model is depicted in Fig. 1, here y and x are Layer 1's inputs, and 1i is node I's output, which would be determined as follows:

$$\mathrm{Out}_{1i} = \mu_{M_i}(X), \quad \text{where } i = 1, 2, \quad \mathrm{Out}_{1i} = \mu_{N_i - 2}(Y) \quad \text{where } i = 3, 4 \tag{1}$$

$$\mu(X) = \frac{1}{e^{-\left(\frac{X-\rho_i}{\alpha_t}\right)^2}}, \tag{2}$$

Its generalized Gaussian membership functions are denoted by the letter μ. The membership values of m are defined by M_i and N_i. Its assumption parameters α_t as well as ρ_t are set.

Layer 2's output is computed as follows:

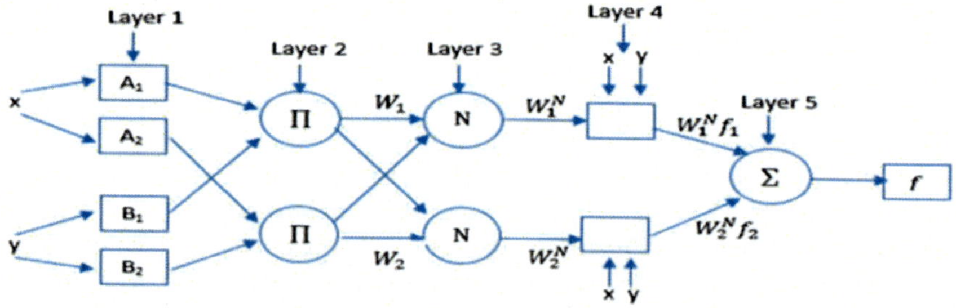

Fig. 1 Structure of the ANFIS model

$$\text{Out}_{2i} = \mu_{M_i}(X)^* \mu_{N_{i-2}}(Y) \tag{3}$$

Similarly, Layer 3's output is calculated as follows:

$$\text{Out}_{3i} \frac{W_i}{\sum_{i=1}^{2} W_i} \tag{4}$$

Layer 4's output is computed as follows:

$$\text{Out}_{4i} = W_i(A_i(X) + B_i(Y) + €) \tag{5}$$

here A_i, B_i, and € are the node i's subsequent properties. Layer 5 has only one node, and its output is as follows:

$$\text{Out}_{5i} = \sum_i W_i f_i \tag{6}$$

2.2 *Algorithm for Flower Pollination (FPA)*

Yang [30] suggested the flower pollination algorithm as an optimization tool. It mimics pollinator transmission of pollen from natural flowers. In this algorithm, two pollination strategies are used. In self-pollination, pollination occurs without the presence of pollinators; however, pollen is transmitted from one plant to another during cross-pollination. Most specifically, cross-pollination is known as universal pollination, whereas self-pollination is known as local pollination.

This following is a mathematical representation of universal pollination or cross-pollination:

$$X_i^{t+1} = X_i^t + P\left(X_i^t + F\right) \tag{7}$$

here X_i^t signifies pollen i at iteration t, and P denotes pollination vigor or step size. The optimal solution or target position is F.

2.3 *Swarm Algorithm by Salp (SSA)*

Introduced the SSA optimization approach. It mimics the salps' natural behavior. The salp chain is the name for this type of activity. SSA's mathematical model begins by dividing its population is divided into two groups: leaders and follower. The whole front salp is the leader, while the other salps are indeed the followers. Using integer variables, the space defined in Eq. (10) is used to update the positions of the salps.

$$X_i^t = F_j + c_1\big((u_{bj} - l_{bj})c_2 + l_{bj}\big)c_3 \leq 0 \quad (8)$$

where X_i^t signifies the position of the leader in the jth dimension. The target position is F_j. The max and min boundaries are represented by u_{bj} and l_{bj}, respectively. Random numbers in the range [0, 1] are denoted by c_2 and, c_1 is parameter.

3 The Proposed Approach

Our suggested FPASSA-ANFIS approach is discussed in this section. As shown in Fig. 2, it is a time series approach for anticipating verified COVID-19 cases.

An enhanced FPA is used by the FPASSA-ANFIS for train its optimizing the parameters of the ANFIS model. As with the standard ANFIS model, the FPASSA-ANFIS has five layers. All input variables are found on layer one. Layer 5 generates

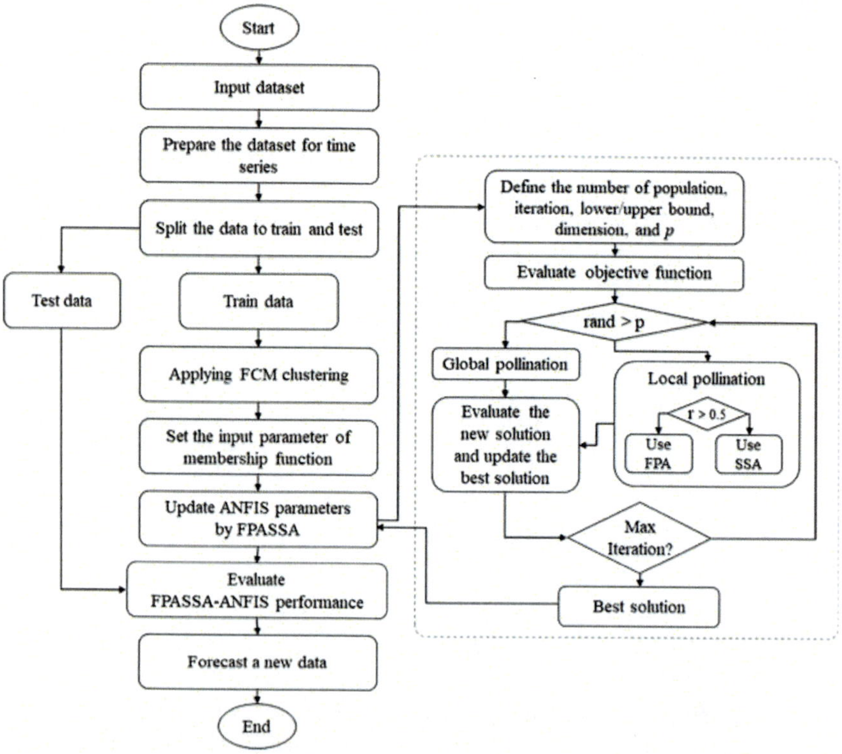

Fig. 2 FPASSA-ANFIS is a method proposed by FPASSA

the expected values. It FPASSA is being used to pick layers with the best weights 4 and 5 during the process of learning.

It FPASSA-ANFIS begins with transforming its input data into a time series. The autocorrelation function (ACF) was used in our scenario. One of the strategies used to discover patterns in the data is ACF, which graphical interface on the correlation between samples separated over various time lags. As a result, factors with ACF bigger than 0.4, i.e., 5-lags, are considered in this study. Furthermore, the training set accounts for 85% of the dataset, while the testing data accounts for 15%. The clusters in the ANFIS model are determined via the fuzzy c-mean (FCM) approach.

The FPASSA method prepares the parameters for the ANFIS model. The computation error [as in Eq. (9)] between both the real data and the anticipated information is used to verify the parameters' accuracy throughout the training phase.

$$\mathrm{MSE} = \frac{1}{L}\sum_{i=1}^{l}(T_i - q_i) \tag{9}$$

where T denotes the actual data and q denotes the expected data. The sample length is L. The goal function's smaller values imply good performance the ANFIS parameter.

4 Experiments

This section includes an explanation of the dataset that was utilized, as well as performance measures, parameterization including all algorithms, experiment results, and comments.

4.1 Description of Datasets

The COVID-19 dataset is the study's core dataset. It came from the World Health Organization's website (https://www.who.int/emergencies/diseases/novel-coronavirus-2019/situation-reports/). According indicated in Table 1, this comprises current confirmed cases on a daily basis in China since January 20, 2021 to February 21, 2021. Researchers used 85% of the remaining data will be used to train the model and 15% to be used to test it.

Furthermore, researchers assessed our suggested method's effectiveness utilizing datasets of monthly virus confirmed cases. The first is known as DS1 and came from the CDC (https://www.cdc.gov/flu/weekly/). This begins in week 40 of 2015 and lasts through week 6 of 2021. The second is referred to as DS2. That one was collected from the WHO website (https://www.who.int/influenza). It covers information on confirmed influenza infections throughout China, starting this week one in 2016 to week eight in 2021.

Table 1 COVID-19 dataset

Date (D/M/Y)	Confirmed	Date (D/M/Y)	Confirmed	Date (D/M/Y)	Confirmed
20/01/2020	278	31/1/2020	9720	10/2/2020	40,554
22/1/2020	309	1/2/2020	11,821	11/2/2020	42,708
23/1/2020	571	2/2/2020	14,411	12/2/2020	44,730
24/1/2020	830	3/2/2020	17,283	13/2/2020	46,550
25/1/2020	1297	4/2/2020	20,471	14/2/2020	48,548
26/1/2020	1985	5/2/2020	24,363	15/2/2020	50,054
27/1/2020	2741	6/2/2020	28,060	16/2/2020	51,174
28/1/2020	4537	7/2/2020	31,211	17/2/2020	70,635
29/1/2020	5997	8/2/2020	34,598	18/2/2020	72,528
30/1/2020	7736	9/2/2020	37,251		

Measures of Performance

Our suggested technique's quality is evaluated that use the following performance measures:

$$\text{Root Mean Square Error (RMSE)}: \sqrt{\frac{1}{N}\sum_{I=1}^{N}(YYP_i - Y_i)} \tag{10}$$

where it projected and original values are YP and Y, correspondingly.

$$\text{MAE} = \frac{1}{N}\sum_{I=1}^{N}(YYP_i - Y_i) \tag{11}$$

Mean absolute percentage error (MAPE):

$$\text{MAPE} = \frac{1}{N}\sum_{I=1}^{N}\frac{YP_i - Y}{YP_i}$$

4.2 Parameter Configuration

One goal of this research would be to compare its FPASSA's capacity to forecast a COVID-19 to other methods, like the ANFIS as well as trained ANFIS models employing FPASSA, PSO, ABC,GA, ABC and FPA. Table 2 shows how to establish the parameters of these models [31, 32].

Some normal life expectancy, including such population size, is set to 30 iterations. Furthermore, the chosen parameters were each algorithm is run for 35 times to ensure comparable situations. The parameters chosen were provided good results in prior studies [33–36].

Algorithm	Parameters setting
ANFIS	Maximum epochs = 90, Error = 0, The first step is to = 0.02, Reduce the rate = 0.8, Boost the rate = 2.2
GA-ANFIS	Type of crossover = 2,
PSO-ANFIS	wight Maximum = 0.8, weight Minimum = 0.3, $C1 = 3$, $C2 = 3$ cp = 2, mp = 0.03
ABC-ANFIS	$a = 2$, bees on the job = $N/2$, bees on the lookout = $N/2$ FPA-ANFISStandard gamma = 1.5, Swich probablity = 0.8
FPASSA-ANFIS	
Standard gamma = 1.5, Swich probablity = 0.3, $C_2 \in [0, 1]$, $C_3 \in [0, 1]$	

4.3 FPASSA's Ability to Forecast DS1

Its performance of proposed FPASSA at predicting both DS1 and DS2 is discussed in this section. Table 2 shows that the FPASSA excelled the comparable approaches on any and all criteria, while the FPA came in second. It FPASSA was rated first in terms of RMSE, and MAPE followed by the FPA, GA, and lastly the ABC (Fig. 3a–f). These findings suggest that the suggested strategy may successfully optimize the ANFIS model's parameters and deliver good outcomes in terms of performance metrics [36].

Table 2 Performance of proposed FPASSA

Dataset	Method	RMSE	MAE	MAPE
DS1	ANFIS	850	560	86.16
	PSO	700	570	45.23
	GA	865	560	23.45
	ABC	913	568	34.80
	FPA	760	540	45.67
	FPASSA	**608**	**340**	**34.23**

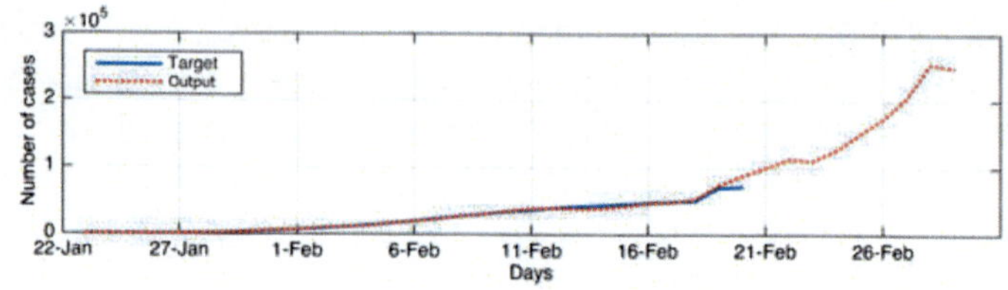

(a) ANFIS

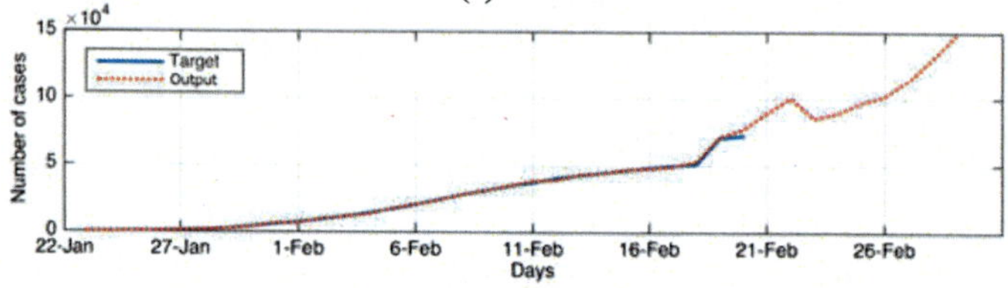

(b)PSO

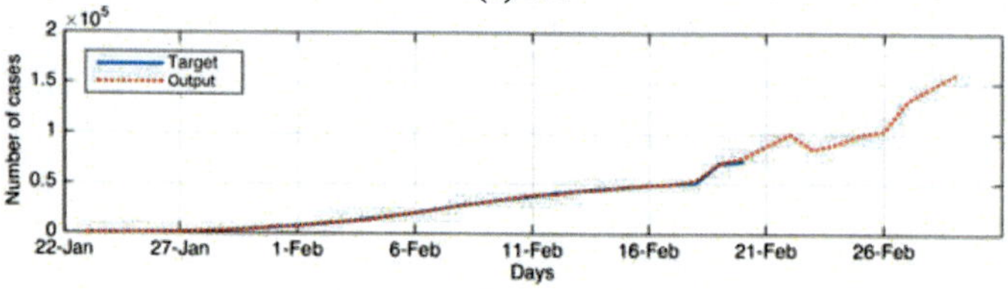

(c) GA

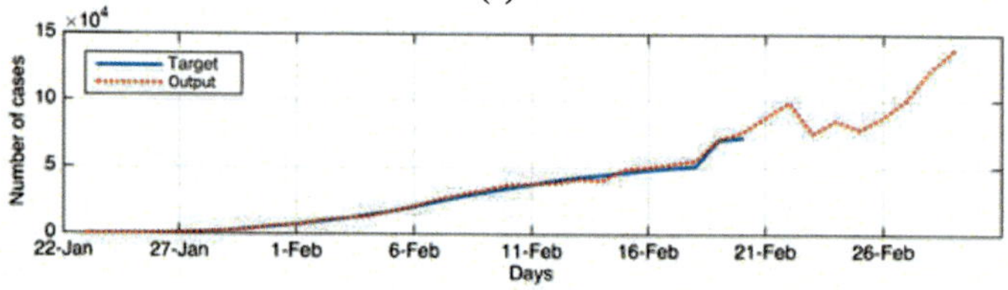

(d) ABC

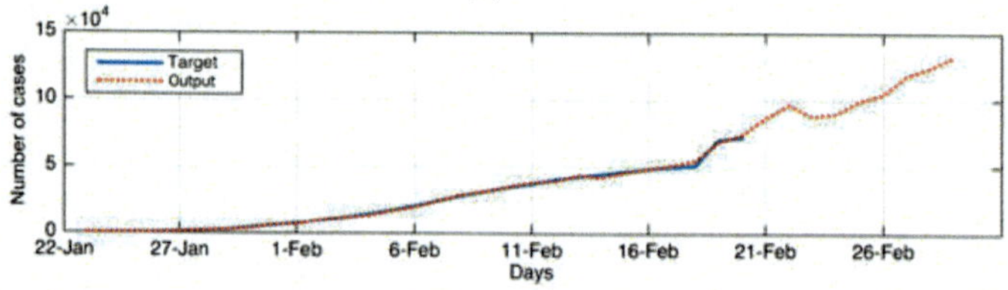

(e) FPA

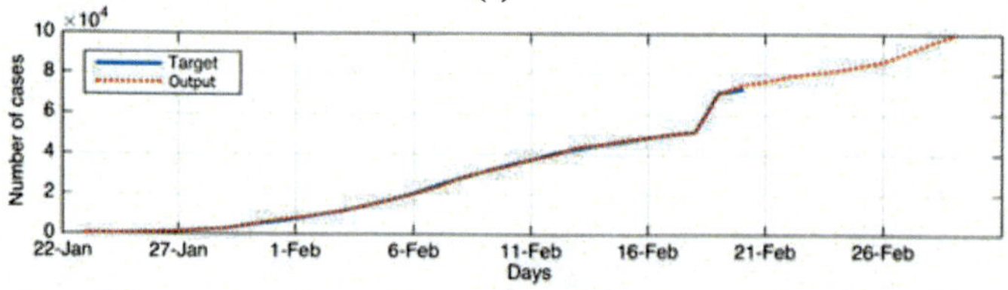

(f) FPASSA

Fig. 3 With all techniques, the real data (target) is compared to the anticipated data output)

5 Conclusions

Using the salp swarm algorithm, this research offered a tweaked version of the pollination algorithm for flowers (FPA). The improved version, known as FPASSA, is used to enhance the ANFIS' efficiency by deciding the best solution for its variables. FPASSA-ANFIS, a newly FPASSA-ANFIS, was created. Suggested within ten days, our FPASSA-ANFIS model does have a ability to accurately anticipate the total number of confirmed instances. FPASSA-ANFIS also beats in terms of other forecasting models MAE, RMSE, and MAPE. Moreover, this suggested method was evaluated using datasets containing weekly pandemic confirmed cases from the USA and China, as well as the results indicated that it performed well. Our suggested FPASSA-ANFIS, based on its promising outcomes, can be used in a variety of forecasting applications.

References

1. Chen Y, Liu Q, Guo D (2020) Emerging coronaviruses: genome structure, replication, and pathogenesis. J Med Virol
2. Siddiqui ST, Singha AK, Ahmad MO, Khamruddin M, Ahmad R (2022) IoT devices for detecting and machine learning for predicting COVID-19 outbreak. In: Recent trends in communication and intelligent systems (pp 107–114). Springer, Singapore
3. Singha AK, Zubair S (2022) Machine learning for hypothesis space and inductive bias: a review. AIJR Abstracts, p 70
4. Cauchemez S, Van Kerkhove M, Riley S, Donnelly C, Fraser C, Ferguson N (2013) Transmission scenarios for middle east respiratory syndrome coronavirus (MERS-CoV) and how to tell them apart. Euro Surveill Bull Eur Sur Les Mal Transm Eur Commun Dis Bull 18, 20503
5. WHO Organization. Novel Coronavirus (2019-nCoV) 2020. Available online: https://www.who.int/. Accessed on 27 Jan 2020
6. Lu R, Zhao X, Li J, Niu P, Yang B, Wu H, Wang W, Song H, Huang B, Zhu N et al (2020) Genomic characterisation and epidemiology of 2019 novel coronavirus: implications for virus origins and receptor binding. Lancet 395:565–574
7. Cheng ZJ, Shan J (2019) Novel coronavirus: where we are and what we know. Infection
8. Guan WJ, Ni ZY, Hu Y, Liang WH, Ou CQ, He JX, Liu L, Shan H, Lei CL, Hui DS et al (2020) Clinical characteristics of 2019 novel coronavirus infection in China. medRxiv. Available online: https://www.medrxiv.org/content/early/2020/02/09/2020.02.06.20020974.full.pdf. Accessed on 9 Feb 2020
9. Siddiqui ST, Ahmad MO, Khamruddin M, Gupta AK, Singha AK (2022) Blockchain and IoT for educational certificates generation and verification. In: 2022 2nd international conference on computing and information technology (ICCIT), pp 298–303. IEEE
10. Pathak N, Siddiqui ST, Singha AK, Mohamed HG, Urooj S, Patil AR, Smart quarantine environment privacy through IoT gadgets using blockchain. Intell Autom Soft Comput (Accepted)
11. Tang B, Wang X, Li Q, Bragazzi NL, Tang S, Xiao Y, Wu J (2020) Estimation of the transmission risk of the 2019-nCoV and its implication for public health interventions. J Clin Med 9:462
12. Thompson RN (2020) novel coronavirus outbreak in Wuhan, China, 2020: intense surveillance is vital for preventing sustained transmission in new locations. J Clin Med 9:498
13. Jung SM, Akhmetzhanov AR, Hayashi K, Linton NM, Yang Y, Yuan B, Kobayashi T, Kinoshita R, Nishiura H (2020) Real time estimation of the risk of death from novel coronavirus (2019-nCoV) infection: inference using exported cases. J Clin Med 9:523

14. DeFelice NB, Little E, Campbell SR, Shaman J (2017) Ensemble forecast of human West Nile virus cases and mosquito infection rates. Nat Commun 8:1–6
15. Shaman J, Karspeck A (2012) Forecasting seasonal outbreaks of influenza. Proc Natl Acad Sci USA 109:20425–20430
16. Singha AK, Singla A, Pandey RK (2016) Study and analysis on biometrics and face recognition methods. EPH-Int J Sci Eng (ISSN: 2454-2016). 2(6):37–41
17. Singha AK, Zubair S et al, Design of ANN based non-linear network using interconnection of parallel processor. Intell Autom Soft Comput (Accepted)
18. Massad E, Burattini MN, Lopez LF, Coutinho FA (2005) Forecasting versus projection models in epidemiology: the case of the SARS epidemics. Med Hypotheses 65:17–22
19. Singha AK, Zubair S et al, An efficient integrated optimize method based on adaptive meta optimizer. Intell Autom Soft Comput (Accepted)
20. Nah K, Otsuki S, Chowell G, Nishiura H (2016) Predicting the international spread of Middle East respiratory syndrome (MERS). BMC Infect Dis 16:356
21. Jang JS (1993) ANFIS: adaptive-network-based fuzzy inference system. IEEE Trans Syst Man Cybern 23:665–685
22. Singha AK, Pathak SN, Sharma N, Gandhar A, Urooj S, Zubair S, Sultana J, Nagalaxmi G (2022) An experimental approach to diagnose covid-19 using optimized CNN. Intell Automa Soft Comput 34(2):1066–1080
23. Pousinho HMI, Mendes V, Catalão JPdS (2012) Short-term electricity prices forecasting in a competitive market by a hybrid PSO–ANFIS approach. Int J Electr Power Energy Syst 39:29–35
24. Svalina I, Galzina V, Lujić R, ŠImunović G (2013) An adaptive network-based fuzzy inference system (ANFIS) for the forecasting: the case of close price indices. Expert Syst Appl 40:6055–6063
25. Ekici BB, Aksoy UT (2011) Prediction of building energy needs in early stage of design by using ANFIS. Expert Syst Appl 38:5352–5358
26. Singha AK, Kumar A, Kushwaha PK (2018) Speed predication of wind using artificial neural network. EPH-Int J Sci Eng 1(1):463–469 (ISSN: 2454–2016)
27. Zubair S, Singha AK (2021) Network in sequential form: combine tree structure components into recurrent neural network. In: IOP conference series: materials science and engineering, vol 1017, No 1. IOP Publishing
28. Sultana J, Singha AK, Siddiqui ST, Nagalaxmi G, Sriram AK, Pathak N (2022) COVID-19 pandemic prediction and forecasting using machine learning classifiers. Intelli Autom Soft Comput 2022:1007–1024
29. Cheng CH, Wei LY (2010) One step-ahead ANFIS time series model for forecasting electricity loads. Optim Eng 11:303–317
30. Zubair S, Singha AK (2020) Parameter optimization in convolutional neural networks using gradient descent. Microservices in big data analytics, pp 87–94. Springer, Singapore
31. Singha AK, Kumar A, Kushwaha PK (2018) Patient cohort approaches to data science using biomedical field. EPH-Int J Sci Eng (ISSN: 2454-2016) 1.1:457–462
32. Singha AK, Kumar A, Kushwaha PK (2018) Classification of brain tumors using deep encoder along with regression techniques. EPH-Int J Sci Eng 1(1):444–449 (ISSN: 2454–2016)
33. Catalão JPDS, Pousinho HMI, Mendes VMF (2010) Hybrid wavelet-PSO-ANFIS approach for short-term electricity prices forecasting. IEEE Trans Power Syst 26:137–144
34. Bhattacharya A, Ghosh G, Mandal R, Ghatak S, Samanta D, Shukla VK, Mukherjee S, Dutta S, Mandal A. Predictive analysis of the recovery rate from corona virus (Covid-19)
35. Singha AK, Kumar A, Kushwaha PK (2018) Recognition of human layered structure using Gradient decent model. EPH-Int J Sci Eng 1(1):450–456 (ISSN: 2454–2016)
36. Singha AK, Zubair S (2020) Enhancing the efficiency of the stochastic method by using non-smooth and non-convex optimization. J Univ Shanghai Sci Technol 22(10) (ISSN: 1007–6735)

37. Ture M, Kurt I (2006) Comparison of four different time series methods to forecast hepatitis A virus infection. Expert Syst Appl 31:41–46
38. Ho YC, Tsai CT (2011) Comparing ANFIS and SEM in linear and nonlinear forecasting of new product development performance. Expert Syst Appl 38:6498–6507
39. Singha AK, Zubair S (2020) Enhancing the efficiency of the stochastic method by using non-smooth and non-convex optimization. J Univ Shanghai Sci Technol 22(10) (ISSN: 1007–6735)

Predictive Analysis of COVID-19 Data Using Two-Step Quantile Regression Method

K. Lavanya, G. V. Vijay Suresh, and Anu Priya Koneru

Abstract In the year 2019, research community began with new challenge called novel coronavirus disease (COVID-19) and has opened up new challenges for the research community. From the report of the World Health Organization (WHO), the new virus COVID-2019 (World Health Organization (2020) Coronavirus disease 2019 (COVID-19): situation report, p 67) causes dangerous illness to the concerned person, and it spread to other peoples with huge rate through contact. Such kind of pandemic analysis needs efficient methods to predict data and also helped further to analyze such epidemic risks. These kinds of analyses are used to handle and control the epidemic kind of diseases. Regression analysis is kind of ML methods and is worked well to analyze such kind of epidemic data. The work in this paper about analysis of COVID-19 data especially focused in the state of Andhra Pradesh. First, the data are collected from the website (i.e., https://prsindia.org/). Next, we applied various regression techniques like linear, multi-linear and quantile regression for COVID-19 data for the prediction of cases. Further extended work to derive penalized quantile results using lasso. The results shows that the two-step quantile regression (TSQR) has been shown to be a better predictive method for predicting confirmed cases compared to linear and multi-linear regressions in terms of MSE and R-Score parameters.

Keywords COVID-2019 · Machine learning · Predictive analysis · Regression analysis

K. Lavanya (✉) · A. P. Koneru
Department of IT, Lakireddy Bali Reddy College of Engineering, Mylavaram, India
e-mail: lavanya.kk2005@gmail.com

G. V. Vijay Suresh
Department of CSE, Lakireddy Bali Reddy College of Engineering, Mylavaram, India

P. Dutta et al. (eds.), *Emerging Technologies in Data Mining and Information Security*, Lecture Notes in Networks and Systems 490,
https://doi.org/10.1007/978-981-19-4052-1_69

1 Introduction

Today, most of the researchers in the entire world panic about the issue of coronavirus (COVID-19) and its consequences [1]. Basically, the virus is discovered in the city of Wuhan, China, and with some reports from the social media, it was created in the China Lab in the month of the November 2019 [2, 3]. Normally, it takes approximately 14 days to showing symptoms of a person affected with COVID-2019. The general symptoms shown in Fig. 1 include fever, cough, issue of breathing, kidney failure, etc. [1]. With reports, most of the countries affected with this disease, and it affects entire economy of world. As per the last year 2020, World Health Organization (WHO) report, a total of 100,075 deaths out of 15,225,252 active cases occurred in the world. Later, the disease spread rapidly across 172 countries from December 1, 2020 to today in the first wave. Recently, second wave of COVID-2019 started and experts suggested to be more conscious.

The study considered the local state analysis (i.e., Andhra Pradesh) of COVID-19 and after studying various reports from the derivative analysis of media from the past to the present date since 05/09/2020 and is illustrated in Fig. 2. The machine learning is an alternative technology where automated data analysis is done by the machines rather than humans. Regression analysis is kind of ML methods and is worked well to analyze realtime data. A number of regression techniques available include linear, multi-linear, and quantile. Today, most of the real-time applications predicted the data with regression analysis. Also these methods were proven their effectiveness in predicting health data [4, 5] in recent days. The study about this paper is done

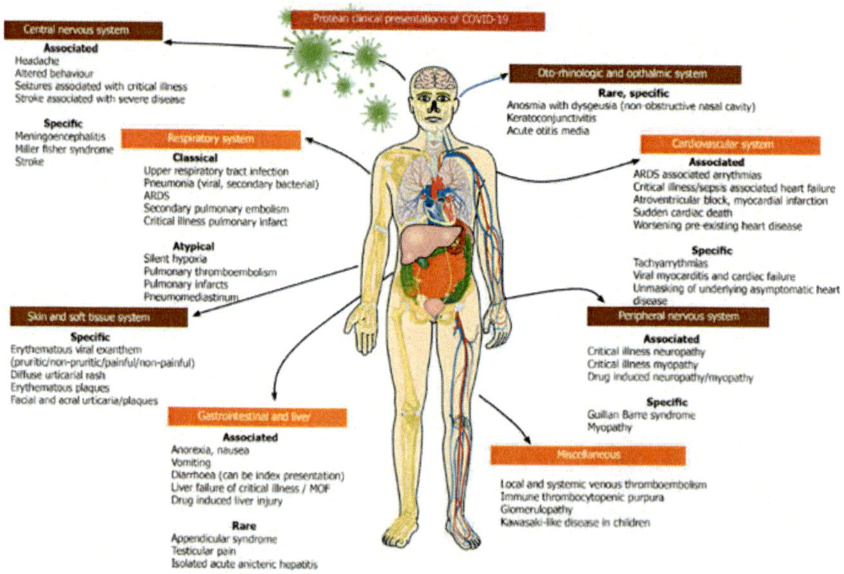

Fig. 1 Typical symptoms caused by COVID-19 artificial intelligence

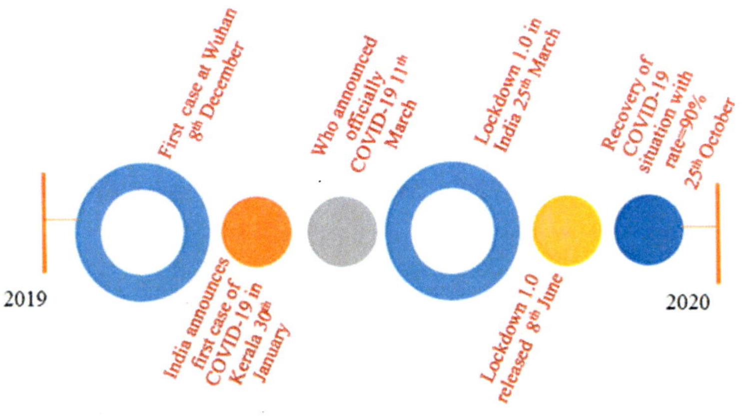

Fig. 2 Timeline diagram of COVID-19

to analyze COVID-19 data especially in the state of Andhra Pradesh. We applied various regression techniques like linear, multi-linear, and quantile regression for COVID-19 data for the prediction of active cases. Table 1 describes the state of Andhra over the past week and shows how the condition is becoming more critical day by day in the event of COVID-19.

It is observed daily that about more than 10,000 cases occur by day. We also notice that from the next few days, the condition will be more critical or become to be quite normal. The complete timeline of the COVID-19 from 2019 to 2020 is shown in Fig. 2.

The main objective of this is study is often estimation of predictive analysis of corona cases with the relationship between confirmed cases with time to death and other independent variables which includes cured number, active cases, and mortality ratio. It is observed that under the normality assumption on the confirmed cases in most of the times in dataset collection leads to non-normality showed only Andhra

Table 1 Confirmed cases in Andhra Pradesh past one year

Days	Confirmed	Active	Cured	Deaths
100	5070	2191	2802	77
125	16,097	8586	7313	198
152	217,040	85,486	129,615	1939
175	414,164	99,129	312,687	3796
223	734,427	49,513	672,479	6086
260	852,995	19,757	826,344	6854
303	881,599	3700	870,342	6976
335	887,720	1289	879,131	7152

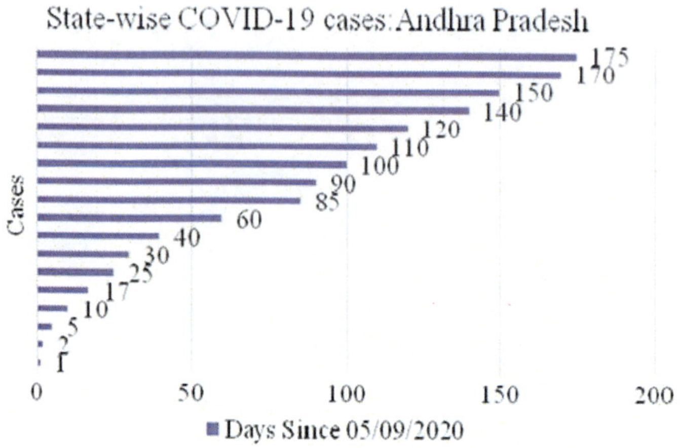

Fig. 3 Number of cases compared to the number of days from March 12, 2020 to 09-05-2020

Pradesh situation in Fig. 3. To solve this, quantile regression is considered and is applicable to many fields like finance, sociology, and immunology.

The advantage of method is need not required special distribution to estimate outcome (i.e., confirmed cases) and also fits for bunch of curves rather than single curve. In this work, a two-step quantile regression (TSQR) approach is introduced for predicting the confirmed corona cases and applied it into the generated dataset. The work is compared with standard linear and quantile regression in terms of MSE and bias and produced promising results by the proposed method.

2 Materials and Methods

2.1 *Data Set*

Work in this paper collected the data of the COVID-19 especially Andhra Pradesh from the website (https://prsindia.org/). Later, we generated dataset with various features which include confirmed, active, cured and death cases from the period from March 12, 2020 to January 30, 2021.

2.2 *Regression Analysis*

Regression analysis is kind of ML methods and is worked well to analyze real-time data [6, 7]. The analysis is a combination of two variables which include predicted and outcome variables. The intention of this method is to establish linear relation among an outcome variable (Y) with many of predictor variables (X's). Similarly, consider the COVID-19 data where confirmed cases are predicted with observed variables which include active, cured, and death cases [8]. The work [9] applied regression analysis to predict confirmed cases of COVID-19 [9]. The regression equation with observed and predictor variables X and Y are to be represented as:

$$Y = aX + b \tag{1}$$

From Eq. (1), it is represented as the intercept on the axis and slope of the regression line is denoted as the and is shown in above equation. Also these parameters are to be considered as regression parameters used for predictive analysis. Hence, it is required suitable learning mechanisms to estimate such parameters. The regression analysis is further defined into various other models using different degree of polynomials which includes exponential, two-degree, three-degree, and fourth-degree polynomials. The regression models with various polynomials include:

$$Y = ae^{bx} \tag{2}$$

$$Y = aX^2 + bX + c \tag{3}$$

$$Y = aX^3 + bX^2 + cX + d \tag{4}$$

$$Y = aX^4 + bX^3 + cX^2 + dX + e \tag{5}$$

From Eqs. (2)–(5) a, b, c, d, and e are called the parameters of the regression analysis with polynomials.

3 Two-Step Quantile Regression (TSQR)

The quantile regression is advanced to linear regression and also never finds any kind of multicollinearity in the predicted variable. The work in this paper introduced a two-step quantile regression method which is shown in Fig. 4. In the first stage, data is shifted into log transformation. Later, we applied correlation among both predicted and observed data to extract optimal features. Then, selected features were applied to quantile.

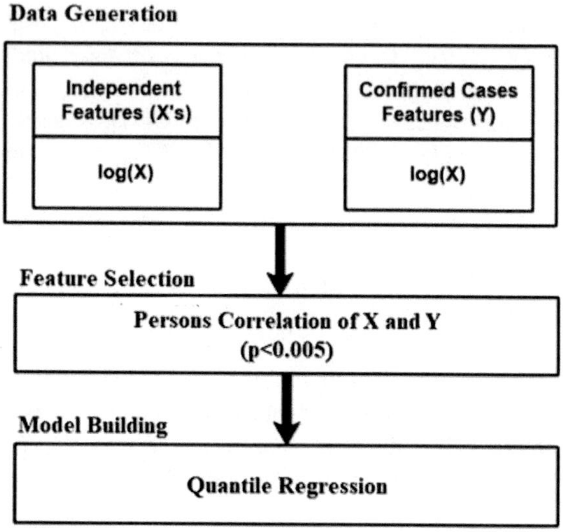

Fig. 4 Two-step quantile regression method

3.1 *Quantile Regression*

Quantile regression is an extended version of the linear regression, and it can be applicable to cases which include linearity, normality, and the independence. The method which is used to evaluate the conditional quantile functions in terms of the linear combination of the predictors. Quantile regression with lasso and ridge regularization is used for lasso and ridge coefficients to build quantile regression model. Solution of lasso coefficient can be written in Lagrangian form as shown below:

Proposed algorithm: two-step quantile regression with lasso
1. For $m = 1$ to M do Select folds and assign observations using cross validation Apply the any Regularization technique using cross fold validation
2. Let $\hat{\lambda}_m$ denote the optimal tuning parameter (regularized coefficient) obtained from either lasso/ridge operator
3. Let $\Lambda = (\hat{\lambda}_1, \hat{\lambda}_{100})$ denote the 100 values of $\hat{\lambda}_m$ with cross-validation for each $\hat{\lambda}$
4. Compute $\lambda(\theta)$, the θ − percentile of $\Lambda = (M)$
5. The computed value is replaced as a solution for lasso/ridge, i.e., $\lambda = \lambda(\theta)$

4 Experimental Results and Discussion

The proposed method is a suitable method for the prediction of confirmed cases in the pandemic situation in Andhra Pradesh.

The work in this paper is applied in various regression methods for the analysis of COVID-19 data. The work is mainly focused about the local state and collected data between March 12, 2020 and February 30, 2021 approx. 1 year data. In this study, five models based on regression analysis are used, namely linear and quantile regression, respectively. In this study, linear and quadratic regression results give unsatisfactory results, while TSQR with lasso give better and satisfactory results on COVID-2019 in Andhra Pradesh and presented in Table 3.

From Fig. 5, it shows that the residual of the proposed fit gives strongly patterned behavior with respect to linear and quantile. Regarding the best fit of the proposed models, the RMSE and MAPE values are calculated for the proposed models and shown result in Table 3. From Table 3, it is shown that minimum error values (i.e., 0.204 and 0.109) in 75% proposed quantile produced are better than all methods. The comparison of the confirmed case (actual result) with existing regression models over proposed model (expected results) for the COVID-2019 Andhra Pradesh dataset is shown in Fig. 5 and Table 2. Therefore, the proposed method is very useful for the future prediction of the COVID-2019 outbreak to the next 100 days from the current date. Moreover, this kind of analysis helps to control the second wave of the corona.

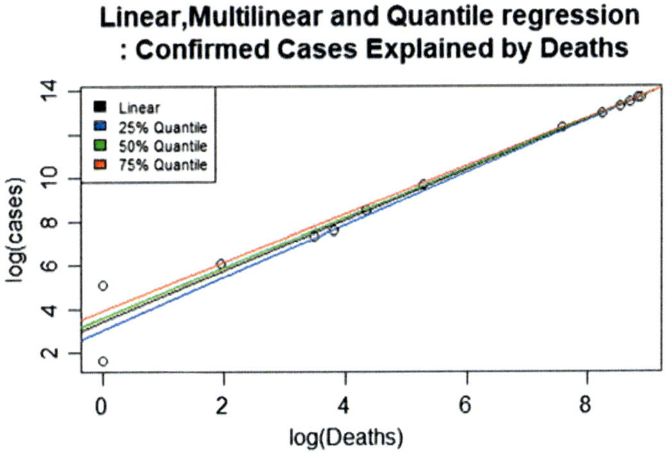

Fig. 5 Regression analysis of COVID-19: Andhra Pradesh case study

Table 2 Regression analysis result on COVID-19

S. No.	Regression model	RMSE	MAPE
1	Linear	0.790	7.820
2	Multi-linear	0.687	7.526
3	25% Quantile	0.703	9.148
4	50% Quantile	0.675	0.569
5	75% Quantile	0.517	0.413
6	25% Two-step quantile	0.6020	0.560
7	50% Two-step quantile	0.489	0.327
8	75% Two-step quantile	0.204	0.109

Table 3 Number of cases predicted using regression analysis on COVID-19 in Andhra Pradesh

Days	Confirmed	Linear	Multi-linear	Quantile	Proposed
10	1.61	3.42	2.95	3.03	1.59
24	5.08	3.42	3.68	3.03	5.06
33	6.07	5.68	5.98	5.37	6.05
62	7.33	7.49	7.46	7.24	7.3
71	7.61	7.84	7.70	7.61	7.61
100	8.53	8.46	8.35	8.25	8.52
125	9.69	9.56	9.70	9.39	9.6
152	12.29	12.20	12.62	12.13	12.26
175	12.93	12.98	13.35	12.94	12.93
193	13.31	13.34	13.64	13.31	13.34
223	13.51	13.52	13.67	13.50	13.51
260	13.66	13.66	13.59	13.65	13.6
303	13.69	13.69	13.26	13.67	13.52
335	13.70	13.72	13.07	13.71	13.68
350	13.70	13.72	13.07	13.71	13.65

5 Conclusions

The work in this paper focused about pandemic analysis for COVID-19 data. Presently, most of the researchers demand for the efficient methods to predict data and also helped further to analyze such epidemic risks. Regression analysis is type of machine learning method and is worked well to analyze such kind of epidemic/pandemic data. The analysis of COVID-19 data especially limited to the state of Andhra Pradesh and in future study will focus other starts too. First, information was collected from the website (i.e., https://prsindia.org/) and generated dataset. Next, various regression techniques are applied like linear, multi-linear, and quantile regression for COVID-19 data to result confirm cases prediction. The results show that the two-step quantile regression has been shown to be a better in predicting confirmed cases compared to linear and multi-linear regressions in terms of MSE and R-Score values.

References

1. Wu Z, McGoogan J (2020) Identification key features and lessons of the 2019 coronavirus (COVID-19) outbreak in China: summary of a report of 72,314 cases from the China Center for Disease Control and Prevention
2. Pirouz B, ShaffieeHaghshenas S, Shaffiee Haghshenas S, Piro P (2020) Investigation of a serious challenge in the process of sustainable development: analysis of confirmed cases of COVID-19 (new type of coronavirus) through a binary classification using intelligence artificial and regression analysis. Durability 12(6):2427
3. Jiang F, Deng L, Zhang L, Cai Y, Cheung CW, Xia Z (2020) Review of clinical features of coronavirus disease 2019 (COVID-19). J Gen Int Med 35:1545–549. https://doi.org/10.1007/s11606-020-05762-w
4. Benvenuto D, Marta G, Lazzaro V, Silvia A, Massimo C (2020) Application of the ARIMA model on the COVID-2019 epidemic dataset. Data in focus 105340. MedRxiv. https://www.medrxiv.org/content/10.1101/2020.03.26.20044511v1
5. Chen M, Hao Y, Hwang K, Wang L, Wang L (2017) Machine learning disease prediction on big data health care communities. IEEE Access 5:8869–8879
6. Yadav RS (2020) Analyzing data from the COVID-2019 outbreak using machine learning methods: a case study from India. Int J Inf Tecnol. https://doi.org/10.1007/s41870-020-00484-y
7. Kavadi DP, Patan R, Ramachandran M, Gandomi AH (2020) Partial derivative nonlinear global pandemic machine learning prediction of COVID 19. Chaos, Solitons & Fractals
8. Kalipe G, Gautham V, Behera RK (2018) Predicting the malaria epidemic using a machine learning and deep learning approach: a review and analysis. In: 2018 International conference on information technology (ICIT). IEEE, pp 33–38
9. Fang Y, Xu P, Yang J, Qin Y (2018) A quantile regression forest based method to predict drug response and assess prediction reliability. PLOS ONE
10. World Health Organization (2020) Coronavirus disease 2019 (COVID19): situation report, p 67

Study on Optimizing Feature Selection in Hate Speech Using Evolutionary Algorithms

Harsh Mittal, Kartikeya Singh Chauhan, and Prashant Giridhar Shambharkar

Abstract Hate speech is an important problem while dealing with user-generated content on online social media platforms. The huge amount of data generated makes it nearly impossible to manually moderate hate speech content and take appropriate measures. In this paper, we utilize various optimization algorithms to enhance the feature extraction and vectorization, of various techniques like TF-IDF, Word2Vec, and Bag of Words and appertain on the machine learning models for two-fold classification. We gauge and visualize the conclusion of the propounded methodology of the hate speech problem about Twitter tweets. We examine our suggested technique on three datasets; out of which, two of the datasets were highly unbalanced, and SMOTE was used for class balance. Our experiments indicate the random behavior of particle swarm optimization and genetic algorithm and the decrease in accuracy when applied individually to the experiments. The results also indicate that the accuracy can be achieved back by applying particle swarm optimization and genetic algorithm parallels, countering their random behavior.

Keywords Particle swarm optimization · Genetic algorithm · Natural language processing · Machine learning · Hate · Cyberbullying

H. Mittal (✉)
Department of Engineering Physics, Delhi Technological University, Bawana Road, New Delhi, Delhi 110042, India
e-mail: 24mithar@gmail.com

K. S. Chauhan (✉)
Department of Electronics and Communications, Delhi Technological University, Bawana Road, New Delhi, Delhi 110042, India
e-mail: ksc13dec@gmail.com

P. G. Shambharkar
Department of Computer Science and Engineering, Delhi Technological University, Bawana Road, New Delhi, Delhi 110042, India
e-mail: prashant.shambharkar@dtu.ac.in

P. Dutta et al. (eds.), *Emerging Technologies in Data Mining and Information Security*, Lecture Notes in Networks and Systems 490,
https://doi.org/10.1007/978-981-19-4052-1_70

1 Introduction

The invention of social media is a global revolution. It is an inexpensive mode of communication that attracts the masses in a very short period. Social media has the advantage of the global reach of access. But, along with this, social media brings with it an important challenge. Mass or global access to these platforms gives a chance for people to post content that is offensive or hateful toward another community or a group of people. Hate speech is described by many dictionaries as; speech addressed to the people, demonstrates hate or emboldens acts of violence against an individual or coterie based on arguments such as caste creed, culture, gender, or sexual orientation. The laws of some nations define hate speech as speech, gesticulations, demeanor, or exhibits that encourage vehemence or detrimental influence against individuals or community based on their belonging in the group, or which belittles or scares and terrorizes a group of individuals based on their belonging in the group. In some nations, hate speech has not yet been added as a legal term, whereas, in other countries, most of it safeguards the citizens constitutionally. In other nations, the sufferer of hate speech may seek remedy under criminal, civil, or both laws [1]. Hate speech in general on social media can cause a lot of harm to any victims who are targeted to fabricate a sense of exclusion among their communities, to toxify public opinion, and coax other forms of radical and detestable behavior. Accordingly, there is a need for an automated mechanism to distinguish hate speech at an exhaustive scale, thereby enabling analyses of large textual datasets collected from social media [2–4]. We can battle the problem of hate speech and cyberbully by promoting awareness of the problem, sensitizing the people about the problem, and reporting hate ourselves when we find it online. The best way of fighting the problem is by educating society in general and creating an environment in a community where such deeds are considered inappropriate. Everyone has to play their part in attempting to make society think about the concomitant of comments they make, and this might deter them from making hateful or abusive comments in the first place [5]. Hate speech is widespread in social media. A glance through the comments section of any social media platform demonstrates how prevalent the penetrating problem is [6, 7]. Social media platforms provide an economical communication medium that gives everyone access to millions of users. Consequently, in these platforms, any kind of content can be published and accessed by anyone, representing a groundbreaking revolution in our society [8]. Safety and security in social media have become a concerning topic, and thus, research work in this field has grown substantially in the last decade. The most impactful aspect of this work is detecting and averting the use of various forms of hate speech [9]. Across all social media platforms, recreation and innovation are continuously transformed by humans. In this heed, social media companies play a major role, mainly the US and Chinese firms which have grown into universal giants [10]. The availability of a large elucidated troupe of social media content and the elaboration of powerful classification vantage points, such as the algorithms we have used in our research, have contributed in an unparalleled way to contrivance the challenge of examining user's statements, comments, and sentiment in online social platforms across time [11].

The rest of the paper is organized as follows: Section 2 gives a brief overview of similar research done in the field of hate speech detection and classification. Section 3 gives a detailed explanation of the methodology used, dataset collection, and how particle swarm optimization and genetic algorithm were implemented on these datasets. Section 4 gives a tabular representation of the result of deploying the aforementioned algorithms on the datasets. A graphical description of the same has also been given. Section 5 gives the limitations of our experimental work. Finally, Section 6 gives the conclusion and future scope for work in this field.

2 Related Work

A review of analogous work was done to collect analyses and create a solution for hate speech on social media. Joshua Uyheng et al. [12] made a dynamic lattice frame to identify hate coterie, emphasizing Twitter tweets are kindred to the COVID epidemic in the United States of America and the Philippines. The data were collected using Twitter's Rest API. In the end, the dataset contained 15 million Twitter tweets, containing 1 million users from the Philippines, and 12 million conversations containing 1.6 million users. The random forest classifier was exercised to enforce decipherability and scalability. The Net mapper software, selected from each conversation; multiple lexical considerations of the use of pronouns, derogatory words, exclusive terms, absolutist words, and specific names among others. Pairwise products of every linguistic measure were used as additional features to extract the ways they occur simultaneously within the Twitter tweets. A study by Araque et al. [13] presented two novel feature extraction algorithms, which use previous Sentic computing resources, affective Space and Sentic Net. It also gave a machine learning framework using an ensemble of two different features to enhance the overall classification performance. Classifiers like support vector machine (SVM), logistic regression (LR), deep learning-based convolutional neural network (CNN), and bidirectional encoder representations from transformers (BERT) pretrained models were used. The weighted F1-score was 0.64 and 0.62 on Trac Facebook English and Hindi datasets, while on Twitter English and Twitter Hindi, the weighted F1-score was 0.58 and 0.50, respectively. The dataset was collected in code—Hindi and English languages from Facebook and Twitter user comments of celebrity posts. Modha et al. [14] suggested BERT model pretrained with Fast Text word embedding; the model marginally outperformed the BOW and term frequency-invert document frequency (TF-IDF) feature extraction techniques. Alsafari et al. [15] used feature extraction techniques like random embedding where every word is converted into a feature vector having a dimension of 300. The vectors are arbitrarily initialized, and their training is a part of the model training. The study in [15] also includes a pretrained word embedding, Fast Text, which is trained on Arabic Wikipedia corpus by Facebook. Miok et al. [16] used the BERT model along with Monte Carlo dropout method and provided a reliable estimation in the transformer network. Wullach et al. [17] used TF-IDF as word embedding, and they also used generative adversarial network (GANs) to scale up the dataset. Salminen et al. [18] proposed an online hate detection by using multi-platform data, where 197,566 comments from YouTube, Wikipedia, Reddit, and Twitter were taken, with 80% of the data docketed as not hate, and the left 20% of the comments was docketed as hateful. Then, experiments with several classification algorithms like support vector machines (SVMs), logistic regression, XGBoost, Naive Bayes, and neural networks were performed. The accentuate delineation has been done using BOW, TF-IDF, Word2Vec, BERT, and their combinations. All the models surpass the keyword-based baseline classifier, XGBoost, plying all foregrounds with an F1-score of 0.92. BERT features were the most impactful. Here, multi-platform training of data has been done, as the uni-platform emphasis is taxing because there is no guarantor that the models developed discern well across other platforms. This helps to overcome the problem of reinventing the wheel. The model performs decently across numerous platforms, using bidirectional encoder representation from transformer (BERT).

3 Methodology

In our analysis, we sunder our experimentation into the subsequent parts: Dataset Collection, Preprocessing, Optimization and classification.

(a) **Dataset Collection**

The data were collected from various public-access hate speech datasets, which are fundraised and annotated by multiple individuals, which are as follows:

- **New Combined Dataset**, which is formed by merging Davidson [19] and personal attack dataset, makes it a highly balanced and reliable dataset [20]. The new combined dataset has a binary output and classifies the tweets as either hate or not hate. It had 37,521 hate and 45,045 not hate entries.
- **TweetBLM** [21] is an open-source dataset, which contains tweets relating to Black Lives Matter. It had 3048 hate and 6081 not hate entries.
- **Kaggle Dataset** [22] is an open-source dataset, available on Kaggle, which contains Twitter tweets, and has binary classification. It had 2,242 hate and 29,720 not hate entries.

(b) **Preprocessing**

Our data consisted of raw tweets, and they had to be preprocessed so that they could be fed to the machine learning model for classification. The ensuing task was taken to preprocess the data: removal of URLs, emoji, Twitter handles, hyperlinks, hashtags, stop words, numbers, converting the tweets into lower case, lemmatization of words.

(c) **Optimization**

For optimization of the feature selection process, two algorithms were deployed, particle swarm optimization and genetic algorithm. We also tried out the combination of the two algorithms when executing one after the other.

Following is the description of the two algorithms which we have used to optimize the feature selection process, namely particle swarm optimization and genetic algorithm.

(d) **PSO (Particle Swarm Optimization)**

PSO is a meta-heuristic optimization algorithm. The foundation of this powerful optimization algorithm lies in artificial life and swarming theory. Its confederation can be traced to genetic algorithms and stochastic processes such as evolutionary programming. The algorithm is computationally inexpensive and involves the use of simple mathematical operators [23]. It looks for its inspiration from the swarm behavior observed in nature such as a swarm of birds or a shoal of fish. The algorithm seeks to simulate and efficiently implement the aforementioned behavior of these processes in nature. A bird's observation range is limited. But, a swarm of birds enables them to be aware of a larger area. This is an example of communal behavior and information transmission which the PSO aims to implement [24]. A velocity vector is delineated to each particle in the swarm to update its current position. These particles have memory, and they have been given the ability to update their memory on their own decision. These particles then update their position based upon the collective behavior of the population that is the swarm. The nature of this process is stochastic. The memory of each particle and the knowledge gained by the swarm is used in the next iterations.

$$x_{k+1}^{i} = x_{k}^{i} + v_{k+1}^{i}\ t \tag{1}$$

Here, x_{k+1}^{i} denotes the position of particle i at iteration $k + 1$. v_{k+1}^{i} is the corresponding velocity vector of a particle. t is the unit time step function used in basic PSO a common scheme for updating the velocity vector of each particle introduced by Shi and Eberhart [25] is

$$V_{k+1}^{i} = v_{k+1}^{i} + c_1 r_1 \left(p^{i} - x_{k}^{i}\right)/t + c_2 r_2 \left(p_{k}^{g} - x_{k}^{i}\right)/t \tag{2}$$

Here, r_1 and r_2 are random independent numbers between 0 and 1. P^i is the best position found by particle i so far, whereas p_k^g is the best position in the swarm at time k. t is the unit time step function that denotes the inertia of each particle. This variable is responsible for controlling the exploration properties of the PSO algorithm. Large values mean that the algorithm will show a more global behavior, whereas a small value means that the algorithm's behavior will be more local. Parameters c_1 indicate how much confidence a particle has in itself, whereas c_2 denotes how much confidence the particle has in the swarm.

Initialization consists of creating a swarm by choosing its size and a random distribution of the particles throughout the design space. Each particle is then given a random velocity. In each iteration, the particle would then update the position of each particle, and a distinction would be made between a particle's memory and the memory of the best particle in the neighborhood.

$$x_o^i = x_{\min} + r_3(x_{\max} - x_{\min}) \tag{3}$$

$$v_o^i = [x_{\min} + r_4(x_{\max} - x_{\min})]/t \tag{4}$$

In Eqs. (3) and (4), v_o^i is the initial velocity vector of a particle; x_o^i is the initial position vector of the particle. r_3 and r_4 are independent random numbers between 0 and 1; $x_{\min}$ is the vector of lower bounds, and $x_{\max}$ is the vector of upper bounds [26]. The basic PSO algorithm was designed for unconstrained problems only. When it is used for optimization on the training set, it performs miserably. It has its demerits in the form that it may converge prematurely, and its random behavior is not suited when implemented on the training set.

2. **GA (Genetic Algorithm)**

GA is a heuristic adaptive search algorithm that belongs to a large category of evolutionary algorithms. It has been inspired by the theories revolving around evolution, natural selection, and genetics. "Survival of the fittest" also known as the natural selection which has already been mentioned above is the main ideology behind this algorithm. These algorithms are used to obtain accurate solutions for search and optimization problems through random search backed by previous data experience for better efficiency and performance in solution space. Genetic algorithms are generally regarded as function optimizers.

- The implementation of the genetic algorithm starts with a population of individuals analogous to a chromosome (represented as an integer/float/ string) which represents a point in the search space and a possible solution. These chromosomes are made up of several variable components known as genes.
- The selection of individuals (parents) is done for mating by the application of crossover and mutation operators on them. This generates new offspring (chromosomes).
- The problem function after a proper assessment is then given generative scope in such a way that the chromosomes which give a superior result are given more chances to "reproduce" as opposed to chromosomes representing inferior solutions.
- This is done by allocating each individual a fitness score. A higher score would imply a better solution and hence a higher probability of "reproduction."
- Since the population is static, the search space will be replaced by newer generations of individuals by causing the death of older individuals.
- This will consequently create new generations which represent a better partial solution set. Each new generation would have better partial solutions than the previous one.

- The aim is to converge the algorithm to a set of optimal solutions for the problem function. This happens when there is no significant difference in the solution set between multiple generations.

The advantages of the genetic algorithm will include its robustness as they do not break on a slight change in input. Moreover, they also optimize over a large space state. Constrained optimization problems can be troublesome for the genetic algorithm as crossover, and mutation operators might give infeasible solutions. Many mechanisms can be employed to overcome this problem, but the discussion of those methods is beyond the scope of our research [27].

3. **Particle Swarm Optimization + Genetic Algorithm**

Lastly, particle swarm optimization and genetic algorithm were deployed in series, to counter the randomness, which was earlier supported by the convex loss function. Firstly, particle swarm optimization was used, and the modified training set was then moved down to the genetic algorithm, which modified the training set again, which were finally passed to various classifiers for binary prediction.

4. **Classification**

For classification, various machine learning techniques like logistic regression, random forest classifiers, and Naive Bayes were used, to produce a binary prediction of hate and not hate.

4 Implementation and Results

We have performed various experiments on baseline classifiers and experimented with different feature extraction techniques on the given datasets.

- In step 1, we deployed machine learning models, logistic regression, random forest classifier, Naive Bayes on feature extraction techniques of TF-IDF, BOW, Word2Vec, on three different datasets, new combined datasets, TweetBLM, and Kaggle dataset.
- In step 2, we deployed machine learning models, logistic regression, random forest classifier, Naive Bayes, on feature extraction techniques of TF-IDF, BOW, Word2Vec, which were optimized with particle swarm optimization, on three different datasets: new combined datasets, TweetBLM, and Kaggle dataset.
- In step 3, we deployed machine learning models, logistic regression, random forest classifier, Naive Bayes, feature extraction techniques of TF-IDF, BOW, Word2Vec, which were optimized with genetic algorithm, on three different datasets, new combined datasets, TweetBLM, and Kaggle dataset.
- In step 4, we deployed machine learning models, logistic regression, random forest classifier, Naive Bayes, feature extraction techniques of TF-IDF, BOW, Word2Vec, which were optimized with particle swarm optimization, followed by

genetic algorithm, on three different datasets, new combined datasets, TweetBLM, and Kaggle dataset.

In Table 1, we have presented all the results which were conducted on the new combined dataset because the new combined dataset is the most authentic and balanced dataset as no sampling technique has been utilized.

After experimenting with various feature extraction techniques, TF-IDF, Word2Vec, and Bag of Words, we concluded that TF-IDF outperforms all other feature extraction techniques. Out of all datasets used, the new combined dataset is the most authentic, and balanced dataset as no sampling technique has been utilized. F1-score achieved with new combined dataset was 77.5688 percent using random forest with particle swarm optimization and genetic algorithms on TF-IDF the F1-score for tweetBLM using random forest on particle swarm optimization + genetic algorithm on TF-IDF was 78.3448. For the other two datasets, Kaggle dataset, and TweetBLM, SMOTE was used for handling class imbalance. The best results are achieved while using TF-IDF on random forest classifier on the Kaggle dataset. Overall, there is no effect of utilizing particle swarm optimization and genetic algorithms on TF-IDF, and when particle swarm optimization and genetic algorithms are applied, they indicate the random behavior of particle swarm optimization and genetic algorithm, as performing the same experiment multiple times produces different results. Figures 1, 2, and 3 show the results of applying logistic regression on Kaggle dataset, TweetBLM dataset and new combined dataset, respectively. Figure 4 displays F1-scores on Kaggle dataset using Bag of Words, TF-IDF, and Word2Vec respectively. Figure 5 displays F1-scores on TweetBLM dataset using bag of words, TF-IDF, and Word2Vec respectively. Figure 6 displays F1-scores on new combined dataset using bag of words, TF-IDF, and Word2Vec respectively.

5 Limitations

Our experimental work so far has been restrained to predicting a binary output. As we are acquainted with the fact that hate speech can be broadly broken down into further subcategories of sexism, abuse, racism, hate and spam. The experimentation can be further expanded to a multiclass-classification problem to examine how competently the technique propounded can classify within those subcategories of hate. Additionally, the task of classification is very subjective and based on personal bias, because the decision on whether some text is termed as hate or not, also depends a lot on the standpoint of the annotator, the culture and customs of a nation as well as the maturity and experience of the annotator. This upheaves the questions concerning the legitimacy and ability of the annotator to precisely identify the user-generated content. This affects the accuracy and precision of our machine learning model and is, therefore, a shortcoming. On the top of that, the sarcastic comments are difficult to discern from hate comments. Since we do not explicitly handle sarcastic tweets, some of the false positives generated by our model that were falsely classified as hate

Table 1 Experimental results

Technique	Features	Model	Accuracy	Recall	Precision	F1-Score
Baseline models	Bag of words	Logistic regression	0.93097	0.9568	0.90989	0.93276
		Naive Bayes	0.77896	0.60856	0.92361	0.73369
		Random forest	0.93531	0.95476	0.91909	0.93658
	TF-IDF	Logistic regression	0.91792	0.9351	0.90415	0.91936
		Naive Bayes	0.8645	0.83709	0.88582	0.86076
		Random forest	0.95891	0.97809	0.94201	0.95971
	Word2Vec	Logistic regression	0.86664	0.84422	0.88403	0.86366
		Naive Bayes	0.80715	0.762	0.83789	0.79814
		Random forest	0.92802	0.8705	0.98377	0.92368
PSO	Bag of words	Logistic regression	0.88484	0.9351	0.84981	0.89042
		Naive Bayes	0.69413	0.66184	0.70786	0.68408
		Random forest	0.90441	0.95293	0.86875	0.9089
	TF-IDF	Logistic regression	0.85313	0.89924	0.82347	0.85969
		Naive Bayes	0.79104	0.8971	0.74029	0.81118
		Random forest	0.91583	0.92084	0.91182	0.91631
	Word2Vec	Logistic regression	0.85711	0.85922	0.85441	0.85681
		Naive Bayes	0.79823	0.80668	0.78482	0.7956
		Random forest	0.93449	0.98411	0.88334	0.93101
GA	Bag of words	Logistic regression	0.88989	0.95273	0.84647	0.89646
		Naive Bayes	0.63132	0.32909	0.83312	0.47181
		Random forest	0.90834	0.96556	0.86651	0.91336

(continued)

Table 1 (continued)

Technique	Features	Model	Accuracy	Recall	Precision	F1-Score
	TF-IDF	Logistic regression	0.84926	0.90066	0.81685	0.85671
		Naive Bayes	0.80929	0.746	0.85438	0.79652
		Random forest	0.89636	0.97066	0.84519	0.90359
	Word2Vec	Logistic regression	0.83715	0.91245	0.78935	0.84645
		Naive Bayes	0.80518	0.76607	0.82535	0.79461
		Random forest	0.70693	0.40698	0.99332	0.57739
PSO + GA	Bag of words	Logistic regression	0.93327	0.96679	0.90613	0.93548
		Naive Bayes	0.78099	0.61864	0.91653	0.73869
		Random forest	0.94561	0.97239	0.92302	0.94706
	TF-IDF	Logistic regression	0.91777	0.94254	0.89816	0.91981
		Naive Bayes	0.86868	0.85135	0.88209	0.86645
		Random forest	0.95917	0.97759	0.9429	0.95993
	Word2Vec	Logistic regression	0.87113	0.90698	0.84298	0.87381
		Naive Bayes	0.8072	0.75718	0.83547	0.7944
		Random forest	0.79374	0.58755	0.98849	0.73702

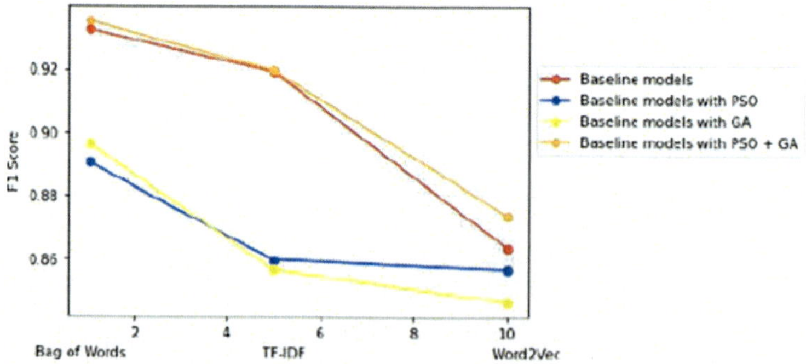

Fig. 1 Logistic regression on Kaggle dataset

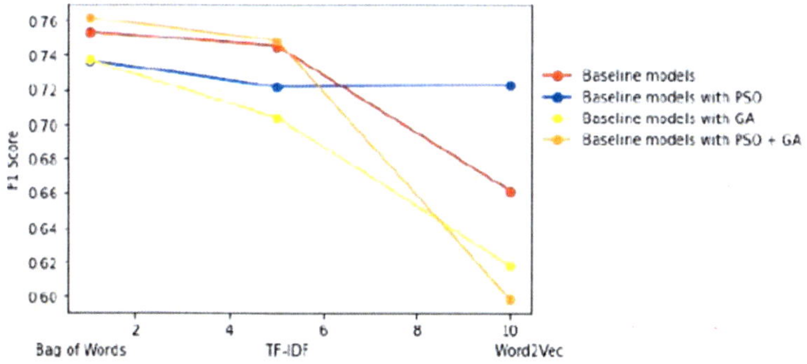

Fig. 2 Logistic regression on TweetBLM

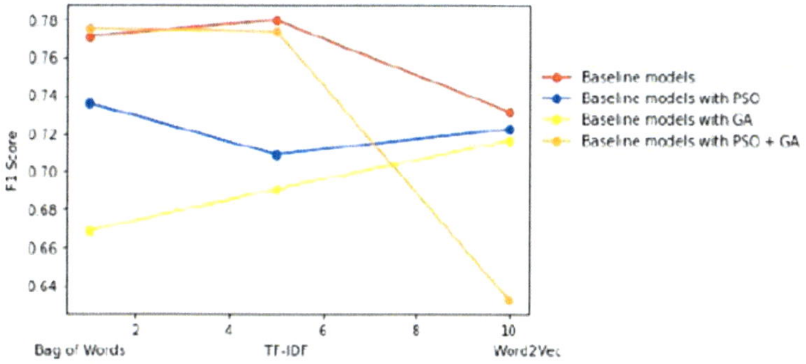

Fig. 3 Logistic regression on new combined dataset

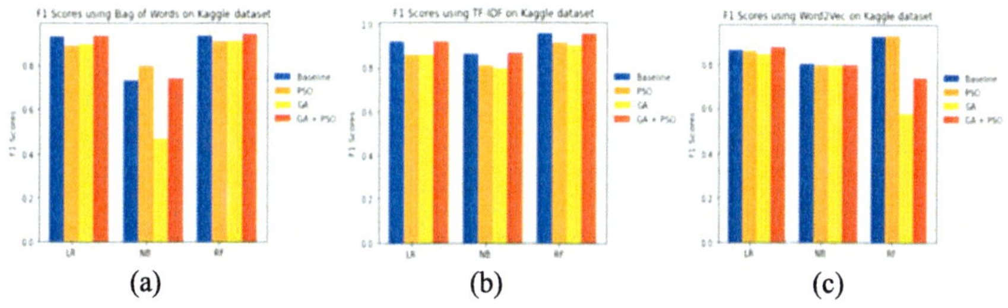

Fig. 4 F1-scores on Kaggle dataset using: **a** Bag of Words, **b** TF-IDF, **c** Word2Vec

were sarcastic tweets, without the intention to hurt or oppose a community. Other than that, many of these tweets that have the hate label, those which were wrongly classified are those which did not necessarily contain any indecent or abusive term, which has a strong opinion against a particular section of the society. Another major issue in this experimentation can be attributed to the way SMOTE works, which

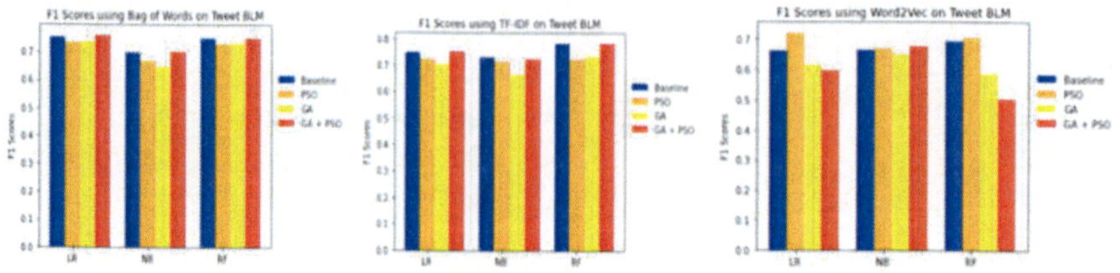

(a) (b) (c)

Fig. 5 F1-scores on TweetBLM dataset: **a** Bag of Words, **b** TF-IDF, **c** Word2Vec

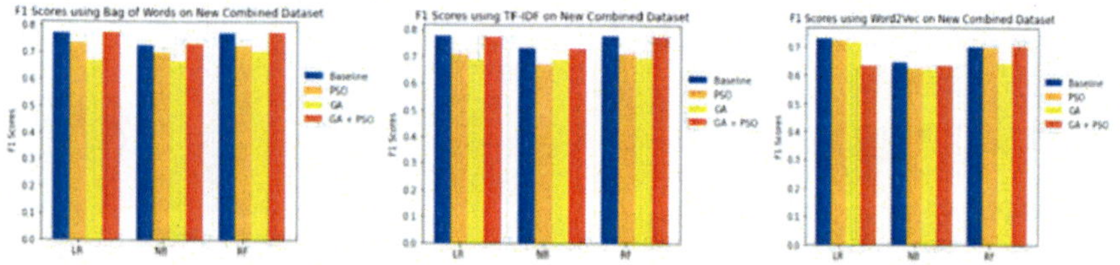

(a) (b) (c)

Fig. 6 F1-scores on new combined dataset: **a** Bag of Words, **b** TF-IDF, **c** Word2Vec

definitely is questionable because either reduction of dataset or multiple accounting of same entries.

6 Conclusion and Future Scope

Combining text with offensive, sarcastic, or innocent words with the strong sentiment prompts the cyberbullying and hate speech detection algorithms to infelicitously flag the tweets and posts. Not rectifying this issue may imply significant dissentious effects such as users' defamations. To enhance the efficiency and precision of a classification mechanism and generalize it to newer datasets, we suggest more emphasis on deep learning-based techniques and improved architectures, rather than working on enhancing feature extraction, using the genetic and evolutionary algorithms because these algorithms are firstly highly random in nature and use convex loss function, which makes the algorithm even more irrelevant in this context. Thus, working on GANs and LSTM might help in solving the problem.

References

1. Hate speech—Wikipedia. https://en.wikipedia.org/wiki/Hate_speech
2. Vidgen B, Yasseri T (2020) Detecting weak and strong Islamophobic hate speech on social media. J Inf Technol Polit 17:66–78. https://doi.org/10.1080/19331681.2019.1702607
3. Guiora A, Park EA (2017) Hate speech on social media. Philos (United States) 45:957–971. https://doi.org/10.1007/s11406-017-9858-4
4. Mathew B, Dutt R, Goyal P, Mukherjee A (2019) Spread of hate speech in online social media. WebSci 2019—In: Proceedings of 11th ACM conference on web science, pp 173–182. https://doi.org/10.1145/3292522.3326034
5. Uyheng J, Carley KM (2021) Characterizing network dynamics of online hate communities around the COVID-19 pandemic. Appl Netw Sci 6. https://doi.org/10.1007/s41109-021-00362-x
6. Ring C (2013) Hate speech IN social media: an exploration of the problem and its proposed solutions. J Chem Inf Model 53:1689–1699
7. Schofield A, Davidson T (2017) Identifying hate speech in social media. XRDS Crossroads, ACM Mag Students 24:56–59. https://doi.org/10.1145/3155212
8. Mondal M, Silva LA, Benevenuto F (2017) A measurement study of hate speech in social media. In: HT 2017—Proceedings of 28th ACM conference hypertext social media, pp 85–94. https://doi.org/10.1145/3078714.3078723
9. Malmasi S, Zampieri M (2017) Detecting hate speech in social media.In: International conference recent advances in natural language processing, pp 467–472. https://doi.org/10.26615/978-954-452-049-6-062
10. Matamoros-Fernández A, Farkas J (2021) Racism, hate speech, and social media: a systematic review and critique. Telev New Media 22:205–224. https://doi.org/10.1177/1527476420982230
11. Florio K, Basile V, Polignano M, Basile P, Patti V (2020) Time of your hate: the challenge of time in hate speech detection on social media. Appl Sci 10. https://doi.org/10.3390/APP10124180
12. Uyheng J, Carley KM (2020) Bots and online hate during the COVID-19 pandemic: case studies in the United States and the Philippines. J Comput Soc Sci 3(2):445–468.
13. Araque O, Iglesias CA (2021) An ensemble method for radicalization and hate speech detection online empowered by sentic computing. Cognit Comput. https://doi.org/10.1007/s12559-021-09845-6
14. Modha S, Majumder P, Mandl T, Mandalia C (2020) Detecting and visualizing hate speech in social media: a cyber watchdog for surveillance. Expert Syst Appl 161:113725. https://doi.org/10.1016/j.eswa.2020.113725
15. Alsafari S, Sadaoui S, Mouhoub M (2020) Hate and offensive speech detection on Arabic social media. Online Soc Netw Media 19:100096. https://doi.org/10.1016/j.osnem.2020.100096
16. Miok K, Škrlj B, Zaharie D, Robnik-Šikonja M (2021) To BAN or not to BAN: Bayesian attention networks for reliable hate speech detection. Cognit Comput. https://doi.org/10.1007/s12559-021-09826-9
17. Wullach T, Adler A, Minkov E (2021) Towards hate speech detection at large via deep generative modeling. IEEE Internet Comput 25:48–57. https://doi.org/10.1109/MIC.2020.3033161
18. Salminen J, Hopf M, Chowdhury SA, Jung S, Almerekhi H, Jansen BJ (2020) Developing an online hate classifier for multiple social media platforms. Human-centric Comput Inf Sci 10:1–34. https://doi.org/10.1186/s13673-019-0205-6
19. Davidson T, Warmsley D, Macy M, Weber I. Automated hate speech detection and the problem of offensive language *
20. Charitidis P, Doropoulos S, Vologiannidis S, Papastergiou I, Karakeva S (2020) Towards countering hate speech against journalists on social media. Online Soc Netw Media 17:1–15. https://doi.org/10.1016/j.osnem.2020.100071
21. Pranesh RR, Kumar S, Shekhar A (2020) TweetBLM: a hate speech dataset and analysis of blacklivesmatter-related microblogs on Twitter. https://doi.org/10.5281/ZENODO.4000539

22. Hate Speech and Offensive Language Dataset | Kaggle, https://www.kaggle.com/mrmorj/hate-speech-and-offensive-language-dataset
23. Okwu MO, Tartibu LK (2021) Particle swarm optimisation. Stud Comput Intell 927:5–13. https://doi.org/10.1007/978-3-030-61111-8_2
24. Marini F, Walczak B (2015) Particle swarm optimization (PSO). Tutorial Chemom Intell Lab Syst 149:153–165. https://doi.org/10.1016/j.chemolab.2015.08.020
25. Shi Y, Eberhart RC (1998) A modi ed particle swarm optimizer. In: Proceedings of the international conference on evolutionary computation, Institute of Electrical and Electronics Engineers, Piscataway, NJ, pp 69–73
26. Hassan R, Cohanim B, De Weck O, Venter G (2005) A comparison of particle swarm optimization and the genetic algorithm. In: Collection of technical papers—AIAA/ASME/ASCE/AHS/ASC structures, structural dynamics and materials conference vol 2, pp 1138–1150. https://doi.org/10.2514/6.2005-1897
27. Whitley D (1994) A genetic algorithm tutorial. Stat Comput 4:65–85. https://doi.org/10.1007/BF001753

Data Mining Approaches for Healthcare Decision Support Systems

Sabyasachi Pramanik, Mohammad Gouse Galety, Debabrata Samanta, and Niju P. Joseph

Abstract Data mining is a user-friendly approach to locating previously unknown or hidden information in data. The employment of data mining technologies in the healthcare system may result in the finding of relevant data. Data mining is used in healthcare medicine to construct learning models that predict a patient's condition. Data mining technologies have the potential to benefit all stakeholders in the healthcare industry. For example, data mining may aid health providers in detecting theft and fraud, medical organizations in making customer service management decisions, physicians in discovering effective therapies and best practices, and customers in obtaining suitable and less expensive healthcare. Contemporary systems, due to their complexity and size, are unable to control and analyze the huge amounts of data generated by healthcare operations. Data mining is a technique and mechanism for converting a large amount of data into useful information. The fundamental purpose of this research is to look at what makes clinical data mining unique, to give an overview of existing clinical decision support systems, to identify and select the most common data mining algorithms used in modern Health and Demographic Surveillance System (HDSS), and to compare different data mining algorithms.

Keywords Bayes · C4.5 · Clinical decision support system · HDSS · WEKA

S. Pramanik
Department of CSE, Haldia Institute of Technology, Haldia, West Bengal, India

M. G. Galety
College of Information Technology and Computer Science, Catholic University in Erbil, Erbil, Iraq

D. Samanta (✉) · N. P. Joseph
Department of Computer Science, CHRIST University, Bangalore, Karnataka, India
e-mail: debabrata.samanta369@gmail.com

P. Dutta et al. (eds.), *Emerging Technologies in Data Mining and Information Security*, Lecture Notes in Networks and Systems 490,
https://doi.org/10.1007/978-981-19-4052-1_71

1 Introduction

Computational science is increasingly being used in medicine and the health sciences. AI is a field of computational science that is increasing progressively and effectively engaged in the medical field. AI has been used to create a variety of healthcare decision support systems [1]. In hospitals and clinics, these systems are increasingly extensively utilized. They have shown to be very beneficial to both patients and medical professionals in making choices. For the creation of such systems, several methods are used. Various methods have different ways of collecting input data and presenting result information. The term "healthcare decision support system" refers to any computer software that assists professionals in making healthcare decisions. Artificial Intelligence has the ability to assist both the production and use of healthcare knowledge, which is a key feature. This paper's primary goal is to describe current developments in medical decision support systems.

The fundamental stage is data mining, which leads to the identification of hidden and predictive information in huge datasets. Data mining denotes the ordinary extraction of implicit, formerly undisclosed, and possibly valuable information about data, according to a proper description of knowledge discovery in databases. Six types of tasks are related to data mining [2]. Anomaly detection identifies odd data information that may be intriguing or data mistakes that need to be investigated more. Exploration for links among variables utilizing association rule learning. A supermarket, for e.g., can gather information on consumer purchasing patterns. The store can use association rule learning for identifying the items that are frequently purchased jointly and utilize that knowledge for marketing reasons. Market basket analysis is a terminology utilized to specify this process.

Clustering [3] denotes the process of identifying groupings and structures in data that are "similar" in a few ways without depending on existing data structures. The effort of generalizing the current structure to adapt to newer data is known as classification. Email software, for example, may try to categorize an email as "genuine" or "spam". Regression [4] Looks for a function that analyzes the data only having the lowest amount of error. Presenting the data set in a more compact form, including visualization and report creation. Data mining is a user-friendly method to discover new and concealed patterns in data. The newly found information may be utilized by hospital management to enhance service enhancement. Medical professionals may utilize the newly found information to minimize the number of untoward drug doses and to propose lesser costly healing comparable options. The below are a few key domains where data mining methods may be very beneficial in healthcare administration.

1. Modeling data for healthcare applications is one of the first things that come to mind.
2. Healthcare Executive Information System.
3. Predicting treatment costs and resource requirements.
4. Predicting a patient's future behavior based on their past behavior.

5. Informatics in Public Health.
6. Healthcare e-governance frameworks.
7. Medical Coverage.

2 Techniques for Data Mining

The most significant method utilized in data mining is knowledge discovery in databases (KDD). In a data mining project, several methods are used. Decision trees, Bayesian networks, Naive Bayes [5], and neural networks [6] are examples of these. The decision tree is a popularly utilized data analysis method. It is utilized in assigning data to the appropriate class and may be utilized in regression and association tasks. Decision trees are used in the medical profession to define the order of characteristics. Nodes provide conditional characteristics, such as symptoms

$$Y = y_1, y_2, \ldots, y_k. \tag{1}$$

branches display the estimates of A^{ih}, i.e., the h^{ih} range for the i^{ih} symptom, and discards current choices and their binary values

$$Z_{dk} = 0, 1. \tag{2}$$

$$X = x_1, x_2, \ldots, x_k. \tag{3}$$

Figure 1 shows the pictorial representation of a decision tree.

The NB classifier is a basic probabilistic classifier that relates to the premise of attribute mutual independency. The Bayes Rule is used to compute the probabilities used in the Naive Bayes algorithm. For example, the probability of hypothesis J may be measured using the hypothesis J and evidence regarding hypothesis K using the following equation:

$$P(J|K) = P(K|J) * P(J)/P(K) \tag{4}$$

In medicinal prognosis, the inputs to the framework in the neural network are the sufferer's symptoms, which are represented by the set X, while the outcome of the diagnostic is represented by the set Y. In NNs, there are three layers: input, hidden, and output layer. The input layer results are hidden in the hidden layer. Weights are given to the conditions that exist between neurons. The back propagation method [7] is used to determine their values. Some nonlinear characteristics are inserted into the network in hidden layers. There may be many output nodes in the out layer that forecast various illnesses. Multiple input layers and one output layer is present in a single neuron. Combination and activation functions are used to generate input and output values.

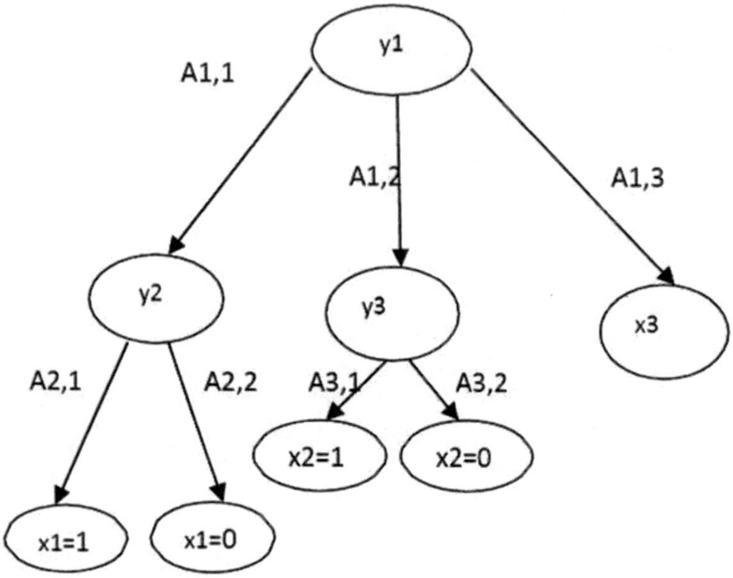

Fig. 1 A decision tree for use in healthcare

2.1 Data Mining Benefits

1. Anticipate forthcoming trends and consumer purchasing patterns,
2. Assist with decision-making,
3. Increase business income and save expenses,
4. Fraud revelation.

2.2 Disadvantages

1. Huge expenditure at installation step,
2. Probable information misuse,
3. Data incorrectness.

3 Data Mining in Clinical Application

Healthcare DSSs and strategy administration/decision-taking are two areas where data mining applications are presently being used in healthcare and medicine.

3.1 A Decision-Making Assistance System for Healthcare

HDSS is Decision Support System (DSS) software that is intended to help doctors and various healthcare providers have decision-taking activities like deciding patient data diagnosis. Modern HDSS is primarily designed to assist physicians while taking care of patients. It implies that a physician will work with an HDSS in assisting with the prognosis, analysis, and other aspects of a patient's data [8]. It is a DSS that provides workers with comprehensive, objective, customized, and up-to-date information on all healthcare issues. Integrated online, phone, and print-based materials provide employees with the knowledge, tools, and assistance they need. This enables workers to make better healthcare choices in collaboration with their own doctors. HDSS is divided into two categories based on knowledge.

3.2 Non-knowledge-Based HDDS

DSS is a paradigm of HDSS, and it denotes how an HDSS can be utilized by a physician. A DDSS will analyze the patient's information and recommend various diagnoses. The physician then examines the DDSS output along with determining the items that are significant and which are not. The timing of the usage of an HDSS is another significant categorization. Doctors utilize the systems while taking care to assist them in handling patients, with the time of usage varying from pre-diagnosing, diagnosing, and post-diagnosing. Pre-diagnosing HDSS technologies are utilized in assisting doctors in making diagnoses. HDSS is a diagnostic tool that helps physicians evaluate and filter their early diagnostic options in order to enhance their final findings. HDSS approaches are also utilized post-diagnosis in mining data in order to draw links among patients with their previous medical records and anticipate further occurrences [9].

3.3 The Characteristics of a Knowledge-Dependent HDSS

The knowledge base, communication approaches, and inference engine are all included in most HDSS. The IF-THEN rules are stored in the knowledge base. The knowledge base rules are merged with the patient's data via the inference engine. The transmission method must enable the framework to display the user the findings as well as receive input. Features of an HDSS that is not based on knowledge Neural networks (NNs) and genetic algorithms are two examples of non-knowledge-based frameworks. NNs find out patterns in a victim's data and generate correlations between symptoms and diagnoses utilizing nodes and weighted connections between them. To obtain optimum HDSS outcomes, GAs are related to simple evolutionary approaches utilizing directed alternatives [10].

3.4 Healthcare Decision Support System Characteristics

HDSSs are computer frameworks that help doctors and medical personnel in making healthcare decisions. Diagnostic aid, treatment critiquing and planning, medicines prescription, information retrieval framework (for example, creating correct clinical questionnaire), along with picture identification and explanation are all included in most healthcare decision support systems (HDSSs) (X-rays, CT, MRI scans), ML systems that are efficient in generating newer medical knowledge are an interesting example of HDSSs [11]. A HDSS may provide a comprehensive representation of input characteristics having the alone feature in clinical situations by studying healthcare cases. It may be helpful in detecting changes in a patient's health state. These technologies have the potential to enhance patient safety by decreasing diagnostic mistakes. They may also aid in the ordering of medicines and tests. Furthermore, as the amount of time doctors spend with a patient increases, the quality of treatment improves. It may be due to the use of appropriate guidelines, current healthcare evidence, and better recordkeeping.

3.5 HDSS Examples

There is a variety of Healthcare Decision Support Systems (HDSSs) available. They aid in the identification of illnesses at an early stage. A handful of the most significant systems are covered in this overview. In hospitals, they are used. Three instances of HDSS are given to demonstrate the concept: HELP, DX plain, and ERA. HELP is the name of a prominent and sophisticated HDSS. It assists doctors in analyzing healthcare data, diagnosing patients' diseases, managing healthcare procedures, and doing other duties. HELP II, a new version, was launched in 2003. Antibiotic assistant and pneumonia diagnostic assistant are two assistants included in this system. The former's goal is to identify the microorganisms that are causing the illness and provide the most cost-effective treatment for individuals with allergies or kidney problems.

3.6 Simple DX

It's a web-dependent HDSS that helps physicians by producing stratified diagnoses depending on user inputs of patient indications and symptoms, lab data, and various medical discoveries. Every clinical uncovering input into DX plain is analyzed to determine its significance and how well it assists a particular prognosis for every illness in the knowledge base [11]. DX simple produces ranked differential diagnoses based on these criteria, with the most probable illnesses receiving the lowest score. Figure 2 shows a neural network used in diagnosis of a medical condition.

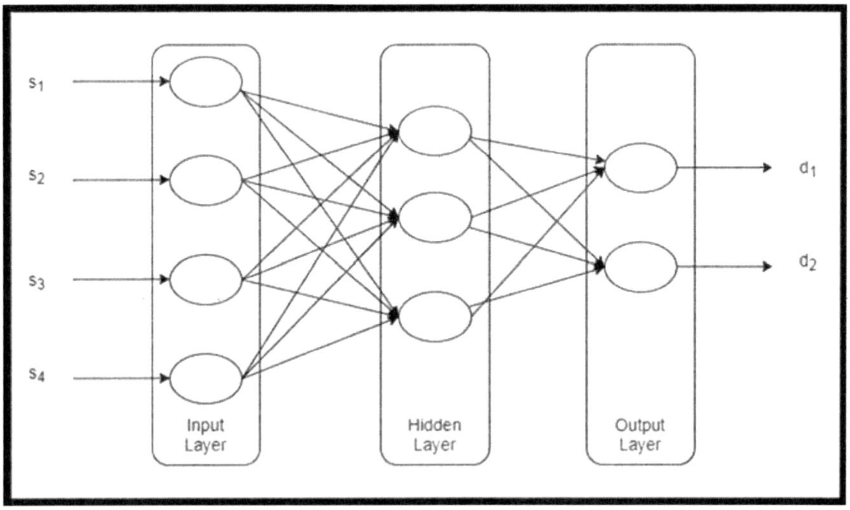

Fig. 2 A neural network used in diagnosis of a medical condition

3.7 ERA

It is an acronym for early referrals application. The current and having the best capability, ERA is devoted to the early identification of many kinds of malignancies. Since 2001, the application has been developed in the United Kingdom by GPs affiliated with the Leicester NHS Trust's academic hospitals.

4 The Value of Healthcare

These are some of the most essential aspects of healthcare.

- access all patient information and identify abnormalities quickly;
- analysis of data utilizing an automated method, that is helpful in the case of severe and recurring abnormalities;
- enhance production and care standard with more frequent, shorter time span, and remote dialogs;
- using technologies shared by the primary care physician and the nurses in charge of everyday patient observation, interact swiftly and simply in an organized manner.

Provide motivational assistance to patients who want it. Using the tool's healthcare database, you may contribute to biomedical research.

4.1 Data Mining in the Healthcare Industry

In terms of using data mining to extract information from data, businesses and marketing companies can be forward when compared to clinical applications. This is rapidly changing. In the healthcare industry, successful mining applications have been developed, three of which are detailed here.

4.2 Infection Control in Hospitals

In the United States, no monomial infections afflict 3,000,000 patients every year, and the incidence of drug-resistant infections has risen to record levels. Proactive monitoring is required to detect outbreaks and develop resistance early [12]. The goal of computer-aided surveillance experiments has been to identify higher-threatening patients, expert systems, and potential instances, as well as to detect variations in the matters of specified events.

4.3 Hospitals Are Ranked

Hospitals and healthcare plans are ranked by organizations based on information provided by healthcare providers. Although consistent reporting is assumed, research indicates that consistency may be improved.

4.4 Tracking Patients at High Risk

Diabetes disease management services are provided by American Health Ways to hospitals and health centers for improving the standards along with the reduction of the expenditure of diabetes care. American Health utilizes predictive modeling technologies to improve the company's capacity to detect high-risk patients in the future.

4.5 Management of Customer Relationships

While CRM [13] is a key component of administering communications between trading companies and the clients (usually banks and merchants), it is essential in the healthcare industry. Call centers, doctors' chambers, billing departments, indoor facilities, and ambulances are all possible places for customers to engage.

5 WEKA Experimentation and Analysis of Algorithms

In the WEKA environment, data mining techniques that have been identified as being highly prevalent in MDSS are accomplished. The goal of this paper is to acquaint the reader with the WEKA approaches' executions, as well as to explain key parameters and demonstrate several methods of presenting the results.

5.1 C4.5 Algorithm

The method C4.5 [14] is known as J48 [15] in the WEKA environment, and it denotes the most recent type of the approach's executions. The C4.5 method has an option that enables you to change the confidence threshold for tree pruning, which is the lowest number of occurrences that may be allowed at a leaf. The tree may also be seen in the graphical (more intuitive) form. The decision tree leaves contain values in brackets, such as (15.0/1.0), indicating that 15 examples successfully followed the algorithm, and one was categorized.

5.2 Benefits and Drawbacks

The C4.5 has the following advantages:

- Creates clearly interpretable models;
- Is simple to apply;
- Can utilize both categorical and continuous variables;
- Handles background noise.

The disadvantages are as follows:

- Minor variations in data may result in various decision trees (particularly when the variables are near in value);
- does not function better on a short training set.

It is the most often utilized method for decision trees construction in classification tasks. Because it works with numeric characteristics and missing data, it is appropriate for real-world situations. The method may be used to create smaller or bigger, more perfect decision trees, and it is also very fast.

5.3 Naïve Bayes

In the WEKA environment, the Naive Bayes classifier offers a very simple interface. It enables you to utilize Supervised Discretization instead of a normal distribution when transforming numeric characteristics to conventional ones. The Naive Bayes classifier's output is in text format. The benefits and drawbacks of Naïve Bayes are as follows: The simplicity, computational economy, and excellent classification performance of the Naive Bayes classifier are its greatest assets. However, three points must be kept in mind. To begin with, obtaining excellent results with the Naive Bayes classifier needs a huge number of data. Second, NB predicts that if a predictor class is not really part of the training data, a new record containing that predictor category has zero probability. If this uncommon predictor value is significant, this may be a problem. Consider the predictor category own boat and the target variable purchased higher-value life insurance. If the training data contains no records with owns boat = 1, NB will allocate a probability value 0 to the target variable purchased higher-value life insurance for any new records with owns yacht = 1. There is no data mining approach that can integrate this likely significant feature into the classification algorithm if there are no training samples with owning boat = 1. The lack of this predictor, on the other hand, "outvotes" all other information in the record to give a 0 to the target value in NB (when, here, it has a relatively fair possibility of being a 1). A larger training set (and, if necessary, careful binning of continuous variables) may help minimize this impact.

The following are the drawbacks:

A minor flaw (or "disadvantage") with Naive Bayes is that the frequency-based probability estimate will be 0 if there are no instances of a class label and a specific attribute value combined (e.g., class="nice," shape="sphere"). Given the conditional independence assumption of Naive Bayes, once all the probabilities are multiplied, the result is zero, which has an impact on the posterior probability estimate. When we take samples from a population, and the drawn vectors are not completely representative of the population, we run into this issue.

5.4 Neural Network

NN denotes a non-knowledge-dependent adaptive HDSS that employs machine learning, a kind of AI that enables computers to learn from previous experiences along with identifying patterns in medical data. It is made up of neuron nodes and weighted connections which send messages back and forth between the neurons in a forward or looping manner. It is divided into three layers: Input, which receives data. With established outcomes for huge quantities of data, the system becomes more efficient.

6 Benefits and Drawbacks

The benefits of NN also include the removal of the requirement to program the systems and the inclusion of expert advice. Because of its adaptive system learning, the HDSS can handle partial data by generating informed guesses regarding incomplete information and improves with each usage. Furthermore, NN frameworks do not need huge databases to hold outgoing data and probabilities. A NN is capable of completing tasks that a linear program is incapable of. Due to the parallel framework of the NN, it may carry out to perform even though its single elements fail. Some drawbacks include the fact that the training procedure can be time-intensive, causing users to underutilize the technology. The NN frameworks develop their individual formulae for weighting [16] and combining input based on statistical recognition trends over time that can be really hard to understand, making the framework's trustworthiness questionable. To function, the NN requires training. Because the structure of a NN differs from that of a microprocessor, it must be mimicked. Appendicitis, back discomfort, myocardial infarction, mental crises, and skin diseases are only a few examples. In other instances, the NN's diagnostic predictions [17] of pulmonary diseases were superior to doctors' forecasts. Patients' records, illness data sets, and anthropometry data sets are among the many kinds of data sets utilized.

7 Datasets

Hepatitis, heart disease, dermatological disease, diabetes, and lung cancer are four UCI medical datasets. According to the latest figures, heart disease was the top cause of death in 2007. Coronary heart disorder, ischemic heart disorder, cardiovascular disorder, cor pulmonale, hereditary heart disorder, hypertensive heart disorder, and valvular heart disorder are the most prevalent heart illnesses. The illness may manifest itself in a variety of ways. Finding trends in heart disease data may aid in the diagnosis of future instances. The V.A. Medical Center in California and the Cleveland Clinic Foundation compiled the heart disease database [18].

It originates from Yugoslavia's Jozef Stefan Institute. The information was collected in 1988. Hepatitis is caused by the hepatitis B virus, which is a deadly infection (HBV). In 15% of instances, chronic hepatitis develops if the illness is not eradicated during the first infection. Diabetes database-Diabetes may present with a wide range of symptoms. A plasma glucose level is tested during diagnosing. This kind of testing determines whether or not the patient is a danger of developing diabetes. Diabetic patients must be diagnosed as soon as feasible.

8 Discussions

Here, we examine how physicians are increasingly unable to identify diseases such as cancer and tumors, resulting in an increase in the mortality rate. Heart disorder is the most prevalent illness in patients and very harmful; thus, we are utilizing contemporary technologies such as data mining and data warehousing. Utilizing this approach, one may quickly uncover the disease's concealed information. We use classification, NB, NN, multilayer perceptron, and other data mining techniques. In this survey, we covered three techniques: the C4.5 algorithm, the NB algorithm, and the NN [19]. It is possible to change the confidence level for tree pruning in the C4.5 algorithm, as well as the least amount of occurrences that are allowed at a leaf. It is desirable to view the tree in graphical form; here, the tree constructs models that are readily understood, are simple to apply, and can utilize both categorical and continuous variables; however, it does not perform well on a short training set. The simplicity, computational economy, and excellent classification performance of the Naive Bayes classifier are its greatest assets. The Naive Bayes classifier, for example, requires a huge number of data in order to provide excellent results. Second, Naive Bayes assumes that if a predictor category is not a part of the training data, a new record containing that predictor category has zero probability. When it classifies, there is no discernible increase in performance. In a NN, there are three layers: the input layer, the hidden layer, and the output layer. The input layer's results are hidden in the hidden layer.

9 Conclusion

The conditions that exist between neurons are assigned weights. Their values are determined using the back propagation approach. In buried layers, several nonlinear features are established for the network. There could be a lot of output nodes in the out layer that predict different ailments. Using this method, the forecasts are difficult to interpret. In this survey, we used a cutting-edge approach to assist doctors and other healthcare providers with decision-making tasks such as analyzing patient data diagnosis. The Healthcare Decision Support System is what it is called HDSS. C4.5 outperforms the Nave Bayes technique in this scenario since there is a thorough outline of the data as well as the necessary preparatory processes. C4.5 offers exceptionally accurate findings after a few folds of cross-validation when analyzing the feature with high-performance gain for classification, however NB classification performance does not show any meaningful improvement. The main goal of this research was to identify the most commonly used data mining algorithms in modern HDSS and to evaluate their performance on a variety of medical datasets. The algorithms employed were C4.5, multilayer perceptron, and Naive Bayes, and multiple sickness databases were used. Many healthcare decision support systems are used in medical facilities all around the world.

References

1. Orte S, Subías Beltráin P, Vargiu E, Miralles F (2021) Perspective and experiences of decision support systems (DSS) and AI in prevention and care for older adults, pp 115–128
2. Eom S (2020) DSS, BI, and data analytics research: current state and emerging trends (2015–2019). In: Moreno-Jiménez JM, Linden I, Dargam F, Jayawickrama U (eds) Decision support systems X: cognitive decision support systems and technologies. Lecture notes in business information processing. Springer, Cham, pp 167–179
3. Comparative analysis of data mining techniques to predict heart disease for diabetic patients
4. Dixon BE, Kharrazi H, Papagari Sangareddy SR (2020) Public health decision support systems. In: Magnuson JA, Dixon BE (eds) Public health informatics and information systems. Health informatics. Springer, Cham, pp 349–365
5. Gupta A, Pramanik S, Bui HT, Ibenu NM (2021) Machine learning and deep learning in steganography and steganalysis
6. Muhammad LJ, Islam MM, Usman SS, Ayon SI (2020) Predictive data mining models for novel coronavirus (COVID-19) infected patients' recovery. SN Comput Sci 1(4):206
7. Meslie Y, Enbeyle W, Pandey BK, Pramanik S, Pandey D, Dadeech P, Belay A, Saini A (2021) Machine intelligence-based trend analysis of COVID-19 for total daily confirmed cases in Asia and Africa
8. Mandal A, Dutta S, Pramanik S (2021) Machine intelligence of Pi from geometrical figures with variable parameters using SCILab
9. Mansour NA, Saleh AI, Badawy M, Ali HA (2021) Accurate detection of Covid-19 patients based on feature correlated Naïve Bayes (FCNB) classification strategy. J Ambient Intell Hum Comput 1–33
10. Narin A, Kaya C, Pamuk Z (2021) Automatic detection of coronavirus disease (COVID-19) using X-ray images and deep convolutional neural networks. Pattern Anal Appl 24(3):1207–1220. arXiv:2003.10849
11. Xinzi W, Kejia W, Jiamu D, Xinqi C, Yi L, Wenlong Z (2021) Predicting water quality during urbanization based on a causality-based input variable selection method modified back-propagation neural network. Environ Sci Pollut Res 28(1):960–973
12. Caixia C, Liwei G, Sheng Z (2021) Design and implementation of bank CRM system based on decision tree algorithm. Neural Comput Appl 33(14):8237–8247
13. Tanyu BF, Abbaspour A, Alimohammadlou Y, Tecuci G (2021) Landslide susceptibility analyses using Random Forest, C4.5, and C5.0 with balanced and unbalanced datasets. CATENA 203:105355
14. Natarajan Y, Arshath Raja R, Kousik NV (2020) Privacy preservation between privacy and utility using ECC-based PSO algorithm, pp 567–573
15. Pramanik S, Singh RP (2017) Role of steganography in security issues
16. Pramanik S, Raja S (2020) A secured image steganography using genetic algorithm. Adv Math: Sci J 9:4533–4541
17. Dutta S, Bandyopadhyay S, Pramanik S (2021) Prediction of weight gain during covid-19 for avoiding complication in health
18. Jumin E, Basaruddin FB, Yusoff YBM, Latif SD, Ahmed AN (2021) Solar radiation prediction using boosted decision tree regression model: a case study in Malaysia. Environ Sci Pollut Res 28(21):26571–26583
19. Sreedhar Kumar S, Ahmed ST, Vigneshwaran P, Sandeep H, Manjunath Singh H (2021) Two phase cluster validation approach towards measuring cluster quality in unstructured and structured numerical datasets. J Ambient Intell Hum Comput 12(7):7581–7594

A Hybrid Gray Wolf Optimizer for Modeling and Control of Permanent Magnet Synchronous Motor Drives

Souvik Ganguli, Gagandeep Kaur, and Prasanta Sarkar

Abstract The model order diminution and controller design of the permanent magnet synchronous motor drive, which is commonly called PMSM, are performed in the complex delta framework using a hybrid metaheuristic algorithm. Two fundamental algorithms, viz., the firefly technique and the gray wolf optimizer (GWO) are combined to develop a new topology termed as the hybrid gray wolf optimizer (HGWO). A PMSM drive made up of both the speed and current controllers was originally generated, by using an identification method used for signal processing, to yield a lower-order model. The second-order model in cascade with an appropriate controller is thus compared with a chosen reference to estimate approximately the control parameters which are not known beforehand. The continuous parameters almost imitated the parameters set with the delta operator. Thus, for the drive, thereby, a unified control system is built. Therefore, both for order reduction and for calculating the control parameters of PMSM drives, the proposed algorithm is successfully used. Besides, a case study for the control of such special motor drives can also be investigated in the future for many household as well as commercial purposes.

Keywords Permanent magnet motor · Model matching · Delta domain method · Firefly algorithm (FA) · Gray wolf optimizer (GWO) · Hybrid gray wolf optimizer (HGWO)

S. Ganguli (✉) · G. Kaur
Department of Electrical and Instrumentation Engineering, Thapar Institute of Engineering and Technology, Patiala, Punjab 147004,, India
e-mail: souvik.ganguli@thapar.edu

G. Kaur
e-mail: gagandeep@thapar.edu

P. Sarkar
Department of Electrical Engineering, National Institute of Technical Teachers' Training and Research, Kolkata, West Bengal 700106,, India
e-mail: psarkar@nitttrkol.ac.in

P. Dutta et al. (eds.), *Emerging Technologies in Data Mining and Information Security*, Lecture Notes in Networks and Systems 490,
https://doi.org/10.1007/978-981-19-4052-1_72

1 Introduction

Mathematical models of higher order are generated by some of the popular control area, for example, fuel injection system, fighter planes, automatic spark timer models, etc., to name a few. These models are hard to manage, too. Therefore, it is preferable to diminish its model, which contributes to the reduction of the computational burden and problems in the implementation of higher-order controllers. Although intelligent control techniques for PMSM are mentioned in the literature [1, 2] due to its incompliant structure and strength in a variety of operations, the PI control system continues to attract industrial applications. The following is a discussion of some important research work in the literature that generates researchers' interest in the speed control of the PMSM drives.

Pillay and Krishnan [3] described the modeling, analysis, and simulation of a controlled PMSM drive system. The state-space models of the controller, the motor was included, and the real-time functioning of inverter switches was examined. The difference in the functioning of the system due to the usage of hysteresis current and PWM was included. A robust-natured current controller of PMSM was designed using adaptive parameter estimation by Kim et al. [4] which was built on the model reference-adaptive method. The efficacy of the roust controller was verified by cross-examining it with the conventionally available schemes at that time. Shyu et al. [5] designed a controller based on variable-structure and linear quadratic-controlled methods for robust control of PMSM. The controller consisted of design flexibility in the conventional feedback methods and was used for controlling the position of PMSM. The results obtained from simulations and experiments were also studied. Mohamed [6] proposed a speed control method for the PMSM drive system and included dead time. Firstly, a self-tuning proportional-integral MFC (ST-PI-MFC) was proposed which was a dead-time compensator. It worked based on the affine projection algorithm (AFM). Secondly, an observer for controlling disturbance was proposed which was then used with the motor and was used for the estimation of the motor. The parameters obtained then were used for updating the other parameters using the AFM. Haque and Rahman [7] presented the modeling, analysis, and control of trajectories of the direct torque control scheme (DTC) of IPMSM. The scheme was used for providing constant torque, and the experimental results were presented. Hakiki et al. [8] investigated the speed control of PMSM using fuzzy logic controller (FLC) using voltage inverters (VSC). The adaptive calculations which include the calculation of error value and generating a voltage based on that were done using FLC, and FLC was also equipped with an error-driven adaptive mechanism for robust control. Liu et al. [9] employed the support vector machine-generalized inverse (SVMGI) with PMSM and developed the pseudo-linear model for PMSM. Then, on this pseudo-linear model, internal model control (IMC) theory was applied for the design of controllers for vigorous speed control of the drive. The merits of SVMGI and IMC were incorporated in the paper, and the results were verified through experiments as well as simulations. Wu and Qi [10] used active disturbance rejection control (ADRC), loop control, and speed control of PMSM.

An extended-state observer (ESO) was employed with PMSM drive which helped in ensuring its high performance and steady-state stability and accuracy. ADRC used along with ESO also determined some of the disturbances and suppressed them. It also provided compensation in the feed-forward system for accurate speed control. Lin et al. [11] proposed a recurrent fuzzy neural cerebellar model articulation network (RFNCMAN) for position and torque control of the permanent magnet-drive system. The method first generated an ideal position torque controller. After that based on the error-generated RFNCMAN tried to mimic the ideal controller by providing compensation for the system. Yu et al. [12] proposed a fractional-order model for the PMSM servo system. This model of the PMSM servo system was calculated theoretically, and then, its parameters were identified using improved Levy method with weights. Two H_∞ norm stabilizers were designed for the feedback control of the system using an algorithm used for the parameter identification of the fractional-order system. Chen et al. [13] presented a model predictive control for controlling the current of twelve-phase PMSM. The total number of switching phases as calculated was 4096 and the paper divided the largest twenty-four available voltage vectors into six vectors and carried out further calculations. Prabhakaran and Karthikeyan [14] enhanced the performance of sensorless surface-mounted PMSM (SSMPMSM) at a standstill and low-speed operating regions using model reference adaptive system speed calculator cascaded with the pseudo-derivative feedback controller. Small signal analysis was employed for the sensitivity and stability analysis of the system. The results obtained were quite promising.

It can be rightly said that almost all of the discrete systems are derived from continuous system sampling. These systems are modeled, examined, and controlled by making use of either a time domain approach or a complex domain applying z-transform analysis. These systems do not, therefore, take into consideration an exceptionally high sampling rate and are not stabilized at close range due to pole crowding near (1,0). This can be achieved if the Middleton and Goodwin-created delta operator is used to formulating the discrete-time systems [15]. Simulating these systems using the delta transform theory helps to construct continuous-time systems at a high sampling frequency and simultaneously increases computational control over the conventional shift operator-based discrete-time models. Delta operator-based systems are thus commonly maneuvered for research and design of control systems [2, 16–18].

The Truxal approach is a synthesis-based approach in the literature of classical control systems [19]. There are two ways by virtue of which the controller synthesis is executed. The first method pertains to the precise model matching, in which a reference is configured to satisfy some specified time and frequency domain parameters. After that, the controller parameters are determined when the closed-loop controlled system coincides with the reference model. The most critical demerit of the exact matching model (EMM) [20] is that the design does not ensure that the controller is physically implemented. Approximate model matching (AMM) [21] is an alternative approach to solve this issue and could be extended to the design control system in the combined domain of analysis. Ganguli et al. [22] developed a hybrid gray wolf optimizer (HGWO) for the order reduction of systems in the integrated delta framework.

This algorithm has been extended to minimize not only the third-order permanent magnet drive system but also in synthesizing the PI controller corresponding to the lower-order drive model in the combined domain, using a model-matching method fitting approximately. The remainder of the article is constructed depicted as follows. The working technique is discussed in Sect. 2; in Sect. 3, the simulation results are narrated, while in Sect. 4, the main conclusions are addressed.

2 An Overview of the Proposed Method

Two important aspects are discussed in this article. In the first section, the PMSM drive model order reduction is achieved using a hybrid technique [22]. In the second part, a PI controller is developed using the same heuristic method. In the model reduction process, the HGWO is used to calculate the numerator and denominator polynomials. The hybridization of the firefly algorithm (FA) [23] and gray wolf optimizer (GWO) [24], coined as HGWO, was utilized to diminish the higher-order single-input single-output models in the combined domain. The adjustment between intensification and diversification features was realized using the firefly algorithm globally, whereas the gray wolf optimization technique conducted a local search mechanism for validation of the exploitation abilities. The quest began with the firefly technique initialized with the help of some random entities. The calculation carried on for a specific tally of iterations for the purpose of achieving the best position in the requisite solution space. Therefore, the outcome produced by the firefly technique was tied-up as initial point for the GWO algorithm. The search method was then diverted to GWO to accelerate the convergence activity toward finding the optimal deliverable finally. In such a manner, the hybrid solution has been able to reach an optimum correctly.

The sum of square errors (SSEs) is reduced by the HGWO method to evaluate the coefficients of the numerator and the denominator. Unlike the step response match, the pseudo-random binary sequence (PRBS) is used as an input signal to get the unknowns in the numerator and denominator. Our method preserves even the correspondence of the DC gain and retains the minimum phase characteristic of the original test system. In addition, features of the parent system are retained both in terms of time and frequency domains. Moreover, it has maintained the dominant pole position. In comparison with the proposed technique, the following methods are used, namely the particle swarm optimization (PSO), genetic algorithm (GA), ant-lion optimization, sine–cosine algorithm (SCA), multi-verse optimizer (MVO), and Salp swarm algorithm (SSA) [25–30], apart from the parent techniques. A feasible low-order PI controller for which the control synthesis is carried out in this paper is one of the purposes of the model order reduction technique. Conventionally, either auto-tuning or standard old PID tuning rules are used to obtain PID parameters [31]. The Truxal method [19] is also a method of accurate model matching control (EMM) synthesis [20], in which the test system coefficients are computed to satisfy the intended output, and controller parameters are measured in a manner that closely resembles the complete closed-loop system. This EMM technique has the main disadvantage that the hardware application of the controller is not guaranteed. Approximate model matching (AMM) [21] will serve as a replacement even for design schemes to overcome this restriction of the EMM process. The AMM technology is therefore used to fetch the controller's coefficients using the HGWO method. Similarly, the HGWO technique is used for checking the effectiveness of the strategy with a certain number of other algorithms.

3 Results and Discussions

The transfer function of the PMSM drive [32, 33] developed from the modeling concept is represented with the help of

$$G_{\mathrm{HOS}}(s) = \frac{7.94s + 13.23}{9.6 \times 10^{-7}s^3 + 3.99 \times 10^{-3}s^2 + 6.97s + 34.63} \tag{1}$$

The objective of this investigation is to produce an appropriate controller for this permanent magnet motor drive. From Eq. (1), we find that the drive produces higher-order terms. So, before developing the right controller for the drive, the 3rd-order model has to be reduced although retaining the inherent features of the original drive model. A PI controller is then designed for the lower-order model. The purpose of using a PI controller over a regular PID is to escape considerable disruptions and noises during the operation of the drive. The transfer function of the drive narrated in Eq. (1) is converted in the combined discrete-time representation at a sampling time $\Delta = 0.001$ s given by

$$G_{\mathrm{HOS}\delta}(\gamma) = \frac{100\gamma^2 + 1200300\gamma + 1995900}{\gamma^3 + 2040\gamma^2 + 1061200\gamma + 5.224200} \tag{2}$$

The reduced models are assumed to be represented in the delta domain with appropriate population sizes and maximum iterations as 20 and 100. As reported in the literature, standard parameters (proposed, parent, and standard heuristics) of all algorithms are considered. The second-order model obtained with the help of the proposed method is given as

$$G_r(\gamma) = \frac{990.5\gamma + 1436.0144}{\gamma^2 + 843.3\gamma + 3759.2} \tag{3}$$

This model also yields the least fitness function when compared with the parent techniques as well as standard heuristic techniques as seen from the graph shown in Fig. 1.

The hybrid gray wolf optimizer (HGWO) not only superseded the constituent methods but also proved better than some of the popular heuristic techniques available

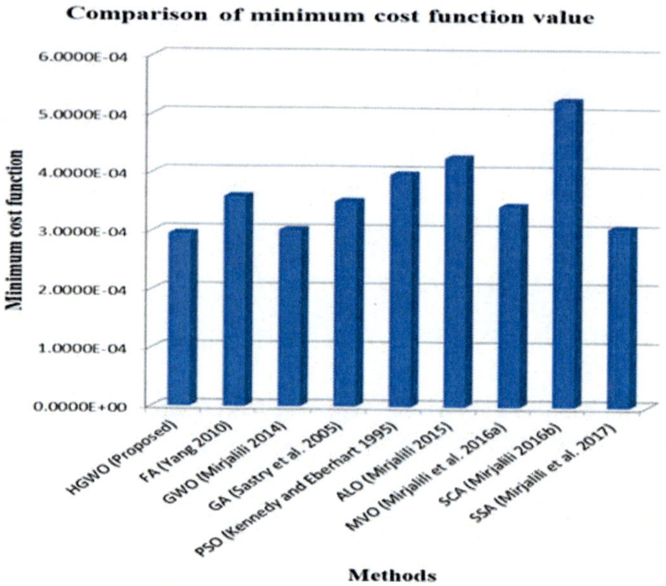

Fig. 1 Comparison of minimum cost function value

in the literature. The proposed HGWO technique also produces the minimum fitness value as evident from Fig. 1. In addition, the transient parameters for the reduced models are also calculated whose results are showcased in Fig. 2.

The integrated approach reflects the closest match with the original system in comparison to the methods taken up for the investigation. Moreover, the error metrics commonly used in the theory of systems and control are also assessed for the reduced-order models as depicted in Fig. 3.

The results of Fig. 3 suggest that the proposed method has outperformed many of the recent heuristic approaches. The last move is to obtain a realize PI controller in the integrated discrete-delta representation. The approximate model matching framework is used to develop the controller. The parameters of the controller are thus obtained heuristically applying the suggested algorithm. The unified controller is thus written as

$$G_c(\gamma) = 1.8679 + \frac{1.4967e - 05}{\gamma} \tag{4}$$

Further, the fitness values of the different methods are also used for comparison with the suggested technique; the results of which are presented in Fig. 4.

The proposed hybrid algorithm reaches the lowest optimum value in comparison to some of the current heuristic techniques. Therefore, the suggested technique could successfully generates PMSM models of a lower order, as well as use the approximate model matching technology to yield a PI controller with the lowest fitness function in contrast to the parent and other standard techniques.

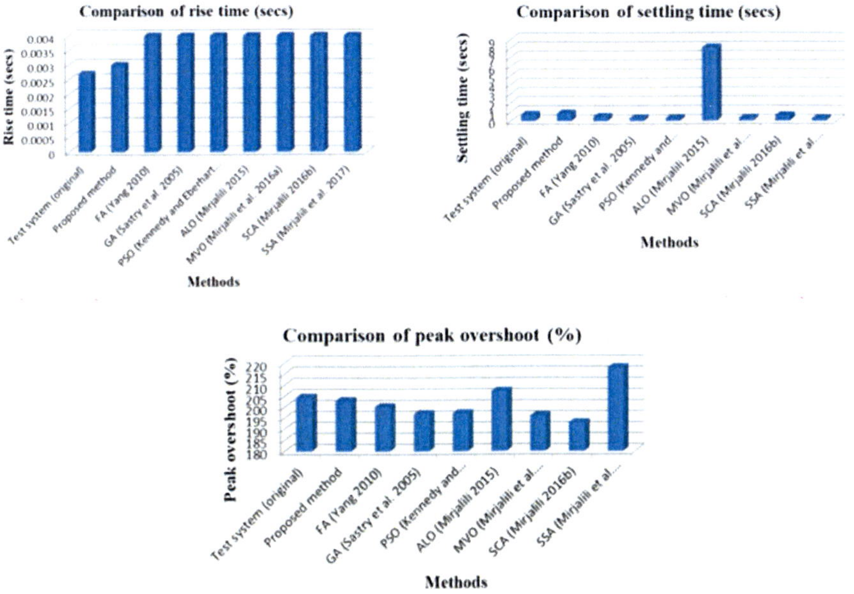

Fig. 2 Comparison of the time-domain metrics of the lower-order permanent magnet drive model applying unified domain approach

4 Conclusions

In this paper, the topic of modeling and development of new controller architecture in the delta framework is addressed for the permanent magnet motor drive. A constrained global optimization technique is employed to determine a new lower-order model for the drive. A hybrid topology combining the gray wolf optimizer with the firefly algorithm has been successfully employed for the purpose. Moreover, the paper also uses the model matching method concepts to yield the PI controller synthesis. The benefit of this AMM approach is that it uses the feedback from the drive output. The control system is so designed that it closely follows the desired model. This paper thus discusses efficiently one important application for modeling and speed control of the permanent magnet drive in the integrated domain. Several new motors such as the switched reluctance motor (SRM) as well as the brushless direct current (BLDC) motor can similarly be investigated by applying this technique.

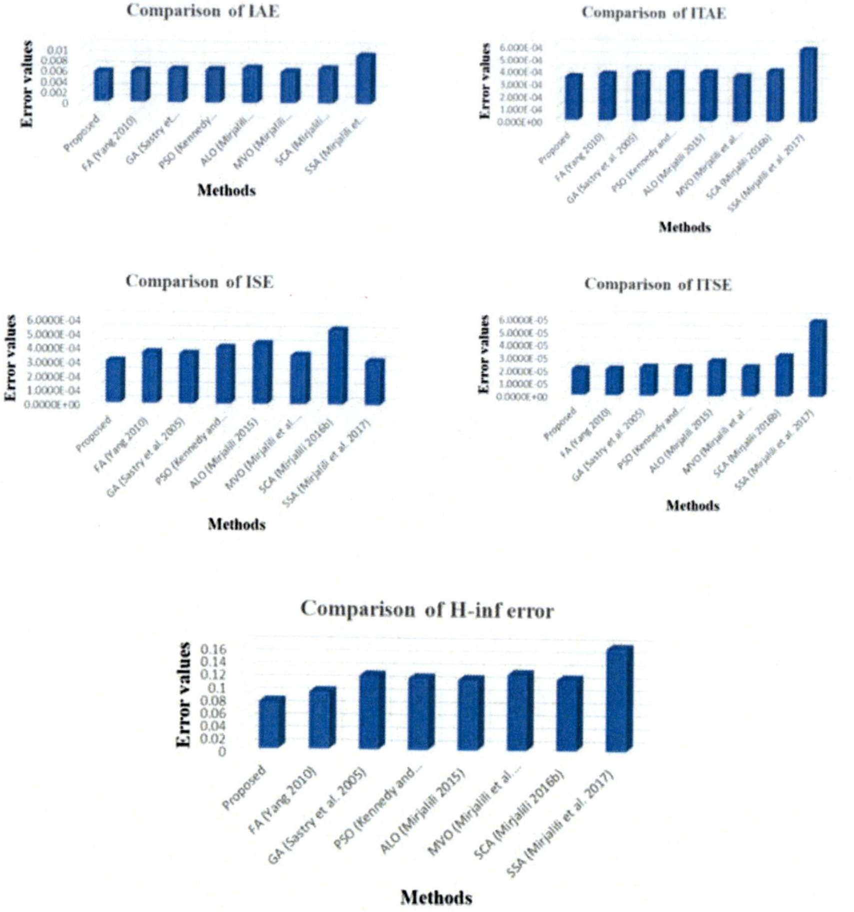

Fig. 3 Comparison of some common performance measures of the lower-order drive in the delta domain

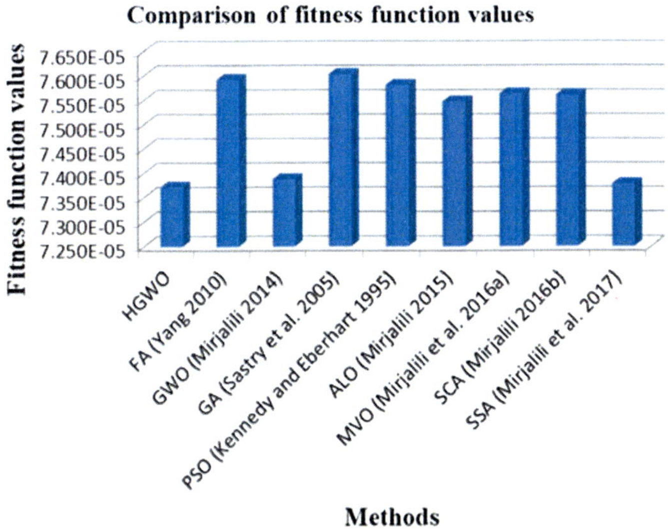

Fig. 4 Comparison of fitness function values

References

1. Ibrahim Z, Levi E (2002) A comparative analysis of fuzzy logic and PI speed control in high-performance AC drives using experimental approach. IEEE Trans Ind Appl 38(5):1210–1218
2. Yu J, Shi P, Dong W, Chen B, Lin C (2014) Neural network-based adaptive dynamic surface control for permanent magnet synchronous motors. IEEE Trans Neur Netw Learn Syst 26(3):640–645
3. Pillay P, Krishnan R (1989) Modeling, simulation, and analysis of permanent-magnet motor drives. I. The permanent-magnet synchronous motor drive. IEEE Trans Ind Appl 25(2):265–273
4. Kim KH, Chung SK, Moon GW, Baik IC, Youn MJ (1995) Parameter estimation and control for permanent magnet synchronous motor drive using model reference adaptive technique. In: Proceedings of IECON'95–21st annual conference on IEEE industrial electronics, vol 1, pp 387–392. IEEE
5. Shyu KK, Lai CK, Tsai YW, Yang DI (2002) A newly robust controller design for the position control of permanent-magnet synchronous motor. IEEE Trans Industr Electron 49(3):558–565
6. Mohamed YARI (2006) Adaptive self-tuning speed control for permanent-magnet synchronous motor drive with dead time. IEEE Trans Energy Convers 21(4):855–862
7. Haque ME, Rahman MF (2009) Incorporating control trajectories with the direct torque control scheme of interior permanent magnet synchronous motor drive. IET Electr Power Appl 3(2):93–101
8. Hakiki K, Meroufel A, Cocquempot V, Chenafa M (2010) A new adaptive fuzzy vector control for permanent magnet synchronous motor drive. In: 18th Mediterranean conference on control and automation, MED'10, pp 922–927. IEEE
9. Liu G, Chen L, Zhao W, Jiang Y, Qu L (2012) Internal model control of permanent magnet synchronous motor using support vector machine generalized inverse. IEEE Trans Industr Inf 9(2):890–898
10. Wu C, Qi R (2013) The simplified active disturbance rejection control for permanent magnet synchronous motor drive system. In: Proceedings of the 32nd Chinese control conference, pp 4172–4176. IEEE

11. Lin FJ, Yang KJ, Sun IF, Chang JK (2015) Intelligent position control of permanent magnet synchronous motor using recurrent fuzzy neural cerebellar model articulation network. IET Electr Power Appl 9(3):248–264
12. Yu W, Luo Y, Chen Y, Pi Y (2016) Frequency domain modelling and control of fractional-order system for permanent magnet synchronous motor velocity servo system. IET Control Theory Appl 10(2):136–143
13. Chen B, Lv J, Jiang X (2019) Simplified model predictive control of a twelve-phase permanent magnet synchronous motor. In: IECON 2019–45th annual conference of the IEEE industrial electronics society, vol 1, pp 1345–1350. IEEE.
14. Prabhakaran KK, Karthikeyan A (2020) Electromagnetic torque-based model reference adaptive system speed estimator for sensorless surface mount permanent magnet synchronous motor drive. IEEE Trans Industr Electron 67(7):5936–5947
15. Middleton RH, Goodwin GC (1990) Digital control and estimation: a unified approach. Prentice Hall Professional Technical Reference
16. Huang C, Naghdy F, Du H (2017) Delta operator-based fault estimation and fault-tolerant model predictive control for steer-by-wire systems. IEEE Trans Control Syst Technol 26(5):1810–1817
17. Kumari K, Bandyopadhyay B, Kim KS, Shim H (2019) Output feedback based event-triggered sliding mode control for delta operator systems. Automatica 103:1–10
18. Ramírez-Neria M, Luviano-Juárez A, Lozada-Castillo N, Ochoa-Ortega G, Madonski R (2020) Discrete-time active disturbance rejection control: a delta operator approach. In: Advanced, contemporary control. Springer, Cham, pp 1383–1395
19. Truxal JG, Weinberg L (1955) Automatic feedback control system synthesis. Phys Today 8(8):17
20. Tzafestas SG, Pimenides TG (1982) Exact model-matching control of three-dimensional systems using state and output feedback. Int J Syst Sci 13(11):1171–1187
21. Wang SS, Chen BS (1987) Model matching: optimal approximation approach. Int J Control 45(6):1899–1907
22. Ganguli S, Kaur G, Sarkar P (2020) A new hybrid algorithm for identification in the unified delta framework. In: AIP conference proceedings, vol 2207, No 1. AIP Publishing LLC, p 040002
23. Yang XS (2010) Firefly algorithm, stochastic test functions and design optimisation. Int J Bio-ins Comput 2(2):78–84
24. Mirjalili S, Mirjalili SM, Lewis A (2014) Grey wolf optimizer. Adv Eng Softw 69:46–61
25. Kennedy J, Eberhart R (1995) Particle swarm optimization. In: Proceedings of ICNN'95-international conference on neural networks, vol 4, pp 1942–1948. IEEE
26. Mirjalili S (2015) The ant lion optimizer. Adv Eng Softw 83:80–98
27. Mirjalili S, Mirjalili SM, Hatamlou A (2016) Multi-verse optimizer: a nature-inspired algorithm for global optimization. Neural Comput Appl 27(2):495–513
28. Mirjalili S (2016) SCA: a sine cosine algorithm for solving optimization problems. Knowl-Based Syst 96:120–133
29. Mirjalili S, Gandomi AH, Mirjalili SZ, Saremi S, Faris H, Mirjalili SM (2017) Salp swarm algorithm: a bio-inspired optimizer for engineering design problems. Adv Eng Softw 114:163–191
30. Sastry K, Goldberg D, Kendall G (2005) Genetic algorithms. In: Search methodologies. Springer, Boston, MA, pp 97–125
31. O'dwyer A (2009) Handbook of PI and PID controller tuning rules. World Scientific
32. Ayyar K (2013) Investigations on model order reduction and design of speed controller for electric drive systems (Doctoral dissertation)
33. Krishnan R (2001) Electric motor drives: modeling, analysis, and control. Pearson

Ensemble Method of Feature Selection Using Filter and Wrapper Techniques with Evolutionary Learning

Sabyasachi Mukherjee, Soumi Dutta, Sushmita Mitra, Soumen Kumar Pati, Farooq Ansari, and Arpit Baranwal

Abstract The improvement in data collection and mining methods has expanded the range of dimensionality or features in the data, which brings about an obstacle to many existing feature selection methodologies. This paper brings forth a fresh feature selection methodology rooted in particle swarm optimization (PSO) as wrapper method and an ensemble method to merge the results of the different filter techniques (chi-square, F-regression, and mutual information) to find an optimal feature set that covers most of the key variables of the dataset. The local search is executed on the global best and makes use of a filter-based method, which then intends to take the advantage of the filter and wrapper methods. Our results exhibit that the proposed methodology can be successfully used to select fewer features and, at the same time, increase the classification efficiency over using all features. The proposed methodology also shows how well an evolutionary learning algorithm like the particle swarm optimizer can be used for search optimization of optimum features in the dataset.

Keywords Evolutionary learning · Machine learning · Feature engineering · Feature selection · Particle swarm optimization · Filter method · Wrapper method

1 Introduction

It is incredibly tricky to diagnose diseases because a proper diagnosis is dependent upon several, different factors [1]. Needless to say, it gets trickier if we try to do

S. Mukherjee (✉) · F. Ansari · A. Baranwal
Asansol Engineering College, Asansol, West Bengal, India
e-mail: sabyasachi.mukherjee24@gmail.com

S. Dutta
Institute of Engineering and Management, Kolkata, India

S. Mitra
Indian Statistical Institute, Kolkata, India

S. K. Pati
Maulana Abul Kalam Azad University of Technology, Nadia, West Bengal, India

P. Dutta et al. (eds.), *Emerging Technologies in Data Mining and Information Security*, Lecture Notes in Networks and Systems 490,
https://doi.org/10.1007/978-981-19-4052-1_73

it automatically with the aid of cutting-edge machine learning paradigms, as we observed in the works of [2, 3], and [4] to name a few. Heart disease is and has been throughout history, one of the major causes of fatality in humans.

The term 'heart disease' itself is an umbrella for various different types of diseases such as coronary heart disease (CAD), hypertension, and cardiac arrest. Most of these diseases are specific to the arteries that provide blood to the heart. Its diagnosis takes in a number of factors that include age, gender, smoking habits, depression, genetic factors, and various metrics like ECG reports, cholesterol level, unhealthy diet, high blood pressure, and so on. In this paper, we have used a dataset with the diagnosis information corresponding to the aforementioned (and more) factors and tests. The dataset is anonymously populated to preserve the privacy of the patients. According to World Health Organization (WHO), 17.9 million people died due to heart disease in 2019, and it is estimated to increase by 2020 [5]. Evolutionary learning has been at the forefront of research in the better part of the last decade. In theory, it involves trying to understand the methodologies and techniques used in natural evolutionary patterns and translating them into repeatable empirical algorithms. There have been many instances of using a genetic algorithm [6], particle swarm optimization [7], or neural networks [8] to diagnose diseases. This approach helps us in this context because the selection of the best features can also be understood as an optimization problem.

As mentioned, particle swarm optimization is one such evolutionary learning algorithm. It relates to understanding and converting evolutionary computations used by birds, fishes, and different insects to guide their traveling group into a definitive step-by-step algorithm that can be used in the field of optimization. The fitness function used in the proposed methodology is a support vector machine classifier. When we are faced with a multi-dimensional dataset, feature selection comes to the fore and makes itself a necessity. To truly understand what features affect the diagnosis the most, we need to separate the grain from the stalk. It is an important step in data preparation as it makes the machine learning model more robust. The reduction in dimension also helps with the computational load required to run the machine learning models.

There are different ways to select the best features within a dataset. Among them, the *filter* approach has been popular and explored by authors [9, 10], and [11] in their works. It does not involve the use of machine learning models and is dependent on the nature of the data itself. The features are shortlisted depending on statistical computations of their correlation with the target variable.

Following are some of the correlation co-efficient calculation methods that were used in this paper:

(a) *Mutual Information*: This calculates the amount of information each of the features holds on them by comparing them in between each other. This removes the redundancy of features.

(b) *F-regression*: This calculates the correlation of correct target variable prediction with respect to each selected feature. It is a repetitive process that outputs the top features that directly affect the target variable.

(c) *Chi-square*: This method is similar to mutual information. It determines the dependency of two variables with each other, which are in this case, an individual feature and the target variable. This is done for all of the features.

In contrast, the *wrapper* method of feature engineering involves the adoption of machine learning models in the feature selection process. We try out different combinational subsets of features with a model and then compare the results, to find the best possible subset. This method has also been widely used by researchers for feature engineering [12–14]. One drawback of this approach is that it is computationally heavier than its corresponding filter methods. In this paper, we explore an ensemble method that uses both of these feature selection techniques.

The rest of the paper is organized thusly. A brief literature survey on machine learning methodologies of heart disease datasets is presented in Sect. 2. Section 3 describes the dataset. The data preprocessing methods used in the pipeline are described in Sect. 4, while Sect. 5 explains the proposed methodology. Section 6 examines the comparative analysis of the results of the various feature selection algorithms. Section 7 wraps up the paper with a few probable future research directions.

2 Related Work

Many researchers have worked on the problem of predicting heart disease in the medical field through data science, but the performance of prediction still has room to grow. The dataset being a popular one has attracted various researchers from time to time. Sun et al. [15] used the rapid miner as an aid followed by decision tree, SVM, logistic regression, Naïve Bayes, and random forest algorithms which are used as feature selection algorithms, thus comparing the outputs of all the combinations to calculate the result.

Reddy et al. [16] showed how a random forest algorithm can be used for classification and feature selection. This model uses a minimum of 8 and 6 selected features for its better performance, and it also shows dropping the number of features will affect the accuracy of the model.

Gárate-Escamila et al. [17] used a combination of six machine learning classifiers out of which chi-square and principal component analysis (CHI-PCA) with random forests (RF) obtained the best performance. They derived a few features pertaining to ST depression and heart vessels such as cholesterol, peak heart rate, and chest pain which are relevant to anatomy and physiology using ChiSqSelector.

Jayaraman et al. [18] described how optimized and hybridized methodology can be used to recognize heart disease in the early stage, for this, it uses an artificial gravitational cuckoo search algorithm which is then optimized by a particle bee optimizer. Associative memory neural network is also used for managing the features which were obtained in earlier classification systems.

Table 1 UCI heart disease dataset [21]

No.	Feature name	Feature description
1	AGE	Represented in years
2	SEX	Sex
3	CP	Type of chest pain
4	TRESTBPS	Blood pressure (resting)
5	CHOL	Serum cholesterol
6	FBS	Blood sugar (fasting)
7	RESTECG	Result of rest electrocardiography
8	THALACH	Peak heart rate obtained
9	EXANG	Exercise-induced angina
10	OLDPEAK	Exercise-induced ST depression
11	SLOPE	Slope of the exercise ST segment in peak
12	CA	Count of major vessels
13	THAL	Thallium test
14	NUM	Target column denoting whether the patient has strong chances of heart failure

In the field of feature selection, Jixiong et al. [19] have tried to combine the filter and wrapper methods and came up with a new distance-based evaluation function to determine the optimal feature subset.

Marc et al. [20] combined the filter and wrapper method and also took the help of an Minimum Spanning Tree (MST) data structure while exploiting a statistical test of *relative certainty gain.*

3 Data Overview

The dataset used in this research is collected from the UCI machine learning repository [21]. It is a multivariate classification dataset about heart disease. It is aggregated from different sources and has 75 features, only 14 of which are primarily used by researchers. There is a total of 303 data points within the dataset, all of which are anonymous patients. Details are given in Table 1.

4 Data Preprocessing

Data preprocessing methods are used to prepare the dataset for machine learning models [22, 23]. It is an essential step required for any experiment in the fields of machine learning and data science. First, to solve the problem of missing values within some columns of the dataset, the help of *imputation* was employed and the missing values were calculated to make the dataset complete. The method used for imputation was that the missing value was replaced by the mean value of that feature in that dataset.

Then, we standardized the different feature values to be within a pre-determined range as they differed in terms of their minimum and maximum possible values. This step ensured that the machine learning model would treat each feature described in Table 1 equally, and there would be no inherent biases in the data.

It can be represented as:

$$z = \frac{x - \mu}{\sigma} \tag{1}$$

Here in Eq. 1, z represents the standardized value, x is the original value, μ is the mean of all the values of that feature, and σ represents the standard deviation of all of the values of that feature. This z-value is calculated for every single value in the dataset.

5 Proposed Methodology

In order to make an efficient pipeline that can churn out maximum performance out of even basic machine learning models; we came up with a novel process flow that incorporated filter and wrapper methods in an ensemble [24–27] way that takes the help of evolutionary learning for optimization. We use an ensemble method to merge the results of the filter techniques (chi-square, F-regression, and mutual information) to find an optimal feature subset.

Since the evolutionary techniques are based on the natural phenomenon of the triumph of superior performance among the mediocre in the sample space, we felt the task of choosing the optimum set of features among many is an appropriate problem statement that could be greatly benefitted by such a technique.

Among the evolutionary learning techniques, the concept of optimization through information exchange between data points of search-space behind particle swarm optimizer greatly resembles the hunt for optimum performance through feature selection.

As depicted in Algorithm 1, Let $F_{chi\text{-}pso}$, F_{reg}, and $F_{info\text{-}pso}$ be the feature matrix obtained from using the *filter* methods chi-square, F-regression, and mutual information with particle storm optimizer wrapper, respectively. A feature would be selected for the *optimal* feature set *if and only if* that feature has appeared and selected in either,

(a) Both $F_{chi\text{-}pso}$ and F_{reg}
(b) Both $F_{chi\text{-}pso}$ and F_i
(c) Both $F_{info\text{-}pso}$ and F_{reg}

(d) All in $F_{chi\text{-}pso}$ and F_{reg} and $F_{info\text{-}pso}$

After finding this optimal subset of features, particle swarm optimization is again utilized for the *wrapper* stage and the performance of this model is calculated.

Algorithm 1: Ensemble feature selection with filter and wrapper methods and evolutionary learning

---

Input: Dataset
Output: Model Evaluation
Split the input dataset D into the following parts for the prediction model;
Preprocess D to impute the missing values and standardize the data points;
Initialize particle storm optimizer P;
Initialize master feature matrix M;
Use chi2 filter method on D to select 9 best features out of possible 14 into Fchi:
Call optimizer P with feature matrix F_{chi} as $F_{chi\text{-}pso} \leftarrow$ optimize (P, F_{chi});
Repeat the process for F-regression and Mutual Info. filters to create F_{reg} and F_{info};
for each feature *f* in $F_{reg} \cap F_{chi\text{-}pso}$ **do**
  add *f* to M;
end for
for each feature f in $F_{reg} \cap F_{info\text{-}pso}$ **do**
  add *f* to M;
end for
for each feature *f* in $F_{chi} \cap F_{info\text{-}pso}$ **do**
  add *f* to M;
end for
for each feature *f* in $F_{chi} \cap F_{info\text{-}pso} \cap F_{reg}$**do**
  add *f* to M;
end for
Use M to create ensemble master dataset D_{master};
for each model L in {Logistic Regression, Random Forest, MLP, SVM Classifiers} **do**
  evaluate (D_{master}, L);
end for

---

6 Results

In order to accentuate and analyze the effects of filter and wrapper techniques in tandem with one another, we shall be evaluating machine learning models at different stages leading up to the model that we set out to create. For clarity, we have also evaluated different feature selection methods separately. This enables us to better understand the effects that our pipeline (see Fig. 1) has on the optimization process by comparison.

For each of them, we calculated the accuracy, F1-Score, specificity, and sensitivity metrics for some of the popular machine learning models. *Accuracy* is the percentage ratio of the total number of correct predictions over all of the predictions. *Specificity or Precision* refers to the percentage ratio of the truly positive diagnosis over every positive diagnosis. *Sensitivity or Recall* is the percentage ratio of correct positive

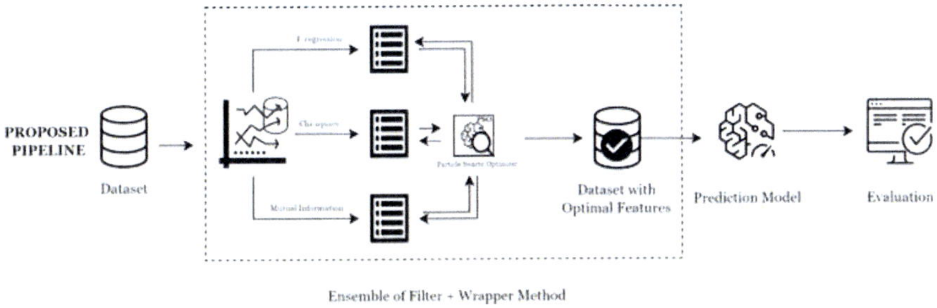

Fig. 1 Illustration [28] of the flow of execution in different phases

diagnosis over all the correct diagnoses. *F1-score* is the harmonic mean of specificity and sensitivity metrics.

As we can observe in Table 2, the metrics improve slightly with the *Wrapper* method depicted in detail in Fig. 3. This is due to the particle swarm optimization that is used for feature selection. As we can see in Fig. 5, there were also some notable results obtained by using Mutual Information Gain as a filter method. The effects of feature selection could be observed in Fig. 2, where the MLP Classifier extracted less than favorable results from the dataset. But the real peak was achieved in *Proposed Methodology* (see Fig. 6), where the specificity metric for the support vector classifier is 100%. This implies that the model has correctly predicted every single patient that would go on to have heart diseases. Figure 4 shows results with just filter method (with F-regression). Figure 7 represents comparison of results on the same dataset of various other works.

Table 2 Comparison of different filter and wrapper methods (*Percentage specificity or precision as a metric is used*)

Classifiers	No feature selection	Wrapper method (PSO)	Filter method (F-reg.)	Filter method (Mutual Info.)	Proposed methodology
Logistic regression	83.93	84.48	81.03	81.03	72.41
Random forest classifier	84.48	70.69	82.76	82.76	74.14
MLP classifier	43.1	72.41	62.07	82.76	58.62
SVM classifier	87.93	87.93	84.48	82.76	100

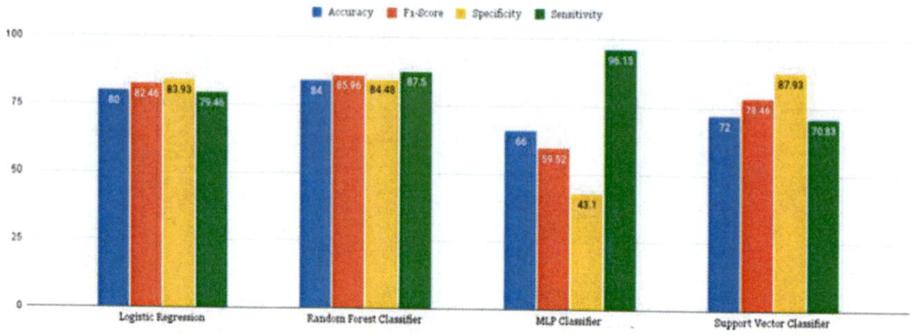

Fig. 2 Results without any feature selection

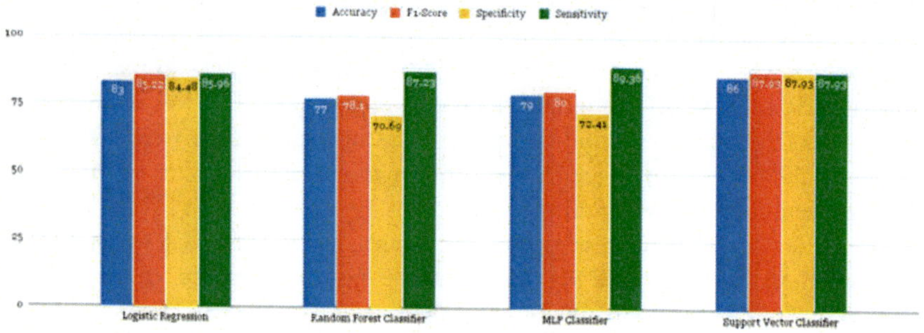

Fig. 3 Results with just wrapper method (with PSO)

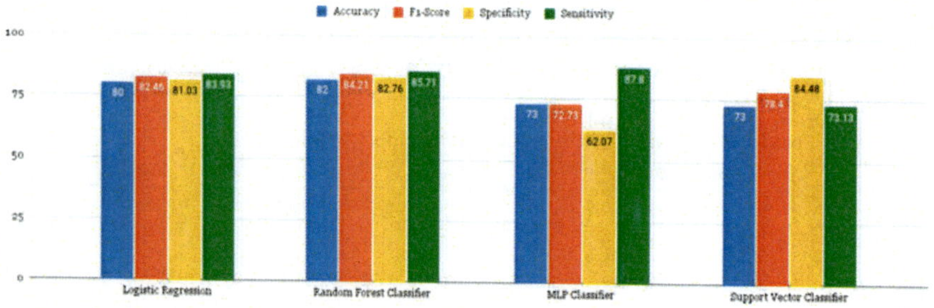

Fig. 4 Results with just filter method (with F-regression)

7 Conclusion

This work used clinical data about heart disease obtained from several different health institutes to provide a methodology to correctly predict a diagnosis of it in individual patients. We fundamentally focused on the selection of features or factors among the ones present in the dataset. The primary novelty offered by the paper is the ensembled and hybrid way of combining *filter* and *wrapper* methods of feature

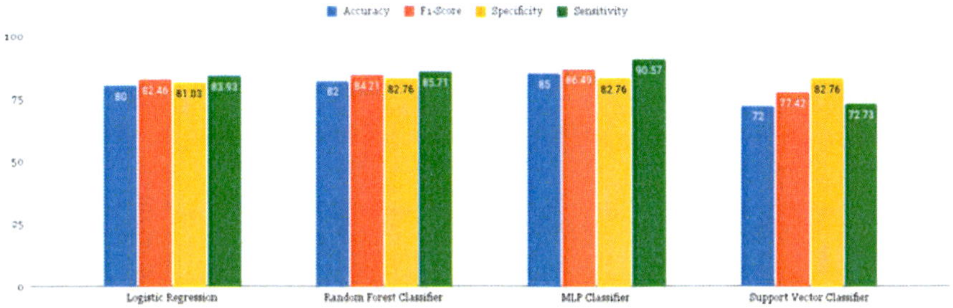

Fig. 5 Results with just filter method (with Mutual Info. Gain)

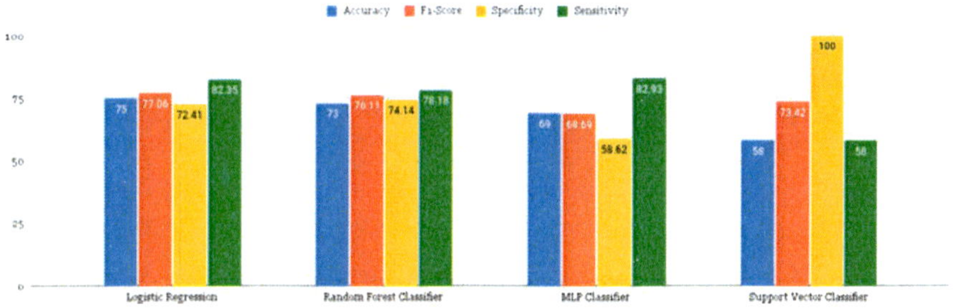

Fig. 6 Results with the proposed methodology

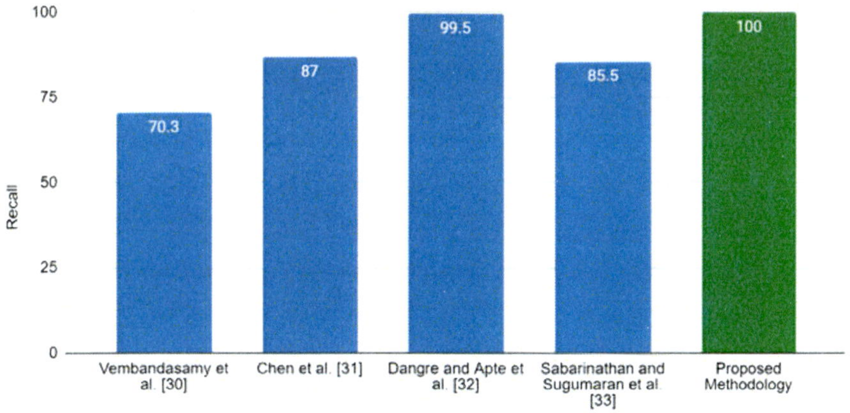

Fig. 7 Comparison of results on the same dataset of various other works

selection with evolutionary learning models and improving the prediction efficacy of any of the popular machine learning models by cleaning the dataset efficiently. The medical field has seen tremendous effort in recent years when it comes to the diagnosis of diseases using machine learning and artificial intelligence paradigms [29, 30]. At the peak of the age of data, where the inflow of data makes it abundant, we need more efficient methods of filtering through the noise and grabbing onto the key variable that affects the prediction models. In these circumstances, feature engineering and feature selection become essential, and in the future, it could be worth exploring the use of methods introduced in this paper on different datasets and various other prediction problems, such as Alzheimer's disease [31], liver disease [32], and other cardiovascular diseases [33]. It is evident through our work that an ensemble and hybrid method of feature selection and evolutionary optimization resulted in a favorable performance from various machine learning models. It illustrates the effectiveness of proper data cleaning and feature selection. This paper also shows how evolutionary learning paradigms alone can be used for optimization purposes of many kinds, including the feature selection problem described by the differences between different *flow* results. In the future, we would like to try and combine the filter and wrapper methods in novel ways to make them work better in tandem with one another.

References

1. Sun R, Liu M, Lu L et al (2015) Congenital heart disease: causes, diagnosis, symptoms, and treatments. Cell Biochem Biophys 72:857–860. https://doi.org/10.1007/s12013-015-0551-6
2. Mohan S, Thirumalai C, Srivastava G (2019) Effective heart disease prediction using hybrid machine learning techniques. IEEE Access 7:81542–81554. https://doi.org/10.1109/ACCESS.2019.2923707
3. Patel J, Tejalupadhyay D, Patel S (2016). Heart disease prediction using machine learning and data mining technique. 10.090592/IJCSC.2016.018
4. Ramalingam VV, Dandapath A, Raja M (2018) Heart disease prediction using machine learning techniques: a survey. Int J Eng Technol 7:684. https://doi.org/10.14419/ijet.v7i2.8.10557
5. https://www.who.int/news-room/fact-sheets/detail/cardiovascular-diseases-(cvds)
6. Parthiban L, Subramanian R (2008) Intelligent heart disease prediction system using CANFIS and genetic algorithm. Int J Biol Med Sci 3(3):2008
7. Khourdifi Y, Bahaj M (2019) Heart disease prediction and classification using machine learning algorithms optimized by particle swarm optimization and ant colony optimization. Int J Intell Eng Syst 12. https://doi.org/10.22266/ijies2019.0228.24
8. Sadek RM, Mohammed SA, Abunbehan ARK, Ghattas AKHA, Badawi MR, Mortaja MN, Abu-Naser SS (2019) Parkinson's disease prediction using artificial neural network
9. Sánchez-Maroño N, Alonso-Betanzos A, Tombilla-Sanromán M (2007) Filter methods for feature selection—a comparative study. In: International conference on intelligent data engineering and automated learning. Springer, Berlin, Heidelberg
10. Yu L, Liu H (2003) Feature selection for high-dimensional data: a fast correlation-based filter solution. In: Proceedings of the 20th international conference on machine learning (ICML-03)
11. Chandrashekar G, Sahin F (2014) A survey on feature selection methods. Comput Electr Eng 40(1):16–28
12. Maldonado S, Weber R (2009) A wrapper method for feature selection using support vector machines. Inf Sci 179(13):2208–2217
13. Chen G, Chen J (2015) A novel wrapper method for feature selection and its applications. Neurocomputing 159:219–226
14. Anam M et al (2017) Wrapper method for feature selection to classify cardiac arrhythmia. In: 2017 39th annual international conference of the IEEE engineering in medicine and biology society (EMBC). IEEE

15. Bashir S, Khan ZS, Hassan Khan F, Anjum A, Bashir K (2019) Improving heart disease prediction using feature selection approaches. In: 2019 16th international Bhurban conference on applied sciences and technology (IBCAST), pp. 619–623. https://doi.org/10.1109/IBCAST.2019.8667106
16. Reddy N, Chandra S et al (2019) Classification and feature selection approaches by machine learning techniques: heart disease prediction. Int J Innov Comput 9(1)
17. Gárate-Escamila AK, El Hassani AH, Andrès E (2020) Classification models for heart disease prediction using feature selection and PCA. Inf Med Unlocked 19:100330, ISSN 2352-9148. https://doi.org/10.1016/j.imu.2020.100330
18. Jayaraman V, Sultana HP (2019) Artificial gravitational cuckoo search algorithm along with particle bee optimized associative memory neural network for feature selection in heart disease classification. J Ambient Intell Human Comput. https://doi.org/10.1007/s12652-019-01193-6
19. Zhang J, Xiong Y, Min S (2019) A new hybrid filter/wrapper algorithm for feature selection in classification. Analytica Chimica Acta 1080:43–54, ISSN 0003–2670. https://doi.org/10.1016/j.aca.2019.06.054
20. Sebban M, Nock R (2002) A hybrid filter/wrapper approach of feature selection using information theory. Pattern Recog 35(4):835–846. ISSN 0031-3203. https://doi.org/10.1016/S0031-3203(01)00084-X
21. UCI Machine Learning Repository, Heart Disease Data Set, https://archive.ics.uci.edu/ml/datasets/heart+disease
22. Kotsiantis SB, Kanellopoulos D, Pintelas PE (2006) Data preprocessing for supervised learning. Int J Comput Sci 1(2):111–117
23. Ahmad T, Aziz MN (2019) Data preprocessing and feature selection for machine learning intrusion detection systems. ICIC Express Lett 13(2):93–101
24. Khullar V, Tiwari RG, Agarwal AK, Dutta S (2022) Physiological signals based anxiety detection using ensemble machine learning. In: Tavares JMRS, Dutta P, Dutta S, Samanta D (eds) Cyber intelligence and information retrieval. Lecture notes in networks and systems, vol 291. Springer, Singapore. https://doi.org/10.1007/978-981-16-4284-5_53
25. Dutta S, Chandra V, Mehra K, Das AK, Chakraborty T, Ghosh S (2018) Ensemble algorithms for microblog summarization. IEEE Intell Syst 33(3):4–14. https://doi.org/10.1109/MIS.2018.033001411
26. Dutta S, Ghatak S, Das AK, Gupta M, Dasgupta S (2019) Feature selection-based clustering on micro-blogging data. In: Behera H, Nayak J, Naik B, Abraham A (eds) Computational intelligence in data mining. Advances in intelligent systems and computing, vol 711. Springer, Singapore. https://doi.org/10.1007/978-981-10-8055-5_78
27. Hattacharya A, Goswami RT, Mukherjee K, Nguyen NG (2019) An ensemble voted feature selection technique for predictive modeling of malwares of android. Int J Inf Syst Model Des (IJISMD) 10(2):46–69. https://doi.org/10.4018/IJISMD.2019040103
28. Images and clip arts were taken from and made using draw.io web-tool https://app.diagrams.net/
29. Prakash KB et al (2020) Analysis, prediction and evaluation of covid-19 datasets using machine learning algorithms. Int J 8(5)
30. Houssein EH et al (2021) Deep and machine learning techniques for medical imaging-based breast cancer: a comprehensive review. Expert Syst Appl 167:114161
31. Chen H et al (2019) A machine learning method for identifying critical interactions between gene pairs in Alzheimer's disease prediction. Front Neurol 10:1162
32. Rahman, AKM Sazzadur, et al (2019) A comparative study on liver disease prediction using supervised machine learning algorithms. Int J Sci Technol Res 8(11):419–422
33. Maini E, Venkateswarlu B, Gupta A (2018) Applying machine learning algorithms to develop a universal cardiovascular disease prediction system. In: International conference on intelligent data communication technologies and internet of things. Springer, Cham

Author Index

P. Dutta et al. (eds.), *Emerging Technologies in Data Mining and Information Security*, Lecture Notes in Networks and Systems 490,
https://doi.org/10.1007/978-981-19-4052-1

S

T

U

V

Y